CHEMISTRY

STRUCTURE AND REACTIONS

HOLT, RINEHART AND WINSTON, INC.

New York Chicago San Francisco London Toronto

MILTON K. SNYDER *Colorado College*

Chemistry
Structure and Reactions

August, 1966

Library of Congress Catalog Card Number: 66–13302

28022–0116

Printed in the United States of America

Cover photograph by George Recsch,
Fundamental Photographs

Preface

Energy change as a criterion for the nature of chemical structure and reactions is the central theme of this book. Beginning with the concepts of force and energy, the tendency to reach the state of lowest potential energy is used to explain atomic and molecular structure as well as the structure of matter in its three physical states. Chemical reactions are viewed from a macroscopic thermodynamic point of view which is used to introduce the concept of chemical equilibrium and which is applied to solubility, acid-base, and oxidation-reduction reactions.

The first ten chapters introduce the structure of matter with bonds and geometries discussed in terms of the lowest potential energy state attainable. Use is made of nuclear binding energies, bond energies, ionization potentials, electron affinities, electronegativities, lattice energies, and heats of fusion and vaporization as well as the ideas of potential energy levels and states. Internucleonic, ionic, nonpolar and polar covalent, coordinate covalent, van der Waals, hydrogen, and metallic bonds are discussed in these energy terms. Use is made of the Valence Shell Electron Pair Repulsion Theory proposed by Gillespie in discussing bonding and geometries within molecules, ions, and crystals.

The final eight chapters emphasize the dynamics of chemical change using thermodynamics and reaction kinetics as the major tools. Although an accurate but brief development of the fundamental concepts of thermodynamics is given, the emphasis is on the application of thermodynamics to chemical systems. Equilibrium constants, half-cell potentials, and reaction EMF's are introduced as thermodynamic functions which provide information about the driving forces behind chemical reactions. Chemical kinetics is interpreted in terms of reaction mechanisms which may serve as a practical control of thermodynamically feasible reactions through high free energies of activation.

Structure provides the thread of continuity between the first and second parts of the book. Enthalpy and entropy changes accompanying chemical reactions are interpreted in terms of the structures and energy states of the reactants and products.

As this suggests, the book is written in sequential order with later parts depending upon earlier material. This is apparent from the implicit use of a modular approach to structure in the first ten chapters. Protons, neutrons, and electrons are the smallest modules from which atoms are constructed. Then atoms become the modules for the construction of molecules. This culminates in a study of the three states of matter, where atoms, molecules, and ions are the modular units in these more gross structures. In particular, the solid state is viewed as the most economical way, in terms of energy to pack large numbers of modules together.

When the emphasis changes from static structures to dynamic reactions, the reactions are interpreted in terms of the previously developed ideas of structure. Thermodynamics, the main theme of this latter portion of the book, is used throughout the final chapters. For example, dissolving processes are discussed in terms of enthalpy and entropy changes which are related to the structures of the solute, solvent, and solution. Reaction kinetics and reaction mechanisms are introduced to explain why thermodynamically feasible reactions may not occur in practice. Acid-base reactions are then used to illustrate a group of reactions with relatively simple kinetics which are thermodynamically controlled and in which the equilibrium constant adequately describes the system since the reactions tend to be rapid. Oxidation-reduction reactions are used to show a group of reactions which may be controlled either thermodynamically or kinetically, so that the reactions which appear to be most possible thermodynamically may not occur because of the high free energy of activation in one of the steps of the mechanism.

In addition, this book has been written so that theory and applications of theory are integrated. Thus as major concepts or ideas are discussed, they are applied and related to the periodic table. For example, in Chapter 5, which follows the discussion of atomic structure, the properties of ionization potential, electron affinity, atomic and ionic sizes, electronegativity, and electrostatic and magnetic polarization are discussed. In this way, students are reminded that the more abstract ideas of the previous chapter pertain to real elements and at the same time they gain a better idea of the nature of atoms.

Immediately after developing some of the concepts of bonding and the states of matter in Chapters 6 and 7, Chapters 8, 9, and 10 are devoted to the application of these concepts to various types of species which the elements form. Similarly, within Chapter 13 on solutions and colloids, a survey is made of the trends in solubility in water and these are related to the thermodynamic properties discussed in Chapter 12. Likewise, within Chapter 15 on acids and bases, a survey is made of the trends in acidic and basic properties of various types of species. Again this survey is based on the periodic table and employs thermodynamic relationships, via structural features, to see the reasons for the observed trends in acid and base strengths. Finally, after an introduction to oxidation-reduction reactions in Chapter 17, all of Chapter 18 is devoted to a discussion of the oxidation states of the elements, which applies not only the principles developed in the previous chapter but also the principles of solubility and acid-base equilibria.

This integration has been used because I feel that students have difficulty in seeing the relationship between theoretical concepts and actual chemical systems. For the most part, students tend to have difficulty in applying principles to specific problems. As a result more time than usual is spent in the first development of a set of concepts to allow the student sufficient time to assimilate and practice using the major principles. For this same reason, the questions and problems at the ends of the chapters have been chosen to emphasize relationships, the meaning of concepts, and the applications of these concepts to actual problems.

Some results of this approach can be seen in Chapter 1 where energy is introduced through a discussion of gravitational, electrostatic, and magnetic forces. It is hoped that the student will come to grasp the types of situations which lead to increases and decreases in potential energy and how these changes might be made manifest. How else can a student have any grasp for potential energy levels in atoms and their nuclei, gamma radiation, bond energies, spectra, lattice energies, or free energy? We must also realize that not all students in first year chemistry have had physics and a substantial introduction to these concepts is essential.

At the same time, Chapter 1 provides the students with ample practice in the use of direct proportionalities, unit conversions, the development of unit systems, the algebraic manipulation of mathematical relationships, and some graphical interpretations. This has been done with the thought that these techniques have more meaning within the context of some scientific subject matter than as prime subject matter themselves.

Similarly, the subject of chemical bonds, solutions and colloids, acids and bases, and redox reactions have been handled in some detail at the outset. After introducing the concepts they are applied to surveys of the elements and the chemical species which they form. In some cases additional practical applications are included as well.

Although many examples of chemical species used in the book are inorganic, organic species have also been included. However, they have been introduced in a natural way as though they fit into the scheme of things and not as though they are somehow to be considered in a separate category. Thus the hydrocarbons, alkyl halides, alcohols, ethers, ketones, nitro compounds, carboxylic acids, esters, acid halides, amines, and amides are introduced structurally in Chapters 8 and 9. Some organic ligands are included in the discussions of coordination complexes in Chapter 10. Similarly organic solvents and solutes are used as illustrations in Chapter 13. Organic species also have been included in the discussions of acids and bases and redox reactions.

Sufficient background is provided for discussions of typical quantitative analysis procedures as desired. The treatment of equilibrium is thorough, with many examples in solubility, acid-base, and redox equilibria. The treatment of solutions includes some of the typical problems of gravimetric determinations through colloid formation and the absorption of foreign ions on precipitates. In the acid-base and redox reaction chapters, applications to titrations are indicated, including the techniques for plotting titration curves, the choice of color indicators, and typical computations from titration data.

Consciously, I have been developing this book over a period of three years and I want to express my thanks to the students in Chemistry 103,104 at The Colorado College who have suffered through the use of mimeographed trial editions during that time and who appear to have survived the deluge of errata and correction sheets which accompanied these early efforts. Also my thanks go to Mrs. Ruth Weber who patiently and expertly undertook the typing and assembling of mimeographed copies for class use on top of her many other duties.

I am also indebted to the staff members of the Chemistry Department of The Colorado College who bailed me out of difficulty as I sank into the morass of unmet deadlines and long days and nights of struggling with the writing of the manuscript. Especially I want to thank Professor Lester A. Michel, chairman of the department, for his encouragement and constant vigilance to see that teaching loads could be worked out to provide maximum writing time. In addition I want to thank Professors William C. Champion and Eldon T. Hitchcock who used a trial edition of the text while I was on leave for a year and who returned valuable suggestions from this experience.

Also I want to thank Jay A. Young of Kings College in Wilkes-Barre, Pennsylvania, for the thorough and extensive criticisms which he made of the text. He undertook the burden of using a trial edition with his class with enthusiasm and vigor and his ability to see the book through the eyes of the students is as remarkable as it was helpful to me.

Finally, I want to acknowledge the understanding of my wife, Barbara, and sons Glenn and Clark who have been widowed and orphaned for long periods of time. My special thanks go to my wife, who has taken time from her busy duties to put her own background in chemistry to good use in reading proofs and making timely suggestions for clarifications.

Milton K. Snyder

Colorado Springs, Colorado
January, 1966

Contents

CHEMISTRY

STRUCTURE AND REACTIONS

PROLOGUE

The methods of science are as diverse as the scientists practicing them. However, it is useful to consider some of the more common methods as background for the study of chemistry.

Natural science, as the name implies, originates in observations of nature. These observations may be made on nature as we find it, or as we have contrived it for our study. Observations of the latter type are made on nature's response to stimuli supplied by the scientist. Thus this stage of the scientific enterprise consists of collecting facts about the behavior of nature.

Most scientists agree, however, that mere observation and fact collecting are not the essence of science. Of greater importance is the mental activity applied to a collection of facts to show their interrelations. Thus the exciting activity of science is the construction of consistent models of nature based upon these interrelated observations.

To many persons, science and technology are interchangeable words. For our purposes, however, it is convenient to separate them in our thinking, relegating the practical applications of science to the realm of technology.

Although the foregoing outline is an oversimplification of a complex enterprise, it will serve to indicate the aim of this text. That is, the presentation in this book is the core of the science of chemistry. That is to say, we shall be concerned less with the gathering of facts and their practical applications and more with the interpretation of these facts.

One of the fundamental tools of the chemist is measurement, the fabric upon which his observations are made. More than this, it is a vocabulary for classifying and interpreting observations. As a fabric, it can be operated upon and fashioned into the desired end products, the models.

Mathematical technology is the language used to put the measurement vocabulary into meaningful structures. We must admit that scientists usually are mathematical technologists. Mathematicians view mathematics in much the same way that scientists view science. They consider mathematics to be a logical system in its own right, regardless of its applicability to practical situations. In this respect, scientists are technologists applying mathematics to their observations of nature.

Therefore, to set the scene for what is to come, we must take a cursory look at some aspects of measurement and the application of mathematical technology to measurement.

Measurement

To say that the distance between two points is 35 in. is to express the result of a measurement. However, there is uncertainty in this measurement, for at best the number indicates that the distance is less than 36 and more than 34 in. This measurement may be satisfactory for some purposes, but not for others. To express a more exact measurement, the distance might be designated as 35.2 in. However, there is still some uncertainty, since this lies between 35.1 in. and 35.3 in. The uncertainty is less, to be sure, but it is still there and it must always be present in any measurement. Exact measurements are not possible.

How the distance is expressed depends upon the way the measurement is made and the way it is to be used. In numbers that express measurements, a statement is built in to indicate how well the measurement was made. Thus the distance would not be written as 35.194 in. if the measurement was made only to the closest inch. Neither would it be written as 35 in. if care was taken to measure it to the nearest thousandth of an inch.

For these reasons, it is agreed that numbers expressing measurements shall contain a number of digits that represents the degree of care taken in making the measurement. These digits are called *significant figures.* Furthermore, it is agreed that since there is some uncertainty in any measurement, the uncertainty in the measurement shall be expressed in the last digit of the number. Thus a distance written as 35.194 in. indicates uncertainty in the fifth digit; the distance lies somewhere between 35.193 and 35.195 in. The distance has been measured to five significant figures.

Significant figures play an important role in computations. In adding 34.2 to 2.21, one is first inclined to say that the answer is 36.41. However, since the first digit beyond the decimal in 34.2 is uncertain, the first digit beyond the decimal in the answer must be uncertain also. As a result there is no value in carrying the answer to more than three significant figures. Therefore the answer is expressed as 36.4. The general rule for addition and subtraction is as follows: the result of addition or subtraction should not contain more than one uncertain digit. This amounts to saying that *the number of digits to the right of the decimal point in a sum or difference should not exceed that of the number having the fewest significant figures to the right of the decimal point.*

For multiplication and division, the determination of the number of significant figures to be retained in the product or quotient is a bit more involved. Suffice it to say here that as a general rule, *the number of significant figures in a product or quotient should not exceed that in the factor, dividend, or divisor with the fewest number of significant figures.*

Some confusion arises when zeros are used in measurements, since zeros may be measured or may be inserted only to locate the decimal point. A distance expressed as 30 in. leaves some doubt about the significance of the second digit. If it is insignificant, then the measurement indicates that the distance lies between 20 and 40 in. If it is significant, then the measurement indicates that the distance lies between 29 and 31 in. However, if the number is expressed as 30.0 in., there is no doubt that the measurement was made to three significant figures, since the last zero is unnecessary to locate the decimal. It is clear that this distance lies between 29.9 and 30.1 in.

Ambiguities of this kind can be avoided by the use of exponential notation, in which numbers are represented as the product of two factors, one of which represents the significant figures and the other the decimal location. In this way, 30 may be written as 3×10^1 or as 3.0×10^1. Here the number of significant figures is one in the first example and two in the second. In writing numbers in *standard exponential form,* the first factor is expressed with one digit to the left of the decimal and the second factor is written as 10 to a power to locate the decimal.

Incidentally, this exponential number notation has several advantages. First, it is convenient to use in expressing very large and very small numbers. A large number such as 4,025,000,000 may be written as 4.025×10^9 to four significant figures. A small number such as 0.0000008297 obviously has four significant figures and may be written as 8.297×10^{-7}. In the exponential form, the positive exponent indicates the number of places to move the decimal to the right and the negative exponent indicates the number of places to move the decimal to the left.

A second advantage of exponential notation is the simplification of some types of computations and the acceleration of the estimation of answers to rather complex problems. The multiplication of exponential numbers may be accomplished simply by adding exponents. Thus $10^2 \times 10^2$ equals 10^4. The division of exponential numbers may be accomplished by subtracting the sum of the exponents in the denominator from the sum of the exponents in the numerator to obtain the exponent in the quotient. Thus $10^4/10^2 = 10^2$ and $10^2/10^6 = 10^{-4}$.

A serial multiplication and division problem may be set up as follows:

$$\frac{39.6 \times 5280 \times 0.601 \times 0.000952}{0.00181 \times 454 \times 95.8}$$

or as an alternative, it might be set up in a factor form with exponential numbers as follows:

$$\frac{3.96 \times 10^{1}}{1.81 \times 10^{-3}} \times \frac{5.28 \times 10^{3}}{4.54 \times 10^{2}} \times \frac{6.01 \times 10^{-1}}{9.58 \times 10^{1}} \times 9.52 \times 10^{-4}$$

From a relatively quick glance over this second formulation, it is possible to estimate that the answer will be near 15×10^{-1}. When the problem is worked out the answer is 1.52. In this way there is a quick check for the correctness of a computed result, and for slide rule computations, a relatively quick method for decimal location.

It should be evident, but sometimes it is forgotten, that the units in which measurements are made are important and must accompany the numbers representing the measurements. It is absurd to say that a distance is 35. The distance might be 35 mi or 35 in. Thus any measurement must be accompanied by the units in which the measurement was made; in fact, it is well to consider the number and the unit as inseparable parts of the measurement.

Direct Proportionality

The concept of proportionality finds wide use in chemistry. A few examples will clarify the concept.

Suppose that you are buying apples at \$.20 per pound. It is clear that the amount you must pay for apples depends upon how many pounds you decide to buy. If you buy 1 lb, the cost will be \$.20; if you buy 2 lb the cost will be \$.40; if you buy 4 lb the cost will be \$.80. Thus the cost and the weight of apples are interrelated variables; they are related through the invariant cost per pound. It is clear that the cost C equals the product of the rate (cost per pound) r multiplied by the weight in pounds W:

$$C = rW$$

This is an example of direct proportionality, which is characterized by an increase in the value of one variable when the other variable increases, a doubling in value of one variable when the other value doubles, a halving of one variable when the other value is halved, and a value of zero for one variable when the other variable is zero. Mathematically, to say that y is proportional to x is to say that y equals x multiplied by a constant, that is, $y = kx$, where k is the constant called the *proportionality constant.*

To see if two variables are directly proportional, two tests may be used. The first comes directly from the equation $y = kx$. If this is solved for k, it is clear that $k = y/x$, and the numerical test is the constancy of the ratio y/x. Thus if several values for the two variables are available, the constancy of their quotient shows that they are directly proportional.

The second test is graphical. If y is plotted as the ordinate and x as the abscissa, the graph is a straight line passing through the origin. Figure P.1 shows such a

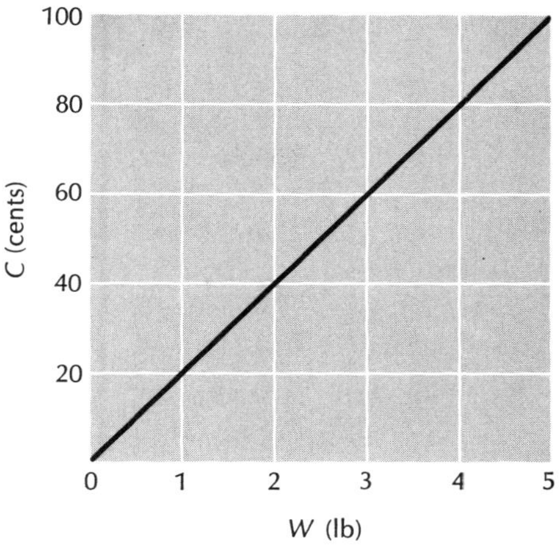

Fig. P.1. Plot of cost against weight

plot for $C = rW$, where C is the cost of apples in cents, r is the rate in cents per pound, and W is the number of pounds of apples. If a right triangle is constructed so that its hypotenuse lies on the graphed line, and it includes the angle made by the line and the horizontal axis of the graph, as shown by the dotted line in Fig. P.1, the ratio of the length of the side parallel to the vertical axis in the plotted units to the length of the side parallel to the horizontal axis in the plotted units is called the *slope* of the line. The slope is the value of the proportionality constant. This is just another way of saying that the slope of the line is the ratio C/W, which is the value for the proportionality constant r.

Now consider an illustration from chemistry. Table P.1 shows the volumes of several samples of mercury along with the corresponding mass in grams at 25°C. In the third column of the table, the first test for proportionality is shown in the record of the ratio of mass to volume. The fact that this ratio is constant indicates that the mass is directly proportional to the volume, and the constant value for the ratio, namely 13.53 grams (g) per milliliter (ml) is the value for the proportionality constant. Thus the mathematical relationship may be written as $m = 13.53V$, where m is the mass in grams and V is the volume in milliliters for mercury at 25°C.

The same result is obtained from the graphical test. In Fig. P.2 the mass and volume data from Table P.1 have been plotted. The straight line through the origin indicates the direct proportionality. The slope of the line is computed to be 13.53 g per ml, the same as the constant ratio shown in Table P.1.

TABLE P. 1
Mass and Volume of Mercury at 25°C

MASS (g)	VOLUME (ml)	MASS/VOLUME (g/ml)
81.18	6.000	13.53
135.3	10.00	13.53
175.9	13.00	13.53
223.2	16.50	13.53
280.1	20.70	13.53

Several comments should be made about this example. The volume samples of mercury in this listing did not go below 6.000 ml nor above 20.70 ml. Therefore, from this data, the only certain part of the straight line in Fig. P.2 is the solid part. We might think that the straight line should continue in both directions, but we cannot be sure that it will. For this reason, the extensions of this line, called *extrapolations*, are shown as dotted lines. The fact that the extension passes through the origin shows that the mass is directly proportional to the volume within the measured region. Of course additional measurements could be made on smaller and larger samples to show whether or not the extrapolations are valid.

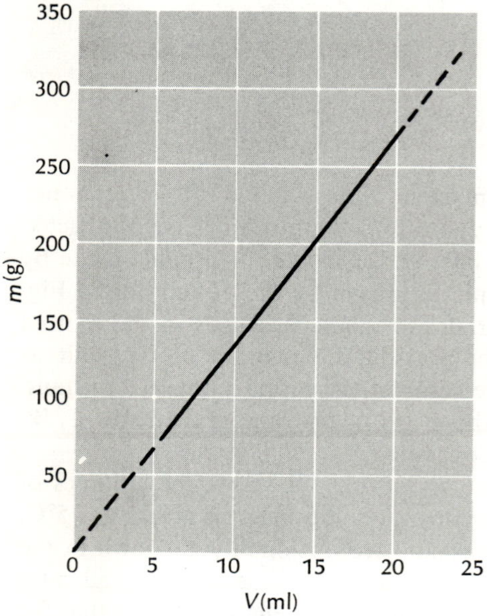

Fig. P.2. Plot of mass against volume of mercury

Also it should be noted that the proportionality constant is an important property of mercury, relating the volume of mercury to its mass. This proportionality constant is called the *density*. Often proportionality constants are significant, and usually they are valuable because they are constant. Many values recorded in handbooks are proportionality constants.

In addition, it should be understood that the use of numerical values for proportionality constants in mathematical relationships clearly fixes the units in which the variables are expressed. In this case, the use of 13.53 g/ml fixes the volume units to milliliters and the mass units to grams. However, these two variables could have been measured in other units, gallons and kilograms, for example. For this reason, it is necessary to keep the units with the numbers. When the density of mercury is given as 13.53 it must be given as 13.53 g/ml, where the mass units have been divided by the volume units.

Suppose that it is desirable to express the density of mercury in other units, say in kilograms per gallon. To say that the density of mercury is 13.53 g/ml is to say that 1.000 ml has a mass of 13.53 g. If the number of kilograms in 13.53 g and the number of gallons in 1 ml can be computed, the two numbers can be divided to provide the density in kilograms per gallon.

Conversions of units of this type involve direct proportionalities. Thus the mass in kilograms of any object is directly proportional to its mass in grams, and the proportionality constant is either remembered or found in a published table. In this case the proportionality constant is given by the relationship 1.00 kg = 1000 g. The number of kilograms in 13.53 g may be found by multiplying the 13.53 g by the correct proportionality constant as follows:

$$13.53\text{ g} \times \frac{1.00\text{ kg}}{1000\text{ g}} = 0.01353\text{ kg} \quad \text{(P.1)}$$

Notice that in carrying out the multiplication of the two factors in Equation (P.1), the units are divided and multiplied just like the numbers. Thus grams/grams is unity, and the unit in the answer is kilograms.

Similarly, gallons are proportional to quarts, quarts are proportional to liters, and liters are proportional to milliliters. The number of gallons in a milliliter may be found as follows:

$$1.00\text{ ml} \times \frac{1.00\text{ l}}{1000\text{ ml}} \times \frac{1.06\text{ qt}}{1.00\text{ l}} \times \frac{1\text{ gal}}{4\text{ qt}} = 2.65 \times 10^{-4}\text{ gal} \quad \text{(P.2)}$$

The density of mercury may be expressed by dividing the result of Equation (P.1) by the result of Equation (P.2), as follows:

$$\text{Density} = \frac{1.35 \times 10^{-2}\text{ kg}}{2.65 \times 10^{-4}\text{ gal}} = 50.9\text{ kg/gal} \quad \text{(P.3)}$$

Note again that in Equations (P.2) and (P.3) the units are divided along with the numbers. The canceling of units provides a convenient guideline for the solution of problems such as this. In fact, the total problem may be set up in a single series of steps that may be solved easily by a single computation on a slide rule or with logarithms. To discover the density of mercury in kilograms per gallon, the original expression for the density is set down as the starting point:

$$\frac{13.53\text{ g}}{1.00\text{ ml}}$$

Then it is noted that the mass units are to be converted from grams to kilograms and in accord with the proportionality relationship, the original expression is multiplied by the appropriate factor:

$$\frac{13.53\text{ g}}{1.00\text{ ml}} \times \frac{1.00\text{ kg}}{1000\text{ g}}$$

If this multiplication is carried out, the answer will be the density of mercury in kg/ml. However, the milliliters must be converted to gallons using the facts that 1.00 l is 1.06 qt and also 1000 ml, and 4.00 qt equal 1.00 gal.

$$\frac{13.53\ \text{g}}{1.00\ \text{ml}} \times \frac{1.00\ \text{kg}}{1000\ \text{g}} \times \frac{1000\ \text{ml}}{1.00\ \text{l}} \times \frac{1.00\ \text{l}}{1.06\ \text{qt}} \times \frac{4.00\ \text{qt}}{1.00\ \text{gal}} = 50.9\ \text{kg/gal}$$

This representation has the advantage of setting all the steps of the computation into a single series which can easily be checked for correctness and which is in order for an easy computation by slide rule or logarithms.

Inverse Proportionality

Closely related to direct proportionality is inverse proportionality. This may be illustrated by the relationship between the average speed of a race car and the time it takes to travel 100.0 mi. Clearly, if the car averages 100.0 mi/hr, it will take one hour, and if it averages 50.00 mi/hr, it will take two hours. In this relationship one variable increases while the other decreases. For example, when the average speed is doubled, the time required is halved.

Perhaps this relationship is easier to see if we consider that the average speed s is the distance traveled D divided by the time t. This is expressed in Equation (P.4).

$$s = \frac{D}{t} = D\frac{1}{t} \qquad \text{(P.4)}$$

Equation (P.4) may be read as follows: the average speed is directly proportional to the *reciprocal* of the time, where the distance traveled is the proportionality constant. If Equation (P.4) is solved for the constant D, then it is clear, as shown by Equation (P.5), that the *product* of the two variables is constant.

$$st = D \qquad \text{(P.5)}$$

This is to be contrasted to the direct proportionality, in which the *ratio* of the two variables is constant.

This new type of relationship is called an *inverse proportionality*. When one variable is inversely proportional to another, that variable is proportional to the reciprocal of the other and the product of the two variables is constant. The constancy of this product is one test for inverse proportionality.

TABLE P. 2 Speed and Time Data for Cars Racing 100.0 Miles

TIME (HR)	1/TIME (HR^{-1})	SPEED (MI/HR)	SPEED × TIME (MI)
0.50	2.00	200.0	100.0
0.75	1.33	133.3	100.0
1.00	1.00	100.0	100.0
1.25	0.800	80.0	100.0
1.50	0.667	66.7	100.0
1.75	0.571	57.1	100.0
2.00	0.500	50.0	100.0

Table P.2 shows some data for the 100.0-mile race. In Fig. P.3 the average speed is plotted directly against the time to show that the result is not a straight line. In Fig. P.4, the speed is plotted against the reciprocal of the time, and the

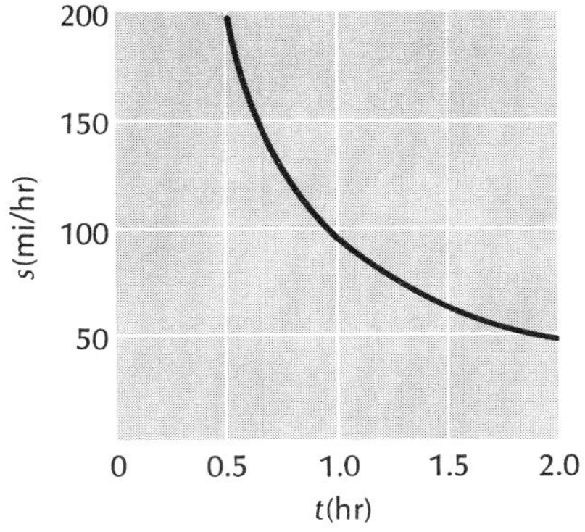

Fig. P.3. Plot of speed against time

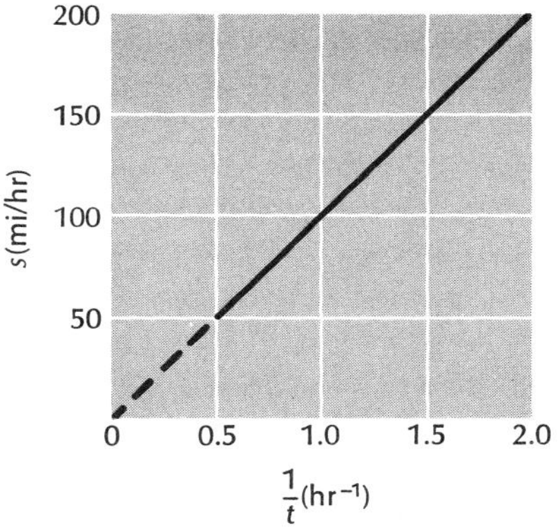

Fig. P.4. Plot of speed against reciprocal time

result is a straight line that can be extrapolated through the origin. Notice that although this line may be extrapolated through the origin, there is no meaningful value for the time at the origin. When the speed is zero, an infinite time is required to travel the 100.0 mi.

Pseudoproportionality

A third kind of proportionality is shown by the relationship between the Fahrenheit and Celsius temperatures. Table P.3 shows some comparable values on the two scales. It is clear that both scales rise and fall together. Thus a higher temperature on the Fahrenheit scale also is a higher temperature on the Celsius scale.

TABLE P. 3
Fahrenheit and Celsius Temperatures

°F	°C
32.0	0.0
68.0	20.0
98.6	37.0
150.0	65.6
212.0	100.0

However, it is clear also that the two scales do not register zero at the same temperature. When the temperature is zero on the Celsius scale the temperature is 32.0° on the Fahrenheit scale.

Figure P.5 shows a plot of Fahrenheit degrees versus Celsius degrees. Obviously the plot is a straight line, but it does not go through the origin. Rather than

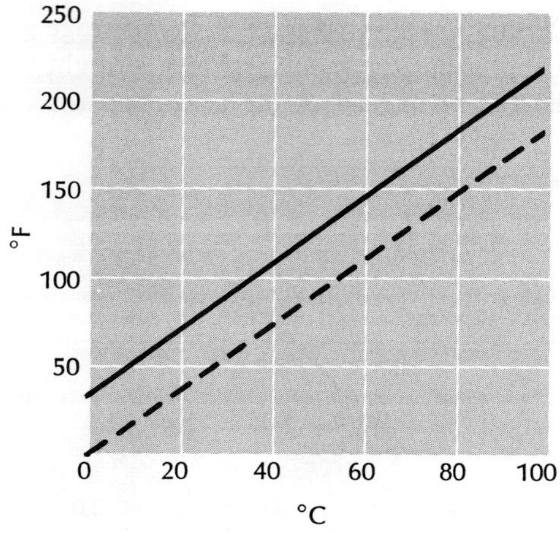

Fig. P.5. Plot of Fahrenheit against Celsius temperature

intersecting the Fahrenheit axis at zero, it intersects at 32.0°. The dotted line in Fig. P.5 is drawn parallel to the solid line but is drawn through the origin. The dotted line, if correct, would represent a direct proportionality between the two temperature scales. It is possible to imagine that the dotted line has been formed by picking up the solid line and moving it 32.0 units downward. Thus every Fahrenheit temperature on the dotted line is 32.0 units below what it ought to be. Thus the temperatures on the dotted line may be represented by $F - 32$ where F is the real Fahrenheit temperature. Then the dotted line is a plot of $F - 32$ against C, where C is the Celsius temperature. Since the dotted line is a plot of a direct proportionality, the relationship between Fahrenheit and Celsius temperatures can be represented by Equation (P.6) where k_1 is the proportionality constant.

$$F - 32 = k_1C \qquad \text{(P.6)}$$

The proportionality constant k_1 in Equation (P.6) is the slope of the dotted line but also the slope of the solid line, since the two lines are parallel. The value for the slope is computed from the graph to be 1.8, and Equation (P.6) may be rearranged to Equation (P.7).

$$F = 1.8C + 32 \qquad \text{(P.7)}$$

We may call this kind of relationship a pseudoproportionality. It resembles the direct proportionality except that it has a second additive constant that is the intercept of the line on the vertical axis. Thus, in very general form, the pseudoproportionality may be written as Equation (P.8).

$$y = k_1x + k_2 \qquad \text{(P.8)}$$

The best test for pseudoproportionality is a graphical one, since there is no simple mathematical test like taking the ratio or the product of two variables.

Consider an example. The density of many substances changes as the temperature changes. The density of mercury is an example. In our earlier consideration of direct proportionality, the density of mercury was taken at constant temperature. The data in Table P.4 show the numerical relationships between density and Celsius temperature for mercury. It is clear that the relation between density and

TABLE P. 4 Density of Mercury Related to Celsius Temperature

DENSITY (g/ml)	TEMPERATURE (°C)	DENSITY × TEMPERATURE (g × °C/ml)
13.60	0.0	0.0
13.55	20.0	271.0
13.50	40.0	540.0
13.45	60.0	807.0
13.40	80.0	1072
13.35	100.0	1335

temperature is not a direct proportionality, since the density decreases as the temperature increases. That it is not an inverse proportionality is shown by the lack of constancy of the product of density and temperature. Figure P.6 shows a

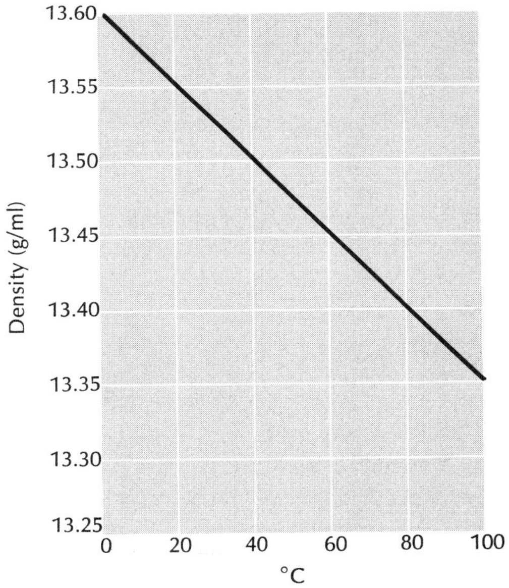

Fig. P.6. Plot of density of mercury against temperature

plot of density versus temperature; it is clear that the relationship is a linear one and that it resembles the pseudoproportionality except that the line slopes in the wrong direction. A change in slope is represented simply by a change in sign, so the density D of mercury is related to the temperature T over the temperature range indicated by Equation (P.9).

$$D = -0.0025T + 13.60 \tag{P.9}$$

The slope used in Equation (P.9) was computed from the graph, or it might have been computed from values in Table P.4.

Finally, consider an example taken from the supermarket as an indication of the kind of observation and treatment of observation discussed here. Our supermarket has eight different sizes of boxes of Brand X detergent powder. The net weights of the contents of these boxes are recorded in Table P.5, both the way they

TABLE P. 5

Dates on Brand X Detergent

NET WEIGHT (LB + OZ)		(LB)	COST (CENTS)
1	3	1.19	41
3	5	3.31	99
4	15	4.94	121
6	1	6.06	133
7	12	7.75	147
9	4	9.25	157
11	7	11.44	168
13	13	13.81	179

appear on the boxes and in a more manageable decimal form. The corresponding prices of the boxes are listed also. An observer wonders what the relationship is between the price and the net weight.

It is obvious that the cost increases as the net weight increases, so this is not an inverse proportionality. The graph of cost against net weight is shown in Fig. P.7. The relationship is not linear. Still undismayed, the observer asks a mathematician friend if the graph has the characteristics of another mathematical function. The mathematician suggests that the graph has some of the characteristics of a plot of

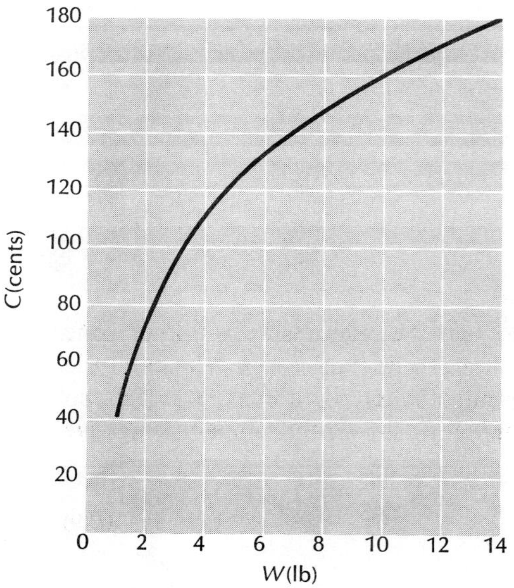

Fig. P.7. Plot of cost against weight for Brand X detergent

y against $\log x$ when $y = \log x$. (The logarithm, of course, is the exponent of 10 when a number is expressed entirely exponentially. Thus $\log 10^5 = 5$ and $\log 10^{0.5} = 0.5$.)

To check this possibility, the observer takes the logarithm of the net weights in Table P.5 and constructs Table P.6, from which he then constructs the graph in Fig. P.8. Within the uncertainties of rounding off costs to the nearest cent, this

TABLE P. 6

Logarithmic Data on Brand X Detergent

WEIGHT (LB)	LOG WEIGHT	COST (CENTS)
1.19	0.076	41
3.31	0.520	99
4.94	0.694	121
6.06	0.782	133
7.75	0.889	147
9.25	0.966	157
11.44	1.057	168
13.81	1.140	179

graph is a straight line with an extrapolated intercept of 31 cents on the vertical axis. The slope of the line is 130. The observer concludes that the relationship between the cost of the detergent and the net weight can be expressed by Equation (P.10), where C is the cost and W is the net weight in pounds.

$$C = 130 \log W + 31 \tag{P.10}$$

In this case the cost of the detergent is pseudoproportional to the logarithm of the net weight. This is an example of a kind of scientific inquiry and an example

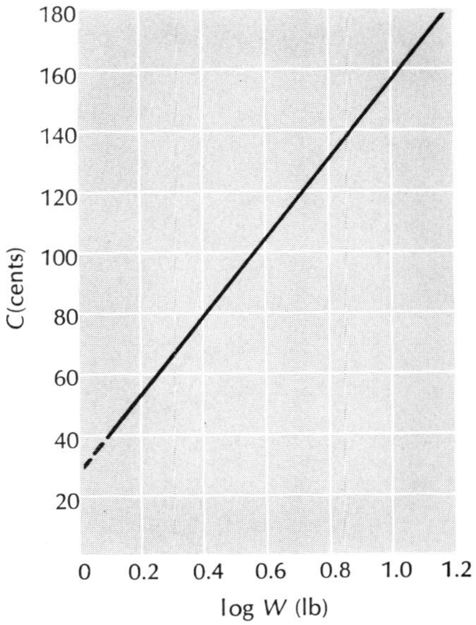

Fig. P.8. Plot of cost against log weight for Brand X detergent

of the application of mathematical technology to practical problems. More and more, scientists are finding practical applications of functions derived by mathematicians. This illustration throws some light upon the activities of chemists who are trying to learn something about the nature of matter and its transformations.

Conclusion

Interesting as the subject may be, we cannot discuss here whether or not the relationship our supermarket observer discovered bears any resemblance to the way in which the prices on the boxes were determined by the store manager. This is a philosophical problem of some current interest. There are those who interpret the results of scientific inquiry as a revelation of mind and plan behind nature. Others feel that the order seen in science is an order imposed by the scientist and methods he uses.

In any event, we may suppose that our supermarket observer came away believing that he had a better understanding of supermarket operation as a result of the inquiry. In a similar way scientists are attempting to come to a better understanding of nature.

In the pages that follow, an attempt is made to show some of the models and patterns chemists have developed and found useful in coming to grips with the problem of the nature of matter.

1

Matter and Energy

INTRODUCTION Chemists are deeply concerned with matter and energy and the relationships between them. In this chapter we shall study matter through its properties of mass, electric charge, and magnetism. Each of these properties is associated with an important force: mass with gravitational force, electric charge with electrostatic force, and magnetism with magnetic force.

Energy will be discussed as kinetic energy, or energy of motion; potential energy, or energy of position; heat; and electromagnetic radiation.

MATTER

Your notion of matter as the "stuff" of which the world and universe is made can serve as a good starting definition. Thus you recognize the solids, liquids, and gases about you as matter. You include as matter the wood and metal of the chair you are sitting in, the air you breathe, and the water you drink. Probably you include yourself and recognize the many different kinds of matter that make up your body.

Mass

Sometimes matter is characterized as that which has mass. Again you have some feeling for the meaning of mass. You recognize that you might push a huge boulder with all your strength for a long time without moving it, but that a similar push will set a smaller rock in motion. Therefore you recognize that the boulder has more mass than the smaller rock. Similarly, if you roll a small snowball in a snow field, it will gather snow and become larger and larger and more difficult to move. If the snowball is rolling downhill it may run out of control and continue to roll at great speed, picking up more and more snow as it goes. *Mass is the property of matter that measures its resistance to being set in motion, or once set in motion, its resistance to a change in that motion.*

Sir Isaac Newton quantified the concept of mass by showing that the force applied to an object is directly proportional to the acceleration produced in that object; mass is the proportionality constant. This may be expressed mathematically by Equation (1.1), where F is the force, a the acceleration, and m the mass.

$$F = ma \tag{1.1}$$

Thus mass is a property of matter. It is clear from Equation (1.1) that if an object has a large mass, a given force will produce a small acceleration, and if the mass is small the same force will produce a larger acceleration.

The units in which forces are measured can be obtained from the relationship shown by Equation (1.1). It follows that if agreement can be reached on the units in which to express any two of the three quantities in Equation (1.1), the units in which to express the third will be determined. For this reason it is necessary to take a brief look at the commonly used systems of measurement units.

The fundamental unit systems normally are defined in terms of units for length, mass, and time. The three systems most used are reviewed in Table 1.1. The two

TABLE 1.1
Some Common Unit Systems

PROPERTY	MKS SYSTEM	CGS SYSTEM	BRITISH SYSTEM
Length	meter(m)	centimeter(cm)	foot(ft)
Mass	kilogram(kg)	gram(g)	slug
Time	second(sec)	second(sec)	second(sec)

metric systems are named by the first letters of the fundamental units employed, namely cgs and mks. The mks system has been recognized by international approval and it is the one that physicists have adopted rather universally. Chemists tend to use the cgs system more, although often they are not consistent in this matter.

Returning to the problem of defining units for force, mass, and acceleration, we see that the unit for mass is defined directly in the three systems in Table 1.1. The unit for acceleration is derived from the fundamental units if it is recalled that acceleration is the rate of change of velocity. Although velocity has two components, namely speed and direction, it is speed we are concerned with here. Therefore, for our purposes, acceleration is defined as the rate of change of speed.

From this point of view, acceleration is speed per unit time. In mks units, speed is given in meters per second, so that acceleration is given as (m/sec)/sec which may be written as m/sec^2. Superscript powers on units have the same meaning as superscript powers on numbers, and sec^2 means sec × sec. The units for speed and acceleration are given in Table 1.2 for the three common unit systems.

TABLE 1.2
Units for Speed and Acceleration

	MKS SYSTEM	CGS SYSTEM	BRITISH SYSTEM
Speed	m/sec	cm/sec	ft/sec
Acceleration	m/sec^2	cm/sec^2	ft/sec^2

The units given in Tables 1.1 and 1.2 now may be applied to Equation (1.1) to define the units of force. This is shown in Table 1.3. The rather complex unit combinations for force are given special names in each of the systems. Thus in the mks

TABLE 1.3
Units of Force Derived from $F = ma$

	MASS	ACCELERATION	FORCE	COMMON NAME
mks system	kg	$\frac{m}{sec^2}$	$\frac{kg \times m}{sec^2}$	newton(nt)
cgs system	g	$\frac{cm}{sec^2}$	$\frac{g \times cm}{sec^2}$	dyne
British system	slug	$\frac{ft}{sec^2}$	$\frac{slug \times ft}{sec^2}$	pound(lb)

system the unit of force is called the newton (nt), and is equivalent to (kg × m)/sec^2. Similarly, in the cgs system the unit of force is the dyne, which is equivalent to (g × cm)/sec^2. In the British system, the unit for force is the pound, which equals the (slug × ft)/sec^2, where the slug is the unit of mass in this system.

GRAVITATIONAL FORCES. One of the important properties associated with mass is the gravitational force that exists between any two objects. This force is always an attractive force and can be related directly to the masses of the objects as well as to the distance between them. The mathematical expression for the relationship is given in Equation (1.2), where F_g is the gravitational force, m_1 and m_2 are the masses of the two objects, and d is the distance between them. Here G is the proportionality constant, called the *universal gravitational constant.*

$$F_g = G\frac{m_1 m_2}{d^2} \tag{1.2}$$

In words, Equation (1.2) may be stated as follows: the gravitational force of attraction between two objects is directly proportional to the masses of the two objects and is inversely proportional to the square of the distance between them.

The numerical value of G depends only on the units in which force, mass, and distance are expressed. Otherwise, it is a true constant. When Equation (1.2) is solved for G, it is clear that G must be expressed in the units (force × $distance^2$)/$mass^2$. In the mks system of units the value of G has been determined to be 6.670 × 10^{-11} (nt × m^2)/kg^2.

One common gravitational system consists of an object on the surface of the earth, and the earth itself. Strictly speaking, the gravitational law is valid only between very small particles, which may be considered to be points. From this point of view, it is necessary to consider the earth to be composed of a very large number of such points, and then to sum up all the effects of each of these points upon any object on the surface. More commonly, the practice is to consider the mass of the earth to be located at a single point, called the center of gravity. For the earth this entails some errors, for the earth is not a uniform sphere, but the result is approximately correct.

WEIGHT AND MASS. On earth, the gravitational force exerted by the earth upon objects near its surface is called the *weight* of those objects. Thus weight is a force and is expressed in force units. The relationship between weight and mass can be derived from Newton's law given as Equation (1.1), where the force (weight) is proportional to the acceleration (of gravity) with mass as the proportionality constant. This is expressed by Equation (1.3) where W is the weight, m is the mass, and g is the acceleration of gravity.

$$W = mg \tag{1.3}$$

It should be noted that g differs from G in Equation (1.2). Whereas G is a constant, g, the acceleration due to gravity, is not. It is true that the differences in the acceleration due to gravity are small on the surface of the earth, and the approximate value of 9.81 m/sec^2 or 32.2 ft/sec^2 is used for approximate purposes.

It follows from Equation (1.3) that weight is measured in force units such as newtons, dynes, or pounds, whereas mass is measured in mass units such as kilograms, grams, or slugs. Also it should be noted that the mass of an object is constant, whereas the weight of an object depends upon its distance from the earth, or in speaking of other planets, upon the mass of the planet upon which the mass is resting. However, it is common practice on earth to express the relationship between weight and mass in similar terms, and often the two are interchanged.

To express this relationship, it is convenient to read Equation (1.3) as follows: The weight of an object is directly proportional to its mass with the acceleration of gravity as the proportionality constant. This is possible since the acceleration of gravity is nearly constant over the surface of the earth. Using this proportionality, it is possible to determine the mass corresponding to a given weight or the weight corresponding to a given mass. Graphically this has been done in Fig. 1.1, where weight in newtons has been plotted against mass in kilograms. The line has the slope g, or 9.81 nt/kg, which is the same as 9.81 m/sec^2. It is possible, given a weight, to read off the corresponding mass directly from the graph.

Mathematically, the conversion may be made as well. For example, the weight, on the earth's surface, of an object with a mass of 3.5 kg can be found as follows:

$$3.5\text{ kg} \times \frac{9.8\text{ nt}}{1.0\text{ kg}} = 34\text{ nt}$$

One of the confusing aspects of measurements reported on earth is that in the British system amounts of substance are reported in weight units (pounds) whereas

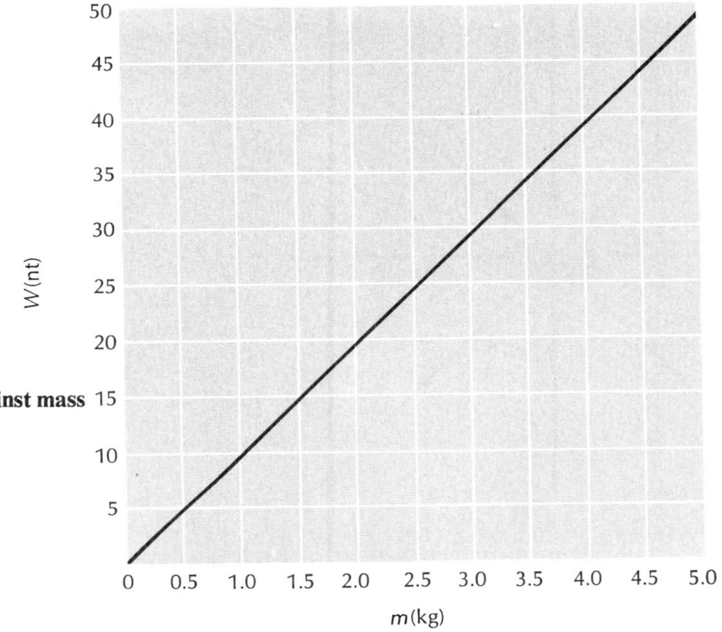

Fig. 1.1. Weight against mass

in the metric system they are expressed in mass units (grams or kilograms). It should be remembered that the conversion units commonly used, such as 1 kg = 2.2 lb or 454 g = 1 lb, are valid only on earth, since they express a relationship between mass and weight which depends upon the acceleration due to gravity on earth. However, such conversion factors may be used in the solution of problems. For example, what is the weight in pounds of 50.0 g of salt? Assuming earth conditions, 1 lb = 454 g, and the result is

$$50.0 \text{ g} \times \frac{1.00 \text{ lb}}{454 \text{ g}} = 0.110 \text{ lb}$$

GRAVITATIONAL FORCE FIELDS. Gravitational forces appear to be exerted through space between separated objects. In modern physics, it has become customary to think about a kind of medium in which the objects in question are immersed and through which the forces may be exerted. This is not a substantive medium. Simply conceived, around any object with mass, the space is altered in some way by the mass, and attractive forces emanate through this space. This altered space is called the *gravitational force field,* or more simply the *gravitational field.*

The gravitational field around a small spherical object is depicted in Fig. 1.2. Forces directed toward the center of the sphere are present throughout the space around the object. This force-containing space is the gravitational field of that object. The forces are represented by arrows pointing in the direction toward which a test object would move when placed in the field. The arrows represent *lines of force.*

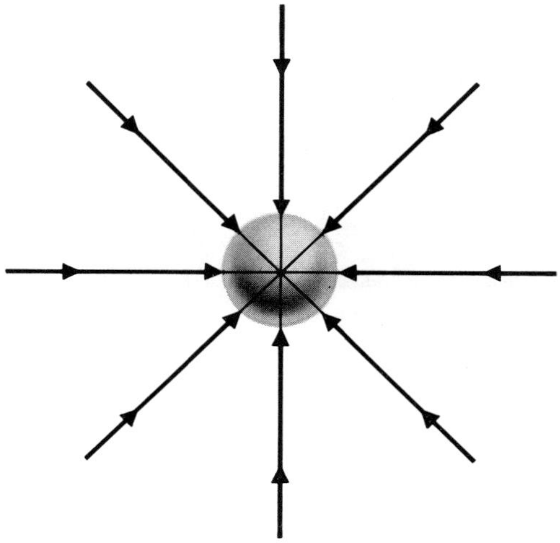

Fig. 1.2. Gravitational lines of force around an isolated object

If two identical objects are set at fixed positions, the two gravitational fields surrounding the two objects will interact in a manner shown in Fig. 1.3. The lines of force associated with the isolated objects are altered, and a new force field is associated with the two objects combined into a single system.

A test mass placed at point *A* in Fig. 1.3 will move downward to a position directly between the two objects, since it is attracted equally by each of them. A test mass placed at point *B* will follow the line of force depicted there. It is attracted by both objects and hence its path is displaced somewhat toward object 1, although it is attracted more strongly to object 2 because it is closer to object 2.

It must be understood that the pictures of gravitational fields and lines of force

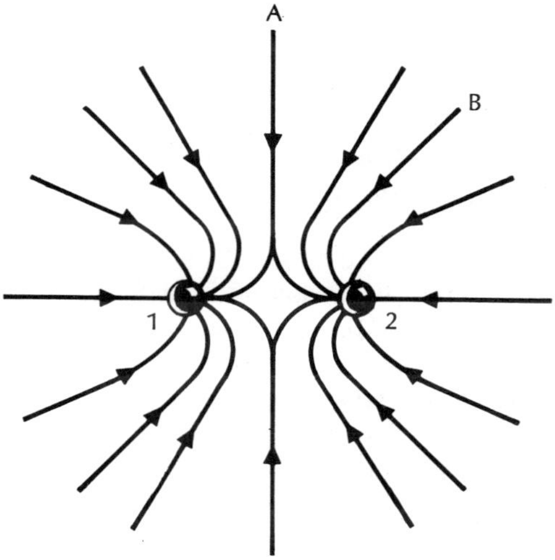

Fig. 1.3. Gravitational lines of force around two fixed objects

are convenient mental constructs useful in the visualization of gravitational forces. Do not be misled into believing that somehow all of space is intertwined with real lines.

Electric Charge

Another force of more general importance to the chemist than gravitational force is electrostatic force. That electrostatic forces exist can be demonstrated by rubbing glass rods with silk and hard rubber rods with animal fur. The glass rods repel each other and the rubber rods repel each other, but the glass rods attract the rubber rods. The rods are said to be electrically charged. Arbitrarily, the charge on the glass rods is called positive charge and the charge on the rubber rods is called negative charge. The repulsive and attractive forces existing between these rods are called *electrostatic forces.* From these observations with glass and rubber rods, it is clear that electrical charges of like type repel, whereas electrical charges of unlike type attract.

COULOMB'S LAW. The effect of distance and charge size upon the magnitude of electrostatic forces is stated in Coulomb's law, which is given mathematically by Equation (1.4).

$$F_e = K\frac{q_1q_2}{d^2} \tag{1.4}$$

In Equation (1.4), F_e is the electrostatic force, K is the proportionality constant, q_1 and q_2 are the magnitudes of the two charges, and d is the distance between the charges. In words, Coulomb's law is as follows: the electrostatic force of attraction or repulsion between two charges is directly proportional to the sizes of the two charges and inversely proportional to the square of the distance between them.

The similarity of Equations (1.4) and (1.2) should be noted. Both are inverse square laws, which means that the force varies inversely as the square of the distance between the objects involved. In addition, it appears that the electrical charge plays the same role in electrostatic forces that mass plays in gravitational forces.

Although the proportionality constant K appears to be similar to G for gravitational forces, it has one very significant difference. Whereas G appears to be a universal constant, K is less constant in that its value depends upon the nature of the medium between the charges. Thus if the charges are immersed in a vacuum, K has one value; if they are immersed in air, K has another value; if they are immersed in water, K has still another value. To express the effect of the medium in the terms usually employed, it is necessary to perform a minor mathematical manipulation. The reasons for this manipulation are not important here; suffice it to say that the method is convenient in other applications of Coulomb's law.

The mathematical manipulation requires substitution of the expression for K given in Equation (1.5) into Equation (1.4).

$$K = \frac{1}{4\pi\epsilon} \tag{1.5}$$

Here π is a constant with the usual value of 3.1416 and ϵ is called the *permittivity constant*. The substitution results in Equation (1.6), the usual form of Coulomb's law.

$$F_e = \frac{q_1 q_2}{4\pi\epsilon d^2} \tag{1.6}$$

Like the value for K, the value for ϵ depends upon the medium between the charges. If the value for ϵ is small, then the electrostatic force between the charges tends to be large. If the value for ϵ is large, then the electrostatic force between the charges tends to be small. Thus a medium with a large permittivity constant tends to reduce the forces between the charges, and a medium with a low permittivity constant tends to magnify the electrostatic force between the charges. The permittivity constant, then, is a measure of the ability of the medium to insulate electrical charges from each other.

In practice, the medium with the smallest permittivity constant is a vacuum. All other media have higher permittivity constants; that is, they are better insulators than a vacuum. In practice the permittivities of media normally are not recorded, but rather they are compared to the permittivity of a vacuum. This is done by taking the ratio of the permittivity of the medium in question to the permittivity of a vacuum. This ratio is called the *dielectric constant* of the medium. Mathematically the dielectric constant, κ, is defined by Equation (1.7) where ϵ_0 is the permittivity of a vacuum and ϵ is the permittivity of the medium.

$$\kappa = \frac{\epsilon}{\epsilon_0} \tag{1.7}$$

An advantage of the dielectric constant is that it is unitless and thus independent of the units in which the permittivity is measured. It should be clear that a high value for the dielectric constant of the medium corresponds to good insulating qualities of the medium, and a low value for the dielectric constant corresponds to poor insulating qualities. A few typical dielectric constants are recorded in Table 1.4. In some circumstances the dielectric constant is an important property of matter.

TABLE 1.4
Some Dielectric Constants of Liquids at 25°C

SUBSTANCE	κ
Hydrogen cyanide	114.9
Water	78.5
Ethylene glycol	37.7
Methyl alcohol	32.6
Chloroform	4.8
Carbon tetrachloride	2.2

Returning to Equation (1.6) for a moment, it is possible to outline the units in which charge and permittivity constants are measured. In the mks system, the unit of charge is the coulomb; in the cgs system it is the statcoulomb, also called the electrostatic unit and abbreviated esu. The values for the permittivity constants corresponding to these charge units are given in Table 1.5. Although the coulomb

TABLE 1.5
Values for the Permittivity of a Vacuum

SYSTEM	CHARGE UNIT	PERMITTIVITY, ϵ_0
mks	coulomb	8.85×10^{-12} coulomb^2/(nt $\times$ m^2)
cgs	statcoulomb	7.96×10^{-2} statcoulomb^2/(dyne $\times$ cm^2)

is a rather large unit of charge and somewhat cumbersome for that reason, it is the preferred unit and will be used almost exclusively throughout this book.

It must be noticed that electrostatic and gravitational forces differ in the way they are affected by the medium. They also differ in that gravitational forces are always attractive whereas electrostatic forces may either attract or repel. The distinction can be made by using the negative sign with negative charges and the positive sign with positive charges. If these are included in Equation (1.6), like charges for q_1 and q_2 will lead to a positive force which must represent a repelling force. Unlike charges for q_1 and q_2 will lead to a negative force which must represent an attractive force.

ELECTROSTATIC FIELDS. Like gravitational forces, electrostatic forces are depicted by a force field. The space around an electrically charged object is altered. This altered space is called the *electrostatic field*, and affects any charged object immersed in it. By convention, the lines of force in an electrostatic field are represented as if the force were brought to bear on a positively charged object immersed in the field. From this point of view, Fig. 1.4 shows the lines of force associated with isolated positive and negative charges. The arrow directions correspond to attraction by the negative charge and repulsion by the positive charge.

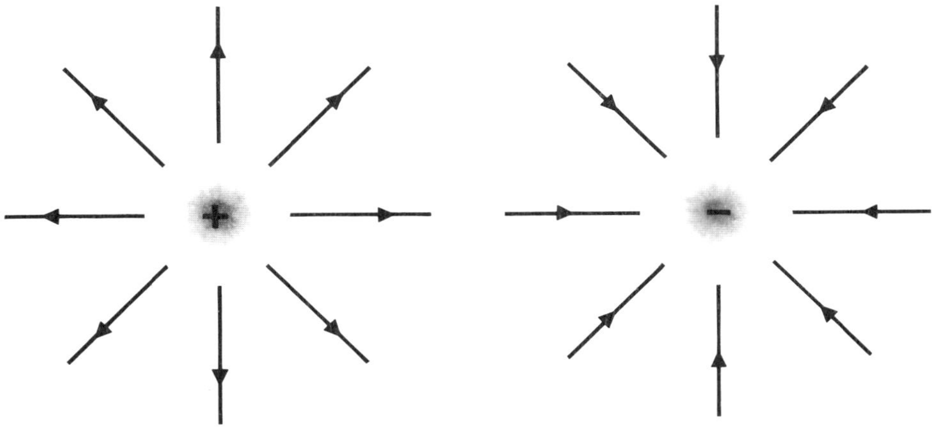

Fig. 1.4. Electrostatic lines of force around isolated positive and negative point charges

Figure 1.5 depicts the fields associated with two charges at fixed distances from each other. The field associated with two negative charges resembles the gravitational field for two objects, since both charges are attracting positive charges. The electrostatic field associated with two positive charges shows that a positive charge tends to be "squeezed out" of the field; that is, it is repelled by both charges. The field associated with a combination of a positive and a negative charge shows the attractive force of the negative charge and the repelling force of the positive charge.

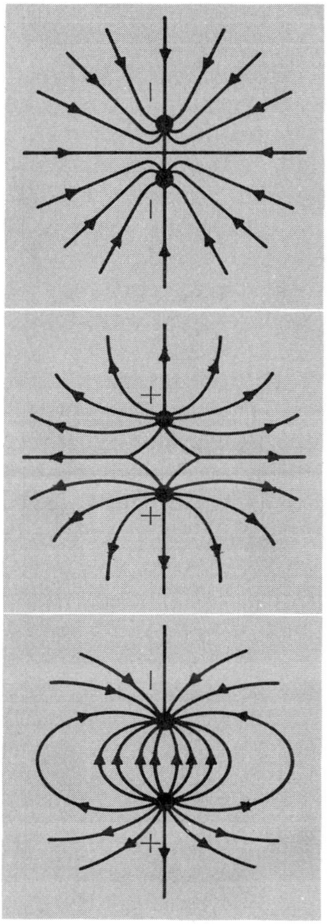

Fig. 1.5. Electrostatic lines of force around various combinations of pairs of point charges

Charge quantization. As indicated earlier, material objects can be charged electrically by friction. In addition, small drops of liquid can be charged electrically by radiating them with x-rays or radiations from radioactive materials. Robert A. Millikan used an apparatus shown schematically in Fig. 1.6, in which electrically charged oil drops were sprayed between two plates on which the electrical charge could be varied, or reversed. Thus the strength of the electrostatic field was made variable and controllable as well as measurable.

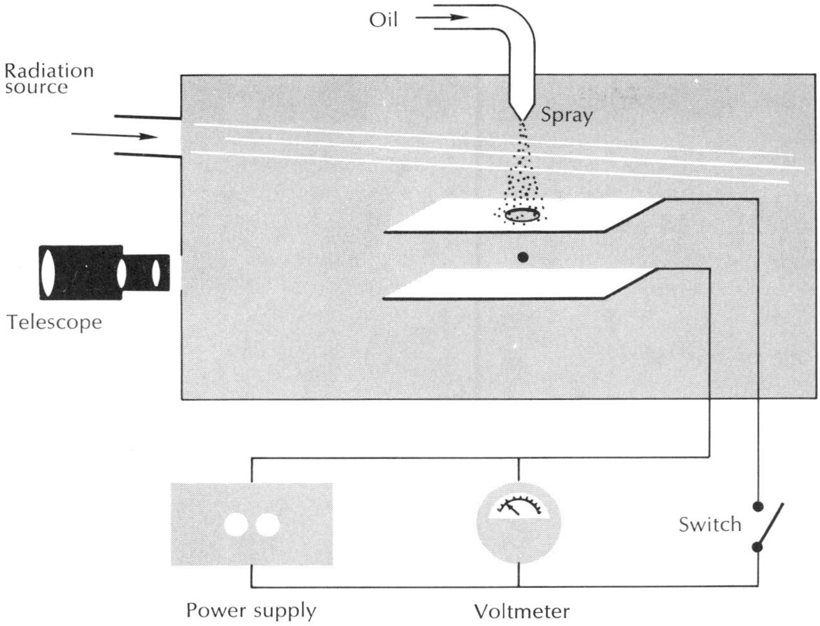

Fig. 1.6. Millikan's oil-drop experiment

Any drop of oil tends to fall downward because of gravitational force. However, if the charge on the top plate is opposite to that of the drop, then the drop is attracted upward, and if the electrostatic field strength is adjusted properly, the drop can be made to remain suspended between the two plates. This occurs when the upward electrostatic force equals the downward gravitational force.

Millikan determined the mass of the oil drops by measuring the radii of the drops and the density of the oil. This, combined with the acceleration of gravity, allowed computation of the downward gravitational force. Then, knowing the strength of the field and using the fact that upward and downward forces must be equal, he computed the charge on the oil drop.

Some typical data obtained by Millikan are recorded in Table 1.6. Astutely he observed that these numbers, along with the many others he obtained, are all inte-

TABLE 1.6
Some Values of Electrical Charge on Millikan's Oil Drops

6.563×10^{-19} coulombs
8.204×10^{-19}
11.50×10^{-19}
13.13×10^{-19}
16.48×10^{-19}
18.08×10^{-19}
19.71×10^{-19}
22.89×10^{-19}
26.13×10^{-19}

gral multiples of 1.641×10^{-19} coulombs. The integral multipliers for the data in Table 1.6 are 4, 5, 7, 8, 10, 11, 12, 14, and 16. From this Millikan concluded that electrical charge is quantized; that is, it exists in discrete quantities. Therefore, any electrical charge must be some integral multiple of a fundamental quantum of charge.

Many experiments carried out since Millikan's show that his work was accurate, and only a slight change has been made in the accepted value for the quantum of charge. That accepted value is 1.602×10^{-19} coulombs or 4.802×10^{-10} esu.

This concept of quantization is an important one and will be encountered again in other contexts.

Magnetism

A flow of electric charge is called a *current*. The magnitude of a current is measured as the rate of flow of charge. This is shown by Equation (1.8), where i is the current and q is the quantity of charge that passes a given point in the time t.

$$i = \frac{q}{t} \tag{1.8}$$

Thus current is measured in charge per unit time; in the mks system this is in coulombs/sec, or *amperes*. In the cgs system it is measured in statcoulombs/sec, or *statamperes*.

The only forces operating between stationary charges are electrostatic. However, it is possible to show that there are additional forces between currents. Using a device schematically represented in Fig. 1.7, where two segments of a current-

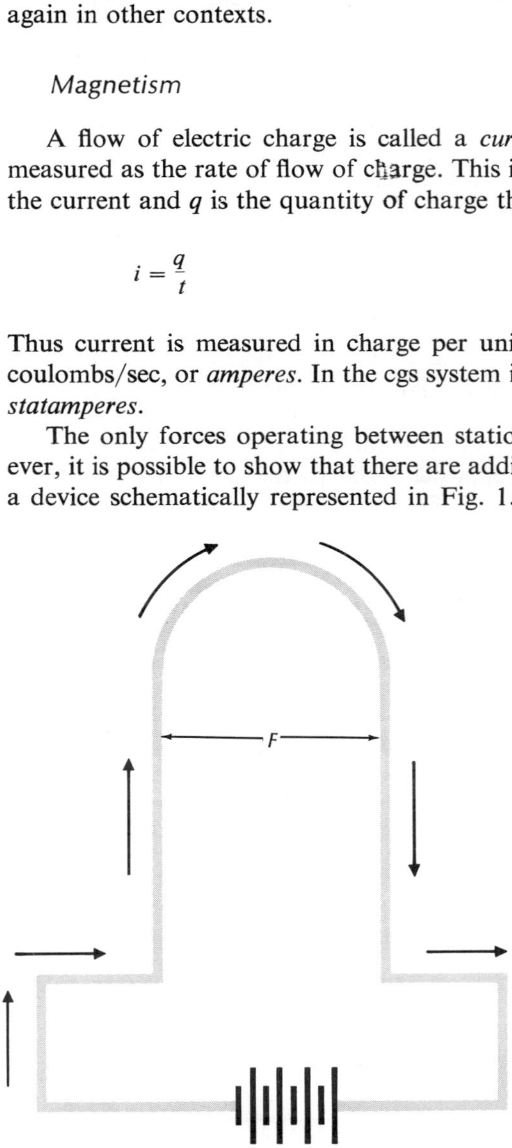

Fig. 1.7. Forces between currents

carrying wire are laid parallel to each other, it can be shown that the two wire segments repel each other. The repulsion is greater than any anticipated electrostatic forces. The arrows *along* the wire indicate the direction of flow of negative charge.

These forces acting between electric currents in addition to any electrostatic forces are called *magnetic forces*. It is an experimental fact that magnetic forces act strongly upon certain metals, of which iron is a good example. Not only are iron objects attracted by magnetic forces, but once introduced into an area around a current carrying wire, they exert what appear to be magnetic forces upon other iron objects. Such objects are said to be *magnetized* or are called *magnets*, and the magnetism is said to have been *induced* in the iron object.

When a magnetized needle is brought near to a current-carrying wire, it will orient itself with respect to that wire. If it is set parallel to the wire, it will turn so that it is perpendicular to the wire as shown in Fig. 1.8. What is more, as indicated in Fig. 1.8, the ends of the magnetized needle can be labeled, and regardless of the way the needle points originally, it always turns so that the ends are directed in the same way. Thus in Figs. 1.8a and 1.8b, end *A* always points to the left. However, if the direction of the current is reversed, as shown in Figs. 1.8c and 1.8d, then end *A* always points to the right.

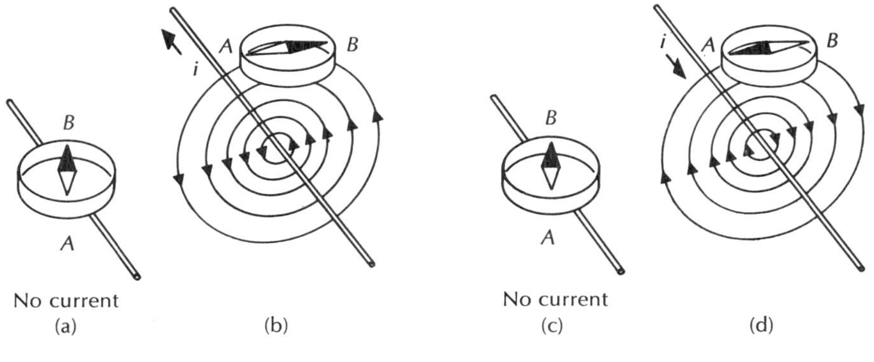

Fig. 1.8. Orientation of magnetized needle with respect to a current-carrying wire

MAGNETIC FIELDS. If a straight current-carrying wire is run through a sheet of paper as indicated in Fig. 1.9, and iron filings are sprinkled on the paper, the iron filings will orient themselves in concentric circles with the wire at the common center of the circles. This pattern strongly resembles a diagram of a force field, and indeed it shows the lines of force as the iron filings line up within the *magnetic field* around the wire.

Thus Fig. 1.9 serves as a pictorial representation of the magnetic field with the concentric circles as the lines of force. Arbitrarily, directions are assigned to the lines of force according to the left-hand rule, which is stated as follows: if the left hand is extended with the thumb out from the hand, the thumb will point in the

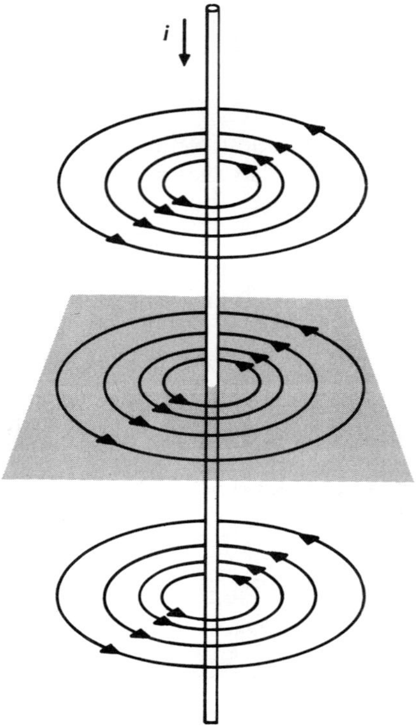

Fig. 1.9. Magnetic lines of force around a straight current-carrying wire

direction of current flow and the natural curl of the fingers will indicate the direction of the lines of force in concentric circles around the current flow.

If a bar magnet is prepared by magnetizing a bar of iron, and if a sheet of paper is placed over this magnet with iron filings sprinkled on the sheet, then the iron filings again assume a pattern of lines as indicated in Fig. 1.10, where the lines of force are shown. Within this pattern there appear to be two very strong centers for the forces near the ends of the bar. These loci are called the *poles* of the magnet, and in comparison to electrostatic fields they appear to be somewhat similar to

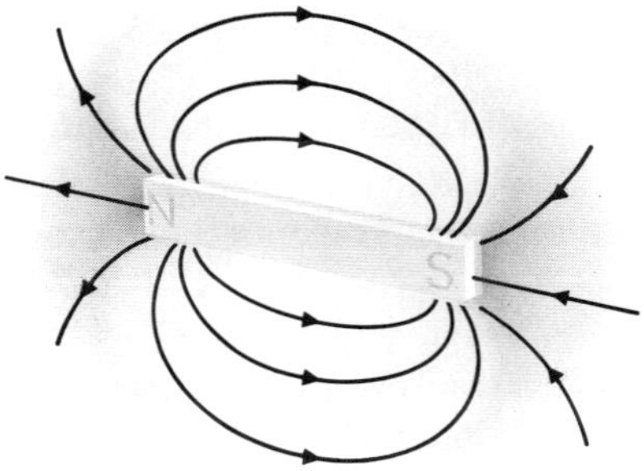

Fig. 1.10. Magnetic lines of force around a bar magnet

electric charge centers. Thus the field shown in Fig. 1.10 is comparable to that shown for the electrostatic field for opposite charges in Fig. 1.5c. If the bar magnet is suspended by a thread, it will align itself in the earth's magnetic field. The pole of the magnet that always points in the northerly direction is called the *north pole* and the pole which always points in the southerly direction is called the *south pole*.

The convention adopted in assigning arrows to the lines of force in Fig. 1.10 is as follows: lines of force are in the direction of the force experienced by a north pole introduced into the magnetic field.

Magnetic fields produced by various combinations of fields are diagrammed in Fig. 1.11. It is interesting to notice how similar these diagrams are to those in Fig. 1.5 for electrostatic fields. However, some care must be exercised in comparing

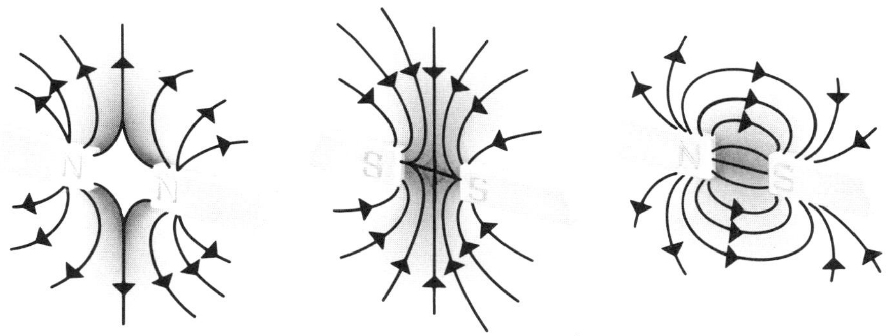

Fig. 1.11. Magnetic lines of force for various combinations of magnetic poles

magnetic poles with electrical charge points. Whereas it is possible to separate electrical charges, so that there are negatively charged objects and positively charged objects, it is not possible to separate magnetic poles. For example, if a bar magnet is cut in half, each of the halves is a new magnet with both a north and a south pole. This is shown in Fig. 1.12.

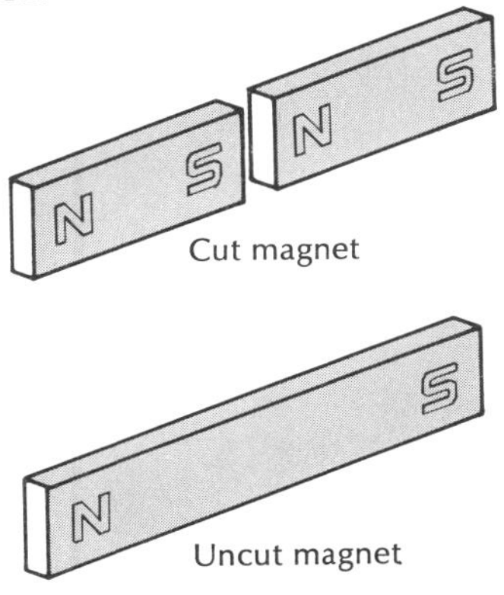

Fig. 1.12. Effect of cutting a bar magnet

Matter in magnetic fields. If, instead of traveling through a straight wire, current travels through a wire coiled around a core, the field can be represented as shown in Fig. 1.13. This field bears a strong resemblance to that of a bar magnet with the two poles near the end of the coil.

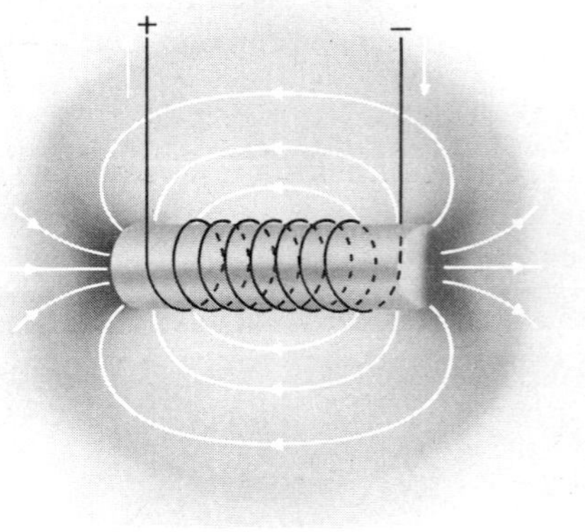

Fig. 1.13. Magnetic lines of force around current loops

Experimentally it is interesting to note what happens to the lines of force of the magnetic field associated with a current coil when various materials are inserted into the core of the coil. If a rod of zinc metal is inserted into the core, the magnetic field becomes less intense than in a vacuum. This is equivalent to saying there is a smaller density of force lines. The effect is shown diagrammatically in Figs. 1.14a and 1.14b. The diagram in Fig. 1.14a represents a vacuum core and the diagram in Fig. 1.14b represents a core containing zinc metal. Types of matter like zinc, which decrease the intensity of a magnetic field, are said to be *diamagnetic*, exhibiting the property of *diamagnetism*.

If aluminum metal is inserted into the core, the intensity of the magnetic field is increased, as indicated in Fig. 1.14c. Any type of matter which exhibits this property of increasing the intensity of a magnetic field is said to be *paramagnetic*, that is, to exhibit the property of *paramagnetism*.

If iron is inserted into the core, the intensity of the magnetic field is increased immensely, as indicated in Fig. 1.14d. Any type of matter which exhibits this property of vastly increasing the intensity of a magnetic field is said to be *ferromagnetic*, that is, to exhibit the property of *ferromagnetism*. Superficially, ferromagnetism appears to be an extreme example of paramagnetism.

Since magnetic poles cannot be separated, the development of quantitative relationships is somewhat more complex than it was for the electrostatic field. Suffice it to say here that the effect of the medium upon magnetic fields is somewhat analogous to the effect of the medium upon electrostatic fields. Thus, comparable to the electrostatic permittivity is the *magnetic permeability*, designated by μ. Now μ is characteristic of the material in the core and when the core of the coil is a vacuum, the permeability is designated by μ_0. If the permeability of the core material is larger than μ_0, the core material is paramagnetic. If the permeability of the core material is less than μ_0, it is diamagnetic.

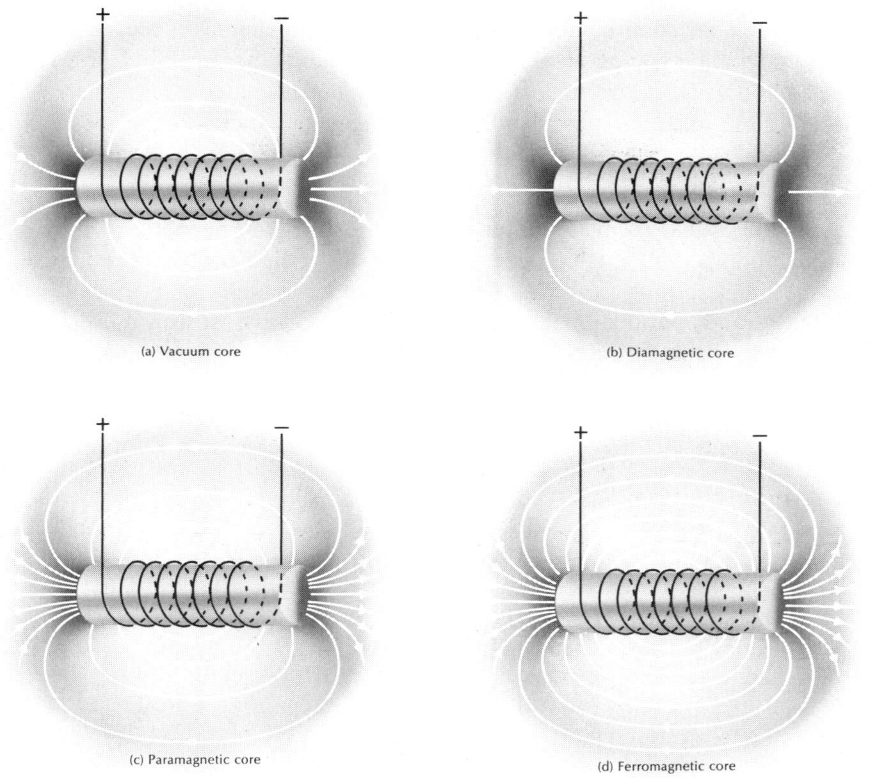

Fig. 1.14. Effects of vacuum, diamagnetic, paramagnetic, and ferromagnetic cores on the magnetic field of a current coil

In describing the effect of the medium upon the electrostatic field, the permittivity of the medium was compared to the permittivity of the vacuum. Similarly in describing the effect of a core material upon a magnetic field, the permeability of the core material is compared to the permeability of a vacuum.

One way in which this comparison is made is identical to the treatment of permittivities; that is, the ratio is taken of the permeability of the material to the permeability of a vacuum. This ratio is called the *relative permeability* and is designated by μ_r. The definition is given by Equation (1.9).

$$\mu_r = \frac{\mu}{\mu_0} \tag{1.9}$$

Like the dielectric constant, the relative permeability is unitless. For paramagnetic materials, μ_r is greater than unity, and for diamagnetic materials, μ_r is less than unity. However, under all circumstances, the relative permeability is positive.

However, there is a second more common way in which to compare the permeabilities. The comparison is made simply by indicating the amount by which the permeability in question is more or less than the permeability of the vacuum. The deviation of a magnetic permeability from the magnetic permeability of a vacuum is called the *magnetic susceptibility*, and usually is designated by χ. The relationship among the magnetic permeabilities and magnetic susceptibility is as follows:

$$\mu = \mu_0 + \chi \tag{1.10}$$

It should be clear from Equation (1.10) that for a paramagnetic core material, μ is larger than μ_0 and χ is positive. Similarly, for a diamagnetic core material, μ is less than μ_0 and χ is negative. Thus diamagnetic materials have negative susceptibilities, and paramagnetic materials have positive susceptibilities. Some typical magnetic susceptibilities are listed in Table 1.7 in the mks units. (The exact nature of the units is not important at this time.)

Notice how small the susceptibilities are in Table 1.7, compared to the permeability of the vacuum, 1.26×10^{-6} mks units. This means that the differences in

TABLE 1.7

Some Magnetic Susceptibilities in mks Units

SUBSTANCE	°C	SUSCEPTIBILITY
Aluminum	18	1.03×10^{-11}
Bismuth	18	-2.12×10^{-11}
Copper	18	-0.14×10^{-11}
Diamond	20	-0.78×10^{-11}
Lead	18	0.19×10^{-11}
Mercury	18	-0.30×10^{-11}
Oxygen gas	20	167×10^{-11}
Oxygen liquid	−219	487×10^{-11}
Silver	18	-0.32×10^{-11}
Zinc	18	-0.25×10^{-11}

permeability are extremely small from one material to another; this is the reason that susceptibilities are preferred to relative permeabilities. The relative permeabilities of the materials in Table 1.7 are so near to unity that they would need to be expressed to many significant figures to show any differences.

ENERGY

In some respects energy is a more elusive term than matter to define. However, the common notion associated with "energetic people" as people who do a lot of work is a good starting point. Indeed, a common but not extremely useful definition of energy is *the capacity to do work.*

Kinetic Energy

To the physicist and chemist, work is done when some exerted force accomplishes a change. Hence a force exerted by a person against a concrete abutment accomplishes no change and thus no work, whereas the same force exerted against a sliding wall may cause the wall to move and in the process accomplish work.

An initial formulation for work is given by Equation (1.11), where w is the work and F is the force acting through distance d.

$$w = Fd \tag{1.11}$$

This equation implies that the application of force must accomplish something, namely a change in position, for if the distance is zero, the work is zero as well.

From the definition in Equation (1.11), the common units for work are derived. Those units are force units multiplied by distance units. These are listed in Table 1.8. Special names are given to the mks and cgs units of work, namely joule and erg respectively.

TABLE 1.8
Work Units

	MKS	CGS	BRITISH
Force	nt	dyne	lb
Distance	m	cm	ft
Work = Fd	nt × m (joule)	dyne × cm (erg)	ft × lb (ft lb)

Since energy is the capacity to do work, energy is measured in terms of the work which it can accomplish. As a result, energy is measured in the same units as work: that is, in joules, ergs, or foot-pounds.

Another relationship for energy of motion may be developed by consideration of a constant force acting upon an object either to set it in motion or change its motion. According to Newton's law in Equation (1.1), an object with constant mass acted upon by a constant force must receive a constant acceleration. The work done on the object by the force is given by the product of the force and the distance through which it acts, as seen in Equation (1.11). Then Equations (1.1) and (1.11) may be combined as follows:

$$w = Fd = mad \tag{1.12}$$

The acceleration in Equation (1.12) is expressed as the change in speed per unit time. The change in speed may be expressed as the difference between the final speed s_f after time t has elapsed, and the initial speed, s_i at the beginning of time period t. Then the acceleration is given by Equation (1.13).

$$a = \frac{s_f - s_i}{t} \tag{1.13}$$

The distance traveled in Equation (1.12) is the product of the average speed s_{av} and the time interval t. The average speed is given by one-half the sum of the final and initial speeds:

$$s_{av} = \frac{s_f + s_i}{2} \tag{1.14}$$

The total distance, d, traversed in time interval t then is given by Equation (1.15), which is derived from Equation (1.14).

$$d = s_{av}t = \frac{s_f + s_i}{2}t \tag{1.15}$$

The expressions in Equations (1.13) and (1.15) may be substituted for the acceleration and distance in Equation (1.12) as follows:

$$w = Fd = mad = m\frac{s_f - s_i}{t} \times \frac{s_f + s_i}{2}t = m\frac{s_f^2 - s_i^2}{2}t$$

This may be rewritten:

$$w = \tfrac{1}{2}ms_f^2 - \tfrac{1}{2}ms_i^2 \tag{1.16}$$

Of course the quantity ms^2 has the units of work or energy. As a result Equation (1.16) may be interpreted as the difference in two energy terms or as a change in energy. Therefore the accomplishment of applied force F is a change in energy. This is energy of motion, since it is related to the speed, and it is called the *kinetic energy*. Each of the terms on the right of Equation (1.16) may be considered to be an expression for kinetic energy; a general expression for kinetic energy may thus be written as Equation (1.17), where E_k is the kinetic energy.

$$E_k = \tfrac{1}{2}ms^2 \tag{1.17}$$

Using the subscripts f for final and i for initial, Equation (1.17) may be used to rewrite Equation (1.16).

$$w = E_{kf} - E_{ki} \tag{1.18}$$

It has become conventional to represent the change in any property from an initial to a final condition by the delta notation, in which Δproperty = property in final condition − property in initial condition. From this point of view the kinetic energy change may be represented by $\Delta E_k = E_{kf} - E_{ki}$, and Equation (1.18) may be rewritten.

$$w = E_{kf} - E_{ki} = \Delta E_k \tag{1.19}$$

From the way in which the delta notation is defined, it should be clear that ΔE_k is positive if the kinetic energy in the final condition is greater than the kinetic energy in the initial condition, that is, if $E_{kf} > E_{ki}$. Under this condition, the applied force acts on the object to increase its acceleration so that the final speed exceeds the initial speed. When ΔE_k is positive, the work is positive, implying that work was done on the object against a resisting force.

Similarly ΔE_k is negative if the kinetic energy in the initial condition is greater than the kinetic energy in the final condition, that is if $E_{ki} > E_{kf}$. Under this condition, the applied force acts on the object to decrease its acceleration so that the final speed is less than the initial speed. When ΔE_k is negative, the work is negative, implying that the object has done work against the applied force.

Thus kinetic energy is energy of motion, and any object in motion possesses kinetic energy. Using the relationships developed here, it is possible to calculate changes in kinetic energy. For example, the change in kinetic energy when a 2.20-kg mass slows from a speed of 50.0 m/sec to a speed of 30.0 m/sec can be found as follows:

$$E_{ki} = \tfrac{1}{2}ms_i^2 = \tfrac{1}{2} \times 2.20 \text{ kg} \times \frac{50.0^2 m^2}{\text{sec}^2} = 2.75 \times 10^3 \text{ joules}$$

$$E_{kf} = \tfrac{1}{2}ms_f^2 = \tfrac{1}{2} \times 2.20 \text{ kg} \times \frac{30.0^2 m^2}{\text{sec}^2} = 0.99 \times 10^3 \text{ joules}$$

$$\Delta E_k = E_{kf} - E_{ki} = (0.99 \times 10^3 \text{ joules}) - (2.75 \times 10^3 \text{ joules})$$
$$= -1.76 \times 10^3 \text{ joules}$$

Potential Energy

Potential energy is the energy an object has because of its position. In a sense it is latent energy that is stored and has the possibility of becoming useful energy. The overworked but pertinent illustration is a rock at the top of a hill; it has potential energy due to its position. When the rock starts to roll down the hill, part of its potential energy is converted into kinetic energy, and its potential energy decreases.

Another illustration is a spring. As a weight is added to a vertical spring so that the spring is stretched out, potential energy increases as the weight does work on the spring, but if the weight is removed, the potential energy decreases as the spring snaps back into position.

Potential energy is stored energy. If work must be done on an object to overcome some attractive or repelling force, energy equivalent to the work put in is stored in the system. For example, if an object is to be lifted from the surface of the earth, work must be done against the attractive gravitational force between the earth and the object. The potential energy is increased by the amount of work done, and if the force holding the object at this distance from the earth is released, the object falls back to earth as the potential energy is converted into kinetic energy.

GRAVITATIONAL POTENTIAL ENERGY. Physicists have shown that the work required to separate two objects of masses m_1 and m_2 from an initial distance d_i to a final distance d_f is given by Equation (1.20), where G is the universal gravitational constant.

$$w = Gm_1m_2\left(\frac{1}{d_i} - \frac{1}{d_f}\right) \tag{1.20}$$

Equation (1.20) may be manipulated algebraically to express it in a more suggestive form as follows.

$$w = Gm_1m_2\left(\frac{1}{d_i} - \frac{1}{d_f}\right) = \frac{Gm_1m_2}{d_i} - \frac{Gm_1m_2}{d_f} = -\frac{Gm_1m_2}{d_f} - \left(-\frac{Gm_1m_2}{d_i}\right) \quad (1.21)$$

The final form of Equation (1.21) looks like typical delta notation in which an initial condition is subtracted from a final condition. Since the work of separating the two objects is stored as potential energy, the work must be the difference of the initial and final potential energies. Therefore the two right-hand terms in Equation (1.21) may be viewed as the final and initial potential energies so that a general expression for the potential energy E_p for an object with respect to another is given by Equation (1.22), where d is the distance of that object from the reference object.

$$E_p = -\frac{Gm_1m_2}{d} \quad (1.22)$$

Using the subscripts f and i for final and initial conditions, Equation (1.22) may be used to rewrite Equation (1.21).

$$\Delta E_p = E_{pf} - E_{pi} \quad (1.23)$$

It must be emphasized that Equation (1.22) is a definition by convention for the *gravitational potential energy*. It is important to realize that actual potential energies cannot be measured. Only differences in potential energies can be measured. In Equation (1.21) the work or potential energy change is represented as a difference in two quantities. The quantities represented there may or may not be the actual potential energies. However, since their difference must be the difference in potential energy, at least they must be proportional to the potential energy. Since the proportionality constant cannot be determined, it is a matter of convenience to assign it the value of one, and in this way define what is meant by the potential energy.

Scrutiny of Equation (1.22) reveals several things. First, all the potential energies due to gravitational attractions are negative. Second, if d is very small, that is, if the two objects are very close, the fraction will be large and the potential energy will be very low, that is, very negative. On the other hand, if d is quite large, so that the objects are well separated, the fraction has a small value and the potential energy is high, that is, it has a small negative value. If d is extremely large—assume it approaches infinity—then the fraction is extremely small, that is, it approaches zero as the distance approaches infinity. Thus by convention the gravitational potential energy is said to be zero when the objects are infinitely far apart. Then as the objects approach each other, the potential energy decreases; that is, it becomes more negative.

As an example, consider two 1.00-kg masses approaching each other. When they are 10.0 m apart, the potential energy may be calculated as follows:

$$\begin{aligned} E_p &= -Gm_1m_2\left(\frac{1}{d}\right) \\ &= \frac{-6.67 \times 10^{-11}\ \text{nt} \times \text{m}}{\text{kg}^2} \times 1.00\ \text{kg} \times 1.00\ \text{kg} \times \frac{1}{10.0\ \text{m}} \\ &= -0.67 \times 10^{-11}\ \text{joule} \end{aligned}$$

Similarly, potential energies may be computed at other intervals as the two 1.00-kg masses approach. Some values are recorded in Table 1.9. These have been

TABLE 1.9
Potential Energy for Two 1.00 kg Masses

DISTANCE (m)	POTENTIAL ENERGY (JOULE)
10.0	-0.67×10^{-11}
9.00	-0.74×10^{-11}
8.00	-0.83×10^{-11}
7.00	-0.95×10^{-11}
6.00	-1.11×10^{-11}
5.00	-1.33×10^{-11}
4.00	-1.66×10^{-11}
3.00	-2.22×10^{-11}
2.00	-3.33×10^{-11}
1.00	-6.67×10^{-11}

plotted in Fig. 1.15. From Fig. 1.15 it is clear that the potential energy decreases as the objects approach each other, that is, as the distance becomes smaller.

It should be clear then that to separate two objects requires work against the attractive gravitational force operating between them. This work is stored as potential energy, and if the separating force is removed, the two objects will move toward each other with a decrease in gravitational potential energy. Without the

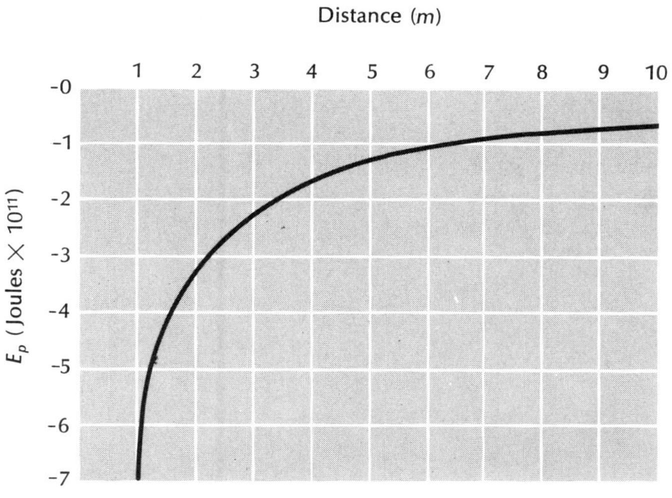

Fig. 1.15. Potential energy change with distance for two kilogram masses that are approaching

intervention of any other forces, *the spontaneous reaction between two objects occurs with a decrease in potential energy due to the attracting gravitational forces which tend to bring the two objects together.* Thus in Fig. 1.15 the spontaneous action is represented along the curve down and to the left.

ELECTROSTATIC POTENTIAL ENERGY. Changing the distance between two electrical charges also produces a change in potential energy. If two opposite charges are moved apart, work must be done against the attractive electrostatic force. If the separating force is removed, the charges move together again. The work done to separate the charges is stored as potential energy, which is converted to kinetic energy when the charges come together again.

Similarly, if two like charges are moved together, work must be done against the repelling electrostatic force. If the compressing force is removed, the charges separate. Again the work done in bringing these charges together is stored as potential energy, which is converted to kinetic energy when the charges separate again.

Notice once more that in both of these cases *the spontaneous reaction occurs with a decrease in potential energy.*

Physicists have shown that the work required to separate two charges q_1 and q_2 from an initial distance d_i to a final distance d_f is given by Equation (1.24).

$$w = -\frac{q_1 q_2}{4\pi\epsilon}\left(\frac{1}{d_i} - \frac{1}{d_f}\right) \tag{1.24}$$

This equation resembles Equation (1.20), and like Equation (1.20), it may be rearranged as follows:

$$w = -\frac{q_1 q_2}{4\pi\epsilon}\left(\frac{1}{d_i} - \frac{1}{d_f}\right) = -\frac{q_1 q_2}{4\pi\epsilon d_i} + \frac{q_1 q_2}{4\pi\epsilon d_f} = \frac{q_1 q_2}{4\pi\epsilon d_f} - \frac{q_1 q_2}{4\pi\epsilon d_i} \tag{1.25}$$

The final form of Equation (1.25) looks like typical delta notation in which an initial condition is subtracted from a final condition. Since the work of separating the two charges is stored as potential energy, the work must be the difference of the initial and final potential energies. Therefore, the two right-hand terms in Equation (1.25) may be viewed as the final and initial electrostatic potential energies; a general expression for the electrostatic potential energy E_p for one charge with respect to another, is given by Equation (1.26), where d is the distance between the charges.

$$E_p = \frac{q_1 q_2}{4\pi\epsilon d} \tag{1.26}$$

Using the subscripts f and i for final and initial conditions, Equation (1.26) may be used to rewrite Equation (1.25).

$$\Delta E_p = E_{pf} - E_{pi} \tag{1.27}$$

Of course Equation (1.27) is identical to Equation (1.23). In one case the derivation was through gravitational potential energy and in the other through electrostatic potential energy. Also, it should be noted that Equation (1.26) is very similar to Equation (1.22). Equation (1.26) is the defining equation for the convention regarding electrostatic potential energies. As seen in this equation, when the two charges are of different signs, the potential energy is negative. This situation is exactly comparable to that of gravitational potential energy. As the size of d becomes small, that is, as the charges come closer, the potential energy decreases, that is, it becomes more negative. As the distance increases, the potential energy becomes less negative, that is, it increases. The limiting case in which d is infinitely large leads to zero potential energy.

When the two charges have the same sign, potential energies are positive. When d is small, the potential energy is large, that is, highly positive. When d is large, the potential energy is small. In the limiting case when d is infinitely large, the potential energy is zero.

The convention adopted for gravitational potential energy is retained for electrostatic potential energy. Thus when charges are infinitely far apart, regardless of charge, the potential energy is zero.

Consider a typical problem. Suppose that two negative charges, each of one quantum, are approaching each other in a vacuum. The potential energy when they are 10.0 m apart may be computed as follows:

$$
\begin{aligned}
E_p &= \frac{q_1 q_2}{4\pi\epsilon}\left(\frac{1}{d}\right) \\
&= \frac{(-1.60 \times 10^{-19}\ \text{coulomb})(-1.60 \times 10^{-19}\ \text{coulomb}) \times \text{nt} \times \text{m}^2}{4 \times 3.14 \times (8.85 \times 10^{-12}\ \text{coulomb}^2)} \times \frac{1}{10.0\ \text{m}} \\
&= 2.31 \times 10^{-29}\ \text{joule}
\end{aligned}
$$

The results of similar computations are summarized in Table 1.10 for the approach of charges of the same sign as well as charges of the opposite sign. The

TABLE 1.10 Electrostatic Potential Energies for Quantum Charges

DISTANCE (m)	UNLIKE CHARGES (JOULE)	LIKE CHARGES (JOULE)
10.0	-2.31×10^{-29}	$+2.31 \times 10^{-29}$
9.00	-2.56×10^{-29}	$+2.56 \times 10^{-29}$
8.00	-2.89×10^{-29}	$+2.89 \times 10^{-29}$
7.00	-3.29×10^{-29}	$+3.29 \times 10^{-29}$
6.00	-3.84×10^{-29}	$+3.84 \times 10^{-29}$
5.00	-4.61×10^{-29}	$+4.61 \times 10^{-29}$
4.00	-5.77×10^{-29}	$+5.77 \times 10^{-29}$
3.00	-7.69×10^{-29}	$+7.69 \times 10^{-29}$
2.00	-11.2×10^{-29}	$+11.2 \times 10^{-29}$
1.00	-23.1×10^{-29}	$+23.1 \times 10^{-29}$

data from Table 1.10 are plotted on a single graph in Fig. 1.16. The upper curve represents the change in potential energy of charges of the same sign with distance, and the lower curve represents the change in potential energy of charges of opposite signs with distance.

For charges with opposite sign, *the spontaneous reaction occurs with a decrease*

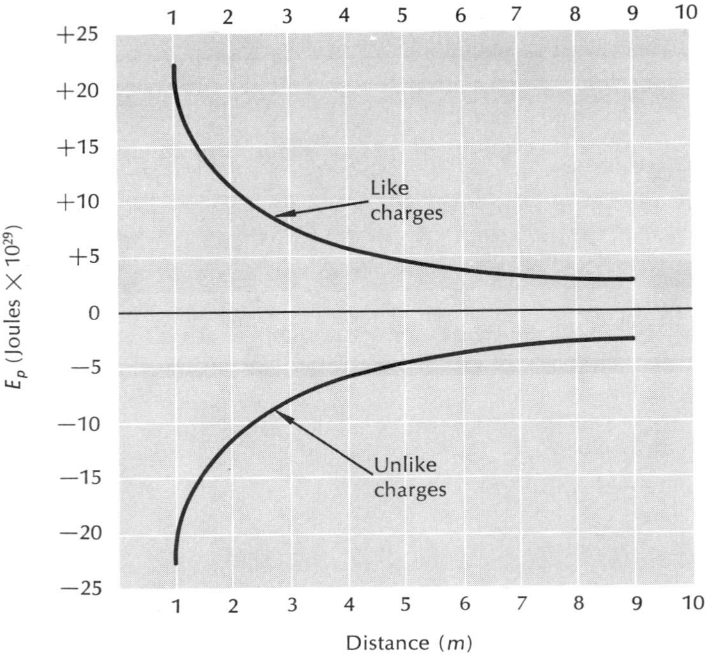

Fig. 1.16. Electrostatic potential energy against distance for quantum charges

in potential energy. This is due to the force of attraction between them. Thus the spontaneous reaction is the movement toward each other which occurs with a decrease in potential energy. In Fig. 1.16, this corresponds to a movement down and to the left along the lower curve.

For charges with the same sign, *the spontaneous reaction occurs with a decrease in potential energy*. This is due to the force of repulsion between them. Thus the spontaneous reaction is the movement away from each other which occurs with a decrease in potential energy. In Fig. 1.16, this corresponds to a movement down and to the right along the upper curve.

Often potential-energy curves such as those shown in Fig. 1.16 are considered to represent potential-energy hills. The spontaneous trend for a system represented by such a diagram is to roll down hill. For example, a rock at the top of a hill also is at the top of a gravitational potential energy hill, and it will roll down the hill to a position of lower potential energy. Similarly, a system composed of two charges of like sign close together is at the top of a potential-energy hill, and the spontaneous trend is for the system to roll down this hill to a position of lower potential energy. This it does by allowing the charges to separate. Also, two unlike charges separated from each other are at the top of a potential-energy hill, and the spontaneous action is for this system to roll down the potential-energy hill by allowing the charges to come together.

It is important to the development of chemical ideas that you have a firm grasp of the idea that *spontaneous change tends to occur in the direction of a decrease in potential energy*.

Heat

When a ball is thrown outward from the earth it leaves the surface of the earth at some initial kinetic energy, and as it moves away from the earth, it does work against the gravitational force of the earth; in so doing it converts kinetic energy to potential energy. When the ball reaches its apex, that is, when it has reached zero speed, it has no kinetic energy but it does have maximum potential energy. Then as it falls back to the earth, its potential energy decreases and its kinetic energy increases. If frictional losses to the atmosphere are ignored, when the ball returns to the earth it has the same kinetic energy it had when it left, and this kinetic energy equals the potential energy it had at the apex of its flight.

This is a simple illustration of the law of *conservation of energy*. Energy is neither created nor destroyed, but it may be converted from one form to another. In the example just cited, the conversions were from kinetic to potential and back to kinetic again.

However, in the instant in which the ball returns to the earth it hits the solid earth and comes to a complete halt. An instant before the ball had much kinetic energy and in another instant it has none, since its speed is zero. The potential energy is the same as it was before it left the earth and very nearly the same as it was just an instant before it came to a halt. The question raised is this: what has happened to the kinetic energy? Has it been destroyed?

Careful studies made by dropping objects through a distance to the earth show that the matter of the objects at the point of impact increases in temperature.

In the previous example with the ball, frictional losses to the atmosphere were ignored, but the fact is that if the ball is thrown out from the earth in the atmosphere, when it returns to earth it has less kinetic energy than it had when it left. Careful experiments again have shown that the air in the path of the ball has increased in temperature.

What is it that causes these temperature changes? By definition, the answer is heat, which is defined as that which when added to matter increases its temperature or when removed from matter decreases its temperature, provided no change of state occurs. (This latter condition is added to make the definition correct. Its significance will be understood later.)

In accordance with the law of conservation of energy, heat is considered to be a form of energy. The apparent losses in kinetic energy in the previous examples can be attributed to the conversion of kinetic energy to heat. Thus when the ball makes an impact on the earth, the kinetic energy is converted to heat. Or when the ball travels through the air, some of its kinetic energy is converted to heat, so the ball returns to the earth with less kinetic energy than it had when it left.

Although it is not possible at this point to define heat any more clearly than this, some of its properties can be indicated. For example, if a cold object is placed in contact with a hot object, the cold object becomes warmer and the hot object becomes cooler, until the two reach the same temperature. *It is a fundamental property of heat that it flows from a hotter to a cooler object spontaneously.*

The fact that heat flow continues until the temperatures of the two objects are the same means that temperature is a measure of the tendency of heat to flow, or stated differently, *the difference in temperature between two objects is a measure of the tendency of heat to flow between them.*

Hence temperature is a measure of heat intensity. However, different kinds of matter have different capacities for holding or storing heat. Perhaps a crude analogy to water storage tanks will be useful in making the distinction between heat and temperature. Consider two water storage tanks, *A* and *B*, connected from beneath by a pipe as shown in Fig. 1.17. Tank *A* is short and fat, whereas tank *B* is tall and thin. Experience shows that with the pipe open, the two levels of water will equalize, despite the fact that tank *A* contains more water than tank *B*. It is the pressure of the water in the two tanks which determines the flow of water

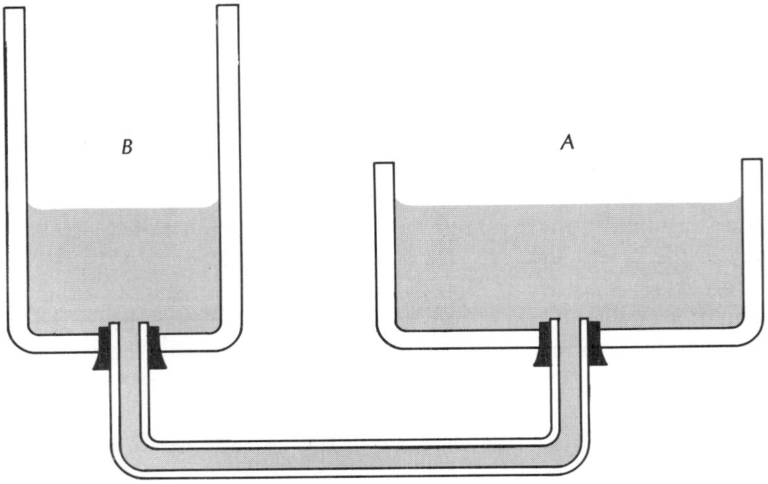

Fig. 1.17. Equalization of levels in two connected water tanks

and not the amount of water. Equalization of the water levels is equivalent to equalization of water pressure.

In these terms, temperature is a measure of heat pressure. Some types of matter are short, fat reservoirs, so to speak, requiring the addition of much heat before the temperature increases very much. In this respect they are like tank *A*, to which much water must be added if the height of the water level is to be increased by very much. Other kinds of matter are tall, thin reservoirs, requiring the addition of very little heat to increase the temperature significantly. In this respect they are like tank *B*, to which very little water must be added to increase the water pressure.

To measure quantities of heat it is convenient to choose a specific quantity of a specific type of matter and use its capacity to store heat as the basic unit for the amount of heat. The standard chosen is one gram of water and the fundamental unit of heat is the *calorie*, which is *the amount of heat required to raise the temperature of one gram of water from 14.5°C to 15.5°C.* The very specific temperature range is necessary since the capacity of water to store heat varies slightly with temperature changes.

With this definition, it is possible to compare the heat storage capacities of various kinds of matter. Often this is given as the *specific heat*, which is defined as *the number of kilocalories of heat required to raise the temperature of one kilogram of matter by one degree Celsius.*

Some specific heats are listed in Table 1.11. An approximate temperature needs to be specified since specific heats vary somewhat with temperature. As an example, the specific heat of water is given at various temperatures in Table 1.12.

TABLE 1.11

Specific Heats in kcal/(kg x °C) Near Room Temperature

Air	0.24	Gold	0.03
Aluminum	0.21	Granite	0.19
Asbestos	0.20	Paraffin	0.69
Brass	0.09	Quartz	0.19
Carbon dioxide	0.18	Silver	0.06
Chalk	0.21	Sugar	0.27
Chloroform	0.15	Water	1.00
Copper	0.09	Wood	0.42

As indicated earlier, when objects are dropped to earth their kinetic energies are converted to heat upon impact. If the conditions are controlled so that the kinetic energy is known at impact, and if the specific heats of all materials involved in the impact are known, the temperature increase can be used as a measure of the heat energy produced, which must equal the kinetic energy at impact. In this way it is possible to determine the relationship between the calorie and the other energy units. Thus it has been found that 1.000 calorie is 4.186 joules.

TABLE 1.12

Specific Heats of Liquid Water

°C	kcal/(kg × °C)	°C	kcal/(kg × °C)
0	1.0074	50	0.9985
5	1.0037	55	0.9990
10	1.0013	60	0.9994
15	1.0000	65	1.0000
20	0.9988	70	1.0007
25	0.9983	75	1.0014
30	0.9980	80	1.0023
35	0.9980	85	1.0033
40	0.9980	90	1.0044
45	0.9983	95	1.0056
		100	1.0070

Water has the highest specific heat listed in Table 1.11, and a characteristic of water is its high specific heat. There are few materials which have a higher specific heat than water.

Electromagnetic Radiation

In the previous discussion of magnetism it was pointed out that a current passing through a wire coil generates a magnetic field. It is equally true that a current may be generated in another coil if it is passed through this magnetic field. A pulsating magnetic field can be imagined if the current in the wire coil is turned on and off or reversed so that magnetic lines of force would tend to spread out from the coil and then recede again. As these fluctuating lines of force cut across another wire coil, a fluctuating electric current is set up in that coil. The net result is that interrelated pulsating magnetic and electric fields are produced.

Electronically, continually varying electric fields may be produced. The plot of field strength E against time is shown for a typical varying field in Fig. 1.18. The positive and negative values for E correspond to opposite directions which the field may have. This type of field strength variation is called *sinusoidal oscillation*, since the curve in Fig. 1.18 resembles a plot of the mathematical sine function. Figure 1.18 is a picture of the way the electric field varies at a given point in space as time passes.

However, the electric field does not exist only at a single point in space. The electric field is all of the space around the oscillator that produces it. Therefore, a picture may be drawn to represent the variation of the electric field with distance from the oscillator at a single instant of time. This has been done in Fig. 1.19. Again a sinusoidal curve results.

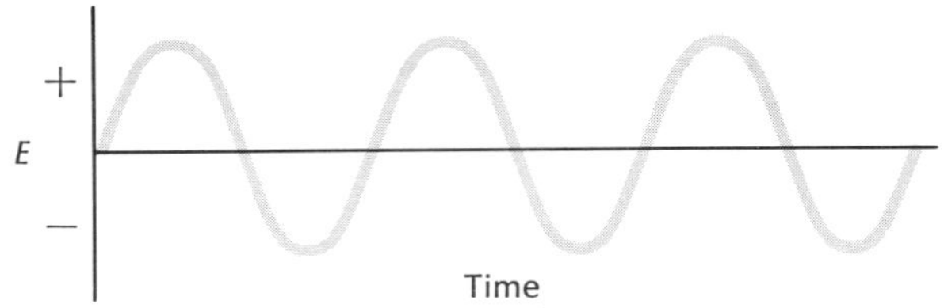

Fig. 1.18. Variation of electric field strength with time from a sinusoidal oscillator

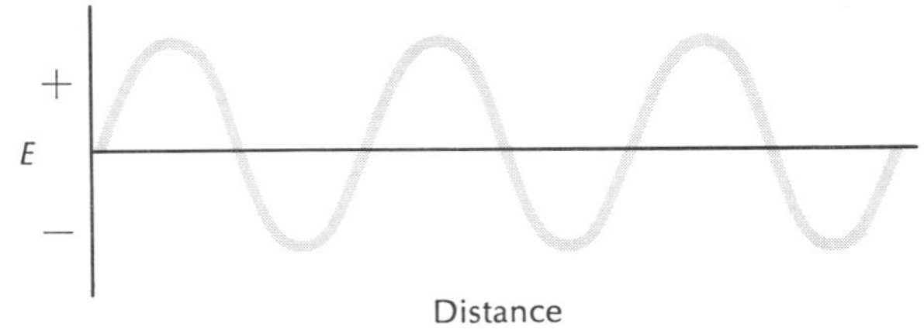

Fig. 1.19. Variation of electric field strength with distance at an instant of time

If an observer were to look at the variation of electric field strength with distance at two successive instants of time, he would see the variation shown in Fig. 1.20. Here the solid curve shows the variation in field strength with distance from the oscillator at the first instant of time, and the dotted curve shows the same re-

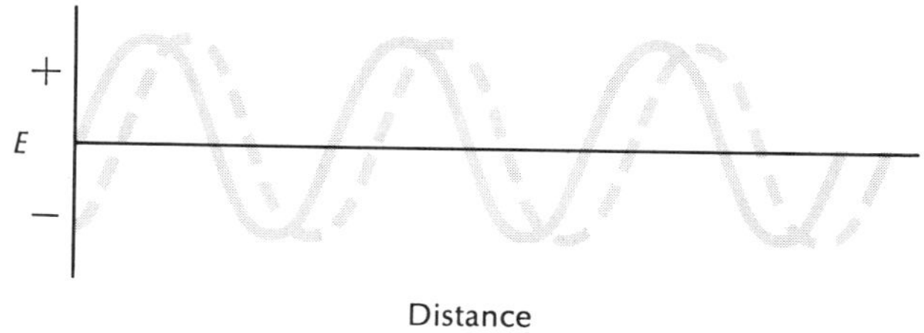

Fig. 1.20. Variation of electric field strength with distance at two instants of time

lationship at the second instant of time. It would appear to the observer that as time passes, the maximum positive field strengths are moving to the right.

Similarly, if the observer focused his attention on a single point in space and watched the electric field go by, so to speak, he would observe a gradual decrease in positive field strength, followed by a reversal of field direction and an increase in negative field strength followed by a decrease in negative field strength, followed by a reversal and an increase in positive field strength to a maximum, followed again by a gradual decrease in positive field strength, and so on. Thus the observer would note a repeating cycle of increasing and decreasing field strengths and reversal of field directions.

From the point of view of Fig. 1.20, the electric field appears to move to the right, and its motion is described as wave motion. The simple wave is reproduced in Fig. 1.21 where several features are named. The distance between any two identical points on adjacent repeating cycles is called the *wavelength* and is denoted by λ. The maximum displacement of the field strength from zero in either direction

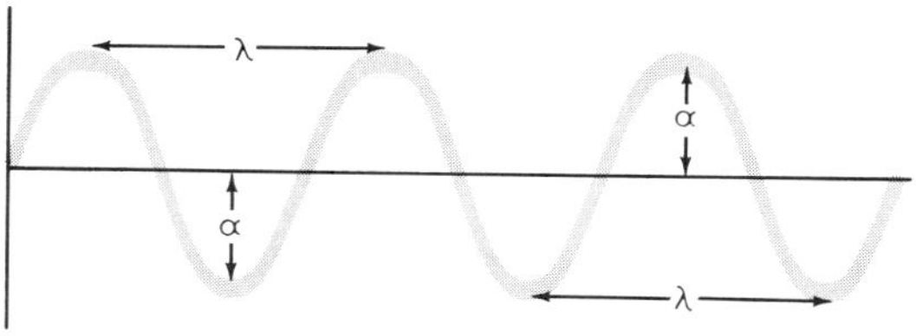

Fig. 1.21. Wavelength and amplitude for a sine wave

is called the *amplitude*, α. The simple repeating unit in the wave consisting of one crest and one trough is called a *cycle*. The number of cycles passing a given point per unit of time is called the *frequency*, ν. Usually the frequency is measured in reciprocal seconds, sec^{-1}, which means cycles per second. The wavelength and the frequency are related to each other through the speed with which the wave appears to be moving. The speed c must be the product of the wavelength λ, which is the distance per cycle, and the frequency ν, which is the number of cycles per second. This is shown in Equation (1.28).

$$c = \lambda\nu \tag{1.28}$$

As the strength and direction of the oscillating electric field vary, the accompanying magnetic field varies also. Thus the magnetic field strength variation may be represented by the graph in Fig. 1.22 where the magnetic field strength is represented by H.

Since the lines of force of a magnetic field are perpendicular to the lines of force of the electric field producing them, the combination of the electric and mag-

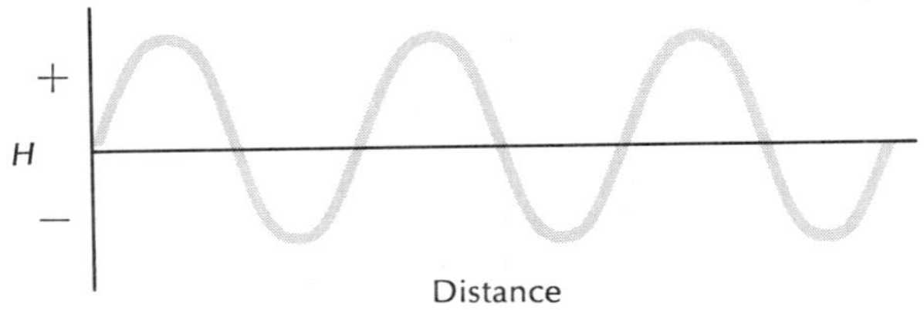

Fig. 1.22. Variation of magnetic field strength with distance at an instant of time

netic fields may be represented graphically by plotting the field strength variations in mutually perpendicular planes. This is done in Fig. 1.23, where the electric field strength variation is plotted in the vertical plane with the magnetic field strength variation in the horizontal plane. As shown in Fig. 1.23, the electric and magnetic fields vary together so that they both change direction at the same time. Thus they both have the same wavelength and they travel at the same speed so they are always in phase.

This kind of combination of pulsating electric and magnetic fields which travels through space without the aid of any material medium is called *electromagnetic radiation.*

In the early part of the nineteenth century, James Clerk Maxwell predicted the existence of this type of radiation, but it was not until after his death that such radiation was produced and detected in a laboratory experiment using an oscillating electric current. Nevertheless, in working out the details of his predictions, Maxwell was able to show that the velocity of such radiation should be constant in a vacuum, regardless of the wavelength (or frequency) and that this velocity should be 3.00×10^8 m/sec. Having made the calculation, Maxwell recognized that this value is the value of the speed of light.

As early as 1675, the astronomer Römmer in France had deduced the speed of light from astronomical measurements. His value was 2×10^8 m/sec. Similarly, Bradley in England had determined the speed of light from astronomical measurements in 1729. His value was 3.04×10^8 m/sec. During Maxwell's lifetime, the French physicists Fizeau and Foucault carried out laboratory measurements of the speed of light, and their values were 3.13×10^8 m/sec and 2.98×10^8 m/sec respectively. In 1876, just three years before Maxwell's death, the Frenchman Cornu also determined the speed of light; his value was 2.9999×10^8 m/sec.

Recognizing the similarity between his calculated value for the speed of electromagnetic radiation in a vacuum and the experimentally determined values for the speed of light, Maxwell proposed that light itself is electromagnetic radiation. This suggestion was the basis for a new view of the nature of light that had not been possible before. Subsequent experimentation showed that many of the phenomena associated with light could be explained in these terms.

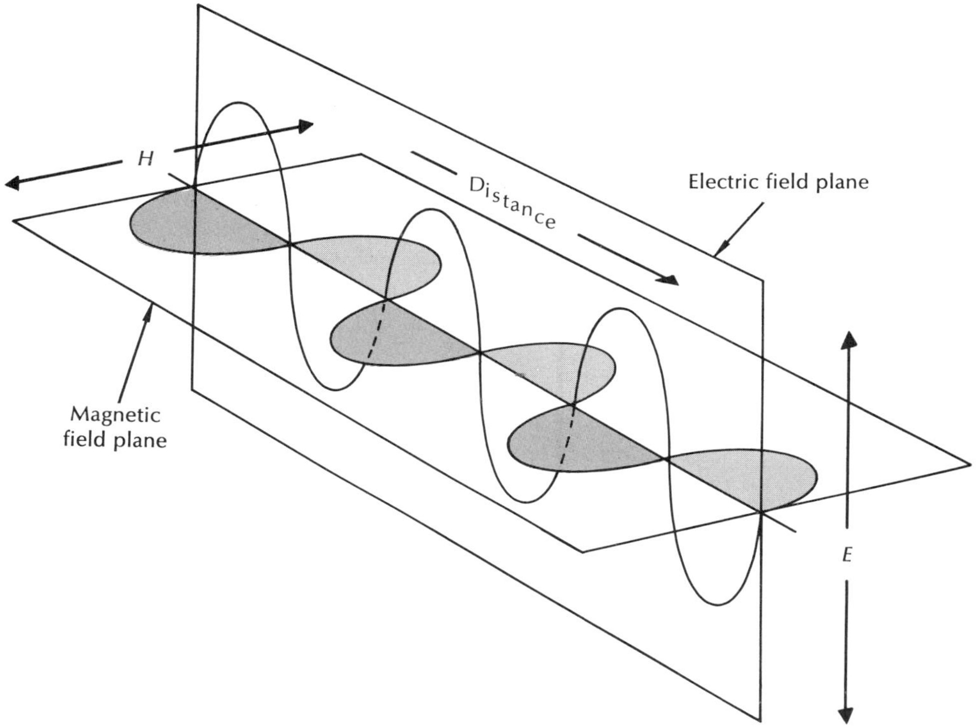

Fig. 1.23. Combined magnetic and electric field strength variation with distance at an instant of time

In 1890, Hertz was able to produce electromagnetic radiation in the laboratory to verify Maxwell's prediction. However, his electromagnetic radiation was not visible. Essentially, the difference between the waves which Hertz produced and visible light was the wavelength.

Since the time of Hertz, it has been shown that electromagnetic radiation exists in a wide variety of wavelengths, only a few of which are detectable by the human eye. The whole range of these radiations is called the *electromagnetic spectrum*, outlined in Table 1.13.

In the table, the values for wavelengths are indicated on several scales representing the various commonly used units of length. Because the wavelengths vary so much from the maximum to the minimum, various length units are used in different regions of the spectrum. The meter, centimeter, and millimeter already are familiar. In addition the following are used also: the micron, μ, which is 10^{-6} m; the millimicron, mμ, which is 10^{-3} μ, and thus 10^{-9} m; the angstrom, Å, which is 10^{-10} m.

Within the small visible region of the electromagnetic spectrum, the various colors of light, as we know them, are determined by the wavelength. The red colors are at the long wavelengths and the blue colors are at the short wavelengths.

TABLE 1.13

The Electromagnetic Spectrum

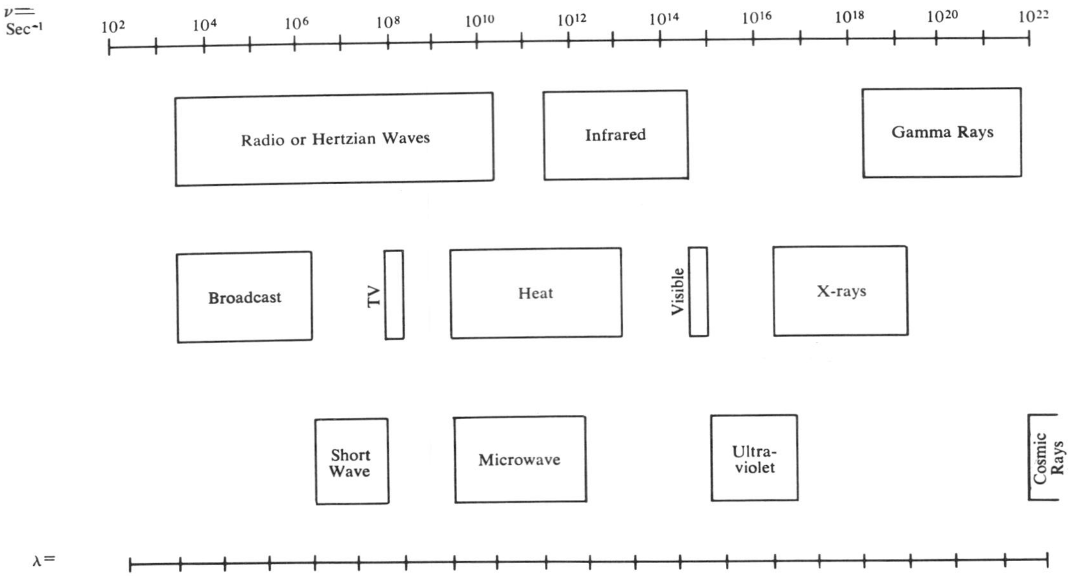

m	10^6	10^4	10^2	10^0	10^{-2}	10^{-4}	10^{-6}	10^{-8}	10^{-10}	10^{-12}	10^{-14}
cm	10^8	10^6	10^4	10^2	10^0	10^{-2}	10^{-4}	10^{-6}	10^{-8}	10^{-10}	10^{-12}
mm	10^9	10^7	10^5	10^3	10^1	10^{-1}	10^{-3}	10^{-5}	10^{-7}	10^{-9}	10^{-11}
μ	10^{12}	10^{10}	10^8	10^6	10^4	10^2	10^0	10^{-2}	10^{-4}	10^{-6}	10^{-8}
mu	10^{15}	10^{13}	10^{11}	10^9	10^7	10^5	10^3	10^1	10^{-1}	10^{-3}	10^{-5}
A	10^{16}	10^{14}	10^{12}	10^{10}	10^8	10^6	10^4	10^2	10^0	10^{-2}	10^{-4}

Adjacent to the visible region on the short wavelength side is the ultraviolet region, and adjacent to the visible region on the long wavelength side is the infrared region. Some approximate ranges for various colors in the visible spectrum are given in Table 1.14.

The breakdown of the entire electromagnetic spectrum in Table 1.13 is very approximate. Near the boundaries between regions often there is no clear break. Indeed the boundaries may be set at different places as a matter of convenience, and some of the overlapping appears in Table 1.13. Sometimes the nature of the detectors used for the radiation define the boundaries due to their mechanical limitations. Sometimes the boundaries are set as a function of the source of the radiation.

It is interesting to note that heat is included in the electromagnetic spectrum. It includes much of the infrared region and most of the microwave region of the

TABLE 1.14
Visible Spectrum Wavelengths in Angstroms

Red	6200–7500
Orange	5900–6200
Yellow	5700–5900
Green	5000–5700
Blue	4500–5000
Violet	4000–4500

spectrum. This accounts for the ability of heat to be transferred without recourse to a material medium. Heat radiation from the sun is included in this category.

SUMMARY

Some properties of matter have been discussed. Mass is the measure of the quantity of matter and a fundamental determinant of gravitational forces. In addition, mass is the property which determines the ability of a moving object to do work, that is, the kinetic energy. In turn, the kinetic energy is a property of moving objects that is converted easily into potential energy. The potential energy is energy of position, which results from forces which act through space and thus change as position changes. Because of these forces, which are related to gravitational and electrostatic forces, natural spontaneous processes are accompanied by a decrease in potential energy.

Electric charge is a property that is easily induced in matter by friction and is the source of electrostatic forces. Electric charge in motion produces new forces called magnetic forces. Both electrostatic and magnetic forces interact with matter to be either enhanced or reduced in the process. Oscillating electrical forces produce concurrent oscillating magnetic forces, which together are called electromagnetic radiation. Electromagnetic radiation is a form of energy and includes heat, which is easily produced from kinetic energy. Throughout all of the natural processes discussed, the law of conservation of energy pervades; that is, energy can neither be created nor destroyed, but it may be converted from one form to another.

QUESTIONS AND PROBLEMS

1.1 Is it true that an object undergoing zero acceleration has zero mass? Explain.
1.2 Define the mks unit of weight.
1.3 In terms of force, length, and time, what are the dimensions of mass?
1.4 An object has a mass of 14.7 g. Express its weight on the surface of the earth in both dynes and pounds.
1.5 Derive the conversion factor necessary to convert weight expressed in pounds to weight expressed in dynes.

1.6 An object acted upon by a force of 220 dynes undergoes an acceleration of 15.0 in./sec^2. What is the mass of the object?

1.7 On the surface of the earth, what is the mass in slugs of an object weighing 175 lb?

1.8 What interpretation can be given to an acceleration expressed as 10(mi/hr)/sec? What is this in mi/hr^2?

1.9 Carry out the following conversions for conditions on the surface of the earth:
a) 200 lb to kilograms
b) 347 g to ounces
c) 4.5×10^7 mg to tons

1.10 The average radius of the earth is 3.96×10^3 mi.
a) What is the approximate mass of the earth?
b) Assuming the earth to be a sphere, what is its average density?

1.11 The mass of the moon is about 7.36×10^{22} kg, and its average radius is about 1.08×10^3 mi.
a) What is the average value for the acceleration of gravity on the surface of the moon?
b) What is the weight in pounds for a 2.20 kg mass on the moon?
c) What is the average density of the moon?

1.12 Distinguish between g in Equation (1.3) and $\mathbf{G}$ in Equation (1.2).

1.13 What justification is there for using a constant acceleration for gravity in solving earth-bound problems?

1.14 Two 1.00-quantum negative charges are 1.00 yd apart in a vacuum.
a) What is the force of repulsion between these two charges?
b) What masses must be associated with these charges so that the gravitational attraction equals the electrostatic repulsion? (Assume equal masses with the two charges.)
c) What is the force of repulsion between these two charges if they are immersed in water, which has a dielectric constant of 78.5?

1.15 Two identical charges are separated by a fixed distance.
a) What is the direction of the force between them?
b) What happens to the force if
1) one of the charges is doubled?
2) both of the charges are doubled?
3) the distance between the charges is cut in half?
4) both charges and the distance are doubled?
5) the old medium is replaced by a new medium of higher dielectric constant?

1.16 Water and carbon tetrachloride have dielectric constants of 78.5 and 2.2 respectively. What differences does this imply between these liquids?

1.17 The force of attraction between two oppositely charged objects is 5.91×10^{-25} nt when they are immersed in a medium with a dielectric constant of 35.0.
a) What is this force in dynes?
b) What is the force of attraction if the medium is replaced by a vacuum?
c) What is the force of attraction if the medium is replaced by one with a dielectric constant of 55.0?

1.18 What dielectric constant is required of a medium if the force of repulsion between two charged objects is to be one-half of what it is in a vacuum?

1.19 How would the force fields diagrammed in Figs. 1.4 and 1.5 differ if they were drawn to represent what happens to a negative test charge placed in the field?

1.20 Is the following statement true or false, and why? It is not possible to have an object with an electric charge of -4.00×10^{-19} coulombs.

1.21 How many quanta of charge pass a given point per minute in a 5.00-amp current?

1.22 If a 12.0-amp current passes through a wire for 5.00 hr, how many coulombs pass through the wire? How many quanta?

1.23 Distinguish among the following terms: a) diamagnetism; b) paramagnetism; c) ferromagnetism.

1.24 For each of the materials listed in Table 1.7, calculate the magnetic permeability.

1.25 Which of the substances in Table 1.7 are paramagnetic?

1.26 Under some conditions, some substances have magnetic susceptibilities of about $+7 \times 10^{-3}$ in the same mks units used in Table 1.7. What kinds of substances are these?

1.27 From the fundamental unit definitions, show that 1.00 joule = 10^7 ergs.

1.28 A 1.50-ton automobile is traveling at 60 mi/hr.

a) Express its kinetic energy in ergs, joules, foot pounds, calories, and electron-volts. (See Appendixes for values for any unfamiliar units.)
b) How fast must a 150-lb man run in order to have the same kinetic energy as the automobile?
c) At this speed, how long would it take the man to run a mile?

1.29 A tiny object of mass 9.11×10^{-28} g is traveling at 3.00×10^7 cm/sec.

a) What is the speed of this object in mi/hr?
b) Express the kinetic energy of this object in ergs, joules, calories, and electron volts.
c) How fast must a 150-lb man run in order to have the same kinetic energy as the object?
d) How fast must a 1.50-ton automobile travel in order to have the same kinetic energy?
e) How many of these tiny objects traveling at this speed are needed in order to have a combined kinetic energy equivalent to the 1.50-ton automobile traveling 60 mi/hr, as found in Problem 1.28?

1.30 Starting at rest, how much work (in ergs, joules, calories, and electron volts) is required to increase the speed of

a) a 1.50-ton automobile by 25 mi/hr?
b) a 1.50-ton automobile by 2.00×10^6 cm/sec?
c) a 9.11×10^{-28} g object by 25 mi/hr?
d) a 9.11×10^{-28} g object by 2.00×10^6 cm/sec?

1.31 Pressure is defined as force per unit area where area is the square of an appropriate length unit, for example, square centimeters, square feet, etc. Volume is defined as the cube of an appropriate length unit, for example, cubic centimeters, cubic feet, etc. Show what units result from the multiplication of pressure by volume.

1.32 Imagine a crane working at a construction site. The crane lifts a 3.0-ton concrete beam 50.0 feet off the ground, then swings it horizontally 30.0 feet so that it is over its final position, and then drops it 20 feet to its final resting place in the new building.

a) What is the change of gravitational potential energy in each step of this process?
b) What is the total change in potential energy from the time the beam leaves the ground until it is in position in the building?

1.33 Two 3.0-ton spheres are suspended so that they are 50 ft above the ground and 10 ft apart, center to center.
a) What is the gravitational potential energy of these spheres with respect to each other?
b) How much work must be done to separate them horizontally until they are 75.0 ft apart?

1.34 Verify the calculated gravitational potential energies in Table 1.9.

1.35 What is the electrostatic potential energy for two 1.00 coulomb charges of opposite sign which are 325 mm apart in a vacuum?

1.36 How much work must be done to bring two −1.00-coulomb charges from a separation of 2.50 m to 0.75 m?

1.37 Verify the calculated electrostatic potential energies in Table 1.10.

1.38 Fuel values for foods are stated in terms of the amount of energy which can be gotten from a specified quantity of a particular food. A commonly stated daily diet requirement is about 3000 kcal. If this amount of heat is added to each of the following at 25°C, what will be the approximate final temperature of each? (See Table 1.11.)
a) 100.0 kg water
b) 1000 kg gold
c) 100.0 kg paraffin

1.39 What is the specific heat of chloroform in joule/(oz × °F)?

1.40 A 50.0-g piece of metal at 50.0°C was placed in 50.0 g of water at 24.0°C. The temperature of the water rose to 33.0°C. Ignoring the container of the water, what is the approximate specific heat of the metal?

1.41 An insulated cup has a mass of 125.6 g and contains 50.0 g of water, and both are at 21.4°C. Then 76.4 g of water at 84.8°C is poured into the cup. When the water is well mixed, the temperature is 47.3°C. What is the specific heat of the cup?

1.42 What role does the law of conservation of energy play in the solution of Problems 1.40 and 1.41?

1.43 A 25.0-lb metal ball is dropped from a height of 250 m into a pool containing 25.0 m^3 of water at the same temperature as the ball.
a) What is the kinetic energy of the ball when it strikes the water?
b) If this kinetic energy is transformed to heat, how many kcal is this?
c) If all of the heat were absorbed by the water by how many degrees would the temperature of the water increase?

1.44 Compute frequencies in sec^{-1} for radiation of the following wavelengths and suggest the regions of Table 1.13 into which each will fall.
a) 2 mi; b) 2 yd; c) 2 ft; d) 2 in.; e) 2 mm; f) 2 Å; g) 2μ; h) 2 mμ.

1.45 Evaluate the validity of each of the following statements:
a) The speed of an object is inversely proportional to its kinetic energy.
b) The gravitational force is inversely proportional to the square of the distance between two masses.

c) The dielectric constant is the proportionality constant between the electrostatic force and the product of the charges.
d) Frequency is directly proportional to the reciprocal of the wavelength.
e) The earth's magnetic pole near the geographical north pole really is a south pole.

2

Fundamental and Composite Particles

INTRODUCTION That matter is electrical in nature has been demonstrated in many ways. Most of the work leading to this conclusion was carried out during the nineteenth century and the early part of the twentieth century. Among the early observations were those of Michael Faraday in the early part of the nineteenth century. Faraday not only observed that matter may be decomposed by an electric current, but also formulated the quantitative relationship between the amount of charge passed through a system and the amount of matter decomposed.

This work eventually resulted in the particle view of matter that is still being elaborated today. Although in many respects the atom is a fundamental particle of matter for the chemist, the atom is best understood as a composite particle of more elementary particles: the electron, the proton, and the neutron. This chapter consists of a brief review of the historical developments that have led to our current understanding of the elementary particles of matter. Included is a discussion of the way this understanding permits a description of the gross atomic properties: atomic number and atomic weight.

VACUUM TUBE EXPERIMENTS

As early as 1705 it was known that a glow of light is produced when an electrically charged object is placed in a container from which some of the air has been removed, that is, in a partial vacuum. The phenomenon occurs only at low air pressures. Faraday studied this observation, but his work was hindered by a lack of good vacuum pumps.

Cathode Rays

By the middle of the nineteenth century, vacuum-pump and glass-blowing techniques permitted the construction of glass tubes which could be evacuated. A typical tube is shown in Fig. 2.1. The two wires and plates on which the electrical charges are placed are called *electrodes*. The positively charged electrode is the *anode* and the negatively charged electrode is the *cathode*. If the electrodes are charged oppositely and the tube slowly evacuated, a glow develops around the cathode. As the evacuation continues, a discharge occurs between the two electrodes and the tube begins to glow. If an object is placed between the electrodes it casts a shadow, suggesting that there is radiation emanating from the cathode toward the anode.

William Crookes studied this radiation and concluded that the radiation, often called *cathode rays*, consists of negatively charged particles emanating from the

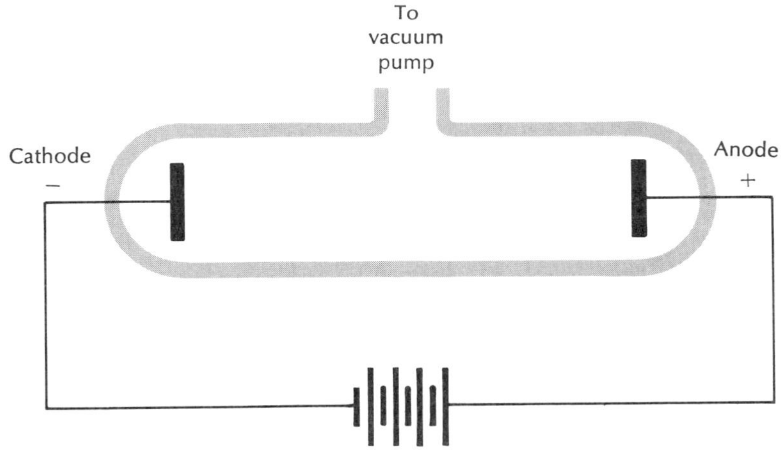

Fig. 2.1. Cathode-ray tube

cathode at very high speed. J. J. Thomson then showed that this beam can be deflected by an electrostatic field as well as a magnetic field. Although these rays seem to travel in a straight line, their paths are bent in an electrostatic or a magnetic field.

Figure 2.2 is a schematic representation of a particle moving through a magnetic field. The force acting on the particle is proportional to the charge q on the particle, and the speed s of the particle. This is shown in Equation (2.1) where F_m is the force acting on the particle, and H is the proportionality constant called the *magnetic field strength*.

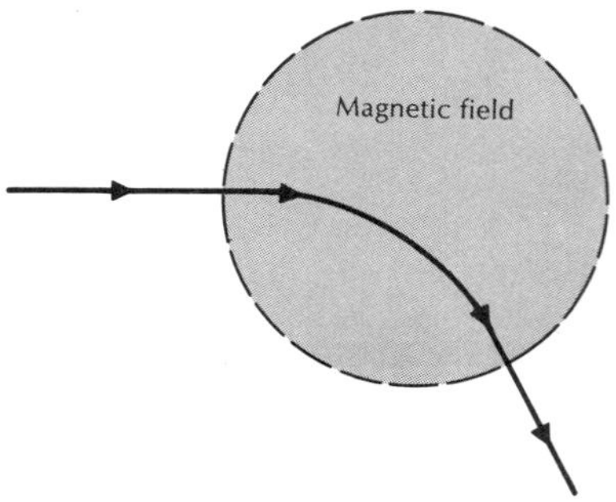

Fig. 2.2. Path of a moving charged particle in a magnetic field

$$F_m = Hqs \tag{2.1}$$

It has been shown that the path of a charged particle in a magnetic field is circular. Thus the force exerted on the particle by the magnetic field must be balanced by the centrifugal force, F_c, which is proportional to the square of the speed s of the particle and inversely proportional to the radius r of the circular path, with the mass m of the particle as the proportionality constant. This is shown in Equation (2.2).

$$F_c = \frac{ms^2}{r} \tag{2.2}$$

If the magnetic force is balanced by the centrifugal force, then Equations (2.1) and (2.2) may be combined into Equation (2.3), and then rearranged into Equation (2.4).

$$Hqs = \frac{ms^2}{r} \tag{2.3}$$

$$\frac{q}{m} = \frac{s}{Hr} \tag{2.4}$$

Experimentally the value of r is found by tracing the curvature of the path of the particle in the magnetic field. However, the values for q, m, and s remain unknown. Thomson eliminated one of these unknowns by devising a technique for measuring s. He superimposed an electrostatic field on the magnetic field so that the path of the charged particle remained a straight line. It was known that the force, F_e on a charged particle in an electrostatic field is proportional to the charge q on the particle, with the proportionality constant E, called the *electrostatic field strength*, as follows:

$$F_e = Eq \tag{2.5}$$

Then if the electrostatic force balances the magnetic force on the particle, Equations (2.1) and (2.5) may be combined.

$$Hqs = Eq \tag{2.6}$$

Then Equation (2.6) may be rearranged to Equation (2.7), showing that the speed of the particle is just the ratio of the strength of the electrostatic field to the strength of the magnetic field.

$$s = \frac{E}{H} \tag{2.7}$$

If this value for s is placed in Equation (2.4), the result is Equation (2.8), which shows how Thomson determined the q/m ratio for the particles in the cathode rays.

$$\frac{q}{m} = \frac{E}{H^2r} \tag{2.8}$$

Thomson's results showed that cathode rays travel at about 3×10^9 cm/sec, or about 10 percent of the speed of electromagnetic radiation. His values for q/m consistently approached 6.6×10^7 coulomb/g. Also he noticed that his results were unaffected by the nature of the material from which the electrodes were constructed, or by the nature of the residual gas remaining in the partially evacuated tube.

Thomson hoped to be able to determine the charge on these particles by independent means, so that his q/m values could be used to compute the masses of the cathode-ray particles. However, he was unable to devise such a method immediately, and for the time being he turned his attention to another area. However, other developments did lead to a value for the charge on the cathode-ray particles.

In 1887, Heinrich Hertz observed that ultraviolet light enhances the electrical discharge between two electrodes. By the time Thomson turned his attention to this *photoelectric effect*, it was known that rays are emitted when ultraviolet light impinges upon certain metals. Thomson showed that these rays are identical to cathode rays; his determined q/m values were identical to the values he had found for cathode rays.

C. T. R. Wilson showed that when the radiation from the photoelectric effect is passed through air saturated with water vapor, the water vapor condenses into liquid droplets. Thomson concluded that the water droplets must be charged electrostatically. Thus he set out to study these water droplets. From the total mass of condensed water, the average radius of the water droplets, the density of water, and total charge on the condensed water, he computed the average charge associated with each drop. He found it to be 1.6×10^{-19} coulombs.

As reported in Chapter 1, Millikan used a refinement of this technique with oil drops. The agreement of Thomson's value with the value for the quantum of charge determined by Millikan led Thomson to the conclusion that the charge on the particles from the photoelectric effect, and therefore the cathode rays, is the quantum of charge, which now is accepted as 1.602×10^{-19} coulomb.

Repetition of Thomson's work and determinations with other techniques have led to the presently accepted value of 1.759×10^{8} coulombs/g for q/m for cathode-ray particles. Using the accepted value for the quantum of charge, the mass of these particles is computed to be 9.108×10^{-28} g. These particles, which bear one quantum of negative electrostatic charge and which have a mass of 9.108×10^{-28} g, are called *electrons*.

Positive Rays

In the meantime, E. Goldstein designed a cathode-ray tube with a large portion of the tube behind the cathode, as shown in Fig. 2.3. In addition, the cathode was perforated. He noted that radiation came through the perforations in the cathode into the area behind the cathode. Initially these were called *canal rays*, but after W. Wien showed that these rays are electrostatically positive, J. J. Thomson proposed the name *positive rays*.

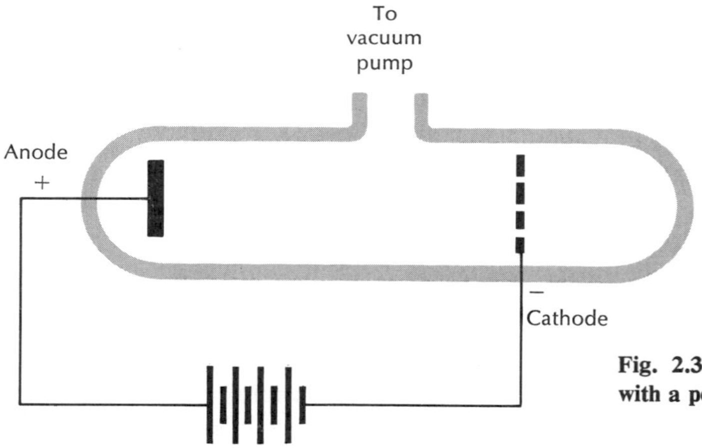

Fig. 2.3. Cathode-ray tube with a perforated cathode

Using Thomson's technique, Wien determined the q/m value for these rays and found that this value depends upon the nature of the residual gas in the partially evacuated tube. The largest value for q/m is obtained when hydrogen is used as the residual gas, and this value is 9.580×10^{4} coulomb/g. If we speculate that the mass of these particles is about the same as the mass of the electrons, the charge must be given as follows:

$$q = \frac{9.580 \times 10^{4}\ \text{coulomb}}{\text{g}} \times 9.109 \times 10^{-28}\ \text{g}$$
$$= 8.725 \times 10^{-23}\ \text{coulomb}$$

This value is much smaller than the quantum of charge that is the smallest permissible charge according to Millikan's work. This suggests that the mass of the positive-ray particles must be much larger than the mass of an electron.

On the other hand, if we assume that these rays have one quantum of charge, that is, 1.602×10^{-19} coulomb, the mass may be calculated as follows from the q/m value:

$$m = \frac{1.602 \times 10^{-19} \text{ coulomb} \times \text{g}}{9.580 \times 10^{4} \text{ coulomb}} = 1.672 \times 10^{-24} \text{ g}$$

Further, this assumption makes sense in the light of some of Faraday's data on the electrical decomposition of substances containing hydrogen. He found that approximately 96,500 coulombs of electrical charge interact with about 1.01 g of hydrogen. When hydrogen is used as a residual gas in a vacuum tube, the mass of the positive rays associated with 96,500 coulombs also is found to be about 1.01 g:

$$96{,}500 \text{ coulombs} \times \frac{1.67 \times 10^{-24} \text{ g}}{1.60 \times 10^{-19} \text{ coulomb}} = 1.01 \text{ g}$$

This confirms the assignment of charge and mass to the positive rays when hydrogen is the residual gas. The positive-ray particles, which carry one quantum of positive charge and have a mass of 1.672×10^{-24} g, are called *protons*.

RADIOACTIVITY

During the time when positive rays were being studied, Henri Becquerel discovered that emanations from substances containing uranium expose photographic plates. He found that these rays penetrate paper and thin metal foils and still expose the plates. Soon after, it was discovered that substances containing thorium, polonium, and radium produce the same effect. This phenomenon of natural radiation is called *radioactivity*.

Becquerel showed that these radioactive radiations can be deflected by a magnetic field and in the same direction as cathode rays. Ernest Rutherford studied the penetrating ability of these radiations and found that there are two different radiations, one more able to penetrate metal foils than the other. He named the radiation with the *lesser* penetrating ability *alpha radiation*, and the other *beta radiation*.

Pierre and Marie Curie showed that beta radiation is deflected easily in a magnetic field, whereas the alpha radiation is not. F. E. Dorn showed that beta radiation is deflected by an electrostatic field also. Then Becquerel determined the speed and q/m ratio to show that beta radiation and cathode rays are identical. Thus beta radiation probably consists of electrons.

Despite the claim by the Curies that alpha radiation is not easily deflected by a magnetic field, Rutherford showed that deflection occurs if the magnetic field is strong enough. He also showed that the deflection occurs in the direction opposite to that of electrons, indicating that alpha radiation is positively charged.

Rutherford measured the q/m value for alpha radiation and found it was approximately 5×10^4 coulomb/g. This is about one-half of the value for the proton. The measurement indicates that if alpha radiation carries a one-quantum charge, the mass of the particles is twice that of protons; if the radiation carries a two-quantum charge, the mass of the particles is four times that of protons; and so on.

Rutherford and Geiger in England and Regener in Germany set out to determine the charge on alpha particles. Using counting instruments then available for the study of radioactive materials, and electrometers to indicate the total charge on emitted rays, they estimated the charge on the alpha particles to be about 3.2×10^{-19} coulombs, or just twice that of the quantum charge. Thus alpha particles have masses four times that of protons and carry positive charges twice those on protons.

Actually a third type of radiation was detected from radioactive sources by P. Villard. These are gamma rays, very high frequency electromagnetic radiations similar to x-rays.

WAVE-PARTICLE DUALITY

Throughout this discussion the concepts of particles and radiation have been intermixed. Thus emanations have been described as radiations as well as moving particles. This mixture of terminology has been intentional, for it suggests the modern train of thought on the nature of fundamental particles and radiations.

By the end of the nineteenth century, the photoelectric effect was well known. As noted above, if ultraviolet light is allowed to fall on certain metals such as zinc, electrons are emitted. This suggests that electrons are embedded in these metals and that the ultraviolet light imparts sufficient energy to the electrons to cause them to fly out of the metal. It was known also that if the intensity of the light is increased, the speed of the emitted electrons remains unchanged, but the rate of emission of the electrons increases. Similarly, it was known that if the wavelength of the light is changed, the speed of the emitted electrons changes. Since the speed of the electron is a measure of its kinetic energy, the wavelength of the light must be related to the energy that the light is capable of imparting to the electron.

Specifically, it is found that the longer the wavelength of the light, the slower the speed of the emitted electrons; and the shorter the wavelength of the light, the greater the speed of the emitted electrons. Thus there appears to be an inverse relationship between the energy of the light and its wavelength, or perhaps a direct relationship between the energy of the light and its frequency. Thus as the frequency increases, the energy of the light increases, and as the frequency of the light decreases, the energy decreases as well. What is more, the fact that the increase in intensity does not increase the energy of the electron suggests that light is absorbed by the electrons in packets; that is, the energy cannot be absorbed continuously. Therefore, the increase in the frequency of the light amounts to an increase in the size of the energy packet, whereas an increase in intensity of light amounts to an increase in the number of packets striking the metal per unit time.

These observations, along with others on the absorption and emission of radiation, led Max Planck to suggest that radiation is absorbed and emitted in discrete units; that is, emitted and absorbed energy is quantized. According to Planck, the energy content of these quanta is directly proportional to the frequency of the radiation. This is shown in Equation (2.9), where E is energy, ν is frequency, and h is the proportionality constant known as Planck's constant.

$$E = h\nu \tag{2.9}$$

The best available value for h is 6.6256×10^{-27} erg $\times$ sec.

Planck startled the scientific world with his suggestion. For many years the wave picture for the propagation of electromagnetic radiation had been accepted. Planck suggested that this picture is not correct, and that electromagnetic radiation is best viewed as particles. Thus in Equation (2.9), $h\nu$ represents an energy quantum often called a *photon*, which is a particle or packet of energy. However, Planck's suggestion was made only for the processes of emission and absorption of energy. It was Albert Einstein who took the final step to propose that radiation is propagated through space in quantized form. This represented a return to the corpuscular theory of light first proposed by Sir Isaac Newton more than 200 years earlier.

It is customary now to talk about the dual nature of radiant energy, which behaves both as a wave and as a particle. Perhaps it is better to think of some properties of radiation in terms of the particle picture and to think of other properties of energy in terms of the wave picture.

Noting the dual nature of radiant energy, and also noting the somewhat radiative character of moving particles such as electrons, Louis de Broglie, in 1923, suggested that any moving particle has wave characteristics associated with it. In fact, he proposed an equation to show the relationship of the mass m, the speed s, and the wavelength λ. This is given as Equation (2.10), where h is Planck's constant.

$$\lambda = \frac{h}{ms} \tag{2.10}$$

On the basis of this prediction, it was suggested that particles such as electrons and protons should exhibit properties characteristic of wave motion. One of the important properties attributed to the wave characteristics of light is *diffraction*, the ability of light to bend around corners or change direction when it passes through a slit that is very narrow compared to its wavelength. In fact, white light, which is the combination of all of the colors in the visible spectrum, may be split into its many wavelengths with a diffraction grating, a device consisting of a series of closely spaced, narrow slits. The angle of bending depends upon the wavelength, so that the various colors are bent at different angles and thus separated.

In 1928, Davisson and Germer actually showed that a beam of electrons can be diffracted like a beam of light. In later years it was shown that protons, alpha particles, and other particles can be diffracted as well. The experimentally determined wavelengths for these particles are in agreement with de Broglie's suggestion in Equation (2.10).

This suggests that material particles have a dual nature also. They exhibit characteristic wave-like properties as well as particle-like properties. The inevitable conclusion appears to be that there is a very fundamental similarity between matter and energy, and in fact they may be different manifestations of the same thing.

With an inversion of history, it is possible to derive Einstein's equation which was derived before de Broglie's equation. Thus Equation (1.28) may be rewritten as Equation (2.11), where the speed of electromagnetic radiation is represented by c.

$$\nu = \frac{c}{\lambda} \tag{2.11}$$

Using this value for ν, Equation (2.9) may be rewritten as follows:

$$E = \frac{hc}{\lambda} \tag{2.12}$$

Then the expression for λ in Equation (2.10) may be substituted into Equation (2.12) to obtain Equation (2.13).

$$E = mc^2 \tag{2.13}$$

Equation (2.13) shows the fundamental relationship between matter and energy. Energy has a mass equivalent, so mass is a property of both energy and matter. Again this indicates the fundamental similarity between matter and energy.

Thus an object in motion has more mass than it has at rest because of its kinetic energy, which has a mass equivalent. Experiments in the determination of q/m values for electrons moving at different speeds show a change attributed to an increase in mass due to the speed. Similarly, an increase in potential energy leads to an increase in mass. However, in most commonly encountered situations the mass change is not detectable because it is very small. In Equation (2.13) it is obvious that the value for c, the speed of electromagnetic radiation, is very large, so c^2 must be extremely large. In order for m, the mass equivalent of the energy, to be large, E, the energy, must be enormous.

Figure 2.4 shows graphically how the mass of an object varies with its speed. In this graph m/m_0 is plotted against s/c, where m is the mass at speed s, m_0 is the rest mass, that is, the mass of the object at rest, and c is the speed of light. From this graph it is clear that the increase in mass due to the speed of the object does not become appreciable until its speed exceeds half the speed of electromagnetic radiation. This means that the mass increase is not appreciable until the object is moving at a speed in excess of 1.50×10^{10} cm/sec, or about 3.35×10^8 mi/hr. Beyond this point, as the speed approaches that of electromagnetic radiation, the

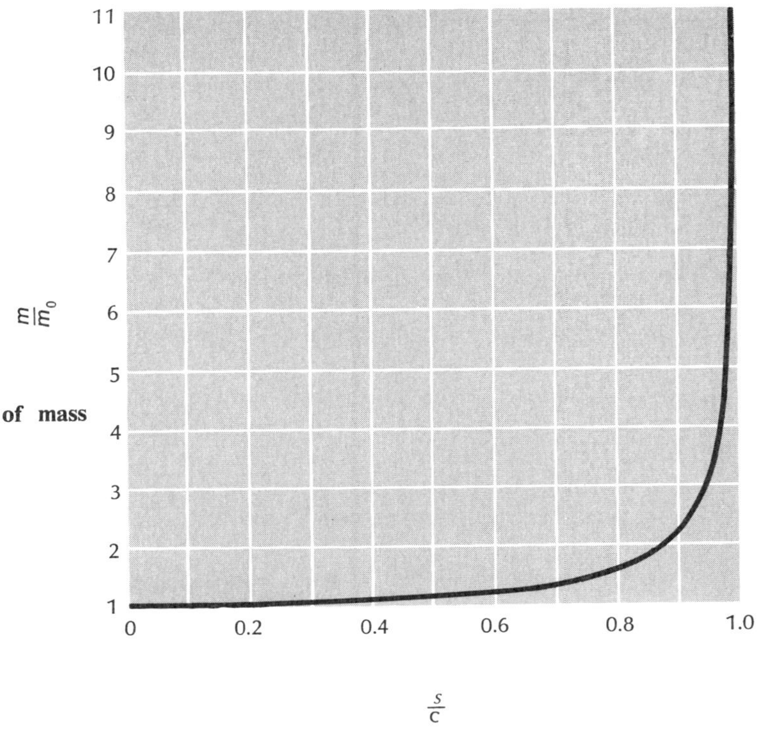

Fig. 2.4. Change of mass with speed

mass rises very rapidly and approaches infinity. Therefore it appears that objects cannot attain the speed of electromagnetic radiation, apparently the maximum in attainable speeds.

It should be noted that particles such as electrons often do travel at speeds that affect their masses measurably. For this reason, the masses quoted thus far in our discussion, and those to be quoted in the future for such particles, are their rest masses.

Although matter and energy may be considered to be composed of particles, there is a fundamental difference between a photon of radiant energy and a particle such as an electron. It has been noted that electromagnetic radiation travels at the very high speed of 3.00×10^{10} cm/sec, whereas the speed of a particle cannot be that great, as shown in the graph in Fig. 2.4. The mass of an electron, for example, approaches infinity as its speed approaches that of electromagnetic radiation.

Furthermore, in the case of an electron, or any other material particle, there is an intrinsic energy associated with the particle at rest. This energy, E_0, is represented in Equation (2.14), a modified version of Equation (2.13) in which the subscripts identify the rest state.

$$E_0 = m_0c^2 \tag{2.14}$$

If the electron assumes a speed s, it reaches a state of higher energy and correspondingly has a higher mass. At rest, the total energy is E_0, whereas at some speed s, the total energy is E_0 plus the kinetic energy E_k. This is shown in Equation (2.15), where E_t is the total energy.

$$E_t = E_0 + E_k \tag{2.15}$$

The relationship between the rest mass m_0 and the mass m at speed s is given by Equation (2.16).

$$m_0 = m\sqrt{1 - \frac{s^2}{c^2}} \tag{2.16}$$

This is the equation from which the graph in Fig. 2.4 can be plotted. In the case of the photon, $s = c$ and $1 - s^2/c^2 = 0$, so the rest mass m_0 is zero. This indicates that a photon has a rest mass of zero and in a sense is a nonmaterial particle, the mass of which is associated entirely with its speed. This serves as a distinction between radiant energy photons, which have zero rest mass, and material particles, which have a real rest mass.

Now it should be apparent that as energy changes occur in various systems, there are corresponding mass changes as well. When a rock rolls from the top of a hill to the foot of the hill, from a state of high potential energy to a state of low potential energy, it undergoes a decrease in mass. When work is done on the rock to place it back at the top of the hill, its mass increases. When two electrostatically charged bodies are brought together by electrostatic attraction, they attain a state of lower potential energy and consequently a state of lower mass.

ATOMS

The history of the atomic theory goes back to the Greek philosophers who pondered the continuity of matter. The problem may be illustrated by a piece of copper wire that is reduced successively in size by one-half. The wire is cut in half and one half is discarded. The remaining half is cut in half and one half is discarded. This is repeated many times. The question is: can this process go on indefinitely or is a point reached at which further division is impossible? One school concluded that the copper consists of small, discrete units which are the smallest units that still have the properties of copper. It was Democritus who coined the word *atoma* for these particles and from this we have derived our word *atom.*

In the first decade of the nineteenth century, John Dalton formalized a picture of atoms as the primary particles of which all matter is constructed. With this picture, he was able to describe the behavior of matter in a more complete and quantitative way than was possible before. His picture of tiny, discrete particles of matter was well accepted by the late nineteenth century.

Ernest Rutherford was interested in the penetrating ability of alpha particles. In his work he collimated alpha particles from a radioactive source through a small slit to form a beam, which was focused on a photographic plate as shown schematically in Fig. 2.5a. He found a sharp exposure line on the photographic plate.

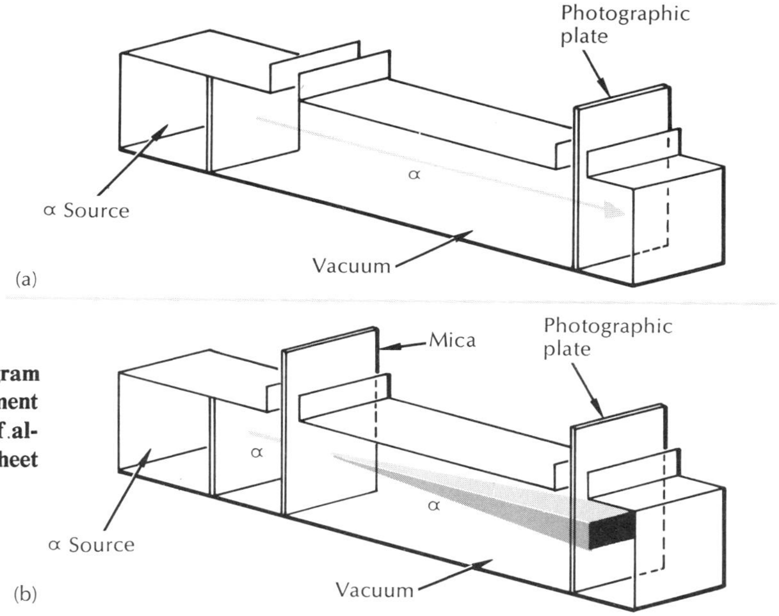

Fig. 2.5. Schematic diagram of Rutherford's experiment showing the scattering of alpha particles by a thin sheet of mica

However, when he inserted a sheet of thin mica into the beam, as shown in Fig. 2.5b, the exposure was more diffuse and less sharp. He was able to show that apparently some of the alpha particles were deflected through angles up to 2° as they passed through the mica.

Working with Rutherford, Geiger used a similar technique, employing thin sheets of metal foil in place of the mica and a phosphorescent screen instead of the photographic plate, as shown in Fig. 2.6. When alpha particles hit the screen they emitted flashes of light which he counted. In this way, he was able to show that in a good vacuum and in the absence of any foil, the alpha particles were deflected but little, whereas they were deflected appreciably in passing through the gold foil. Typical data are plotted in Fig. 2.7.

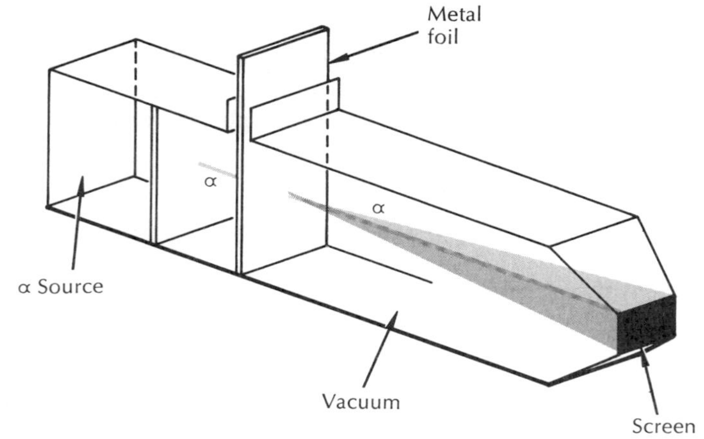

Fig. 2.6. Schematic diagram of Geiger's experiment showing the scattering of alpha particles by thin metal foils

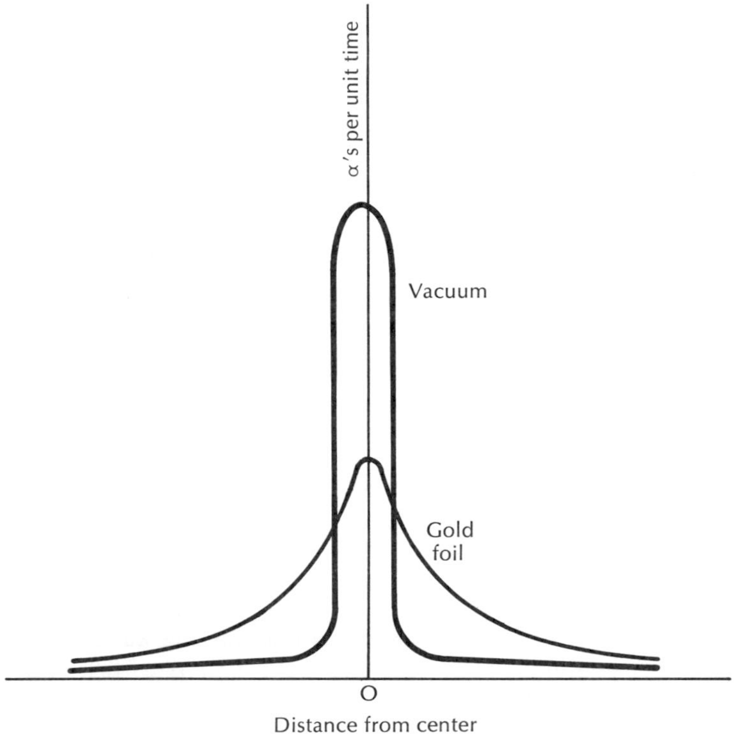

Fig. 2.7. Graph showing distribution of scattered end particles

Geiger and Marsden used high intensity alpha radiation to show that some alpha particles are deflected through 90° and more when they bombard metal foils. The apparatus and results are shown schematically in Fig. 2.8. This was a rather startling discovery. It was shown that for such deflections to occur, the alpha particles must encounter very strong electrostatic fields or centers of high mass or both. In 1911, Rutherford summarized these observations by suggesting that the mass of the atom is concentrated in a small center, or *nucleus*. By comparing the numbers of undeflected alpha particles with those that were deflected, the diameter of the nucleus was estimated to be about 10^{-4} times that of the entire atom.

By analysis of the angles of deflection, Rutherford and his coworkers showed that there are strong electrostatic interactions between the nucleus and the bombarding alpha particles. Rutherford proposed that the nucleus is electrostatically charged, and moreover that it is charged positively. This is consistent with the vacuum-tube observations that negative rays are electrons with small masses.

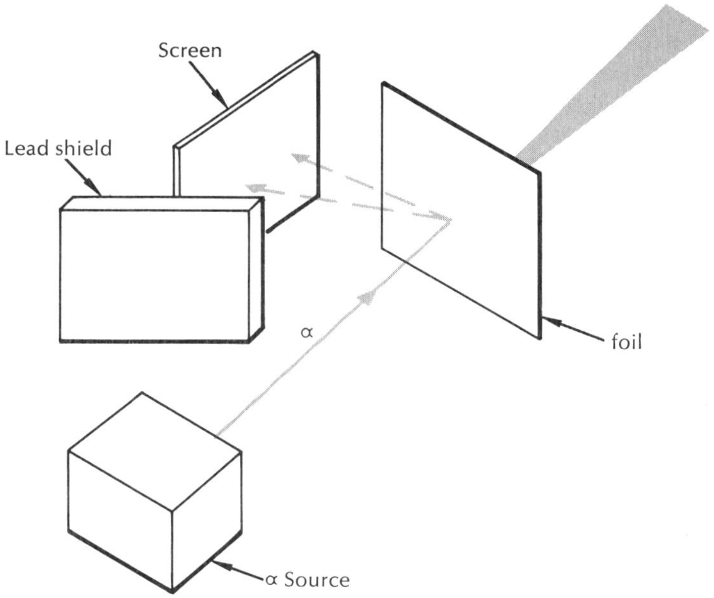

Fig. 2.8. Schematic diagram of the experiment by Geiger and Marsden to show the deflection of alpha particles through more than 90°

However, nearly all of the mass appears to be associated with the positive rays. For example, hydrogen provides the lighest positive-ray particles, and these are nearly 2000 times the mass of the electron that bears the negative charge. Therefore it is reasonable that the positive charge be associated with most of the mass of the atom, and the nuclei probably are positively charged.

Thus Rutherford was responsible for the development of the picture of the nuclear atom. He proposed that atoms consist of tiny nuclei of high mass and positive charge surrounded by a cloud of negative electrons which have almost negligible mass in comparison to the nuclei.

Atomic Numbers

Still, the charge on atomic nuclei was a matter of concern. There were unsuccessful attempts to evaluate the nuclear charge on the basis of alpha particle deflections, but the results were inconsistent and inconclusive. Another approach led to the solution of this problem.

Other chemical developments prior to this time had led to the identification of many substances as elements. Elements were defined as substances which cannot be decomposed by normal chemical means. As early as 1868, the Russian Dmitri Mendeleef, had arranged the elements in a sequential order according to the gradations in various properties. This arrangement was called the *periodic table.* Each element was assigned an ordinal number to designate its place in the sequence.

Rutherford was the first to suggest the possibility that the ordinal numbers assigned to the elements in the periodic table might be the number of quanta of positive charge on the nuclei of the atoms of the element. It was one of Rutherford's students, H. G. J. Moseley, who convinced the scientific world that Rutherford's insight was correct. He did this by analyzing the characteristic x-ray emissions of the elements.

In principle, x-ray tubes are cathode-ray tubes in which a substance is placed in the electron beam. Upon bombardment with electrons, x-rays are emitted from the target. By the time Moseley began his work, it was known that some of the x-rays emitted are characteristic of the energy of the electron beam, whereas other x-rays emitted are characteristic of the target material. In Moseley's work, each element used as a target emitted two wavelengths of x-rays which were characteristic of the element. From his results, Moseley showed that the square root of the frequency of these characteristic x-rays is pseudoproportional to the ordinal numbers of the elements as they appear in the periodic table. Figure 2.9 shows a plot of some of his data.

Moseley reasoned that the ordinal number is the positive charge on the atomic nuclei. As the nuclear charge increases, the frequency (ν), and consequently the energy, of the emitted x-rays increases as well. Each quantum increase in nuclear charge leads to a characteristic increase in the square root of the frequency of the emitted x-rays. *Atomic number* is the name given to the quantum positive charges on the nuclei of atoms.

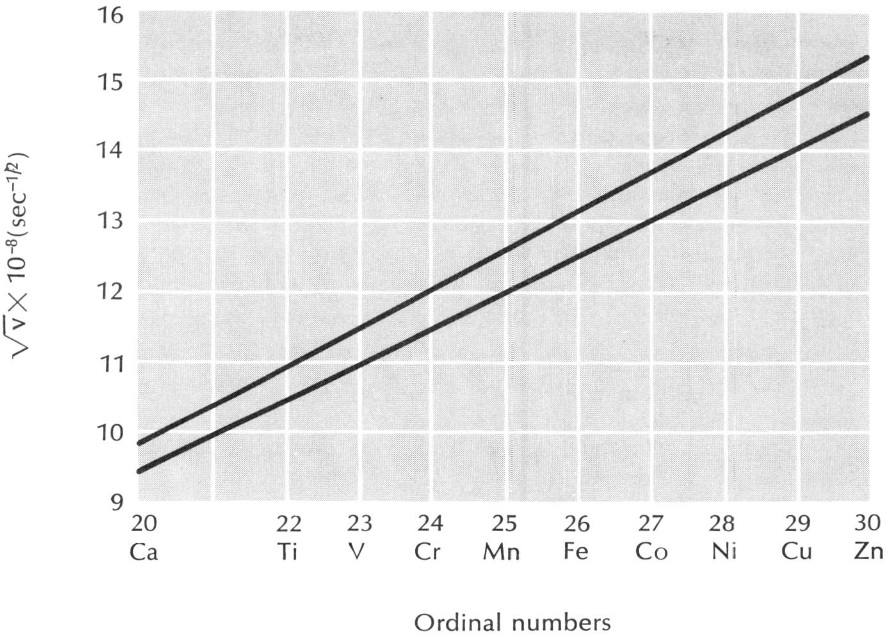

Fig. 2.9. Moseley's frequency function variation with ordinal number

The Elements

From the studies of all these men, a picture of the atom began to emerge. In general, atoms are composed of small, dense, positively charged nuclei surrounded by a cloud of negative electrons. The elements differ from each other in the number of positive quantum charges of their atomic nuclei. Since matter appears to be electrostatically neutral under normal conditions, the number of electrons outside the nucleus must equal the number of quantum charges on the nucleus. In other words, the atomic number is the number of electrons outside the nucleus as well as the number of positive quantum charges on the nucleus of a neutral atom. Also, all atoms with the same atomic number are atoms of the same element. In fact it is possible to define an element as a substance all of whose atoms have the same atomic number.

An alphabetical list of the known elements is given in Table 2.1. These same elements are listed in order of increasing atomic number in Table 2.2.

TABLE 2.1
Alphabetical Listing of the Elements

NAME	SYMBOL	ATOMIC NUMBER	ATOMIC WEIGHT
Actinium	Ac	89	(227)†
Aluminum	Al	13	26.98154
Americium	Am	95	(243)
Antimony	Sb	51	121.75
Argon	Ar	18	39.948
Arsenic	As	33	74.92158
Astatine	At	85	(210)
Barium	Ba	56	137.34
Berkelium	Bk	97	(247)
Beryllium	Be	4	9.01219
Bismuth	Bi	83	208.980
Boron	B	5	10.811
Bromine	Br	35	79.909
Cadmium	Cd	48	112.40
Calcium	Ca	20	40.08
Californium	Cf	98	(251)
Carbon	C	6	12.01115
Cerium	Ce	58	140.12
Cesium	Cs	55	132.9051
Chlorine	Cl	17	35.453
Chromium	Cr	24	51.996
Cobalt	Co	27	58.93319
Copper	Cu	29	63.54
Curium	Cm	96	(247)
Dysprosium	Dy	66	162.50
Einsteinium	Es	99	(252)
Erbium	Er	68	167.26
Europium	Eu	63	151.96
Fermium	Fm	100	(253)
Fluorine	F	9	18.99840
Francium	Fr	87	(223)
Gadolinium	Gd	64	157.25
Gallium	Ga	31	69.72
Germanium	Ge	32	72.59
Gold	Au	79	196.96655
Hafnium	Hf	72	178.49
Helium	He	2	4.00260
Holmium	Ho	67	164.9303
Hydrogen	H	1	1.00797
Indium	In	49	114.82
Iodine	I	53	126.90447
Iridium	Ir	77	192.2
Iron	Fe	26	55.847
Krypton	Kr	36	83.80
Lanthanum	La	57	138.91

TABLE 2.1 (cont.)

NAME	SYMBOL	ATOMIC NUMBER	ATOMIC WEIGHT
Lawrencium	Lw	103	(257)
Lead	Pb	82	207.19
Lithium	Li	3	6.939
Lutetium	Lu	71	174.97
Magnesium	Mg	12	24.312
Manganese	Mn	25	54.93805
Mendelevium	Md	101	(256)
Mercury	Hg	80	200.59
Molybdenum	Mo	42	95.94
Neodymium	Nd	60	144.24
Neon	Ne	10	20.183
Neptunium	Np	93	(237)
Nickel	Ni	28	58.71
Niobium	Nb	41	92.9060
Nitrogen	N	7	14.0067
Nobelium	No	102	(256)
Osmium	Os	76	190.2
Oxygen	O	8	15.9994
Palladium	Pd	46	106.4
Phosphorus	P	15	30.97376
Platinum	Pt	78	195.09
Plutonium	Pu	94	(244)
Polonium	Po	84	(209)
Potassium	K	19	39.102
Praseodymium	Pr	59	140.90739
Promethium	Pm	61	(145)
Protactinium	Pa	91	(231)
Radium	Ra	88	(226)
Radon	Rn	86	(222)
Rhenium	Re	75	186.2
Rhodium	Rh	45	102.90551
Rubidium	Rb	37	85.47
Ruthenium	Ru	44	101.07
Samarium	Sm	62	150.35
Scandium	Sc	21	44.9559
Selenium	Se	34	78.96
Silicon	Si	14	28.086
Silver	Ag	47	107.870
Sodium	Na	11	22.98977
Strontium	Sr	38	87.62
Sulfur	S	16	32.064
Tantalum	Ta	73	180.948
Technetium	Tc	43	(97)
Tellurium	Te	52	127.60
Terbium	Tb	65	158.9250

TABLE 2.1 (cont.)

NAME	SYMBOL	ATOMIC NUMBER	ATOMIC WEIGHT
Thallium	Tl	81	204.37
Thorium	Th	90	232.038
Thulium	Tm	69	168.934
Tin	Sn	50	118.69
Titanium	Ti	22	47.90
Tungsten	W	74	183.85
Uranium	U	92	238.03
Vanadium	V	23	50.942
Xenon	Xe	54	131.30
Ytterbium	Yb	70	173.04
Yttrium	Y	39	88.9054
Zinc	Zn	30	65.37
Zirconium	Zr	40	91.22

†Parenthetical values are mass numbers of the most stable nuclides.

TABLE 2.2

The Elements in Order of Increasing Atomic Number

ATOMIC NUMBER	SYMBOL	NAME	ATOMIC WEIGHT
1	H	Hydrogen	1.00797
2	He	Helium	4.00260
3	Li	Lithium	6.939
4	Be	Beryllium	9.01219
5	B	Boron	10.811
6	C	Carbon	12.01115
7	N	Nitrogen	14.0067
8	O	Oxygen	15.9994
9	F	Fluorine	18.99840
10	Ne	Neon	20.183
11	Na	Sodium	22.98977
12	Mg	Magnesium	24.312
13	Al	Aluminum	26.98154
14	Si	Silicon	28.086
15	P	Phosphorus	30.97376
16	S	Sulfur	32.064
17	Cl	Chlorine	35.453
18	Ar	Argon	39.948
19	K	Potassium	39.102
20	Ca	Calcium	40.08
21	Sc	Scandium	44.9559
22	Ti	Titanium	47.90
23	V	Vanadium	50.942

TABLE 2.2 (cont.)

ATOMIC NUMBER	SYMBOL	NAME	ATOMIC WEIGHT
[illegible]	Cr	Chromium	51.996
[illegible]	Mn	Manganese	54.93805
26	Fe	Iron	55.847
27	Co	Cobalt	58.93319.
28	Ni	Nickel	58.71
29	Cu	Copper	63.54
30	Zn	Zinc	65.37
31	Ga	Gallium	69.72
32	Ge	Germanium	72.59
33	As	Arsenic	74.92158
34	Se	Selenium	78.96
35	Br	Bromine	79.909
36	Kr	Krypton	83.80
37	Rb	Rubidium	85.47
38	Sr	Strontium	87.62
39	Y	Yttrium	88.9054
40	Zr	Zirconium	91.22
41	Nb	Niobium	92.9060
42	Mo	Molybdenum	95.94
43	Tc	Technetium	(97)†
44	Ru	Ruthenium	101.07
45	Rh	Rhodium	102.90551
46	Pd	Palladium	106.4
47	Ag	Silver	107.870
48	Cd	Cadmium	112.40
49	In	Indium	114.82
50	Sn	Tin	118.69
51	Sb	Antimony	121.75
52	Te	Tellurium	127.60
53	I	Iodine	126.90447
54	Xe	Xenon	131.30
55	Cs	Cesium	132.9051
56	Ba	Barium	137.34
57	La	Lanthanum	138.91
58	Ce	Cerium	140.12
59	Pr	Praseodymium	140.90739
60	Nd	Neodymium	144.24
61	Pm	Promethium	(145)
62	Sm	Samarium	150.35
63	Eu	Europium	151.96
64	Gd	Gadolinium	157.25
65	Tb	Terbium	158.9250
66	Dy	Dysprosium	162.50
67	Ho	Holmium	164.9303
68	Er	Erbium	167.26
69	Tm	Thulium	168.934

TABLE 2.2 (cont.)

ATOMIC NUMBER	SYMBOL	NAME	ATOMIC WEIGHT
70	Yb	Ytterbium	173.04
71	Lu	Lutetium	174.97
72	Hf	Hafnium	178.49
73	Ta	Tantalum	180.948
74	W	Tungsten	183.85
75	Re	Rhenium	186.2
76	Os	Osmium	190.2
77	Ir	Iridium	192.2
78	Pt	Platinum	195.09
79	Au	Gold	196.96655
80	Hg	Mercury	200.59
81	Tl	Thallium	204.37
82	Pb	Lead	207.19
83	Bi	Bismuth	208.980
84	Po	Polonium	(209)
85	At	Astatine	(210)
86	Rn	Radon	(222)
87	Fr	Francium	(223)
88	Ra	Radium	(226)
89	Ac	Actinium	(227)
90	Th	Thorium	232.038
91	Pa	Protactinium	(231)
92	U	Uranium	238.03
93	Np	Neptunium	(237)
94	Pu	Plutonium	(244)
95	Am	Americium	(243)
96	Cm	Curium	(247)
97	Bk	Berkelium	(247)
98	Cf	Californium	(251)
99	Es	Einsteinium	(252)
100	Fm	Fermium	(253)
101	Md	Mendelevium	(256)
102	No	Nobelium	(256)
103	Lw	Lawrencium	(257)

†Parenthetical values are mass numbers of the most stable nuclides.

Each of the elements has a name usually given to it by the person discovering the element. These names are abbreviated into *symbols* which provide a shorthand notation by which to refer to the elements. In some cases the symbols given in Tables 2.1 and 2.2 do not appear to be derived from the names given there. With the exception of tungsten which has the symbol W derived from the German name wolfram, these symbols are derived from the Latin names for the elements. A list of these is given in Table 2.3.

TABLE 2.3
Symbols Derived from Latin

PRESENT NAME	SYMBOL	LATIN NAME
Antimony	Sb	Stibium
Copper	Cu	Cuprum
Gold	Au	Aurum
Iron	Fe	Ferrum
Lead	Pb	Plumbum
Mercury	Hg	Hydrargyrum
Potassium	K	Kalium
Silver	Ag	Argentum
Sodium	Na	Natrium
Tin	Sn	Stannum

From the lists in Tables 2.1 and 2.2, it is obvious that 103 elements are known, corresponding to atoms with atomic numbers from 1 to 103. Of these, only 88 have been discovered occurring naturally on earth. The rest are known as products of laboratory research.

To understand the meaning of the atomic weights listed in Tables 2.1 and 2.2, it is necessary to consider the discovery of still another particle.

Neutrons

The development of the picture of the atom permitted a better explanation of the nature of cathode and positive rays. Apparently the electrostatic field in the vacuum tube attracts electrons from the clouds of electrons outside the nuclei of the atoms of the gas in the tube. The electrons appear as cathode rays. The atoms from which electrons have been removed are no longer electrostatically neutral, since some of the negative charge has been removed with the electrons; such atoms are called *ions*. These positively charged ions constitute the positive rays.

Studies on the deflections of positive rays in magnetic and electrostatic fields led to determinations of the masses of the atoms. These determinations suggested that the masses of many of the atoms of low atomic number are nearly twice that of the atomic number of protons. Thus a typical helium atom with atomic number 2 has a mass that is about that of 4 protons. Similarly, a typical nitrogen atom with atomic number 7 has a mass that is about that of 14 protons, and a typical oxygen atom with atomic number 8 has a mass that is about that of 16 protons.

To reconcile these observations, it was postulated that the positive quantum charges on the nucleus are due to the presence of protons in the nucleus, and the excess mass of the atom is due to the presence of yet undiscovered particles without electrostatic charge. These new particles were named *neutrons* and were thought to be combinations of protons and electrons. Thus the mass of the neutron would be near that of the proton but the positive charge of the proton would be neutralized by the negative charge of the electron.

Once the neutron was postulated, in 1920, the search began for this elusive particle. The first break came in 1930 when Bothe and Becker reported that if certain substances such as beryllium, boron, or lithium are radiated with alpha particles, a new kind of highly penetrating radiation is emitted. In 1932 Irene Joliot-Curie and her husband, Frederic Joliot, repeated these experiments and also noted that high-speed protons are ejected from paraffin placed in the path of this new radiation.

Later in the same year, another of Rutherford's students, James Chadwick, proposed that these observations could be explained on the assumption that this new radiation consisted of neutrons. Thus a particle of about the same mass as the proton could impart the necessary kinetic energy to a proton to give it the observed speed. It was on this basis that the neutron was discovered.

The mass of the neutron is difficult to measure since neither magnetic nor electrostatic fields affect it. To obtain the mass of the neutron, it is possible to study the effect of the neutron upon other particles. It was mentioned earlier that charged particles traveling through air saturated with water vapor cause the water vapor to condense. Such condensation is not observed when neutrons travel through the moisture-laden air. Thus it is possible to devise a *cloud chamber*, in which the impact of neutrons is used to propel protons, which then form a stream of condensation in their path. In this way, the effect of the neutron upon the proton can be measured and the mass of the neutron can be determined.

The best current value for the mass of the neutron is 1.675×10^{-24} g. This is considerably more than the mass of the proton. In fact, the difference between the mass of the neutron and the mass of the proton is more than twice the mass of the electron. This, along with theoretical considerations, suggests that the neutron is not a simple composite particle consisting of an electron and a proton.

With the addition of the neutron to the list of particles, there is sufficient information available to explain in elementary form the modern ideas of atomic structure. The properties of the particles discussed so far are listed in Table 2.4.

TABLE 2.4 Subatomic Particles

PARTICLE	SYMBOLS	CHARGE (QUANTA)	REST MASS (GRAMS)
Electron	${}_{-1}^{0}e$, e, e^-, β, β^-	-1	9.1091×10^{-28}
Proton	p, p^+, H^+, ${}_{1}^{1}H$	$+1$	1.6725×10^{-24}
Alpha	α, ${}_{2}^{4}He$	$+2$	6.6455×10^{-24}
Neutron	n, ${}_{0}^{1}n$	0	1.6748×10^{-24}

Included in this table are the symbols commonly used for each of the particles; in some cases these symbols will become more understandable later in this chapter. The electrostatic charge is given as the number of quanta of charge on the particle, using the + and − signs to indicate the type of charge. Rest masses are given in this reference table to one more significant figure than those used in the body of the text.

Mass Numbers and Atomic Weights

It has been noted already that most of the mass of an atom is concentrated in its nucleus. This mass is distributed among the protons and neutrons in the nucleus.

Since the masses of the protons and the neutrons are *roughly* the same, often they are assigned an approximate, arbitrary value of 1 mass unit. Often for approximate purposes it is not necessary to distinguish between the protons and neutrons in the nucleus and when this distinction is not needed, the protons and neutrons together are called *nucleons*. In an approximate way, then, the mass of an atom is referred to as the number of nucleons in the atom and this number of nucleons is called the *mass number* of the atom. Thus the *mass number* of an atom is the sum of the number of protons and neutrons in the nucleus.

Although all atoms of a given element have the same atomic number, it is not necessary that all atoms of the same element have the same mass numbers. Thus there are atoms of the same element with different mass numbers. To say it still differently, there are atoms with the same number of protons but different numbers of neutrons. These are all equivalent definitions of *isotopes*. Thus isotopes are two or more different kinds of atoms of the same element. Each specific kind of atom of an element with a specific mass number is called a *nuclide*. Thus hydrogen has three isotopes. These consist of the nuclide of mass number 1 called hydrogen-1, the nuclide of mass number 2 called hydrogen-2, and the nuclide of mass number 3 called hydrogen-3. Since all hydrogen atoms have an atomic number of 1, they all contain a single proton in the nucleus. It should be clear from the definitions, then, that hydrogen-2 atoms contain one neutron and hydrogen-3 atoms contain two neutrons in their nuclei in addition to the single proton.

It should be clear that in writing out the names of the nuclides, the mass number follows the name of the element and is separated from it by a hyphen. When the symbols for the elements are used, it is approved practice to use the mass number as a superscript to the left of the symbol and the atomic number as a subscript to the left of the symbol. Thus hydrogen-1 is written as ${}^{1}_{1}H$; hydrogen-2 is written as ${}^{2}_{1}H$; hydrogen-3 is written as ${}^{3}_{1}H$. (It should be noted that older practice places the mass number as a superscript to the right of the symbol.)

Where more exact masses must be used, it is common practice to use a comparative mass scale for atoms called the *atomic weight scale*. It would be better to call this the atomic mass scale, but a suggestion to make this change was defeated by the International Union of Pure and Applied Chemistry in 1961. So, officially it is still called the atomic weight scale.

On this scale, the mass of some atom is chosen arbitrarily as the standard to which the masses of all other atoms are compared. Historically several standards have been used. However, the current standard is carbon-12, which has been assigned a mass of exactly 12 *atomic mass units*. Often the atomic mass unit is abbreviated to amu. Thus by definition 1 atomic mass unit is $\frac{1}{12}$ of the mass of the carbon-12 atom.

Since a carbon-12 atom contains 12 nucleons, this is equivalent to assigning a mass of 1 amu to each nucleon. In this way the mass of any nuclide should correspond rather closely to its mass number. In fact, however, there are variations, so that the atomic weights of most nuclides are not integral numbers.

The most interesting reason for this deviation from integral numbers for atomic weights stems from an observation that will be outlined briefly here and discussed in more detail in Chapter 3. The mass of the carbon-12 atom has been determined rather accurately to be 1.9924×10^{-23} g. From this it can be seen that 1.0000 amu = 1.6603×10^{-24} g, which is $\frac{1}{12}$ of the mass of the carbon-12 atom. When the mass value is compared with the values in Table 2.4, it is obvious that 1 amu is less than the mass of either a proton or a neutron. Putting it in other terms, *the actual mass of the carbon-12 atom is less than the sum of the rest masses of the protons, neutrons, and electrons of which it is composed.* This is a very important observation, suggesting that the particles within the atom must be at a state of lower energy (and consequently lower mass) than the same particles are when they are isolated *and* at rest. However, this means that depending upon the energy states within individual atoms, there will be mass differences, and these differences will lead to atomic weights that deviate more or less from the mass numbers.

With the gram equivalent of the atomic mass unit established, it is possible to express the masses of the subatomic particles in atomic mass units. Since these are used often, they have been tabulated in Table 2.5.

TABLE 2.5
Rest Masses in Atomic Mass Units

Electron	5.4860×10^{-4}	amu
Proton	1.007277	amu
Neutron	1.008665	amu

For the day-to-day work in chemistry, the atomic weights of greatest use are those of the elements as they are found in nature. The elements usually are found as mixtures of nuclides, and the atomic weights used are the weighted averages of the actual nuclides found in natural mixtures.

For example, it has been found that naturally occurring boron is a mixture of boron-10, which has an atomic mass of 10.013 amu and boron-11, which has an atomic mass of 11.009 amu. The *atom* percent is 19.923 for boron-10 and 80.077 for boron-11. This means that on the average, 100 atoms of naturally occurring boron contain 19.923 atoms of boron-10 and 80.077 atoms of boron-11. Thus the average mass of an atom of naturally occurring boron can be found by taking a weighted average of these two masses. This mass, calculated below, is called the atomic weight of boron.

$$1 \text{ atom B}\left(\frac{19.923 \text{ atoms } {}^{10}\text{B}}{100 \text{ atoms B}}\times\frac{10.013 \text{ amu}}{1 \text{ atom } {}^{10}\text{B}}+\frac{80.077 \text{ atoms } {}^{11}\text{B}}{100 \text{ atoms B}}\times\frac{11.099 \text{ amu}}{1 \text{ atom } {}^{11}\text{B}}\right) = 10.811 \text{ amu}$$

This indicates that 19.923 percent of the atomic mass of boron-10 is added to 80.077 percent of the atomic mass of boron-11 to get the atomic weight of boron.

In this context a distinction can be made between the terms *atomic mass* and *atomic weight.* The atomic weight is the weighted average of the atomic masses of the elements as found in nature. The atomic mass is the mass of a particular nuclide measured on the atomic weight scale.

In normal laboratory work, it is difficult to deal in atomic quantities. It is necessary to work with quantities that may be weighed and seen. Such quantities contain very large numbers of atoms. For this reason, a new quantity is defined which is a specified large number of atoms to be used as a standard. This number is called the *mole* and has the value of 6.0225×10^{23}. In a sense, the mole is the chemist's dozen. The mole is a widely used quantity which may be applied to all objects. Thus a mole of atoms is 6.0225×10^{23} atoms; a mole of electrons is 6.0225×10^{23} electrons; and a mole of alpha particles is 6.0225×10^{23} alpha particles.

At first sight, the choice of 6.0225×10^{23} as a standard number might seem a bit strange. However, there are certain advantages to be gained by using this particular number. It was noted previously that the mass of the carbon-12 atom is 1.9924×10^{-23} g. Therefore, the mass of a mole of carbon-12 atoms is

$$1 \text{ mole } {}^{12}\text{C atoms} \times \frac{6.0225 \times 10^{23}\ {}^{12}\text{C atoms}}{1 \text{ mole } {}^{12}\text{C atoms}} \times \frac{1.9924 \times 10^{-23}\text{ g}}{1\ {}^{12}\text{C atom}} = 12.0000\text{ g}$$

Thus, by choosing the value 6.0225×10^{23} as the mole, the mass of a mole of carbon-12 atoms is numerically equal to the atomic mass expressed in grams. Consequently Tables 2.1 and 2.2 become even more useful. A mole of atoms of any element has a mass equal to the atomic weight expressed in grams. Thus a mole of naturally occurring oxygen atoms has a mass of 15.9994 g; a mole of naturally occurring silver atoms has a mass of 107.870 g; a mole of naturally occurring uranium atoms has a mass of 238.03 g.

It is convenient to associate a specific quantity of an element with the symbol for that element. One obvious quantity is the single atom. Thus the symbol Ne represents not only neon in the general sense, but specifically one atom of neon. Similarly, the symbol Cr represents one atom of chromium and the symbol K represents one atom of potassium. However, since it is more convenient, in many cases, to think in larger terms, the symbol for an element may represent one *mole* of atoms of that element as well. Thus the symbol Ne may represent either one atom of neon or one mole of neon atoms. Similarly, the symbol Cr may represent either one atom of chromium or one mole of chromium atoms, and the symbol K may represent one atom of potassium or one mole of potassium atoms. The interpretation to be given to the symbol normally is clear within the context of its usage.

SUMMARY

Matter is pictured as a composite of many small particles. For the chemist, the most significant of these is the atom. However, the atom is a composite of several particles, the most important of which are the electron, the proton, and the neutron. The more massive protons and neutrons are concentrated in the central nucleus which is surrounded by a cloud of electrons. The fundamental particles are closely related to electromagnetic radiation. They differ from the photon, which is a particle of electromagnetic radiation, by having a rest mass, whereas the photon has zero rest mass. They also differ in not being able to travel at the speed of electromagnetic radiation. However, electromagnetic radiation has a mass equivalent, and the fundamental particles have an energy equivalent due to their mass. As a result, the attractions existing among the fundamental particles within an atom account for the atom's existing in a low state of potential energy, in which the mass of the atom is less than the sum of the masses of its component parts.

Therefore, rather than measure the masses of the atoms in terms of the sum of the masses of the particles of which they are composed, it is more convenient to compare the masses of atoms to the mass of a single standard atom on the atomic weight scale. Such atomic weights really are averages of the masses of the different nuclides for each element, since atoms of a given element may have different masses depending upon the number of neutrons in the nuclei of the atoms. However, regardless of the masses of these nuclides, the atoms of a given element always have the same atomic number; that is, they have the same number of protons in the nucleus.

QUESTIONS AND PROBLEMS

2.1 What are the mks units for the following:
a) Magnetic field strength, H, in Equation (2.1)?
b) Electrostatic field strength, E, in Equation (2.5)?
c) The ratio, E/H?

2.2 What was the relationship between the work of Millikan and that of Thomson?

2.3 For positive rays, the highest q/m value is obtained when hydrogen is used as the residual gas in the cathode-ray tube. What does a high value for q/m imply?

2.4 An automobile has a rest mass of 1.50 tons.
a) What is its mass when it is traveling 50.0 mi/hr?
b) At what speed must the car travel to increase its mass by 1.00 percent of its rest mass?

2.5 A man has a rest mass of 70.0 kg.
a) What is the man's mass when he is traveling 700 mi/hr in a jet airplane?
b) At what speed must the jet airplane travel in order to increase the man's mass by 1.00 kg over his rest mass?

2.6 An electron has a rest mass of 9.11×10^{-28} g.
a) What is the mass of the electron when it is traveling at 10.0 percent of the speed of electromagnetic radiation?
b) At what speed must the electron travel in order to have a mass that is double its rest mass?

2.7 Outline Rutherford's thinking for the postulation of the nuclear atom. Why did he propose a positive nucleus rather than a negative nucleus?

2.8 From the graph in Fig. 2.9 write the general form of the mathematical equation for each of the straight lines and evaluate the constants in each of these expressions.

2.9 One critic of the concept of the nuclear atom has claimed that Rutherford erred in postulating that the deflections of the alpha particles were due to the atomic nucleus. The critic maintains that these deflections were caused by the atoms themselves, and thus what Rutherford called the atomic nucleus should be considered to be the entire atom.

a) What requirement does this view impose upon the packing of the atoms in the metal foil which Rutherford used?

b) How would this view affect the definition of atomic number?

2.10 Often the names for the elements are derived from words or names which depict some significant property of the element or which indicate something about the discovery of the element. In the library, look up the meanings of the following names of the elements:

a) hydrogen
b) helium
c) oxygen
d) neon
e) silicon
f) magnesium
g) phosphorus
h) argon
i) potassium
j) chlorine
k) arsenic
l) selenium
m) bromine
n) krypton
o) rubidium
p) technetium
q) tellurium
r) iodine
s) xenon
t) praseodymium
u) gadolinium
v) dysprosium
w) tungsten
x) radium

2.11 Compute the number of protons and neutrons in the nuclei of each of the following atoms:

a) iodine-127
b) neon-22
c) phosphorus-31
d) $^{26}_{12}Mg$
e) $^{10}_{5}B$
f) $^{27}_{13}Al$

2.12 How would the atomic weights of the elements listed in Tables 2.1 and 2.2 differ if naturally occurring magnesium were taken as the standard at 24.0000 amu? Compute the atomic weights of hydrogen, carbon, and oxygen using this standard.

2.13 Using the data in Table 2.4, compute the q/m value for each particle listed.

2.14 Although the charge on the alpha particle is twice that on the proton, the q/m value for the alpha particle is about half that for the proton. Why?

2.15 Suggest feasible descriptions for particles with the following q/m values:

a) $+1.76 \times 10^8$ coulombs/g
b) -9.58×10^7 coulombs/kg

2.16 Distinguish among the following terms: atomic number, mass number, and atomic weight.

2.17 Why is the atomic weight of carbon listed as 12.01115 instead of 12.00000, since carbon is the standard by which comparative atomic weights are measured?

2.18 Elemental fluorine exists in nature as a single isotope. Why is its atomic weight listed as a fractional value instead of an integral value?

2.19 Naturally occurring hydrogen and lithium each are composed of two nuclides. From the following experimentally determined nuclidic masses in atomic mass units and the data in Tables 2.1 and 2.2, compute the atom percent composition of each of these naturally occurring elements:

a) hydrogen-1: 1.0078 amu; hydrogen-2: 2.0141 amu
b) lithium-6: 6.0151 amu; lithium-7: 7.0160 amu

2.20 In the following table, the compositions of four naturally occurring elements are given in terms of the mass numbers and the experimentally determined masses of the nuclides in atomic mass units. Over-all composition is indicated in *atom* percent. For each of the elements compute the atomic weight.

	MASS NUMBER	NUCLIDIC MASS (amu)	ATOM PERCENT
a) Silicon	28	27.984	92.21
	29	28.984	4.70
	30	29.981	3.09
b) Chlorine	35	34.975	75.77
	37	36.973	24.23
c) Argon	36	35.975	0.337
	38	37.971	0.063
	40	39.971	99.60
d) Potassium	39	38.972	93.83
	40	39.973	0.012
	41	40.971	6.905

2.21 Express the rest mass of the alpha particle as it is recorded in Table 2.4 in atomic mass units.

2.22 Compute the numbers of atoms and moles of atoms in the following:

a) 1.00 g of sulfur
b) 1.00 g of oxygen
c) 1.00 g of uranium
d) 4.25 lb of aluminum
e) 2.00 tons of iron
f) 10^{-9} kg of chromium
g) 43.7 mg of chlorine
h) 1.00 oz of tin
i) 6.58×10^{-13} g of lead
j) 175 pounds of platinum

2.23 Compute the mass in grams of

a) 8.48×10^{18} atoms of naturally occurring zinc
b) 3.14×10^{27} atoms of naturally occurring sodium
c) 5.81 moles of naturally occurring iodine atoms
d) 0.37 mole of naturally occurring magnesium atoms
e) 1 atom of fluorine (Naturally occurring fluorine contains only one nuclide.)

2.24 Criticize each of the following statements:

a) Since particles such as electrons, protons, and neutrons are derived from matter, they must be component parts of matter.
b) Atoms with the same mass number are atoms of the same element, whereas atoms with different mass numbers are atoms of different elements.

3

Atomic Nuclei

INTRODUCTION How the nucleus of an atom is held together remains one of the mysteries of atomic structure. On the average, the radius of an atomic nucleus has been estimated to be about 10^{-13} cm, which is 10^{-5} Å. Radii of whole atoms are about 10^{-8} cm, or 1 Å. These approximate figures suggest that the whole atom has a radius 100,000 times the radius of the nucleus. If it is assumed that atoms and their nuclei are spherical, the volume of the whole atom is about 10^{15} times the volume of the nucleus. Yet, within this very tiny nucleus most of the atomic mass and all of the positive electrostatic charge must be packed. Considering the repulsions that exist between like charges, it is amazing that within an atom of lead, for example, 82 protons may be packed so tightly.

Much current nuclear research revolves around this central problem. At best, the model which describes the nucleus from data now available is not very fully developed. In this chapter, some aspects of nuclear structure and behavior will be summarized through a study of natural and artificial radioactivity.

NATURAL RADIOACTIVITY

In Chapter 2 the discovery and study of radioactivity was mentioned as one of the historical steps in the discovery of fundamental atomic particles. If the Rutherford model of the nuclear atom is accepted, radioactivity provides evidence of nuclear instability, since the ejected particles must originate in the nucleus. Since reactions which occur spontaneously tend to occur in the direction of increasing stability, the ejection of particles from the nucleus must be done to provide a more stable nucleus. Therefore observations of these spontaneous processes provide some notions about the factors that are important in determining nuclear stability.

The three most common changes in naturally occurring radioactive nuclides are alpha emission, beta emission, and gamma emission. It should be noted that whereas alpha and beta emission rarely if ever occur simultaneously from the same nucleus, gamma emission very often accompanies alpha or beta emission.

Alpha Emission

Alpha particles have a mass number of 4 and an atomic number of 2. Thus if a nuclide undergoes alpha emission, the parent nuclide decomposes or is *transmuted* into a daughter nuclide which has both a smaller mass number and a smaller atomic number than the parent nuclide. This transformation may be represented by a nuclear equation in which the parent nuclide is shown to the left of an equals sign with the daughter and emitted particles and/or radiation shown as products on the right. In this way Equation (3.1) represents the spontaneous decay of uranium-238 into thorium-234, alpha particles, and gamma radiation.

$$^{238}_{92}\text{U} = ^{234}_{90}\text{Th} + ^{4}_{2}\text{He} + \gamma \tag{3.1}$$

Since both the alpha particle and the nucleus of helium-4 contain two protons and two neutrons, the alpha particle in Equation (3.1) is represented as the nucleus of a helium-4 atom. As a matter of record, the element helium is found in the vicinity of any alpha-emitting radioactive nuclide.

Equation (3.1) is typical of all nuclear equations in that both the mass numbers and atomic numbers are conserved. Thus the sum of the mass numbers on the right side of the equation equals the sum of the mass numbers on the left side of the equation. Similarly, the sum of the atomic numbers on the right side of the equation equals the sum of the atomic numbers on the left side of the equation.

Equation (3.1) may be interpreted as follows: The nucleus of an atom of uranium-238 decays to the nucleus of an atom of thorium-234, and the nucleus of an atom of helium-4 (an alpha particle) with the concurrent emission of gamma radiation.

In discussing nuclear transformations, it is customary to ignore the extranuclear electrons. However, in fact, whole atoms are involved, including the electrons outside the nucleus. Since the alpha particle is ejected at great speed, it does not pick up any electrons as it passes through the electron cloud of the ^{238}U nucleus. However, when it reaches the atmosphere outside of the atom, it collides with other atoms. Since it is positively charged, it attracts electrons from these other atoms and eventually becomes an electrically neutral helium atom. However, it does this at the expense of other atoms, which are left as positively charged ions.

When the alpha particle leaves the nucleus of the ^{238}U atom, which originally contained 92 protons and was surrounded by 92 electrons, the nucleus is transformed into the nucleus of a ^{234}Th atom, which contains 90 protons but is surrounded by 92 electrons. However, positive ions in the surrounding atmosphere attract the extra electrons from this negatively charged ion, leaving behind an electrically neutral atom of ^{234}Th. The net result is that the surrounding atmosphere has been enriched by the components of a helium atom, first by the doubly charged nucleus and then by the two electrons.

Understanding that the necessary electron changes are made also, Equation (3.1) may be interpreted as follows: an atom of ^{238}U decays into an atom of ^{234}Th and an atom of ^{4}He with concurrent emission of gamma radiation.

For practical purposes, often it is desirable to interpret nuclear equations in terms of quantities larger than an atom. Thus the mole-sized quantities may be used to give the following equivalent interpretations to Equation (3.1):

1. One mole of ^{238}U atoms decays into 1 mole of ^{234}Th atoms and 1 mole of ^{4}He atoms with concurrent emission of gamma radiation;
2. 6.02×10^{23} atoms of ^{238}U decay into 6.02×10^{23} atoms of ^{234}Th and 6.02×10^{23} atoms of ^{4}He (or 6.02×10^{23} alpha particles) with concurrent emission of gamma radiation;
3. In integral numbers, 238 g of ^{238}U decays into 234 g of ^{234}Th and 4 g of ^{4}He with concurrent emission of gamma radiation.

Such interpretations of nuclear equations provide valuable quantitative information. For example, deposits of helium often are found in pockets in the earth near radioactive materials. Assuming that all of the He in the deposit came from the alpha decay of the radioactive material present, and assuming that none of the He has escaped from the deposit over the years, it is possible to compute how much of the radioactive material was present originally.

For example, a small deposit containing 10^6 g of ^{238}U was found in a pocket of the earth which also contained 17,000 g of ^{4}He. From Equation (3.1), it is clear that 4.00 g of helium is produced for each 238 g of ^{238}U which decays. Thus the amount of ^{238}U which has decayed over the years can be found, assuming that the decay of ^{238}U is the only source of helium.

$$1.7 \times 10^4 \text{ g He} \times \frac{38 \text{ g } {}^{238}_{92}\text{U}}{4.00 \text{ g He}} = 1 \times 10^6 \text{ g } {}^{238}_{92}\text{U decayed}$$

The total amount of ^{238}U originally present in the deposit must be the sum of that now present and that which has decayed, or $10^6 \text{ g} + 10^6 \text{ g} = 2 \times 10^6$ g.

Beta Emission

Beta particles are electrons. The modern nuclear atom model is based on evidence that electrons do not exist within the nucleus. However, it is possible that a nucleus may emit an electron, not because an electron as such exists as an entity within the nucleus, but because it may be generated within the nucleus from existing nuclear matter. This may be compared to producing small pieces of glass by throwing a ball through a glass window. The small pieces of glass produced did not exist as separate entities within the pane of glass before the ball hit it, but they were created from the existing glass when the ball hit the window. Similarly, an electron may be created from existing nuclear material, although the electron does not exist as an entity there. This electron may be emitted and is then called a beta particle.

A typical beta-emission process is depicted in Equation (3.2).

$$^{234}_{90}\text{Th} = {}^{234}_{91}\text{Pa} + {}^{0}_{-1}\text{e} + \gamma \tag{3.2}$$

Again note that both atomic number and mass number are conserved, where the atomic number (electrostatic charge) of the electron is designated as -1.

Despite objections to a model of a neutron as a combination of an electron and a proton, the net effect of beta emission can be described in terms of the transformation of a nuclear neutron into a nuclear proton and an electron which is ejected as a beta particle. Thus when beta emission occurs, the parent and daughter nuclides have the same mass numbers, since the net mass effect in the nucleus has been the conversion of one nucleon into another nucleon. However, the daughter nuclide has one more nuclear proton than the parent nuclide, since one of the parent neutrons was converted into a proton in the daughter nuclide. Thus the atomic number of the parent nuclide is one less than the atomic number of the daughter nuclide.

Recalling the interpretations given to the alpha emission Equation (3.1), Equation (3.2) may be interpreted as follows:

1. An atom of ^{234}Th decays into an atom of protactinium-234 and an electron (beta particle) with concurrent emission of gamma radiation;
2. 6.02×10^{23} atoms of ^{234}Th decay into 6.02×10^{23} atoms of ^{234}Pa and 6.02×10^{23} electrons (beta particles) with concurrent emission of gamma radiation;
3. A mole of ^{234}Th atoms decays into a mole of ^{234}Pa atoms and a mole of electrons (beta particles) with concurrent emission of gamma radiation.
4. In integral numbers, 234 g of ^{234}Th decays into 234 g of ^{234}Pa with concurrent emission of beta particles and gamma radiation.

Gamma Emission

Gamma radiation is high-frequency electromagnetic radiation which accompanies many nuclear transformations, including many alpha and beta emissions. Gamma radiation by itself does not affect the atomic number or mass number and so does not lead to transmutations by which one element is transformed into another element.

Radioactive Disintegration Series

Only a few of the naturally occurring nuclides of elements of low atomic number are radioactive. However, all of the naturally occurring nuclides with atomic numbers greater than 83 are radioactive and may be placed in one of the three radioactive disintegration series. Since the only natural radioactive process occurring among these elements which leads to any change in mass number is alpha emission, and since the mass number always changes by four units during this process, these radioactive disintegration series are defined in terms of the mass numbers of the nuclides in each series. The $4n$ series contains all nuclides for which the mass number is exactly divisible by 4 with no remainder. The $4n + 2$ series contains all nuclides for which the mass number is divisible by 4 with a remainder of 2. The $4n + 3$ series contains all nuclides for which the mass number is divisible by 4 with a remainder of 3.

The $4n + 1$ series is conspicuous by its absence in this list. None of the radioactive nuclides which should fall into this series has been found in nature, although the series has been identified through artificially produced nuclides.

The $4n$ radioactive disintegration series decay scheme is given in detail in Table 3.1. In this table, bismuth-212 provides an example of *branched chain* decay

TABLE 3.1
The 4n Radioactive Disintegration Series

$$
\begin{aligned}
{}^{232}_{90}\mathrm{Th} &= {}^{228}_{88}\mathrm{Ra} + {}^{4}_{2}\mathrm{He} + \gamma \\
{}^{228}_{88}\mathrm{Ra} &= {}^{228}_{89}\mathrm{Ac} + {}^{0}_{-1}\mathrm{e} \\
{}^{228}_{89}\mathrm{Ac} &= {}^{228}_{90}\mathrm{Th} + {}^{0}_{-1}\mathrm{e} + \gamma \\
{}^{228}_{90}\mathrm{Th} &= {}^{224}_{88}\mathrm{Ra} + {}^{4}_{2}\mathrm{He} + \gamma \\
{}^{224}_{88}\mathrm{Ra} &= {}^{220}_{86}\mathrm{Rn} + {}^{4}_{2}\mathrm{He} + \gamma \\
{}^{220}_{86}\mathrm{Rn} &= {}^{216}_{84}\mathrm{Po} + {}^{4}_{2}\mathrm{He} + \gamma \\
{}^{216}_{84}\mathrm{Po} &= {}^{212}_{82}\mathrm{Pb} + {}^{4}_{2}\mathrm{He} \\
{}^{212}_{82}\mathrm{Pb} &= {}^{212}_{83}\mathrm{Bi} + {}^{0}_{-1}\mathrm{e} + \gamma \\
{}^{212}_{83}\mathrm{Bi} \nearrow &= {}^{212}_{84}\mathrm{Po} + {}^{0}_{-1}\mathrm{e} + \gamma \\
{}^{212}_{83}\mathrm{Bi} \searrow &= {}^{208}_{81}\mathrm{Tl} + {}^{4}_{2}\mathrm{He} \\
{}^{212}_{84}\mathrm{Po} &= {}^{208}_{82}\mathrm{Pb} + {}^{4}_{2}\mathrm{He} \\
{}^{208}_{81}\mathrm{Tl} &= {}^{208}_{82}\mathrm{Pb} + {}^{0}_{-1}\mathrm{e} + \gamma
\end{aligned}
$$

in which ${}^{212}Bi$ actually decays by two different paths; some of the nuclei decay by alpha emission and others by beta emission. It should be noted that this branch is closed, since both of the products of these two decay patterns decay to the same nuclide in the next step. Thus ${}^{212}Bi$ decays to both polonium-212 and thallium-208. However, both ${}^{212}Po$ and ${}^{208}Tl$ decay to lead-208. All of the other radioactive disintegration series have similar branches involving other nuclides.

The $4n + 1$ series is the artificial series which begins with plutonium-241 and ends with bismuth-209. The $4n + 2$ series begins with ${}^{238}U$ and ends with ${}^{206}Pb$. The $4n + 3$ series begins with ${}^{235}U$ and ends with ${}^{207}Pb$.

A few radioactive nuclides with atomic numbers less than 83 do occur in nature, but generally they are not very abundant. Some of these are listed in Table 3.2, along with their mode of decay.

TABLE 3.2
Some Naturally Occurring Radioactive Nuclides Among the Elements of Atomic Number Less than 83

NUCLIDE	DECAY	NUCLIDE	DECAY
$^{14}_{6}C$	β	$^{147}_{62}Sm$	α
$^{40}_{19}K$	β	$^{148}_{62}Sm$	α
$^{50}_{23}V$	β	$^{152}_{64}Gd$	α
$^{87}_{37}Rb$	β	$^{176}_{71}Lu$	β
$^{115}_{49}In$	β	$^{174}_{72}Hf$	α
$^{138}_{57}La$	β	$^{187}_{75}Re$	β
$^{142}_{58}Ce$	α	$^{190}_{78}Pt$	α
$^{144}_{60}Nd$	α	$^{192}_{78}Pt$	α

NUCLEAR STABILITY

The nonradioactive, naturally occurring nuclides of the elements are listed in Table 3.3. Careful scrutiny of Table 3.3 shows several significant relationships.

TABLE 3.3
The Naturally Occurring Nonradioactive Nuclides

ATOMIC NUMBER	ELEMENT	MASS NUMBERS
1	Hydrogen	1, 2
2	Helium	3, 4
3	Lithium	6, 7
4	Beryllium	9
5	Boron	10, 11
6	Carbon	12, 13
7	Nitrogen	14, 15
8	Oxygen	16, 17, 18
9	Fluorine	19
10	Neon	20, 21, 22
11	Sodium	23
12	Magnesium	24, 25, 26
13	Aluminum	27
14	Silicon	28, 29, 30
15	Phosphorus	31
16	Sulfur	32, 33, 34, 36
17	Chlorine	35, 37
18	Argon	36, 38, 40
19	Potassium	39, 41
20	Calcium	40, 42, 43, 44, 46
21	Scandium	45
22	Titanium	46, 47, 48, 49, 50
23	Vanadium	51
24	Chromium	50, 52, 53, 54
25	Manganese	55

TABLE 3.3 (cont.)

ATOMIC NUMBER	ELEMENT	MASS NUMBERS
26	Iron	54, 56, 57, 58
27	Cobalt	59
28	Nickel	58, 60, 61, 62, 64
29	Copper	63, 65
30	Zinc	64, 66, 67, 68, 70
31	Gallium	69, 71
32	Germanium	70, 72, 73, 74, 76
33	Arsenic	75
34	Selenium	74, 76, 77, 78, 80, 82
35	Bromine	79, 81
36	Krypton	78, 80, 82, 83, 84, 86
37	Rubidium	85
38	Strontium	84, 86, 87, 88
39	Yttrium	89
40	Zirconium	90, 91, 92, 94, 96
41	Niobium	93
42	Molybdenum	92, 94, 95, 96, 97, 98, 100
43	Technetium	
44	Ruthenium	96, 98, 99, 100, 101, 102, 104
45	Rhodium	103
46	Palladium	102, 104, 105, 106, 108, 110
47	Silver	107, 109
48	Cadmium	106, 108, 110, 111, 112, 113, 114, 116
49	Indium	113
50	Tin	112, 114, 115, 116, 117, 118, 119, 120, 122, 124
51	Antimony	121, 123
52	Tellurium	120, 122, 123, 124, 125, 126, 128, 130
53	Iodine	127
54	Xenon	124, 126, 128, 129, 130, 131, 132, 134, 136
55	Cesium	133
56	Barium	130, 132, 134, 135, 136, 137, 138
57	Lanthanum	139
58	Cerium	136, 138, 140
59	Praseodymium	141
60	Neodymium	142, 143, 145, 146, 148, 150
61	Promethium	
62	Samarium	144, 148, 149, 150, 152, 154
63	Europium	151, 153
64	Gadolinium	154, 155, 156, 157, 158, 160
65	Terbium	159
66	Dysprosium	156, 158, 160, 161, 162, 163, 164
67	Holmium	165
68	Erbium	162, 164, 166, 167, 168, 170

TABLE 3.3 (cont.)

ATOMIC NUMBER	ELEMENT	MASS NUMBERS
69	Thulium	169
70	Ytterbium	168, 170, 171, 172, 173, 174, 176
71	Lutetium	175
72	Hafnium	176, 177, 178, 179, 180
73	Tantalum	180, 181
74	Tungsten	180, 182, 183, 184, 186
75	Rhenium	185
76	Osmium	184, 186, 187, 188, 189, 190, 192
77	Iridium	191, 193
78	Platinum	192, 194, 195, 196, 198
79	Gold	197
80	Mercury	196, 198, 199, 200, 201, 202, 204
81	Thallium	203, 205
82	Lead	204, 206, 207, 208
83	Bismuth	209

These observations were first noted by W. D. Harkins in 1928 and are sometimes called Harkins' rules.

1. Nuclides with even atomic numbers are much more numerous than nuclides with odd atomic numbers. There are no elements known which have odd atomic numbers and more than two stable (nonradioactive) naturally occurring nuclides. There are two elements with atomic numbers less than 83, the existence of which, in nature, has not been proven. Both of these have odd atomic numbers, namely 43 and 61. By actual count, of the 269 nuclides listed in Table 3.3, only 55 have odd atomic numbers whereas the remaining 214 have even atomic numbers.
2. Nuclides with even mass numbers are more numerous than nuclides with odd mass numbers. By actual count in Table 3.3, there are 105 nuclides with odd mass numbers and 164 nuclides with even mass numbers.
3. Nuclides with even numbers of neutrons are more numerous than nuclides with odd numbers of neutrons. By actual count in Table 3.3 there are 60 nuclides with odd numbers of neutrons and 209 with even numbers of neutrons.
4. Among the stable (nonradioactive) naturally occurring nuclides, there is a tendency for nuclei to contain even numbers of both protons and neutrons, and very little tendency for nuclei to contain odd numbers of both protons and neutrons. The odd-even arrangements of nucleons for the nuclides shown in Table 3.3 are given in Table 3.4. Of the five nuclides which have odd numbers of both protons and neutrons, four have atomic numbers of 7 or less.
5. Only hydrogen-1 and helium-3 have more protons than neutrons in their atomic nuclei. Among the nuclides of low atomic number (about 20 or less), the numbers of protons and neutrons tend to be about the same. For nuclides of higher atomic number, the number of neutrons tends to be greater than the number of protons.

TABLE 3.4

Odd-Even Proton and Neutron Distributions Among the Nuclides Shown in Table 3.3

Even number of protons, even number of neutrons	159
Even number of protons, odd number of neutrons	55
Odd number of protons, even number of neutrons	50
Odd number of protons, odd number of neutrons	5

Another illustration of the tendency toward even numbers of nucleons in the nuclei of stable nuclides is given in Table 3.5, which lists the seven nuclides that constitute 95 atom percent of all of the nuclides in the crust of the earth. Note that only two of these seven have odd numbers of protons and odd mass numbers, whereas all of them have even numbers of neutrons. All of them have atomic numbers less than 30 and mass numbers less than 60. Four of these nuclides have equal numbers of neutrons and protons and two of the remaining nuclides have just one more neutron than protons.

TABLE 3.5

The Nuclidic Composition of the Earth's Crust for the Seven Most Abundant Nuclides

NUCLIDE	ATOM PERCENT
$^{16}_{8}O$	62
$^{28}_{14}Si$	19
$^{27}_{13}Al$	6.4
$^{23}_{11}Na$	2.5
$^{40}_{20}Ca$	1.9
$^{56}_{26}Fe$	1.8
$^{24}_{12}Mg$	1.5

One of the early observations concerning nuclear stability involved the relationship between the numbers of protons and neutrons in the nucleus. Figure 3.1 shows a plot of the number of neutrons versus the number of protons in the stable nuclei. This plot illustrates Harkins' fifth rule and shows that the stable nuclides fall into a fairly narrow belt. Sometimes this is called the *belt of stability.* As noted in Harkins' rule, these nuclides lie close to the straight line in Fig. 3.1 which represents equal numbers of protons and neutrons. However, as atomic number increases, the number of neutrons increases more rapidly than the number of protons.

From Harkins' rule and the belt of stability, it has been suggested that nuclear stability is enhanced by approximately equal numbers of neutrons and protons in the nucleus. Stated in another way, stability of the nucleus is to be expected when the ratio of the number of neutrons to protons, often represented by n/p, is nearly unity. Roughly, when this ratio exceeds 1.4 to 1.6, nuclei tend to be unstable and radioactive. Insofar as this is true, the effect of radioactive decay should be to reduce the value of n/p.

Beta emission can be explained in these terms. When a beta particle is emitted

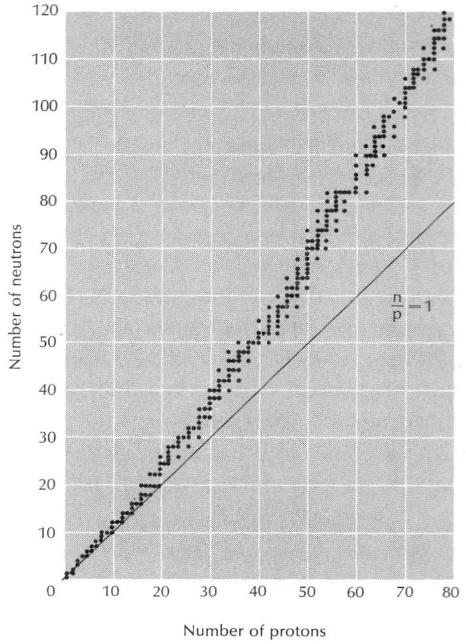

Fig. 3.1. Belt of stability: number of neutrons against number of protons for stable nuclides

the effect is the same as the conversion of a neutron into an electron and a proton. Thus the number of neutrons is decreased by one and the number of protons is increased by one so that the n/p value is decreased.

However, alpha emission cannot be explained in these terms, suggesting that there is more than neutron-to-proton ratio involved in the determination of nuclear stability. In fact, alpha emission usually leads to a higher n/p value. For example, $^{232}_{90}Th$ decays by alpha emission. This parent nuclide contains 90 protons and 142 neutrons, so that $n/p = 142/90 = 1.58$. The daughter nuclide is $^{228}_{88}Ra$, which contains 88 protons and 140 neutrons, so that $n/p = 140/88 = 1.59$, slightly higher than the value for the ratio for the parent.

Nucleon Interactions

Some of the other factors involved in nuclear stability may be understood through consideration of the approach of two protons. Electrostatically these protons repel each other so that the approach is accompanied by an increase in potential energy. This is shown by Fig. 3.2, which is identical to the upper curve in Fig. 1.16. Clearly, the potential energy becomes immense by the time the protons are as close as they must be in an atomic nucleus. Two possible explanations are immediately obvious. If large numbers of protons are to be packed into very small nuclei, either the electrostatic repulsions do not exist under nuclear conditions, or attractive forces within the nucleus exceed the repulsive forces among the protons. The second approach has proved the more successful.

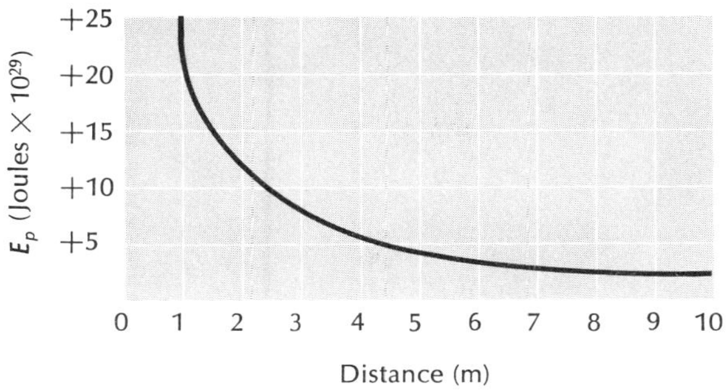

Fig. 3.2. Potential energy change as two protons approach

To understand this point of view, consider the hydrogen-2 nuclide, often called *deuterium* or heavy hydrogen, which has a nucleus composed of a proton and a neutron. This nucleus is called a *deuteron*. Next, imagine a proton approaching the deuteron. Since the deuteron and the proton are positively charged, the two particles repel each other. This means that work must be done to bring them together; that is, the system reaches higher and higher potential energy states as the two particles approach. This is shown graphically in Fig. 3.3, which resembles

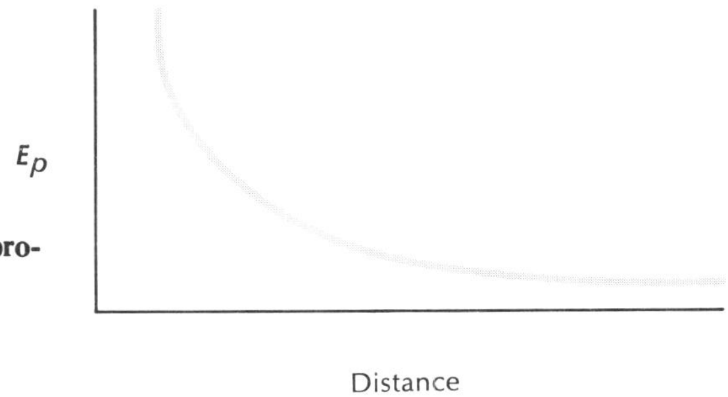

Fig. 3.3. Potential energy change as a proton approaches a deuteron

Fig. 3.2. Since the qualitative picture is to be developed here, numbers have been omitted from the graph in Fig. 3.3.

Figure 3.3 suggests that the potential energy approaches infinity as the distance between the proton and the deuteron approaches zero. However, this does not appear to be the case. The graph in Fig. 3.3 was discontinued before very small distances were considered. Figure 3.4 shows a continuation of Fig. 3.3 at very

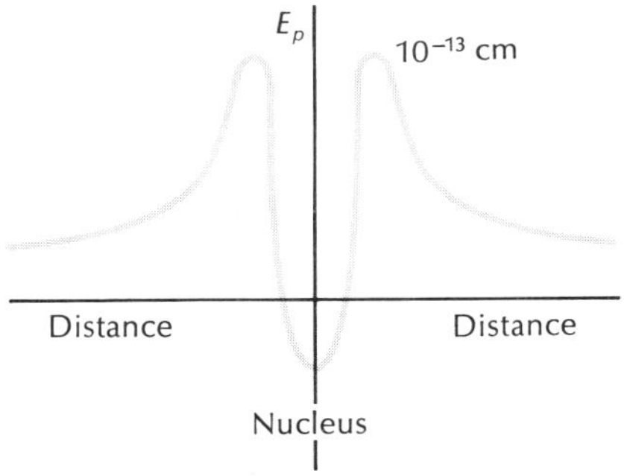

Fig. 3.4. Potential energy change as a proton approaches a deuteron very closely

short distances. When the distance between the proton and the deuteron approaches 10^{-13} cm, very strong attractive forces come into play. The nature of these forces is not understood, but the forces appear to be greater than the repulsive forces between the protons. With the attractive forces operating, the system decreases very markedly in potential energy.

In using the diagram in Fig. 3.4, it has become customary to look at the potential energy change from the point of view of the proton as it approaches the deuteron. Thus the proton climbs a *potential energy hill;* when it reaches the critical distance at which the strong nuclear attractive forces become significant, it is said to fall into a *potential energy well.* To use another related analogy, the change in potential energy for the proton is similar to the change in gravitational potential energy for an animal which climbs the side of a volcano and then suddenly falls into the crater.

Although the exact nature of these strong attractive forces is not well understood, it does appear that they are internucleonic forces independent of electrostatic charges. Considering the two kinds of nucleons, namely protons and neutrons, there can be three different attractive forces defined, specifically those between protons (designated as *p-p* forces), those between neutrons (designated as *n-n* forces), and those between neutrons and protons (designated as *n-p* forces). It should be noted that the *p-p* forces defined here are attractive forces and do not represent the additional repulsive forces due to electrostatic repulsions.

Using this approach, it is possible to begin to explain the n/p value trends. The internucleonic attractive forces appear to be very short range forces, and as nuclei become more complex, the addition of more nucleons leads to smaller and smaller decreases in potential energy. In addition, as the number of protons increases, the potential energy increases due to the electrostatic repulsions. Thus nuclear stability depends upon the balance between the internucleonic attractions and the electrostatic repulsions among the protons.

Since the nuclei of nuclides with low atomic numbers have relatively few protons, their internucleonic attractions are greater than their electrostatic repulsions. However, for nuclides of somewhat higher atomic number, the number of protons is greater and the electrostatic repulsions are greater. As a result, the number of neutrons must be greater to provide additional *n-n* and *n-p* forces to counterbalance this marked increase in repulsion. For nuclides with very high atomic numbers, where there are diminishing returns in attractive forces with increasing numbers of neutrons, the important factor is the protonic repulsions. Consequently, alpha emission is common among these nuclides as a means of decreasing the number of protons and thus of decreasing electrostatic repulsion.

Potential-energy Levels

It still is not clear why such electrostatic repulsions are diminished by alpha emission instead of direct proton emission. Some insight into this problem may be gained by making an improvement in the depiction of the potential-energy well of Fig. 3.4. The nature of gamma emission as discussed in the following section provides some evidence that the potential-energy decrease is not continuous as the proton falls into the potential-energy well. A better representation is given in Fig. 3.5, where the potential-energy decrease is shown as a series of steps down the side of the potential-energy well. Thus the proton approaches the state of lowest potential energy by assuming only those potential energies which correspond to the horizontal steps in Fig. 3.5.

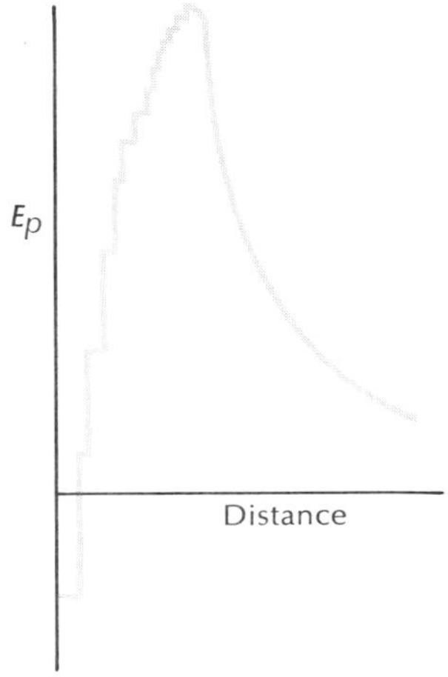

Fig. 3.5. Stepwise potential energy decrease for the close approach of a proton and a deuteron

Each of the horizontal steps shown in Fig. 3.5 is called an *energy level.* The final potential energy of the proton in the nucleus depends upon a number of factors, including the nature of the internucleonic forces. Since spontaneous reactions tend to occur toward states of lowest potential energy, it is expected that the proton will assume the lowest potential energy available to it. However, the key to this expectation is in the word "available." If several nucleons are present in the nucleus, it may not be possible for them all to occupy the same energy level; that is, it may not be possible for all of them to be on the same step. There appear to be limits to the numbers of nucleons that can occupy any given potential-energy level. This means that an approaching nucleon will occupy the lowest energy level available to it.

The capacities of the nuclear energy levels are not well known. There is some evidence that a group of two protons and two neutrons may constitute a kind of closed energy level, that is, a full step. This group of two protons and two neutrons is a potential alpha particle. This may help to explain why alpha-particle emission is so common among the naturally occurring nuclides.

Gamma Emission

As a proton approaches a deuteron and descends into the energy well, it undergoes finite and discrete changes in potential energy. Thus if the proton has potential energy E_1 on a top step and then falls to potential energy E_2 on a lower step, the potential-energy change in this transition may be represented as follows:

$$\Delta E_p = E_2 - E_1 \qquad \textbf{(3.3)}$$

Since there is a decrease in potential energy in the process being described, ΔE_p is a precisely defined and discrete packet of energy which must be lost to the nuclear system. This discrete packet of energy is an emitted photon. For transitions within atomic nuclei, the emitted photons appear as gamma radiation. Since photons of gamma radiation have fairly high frequencies and consequently high energies, the jumps between nuclear energy levels are fairly large.

In natural processes, rarely are particles observed approaching nuclei and falling into potential-energy wells. More frequently, some particle emission is observed with accompanying gamma radiation. In the process of undergoing beta or alpha decay, a daughter nucleus may be produced in a high-energy state called an *excited state*. This means that one or more nucleons are not in the lowest possible potential-energy levels after the particle has been emitted. Then the nucleons in the nucleus shift from their high energy levels to the lower energy levels. During this adjustment, gamma radiation is emitted.

Since the nucleons must occupy specific, discrete potential energy levels, only a finite number of specific transitions are possible in the shift from high to low energy levels. As a result, gamma radiation is observed in only a few specific frequencies. This is true since ΔE_p from Equation (3.3) is the energy E_γ, which is associated with the photon of gamma radiation. However, the energy of the photon of gamma radiation is related to the frequency of the gamma radiation ν_γ through Equation (2.9).

$$E_\gamma = h\nu_\gamma \tag{3.4}$$

Thus each frequency of gamma radiation emitted represents a transition of a nucleon from one energy level to another lower energy level.

This may be made clearer with a specific example. Bismuth-212 decays to thallium-208 by alpha emission with simultaneous gamma emission as represented in Equation (3.5).

$$^{212}_{83}\text{Bi} = {}^{208}_{81}\text{Tl} + {}^{4}_{2}\text{He} + \gamma \tag{3.5}$$

The wavelengths and corresponding energy equivalents are listed in Table 3.6 for the gamma radiation detected in this process. These data have been interpreted to mean that bismuth-212 decays into an excited thallium-208 nuclide. The nucleons

TABLE 3.6
Gamma Radiation Associated with the Alpha Decay of Bismuth-212

WAVELENGTH (microns)	PHOTON ENERGY (kcal)
3.12×10^{-5}	1.53×10^{-18}
4.32×10^{-6}	1.10×10^{-17}
3.79×10^{-6}	1.26×10^{-17}
2.86×10^{-6}	1.66×10^{-17}
2.66×10^{-6}	1.73×10^{-17}
2.63×10^{-6}	1.81×10^{-17}

in this excited thallium-208 nuclide then rearrange themselves by making transitions to lower available energy levels. Each of these transitions is accompanied by a specific wavelength of gamma radiation. The two-step nature of the over-all process is represented by Equation (3.6), where $[^{208}_{81}\text{Tl}]^*$ represents the excited potential energy state for thallium-208.

$$^{212}_{83}\text{Bi} = {}^{4}_{2}\text{He} + [^{208}_{81}\text{Tl}]^* \qquad (3.6)$$
$$\| $$
$$^{208}_{81}\text{Tl} + \gamma$$

The data in Table 3.6 have been interpreted diagrammatically in Fig. 3.6. In this diagram, for the sake of convenience, the lowest of the five energy levels represented has been given an arbitrary potential energy of zero so that all energy values assigned to the other levels represent values above the lowest. With the energy levels numbered from 1 to 5 starting with the lowest, the six transitions observed are from level 2 to level 1; from level 3 to level 1; from level 3 to level 2; from level 4 to level 1; from level 4 to level 2; from level 5 to level 2.

The diagram in Fig. 3.6 is really a front view of a portion of the potential-energy well steps for the thallium-208 nucleus, which from the side resembles Fig. 3.5. The energy levels in Fig. 3.6 are spaced to scale; it is obvious that the levels are not necessarily equally spaced. Similarly, it is obvious that the nucleons may jump through any integral number of levels, and that some levels may be skipped in any single transition. It should be noted also that Fig. 3.6 does not necessarily include all of the potential energy levels in the nucleus, but only those involved in the

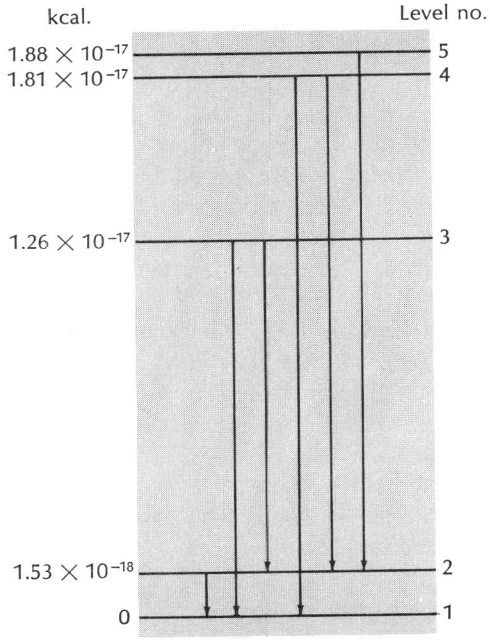

Fig. 3.6. Nuclear energy level diagram for thallium-208

transitions from the excited thallium-208 nucleus left by the departure of an alpha particle from the bismuth-212 nucleus.

Binding Energy

Since gamma radiation is high-energy radiation, nuclear energy changes are large. Considering the small masses of the particles involved, it should not be surprising that the large energy changes are manifested in detectable mass changes as well. These mass changes reveal something about nuclear stability.

For any nucleus, the difference between the sum of the rest masses of the nucleons and the actual mass of the nucleus represents the actual net decrease in potential energy for these nucleons. This can be seen in Figure 3.7, where the idealized potential energy curve is shown. Here potential energy E_A is the zero potential energy characteristic of infinite separation of all nucleons. Potential energy E_B is the aggregate potential energy of the nucleons in the nucleus at its lowest state of potential energy. The net loss in potential energy is represented by $\Delta E_p = E_B - E_A$. The potential energy E_C represents the potential energy at the

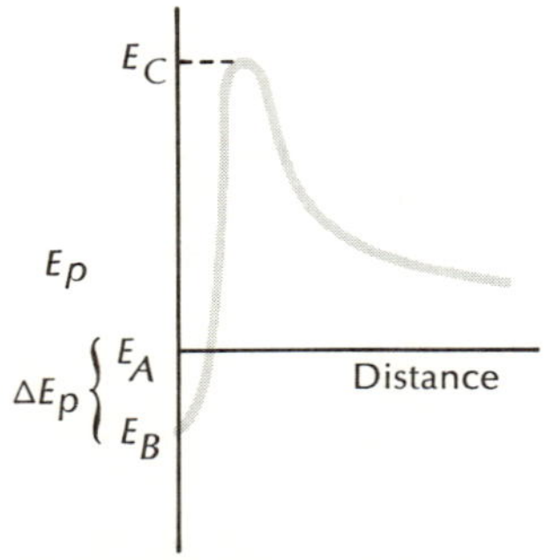

Fig. 3.7. Net potential energy decrease for nucleons forming a nucleus

crest of the potential-energy barrier as the nucleons approach each other. However, any increase in potential energy required to get over this hill is recovered by the large decrease in potential energy within the well. Thus ΔE_p is the net decrease in potential energy, and the mass equivalent of ΔE_p is the observed decrease in mass. This net decrease in potential energy, ΔE_p, is called the *nuclear binding energy*.

The binding energy for any nuclide can be calculated by taking the difference between the sum of the masses of the nucleons at rest and the nuclear mass. In practice, however, with all but the simplest nuclei, it is difficult to measure the exact mass of the nucleus without including the masses of the extranuclear electrons. This is taken into account by using experimentally determined atomic masses from which the rest masses of the electrons, protons, and neutrons are subtracted. This assumes that the electrons lose very little mass in going from their free rest state to the electron cloud outside the nucleus of an atom. This is a fairly valid assumption in most cases.

Consider deuterium as an example. The atom consists of one electron, one proton, and one neutron. The atomic mass of the deuterium atom is 2.014102 amu.

The rest masses of the separate particles are as follows:

Rest mass of one electron	0.000549 amu
Rest mass of one proton	1.007277 amu
Rest mass of one neutron	1.008665 amu
Sum of rest masses	2.016491 amu

The difference between the sum of the rest masses of the particles and the mass of the atom itself is 2.016491 − 2.014102 amu, or 0.002389 amu. The energy equivalent of this mass may be found directly from the energy-mass relationship: $E = mc^2$.

$$E = mc^2 = 2.389 \times 10^{-3}\ \text{amu} \times \frac{1.6602 \times 10^{-24}\ \text{g}}{1.0000\ \text{amu}} \times \frac{(2.9979 \times 10^{10})^2\ \text{cm}^2}{\text{sec}^2}$$

$$= 3.564 \times 10^{-6}\ \text{erg}$$

This is the value of the nuclear binding energy. However, since chemists tend to express energies in calories or kilocalories, ergs may be converted to kilocalories as follows:

$$3.564 \times 10^{-6}\ \text{erg} \times \frac{2.389 \times 10^{-11}\ \text{kcal}}{1.0000\ \text{erg}} = 8.514 \times 10^{-17}\ \text{kcal}$$

Thus the nuclear binding energy for deuterium is 8.514×10^{-17} kcal.

The binding energy is a measure of nuclear stability, since it not only represents the amount of potential energy decrease observed when the nucleons are brought together to form the nucleus, but in reverse fashion, it represents the net amount of energy that must be returned to the nucleus in order to separate it into its individual nucleons.

However, it is unfair to compare nuclei of two different atoms with different numbers of nucleons, since larger numbers of nucleons normally lead to larger total binding energies. Somewhat more meaningful trends can be observed if the total binding energy for a nuclide is divided by the total number of nucleons to obtain the *nuclear binding energy per nucleon.* In a sense this is unfair, too, since this tends to imply that each nucleon is bound into the same nucleus with the same amount of nuclear binding energy, which is not correct. Certainly nucleons in high potential energy levels in the nucleus contribute less to the over-all binding energy than do nucleons in lower potential energy levels.

Figure 3.8 shows a plot of the total binding energy of nonradioactive naturally occurring nuclides versus the mass number. Figure 3.9 shows a similar plot of the nuclear binding energy per nucleon for these same nuclides versus the mass number. It is clear in Fig. 3.8 that total binding energy increases as mass number increases, but that the rate of increase falls off at high mass numbers.

A moderately good, smooth, average curve can be drawn through the points in Fig. 3.9. However, the nuclear binding energies per nucleon are high for helium-4, carbon-12, and oxygen-16, all of which have just the right number of nucleons to form integral numbers of alpha particle configurations in the nucleus. This is

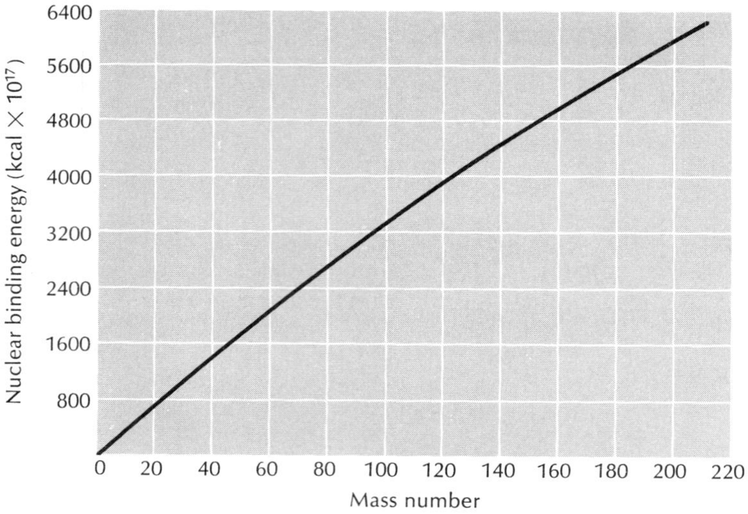

Fig. 3.8. Total binding energy against mass number

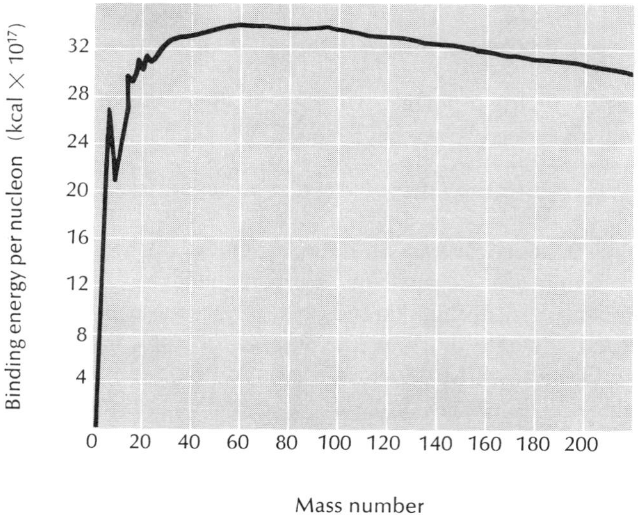

Fig. 3.9. Binding energy per nucleon against mass number

further evidence for the closed nuclear energy levels consisting of two neutrons and two protons. The curve in Fig. 3.9 also shows a maximum value for the nuclear binding energy per nucleon in the vicinity of mass number 56. As the mass number increases, the nuclear binding energy per nucleon rises rapidly when the mass numbers are small. This reflects the large increase in internucleonic attractions when there are relatively few nucleons present. At higher mass numbers, the binding energy per nucleon levels off and eventually decreases. This reflects the diminishing returns obtained in potential energy decreases as nucleons are added to nuclei containing large numbers of nucleons.

In Fig. 3.9 it can be seen that the splitting of the nucleus of a nuclide of mass number 200 into two nuclides with mass numbers of about 100 each is accompanied by an increase in nuclear binding energy per nucleon. This corresponds to moving from a point on the curve to the far right upward to a point further to the left. The increase in nuclear binding energy per nucleon corresponds to a decrease in potential energy, which means that energy is evolved in the process. This is the basis for nuclear *fission* in which nuclides with large mass numbers are split into nuclides with smaller mass numbers. Nuclear fission is the basis for atomic energy.

Also from Fig. 3.9, it is apparent that large amounts of energy might be obtained if nuclei with very small mass numbers could be fused to produce nuclei with larger mass numbers. For example, if two deuterons could be fused to produce a helium-4 nucleus, the steep rise in the nuclear binding energy per nucleon curve suggests that very large amounts of energy would be evolved. This process is called *nuclear fusion*, which is the basis for current research on thermonuclear energy, of which the hydrogen bomb is one aspect.

Radioactive Decay Rates

Although a nuclide may be radioactive, not all of the atoms disintegrate at the same time. Rather, the sample decays gradually over a period of time. For example, thorium-231 is a beta emitter. If a series of identical samples of this nuclide are prepared and allowed to decay so that each sample in the series decays for a different length of time, the amount of thorium-231 remaining after each decay period can be related to the length of the decay period. Some typical data on four sets of samples are given in Table 3.7. The symbol N_0 represents the number of milli-

TABLE 3.7
Decay Rate Data for Thorium-231

t (HR)	N_t = mg OF THORIUM-231 LEFT AFTER TIME t			
	$N_0 = 1.00$ mg	$N_0 = 1.89$ mg	$N_0 = 3.00$ mg	$N_0 = 5.00$ mg
0	1.00	1.89	3.00	5.00
8	0.805	1.52	2.42	4.03
16	0.647	1.22	1.94	3.24
24	0.521	0.984	1.56	2.61
32	0.419	0.791	1.26	2.09
40	0.337	0.637	1.01	1.69
48	0.272	0.513	0.815	1.36
56	0.218	0.412	0.655	1.09
64	0.176	0.332	0.528	0.879
72	0.141	0.267	0.424	0.706

grams of thorium-231 present in the sample initially. The symbol N_t represents the number of milligrams of thorium-231 present in the sample after time t has elapsed.

In Fig. 3.10, N_t is plotted against t in hours for each of these four sets of samples. The decreasing steepness of a single curve as time passes, as well as the greater

TABLE 3.8

Logarithmic Decay rate Data for Thorium-231

	N_o = 1.00 mg			N_o = 1.89 mg			N_o = 3.00 mg			N_o = 5.00 mg		
t (HR)	N_t(mg)	LOG N_t	LOG $\frac{N_t}{N_o}$	N_t(mg)	LOG N_t	LOG $\frac{N_t}{N_o}$	N_t(mg)	LOG N_t	LOG $\frac{N_t}{N_o}$	N_t(mg)	LOG N_t	LOG $\frac{N_t}{N_o}$
0	1.00	0	0	1.89	0.276	0	3.00	0.477	0	5.00	0.699	0
8	0.805	−0.094	−0.094	1.52	0.182	−0.094	2.42	0.383	−0.094	4.03	0.605	−0.094
16	0.647	−0.189	−0.189	1.22	0.087	−0.189	1.94	0.288	−0.189	3.24	0.510	−0.189
24	0.521	−0.283	−0.283	0.984	−0.007	−0.283	1.56	0.194	−0.283	2.61	0.416	−0.283
32	0.419	−0.378	−0.378	0.791	−0.102	−0.378	1.26	0.099	−0.378	2.09	0.321	−0.378
40	0.337	−0.472	−0.472	0.637	−0.196	−0.472	1.01	0.005	−0.472	1.69	0.227	−0.472
48	0.272	−0.566	−0.566	0.513	−0.290	−0.566	0.815	−0.089	−0.566	1.36	0.133	−0.566
56	0.218	−0.661	−0.661	0.412	−0.385	−0.661	0.655	−0.184	−0.661	1.09	0.038	−0.661
64	0.176	−0.755	−0.755	0.332	−0.479	−0.755	0.528	−0.278	−0.755	0.879	−0.056	−0.755
72	0.141	−0.850	−0.850	0.267	−0.574	−0.850	0.424	−0.373	−0.850	0.706	−0.151	−0.850

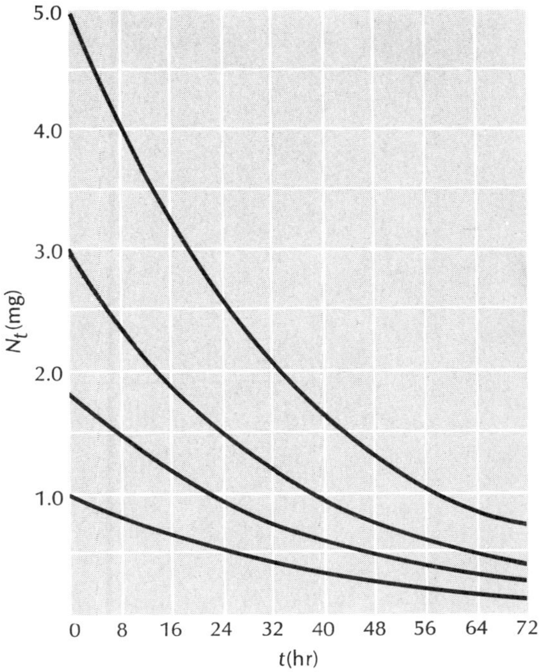

Fig. 3.10. Radioactive decay of thorium-231: N_t against time

steepness of the curve foı $N_0 = 5.00$ mg as compared to the curves when N_0 is less than 5.00 mg, shows that the *rate* at which the decay occurs depends upon the amount of radioactive nuclide present. Thus the greater the amount of nuclide present, the greater the *rate* of decay, expressed in this case as mg/hr.

Although the relationship between N_t and time is not linear, the shape of the curves suggests a linear logarithmic relationship. To test this possibility, the values for log N_t have been recorded in Table 3.8, and these have been plotted against time in Fig. 3.11. From this graph it is obvious that log N_t is pseudopro-proportional to t. Since the four lines are parallel, they have the same slope, but they differ in the vertical axis intercept. The general relationship is expressed by Equation (3.7), where K_1 is the slope, which is the same for all samples, and K_2 is the vertical axis intercept, which has a different numerical value for each sample.

$$\log N_t = K_1 t + K_2 \tag{3.7}$$

Scrutiny of the intercepts on the vertical axis reveals that although the numerical values differ from sample to sample, all of them are the value for log N_0 for the sample. Thus $K_2 = \log N_0$ and Equation (3.7) may be rewritten as Equation (3.8), which may be rearranged to Equation (3.9).

$$\log N_t = K_1 t + \log N_0 \tag{3.8}$$

$$\log N_t - \log N_0 = K_1 t$$

$$\log \frac{N_t}{N_0} = K_1 t \tag{3.9}$$

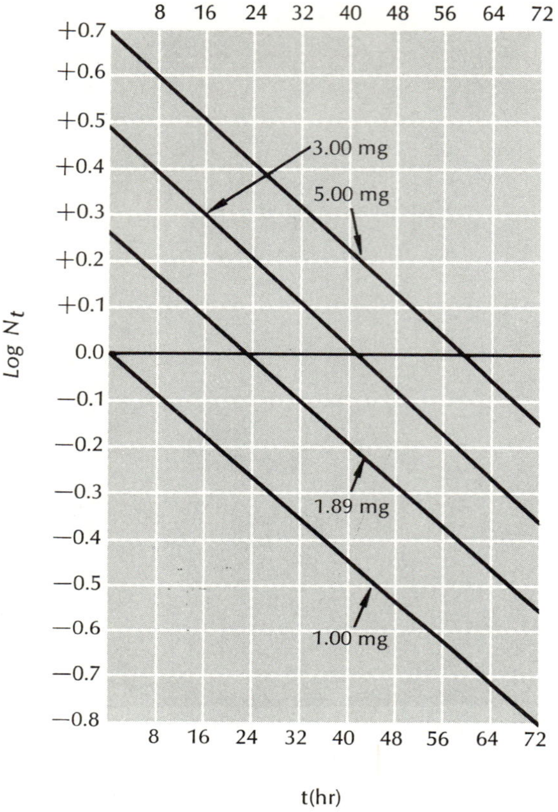

Fig. 3.11. Radioactive decay of thorium-231: log N_t against time

From Equation (3.9) it would appear that log N_t/N_0 must be constant for any time interval t, and thus independent of sample size. This is borne out by the data in Table 3.8. Thus Equation (3.9) and the graph in Fig. 3.12 derived from this equation are descriptions of the decay of thorium-231 regardless of the size of the sample which is chosen. In practice, the proportionality constant K_1 is expressed as the product of two constants:

$$K_1 = -0.434\lambda$$

This permits rewritiag Equation (3.9) as Equation (3.10) ,which is the form usually encountered.

$$\log \frac{N_t}{N_0} = -0.434\lambda t \tag{3.10}$$

The value 0.434 is factored out as a matter of convenience since it is the factor used to convert natural logarithms to the base e to common logarithms to the base 10. The negative sign is included, since the slope of the line in Fig. 3.12 is

Fig. 3.12. Radioactive decay of thorium-231: $\log(N_t/N_o)$ against time

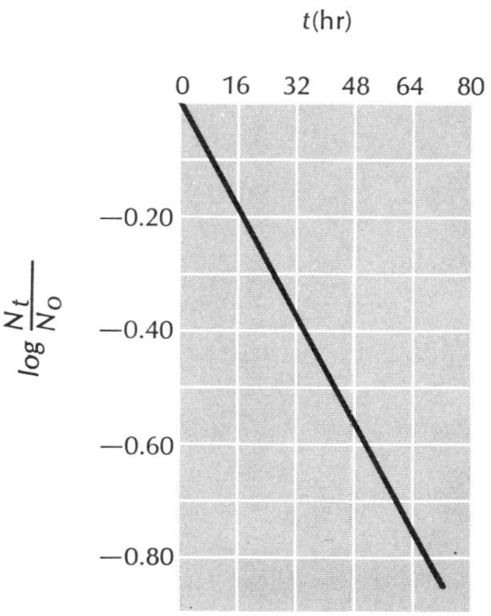

negative. The constant λ is called the *decay constant* and has a value that is characteristic of the nuclide under consideration.

The decay constant λ has the unit of reciprocal time. Thus if the time is expressed in seconds, λ has the units sec^{-1}. In the case of thorium-231, the time has been expressed in hours and λ then is given in hr^{-1}. Since the slope of the line in Fig. 3.12 is −0.0118, $K_1 = -0.0118\ hr^{-1}$ and $\lambda = 0.0272\ hr^{-1}$.

It should be clear that the ratio, N_t/N_0 is the fraction of the original sample which remains after time t has elapsed. Thus Equation (3.10) means that the *fraction* of a sample which disintegrates in a specified time interval is constant regardless of the actual size of the original sample. For example, the fraction of a sample of thorium-231 that remains after disintegration has gone on for five hours may be found as follows by using Equation (3.10).

$$\log \frac{N_t}{N_0} = -0.434 \times 0.0272\ hr^{-1} \times 5.00\ hr$$
$$= -0.0590$$
$$\frac{N_t}{N_0} = 0.873$$

This means that after a period of five hours, 87.3 percent of any sample of thorium-231 should remain; that is, 12.7 percent of the sample should have disintegrated. If the original sample contained 1.00 g of thorium-231, then after five hours 0.13 g should have disintegrated. However, if the original sample contained 0.300 g of thorium-231, then after five hours 0.038 g should have disintegrated. Thus the actual amount of radioactive nuclide that decays in a specified period of time depends upon the amount of the nuclide present at the beginning of the decay period, whereas the percent of the nuclide which decays in a specified period of time is constant for any specific nuclide and is independent of the original amount of nuclide present.

It is customary to express the rate of radioactive decay for various nuclides by the time period required for 50 percent of any sample to decay, rather than by the decay constant. This time period is called the *half-life* of the nuclide and often is represented symbolically by $t_{1/2}$. Obviously, when the half-life period has elapsed, $N_t/N_0 = \frac{1}{2}$. Using this value, Equation (3.10) may be solved for $t_{1/2}$.

$$t_{1/2} = -\frac{\log(1/2)}{0.434\lambda} = \frac{0.693}{\lambda} \tag{3.11}$$

In the case of thorium-231, the half-life is

$$t_{1/2} = \frac{0.693}{0.0272 \text{ hr}^{-1}} = 25.5 \text{ hr}$$

The wide range in values for half-lives is shown with examples in Table 3.9.

TABLE 3.9 *Some Typical Half-Lives*

NUCLIDE	HALF-LIFE	NUCLIDE	HALF-LIFE
$^{204}_{82}Pb$	1.4×10^{17} yr	$^{222}_{86}Rn$	3.8 days
$^{142}_{58}Ce$	5×10^{15} yr	$^{212}_{82}Pb$	10.6 hr
$^{148}_{62}Sm$	1.2×10^{13} yr	$^{228}_{89}Ac$	6.13 hr
$^{238}_{92}U$	4.5×10^{9} yr	$^{212}_{83}Bi$	60.6 min
$^{230}_{90}Th$	8×10^{4} yr	$^{211}_{82}Pb$	36.1 min
$^{226}_{88}Ra$	1.62×10^{3} yr	$^{206}_{81}Tl$	4.20 min
$^{227}_{89}Ac$	21.2 yr	$^{220}_{86}Rn$	56 sec
$^{228}_{88}Ra$	6.7 yr	$^{218}_{85}At$	1.3 sec
$^{234}_{90}Th$	24.1 days	$^{214}_{84}Po$	1.64×10^{-4} sec
$^{223}_{88}Ra$	11.7 days	$^{212}_{84}Po$	3×10^{-7} sec

Radioactive decay is a statistically random process in which it is not possible to predict when a specific atom will decay. Only on the average is it possible to say that some average fraction of the nuclide will disintegrate within a given time period. This statistical approach is valid only when large numbers of atoms are involved. As the number of atoms becomes very small, statistical predictions become very inaccurate. Although the discussion to this point has suggested that it is impossible for all atoms of a particular sample to disintegrate, this is not the case, since the relationships used here are not valid once the number of remaining atoms becomes very small.

In conclusion it should be noted that the terms in the ratio, N_t/N_0, may be measured in units other than mass units, provided both terms are measured in the same units. Any property proportional to the number of atoms is sufficient. Thus N_t and N_0 might be actual numbers of atoms or they might be sample masses or volumes, or the number of particles ejected, or other units. Thus a wide range of possible experimental techniques is available for measuring N_t and N_0.

ARTIFICIAL RADIOACTIVITY

The potential energy diagram in Fig. 3.2 indicates that new nuclei might be formed by adding particles to existing nuclei. In that case a proton would be added to the nucleus of an atom of deuterium. However, that diagram shows also that a sizable potential-energy hill must be climbed if the proton is to get close enough to the deuteron for the internucleonic forces to become effective. This potential-energy hill results from the electrostatic repulsions between the deuteron and the proton. Because this potential-energy barrier exists, much kinetic energy must be imparted to the proton so that it may overcome this potential-energy barrier.

Accelerators

Nuclei of many atoms have been bombarded with a wide variety of particles, including neutrons, protons, deuterons, nuclei of atoms of low atomic number, and electrons. In one respect, neutrons are the simplest particles to use for bombardment, since they do not encounter an electrostatic potential-energy barrier upon approaching an atom. The potential-energy diagram for the approach of a neutron and an atom whose nucleus can capture the neutron is shown quantitatively in Fig. 3.13.

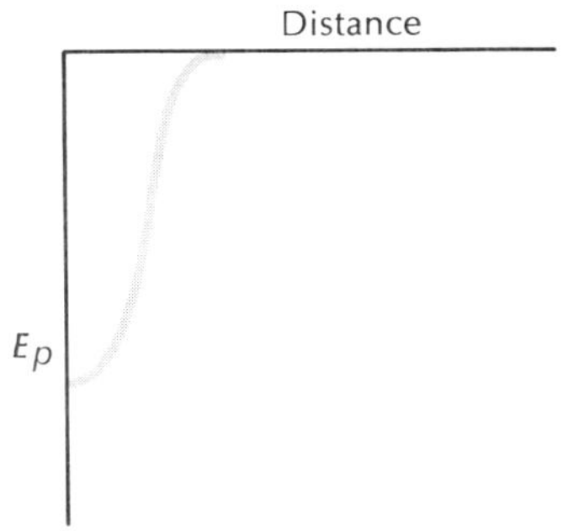

Fig. 3.13. Potential energy change as a neutron approaches a nucleus

As indicated by Fig. 3.2, any positively charged particle encounters a potential-energy barrier. For that reason, it is necessary to devise techniques for accelerating such particles so that they have sufficient kinetic energy to overcome the electrostatic repulsions. A typical linear accelerator is diagrammed schematically in Fig. 3.14.

A beam of positive particles from source *S* is directed into the first of a succession of hollow tubes that are enclosed in a vacuum system and which are charged electrostatically. At any given moment, these tubes are charged alternately: $- + - + - +$, and so on. While a given particle is traveling through tube 1, tubes 1, 3, 5, . . . are charged negatively. As the particle reaches the end of tube 1,

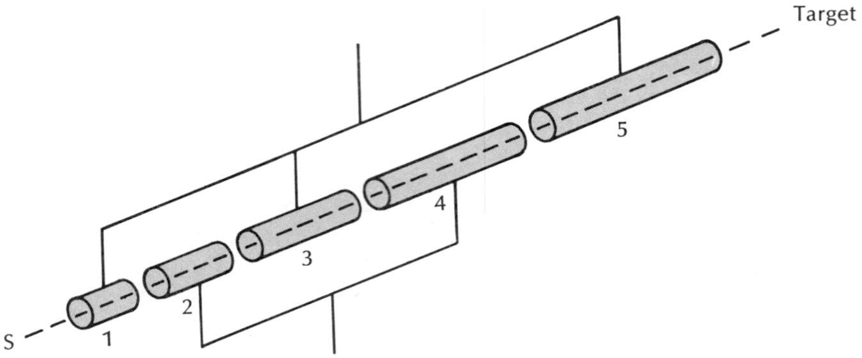

Fig. 3.14. Schematic diagram of a linear accelerator

the charges are reversed so that tubes 2, 4, 6, . . . are charged negatively. Thus the positive particle is accelerated as it jumps from tube 1 to tube 2. When the particle reaches the end of tube 2, the charges are reversed again and the particle is accelerated once more in jumping from tube 2 to tube 3. This process continues throughout all of the tubes of the accelerator and the particle strikes the target nuclide at the end of the last tube.

Since the alternation of charges on the accelerator tubes occurs at constant time intervals, each successive tube must be longer than the previous one. Thus the faster particles near the end of the accelerated path travel a greater distance in the time interval than do those slower particles just starting down the path.

The cyclotron resembles the linear accelerator in principle except that a magnetic field is superimposed over the electrostatic field so that the path of the particle is curved rather than linear. Instead of linear tubes, the cyclotron is constructed of two hollow, flat, semicircular containers, which are called *dees* because of their shape.

Looking down on a cross section of the two *D*'s in Fig. 3.15, we would observe particles emanating from a source *S* between the two *D*'s. These particles are attracted to the negatively charged *D*, D_1. While the particle is inside D_1, the magnetic field causes the particle to move in a semicircular path. When it reaches

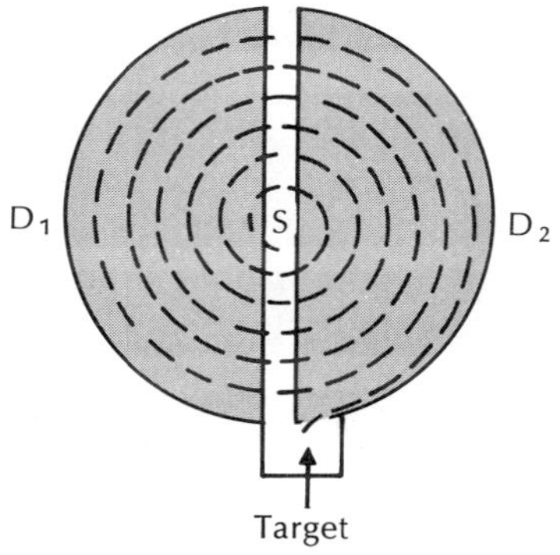

Fig. 3.15. Cross-sectional diagram of particle path in a cyclotron

the gap between the dees, the electrical charges are reversed on the two dees and the particle is accelerated as it jumps to D_2. Because it has been accelerated, it traverses a somewhat larger semicircle in D_2 than it did in D_1. Again at the gap the charges are reversed and the particle is accelerated once more. This continues until the particle has been accelerated sufficiently to leave the side of one of the dees to strike the target nuclide.

The most common negative particles used as projectiles in bombardment processes are electrons. Electrons also face a potential energy barrier in approaching atoms because they are repelled by the clouds of electrons outside the nucleus. As a result, electrons must be accelerated also. Special accelerators are used for electrons to take into account the relatively large mass change which occurs during acceleration. For our purposes, however, the current results of electron bombardment are too diffuse to be considered here.

Nuclear Reactions

When a projectile particle makes a successful attack on a target nucleus, the initial step probably is the capture of the projectile by the target to form a *compound nucleus* which is in a highly excited energy state. In a very short time of about 10^{-16} to 10^{-12} sec, this compound nucleus adjusts to a state of lower energy. In general three types of adjustment are observed.

CAPTURE PROCESSES. The compound nucleus may retain the captured particle and adjust to a lower energy state by emitting gamma radiation.

Neutron capture is very common. This is to be expected since the neutron faces no electrostatic potential-energy barrier upon approaching the target nucleus. Therefore it does not need very high kinetic energy to enter the nucleus. As a low-energy particle, it does not disturb the nucleons in the nucleus very much. Once captured, the neutron does not increase electrostatic repulsions, but does increase the *n-n* and *n-p* forces.

An example of neutron capture by uranium-238 is given in Fig. 3.16. This diagram shows qualitative energy changes with higher positions corresponding to

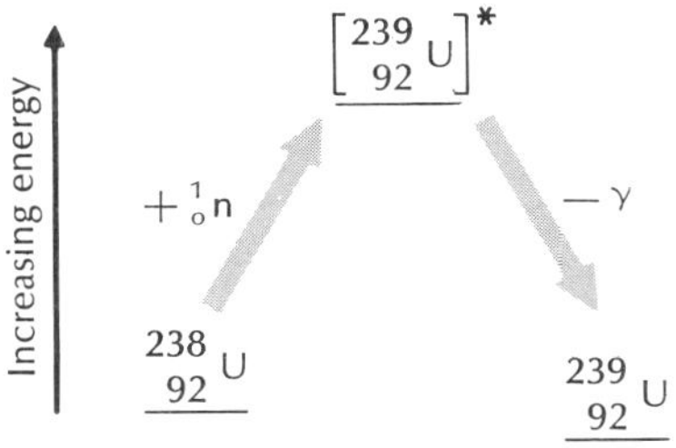

Fig. 3.16. Energy diagram of a typical capture process

higher energies. However, the energy levels are not drawn to scale. This diagram shows the initial capture of the neutron to form the excited compound nucleus of uranium-239, $[^{239}_{92}U]^*$ This excited compound nucleus emits gamma radiation to form the uranium-239 nucleus in its lowest energy state. This over-all process may be represented by the following equation:

$$^{238}_{92}U + ^{1}_{0}n = [^{239}_{92}U]^* = ^{239}_{92}U + \gamma$$

Often in writing the equations for these nuclear reactions, the excited compound nucleus is omitted. Thus the usual representation of neutron capture by uranium-238 is given by Equation (3.12).

$$^{238}_{92}U + ^{1}_{0}n = ^{239}_{92}U + \gamma \tag{3.12}$$

Another example of neutron capture is given by Equation (3.13).

$$^{79}_{35}Br + ^{1}_{0}n = ^{80}_{35}Br + \gamma \tag{3.13}$$

Positive-particle capture is less common, since the positive particles must be highly energetic to overcome the electrostatic potential energy barrier. In addition they increase electrostatic repulsions. Proton capture is somewhat common among the nuclides of low mass number, where the electrostatic repulsions among the protons still is insufficient to overbalance the nuclear attractive forces. An example of proton capture is given by Equation (3.14).

$$^{14}_{7}N + ^{1}_{1}H = ^{15}_{8}O + \gamma \tag{3.14}$$

PARTICLE-PARTICLE PROCESSES. Compound nuclei often return to lower energy states by losing a particle. Usually the only particles lost in this way are alpha particles, protons, deuterons, and neutrons. It should be noted that just as particles must cross a high electrostatic potential energy barrier to enter the nucleus, they must cross this same barrier in the reverse direction to leave the nucleus. This is to say, they must climb the walls of the well in which they reside in order to get out of the nucleus. Therefore,.the loss of positive particles is favored among nuclides of low atomic numbers where the electrostatic potential energy barrier is still low.

Because they do not face a potential-energy barrier, neutrons are good projectile particles. Although neutron capture is common, among the nuclides of low atomic number bombardment with neutrons often leads to the ejection of charged particles. A typical example is diagrammed in Fig. 3.17, which shows an excited compound nucleus of nitrogen-15 formed by capture of a neutron. This compound nucleus then loses a proton to form an excited carbon-14 nucleus, which stabilizes by gamma emission. The over-all process is represented by Equation (3.15).

$$^{14}_{7}N + ^{1}_{0}n = ^{14}_{6}C + ^{1}_{1}H + \gamma \tag{3.15}$$

Another particle emission initiated by neutrons is shown in Equation (3.16).

$$^{6}_{3}Li + ^{1}_{0}n = ^{3}_{1}H + ^{4}_{2}He + \gamma \tag{3.16}$$

The two reactions represented by Equations (3.15) and (3.16) are important ones. The first one is the reaction by which carbon-14 is produced in the atmosphere

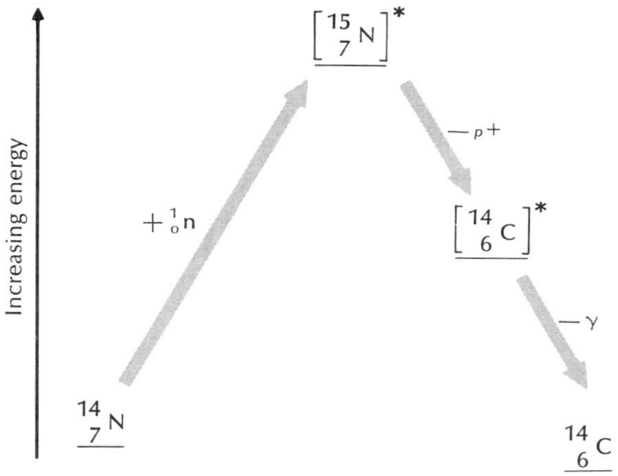

Fig. 3.17. Energy diagram of a typical particle-particle process

by cosmic rays. Carbon-14 has become an important tool in determining the age of old plant materials found in archeological excavations. The second shows a method for the production of hydrogen-3, often called tritium, which is one of the isotopes of hydrogen used in thermonuclear research.

When protons are used as projectiles, neutrons often are lost. When a charged particle enters the nucleus, the neutron is easily ejected since it faces a small potential energy barrier. Also this ejection is favored when the n/p value is high, since the addition of a proton and the loss of a neutron reduce this value considerably. An example of neutron ejection from proton bombardment is given by Equation (3.17).

$$^{18}_{8}O + {}^{1}_{1}H = {}^{18}_{9}F + {}^{1}_{0}n + \gamma \qquad (3.17)$$

A common reaction for deuteron bombardment is the assimilation of the deuteron with the loss of a proton. In effect this amounts to neutron capture. An example of this process is given by Equation (3.18).

$$^{37}_{17}Cl + {}^{2}_{1}H = {}^{38}_{17}Cl + {}^{1}_{1}H + \gamma \qquad (3.18)$$

In general, alpha bombardment produces changes similar to those induced by proton bombardment. Neutron loss is common as illustrated by Equation (3.19).

$$^{27}_{13}Al + {}^{4}_{2}He = {}^{30}_{15}P + {}^{1}_{0}n + \gamma \qquad (3.19)$$

It should be noted in these over-all reactions that the process is one of capture of one particle with the emission of another, hence the name particle-particle processes. Also it should be noted that these particle-particle processes usually are accompanied by gamma emission, representing the additional stabilization required beyond the loss of a particle as illustrated in Fig. 3.17.

Fission

Compound nuclei may adjust to lower energy states by splitting into other medium-sized nuclei. This process is called *fission.* Fission can be induced in many nuclides by bombardment with various projectile particles, but the fission processes of greatest use are those induced by neutrons. Nuclides which undergo neutron fission are uranium-235, uranium-233, and plutonium-239.

A typical neutron-induced fission reaction is represented by Equation (3.20).

$$^{235}_{92}U + ^{1}_{0}n = ^{141}_{56}Ba + ^{92}_{36}Kr + 3^{1}_{0}n + \gamma \tag{3.20}$$

However, it is not possible to write a single equation for a fission process since the target nuclide splits in many different ways into many different fragments, not just two as suggested by Equation (3.20). For example, more than 150 different nuclides have been produced by the fission of uranium-235.

Decay of Artificial Nuclides

It should not be assumed that the nuclides produced by artificial means are nonradioactive. In fact, many of them are radioactive. There is no fundamental difference in principle between the radioactive decay of a naturally occurring nuclide and the radioactive decay of an artificially produced nuclide. In fact, some of the artificially produced nuclides may have occurred in nature at one time, but due to very short half-lives, they have since disappeared. The similarities in the decay processes for natural and artificial nuclides are emphasized in Fig. 3.18

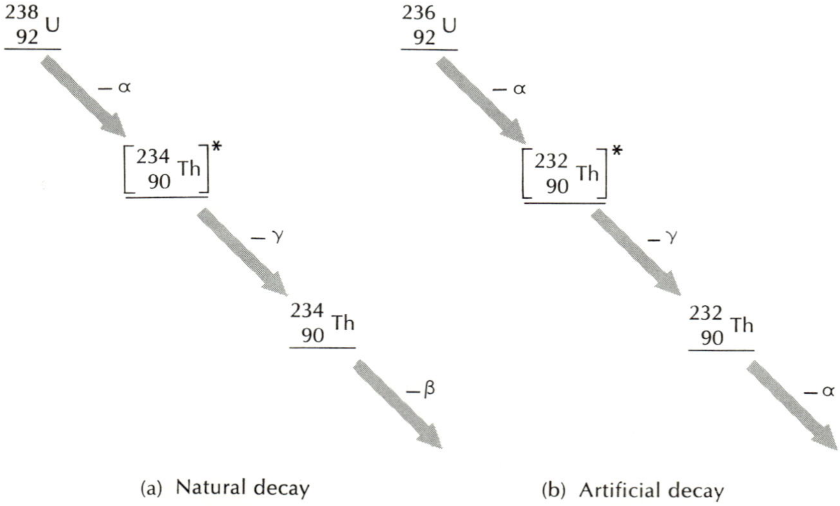

Fig. 3.18. Energy diagram comparison of natural and artificial radioactive decay

where schematic energy diagrams are shown for the artificially produced uranium-236 and the naturally occurring uranium-238. Both are alpha emitters with rather long half-lives, 2.39×10^7 yr and 4.51×10^9 yr respectively. Both decay processes lead to excited thorium nuclides which stabilize by gamma emission. Even the products of these decay processes are radioactive and decay further by alpha emission.

Among the naturally occurring nuclides radioactive decay most frequently occurs through beta or alpha emission. These modes of decay are common for the artificially produced nuclides as well. However, other modes of decay are found commonly also among the artificially produced nuclides. Two of these are included in the following summary.

BETA EMISSION. One of the most common decay processes for artificially produced nuclides is beta emission. This is the only process known whereby a high n/p value can be reduced. This decay process occurs most readily for those nuclides with high n/p values. Two beta decay processes are shown schematically in Fig. 3.19. The simplest of these is the decay of tritium to helium-3. In this case the entire process involves the beta emission. There is no accompanying gamma emission.

The decay of argon-42 represented in Fig. 3.19 involves several steps. The first is beta emission with the production of excited potassium-42 which stabilizes by gamma emission. Potassium-42 is an isolatable nuclide with a measured half-life of 12.4 hours. However, it undergoes another similar process emitting a beta

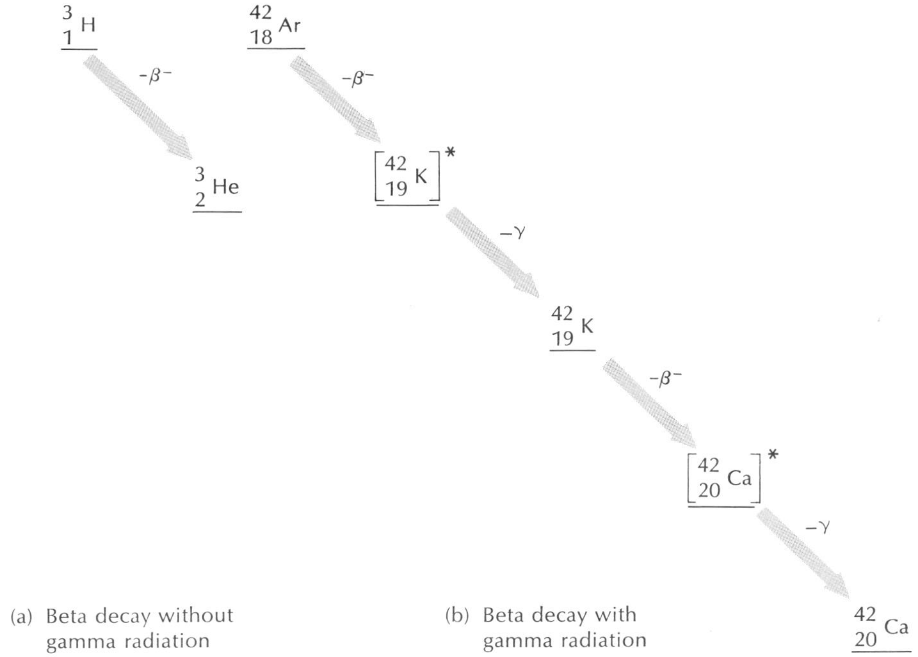

Fig. 3.19. Energy diagrams of typical beta-decay processes

particle to form an excited calcium-42 nuclide which stabilizes by gamma emission. The calcium-42 is nonradioactive.

The equations for the processes shown in Fig. 3.19 are included among the following examples of beta emission from artificially produced nuclides.

$$^{3}_{1}\text{H} = {}^{3}_{2}\text{He} + {}^{0}_{-1}\text{e} \tag{3.21}$$

$$^{42}_{18}\text{Ar} = {}^{42}_{19}\text{K} + {}^{0}_{-1}\text{e} + \gamma \tag{3.22}$$

$$^{42}_{19}\text{K} = {}^{42}_{20}\text{Ca} + {}^{0}_{-1}\text{e} + \gamma \tag{3.23}$$

$$^{70}_{31}\text{Ga} = {}^{70}_{32}\text{Ge} + {}^{0}_{-1}\text{e} + \gamma \tag{3.24}$$

$$^{128}_{53}\text{I} = {}^{128}_{54}\text{Xe} + {}^{0}_{-1}\text{e} + \gamma \tag{3.25}$$

POSITRON EMISSION. Positrons are particles with all of the properties of electrons except that they have a positive electrostatic charge. They are represented by the symbols β^+, e^+, and ${}^{0}_{1}\text{e}$. Positron emission is found most commonly among artificially produced nuclides. It represents one of two ways in which a low n/p value can be increased.

It has been suggested that electrons may be generated within the nucleus by transformation of a neutron into a proton and an electron. In a similar way it is convenient to think of a positron generated by transformation of a proton into a neutron and a positron. Thus positron emission does not change the mass number but does decrease the atomic number.

Two positron emission schemes are shown in Fig. 3.20. Sodium-21 emits a positron to form excited neon-21 which stabilizes by gamma emission. Neon-21 is nonradioactive.

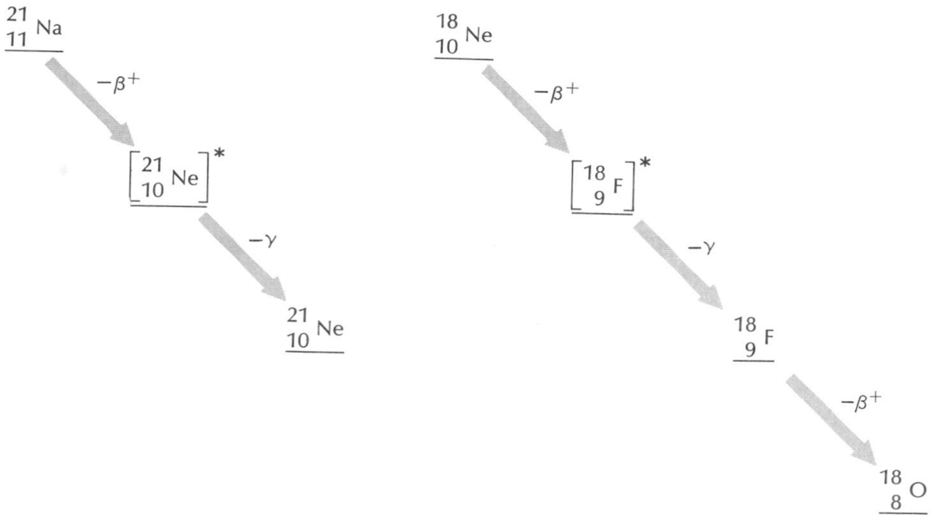

Fig. 3.20. Energy diagrams of typical positron-emission processes

Neon-18, however, is radioactive and emits a positron to form excited fluorine-18, which stabilizes by gamma emission. Fluorine-18 is an isolatable nuclide with a measured half-life of 111 min. However, it too emits a positron to form oxygen-18, which is nonradioactive.

The equations for the processes shown in Fig. 3-20 are included among the following examples of positron emission from artificially produced nuclides.

$$^{10}_{6}C = {}^{10}_{5}B + {}^{0}_{1}e + \gamma \tag{3.26}$$

$$^{18}_{10}Ne = {}^{18}_{9}F + {}^{0}_{1}e + \gamma \tag{3.27}$$

$$^{18}_{9}F = {}^{18}_{8}O + {}^{0}_{1}e \tag{3.28}$$

$$^{21}_{11}Na = {}^{21}_{10}Ne + {}^{0}_{1}e + \gamma \tag{3.29}$$

$$^{59}_{29}Cu = {}^{58}_{28}Ni + {}^{0}_{1}e + \gamma \tag{3.30}$$

$$^{129}_{56}Ba = {}^{129}_{55}Cs + {}^{0}_{1}e + \gamma \tag{3.31}$$

ELECTRON CAPTURE. Electron capture is the process whereby an electron from the cloud of electrons surrounding the nucleus is captured by the nucleus. It is convenient to think of this process as involving the conversion of a proton to a neutron. This is the second process whereby a low value for n/p may be increased. One illustration of electron capture is shown schematically in Fig. 3.21. Note that

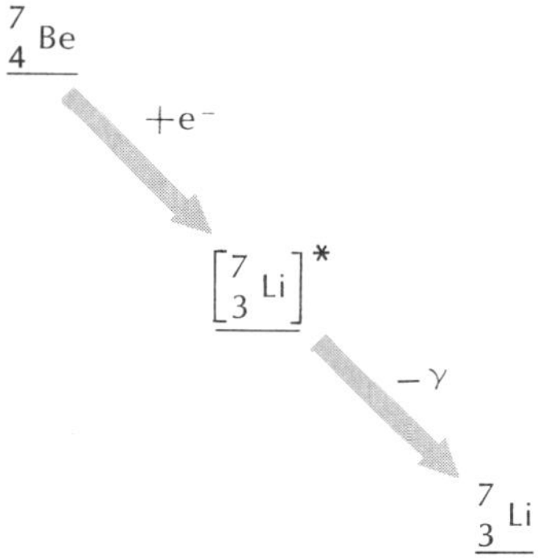

Fig. 3.21. Energy diagram of a typical electron-capture process

the initial effect of electron capture is to increase the energy of the nucleus, which then decays by gamma emission to the stabilized form.

The equation for the process shown in Fig. 3.21 is included among the following examples of electron capture by artificially produced nuclides.

$$^{7}_{4}Be + {}^{0}_{-1}e = {}^{7}_{3}Li + \gamma \tag{3.32}$$

$$^{86}_{40}Zr + {}^{0}_{-1}e = {}^{86}_{39}Y + \gamma \tag{3.33}$$

$$^{190}_{80}Hg + {}^{0}_{-1}e = {}^{190}_{79}Au + \gamma \tag{3.34}$$

ALPHA EMISSION. Alpha emission is found only among nuclides of high atomic number where proton repulsion is dominant. A schematic energy diagram was shown for alpha decay in Fig. 3.18. Three examples of alpha emission among synthetic nuclides are shown by the following equations:

$$^{208}_{84}\mathrm{Po} = {}^{204}_{82}\mathrm{Pb} + {}^{4}_{2}\mathrm{He} + \gamma \tag{3.35}$$

$$^{216}_{86}\mathrm{Rn} = {}^{212}_{84}\mathrm{Po} + {}^{4}_{2}\mathrm{He} \tag{3.36}$$

$$^{221}_{87}\mathrm{Fr} = {}^{217}_{85}\mathrm{At} + {}^{4}_{2}\mathrm{He} + \gamma \tag{3.37}$$

SUMMARY

The structure of atomic nuclei is understandable in terms of strong, short-range internucleonic forces of attraction. Such attractive forces can be used to explain the stability of nuclei containing fairly large numbers of protons, as well as the instability of nuclei containing very large numbers of protons. In addition these forces are helpful in explaining the need for neutrons to provide additional internucleonic attractions to counterbalance the interprotonic repulsions.

The energy-well point of view, which is possible with the postulation of internucleonic attractions, provides some insights into the energetics of nuclear reactions including the prevalence of alpha emission, gamma emission, the prevalence of neutron capture, and the formation and decomposition of compound nuclei in bombardment processes.

Radioactive decay rates are best understood in terms of statistically random processes involving large numbers of nuclei.

QUESTIONS AND PROBLEMS

3.1 Write a nuclear equation for each of the following nuclear processes, including gamma emission in each case:
a) Protactinium-231 undergoes alpha decay.
b) Thorium-227 is produced by decay of actinium-227.
c) Astatine-219 decays to bismuth-215.
d) Francium-223 decays to radium-223.
e) Lead-211 undergoes beta decay.
f) Uranium-235 is produced by alpha decay.

3.2 Hafnium-174 is an alpha emitter. Write the nuclear equation for this process.
a) How many moles of helium-4 atoms will be produced when 4.0 g of hafnium-174 decays?
b) How many atoms of hafnium-174 must decay in order to produce 4.0 g of helium-4?

3.3 Write the equation for the beta decay of hydrogen-3 (tritium).
a) Approximately how many grams of hydrogen-3 must decay in order to produce 4.0 g of helium?
b) How many electrons are emitted during the formation of 7.0 moles of helium atoms?
c) If some material is 15 percent tritium, how many grams of this material would supply enough tritium to produce 8.0×10^{45} atoms of helium-3?

3.4 A deposit of 10^6 tons of platinum is accompanied by a 500-ton deposit of helium-4. Assuming that all of the helium came from the decay of platinum-190, and that none escaped, how much of this isotope of platinum has decayed since it was deposited? How much platinum was in the original deposit? (Assume that there were no other radioactive isotopes of platinum in the deposit.)

3.5 A natural deposit containing 50,000 tons of rubidium is accompanied by a deposit of 5000 tons of strontium-87. Assuming that the strontium-87 came from the decay of rubidium-87, how much rubidium has decayed since it was deposited? How much rubidium was in the original deposit?

3.6 Using the beta emission of thorium-234 as an example, describe what happens to the extranuclear electrons in the atom during the beta decay process.

3.7 The $4n + 3$ radioactive disintegration series begins with uranium-235, from which particles are emitted in the following order (ignoring the relatively unimportant branched chains): alpha, beta, alpha, beta, alpha, alpha, alpha, alpha, beta, alpha, beta.

a) What is the end product of this chain of reactions?

b) Write an equation for each step in the series.

3.8 In Table 3.5,

a) Which two nuclides have odd numbers of protons?

b) Which four nuclides have equal numbers of neutrons and protons?

c) Which two nuclides have just one more neutron than protons?

3.9 For the twenty elements with atomic numbers from 1 to 20, the most abundant nuclides for each of those elements with *even* atomic numbers are: ${}^{4}_{2}He$, ${}^{9}_{4}Be$, ${}^{12}_{6}C$, ${}^{16}_{8}O$, ${}^{20}_{10}Ne$, ${}^{24}_{12}Mg$, ${}^{28}_{14}Si$, ${}^{32}_{16}S$, ${}^{40}_{18}Ar$, ${}^{40}_{20}Ca$. What suggestions about nuclear stabilities does this observation support?

3.10 Figure 3.4 shows how a proton may be captured by a deuteron. What nucleus is formed when a deuteron captures the proton?

3.11 Graphically and verbally describe the energetics of the approach of a proton and a lithium-6 nucleus.

3.12 In which case does a proton experience a higher potential-energy barrier: in approaching a carbon-12 nucleus or in approaching a uranium-238 nucleus? Explain.

3.13 What experimental evidence is there for the energy level model of the nucleus?

3.14 Each of the following transformations occurs with accompanying gamma radiation of the characteristic indicated. For each reaction, write a nuclear equation and include the energy term corresponding to the gamma radiation. Express this in kilocalories and use an atomic interpretation for symbols. For example, "uranium-239 decays by beta emission and the gamma radiation has an energy of 0.074 Mev (million electron volts)" is written as follows:

$$ {}^{239}_{92}U = {}^{0}_{-1}e + {}^{239}_{93}Np + 2.83 \times 10^{-18} \text{ kcal} $$

a) Copper-66 undergoes beta decay and gamma emission with a wavelength of 1.20×10^{-2} Å.

b) Polonium-210 undergoes alpha decay and gamma emission with a frequency of 1.16×10^{21} min^{-1}.

3.15 In Table 3.6 show that the listed photon energies are correct for the wavelengths listed.

3.16 In comparing the diagram in Fig. 3.6 with the data in Table 3.6,

a) Why are some of the photon energies listed in Table 3.6 not assigned as values for potential-energy levels in Fig. 3.6?

b) How is the value of 1.88×10^{-17} kcal for energy level 5 in Fig. 3.6 computed from the data in Table 3.6?

3.17 In kilocalories, what is the energy equivalent of one atomic mass unit?

3.18 Atomic masses for several nuclides are given below in atomic mass units. For each, calculate the total binding energy and the binding energy per nucleon in kilocalories.

a) hydrogen-1: 1.007825 amu
b) lithium-7: 7.01600 amu
c) carbon-13: 13.003354 amu
d) manganese-55: 54.9380 amu
e) iodine-127: 126.9044 amu
f) gold-197: 196.967 amu

3.19 From the following *binding energy per nucleon* values, compute the atomic mass of each nuclide in atomic mass units.

a) beryllium-9: 2.47×10^{-16} kcal
b) aluminum-27: 3.19×10^{-16} kcal
c) rhodium-103: 3.28×10^{-16} kcal
d) thorium-232: 2.91×10^{-16} kcal

3.20 Compute the decay constants for the nuclides listed in Table 3.9.

3.21 Choose any nuclide listed in Table 3.9 and prepare graphs for that nuclide corresponding to Figs. 3.10, 3.11, and 3.12.

3.22 In recognizing that the graph in Fig. 3.10, plotted from data in Table 3.7, resembles a logarithmic relationship between N_t and t, can a linear relationship be defined by plotting N_t versus log t? If so, what is this relationship? If not, why not?

3.23 Suppose that radioactive decay rates for nuclides were to be given in quarter-lives instead of half-lives, where the quarter-life is defined as the time required for one-quarter of the radioactive nuclide to decay.

a) For a given nuclide, is the quarter-life or the half-life the longer?
b) Derive the quantitative relationship between the quarter-life and the decay constant, λ.
c) Derive the quantitative relationship between the quarter-life and the half-life.
d) Compute the quarter-lives of some of the nuclides listed in Table 3.9.

3.24 Thorium-232 is an alpha emitter with a half-life of 1.39×10^{10} yr. A deposit of material containing 1.00×10^{30} g of thorium-232 was discovered in contact with a deposit of 1.72×10^{27} g of helium-4. How long had the deposit been there?

3.25 A sample containing 34.5 mg of phosphorus-32 was allowed to stand for exactly one day. At the end of that time, 32.9 mg of phosphorus-32 remained.

a) What is the half-life of phosphorus-32?
b) What other element was produced? (Phosphorus-32 is a beta emitter.)
c) How much of this other element was produced in this one day?

3.26 Chromium-51 has a half-life of 27 days. If 25 g of a sample of chromium-51 is present now, how much of it was present 10 days ago? How much will be left 10 days from now?

3.27 Carbon-14 is useful in dating old objects derived from wood. Living plants absorb carbon dioxide from the atmosphere, which contains a relatively constant percentage of carbon-14. Thus while the plant lives, it too maintains a constant level of carbon-14. However, when the plant dies, the intake of carbon dioxide and consequently the intake of carbon-14 ceases. Thus the carbon-14 already in the tissues of the plant begins to decay and the carbon-14 content decreases in the sample.

The half-life of carbon-14 is 5760 years. Three wooden archeological artifacts were sampled and the carbon-14 content determined. The percent of the living carbon-14 content is indicated below. Calculate the time elapsed since the death of the trees from which these artifacts were made.

a) Artifact 1 contained 75 percent of the living carbon-14 content.
b) Artifact 2 contained 50 percent of the living carbon-14 content.
c) Artifact 3 contained 25 percent of the living carbon-14 content.

3.28 Draw feasible schematic energy diagrams corresponding to the excited compound nucleus stabilization processes represented by Equations (3.13), (3.14), (3.16), (3.17), (3.18), and (3.19).

3.29 Draw feasible schematic energy diagrams corresponding to the decay of artificial nuclides as represented by Equations (3.24), (3.25), (3.26), (3.30), (3.31), (3.33), (3.34), (3.35), and (3.36).

3.30 Write an equation for each of the following nuclear processes:

a) Samarium-146 undergoes alpha decay.
b) Radon-206 decays to polonium-202.
c) Bismuth-202 decays to lead-202.
d) Barium-129 decays to cesium-129.
e) Xenon-128 is formed by beta emission.
f) Silver-104 is formed by positron emission.
g) Calcium-41 undergoes electron capture.
h) Gold-197 is produced by decay of platinum-197.
i) Carbon-12 is formed by alpha bombardment of beryllium-9.
j) Carbon-12 undergoes proton capture.
k) A proton is lost when sodium-23 is bombarded by deuterons.
l) An alpha particle and nitrogen-16 are formed in a neutron bombardment process.

3.31 In terms of the observations made on nuclei, explain how each of the following transitions leads to increased nuclear stability:

a) alpha decay
b) beta decay
c) neutron capture
d) fission
e) fusion
f) positron emission
g) electron capture

3.32 Suggest explanations for the following observations:

a) Nuclides of low mass number (or atomic number) do not undergo alpha decay.
b) Positron emission is more common among nuclides of low atomic number than among nuclides of high atomic number.
c) Of all of the spontaneous nuclear reactions for both natural and synthetic nuclides, only beta decay leads to a decrease in the n/p value.
d) Electrons and positrons may be emitted from nuclei although they do not exist in these nuclei.
e) The neutron experiences no potential-energy barrier or hill upon approaching nuclei and only a relatively low barrier upon leaving the nucleus.
f) Neutrons are captured more easily than other particles.
g) Charged particles are emitted from the nucleus of an atom of low atomic number more easily than from the nucleus of an atom of high atomic number.

4

Electronic Structure and the Periodic Table

INTRODUCTION The forces acting on electrons outside the nucleus are mostly electrostatic. The negatively charged electrons are attracted to the positively charged nucleus, whereas the electrons repel each other. Observations of the energetics of the behavior of atoms have permitted the development of a useful model for the arrangements of electrons outside nuclei. This model can be correlated to the properties of the elements so that the elements may be classified in a meaningful way through the periodic table. It is the purpose of this chapter to develop some of the significant aspects of the modern view of electronic structure of atoms and to relate this to the periodic table.

THE HYDROGEN ATOM

Because it consists of one proton and one electron, the hydrogen-1 atom is the simplest of all atoms. For this reason the structure of the hydrogen atom has been studied in detail and from these studies a model of the electronic structure of atoms in general has been developed.

Energy Levels

For the moment, imagine the formation of the simple hydrogen atom by allowing an electron to approach a proton from an infinite distance. The anticipated change in potential energy is shown graphically in Fig. 4.1 which is the lower part of Fig. 1.16 (Chapter 1). This suggests that the potential energy should decrease as these particles approach, due to the electrostatic attraction between the positively charged proton and the negatively charged electron.

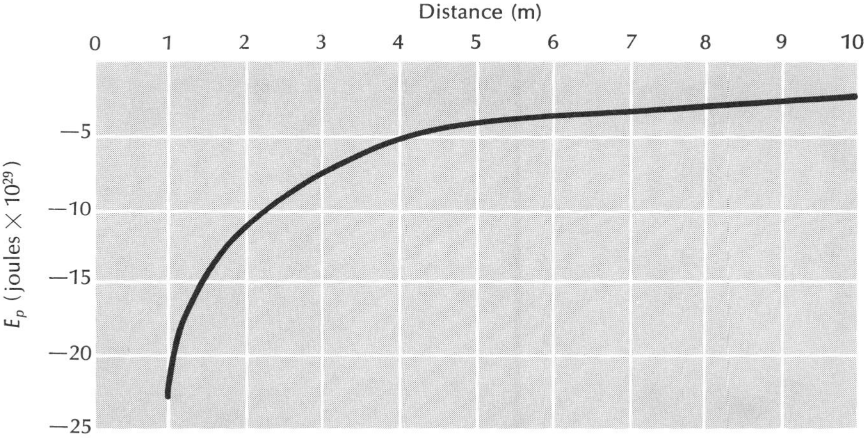

Fig. 4.1. Idealized potential energy change as an electron approaches a proton

Recalling the nature of the potential-energy well encountered in the discussion of the nucleus in Chapter 3, we see that there is good reason to wonder about the nature of the potential-energy decrease here as well. In fact, hydrogen can be made to emit discrete wavelengths of electromagnetic radiation in much the same way that nuclei emit gamma radiation. Thus, in principle, the energy decrease as an electron approaches a proton appears to be the same as that seen when a proton approaches a deuteron; that is, the potential-energy decrease occurs stepwise as suggested in Fig. 4.2.

By measuring the frequency of the electromagnetic radiations emitted from excited hydrogen atoms, it is possible to determine the energy differences between some of the energy levels. The frequencies of three series of radiations are given in Table 4.1. These frequencies are of historical importance in the development of the model of the electronic structure of the hydrogen atom. The series are named for the men who measured them.

A comparison of the values in Table 4.1 with the electromagnetic spectrum given earlier in Fig. 1.13 shows that the frequencies of these radiations are much lower than gamma radiations from nuclear transformations. In fact, these electromagnetic radiations are near the visible region of the spectrum. This suggests that potential-energy steps experienced by the electron approaching the proton are smaller than energy steps experienced by the proton approaching the deuteron.

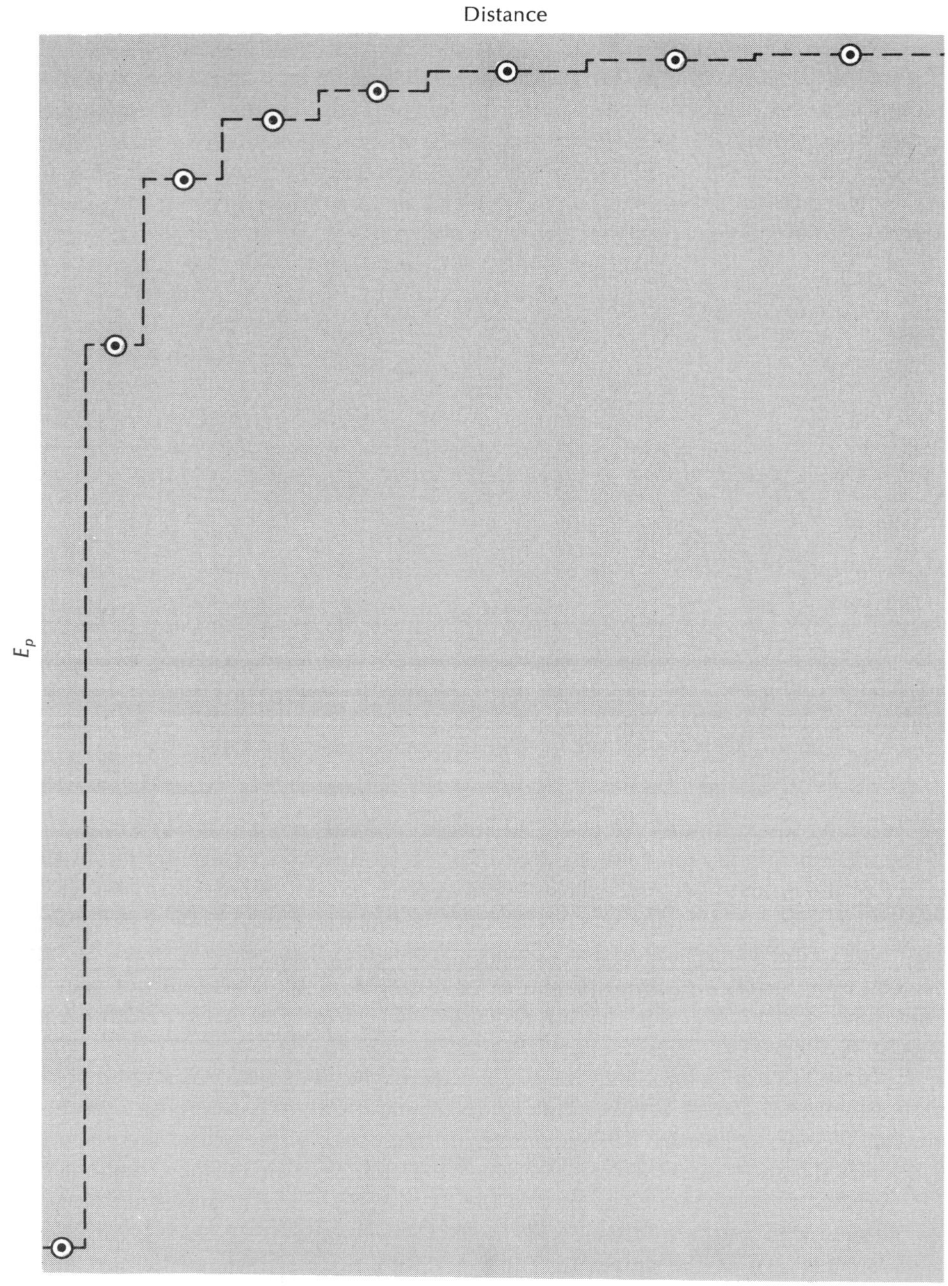

Fig. 4.2. Stepwise potential energy change as an electron approaches a proton

The data in Table 4.1 can be used to sketch a front view of the steps shown from the side in Fig. 4.2. This is done in Fig. 4.3. Each of the steps is called an

TABLE 4.1 *Frequencies of Three Series of Electromagnetic Radiations from the Hydrogen Atom in Sec^{-1}*

LYMAN SERIES	BALMER SERIES	PASCHEN SERIES
2.4657×10^{15}	4.5668×10^{14}	1.5983×10^{14}
2.9227×10^{15}	6.1651×10^{14}	2.3381×10^{14}
3.0825×10^{15}	6.9049×10^{14}	2.7400×10^{14}
3.1565×10^{15}	7.3067×10^{14}	
3.1976×10^{15}		

energy level, and in Fig. 4.3 the energy levels are spaced to scale, with the energies listed in kcal $\times 10^{24}$. This means that the energy values shown on the scale have been multiplied by 10^{24}, and if any value is to be extracted from the diagram, it must be multiplied by 10^{-24}. Thus the potential energy listed for the topmost energy level in Fig. 4.3 is -14.5×10^{-24} kcal.

The Lyman, Balmer, and Paschen series of electromagnetic radiations are sketched into Fig. 4.3 so that each arrow corresponds to the energy-level transition that accounts for one of the observed frequencies. It should be noted also in Fig. 4.3 that energy differences are greatest at the lower potential energy levels. This is shown in Table 4.2, where the potential energy values for the lowest 20 potential

TABLE 4.2 *Potential Energies of the Lowest 20 Energy Levels in the Hydrogen Atom*

LEVEL	POTENTIAL ENERGY kcal $\times 10^{24}$
20	−1.30
19	−1.45
18	−1.61
17	−1.80
16	−2.03
15	−2.33
14	−2.68
13	−3.10
12	−3.64
11	−4.32
10	−5.20
9	−6.45
8	−8.15
7	−10.6
6	−14.5
5	−20.8
4	−32.5
3	−57.8
2	−130.1
1	−520.3

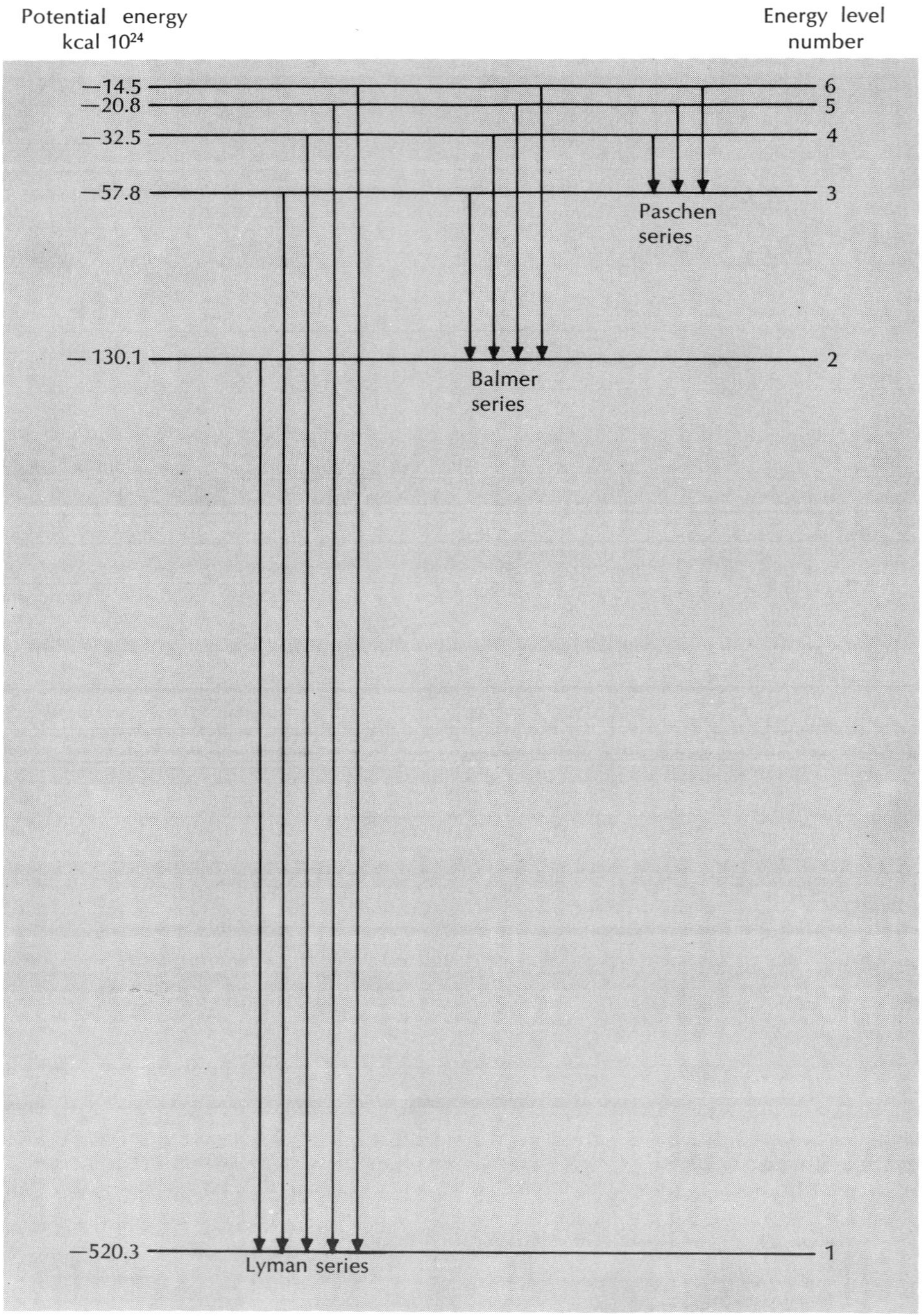

Fig. 4.3. Energy-level transition interpretation for a portion of the hydrogen atom spectrum

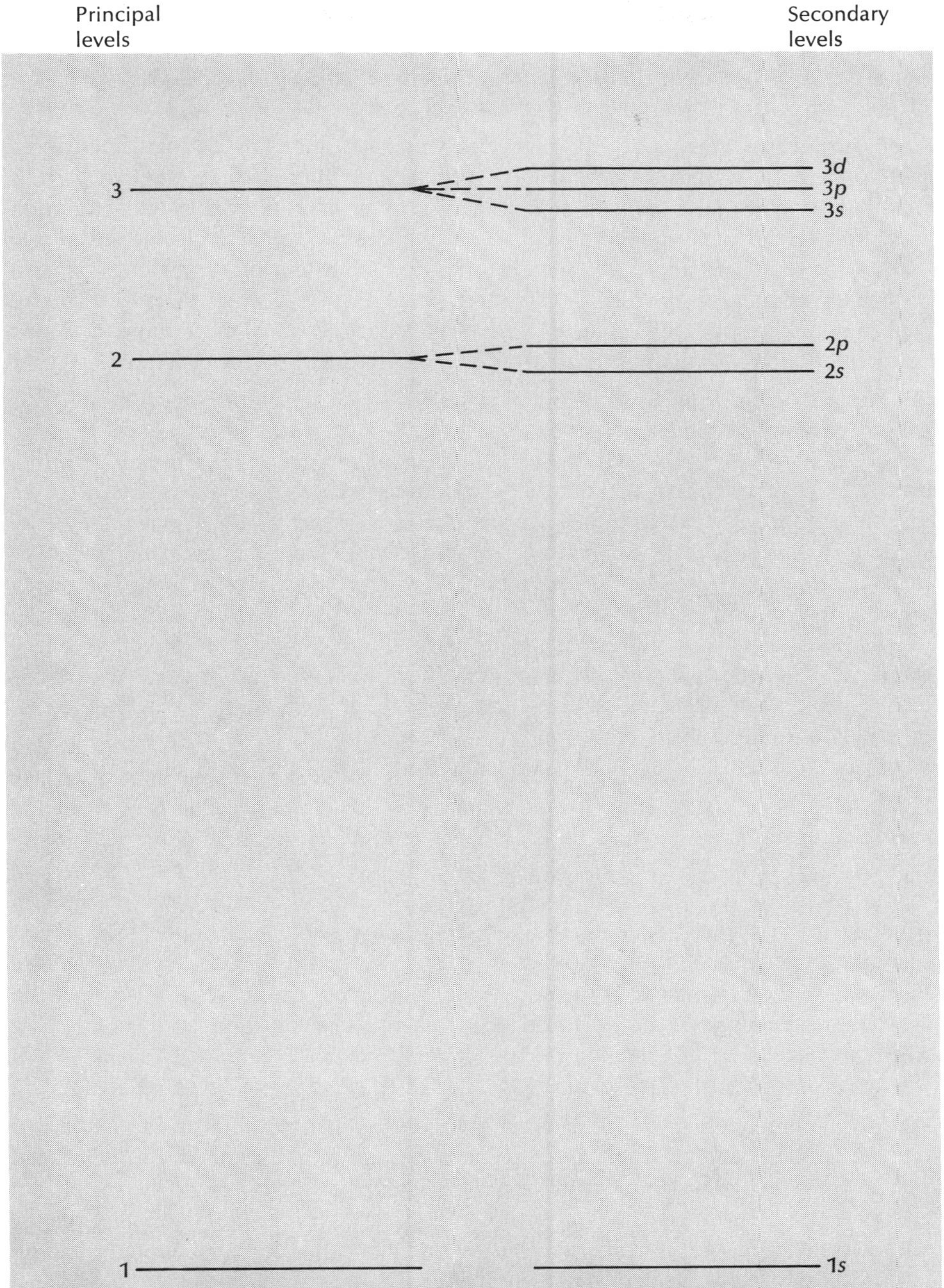

Fig. 4.4. Interpretation of resolved spectral lines for hydrogen

energy levels in hydrogen are listed. By convention, the energy levels are numbered numerically from one, beginning with the lowest energy level, as labeled in Table 4.2. It is apparent that the difference between energy levels is smaller in the high energy region than it is in the low energy region.

By using instruments of very high resolving capabilities to distinguish between electromagnetic radiations which have very small differences in frequency, it is possible to show that what was thought earlier to be a single frequency of radiation often is a cluster of several frequencies. This has been interpreted to mean that the energy levels shown in Fig. 4.3 really consist of clusters of very closely spaced secondary energy levels. This is diagrammed in Fig. 4.4, which is not drawn to scale because of space limitations. The separation between the secondary levels in the hydrogen atom is not as great as might be implied by the diagram.

The gross energy levels shown in Fig. 4.3 are called *principal* energy levels. Thus each of these principal energy levels appears to be subdivided into *secondary* energy levels. As suggested in Fig. 4.4, the *number* of secondary levels within each principal energy level turns out to be equal to the number which has been assigned to the principal energy level. In other words, principal energy level 1 has only one secondary energy level; principal energy level 2 has two secondary energy levels; principal energy level 3 has three secondary energy levels, and so on. This trend appears to continue, so that principal energy level 40 should have forty secondary energy levels. However, the fact is that at high potential energy the principal energy levels differ so little in energy that distinguishing among them is difficult. As a result, distinguishing among the secondary energy levels in this region of high potential energy is virtually impossible.

Just as the principal energy levels have been numbered, the secondary energy levels have been given letter names, some of which have been derived from the spectroscopists' nomenclature to describe spectral observations. Within each principal energy level, the secondary energy level representing the lowest potential energy is called the *s* secondary level. Thus the only secondary energy level in principal energy level 1 is called the 1*s* secondary energy level. Similarly the lowest secondary energy level in principal energy level 2 is called the 2*s* secondary energy level and the lowest secondary energy level in principal energy level 3 is called the 3*s* secondary energy level, and so on. The *s* in this designation is derived from the word *single*.

Within each principal energy level in the hydrogen atom, the secondary energy level immediately above the *s* secondary energy level in potential energy is called the *p* secondary energy level. Since there is no additional secondary energy level above the *s* secondary level in principal energy level 1, there is no 1*p* secondary energy level. However, the secondary energy level immediately above the 2*s* secondary energy level in principal energy level 2 is called the 2*p* secondary energy level. The *p* in these designations is derived from the word *principal*.

In a similar fashion, the successive secondary energy levels above the *p* secondary level within each principal energy level in the hydrogen atom are labeled *d* for *diffuse*, *f* for *fundamental* and *g*, *h*, *i*, etc. alphabetically after *f*. Table 4.3 lists the secondary energy levels for the lowest five of the principal energy levels for hydrogen, with the lowest level at the bottom and highest level at the top.

The principal and secondary energy levels, as just defined, appear to be adequate descriptions of the potential energy states an electron may assume in isolated hydrogen atoms. However, if the excited hydrogen atoms from which electromagnetic radiation is being emitted are placed between the poles of a very powerful magnet, there is a large increase in the number of different frequencies of radiation observed. Again these radiations tend to cluster close to the frequencies assigned to differences between principal and secondary energy levels already discussed.

TABLE 4.3

Secondary Energy Levels for the First Five Principal Energy Levels of Hydrogen

5*g*
5*f*
5*d*
5*p*
5*s*
4*f*
4*d*
4*p*
4*s*
3*d*
3*p*
3*s*
2*p*
2*s*
1*s*

In fact, it is possible to account for these observations upon the assumption that the secondary energy levels are further subdivided into other energy levels by the presence of the magnetic field. These new subdivisions of the secondary energy levels are called *orbitals*. Apparently, in the absence of a magnetic field, the orbitals of any given secondary energy level of hydrogen all represent the same potential energy for an electron, but they do not when the hydrogen is placed in a magnetic field. A schematic representation is shown in Fig. 4.5. In this diagram the unresolved spectral determinations of the first three principal energy levels are shown on the left. With better instrumental resolving power, it can be shown that there

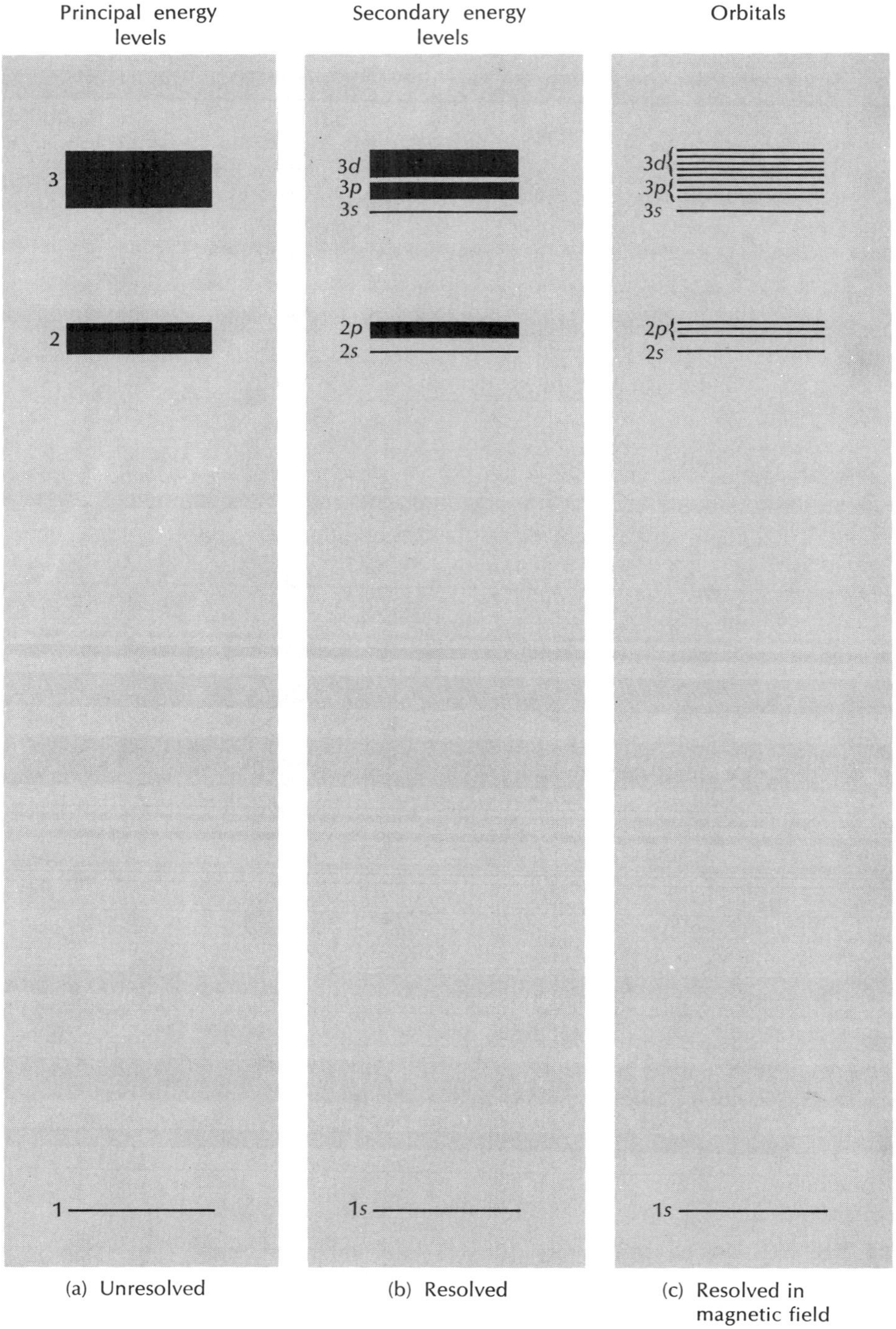

Fig. 4.5. Interpretation of hydrogen spectral lines in a magnetic field

are secondary energy levels as shown in the middle. However, in the presence of a magnetic field, the secondary energy levels appear to split into clusters of closely related energy levels or orbitals.

The diagram in Fig. 4.5 shows that it is necessary to propose that *s* type secondary energy levels do not split in a magnetic field. However, *p* type secondary energy levels appear to be composed of three orbitals, *d* type secondary energy levels appear to be composed of five orbitals, *f* type secondary energy levels appear to be composed of seven orbitals, and so on.

Quantum Numbers

It is apparent that in the simple hydrogen atom which contains just one electron, the pattern of energy levels in which the electron may exist is rather complex. To describe the electron in various energy levels within the hydrogen atom, a series of four *quantum numbers* has been defined. The derivation and significance of the numerical values of these numbers are beyond the scope of this treatment, but they provide an organized description of the potential energy levels which an electron may assume within a hydrogen atom, and eventually within other atoms.

PRINCIPAL QUANTUM NUMBER. The principal quantum number is the integer associated with the identification of the principal energy levels shown in Fig. 4.3 and Table 4.2. The numerical values for this quantum number are the integers 1, 2, 3, 4 . . ., the lower numbers corresponding to the lower potential energy levels. The usual symbol employed for this quantum number is n. Sometimes the numerical designations for the principal quantum number are replaced by capital letters starting with K and continuing alphabetically. Thus the equivalence of letter and numerical values are $K = 1$, $L = 2$, $M = 3$, $N = 4$, and so on.

Recalling Figs. 4.1 and 4.2, which show the decrease in potential energy as the distance between the proton and the electron decreases, we conclude that the principal quantum number n represents the distance between the proton and the electron. Thus, increasing numerical values for n represent increasing distances between the proton and electron, which in turn represent increasing potential energy.

SECONDARY QUANTUM NUMBER. The secondary quantum number identifies the secondary energy levels and usually is designated by the symbol l. The numerical values for this quantum number are all of the integers from zero to $n - 1$, where n is the principal quantum number. Thus when $n = 1$, the only value for l is zero. However, when $n = 2$, there are two values for l, namely 0 and 1. Similarly, when $n = 3$, $l = 0$, 1, or 2.

Each of the numerical values for l has a corresponding letter designation. These letters have been indicated already in Table 4.3, but they are repeated in Table 4.4, which shows the relationship between the numerical value of l and the letter designator.

It has been stated before that the potential energy of an electron in a hydrogen atom is quantized, that is, the electron may assume only certain discrete potential energy values. This was shown in Fig. 4.2. The principal and secondary quantum numbers provide a means for designating those discrete potential energy values. Thus for the hydrogen atom, these values are represented symbolically as 1*s*, 2*s*, 2*p*, 3*s*, 3*p*, 3*d*, and so on.

MAGNETIC QUANTUM NUMBER. The magnetic quantum number identifies the orbitals, which indicate the effect of the magnetic field upon the potential energy levels of the electron in the hydrogen atom. The usual symbol for this quantum number is m_l. The values for m_l are the integers from $-l$ to $+l$ including zero. Thus when $l = 0$, the only value for m_l is zero. This is equivalent to saying that an *s* secondary energy level has one orbital. Similarly, when $l = 1$, the values for m_l are -1, 0 or $+1$. This is equivalent to saying that a *p* secondary energy level has three orbitals.

Whereas the principal and secondary quantum numbers represent the discrete potential energies which the electron may assume in the hydrogen atom, the magnetic quantum number represents the latent ability of these discrete potential energy levels to be altered or subdivided by the application of an external magnetic field. This suggests that the electron has associated with it a magnetic field which

TABLE 4.4

Letter Equivalents for l Values

l VALUE	LETTER
0	*s*
1	*p*
2	*d*
3	*f*
4	*g*
5	*h*
6	*i*

then interacts with an applied magnetic field. Recalling that a magnetic field is associated with the motion of electrostatic charge, this suggests that the electron may be in motion in the atom. Like other magnetic fields, the magnetic field of the electron must be directional, and the fact that an electron in different orbitals interacts in different ways with an applied magnetic field to lead to different potential energies suggests that the magnetic field of the electron may be oriented in different directions. These different orientations of the magnetic field are represented by the orbitals. Thus, in the isolated atom, the different orientations of the magnetic field do not lead to differences in potential energy, but they do lead to differences in potential energy when they interact with an applied magnetic field.

The fact that only certain discrete values are possible for the magnetic quantum number indicates that the directional characteristics of the magnetic field associated with the electron in the hydrogen atom are limited in number, that is the directional characteristics of the magnetic field are quantized. These specific possible directions are represented by orbitals. Thus in the 1*s*, 2*s*, 3*s*, or any other *ns* potential energy state, only one direction of magnetic field is permitted, as indicated by the single orbital for which $m_l = 0$. However, in the 2*p*, 3*p*, 4*p*, or any other *np* potential energy state, three directions of magnetic field are permitted, as indicated by the three orbitals for which $m_l = -1$, 0, and $+1$. Similarly, in the 3*d*, 4*d*, 5*d*, or any other *nd* potential energy state, five directions of magnetic field are permitted, as indicated by the five orbitals for which $m_l = -2$, -1, 0, $+1$, and $+2$.

SPIN QUANTUM NUMBER. From what has been said so far, it would appear that all hydrogen atoms in which the electron is in the 1*s* potential-energy level have the same set of properties. The 1*s* energy level is not split by the presence of a magnetic field. Thus it is to be assumed that there is only one magnetic field orientation associated with an electron in the 1*s* potential-energy level.

However, Fig. 4.6 shows a schematic diagram of an experiment which shows that in a beam of hydrogen atoms there are two different kinds of atoms. In the experiment, a nonuniform magnetic field is produced by means of north and south

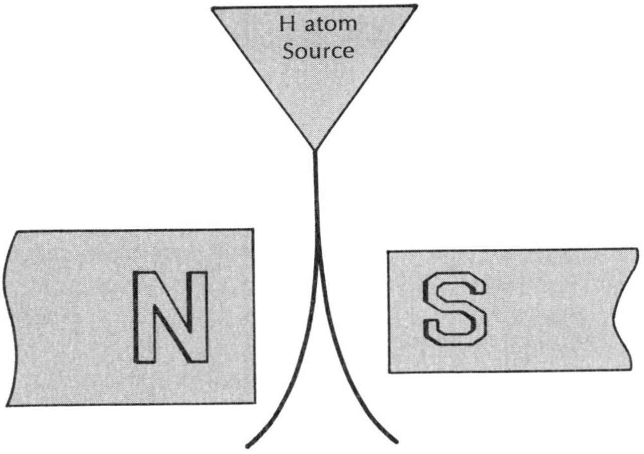

Fig. 4.6. Separation of hydrogen atoms in a nonuniform magnetic field

poles of different strengths. A stream of hydrogen atoms with their electrons in the 1*s* potential-energy state is passed through this field. The stream of hydrogen atoms splits into two paths as it passes through the magnetic field. This suggests that there are two different kinds of hydrogen atoms in the original stream. It also suggests that these two kinds of hydrogen atoms differ in the directional characteristics of the magnetic fields associated with the electron in the atom.

To explain this phenomenon, it has been suggested that there are two components to the magnetic field associated with the electron in the hydrogen atom. One of these is the component due to the translational motion of the electron in the atom; this component is the main contributor to the determination of m_l values. The second component is associated with the spinning of the electron on an axis in much the same way that the earth spins on an axis.

However, the spin appears to be quantized and limited to two directions, which may be thought of as clockwise and counterclockwise. The spin is designated by the spin quantum number, usually represented by the symbol, m_s. The values for m_s are $+\frac{1}{2}$ or $-\frac{1}{2}$.

QUANTUM NUMBER SUMMARY. On the basis of the foregoing discussion, we can consider all of the possible states in which an electron may exist within the hydrogen atom. Some of these states are the same in potential energy but different in the directional characteristics of the magnetic field associated with the electron. The possible states for the electron in a hydrogen atom can be designated by the four quantum numbers. A summary of these is given in Table 4.5.

TABLE 4.5 Quantum Number Summary

n [1,2,...]	l [0 TO $(n-1)$]	m_l [$-l$ TO 0 TO $+l$]	m_s [$-\frac{1}{2}$ OR $+\frac{1}{2}$]
1	0	0	$-\frac{1}{2}$ or $+\frac{1}{2}$
2	0	0	$-\frac{1}{2}$ or $+\frac{1}{2}$
	1	$-1, 0, +1$	$-\frac{1}{2}$ or $+\frac{1}{2}$
	0	0	$-\frac{1}{2}$ or $+\frac{1}{2}$
3	1	$-1, 0, +1$	$-\frac{1}{2}$ or $+\frac{1}{2}$
	2	$-2, -1, 0, +1, +2$	$-\frac{1}{2}$ or $+\frac{1}{2}$
	0	0	$-\frac{1}{2}$ or $+\frac{1}{2}$
4	1	$-1, 0, +1$	$-\frac{1}{2}$ or $+\frac{1}{2}$
	2	$-2, -1, 0, +1, +2$	$-\frac{1}{2}$ or $+\frac{1}{2}$
	3	$-3, -2, -1, 0, +1, +2, +3$	$-\frac{1}{2}$ or $+\frac{1}{2}$

The combination of four quantum numbers designates the state of the electron in the hydrogen atom. For example, the electron in the hydrogen atom is in its lowest potential-energy state when $n = 1$. However, there are two possible states for an electron for which $n = 1$. The two states are the same in potential energy, but they differ in spin. The four quantum numbers representing each of these states are summarized in Table 4.6. The four quantum numbers are represented inside square brackets in the order $[n, l, m_l, m_s]$.

TABLE 4.6 Electron States for Hydrogen When $n = 1$

$[1,0,0,-\frac{1}{2}]$
$[1,0,0,+\frac{1}{2}]$

When $n = 2$, the electron is in a higher potential-energy level than it is when $n = 1$. However, when the electron is in the $n = 2$ energy level, it may assume eight different states at two different potential energies. These eight states are summarized in Table 4.7, where each vertical column represents a single potential

TABLE 4.7 Electron States for Hydrogen When $n = 2$

$[2,0,0,-\frac{1}{2}]$	$[2,1,-1,-\frac{1}{2}]$
$[2,0,0,+\frac{1}{2}]$	$[2,1,-1,+\frac{1}{2}]$
	$[2,1,0,-\frac{1}{2}]$
	$[2,1,0,+\frac{1}{2}]$
	$[2,1,+1,-\frac{1}{2}]$
	$[2,1,+1,+\frac{1}{2}]$

energy. In the second column of Table 4.7 there are three pairs of states, representing the three different orientations of the magnetic field of the electron in the atom. The states within each pair differ only in the spin component of the magnetic field of the electron.

When $n = 3$, the electron is in a still higher potential energy level and may assume eighteen different states corresponding to three different secondary potential energy levels. These are summarized in Table 4.8. In the second column of Table 4.8 there are three pairs of states which are equivalent in potential energy but differ in the directional characteristics of the magnetic field associated with the electron. Within each pair two different spin orientations are indicated. All three of these pairs have the same l value and thus belong to the same secondary energy level. However, each pair differs from the other two pairs in m_l value and thus represents an orbital.

TABLE 4.8

Electron States for Hydrogen When n=3

[3,0,0,$-\frac{1}{2}$]	[3,1,$-$1,$-\frac{1}{2}$]	[3,2,$-$2,$-\frac{1}{2}$]
[3,0,0,$+\frac{1}{2}$]	[3,1,$-$1,$+\frac{1}{2}$]	[3,2,$-$2,$+\frac{1}{2}$]
	[3,1,0,$-\frac{1}{2}$]	[3,2,$-$1,$-\frac{1}{2}$]
	[3,1,0,$+\frac{1}{2}$]	[3,2,$-$1,$+\frac{1}{2}$]
	[3,1,+1,$-\frac{1}{2}$]	[3,2,0,$-\frac{1}{2}$]
	[3,1,+1,$+\frac{1}{2}$]	[3,2,0,$+\frac{1}{2}$]
		[3,2,+1,$-\frac{1}{2}$]
		[3,2,+1,$+\frac{1}{2}$]
		[3,2,+2,$-\frac{1}{2}$]
		[3,2,+2,$+\frac{1}{2}$]

In the third column of Table 4.8 there are five pairs of states which are equivalent in potential energy but differ in the directional characteristics of the magnetic field associated with the electron. Each of these five pairs represents an orbital in the 3*d* secondary potential energy level.

This orderly sequence of quantum numbers describing the various states which an electron may assume in a hydrogen atom permits some generalizations about the number of states in which the electron may exist within the hydrogen atom. Table 4.9 lists the number of different states within each of the first four principal

TABLE 4.9

Number of States of the Hydrogen Atom Related to Principal Quantum Number

n	STATES
1	2
2	8
3	18
4	32

energy levels. Scrutiny of this table reveals that the total number of states possible within any given principal energy level can be found from the formula $2n^2$, where n is the principal quantum number.

It should be noted in Tables 4.6, 4.7, and 4.8 that for any given value of l there is a specific number of states in which the electron may exist at the same secondary potential energy level. These numbers of states are listed in Table 4.10.

TABLE 4.10

Number of States of the Hydrogen Atom Related to Secondary Quantum Numbers

l	LETTER	STATES
0	*s*	2
1	*p*	6
2	*d*	10
3	*f*	14

It is apparent from this table that as the value for l increases by one, the number of states increases by four. The number of states in which the electron can exist within a given secondary energy level of the hydrogen atom can be designated by the formula $2(2l + 1)$, where l is the secondary quantum number.

Uncertainty and Probability

One of the difficulties in studying subatomic phenomena is expressed in the Heisenberg uncertainty principle. In one form, this principle may be stated as follows: *It is impossible to determine accurately both the exact position and exact velocity of an electron at any given time.*

To see the rationality of this statement, consider how the location and velocity of an automobile might be determined. In some way it is necessary to "see" the automobile in order to measure its location and velocity with respect to some standards. The "seeing" may be done with the eyes, with the hands by feeling the car, or with appropriate instrumentation. If we consider seeing the car with our eyes, then we can see the car only if there is light impinging upon the car. If the car is in total darkness, it cannot be seen. The light must affect the car since it exerts a pressure. Thus the light which permits us to see the car changes the location and motion of the car. In the case of a car, these changes are negligible.

To "see" an electron, however, requires a probe such as high-frequency electromagnetic radiation. The effect of this radiation upon the very small electron is not negligible. As a result, by just looking at the electron, we change its position.

For this reason, we must be satisfied with a rather fuzzy notion of electron location in the atom. We must be satisfied with a statistical approach which at best gives us some notion of probable locations. Although we may be reasonably certain that an electron will be in some general location at some specific time, we cannot be absolutely certain, for there remains the possibility that it might be somewhere else.

This statistical model may be compared to school grades. In a large school class, over a period of many years, it is expected that about 60 percent of the students will receive grades of C, about 20 percent will receive each of the grades B and D and about 10 percent will receive each of the grades A and F.

In a random class of some particular year in a given course, we hope and expect that the grades will be distributed in about the same way. However, the certainty of our prediction is less than the certainty of this distribution over a long period of time. It may be a good class in which no one receives a grade of F and only a few receive a grade of D, while most of the students are split evenly among the C's, B's, and A's. Again, it may be a poor class in which no one receives a grade of A.

To try to predict the grade which a specific, randomly chosen student will receive in a particular year is even more risky. Of course we are 100 percent confident that he will receive a grade of A, B, C, D, or F. We may be about 80 percent confident that he will receive a grade of B, C, or D and less than 60 percent confident that he will receive a grade of C.

As we try to become more specific, we lose confidence in our ability to predict the grade. Similarly, our knowledge about the location of an electron in an atom is uncertain. The location of a specific electron at a given instant of time can be stated only in probabilities and in terms of our confidence that it is within a certain unit of space.

Taking de Broglie's original postulates about the wave nature of matter, Erwin Schrödinger was able to derive a number of equations describing the wave characteristics of electrons in atoms. The actual meanings of these equations were somewhat obscure until Max Born suggested that they might be used to state the probability of finding an electron in a particular location.

Some of the results of this approach are summarized graphically in Fig. 4.7. In the graphs in this diagram, probability of finding an electron is plotted against the distance from the nucleus of the hydrogen atom. Distance is plotted in angstrom units and the probability is plotted in somewhat arbitrary units, the nature of which does not concern us here, except that they are the same on all three graphs.

It is interesting to note some of the specific probabilities in Fig. 4.7. The maximum probability for finding an electron in a hydrogen atom is at a distance of about 0.5 Å from the nucleus. This corresponds to the electron in the first principal energy level, that is when $n = 1$.

Next, there are two similar maximum probabilities in the vicinity of 2 to 2.5 Å from the nucleus. These two maxima correspond to the probability of finding the electron in the second principal energy level, that is when $n = 2$. The two separate curves, however, correspond to the two secondary energy levels associated with the second principal energy level, namely $2s$ and $2p$. The curve which corresponds

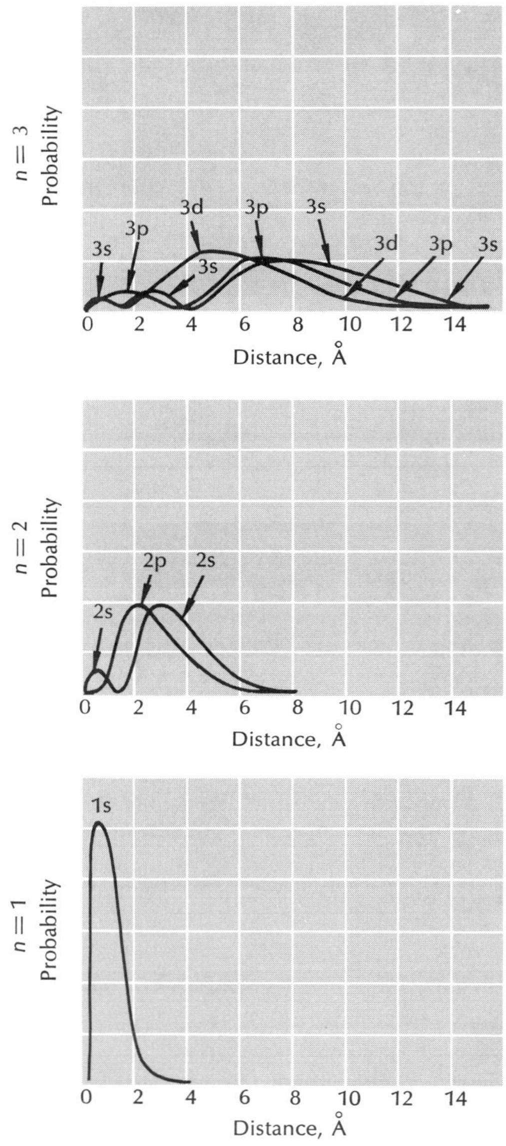

Fig. 4.7. Probability distributions for the secondary energy levels of the first three principal energy levels in the hydrogen atom

to the $2s$ secondary energy level has two maxima: a large one near 2.5 Å and a small one near 0.5 Å. However, the curve corresponding to the $2p$ secondary energy level has a single large maximum near 2 Å. Thus it is said that an electron in the $2s$ secondary energy level has greater *penetration* than an electron in the $2p$ secondary energy level. Higher penetration indicates higher probability for finding the electron near the nucleus. An electron with high penetration comes under the electrostatic influence of the nucleus more readily than an electron with low penetration. Thus, despite the fact that the large maximum probability for locating an electron in the $2p$ secondary energy level is closer to the nucleus than the large maximum probability for locating an electron in the $2s$ secondary energy level, an electron in the $2s$ secondary energy level is at a slightly lower potential energy than an electron in the $2p$ secondary energy level. This is because the electron in the $2s$ secondary energy level has greater penetration than does the electron in the $2p$ secondary energy level.

The topmost graph in Fig. 4.7 shows three broad maxima near 5 to 6.5 Å. These peaks are lower than those in the bottom two graphs corresponding to a lower probability of finding the electron at these distances from the nucleus. The three curves correspond to the $3s$, $3p$, and $3d$ secondary energy levels associated with principal energy level 3. The $3s$ secondary energy level curve shows two minor maxima near 2 Å and 0.5 Å. The $3p$ secondary energy level curve shows a single minor maximum near 1.5 Å. The $3d$ secondary energy level curve shows a single broad maximum with no minor maxima. Thus an electron in the $3s$ secondary energy level has greater penetration than an electron in the $3p$ secondary energy level and an electron in the $3p$ secondary energy level has greater penetration than an electron in the $3d$ secondary energy level.

Although the graphical representation of the probable electron distance from the nucleus of the hydrogen atom, as shown in Fig. 4.7, is useful, it does not distinguish among the various orbitals within a specified secondary energy level. Thus there are three $2p$ orbitals, and an electron in any one of these three states has the same probability distance from the nucleus as an electron in any other of these three states.

In accordance with the observation that orbital potential energy differences occur only in a magnetic field, the orbitals have associated with them directionally oriented, highly probable locations for electrons in the hydrogen atom. Thus it is possible to sketch three-dimensional units of space within which the electron may most probably be found when it is in a state corresponding to an orbital. Such spaces may be interpreted as the spaces within which an electron most probably will be found at any instant of time. In addition, they may be interpreted as the spaces through which the electron moves over a long period of time. From this point of view, this probable location space is the space through which the charge of the electron is distributed over a long period of time. Thus these spaces are called the *charge distributions* for electrons in the orbitals.

The charge distributions for s, p, and d orbitals are shown in Fig. 4.8. Care must be exercised *not* to view these as photographs of real containers for electrons. They are representations of probable electron locations consistent with the spectral observations discussed earlier in this chapter.

Note that the charge distributions for electrons in s-type orbitals are spherical and thus nondirectional in character. Actually, it is more proper to consider the $1s$ charge distribution to be a hollow sphere, with the thickness of the shell approximately corresponding to the thickness of the peak in the bottom graph of Fig. 4.7. The charge distribution for an electron in the $2s$ orbital resembles two concentric

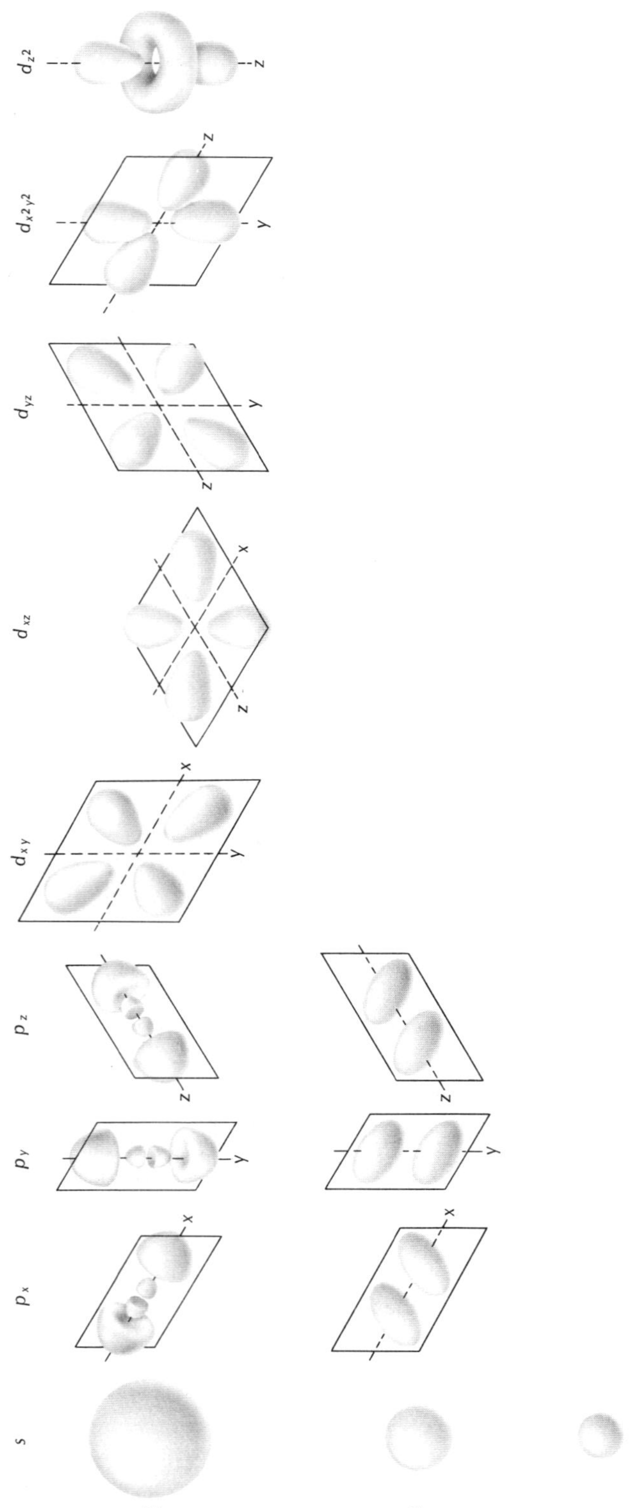

Fig. 4.8. Probable charge distributions for the *s*, *p*, *d* orbitals of the first three principal energy levels in the hydrogen atom

hollow spheres. The outer sphere has a shell approximately as thick as the major 2*s* peak in Fig. 4.7 and the inner sphere has a shell approximately as thick as the minor 2*s* peak in Fig. 4.7.

The charge distributions for electrons in *p* orbitals are directed along the three mutually perpendicular Cartesian axes. The distributions for the 2*p* orbitals differ somewhat from the distributions for the 3*p* orbitals, since it is necessary to show the two maxima from Fig. 4.7 in the model of the 3*p* charge distributions. Thus the 2*p* charge distributions consist of two ellipsoids lying across each of the axes, whereas the 3*p* charge distributions consist of two mushroom-shaped spaces directed along each axis. The cap of the mushroom corresponds to the major maximum in the 3*p* curve of Fig. 4.7 and the stem of the mushroom corresponds to the minor maximum in the 3*p* curve of Fig. 4.7.

Due to the directional characteristics of the *p* orbital charge distributions, they are designated by the superscripts *x*, *y*, and *z* corresponding to the usual designations for the Cartesian axes.

The charge distributions corresponding to the five 3*d* orbitals are more complex. Three of these consist of four lobes, each directed between the axes in the *xy*, *xz*, and *yz* planes respectively. These three are designated as the d_{xy}, d_{xz}, and d_{yz} charge distributions. A fourth charge distribution consists of four lobes directed along the *x* and *y* axes. This is called the $d_{x^2-y^2}$ charge distribution. The final charge distribution consists of two lobes directed along the *z* axis and encircled by a doughnut-shaped lobe in the *xy* plane. This is the d_{z^2} charge distribution.

Hydrogen Atom Summary

Where the charge distribution orientations are not important, an abbreviated schematic diagram is used to represent the state of the electron in the hydrogen atom. The format for this scheme is shown in Fig. 4.9, where boxes represent the orbitals, which are grouped by secondary energy level. The orbitals are arranged according to increasing potential energy for the hydrogen atom from left to right.

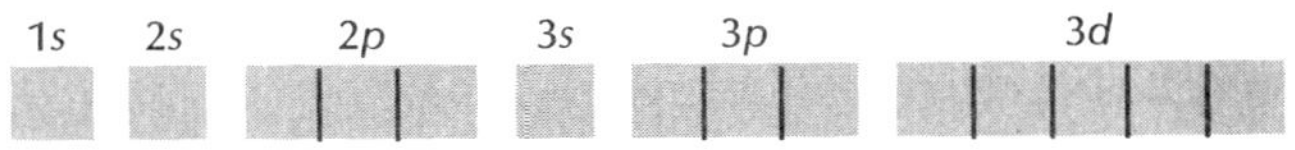

Fig. 4.9. Schematic representation of atomic orbitals

The electron is represented in these diagrams by an arrow which may point upward or downward to represent the two spins an electron may assume. By means of the arrow, several different states of the electron in the hydrogen atom are represented in Fig. 4.10. States *a* and *b* in Fig. 4.10 do not differ in potential energy but only in spin of the electron, since the electron is in the single 1*s* orbital in both cases. However states *b* and *c* in Fig. 4.10 differ in potential energy although they do not differ in spin. States *d* and *e* do not differ in either potential energy or spin, but they differ in orbitals within the 2*p* secondary energy level and thus they differ in the direction of the magnetic field associated with the electron.

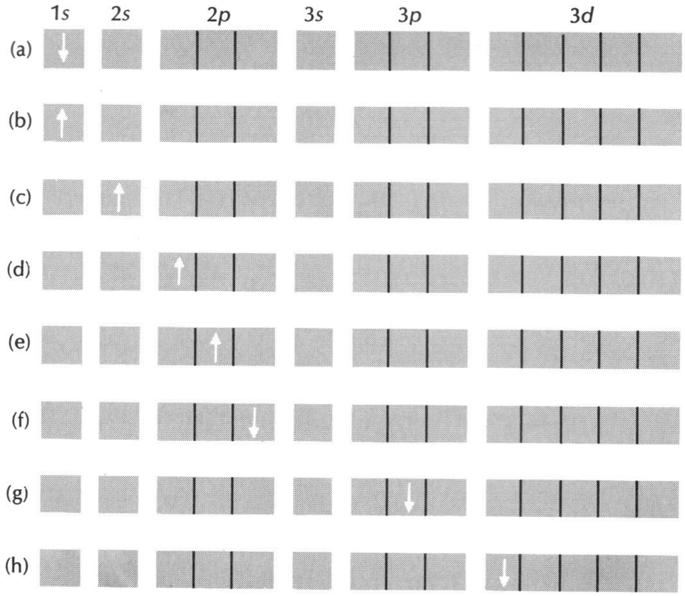

Fig. 4.10. Several different states for the hydrogen atom

ATOMS WITH MORE THAN ONE ELECTRON

Of all the atoms known, the hydrogen atom alone has been studied in rather complete detail. Atoms with more than one electron are more difficult to study and experimental results are more difficult to interpret. Using the model of the H atom that has just been discussed, a reasonable pattern of electronic structures for the other atoms has been developed that is consistent with the experimental facts as they are now known.

The Hydrogen Atom Analogy

The states of the electrons in atoms with more than one electron are described by the same four quantum numbers used for the various states of the single electron in the hydrogen atom. The values for these quantum numbers are the same as those listed in Table 4.5.

One difference between the hydrogen atom and other atoms lies in the number of electrons in each atom which must be described by the quantum numbers. In the hydrogen atom, the states described by the four quantum numbers could be assumed by only one electron at different times. The lowest potential energy state of any atom is called the *ground state* of that atom, whereas other higher states are called *excited states*. Thus for hydrogen in the ground state, the single electron occupies the 1*s* secondary energy level. To have an electron in any higher potential energy state requires excitation of the atom.

Although an atom containing many electrons may be in the ground state, the electrons within it may be in different potential energy states. One of the problems in describing the electronic structures of such atoms is the assignment of the electrons to their various possible states.

EXCLUSION AND MULTIPLICITY. To aid in the assignment of quantum numbers to electrons in atoms containing more than one electron, two rules have been stated. The application of these two rules leads to the assignment of quantum numbers which appears to be consistent with many observations of the behavior of the atoms.

The first rule is known as the *Pauli exclusion principle*. According to this rule, *no two electrons may be in the same state in the same atom*. This is equivalent to saying that no two electrons can have the same set of four quantum numbers in the same atom. The rule implies that this is the only exclusion, however, and that electrons may assume all of the states permitted by the combinations of quantum numbers specified for the hydrogen atom.

The exclusion principle indicates that two electrons may have the same orbital state since those two electrons may have different spins. Thus two electrons in the 1*s* orbital may have two different sets of quantum numbers: $[1,0,0,-\frac{1}{2}]$ and $[1,0,0,+\frac{1}{2}]$.

Reference to Tables 4.6, 4.7, and 4.8 suggests the following observations as well:

1. The first principal energy level ($n = 1$) can contain a maximum ($2n^2$) of two electrons; these will be in the 1*s* secondary energy level and will have opposite spins.
2. The second principal energy level can contain a maximum of eight electrons, with two electrons of opposite spin in the 2*s* secondary energy level and six electrons in the three orbitals of the 2*p* secondary energy level. The two electrons in each of the orbitals must have opposite spins.
3. The third principal energy level can contain a maximum of eighteen electrons. These will consist of two electrons of opposite spin in the 3*s* secondary energy level; six electrons in the 3*p* secondary energy level, with paired electrons in each orbital having opposite spins; and ten electrons in the five 3*d* orbitals, the paired electrons in any orbital having opposite spins.

The second rule employed for assigning quantum numbers to electrons in atoms with more than one electron is known as *Hund's rule of maximum multiplicity*. According to this rule, *multiple electrons within the same secondary energy level are in the lowest potential energy state when they have the same spin*, that is, *when they have the same value for* m_s. For example, if there are three electrons in the 2*p* energy level of an atom, each of these electrons will be in a different orbital, since to be in the same orbital they must have opposite spins. Of course, if there are four electrons in the 2*p* energy level, two of them must be in the same orbital with opposite spins. However, the other two electrons will occupy the other two orbitals singly and will have the same spin.

POTENTIAL ENERGY AND ATOMIC NUMBER. Another significant difference between the hydrogen atom and other atoms is the nucleus. In the hydrogen atom, the nucleus contains only one proton, that is, only one quantum of positive charge. In atoms containing more than one electron, the nuclear charge is higher. The changes in potential energy as an electron approaches a nucleus, as represented in Figs. 4.1 and 4.2, depend upon the nuclear charge, since the forces of attraction become greater as the nuclear charge increases. Similarly, the actual levels of potential energy available to an electron depend upon the nuclear charge.

It is to be expected that the actual potential-energy value for a given atomic secondary energy level decreases as the nuclear charge increases. However, many factors are involved, including electrostatic attractions between nuclei and electrons, electrostatic repulsions among the electrons outside the nucleus, and electronic orbital penetration. Consequently, the exact relationship between the nuclear charge (atomic number) and the potential-energy value for any specified secondary potential-energy level is not a simple one. Mathematical techniques have been devised to estimate the relationship between atomic number and the potential energy of the secondary potential-energy levels. One suggested but *approximate* picture is provided by the graph in Fig. 4.11 where potential energy is plotted against atomic number. For our purposes, the actual units of the potential energy are not significant and so have not been specified.

Several very significant trends can be discerned in Fig. 4.11. Whereas all of the secondary energy levels of a specific principal energy level have very nearly the same potential energy in the hydrogen atom, this is not true in atoms containing more than one electron. For example, there is a sizable difference in potential energy between the 2*s* and 2*p* secondary energy levels in all atoms except hydrogen. Similarly, there is a sizable difference in potential energy-values for 3*s*, 3*p*, and 3*d* secondary levels in atoms of high atomic number.

In addition, it is apparent from Fig. 4.11 that electrons in highly penetrating orbitals are affected much more strongly at low atomic numbers than are the electrons in less highly penetrating orbitals. This is quite obvious in the third principal energy level. At low atomic number 1 for hydrogen, the 3*s*, 3*p*, and 3*d* secondary energy levels are very close in potential energy. However, as the atomic number increases, the most highly penetrating orbital, 3*s*, decreases in potential energy more rapidly than the 3*p* orbitals, which are of intermediate penetration. In turn the 3*p* orbitals decrease in potential energy more rapidly than the less penetrating 3*d* orbitals.

This is not surprising, since the more highly penetrating electrons should be affected more strongly by the increase in nuclear charge. However, the difference leads to some interesting changes in the potential-energy order for the various secondary energy levels.

In the hydrogen atom, secondary energy levels increase in potential energy in the order 1*s*, 2*s*, 2*p*, 3*s*, 3*p*, 3*d*, 4*s*, 4*p*, 4*d*, 4*f*, and so on. However, as the nuclear charge increases, the electrons in highly penetrating orbitals undergo such rapid decreases in potential energy that there are some inversions. For example, the electrons in the highly penetrating 4*s* orbital drop in potential energy so rapidly that at atomic number 7, the potential energy of the electrons in the 4*s* orbital is about the same as the potential energy of electrons in the less highly penetrating 3*d*

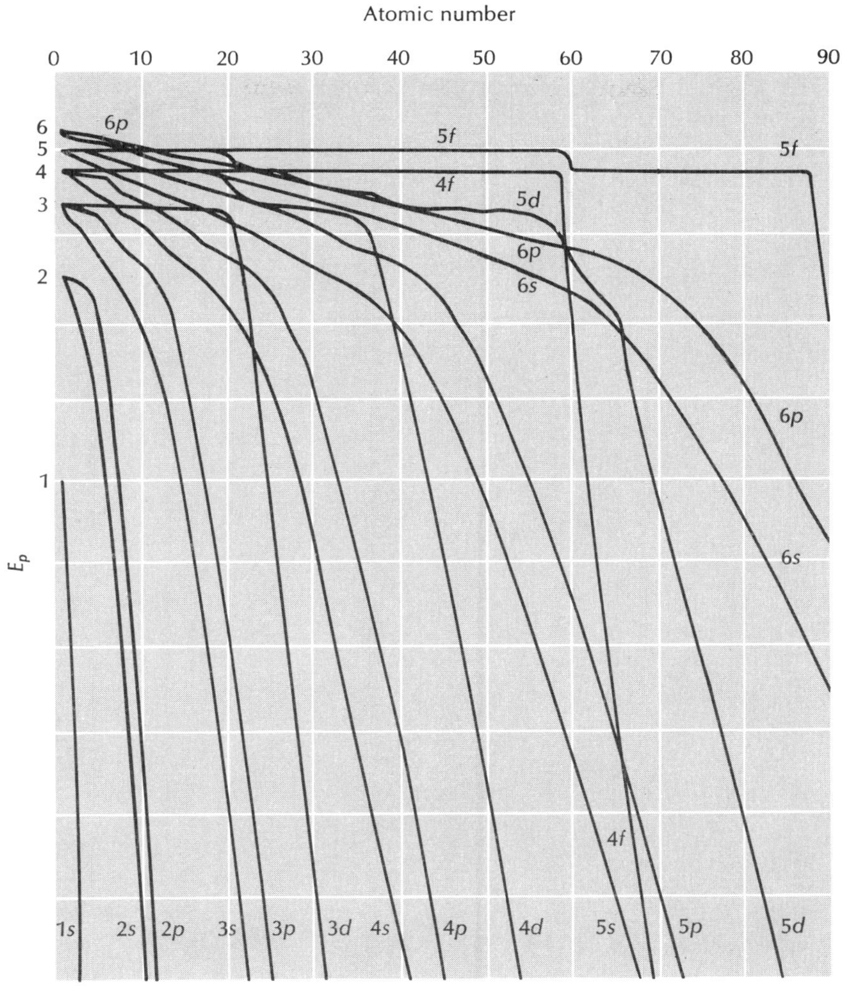

Fig. 4.11. Effect of changing atomic number on the potential energy of the secondary energy levels

orbitals. Thus in an atom of atomic number 8, the order of increasing potential energy is 1*s*, 2*s*, 2*p*, 3*s*, 3*p*, 4*s*, 3*d*, 4*p*, etc. At the same time, electrons in the fairly highly penetrating 4*p* orbitals undergo a rapid decrease in potential energy also. Near atomic number 11, the potential energy of electrons in the 4*p* orbitals is nearly the same as the potential energy of electrons in the 3*d* orbitals. Thus, at atomic number 15 the order of increasing potential energy is 1*s*, 2*s*, 2*p*, 3*s*, 3*p*, 4*s*, 4*p*, 3*d*, etc. However, at atomic number 20, the electrons in the 3*d* orbitals decrease in potential energy rapidly, so that near atomic number 22 the electrons in the 3*d* and 4*p* orbitals have nearly the same potential energy. Similarly, at atomic number

23, electrons in the $3d$ and $4s$ orbitals also have nearly the same potential energy. Consequently, for an atom with atomic number 40, the order of increasing potential energy, once more, is $1s$, $2s$, $2p$, $3s$, $3p$, $3d$, $4s$, $4p$, etc.

The graph in Fig. 4.11 inconveniently terminates at the bottom long before all of the energies can be recorded. As a matter of convenience, to put all of the energies for the first 92 elements on a single graph on a single page, a function of the potential energy is plotted against atomic number in Fig. 4.12. The function

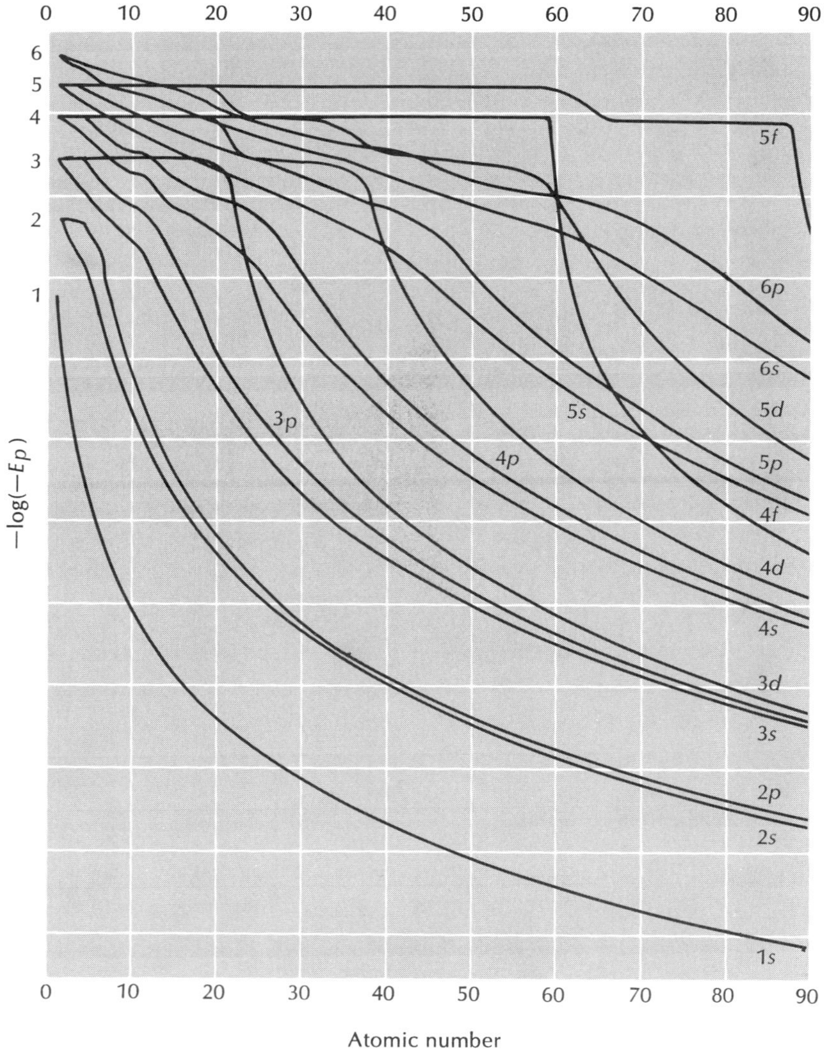

Fig. 4.12. Plot of $-\log(-E_p)$ against atomic number for the secondary energy levels

chosen arbitrarily for convenience is $-\log(-E_p)$. The logarithm is taken of the negative potential energy because the potential energy is negative and the negative potential energy is therefore positive. Logarithms of negative numbers cannot be taken. The negative sign before the logarithm term makes the entire function negative so that it resembles the potential energy in sign. Thus a decrease in the value of this negative function corresponds to a decrease in potential energy.

Figure 4.12 is used in the same way as Fig. 4.11. To see the order of increasing potential energy for secondary energy levels, we scan from bottom to top at the atomic number desired. For example, at atomic number 15, potential energy increases in the order 1*s*, 2*s*, 2*p*, 3*s*, 3*p*, 4*s*, 4*p*, 3*d*, 5*s*, 5*p*, etc.

Electronic Configurations of the Atoms

In the ground state, all electrons in all atoms are in the lowest energy levels consistent with the Pauli exclusion principle and Hund's rule of maximum multiplicity. The quantum number sets describing the states of all electrons for the ground state of the atom are called the *electronic configuration* of the atom. The scheme shown in Fig. 4.9 and 4.10 is used in Table 4.11 to list the electronic configurations of the known atoms. To conserve space the letter symbols for the principal energy levels are used to represent completed principal energy levels. Thus $KLMN_{spd}$ means that principal energy levels 1, 2, and 3 are complete along with the *s*, *p*, and *d* secondary energy levels of principal level 4.

For atoms with many electrons, it is not always possible to make unambiguous assignments of quantum numbers to all electrons. Spectral interpretations are very complex because large numbers of electron transitions occur. Thus approximate assignments are made on the basis of the best evidence available. In some cases, it is not clear what is the best evidence. Therefore, some books and references may list assignments slightly different from those used here.

A somewhat more abbreviated notation often is used in writing electronic configurations. The designations for the secondary energy levels, such as 1*s*, 2*p*, 3*d*, are written in the order of increasing potential energy *as they appear in the hydrogen atom* and not necessarily in the order of increasing potential energy within the atom for which the configuration is written. To each of these designations is added a superscript number corresponding to the number of electrons in that secondary energy level. Thus the notation $1s^2 2s^2 2p^6 3s^1$ indicates that there are two electrons in the 1*s* secondary energy level, two electrons in the 2*s* secondary energy level, six electrons in the 2*p* secondary energy level, and one electron in the 3*s* secondary energy level. The total number of electrons in the atom may be found by adding the superscript numbers. In this case, the total is 11; this is the electronic configuration for sodium, Na, which has an atomic number of 11.

However, this more abbreviated notation does *not* show the distribution of electrons in the orbitals or the spin arrangements of the electrons. In the following discussion, the electronic configurations of the atoms of the elements will be indicated in this more abbreviated notation, which should be checked against Table 4.11 for more detail.

Table 4.11
Electronic Configurations

Atomic No.	Symbol	1*s*	2*s*	2*p*	3*s*	3*p*	3*d*	4*s*	4*p*	4*d*	4*f*	5*s*	5*p*
1	H	[↓]											
2	He	[↓↑]											
3	Li	[↓↑]	[↓]										
4	Be	[↓↑]	[↓↑]										
5	B	[↓↑]	[↓↑]	[↓][ ][ ]									
6	C	[↓↑]	[↓↑]	[↓][↓][ ]									
7	N	[↓↑]	[↓↑]	[↓][↓][↓]									
8	O	[↓↑]	[↓↑]	[↓↑][↓][↓]									
9	F	[↓↑]	[↓↑]	[↓↑][↓↑][↓]									
10	Ne	[↓↑]	[↓↑]	[↓↑][↓↑][↓↑]									
11	Na	[↓↑]	[↓↑]	[↓↑][↓↑][↓↑]	[↓]								
12	Mg	[↓↑]	[↓↑]	[↓↑][↓↑][↓↑]	[↓↑]								
13	Al	[↓↑]	[↓↑]	[↓↑][↓↑][↓↑]	[↓↑]	[↓][ ][ ]							

Table 4.11
Electronic Configurations
(cont.)

Atomic No.	Symbol	1s	2s	2p	3s	3p	3d	4s	4p	4d	4f	5s	5p
14	Si	↓↑	↓↑	↓↑ ↓↑ ↓↑	↓↑	↓ ↓							
15	P	↓↑	↓↑	↓↑ ↓↑ ↓↑	↓↑	↓ ↓ ↓							
16	S	↓↑	↓↑	↓↑ ↓↑ ↓↑	↓↑	↓↑ ↓ ↓							
17	Cl	↓↑	↓↑	↓↑ ↓↑ ↓↑	↓↑	↓↑ ↓↑ ↓							
18	Ar	↓↑	↓↑	↓↑ ↓↑ ↓↑	↓↑	↓↑ ↓↑ ↓↑							
19	K	↓↑	↓↑	↓↑ ↓↑ ↓↑	↓↑	↓↑ ↓↑ ↓↑		↓					
20	Ca	↓↑	↓↑	↓↑ ↓↑ ↓↑	↓↑	↓↑ ↓↑ ↓↑		↓↑					
21	Sc	↓↑	↓↑	↓↑ ↓↑ ↓↑	↓↑	↓↑ ↓↑ ↓↑	↓	↓↑					
22	Ti	↓↑	↓↑	↓↑ ↓↑ ↓↑	↓↑	↓↑ ↓↑ ↓↑	↓ ↓	↓↑					
23	V	↓↑	↓↑	↓↑ ↓↑ ↓↑	↓↑	↓↑ ↓↑ ↓↑	↓ ↓ ↓	↓↑					
24	Cr	↓↑	↓↑	↓↑ ↓↑ ↓↑	↓↑	↓↑ ↓↑ ↓↑	↓ ↓ ↓ ↓ ↓	↓					
25	Mn	↓↑	↓↑	↓↑ ↓↑ ↓↑	↓↑	↓↑ ↓↑ ↓↑	↓ ↓ ↓ ↓ ↓	↓↑					
26	Fe	↓↑	↓↑	↓↑ ↓↑ ↓↑	↓↑	↓↑ ↓↑ ↓↑	↓↑ ↓ ↓ ↓ ↓	↓↑					
27	Co	↓↑	↓↑	↓↑ ↓↑ ↓↑	↓↑	↓↑ ↓↑ ↓↑	↓↑ ↓↑ ↓ ↓ ↓	↓↑					

Table 4.11
Electronic Configurations
(cont.)

Atomic No.	Symbol	1*s*	2*s*	2*p*	3*s*	3*p*	3*d*	4*s*	4*p*	4*d*	4*f*	5*s*	5*p*
28	Ni	↓↑	↓↑	↓↑ ↓↑ ↓↑	↓↑	↓↑ ↓↑ ↓↑	↓↑ ↓↑ ↓↑ ↓ ↓	↓↑	_ _ _	_ _ _ _ _	_ _ _ _ _ _ _	_	_ _ _
29	Cu	↓↑	↓↑	↓↑ ↓↑ ↓↑	↓↑	↓↑ ↓↑ ↓↑	↓↑ ↓↑ ↓↑ ↓↑ ↓↑	↓	_ _ _	_ _ _ _ _	_ _ _ _ _ _ _	_	_ _ _
30	Zn	↓↑	↓↑	↓↑ ↓↑ ↓↑	↓↑	↓↑ ↓↑ ↓↑	↓↑ ↓↑ ↓↑ ↓↑ ↓↑	↓↑	_ _ _	_ _ _ _ _	_ _ _ _ _ _ _	_	_ _ _
31	Ga	↓↑	↓↑	↓↑ ↓↑ ↓↑	↓↑	↓↑ ↓↑ ↓↑	↓↑ ↓↑ ↓↑ ↓↑ ↓↑	↓↑	↓ _ _	_ _ _ _ _	_ _ _ _ _ _ _	_	_ _ _
32	Ge	↓↑	↓↑	↓↑ ↓↑ ↓↑	↓↑	↓↑ ↓↑ ↓↑	↓↑ ↓↑ ↓↑ ↓↑ ↓↑	↓↑	↓ ↓ _	_ _ _ _ _	_ _ _ _ _ _ _	_	_ _ _
33	As	↓↑	↓↑	↓↑ ↓↑ ↓↑	↓↑	↓↑ ↓↑ ↓↑	↓↑ ↓↑ ↓↑ ↓↑ ↓↑	↓↑	↓ ↓ ↓	_ _ _ _ _	_ _ _ _ _ _ _	_	_ _ _
34	Se	↓↑	↓↑	↓↑ ↓↑ ↓↑	↓↑	↓↑ ↓↑ ↓↑	↓↑ ↓↑ ↓↑ ↓↑ ↓↑	↓↑	↓↑ ↓ ↓	_ _ _ _ _	_ _ _ _ _ _ _	_	_ _ _
35	Br	↓↑	↓↑	↓↑ ↓↑ ↓↑	↓↑	↓↑ ↓↑ ↓↑	↓↑ ↓↑ ↓↑ ↓↑ ↓↑	↓↑	↓↑ ↓↑ ↓	_ _ _ _ _	_ _ _ _ _ _ _	_	_ _ _
36	Kr	↓↑	↓↑	↓↑ ↓↑ ↓↑	↓↑	↓↑ ↓↑ ↓↑	↓↑ ↓↑ ↓↑ ↓↑ ↓↑	↓↑	↓↑ ↓↑ ↓↑	_ _ _ _ _	_ _ _ _ _ _ _	_	_ _ _
37	Rb	↓↑	↓↑	↓↑ ↓↑ ↓↑	↓↑	↓↑ ↓↑ ↓↑	↓↑ ↓↑ ↓↑ ↓↑ ↓↑	↓↑	↓↑ ↓↑ ↓↑	_ _ _ _ _	_ _ _ _ _ _ _	↓	_ _ _
38	Sr	↓↑	↓↑	↓↑ ↓↑ ↓↑	↓↑	↓↑ ↓↑ ↓↑	↓↑ ↓↑ ↓↑ ↓↑ ↓↑	↓↑	↓↑ ↓↑ ↓↑	_ _ _ _ _	_ _ _ _ _ _ _	↓↑	_ _ _
39	Y	↓↑	↓↑	↓↑ ↓↑ ↓↑	↓↑	↓↑ ↓↑ ↓↑	↓↑ ↓↑ ↓↑ ↓↑ ↓↑	↓↑	↓↑ ↓↑ ↓↑	↓ _ _ _ _	_ _ _ _ _ _ _	↓↑	_ _ _
40	Zr	↓↑	↓↑	↓↑ ↓↑ ↓↑	↓↑	↓↑ ↓↑ ↓↑	↓↑ ↓↑ ↓↑ ↓↑ ↓↑	↓↑	↓↑ ↓↑ ↓↑	↓ ↓ _ _ _	_ _ _ _ _ _ _	↓↑	_ _ _
41	Nb	↓↑	↓↑	↓↑ ↓↑ ↓↑	↓↑	↓↑ ↓↑ ↓↑	↓↑ ↓↑ ↓↑ ↓↑ ↓↑	↓↑	↓↑ ↑ ↓↑	↓ ↓ ↓ ↓ _	_ _ _ _ _ _ _	↓	_ _ _

Table 4.11
Electronic Configurations
(cont.)

Atomic No.	Symbol	1s	2s	2p	3s	3p	3d	4s	4p	4d	4f	5s	5p
42	Mo	[↓↑]	[↓↑]	[↓↑][↓↑][↓↑]	[↓↑]	[↓↑][↓↑][↓↑]	[↓↑][↓↑][↓↑][↓↑][↓↑]	[↓↑]	[↓↑][↓↑][↓↑]	[↓][↓][↓][↓][↓]	[ ][ ][ ][ ][ ][ ][ ]	[↓]	[ ][ ][ ]
43	Tc	[↓↑]	[↓↑]	[↓↑][↓↑][↓↑]	[↓↑]	[↓↑][↓↑][↓↑]	[↓↑][↓↑][↓↑][↓↑][↓↑]	[↓↑]	[↓↑][↓↑][↓↑]	[↓↑][↓][↓][↓][↓]	[ ][ ][ ][ ][ ][ ][ ]	[↓]	[ ][ ][ ]
44	Ru	[↓↑]	[↓↑]	[↓↑][↓↑][↓↑]	[↓↑]	[↓↑][↓↑][↓↑]	[↓↑][↓↑][↓↑][↓↑][↓↑]	[↓↑]	[↓↑][↓↑][↓↑]	[↓↑][↓↑][↓][↓][↓]	[ ][ ][ ][ ][ ][ ][ ]	[↓]	[ ][ ][ ]
45	Rh	[↓↑]	[↓↑]	[↓↑][↓↑][↓↑]	[↓↑]	[↓↑][↓↑][↓↑]	[↓↑][↓↑][↓↑][↓↑][↓↑]	[↓↑]	[↓↑][↓↑][↓↑]	[↓↑][↓↑][↓↑][↓][↓]	[ ][ ][ ][ ][ ][ ][ ]	[↓]	[ ][ ][ ]
46	Pd	[↓↑]	[↓↑]	[↓↑][↓↑][↓↑]	[↓↑]	[↓↑][↓↑][↓↑]	[↓↑][↓↑][↓↑][↓↑][↓↑]	[↓↑]	[↓↑][↓↑][↓↑]	[↓↑][↓↑][↓↑][↓↑][↓↑]	[ ][ ][ ][ ][ ][ ][ ]	[ ]	[ ][ ][ ]
47	Ag	[↓↑]	[↓↑]	[↓↑][↓↑][↓↑]	[↓↑]	[↓↑][↓↑][↓↑]	[↓↑][↓↑][↓↑][↓↑][↓↑]	[↓↑]	[↓↑][↓↑][↓↑]	[↓↑][↓↑][↓↑][↓↑][↓↑]	[ ][ ][ ][ ][ ][ ][ ]	[↓]	[ ][ ][ ]
48	Cd	[↓↑]	[↓↑]	[↓↑][↓↑][↓↑]	[↓↑]	[↓↑][↓↑][↓↑]	[↓↑][↓↑][↓↑][↓↑][↓↑]	[↓↑]	[↓↑][↓↑][↓↑]	[↓↑][↓↑][↓↑][↓↑][↓↑]	[ ][ ][ ][ ][ ][ ][ ]	[↓↑]	[ ][ ][ ]
49	In	[↓↑]	[↓↑]	[↓↑][↓↑][↓↑]	[↓↑]	[↓↑][↓↑][↓↑]	[↓↑][↓↑][↓↑][↓↑][↓↑]	[↓↑]	[↓↑][↓↑][↓↑]	[↓↑][↓↑][↓↑][↓↑][↓↑]	[ ][ ][ ][ ][ ][ ][ ]	[↓↑]	[↓][ ][ ]
50	Sn	[↓↑]	[↓↑]	[↓↑][↓↑][↓↑]	[↓↑]	[↓↑][↓↑][↓↑]	[↓↑][↓↑][↓↑][↓↑][↓↑]	[↓↑]	[↓↑][↓↑][↓↑]	[↓↑][↓↑][↓↑][↓↑][↓↑]	[ ][ ][ ][ ][ ][ ][ ]	[↓↑]	[↓][↓][↓]
51	Sb	[↓↑]	[↓↑]	[↓↑][↓↑][↓↑]	[↓↑]	[↓↑][↓↑][↓↑]	[↓↑][↓↑][↓↑][↓↑][↓↑]	[↓↑]	[↓↑][↓↑][↓↑]	[↓↑][↓↑][↓↑][↓↑][↓↑]	[ ][ ][ ][ ][ ][ ][ ]	[↓↑]	[↓][↓][↓]
52	Te	[↓↑]	[↓↑]	[↓↑][↓↑][↓↑]	[↓↑]	[↓↑][↓↑][↓↑]	[↓↑][↓↑][↓↑][↓↑][↓↑]	[↓↑]	[↓↑][↓↑][↓↑]	[↓↑][↓↑][↓↑][↓↑][↓↑]	[ ][ ][ ][ ][ ][ ][ ]	[↓↑]	[↓↑][↓][↓]
53	I	[↓↑]	[↓↑]	[↓↑][↓↑][↓↑]	[↓↑]	[↓↑][↓↑][↓↑]	[↓↑][↓↑][↓↑][↓↑][↓↑]	[↓↑]	[↓↑][↓↑][↓↑]	[↓↑][↓↑][↓↑][↓↑][↓↑]	[ ][ ][ ][ ][ ][ ][ ]	[↓↑]	[↓↑][↓↑][↓]
54	Xe	[↓↑]	[↓↑]	[↓↑][↓↑][↓↑]	[↓↑]	[↓↑][↓↑][↓↑]	[↓↑][↓↑][↓↑][↓↑][↓↑]	[↓↑]	[↓↑][↓↑][↓↑]	[↓↑][↓↑][↓↑][↓↑][↓↑]	[ ][ ][ ][ ][ ][ ][ ]	[↓↑]	[↓↑][↓↑][↓↑]

Table 4.11
Electronic Configurations
(cont.)

Atomic No.	Symbol		4f	5s	5p	5d	5f	5g	6s	6p	6d
55	Cs	$KLMN_{spd}$		↓↑	↓↑ ↓↑ ↓↑				↓		
56	Ba	$KLMN_{spd}$		↓↑	↓↑ ↓↑ ↓↑				↓↑		
57	La	$KLMN_{spd}$		↓↑	↓↑ ↓↑ ↓↑	↓			↓↑		
58	Ce	$KLMN_{spd}$	↓ ↓	↓↑	↓↑ ↓↑ ↓↑				↓↑		
59	Pr	$KLMN_{spd}$	↓ ↓ ↓	↓↑	↓↑ ↓↑ ↓↑				↓↑		
60	Nd	$KLMN_{spd}$	↓ ↓ ↓ ↓	↓↑	↓↑ ↓↑ ↓↑				↓↑		
61	Pm	$KLMN_{spd}$	↓ ↓ ↓ ↓ ↓	↓↑	↓↑ ↓↑ ↓↑				↓↑		
62	Sm	$KLMN_{spd}$	↓ ↓ ↓ ↓ ↓ ↓	↓↑	↓↑ ↓↑ ↓↑				↓↑		
63	Eu	$KLMN_{spd}$	↓ ↓ ↓ ↓ ↓ ↓ ↓	↓↑	↓↑ ↓↑ ↓↑				↓↑		
64	Gd	$KLMN_{spd}$	↓ ↓ ↓ ↓ ↓ ↓ ↓	↓↑	↓↑ ↓↑ ↓↑	↓			↓↑		
65	Tb	$KLMN_{spd}$	↓↑ ↓↑ ↓ ↓ ↓ ↓ ↓	↓↑	↓↑ ↓↑ ↓↑				↓↑		
66	Dy	$KLMN_{spd}$	↓↑ ↓↑ ↓↑ ↓ ↓ ↓ ↓	↓↑	↓↑ ↓↑ ↓↑				↓↑		
67	Ho	$KLMN_{spd}$	↓↑ ↓↑ ↓↑ ↓↑ ↓ ↓ ↓	↓↑	↓↑ ↓↑ ↓↑				↓↑		
68	Er	$KLMN_{spd}$	↓↑ ↓↑ ↓↑ ↓↑ ↓↑ ↓ ↓	↓↑	↓↑ ↓↑ ↓↑				↓↑		

Table 4.11
Electronic Configurations
(cont.)

Atomic No.	Symbol		4*f*	5*s*	5*p*	5*d*	5*f*	5*g*	6*s*	6*p*	6*d*
69	Tm	$KLMN_{spd}$	↓↑ ↓↑ ↓↑ ↓↑ ↓↑ ↓↑ ↓	↓↑	↓↑ ↓↑ ↓↑				↓↑		
70	Yb	$KLMN_{spd}$	↓↑ ↓↑ ↓↑ ↓↑ ↓↑ ↓↑ ↓↑	↓↑	↓↑ ↓↑ ↓↑				↓↑		
71	Lu	$KLMN_{spd}$	↓↑ ↓↑ ↓↑ ↓↑ ↓↑ ↓↑ ↓↑	↓↑	↓↑ ↓↑ ↓↑	↓			↓↑		
72	Hf	$KLMN_{spd}$	↓↑ ↓↑ ↓↑ ↓↑ ↓↑ ↓↑ ↓↑	↓↑	↓↑ ↓↑ ↓↑	↓ ↓			↓↑		
73	Ta	$KLMN_{spd}$	↓↑ ↓↑ ↓↑ ↓↑ ↓↑ ↓↑ ↓↑	↓↑	↓↑ ↓↑ ↓↑	↓ ↓ ↓			↓↑		
74	W	$KLMN_{spd}$	↓↑ ↓↑ ↓↑ ↓↑ ↓↑ ↓↑ ↓↑	↓↑	↓↑ ↓↑ ↓↑	↓ ↓ ↓ ↓					
75	Re	$KLMN_{spd}$	↓↑ ↓↑ ↓↑ ↓↑ ↓↑ ↓↑ ↓↑	↓↑	↓↑ ↓↑ ↓↑	↓ ↓ ↓ ↓ ↓			↓↑		
76	Os	$KLMN_{spd}$	↓↑ ↓↑ ↓↑ ↓↑ ↓↑ ↓↑ ↓↑	↓↑	↓↑ ↓↑ ↓↑	↓↑ ↓ ↓ ↓ ↓			↓↑		
77	Ir	$KLMN_{spd}$	↓↑ ↓↑ ↓↑ ↓↑ ↓↑ ↓↑ ↓↑	↓↑	↓↑ ↓↑ ↓↑	↓↑ ↓↑ ↓↑ ↓↑ ↓					
78	Pt	$KLMN_{spd}$	↓↑ ↓↑ ↓↑ ↓↑ ↓↑ ↓↑ ↓↑	↓↑	↓↑ ↓↑ ↓↑	↓↑ ↓↑ ↓↑ ↓↑ ↓			↓		
79	Au	$KLMN_{spd}$	↓↑ ↓↑ ↓↑ ↓↑ ↓↑ ↓↑ ↓↑	↓↑	↓↑ ↓↑ ↓↑	↓↑ ↓↑ ↓↑ ↓↑ ↓↑			↓		
80	Hg	$KLMN_{spd}$	↓↑ ↓↑ ↓↑ ↓↑ ↓↑ ↓↑ ↓↑	↓↑	↓↑ ↓↑ ↓↑	↓↑ ↓↑ ↓↑ ↓↑ ↓↑			↓↑		
81	Tl	$KLMN_{spd}$	↓↑ ↓↑ ↓↑ ↓↑ ↓↑ ↓↑ ↓↑	↓↑	↓↑ ↓↑ ↓↑	↓↑ ↓↑ ↓↑ ↓↑ ↓↑			↓↑	↓	
82	Pb	$KLMN_{spd}$	↓↑ ↓↑ ↓↑ ↓↑ ↓↑ ↓↑ ↓↑	↓↑	↓↑ ↓↑ ↓↑	↓↑ ↓↑ ↓↑ ↓↑ ↓↑			↓↑	↓	

Table 4.11
Electronic Configurations
(cont.)

Atomic No.	Symbol		4f	5s	5p	5d	5f	5g	6s	6p	6d
83	Bi	$KLMN_{spd}$	[↓↑][↓↑][↓↑][↓↑][↓↑][↓↑][↓↑]	[↓↑]	[↓↑][↓↑][↓↑]	[↓↑][↓↑][↓↑][↓↑][↓↑]	[ ][ ][ ][ ][ ][ ][ ]		[↓↑]	[ ][↓][↓]	[ ][ ][ ][ ][ ]
84	Po	$KLMN_{spd}$	[↓↑][↓↑][↓↑][↓↑][↓↑][↓↑][↓↑]	[↓↑]	[↓↑][↓↑][↓↑]	[↓↑][↓↑][↓↑][↓↑][↓↑]	[ ][ ][ ][ ][ ][ ][ ]		[↓↑]	[↓↑][↓][↓]	[ ][ ][ ][ ][ ]
85	At	$KLMN_{spd}$	[↓↑][↓↑][↓↑][↓↑][↓↑][↓↑][↓↑]	[↓↑]	[↓↑][↓↑][↓↑]	[↓↑][↓↑][↓↑][↓↑][↓↑]	[ ][ ][ ][ ][ ][ ][ ]		[↓↑]	[↓↑][↓↑][↓]	[ ][ ][ ][ ][ ]
86	Rn	$KLMN_{spd}$	[↓↑][↓↑][↓↑][↓↑][↓↑][↓↑][↓↑]	[↓↑]	[↓↑][↓↑][↓↑]	[↓↑][↓↑][↓↑][↓↑][↓↑]	[ ][ ][ ][ ][ ][ ][ ]		[↓↑]	[↓↑][↓↑][↓↑]	[ ][ ][ ][ ][ ]

Atomic No.	Symbol		5f	5g	6s	6p	6d	6f	6g	6h	7s
87	Fr	$KLMNO_{spd}$	[ ][ ][ ][ ][ ][ ][ ]		[↓↑]	[↓↑][↓↑][↓↑]	[ ][ ][ ][ ][ ]	[ ][ ][ ][ ][ ][ ][ ]			[↓]
88	Ra	$KLMNO_{spd}$	[ ][ ][ ][ ][ ][ ][ ]		[↓↑]	[↓↑][↓↑][↓↑]	[ ][ ][ ][ ][ ]	[ ][ ][ ][ ][ ][ ][ ]			[↓↑]
89	Ac	$KLMNO_{spd}$	[ ][ ][ ][ ][ ][ ][ ]		[↓↑]	[↓↑][↓↑][↓↑]	[↓][ ][ ][ ][ ]	[ ][ ][ ][ ][ ][ ][ ]			[↓↑]
90	Th	$KLMNO_{spd}$	[ ][ ][ ][ ][ ][ ][ ]		[↓↑]	[↓↑][↓↑][↓↑]	[↓][↓][ ][ ][ ]	[ ][ ][ ][ ][ ][ ][ ]			[↓↑]
91	Pa	$KLMNO_{spd}$	[↓][↓][ ][ ][ ][ ][ ]		[↓↑]	[↓↑][↓↑][↓↑]	[↓][ ][ ][ ][ ]	[ ][ ][ ][ ][ ][ ][ ]			[↓↑]
92	U	$KLMNO_{spd}$	[ ][↓][↓][↓][ ][ ][ ]		[↓↑]	[↓↑][↓↑][↓↑]	[↓][ ][ ][ ][ ]	[ ][ ][ ][ ][ ][ ][ ]			[↓↑]
93	Np	$KLMNO_{spd}$	[↓][↓][↓][↓][ ][ ][ ]		[↓↑]	[↓↑][↓↑][↓↑]	[↓][ ][ ][ ][ ]	[ ][ ][ ][ ][ ][ ][ ]			[↓↑]

Table 4.11
Electronic Configurations
(cont.)

Atomic No.	Symbol		5f	5g	6s	6p	6d	6f	6g	6h	7s
94	Pu	KLMNO$_{spd}$	[↓][↓][↓][↓][↓][↓][ ]		[↓↑]	[↓↑][↓↑][↓↑]	[ ][ ][ ][ ][ ]	[ ][ ][ ][ ][ ][ ][ ]			[↓↑]
95	Am	KLMNO$_{spd}$	[↓][↓][↓][↓][↓][↓][↓]		[↓↑]	[↓↑][↓↑][↓↑]	[ ][ ][ ][ ][ ]	[ ][ ][ ][ ][ ][ ][ ]			[↓↑]
96	Cm	KLMNO$_{spd}$	[↓][↓][↓][↓][↓][↓][↓]		[↓↑]	[↓↑][↓↑][↓↑]	[↓][ ][ ][ ][ ]	[ ][ ][ ][ ][ ][ ][ ]			[↓↑]
97	Bk	KLMNO$_{spd}$	[↓↑][↓↑][↓][↓][↓][↓][↓]		[↓↑]	[↓↑][↓↑][↓↑]	[ ][ ][ ][ ][ ]	[ ][ ][ ][ ][ ][ ][ ]			[↓↑]
98	Cf	KLMNO$_{spd}$	[↓↑][↓↑][↓↑][↓][↓][↓][↓]		[↓↑]	[↓↑][↓↑][↓↑]	[ ][ ][ ][ ][ ]	[ ][ ][ ][ ][ ][ ][ ]			[↓↑]
99	Es	KLMNO$_{spd}$	[↓↑][↓↑][↓↑][↓↑][↓][↓][↓]		[↓↑]	[↓↑][↓↑][↓↑]	[ ][ ][ ][ ][ ]	[ ][ ][ ][ ][ ][ ][ ]			[↓↑]
100	Fm	KLMNO$_{spd}$	[↓↑][↓↑][↓↑][↓↑][↓↑][↓][↓]		[↓↑]	[↓↑][↓↑][↓↑]	[ ][ ][ ][ ][ ]	[ ][ ][ ][ ][ ][ ][ ]			[↓↑]
101	Md	KLMNO$_{spd}$	[↓↑][↓↑][↓↑][↓↑][↓↑][↓↑][↓]		[↓↑]	[↓↑][↓↑][↓↑]	[ ][ ][ ][ ][ ]	[ ][ ][ ][ ][ ][ ][ ]			[↓↑]
102	No	KLMNO$_{spd}$	[↓↑][↓↑][↓↑][↓↑][↓↑][↓↑][↓↑]		[↓↑]	[↓↑][↓↑][↓↑]	[ ][ ][ ][ ][ ]	[ ][ ][ ][ ][ ][ ][ ]			[↓↑]
103	Lw	KLMNO$_{spd}$	[↓↑][↓↑][↓↑][↓↑][↓↑][↓↑][↓↑]		[↓↑]	[↓↑][↓↑][↓↑]	[↓][ ][ ][ ][ ]	[ ][ ][ ][ ][ ][ ][ ]			[↓↑]

ATOMIC NUMBERS 1 AND 2. Since the $1s$ secondary energy level is the lowest potential energy level in all atoms, it is expected that one electron in the hydrogen atom and two electrons in the helium atom are in the $1s$ level in the ground state. The helium atom configuration is the first in which the first principal energy level has been completed, that is, it contains the maximum ($2n^2$) of two electrons. The electronic configurations of the hydrogen and helium atoms in the ground state are shown in Table 4.12.

TABLE 4.12
Ground State Configurations of Atoms with Atomic Numbers 1 and 2

H:	$1s^1$
He:	$1s^2$

ATOMIC NUMBERS 3 THROUGH 10. From Figs. 4.11 and 4.12 it is apparent that after the $1s$ level, the $2s$ and $2p$ secondary energy levels are the next lowest in potential energy, and this is true for all of the atoms. Thus for atoms with atomic numbers 3 through 10, the electrons beyond the first two are in the $2s$ and $2p$ secondary energy levels. The successive configurations of these eight atoms are shown in Table 4.13.

TABLE 4.13
Ground State Configurations of Atoms with Atomic Numbers 3 through 10

Li:	$1s^22s^1$
Be:	$1s^22s^2$
B:	$1s^22s^22p^1$
C:	$1s^22s^22p^2$
N:	$1s^22s^22p^3$
O:	$1s^22s^22p^4$
F:	$1s^22s^22p^5$
Ne:	$1s^22s^22p^6$

In Table 4.11, note should be made of the application of the rule of maximum multiplicity and the Pauli exclusion principle. Carbon is the first element to have more than one electron in the $2p$ secondary energy level; these two electrons are in different orbitals within that level in agreement with the rule of maximum multiplicity. If these two electrons were in the same orbital, then according to the Pauli exclusion principle they would have opposite spins, for otherwise they would have the same set of four quantum numbers.

ATOMIC NUMBERS 11 THROUGH 18. In Figs. 4.11 and 4.12 it is apparent that immediately above the $1s$, $2s$, and $2p$ levels in potential energy are the $3s$ and $3p$ secondary levels, and this is true for all atoms. Thus for atoms with atomic numbers 11 through 18, the electrons beyond the first ten are in the $3s$ and $3p$ levels. The configurations of these atoms are shown in Table 4.14. The configurations listed in Table 4.11 illustrate maximum multiplicity in the $3p$ level.

TABLE 4.14

Ground State Configurations of Atoms with Atomic Numbers 11 through 18

Na:	$1s^22s^22p^63s^1$
Mg:	$1s^22s^22p^63s^2$
Al:	$1s^22s^22p^63s^23p^1$
Si:	$1s^22s^22p^63s^23p^2$
P:	$1s^22s^22p^63s^23p^3$
S:	$1s^22s^22p^63s^23p^4$
Cl:	$1s^22s^22p^63s^23p^5$
Ar:	$1s^22s^22p^63s^23p^6$

ATOMIC NUMBERS 19 THROUGH 36. Careful scrutiny of Figs. 4.11 and 4.12 shows that in the vicinity of atomic number 19 the next lowest energy level after the $3p$ level is the $4s$ level. Consequently the nineteenth and twentieth electrons in the potassium and calcium atoms fit into the $4s$ level, as indicated in Tables 4.11 and 4.15.

TABLE 4.15

Ground State Configurations of Atoms with Atomic Numbers 19 through 36

K:	$1s^22s^22p^63s^23p^64s^1$
Ca:	$1s^22s^22p^63s^23p^64s^2$
Sc:	$1s^22s^22p^63s^23p^63d^14s^2$
Ti:	$1s^22s^22p^63s^23p^63d^24s^2$
V:	$1s^22s^22p^63s^23p^63d^34s^2$
Cr:	$1s^22s^22p^63s^23p^63d^54s^1$
Mn:	$1s^22s^22p^63s^23p^63d^54s^2$
Fe:	$1s^22s^22p^63s^23p^63d^64s^2$
Co:	$1s^22s^22p^63s^23p^63d^74s^2$
Ni:	$1s^22s^22p^63s^23p^63d^84s^2$
Cu:	$1s^22s^22p^63s^23p^63d^{10}4s^1$
Zn:	$1s^22s^22p^63s^23p^63d^{10}4s^2$
Ga:	$1s^22s^22p^63s^23p^63d^{10}4s^24p^1$
Ge:	$1s^22s^22p^63s^23p^63d^{10}4s^24p^2$
As:	$1s^22s^22p^63s^23p^63d^{10}4s^24p^3$
Se:	$1s^22s^22p^63s^23p^63d^{10}4s^24p^4$
Br:	$1s^22s^22p^63s^23p^63d^{10}4s^24p^5$
Kr:	$1s^22s^22p^63s^23p^63d^{10}4s^24p^6$

However, beyond atomic number 20, it is not clear which the next lowest energy level is after the $4s$ level, since the $3d$ and $4p$ levels are very nearly the same. In fact, in the vicinity of atomic number 23, the $3d$ and $4s$ energy levels are at about the same potential energy. Remembering that Figs. 4.11 and 4.12 are approximate, and taking into consideration the behavior of atoms with atomic numbers 21 through 30, note that these atoms are assigned configurations that differ from each other in the number of $3d$ electrons. There appears to be some slight variation at chromium and copper, in which there is one more electron in the $3d$ level and one less electron in the 4*s* level than might otherwise be expected. However, throughout this region of atomic numbers 21 through 30, the 3*d* level is always below the 4*p* level in potential energy, and for the most part it is near or below the 4*s* level as well.

Once the 3*d* and 4*s* levels are filled at atomic number 30, there is no further question about the next lower potential-energy level, which is the 4*p* level. The orderly arrangement of electrons in the 4*p* level as well as the trends in electronic configurations throughout this region of atomic number values can be seen in Tables 4.11 and 4.15.

ATOMIC NUMBERS 37 THROUGH 54. In the region of atomic number 37, Figs. 4.11 and 4.12 suggest that the next higher energy level after 4*p* is 5*s*, although at slightly higher atomic numbers the 4*d* energy level is lower in potential energy than the 5*s* energy level. Thus the configurations assigned to rubidium and strontium have electrons in the 5*s* level, but the succeeding elements, beginning with yttrium and ending with cadmium, differ from each other mainly in the number of 4*d* electrons in the atoms. Although two electrons are assigned to the 5*s* level in strontium, the atoms of the elements with atomic numbers 41 through 47 have fewer 5*s* electrons, palladium having none at all. This is indicative of the lower energy of the 4*d* level into which the electrons tend to cluster. Tables 4.11 and 4.16 show the electronic configurations of these atoms.

TABLE 4.16

Ground State Configurations of Atoms with Atomic Numbers 37 through 54

Rb:	$1s^22s^22p^63s^23p^63d^{10}4s^24p^65s^1$
Sr:	$1s^22s^22p^63s^23p^63d^{10}4s^24p^65s^2$
Y:	$1s^22s^22p^63s^23p^63d^{10}4s^24p^64d^15s^2$
Zr:	$1s^22s^22p^63s^23p^63d^{10}4s^24p^64d^25s^2$
Nb:	$1s^22s^22p^63s^23p^63d^{10}4s^24p^64d^45s^1$
Mo:	$1s^22s^22p^63s^23p^63d^{10}4s^24p^64d^55s^1$
Tc:	$1s^22s^22p^63s^23p^63d^{10}4s^24p^64d^65s^1$
Ru:	$1s^22s^22p^63s^23p^63d^{10}4s^24p^64d^75s^1$
Rh:	$1s^22s^22p^63s^23p^63d^{10}4s^24p^64d^85s^1$
Pd:	$1s^22s^22p^63s^23p^63d^{10}4s^24p^64d^{10}$
Ag:	$1s^22s^22p^63s^23p^63d^{10}4s^24p^64d^{10}5s^1$
Cd:	$1s^22s^22p^63s^23p^63d^{10}4s^24p^64d^{10}5s^2$
In:	$1s^22s^22p^63s^23p^63d^{10}4s^24p^64d^{10}5s^25p^1$
Sn:	$1s^22s^22p^63s^23p^63d^{10}4s^24p^64d^{10}5s^25p^2$
Sb:	$1s^22s^22p^63s^23p^63d^{10}4s^24p^64d^{10}5s^25p^3$
Te:	$1s^22s^22p^63s^23p^63d^{10}4s^24p^64d^{10}5s^25p^4$
I:	$1s^22s^22p^63s^23p^63d^{10}4s^24p^64d^{10}5s^25p^5$
Xe:	$1s^22s^22p^63s^23p^63d^{10}4s^24p^64d^{10}5s^25p^6$

Once the 4*d* and 5*s* levels are filled at atomic number 48, there is no further question about the next lower potential energy level, which is the 5*p* level.

ATOMIC NUMBERS 55 THROUGH 86. As atomic numbers increase and the number of electrons per atom increases, the assigned configurations become less certain. The configurations given here are very problematical and seem to be the most probable.

From Figs. 4.11 and 4.12 it appears that in the vicinity of atomic number 55 the next highest energy level after $5p$ is the $6s$ energy level. However, beyond this, the $6p$, $5d$, and $4f$ levels are rather close in potential energy. The configurations shown in Tables 4.11 and 4.17 reflect this confusion. In an approximate way, the first two elements in this sequence differ from each other in the number of $6s$ electrons per atom; the next 14 or 15 elements differ from each other in the number of $4f$ electrons per atom; the next 9 or 10 elements differ from each other in the number of $5d$ electrons per atom; the last six elements differ from each other in the number of $6p$ electrons per atom.

Atomic numbers 87 through 103. In the vicinity of atomic number 87, Figs. 4.11 and 4.12 suggest that immediately above the $6p$ level is the $7s$ level. However,

TABLE 4.17
Ground State Configurations of Atoms with Atomic Numbers 55 through 86

Cs:	$1s^22s^22p^63s^23p^63d^{10}4s^24p^64d^{10}5s^25p^66s^1$
Ba:	$1s^22s^22p^63s^23p^63d^{10}4s^24p^64d^{10}5s^25p^66s^2$
La:	$1s^22s^22p^63s^23p^63d^{10}4s^24p^64d^{10}5s^25p^65d^16s^2$
Ce:	$1s^22s^22p^63s^23p^63d^{10}4s^24p^64d^{10}4f^25s^25p^66s^2$
Pr:	$1s^22s^22p^63s^23p^63d^{10}4s^24p^64d^{10}4f^35s^25p^66s^2$
Nd:	$1s^22s^22p^63s^23p^63d^{10}4s^24p^64d^{10}4f^45s^25p^66s^2$
Pm:	$1s^22s^22p^63s^23p^63d^{10}4s^24p^64d^{10}4f^55s^25p^66s^2$
Sm:	$1s^22s^22p^63s^23p^63d^{10}4s^24p^64d^{10}4f^65s^25p^66s^2$
Eu:	$1s^22s^22p^63s^23p^63d^{10}4s^24p^64d^{10}4f^75s^25p^66s^2$
Gd:	$1s^22s^22p^63s^23p^63d^{10}4s^24p^64d^{10}4f^75s^25p^65d^16s^2$
Tb:	$1s^22s^22p^63s^23p^63d^{10}4s^24p^64d^{10}4f^95s^25p^66s^2$
Dy:	$1s^22s^22p^63s^23p^63d^{10}4s^24p^64d^{10}4f^{10}5s^25p^66s^2$
Ho:	$1s^22s^22p^63s^23p^63d^{10}4s^24p^64d^{10}4f^{11}5s^25p^66s^2$
Er:	$1s^22s^22p^63s^23p^63d^{10}4s^24p^64d^{10}4f^{12}5s^25p^66s^2$
Tm:	$1s^22s^22p^63s^23p^63d^{10}4s^24p^64d^{10}4f^{13}5s^25p^66s^2$
Yb:	$1s^22s^22p^63s^23p^63d^{10}4s^24p^64d^{10}4f^{14}5s^25p^66s^2$
Lu:	$1s^22s^22p^63s^23p^63d^{10}4s^24p^64d^{10}4f^{14}5s^25p^65d^16s^2$
Hf:	$1s^22s^22p^63s^23p^63d^{10}4s^24p^64d^{10}4f^{14}5s^25p^65d^26s^2$
Ta:	$1s^22s^22p^63s^23p^63d^{10}4s^24p^64d^{10}4f^{14}5s^25p^65d^36s^2$
W:	$1s^22s^22p^63s^23p^63d^{10}4s^24p^64d^{10}4f^{14}5s^25p^65d^46s^2$
Re:	$1s^22s^22p^63s^23p^63d^{10}4s^24p^64d^{10}4f^{14}5s^25p^65d^56s^2$
Os:	$1s^22s^22p^63s^23p^63d^{10}4s^24p^64d^{10}4f^{14}5s^25p^65d^66s^2$
Ir:	$1s^22s^22p^63s^23p^63d^{10}4s^24p^64d^{10}4f^{14}5s^25p^65d^9$
Pt:	$1s^22s^22p^63s^23p^63d^{10}4s^24p^64d^{10}4f^{14}5s^25p^65d^96s^1$
Au:	$1s^22s^22p^63s^23p^63d^{10}4s^24p^64d^{10}4f^{14}5s^25p^65d^{10}6s^1$
Hg:	$1s^22s^22p^63s^23p^63d^{10}4s^24p^64d^{10}4f^{14}5s^25p^65d^{10}6s^2$
Tl:	$1s^22s^22p^63s^23p^63d^{10}4s^24p^64d^{10}4f^{14}5s^25p^65d^{10}6s^26p^1$
Pb:	$1s^22s^22p^63s^23p^63d^{10}4s^24p^64d^{10}4f^{14}5s^25p^65d^{10}6s^26p^2$
Bi:	$1s^22s^22p^63s^23p^63d^{10}4s^24p^64d^{10}4f^{14}5s^25p^65d^{10}6s^26p^3$
Po:	$1s^22s^22p^63s^23p^63d^{10}4s^24p^64d^{10}4f^{14}5s^25p^65d^{10}6s^26p^4$
At:	$1s^22s^22p^63s^23p^63d^{10}4s^24p^64d^{10}4f^{14}5s^25p^65d^{10}6s^26p^5$
Rn:	$1s^22s^22p^63s^23p^63d^{10}4s^24p^64d^{10}4f^{14}5s^25p^65d^{10}6s^26p^6$

immediately beyond this atomic number, there is similarity in potential energy for the $6d$ and $5f$ levels, although the $5f$ level soon is clearly below the $6d$ level. In an approximate manner it can be said that the elements with atomic number 89 through 103 differ from each other in the number of $5f$ electrons per atom. The configurations of these elements are shown in Tables 4.11 and 4.18.

TABLE 4.18

Ground State Configurations of Atoms with Atomic Numbers 87 through 103

Fr:	$1s^22s^22p^63s^23p^63d^{10}4s^24p^64d^{10}4f^{14}5s^25p^65d^{10}6s^26p^67s^1$
Ra:	$1s^22s^22p^63s^23p^63d^{10}4s^24p^64d^{10}4f^{14}5s^25p^65d^{10}6s^26p^67s^2$
Ac:	$1s^22s^22p^63s^23p^63d^{10}4s^24p^64d^{10}4f^{14}5s^25p^65d^{10}6s^26p^66d^17s^2$
Th:	$1s^22s^22p^63s^23p^63d^{10}4s^24p^64d^{10}4f^{14}5s^25p^65d^{10}6s^26p^66d^27s^2$
Pa:	$1s^22s^22p^63s^23p^63d^{10}4s^24p^64d^{10}4f^{14}5s^25p^65d^{10}5f^26s^26p^66d^17s^2$
U:	$1s^22s^22p^63s^23p^63d^{10}4s^24p^64d^{10}4f^{14}5s^25p^65d^{10}5f^36s^26p^66d^17s^2$
Np:	$1s^22s^22p^63s^23p^63d^{10}4s^24p^64d^{10}4f^{14}5s^25p^65d^{10}5f^46s^26p^66d^17s^2$
Pu:	$1s^22s^22p^63s^23p^63d^{10}4s^24p^64d^{10}4f^{14}5s^25p^65d^{10}5f^66s^26p^67s^2$
Am:	$1s^22s^22p^63s^23p^63d^{10}4s^24p^64d^{10}4f^{14}5s^25p^65d^{10}5f^76s^26p^67s^2$
Cm:	$1s^22s^22p^63s^23p^63d^{10}4s^24p^64d^{10}4f^{14}5s^25p^65d^{10}5f^76s^26p^66d^17s^2$
Bk:	$1s^22s^22p^63s^23p^63d^{10}4s^24p^64d^{10}4f^{14}5s^25p^65d^{10}5f^96s^26p^67s^2$
Cf:	$1s^22s^22p^63s^23p^63d^{10}4s^24p^64d^{10}4f^{14}5s^25p^65d^{10}5f^{10}6s^26p^67s^2$
Es:	$1s^22s^22p^63s^23p^63d^{10}4s^24p^64d^{10}4f^{14}5s^25p^65d^{10}5f^{11}6s^26p^67s^2$
Fm:	$1s^22s^22p^63s^23p^63d^{10}4s^24p^64d^{10}4f^{14}5s^25p^65d^{10}5f^{12}6s^26p^67s^2$
Md:	$1s^22s^22p^63s^23p^63d^{10}4s^24p^64d^{10}4f^{14}5s^25p^65d^{10}5f^{13}6s^26p^67s^2$
No:	$1s^22s^22p^63s^23p^63d^{10}4s^24p^64d^{10}4f^{14}5s^25p^65d^{10}5f^{14}6s^26p^67s^2$
Lw:	$1s^22s^22p^63s^23p^63d^{10}4s^24p^64d^{10}4f^{14}5s^25p^65d^{10}5f^{14}6s^26p^66d^17s^2$

THE PERIODIC TABLE

The configurations shown in Tables 4.12 through 4.18 suggest a regularity in the construction of successive atoms and that there are repeating types of configurations when the atoms are arranged in the order of increasing atomic number. For example, the initial entry in each of these tables has been put into Table 4.19,

TABLE 4.19

Configurations of Atoms of the ns^1 Type

H:	$1s^1$
Li:	$1s^22s^1$
Na:	$1s^22s^22p^63s^1$
K:	$1s^22s^22p^63s^23p^64s^1$
Rb:	$1s^22s^22p^63s^23p^63d^{10}4s^24p^65s^1$
Cs:	$1s^22s^22p^63s^23p^63d^{10}4s^24p^64d^{10}5s^25p^66s^1$
Fr:	$1s^22s^22p^63s^23p^63d^{10}4s^24p^64d^{10}4f^{14}5s^25p^65d^{10}6s^26p^67s^1$

where it is apparent that these configurations are similar since they have a final ns^1 arrangement. These differ only in the principal energy level in which the configuration occurs.

Similarly, the final entries of Tables 4.13 through 4.17 are listed in Table 4.20. These configurations are similar in that they end with the ns^2np^6 arrangement.

TABLE 4.20
Configurations of Atoms of the $ns^2\,np^6$ Type

Ne:	$1s^22s^22p^6$
Ar:	$1s^22s^22p^63s^23p^6$
Kr:	$1s^22s^22p^63s^23p^63d^{10}4s^24p^6$
Xe:	$1s^22s^22p^63s^23p^63d^{10}4s^24p^64d^{10}5s^25p^6$
Rn:	$1s^22s^22p^63s^23p^63d^{10}4s^24p^64d^{10}4f^{14}5s^25p^65d^{10}6s^26p^6$

This type of observation suggests that there is a periodic repetition of electronic configurations when the elements are arranged in the order of increasing atomic number. Even before the electron was discovered, many properties of the elements had been shown to recur in a periodic way. From these observations the elements were put into an orderly array called the *periodic table*. It was pointed out in Chapter 2 that the ordinal numbers used to identify the elements in the periodic table arrangement are what we now call the atomic numbers.

As it turns out, the recurrence of similar electronic configurations coincides with the recurrence of similar properties of the elements. Consequently the periodic table may be interpreted in terms of electronic configurations of the atoms.

A modern form of the periodic table is shown as Table 4.21. In this table, the elements are arranged in the order of increasing atomic number. The seven horizontal rows of the table are called *periods* and these are numbered from top to bottom. The vertical columns are called *families* or *groups* and are numbered as indicated across the top of the table. However, as an alternative, the families sometimes are named by the first element in the family. Thus family 4A sometimes is called the carbon family.

Some of the families of elements within the periodic table have common names in addition to those already mentioned. These are listed in Table 4.22. It should be noted also that there are three families grouped under the number 8. These usually are referred to individually as the iron, cobalt, and nickel families.

The two partial periods at the bottom of the table beginning with atomic number 58 and 90 respectively really are parts of periods 6 and 7. They are placed at the bottom to conserve space. The partial period beginning with atomic number 58 and often called the *lanthanide series* belongs in period 6 between atomic numbers 57 and 72. The partial period beginning with atomic number 90 and often called the *actinide series* belongs in period 7 immediately after atomic number 89.

The periodic table may be used as a road map for electronic configurations. It should be recognized that the elements in each of the Tables 4.12 through 4.18 are the elements for each of the periods in the periodic table. Through each of the periods there is a somewhat similar progression of configurations, and similar configurations fall into the vertical columns or families of the periodic table. Once the similarities and regularities are recognized, it is possible to write reasonable electronic configurations for atoms by locating them in the periodic table.

First, it is interesting to note the characteristic clustering of families in the periodic table. At the far left, there are two families (1A and 2A); at the far right there are six families (3A through O); in the middle there are 10 families (3B

TABLE 4.21
The Periodic Table With Atomic Weights

	1A	2A	3B	4B	5B	6B	7B	8	8	8	1B	2B	3A	4A	5A	6A	7A	0
1	1 **H** 1.008																1 **H** 1.008	2 **He** 4.003
2	3 **Li** 6.939	4 **Be** 9.012											5 **B** 10.81	6 **C** 12.01	7 **N** 14.01	8 **O** 16.00	9 **F** 19.00	10 **Ne** 20.18
3	11 **Na** 23.00	12 **Mg** 24.31											13 **Al** 26.98	14 **Si** 28.09	15 **P** 30.97	16 **S** 32.06	17 **Cl** 35.45	18 **Ar** 39.95
4	19 **K** 39.10	20 **Ca** 40.08	21 **Sc** 44.96	22 **Ti** 47.90	23 **V** 50.94	24 **Cr** 52.00	25 **Mn** 54.94	26 **Fe** 55.85	27 **Co** 58.93	28 **Ni** 58.71	29 **Cu** 63.54	30 **Zn** 65.37	31 **Ga** 69.72	32 **Ge** 72.59	33 **As** 74.92	34 **Se** 78.96	35 **Br** 79.91	36 **Kr** 83.80
5	37 **Rb** 85.47	38 **Sr** 87.62	39 **Y** 88.90	40 **Zr** 91.22	41 **Nb** 92.91	42 **Mo** 95.94	43 **Tc** (97)	44 **Ru** 101.1	45 **Rh** 102.9	46 **Pd** 106.4	47 **Ag** 107.9	48 **Cd** 112.4	49 **In** 114.8	50 **Sn** 118.7	51 **Sb** 121.8	52 **Te** 127.6	53 **I** 126.9	54 **Xe** 131.3
6	55 **Cs** 132.9	56 **Ba** 137.3	57 **La** 138.9	72 **Hf** 178.5	73 **Ta** 180.9	74 **W** 183.8	75 **Re** 186.2	76 **Os** 190.2	77 **Ir** 192.2	78 **Pt** 195.1	79 **Au** 197.0	80 **Hg** 200.6	81 **Tl** 204.4	82 **Pb** 207.2	83 **Bi** 209.0	84 **Po** (209)	85 **At** (210)	86 **Rn** (222)
7	87 **Fr** (223)	88 **Ra** (226)	89 **Ac** (227)															
					58 **Ce** 140.1	59 **Pr** 140.9	60 **Nd** 144.2	61 **Pm** (145)	62 **Sm** 150.4	63 **Eu** 152.0	64 **Gd** 157.2	65 **Tb** 158.9	66 **Dy** 162.5	67 **Ho** 164.9	68 **Er** 167.3	69 **Tm** 168.9	70 **Yb** 173.0	71 **Lu** 175.0
					90 **Th** 232.0	91 **Pa** (231)	92 **U** 238.0	93 **Np** (237)	94 **Pu** (244)	95 **Am** (243)	96 **Cm** (247)	97 **Bk** (247)	98 **Cf** (251)	99 **Es** (252)	100 **Fm** (253)	101 **Md** (256)	102 **No** (256)	103 **Lw** (257)

TABLE 4.22
Common Family Names in the Periodic Table

Family 1A	The Alkali Metals
Family 2A	The Alkaline Earths
Family 1B	The Coinage Metals
Family 6A	The Chalcogens
Family 7A	The Halogens
Family O	The Rare or Inert Gases

through 2B); at the bottom, there are two partial periods comprising 14 small families. The numbers 2, 6, 10, and 14 which occur here should be recognized as the maximum numbers of electrons which may be found in the *s*, *p*, *d*, and *f* secondary energy levels. See Table 4.10.

To characterize these clusters of families, it is common practice to represent the highest principal energy level for an electron in the ground state of a given atom by the principal quantum number symbol, n. Then it is possible to refer to the ns level which is the $1s$ level for hydrogen, the $2s$ level for lithium, the $3s$ level for sodium, etc. Usually, for any element, the value for n corresponds to the number of the period of the periodic table in which that element is located. Lower principal energy levels are designated as $(n - 1)$, $(n - 2)$, etc. For example, the np level for bromine is the $4p$ level, the $(n - 1)p$ level is the $3p$ level, and the $(n - 2)s$ level is the $2s$ level.

In this way types of configurations may be designated easily. The atoms of elements in families 1A and 2A differ from each other in the number of electrons in the ns level. The atoms of the elements in families 3A through O differ from each other in the number of electrons in the np level. Collectively, the elements in these eight families, whose atoms differ mainly in the ns and np levels, are called the *representative elements.*

The atoms of the elements in families 3B through 2B differ from each other in the number of electrons in the $(n - 1)d$ level. Collectively the elements in these ten families in the center of the periodic table are known as the *transition elements.* Finally, the atoms of the elements in the lanthanide and actinide series differ from each other mainly in the numbers of electrons in the $(n - 2)f$ level. Collectively, these elements are known as the *inner transition elements.*

Representative Elements

The configurations of the atoms of the representative elements vary in a regular manner and may be written easily with reference to the periodic table. The differences in configuration from family to family among the representative elements are shown in Table 4.23.

Hydrogen and helium are somewhat anomalous due to the absence of a $1p$ secondary energy level. The symbol for hydrogen has been put in two places in Table 4.21. In terms of electronic configuration it appears to belong in family 1A.

TABLE 4.23

Configuration Types for the Representative Elements

Family 1A	ns^1
Family 2A	ns^2
Family 3A	ns^2np^1
Family 4A	ns^2np^2
Family 5A	ns^2np^3
Family 6A	ns^2np^4
Family 7A	ns^2np^5
Family O	ns^2np^6

However, the chemical characteristics of hydrogen suggest similarities to both family 1A and family 7A. Thus the hydrogen symbol appears in two locations of the periodic table. On the other hand, the characteristics of helium definitely are those of an inert gas. Thus the symbol for helium has been put into the inert gas family in the periodic table.

For any of the first twenty elements, the electronic configurations can be read from the periodic table very easily. Consider the element calcium, with atomic number 20. Since it is located in period 4 and family 2A, its configuration must end with $4s^2$. In addition, those energy levels below the fourth energy level which were filled in the meantime for elements of atomic number less than 20, remain filled. These are $1s$, $2s$, $2p$, $3s$, and $3p$ secondary energy levels. Thus the electronic configuration for the calcium atom is $1s^22s^22p^63s^23p^64s^2$.

Transition Elements

The electronic configurations of the transition elements are less regular than those for the representative elements. For these elements the difficulty lies in the fact that the energies of electrons in the $(n - 1)d$ and ns levels are not very different. A regular progression across a given transition series in any period would lead to a regular increase in the number of electrons in the $(n - 1)d$ energy level. This regular progression is shown in Table 4.24 along with the apparent variations which are found among the assignments made in Tables 4.11, 4.15, 4.16, and 4.17.

TABLE 4.24

Configuration Types for the Transition Elements

FAMILY	REGULAR PROGRESSION	ASSIGNED VARIATIONS
3B	$(n-1)d^1ns^2$	none
4B	$(n-1)d^2ns^2$	none
5B	$(n-1)d^3ns^2$	$(n-1)d^4ns^1$
6B	$(n-1)d^4ns^2$	$(n-1)d^5ns^1$
7B	$(n-1)d^5ns^2$	$(n-1)d^6ns^1$
8(Fe)	$(n-1)d^6ns^2$	$(n-1)d^7ns^1$
8(Co)	$(n-1)d^7ns^2$	$(n-1)d^8ns^1$; $(n-1)d^9$
8(Ni)	$(n-1)d^8ns^2$	$(n-1)d^9ns^1$; $(n-1)d^{10}$
1B	$(n-1)d^9ns^2$	$(n-1)d^{10}ns^1$
2B	$(n-1)d^{10}ns^2$	none

In general, the atoms of the transition elements tend to have more $(n-1)d$ electrons than predicted by the regular progression, and this tendency increases as the period number increases. For example, in period 4, the only elements showing variations from the regular progression are chromium in family 6B and copper in family 1B. However, in period 5, elements in families 5B, 6B, 7B, 8(Fe), 8(Co), 8(Ni), and 1B all show variations from the regular progression.

Regular progression configurations may be written easily from the periodic table. For example, nickel is listed in period 4. Thus the configuration of the nickel atom ends in $4s^2$. Nickel also is listed in family 8(Ni), which is the eighth family in the transition element sequence. Thus, more specifically, the configuration of the nickel atom ends in $3d^84s^2$. Looking at the elements listed before nickel in the periodic table, we see that the following energy levels contain the maximum number of electrons: $1s$, $2s$, $2p$, $3s$, $3p$, and $4s$. Therefore the complete electronic configuration for nickel is $1s^22s^22p^63s^23p^63d^84s^2$.

Inner Transition Elements

The electronic configurations for the atoms of the inner transition elements are less well determined than those listed for the representative and transition elements. In general, the atoms of these elements differ from each other in the number of $(n-2)f$ electrons in the atoms. However the $(n-1)d$ and $(n-2)f$ electrons are at about the same potential energy. Thus variations appear in the progression through the lanthanide and actinide sequences.

From the arrangement of the periodic table, lanthanum appears as a transition element; one electron is in the $(n-1)d$ state. It might be expected that the next element, cerium, would be the first to exhibit an electron in the $(n-2)f$ state. Although this appears to be so, the cerium atom does not appear to have an $(n-1)d$ electron. In fact, it is probable that the atoms of most of the inner transition elements do not have electrons in the $(n-1)d$ state. The specifics of the actual configurations of the atoms of these elements are of relatively little importance to us, although it is important to see how these elements fit into the over-all picture as presented by the periodic table.

SUMMARY

Studies of the spectral and magnetic properties of hydrogen lead to the conclusion that the electron in the atom may exist only in specific, discrete states. These states may be described by four quantum numbers. The principal and secondary quantum numbers describe the specific potential energies which the electron may assume in the hydrogen atom. The remaining two quantum numbers describe the magnetic properties of the electron in the hydrogen atom. The magnetic quantum number describes the discrete directions of the magnetic field which may be associated with the translational motion of the electron in the atom. In addition, a wave mechanical treatment of the electron suggests probable electron locations associated with each of these permitted directions. Furthermore, the electron may have either of two spins, which produce a second component of the magnetic field of the electron.

The states of electrons in atoms containing more than one electron are described by the same four quantum numbers used for the possible states of the single electron in the hydrogen atom. However, each electron in the atom has a different set of the four quantum numbers. Thus secondary energy levels may contain more than one electron. According to the rule of maximum multiplicity, electrons preferentially occupy the maximum number of orbitals within a given secondary energy level, and when electrons are unpaired, their spins are the same. However, electrons in the same orbital are paired, that is, their spins are opposite.

The potential energies of electrons in the secondary energy levels are affected by the nuclear charge of the atom. However, not all secondary energy levels are affected equally or in the same way. As a result, electrons may occupy different combinations of secondary energy levels in different atoms. The periodic table conveniently displays the elements in a sequence which reveals the pattern of electronic configurations of the atoms. From the location of the symbol of an element in the periodic table, it is possible to write a reasonable electronic configuration for the atom of that element.

For example, consider lead in period 6 and family 4A of the periodic table. Lead is a representative element whose characteristic configuration term is ns^2np^2, that is, $6s^26p^2$. As seen in the periodic table, the following secondary energy levels are completed: $1s$, $2s$, $2p$, $3s$, $3p$, $4s$, $3d$, $4p$, $5s$, $4d$, $5p$, $6s$, $4f$, and $5d$. Therefore, the complete electronic configuration for the lead atom is $1s^22s^22p^63s^23p^63d^{10}4s^24p^6$-$4d^{10}4f^{14}5s^25p^65d^{10}6s^26p^2$.

QUESTIONS AND PROBLEMS

4.1 Describe a hydrogen atom.

4.2 Correlate Fig. 4.3 with Table 4.1 by computing the values in Table 4.1 from the data in Fig. 4.3.

4.3 Refer to Table 4.2. Compute the frequencies and positions in the electromagnetic spectrum for the electromagnetic radiation which is emitted from the hydrogen atom when the electron makes each of the following transitions:
a) From energy level 20 to energy level 1.
b) From energy level 15 to energy level 10.
c) From energy level 15 to energy level 14.

4.4 Do any of the spectral frequencies in the Lyman, Balmer or Paschen series given in Table 4.1 appear in the visible portion of the electromagnetic spectrum? If any do, what colors do they represent? Which of these radiations appear in the ultraviolet and infrared portions of the spectrum?

4.5 Sodium-containing materials such as glass or table salt produce a yellow flame when introduced into the flame of a bunsen burner. Suggest a reason for this observation.

4.6 The wavelength ranges for the colors of the visible spectrum are shown in Table 1.14.

a) The wavelengths of light emitted by sodium containing materials heated in a bunsen flame are 5896 Å and 5890 Å. What colors are these emissions and to what energy differences in kilocalories do they correspond?

b) The wavelengths of light emitted by lithium-containing materials heated in a bunsen flame are 6708 Å and 6104 Å. What colors are these emissions and to what energy differences in kilocalories do they correspond?

c) The frequency of light emitted by potassium-containing materials heated in a bunsen flame is 7.41×10^{14} sec^{-1}. What color is this emission and to what energy difference does it correspond?

d) Compare the answers to parts a), b), and c). What happens to the sizes of the energy transitions from lithium to sodium to potassium?

4.7 In a bunsen flame the following elements undergo electron transitions of the listed magnitudes. Refer to Table 1.14 to suggest the colors imparted to the flame by each of these elements.

a) barium: 3.62×10^{-12} ergs

b) strontium: 3.06×10^{-12} ergs; 3.11×10^{-12} ergs; 4.33×10^{-12} ergs

c) calcium: 3.21×10^{-12} ergs; 3.49×10^{-12} ergs

4.8 Suggest some commercial applications for emission spectra.

4.9 In the experiment sketched in Fig. 4.6, why must the two magnetic poles be of unequal strengths?

4.10 Summarize the significance of the four quantum numbers.

4.11 Expand Fig. 4.5 to include principal energy level 4.

4.12 Expand Table 4.5 showing the values for l, m_l, and m_s when $n = 5, 6, 7$, and 8.

4.13 Prepare a table like Table 4.8 showing the electron states for the hydrogen atom when a) $n = 4$; b) $n = 5$; c) $n = 6$.

4.14 Expand Table 4.9 for $n = 5, 6, 7$, and 8.

4.15 Expand Table 4.10 for $l = 4, 5$, and 6.

4.16 Distinguish between the Heisenberg uncertainty principle and the Pauli exclusion principle.

4.17 How many *orbitals* are there in each of the following secondary energy levels?

a) 5*s* b) 7*d* c) 6*p* d) 10*f*

4.18 Draw feasible probability-distance curves for $n = 4$, similar to those in Fig. 4.7. Which has the greater penetration, an electron in the 4*s* or 4*p* secondary energy level?

4.19 How do the 4*d* charge distributions differ from the 3*d* charge distributions?

4.20 In Fig. 4.10, indicate the differences in the following pairs of states:

a) states *a* and *d* c) states *d* and *f*

b) states *f* and *g* d) states *e* and *g*

4.21 State Hund's rule of maximum multiplicity.

4.22 Why is an electron in the 1*s* energy level of a carbon atom in a lower state of potential energy than an electron in the 1*s* energy level of a hydrogen atom?

4.23 Would you expect any energy difference between an electron in the 1*s* energy level of ^1_1H and an electron in the 1*s* energy level of ^3_1H? Why?

4.24 What are the maximum numbers of electrons possible in each of the following?

a) *Q* principal energy level b) 10*j* secondary energy level c) a 5*f* orbital

4.25 Using Fig. 4.12, list the secondary energy levels in the order of increasing potential energy for

a) neon b) iron c) tin d) gold

4.26 Criticize the following statement: The calcium atom does not have any 3*d* secondary energy levels.

4.27 Criticize the following two rules which often have been stated as guides for writing electronic configurations:

a) No atom contains more than 8 electrons in its *n* principal energy level.

b) No atom contains more than 18 electrons in its $(n - 1)$ principal energy level.

4.28 Identify the following elements as representative, transition, or inner transition elements:

a) manganese
b) phosphorus
c) technetium
d) uranium
e) radon
f) platinum

4.29 Prepare a table like Tables 4.19 and 4.20 for each of the following configuration types:

a) ns^2
b) ns^2np^4
c) $(n - 1)d^2ns^2$
d) $(n - 1)d^5ns^1$

4.30 The periodic law has been stated as follows: The periodicity of the elements is a function of the atomic number. What does this mean?

4.31 Account for the fact that the third period of the periodic table has eight and not eighteen elements.

4.32 Why are calcium and zinc not placed in the same periodic family since each of their configurations ends in ns^2?

4.33 An atom has 2 electrons in the *K* principal energy level, 8 electrons in the *L* principal energy level, and 7 electrons in the *M* principal energy level. Give its atomic number and the total number of electrons in *s*, *p*, and *d* secondary energy levels.

4.34 For the variations listed in Table 4.24, list the elements whose atoms show each of these variations.

4.35 Which families among the transition elements always show variations from the regular progression as shown in Table 4.24?

4.36 Using only the periodic table, write the complete electronic configurations for the following elements and classify these elements as representative, transition, or inner transition:

a) carbon
b) arsenic
c) xenon
d) iron
e) silver
f) tungsten
g) actinium
h) thorium

4.37 Using only the periodic table, indicate the number of electrons with unpaired spins in the atoms of each of the following elements. Check your answers with Table 4.11.

a) chromium
b) zinc
c) aluminum
d) chlorine
e) krypton
f) cobalt
g) calcium
h) uranium

4.38 For each of the elements with atomic numbers 1 through 7 write the electronic configuration of the atom and assign the numerical values for the four quantum numbers to each electron.

5

Atomic Properties and the Periodic Table

INTRODUCTION It is useful to distinguish between *atomic* and *aggregate* properties of the elements. Quantities of elements large enough to see and handle by ordinary techniques contain large numbers of atoms in close proximity. Each atom has associated with it a magnetic and electrostatic field, each of which may be considered to be the result of individual magnetic and electrostatic fields associated with the positively charged nucleus and the negatively charged cloud of electrons outside of the nucleus. It is impossible for these atoms to be so close together without interaction. For example, the nuclei repel each other; the peripheral electron clouds repel each other; the nucleus of one atom attracts the electrons of other atoms. Many properties of the elements depend upon this mutual interaction of atoms; such properties are *aggregate properties* of the elements.

Although the aggregate properties of the elements depend upon the properties of the individual atoms, some care must be exercised not to confuse aggregate properties with *atomic properties*, the properties associated with the individual atoms. For example, the gross density of an element is a property measured for a large number of atoms and it could be incorrect to assign this same density to individual atoms.

Similarly, the states of matter depend upon interatomic interactions and it is not possible to say that an individual atom is either a solid, a liquid, or a gas under a given set of conditions, since these terms have no meaning for individual atoms. Whether or not an element in the bulk is a solid, liquid, or gas under a given set of conditions depends upon the properties of the individual atoms, although the individual atoms do not exhibit the properties of any of the three states of matter.

In this chapter, a few atomic properties will be considered. For the most part, these will be related directly to the electrostatic and magnetic fields associated with the atoms. However, it is well to remember that the measurement of properties normally must be carried out on large samples. Thus the atomic properties often must be inferred from the results of experiments performed with bulk quantities. Often this means that the data available for atomic properties are more approximate than we might desire. There are arguments about the specific values of some properties. However, usually, there is fair agreement about the relative trends in the properties among the elements. The best approach is to use a set of internally consistent values so that the trends may be observed. In these terms, the periodic table is a useful tool in correlating the trends.

IONIZATION POTENTIAL

In Chapter 4, the potential energy of an electron in the 1*s* level of the hydrogen atom was listed as -520.3×10^{-24} kcal. This represents the amount of energy liberated when an electron approaches the hydrogen nucleus from an infinite distance. To remove the electron from the hydrogen atom, that is, to remove the electron an infinite distance from the nucleus, requires putting 520.3×10^{-24} kcal of energy into the atom. This energy is called the *ionization potential*, if the measurement is made on a bulk quantity of the element in the gaseous state.

For atoms which contain more than one electron, there is more than one ionization potential. Actually there is an ionization potential for each electron in the atom. However, it must be understood that the second ionization potential is the energy required to recove the second electron after the first electron has been removed. Similarly, the third ionization potential is the energy required to remove the third electron after the first two electrons have been removed. When the term *ionization potential* is used without a qualifying adjective, it is understood that the first ionization potential is meant.

Cation Formation

After an electron has been removed from an atom, the product atom no longer is electrically neutral. Rather, it is positively charged, since it contains more protons in the nucleus than it has electrons outside of the nucleus. Any electrically charged atom is called an *ion*, and any positively charged ion is called a *cation*. If enough energy is added to the atom, then, one or more electrons will be removed and the product will be a cation. Numerically, the number of quanta of positive charge on a cation equals the number of electrons removed from the neutral atom.

The positive charge associated with a cation makes the cation different from the parent atom. In the first place, the removal of an electron from an atom decreases the net electrostatic repulsions among the electrons left behind. As a result, the remaining electrons are in slightly lower states of potential energy. This suggests that each successively higher ionization potential for a particular element has a greater value than the one before. Thus the second ionization potential for an atom of an element is higher in value than the first ionization potential, and the third ionization potential is higher in value than the second ionization potential, and so on. Consequently cations with high positive charge are more difficult to produce than cations with low positive charge.

The magnetic and electrostatic fields associated with atoms may be considered to be composites of the magnetic and electrostatic fields of the positive nuclei and the negative electron clouds. The removal of an electron results in a decrease in the contribution of the electron cloud to the total magnetic and electrostatic fields associated with the particle. In this respect, the total magnetic and electrostatic fields associated with a cation are somewhat more like the fields associated with the nucleus than are the fields associated with the neutral atom.

Since ions differ from the atoms from which they are formed, they must be represented with their appropriate electrostatic charges. Symbolically, ions are represented by the symbols of the elements from which they are derived, followed by the quantum electrical charge as a superscript. Thus the hydrogen cation derived from the hydrogen atom is written as H^+ or H^{1+}. The two ions derived from helium atoms, which represent the respective removal of one and two electrons, may be written as He^+ and He^{++} or He^{1+} and He^{2+}.

Using these symbols it is possible to write equations to represent the removal of electrons from atoms. For example, Equation (5.1) represents the removal of an electron from the H atom.

$$H + \text{energy} = H^+ + e^- \tag{5.1}$$

Equation (5.1) may be read in either of two ways:

1. One hydrogen atom absorbs energy to form one H cation and one electron;
2. One mole of H atoms absorbs energy to form one mole of H cations and one mole of electrons.

Since the H cation is just one electron short of being an atom of hydrogen, the mass difference between an H atom and an H cation is negligible for most purposes. Consequently the numerical value for the atomic weight of hydrogen is used also for the H cation, but for the cation it is called the *ionic weight*. Thus the ionic weight of the H cation is given as 1.01, which means that one H cation has a mass of 1.01 amu and one mole of H cations has a mass of 1.01 g.

As an alternative, Equation (5.1) may be written with a numerical value for the energy term. The energy term is the ionization potential. Thus Equation (5.1) may be written as Equation (5.2), where the ionization potential has been inserted for the energy term.

$$H + 5.20 \times 10^{-22}\ \text{kcal/atom} = H^+ + e^- \tag{5.2}$$

Equation (5.2) may be read as follows: One H atom absorbs 5.20×10^{-22} kcal of energy to form one H cation and one electron. The use of a numerical value for the energy fixes the interpretation of the equation. Thus Equation (5.2) must be interpreted in atom quantities and not in mole quantities. However, Equation (5.2) may be rewritten with another energy value so that it may be interpreted in mole quantities. Since 1 mole of atoms is 6.02×10^{23} atoms, and since 1 atom of hydrogen absorbs 5.20×10^{-22} kcal of energy to form the H cation, the number of kcal of energy which 1 mole of H atoms must absorb to form 1 mole of H ions may be found as follows:

$$1 \text{ mole H atoms} \times \frac{6.02 \times 10^{23} \text{ H atoms}}{1 \text{ mole H atoms}} \times \frac{5.20 \times 10^{-22} \text{ kcal}}{1 \text{ H atom}} = 313 \text{ kcal}$$

Using this value, Equation (5.2) may be rewritten as Equation (5.3).

$$H + 313 \text{ kcal/mole} = H^+ + e^- \qquad (5.3)$$

Equation (5.3) may be interpreted only in mole quantities as follows: 1 mole of H atoms absorbs 313 kcal of energy to form 1 mole of H cations and 1 mole of electrons.

Electronic Configurations of Cations

Of course the hydrogen cation contains no electrons. However, most cations do contain electrons and have electronic configurations that can be related to the electronic configurations of the atoms from which they have been formed. As an atom is energized, the first electron to leave the atom is the electron of highest potential energy in the atom. Therefore a review of the potential energies of the various electrons in the atoms of the elements can provide some insights into the electronic configurations of the cations of these elements.

Table 5.1 is a periodic table in which the levels from which the first two electrons are lost are indicated. Among the representative elements, there are no surprises. In all cases the np and ns levels represent the highest potential energies for electrons in these atoms and electrons in these states are removed first.

Therefore, it is possible to write the configurations of cations derived from representative elements by subtracting the appropriate electrons from the atom. For example, the electronic configuration for the fluorine atom is $1s^22s^22p^5$. Since the electrons in the $2p$ level are highest in potential energy, one of these is removed first and the electronic configuration for F^+ is $1s^22s^22p^4$. Similarly, the electronic configuration for F^{++} is $1s^22s^22p^3$. The electronic configurations for ions derived from several other representative elements are shown in Table 5.2.

TABLE 5.1

reriodic Table Showing the Secondary Energy Levels from Which the First Two Electrons are Removed

	1A	2A	3B	4B	5B	6B	7B	8			1B	2B	3A	4A	5A	6A	7A	0
1	1 **H** 1s																1 **H** 1s	2 **He** 1s,1s
2	3 **Li** 2s,1s	4 **Be** 2s,2s											5 **B** 2p,2s	6 **C** 2p,2p	7 **N** 2p,2p	8 **O** 2p,2p	9 **F** 2p,2p	10 **Ne** 2p,2p
3	11 **Na** 3s,2p	12 **Mg** 3s,3s											13 **Al** 3p,3s	14 **Si** 3p,3p	15 **P** 3p,3p	16 **S** 3p,3p	17 **Cl** 3p,3p	18 **Ar** 3p,3p
4	19 **K** 4s,3p	20 **Ca** 4s,4s	21 **Sc** 4s,4s	22 **Ti** 4s,4s	23 **V** 4s,3d	24 **Cr** 4s,3d	25 **Mn** 4s,4s	26 **Fe** 4s,4s	27 **Co** 4s,3d	28 **Ni** 4s,3d	29 **Cu** 4s,3d	30 **Zn** 4s,4s	31 **Ga** 4p,4s	32 **Ge** 4p,4p	33 **As** 4p,4p	34 **Se** 4p,4p	35 **Br** 4p,4p	36 **Kr** 4p,4p
5	37 **Rb** 5s,4p	38 **Sr** 5s,5s	39 **Y** 4d,5s	40 **Zr** 5s,5s	41 **Nb** 5s,4d	42 **Mo** 5s,4d	43 **Tc** 5s,4d	44 **Ru** 5s,4d	45 **Rh** 5s,4d	46 **Pd** 4d,4d	47 **Ag** 5s,4d	48 **Cd** 5s,5s	49 **In** 5p,5s	50 **Sn** 5p,5p	51 **Sb** 5p,5p	52 **Te** 5p,5p	53 **I** 5p,5p	54 **Xe** 5p,5p
6	55 **Cs** 6s,5p	56 **Ba** 6s,6s	57 **La** 6s,5d	72 **Hf** 5d,6s	73 **Ta** 6s,6s	74 **W** 6s,6s	75 **Re** 6s,6s	76 **Os** 6s,6s	77 **Ir** 5d	78 **Pt** 6s,5d	79 **Au** 6s,5d	80 **Hg** 6s,6s	81 **Tl** 6p,6s	82 **Pb** 6p,6p	83 **Bi** 6p,6p	84 **Po** 6p,6p	85 **At** 6p,6p	86 **Rn** 6p,6p
7	87 **Fr**	88 **Ra**	89 **Ac** 6d,7s															
					58 **Ce**	59 **Pr**	60 **Nd**	61 **Pm**	62 **Sm**	63 **Eu**	64 **Gd**	65 **Tb**	66 **Dy**	67 **Ho**	68 **Er**	69 **Tm**	70 **Yb**	71 **Lu**
					90 **Th**	91 **Pa**	92 **U**	93 **Np**	94 **Pu**	95 **Am**	96 **Cm**	97 **Bk**	98 **Cf**	99 **Es**	100 **Fm**	101 **Md**	102 **No**	103 **Lw**

TABLE 5.2
Some Ground State Configurations of Ions Derived from Representative Elements

Al:	KL	$3s^23p^1$	Al^+:	KL	$3s^2$	Al^{++}:	KL	$3s^1$
K:	KL	$3s^23p^64s^1$	K^+:	KL	$3s^23p^6$	K^{++}:	KL	$3s^23p^5$
Kr:	KLM	$4s^24p^6$	Kr^+:	KLM	$4s^24p^5$	Kr^{++}:	KLM	$4s^24p^4$

However, among the transition elements, there are some interesting observations to be made. Since the potential energy of the ns and $(n - 1)d$ electrons is somewhat similar, it is not obvious which should ionize first. From Table 5.1, it is obvious that the ns electron is the first one removed in most cases. The rapid drop in potential energy for $(n - 1)d$ electrons near the atomic number range for the transition elements, as seen in Figs. 4.11 and 4.12, puts the $(n - 1)d$ electrons at a lower potential energy than the ns electrons. The only atoms in which the first electron removed is an $(n - 1)d$ electron are palladium and iridium, neither of which contains an ns electron; and yttrium, hafnium, and actinium, which are close to the crossover point in potential energy between the ns and $(n - 1)d$ levels.

Thus for most of the simple cations derived from the atoms of transition elements, the configurations may be written by subtracting electrons first from the ns level and then the $(n - 1)d$ level of the parent atom. For example, the electronic configuration of the chromium atom is $1s^22s^22p^63s^23p^63d^54s^1$. The electronic configuration for Cr^+ is $1s^22s^22p^63s^23p^63d^5$ and the electronic configuration for Cr^{++} is $1s^22s^22p^63s^23p^63d^4$. Several similar examples are given in Table 5.3.

TABLE 5.3
Some Ground State Configurations of Ions Derived from Transition Elements

Mn:	KL	$3s^23p^63d^54s^2$	Mn^+:	KL	$3s^23p^63d^54s^1$	Mn^{++}:	KL	$3s^23p^63d^5$
Cu:	KL	$3s^23p^63d^{10}4s^1$	Cu^+:	KL	$3s^23p^63d^{10}$	Cu^{++}:	KL	$3s^23p^63d^9$
Pd:	KLM	$4s^24p^64d^{10}$	Pd^+:	KLM	$4s^24p^64d^9$	Pd^{++}:	KLM	$4s^24p^64d^8$

Among the transition elements that have atoms with a configuration of the $(n - 1)d^xns^2$ type, there are four in which the removal of one of the ns^2 electrons leads to the conversion of the other to an $(n - 1)d$ electron. These are listed in Table 5.4.

TABLE 5.4
Electronic Configuration Inversions Accompanying Removal of the First Electron from Some Atoms of Transition Elements

V:	KL	$3s^23p^63d^34s^2$	V^+:	KL	$3s^23p^63d^4$
Co:	KL	$3s^23p^63d^74s^2$	Co^+:	KL	$3s^23p^63d^8$
Ni:	KL	$3s^23p^63d^84s^2$	Ni^+:	KL	$3s^23p^63d^9$
La:	KLMN	$5s^25p^65d^16s^2$	La^+:	KLMN	$5s^25p^65d^2$

Information has not been tabulated here for the inner transition elements, for which the data is spotty at best. Probably the $(n - 2)f$ electrons are not among the first to be removed from these atoms. In all probability the first to be removed are those in the *ns* energy level, followed by those in the $(n - 1)d$ level.

The important thing to remember is that for nearly all atoms, the electrons of highest potential energy are the *np* and *ns* electrons, followed by the $(n - 1)d$ electrons. These are the electrons most easily removed from the atom when energy is applied.

Ionization Potential Trends

The first ionization potentials for the elements are shown in Table 5.5; these values are plotted graphically against atomic number in Fig. 5.1. From the graph it is clear that there are periodic increases and decreases in ionization potential, culminating in maxima at atomic numbers 2, 10, 18, 36, 54, and 86. This corresponds to a general increase in ionization potential across any period of the periodic table from left to right, with the maxima occurring at the rare gases, family 0.

Among the representative elements, the trends are fairly clear. The general trend across a period of the periodic table from left to right is toward increasing ionization potentials. This is attributed to the decrease in potential energy for electrons in any secondary energy level with increasing nuclear charge. Thus the family 2A potentials are larger than the corresponding family 1A potentials, since the potential energy of electrons in the *ns* level decreases as the nuclear charge increases. From family 2A to family 3A there is a change in the energy level from which the first electron is removed: *ns* in family 2A, and *np* in family 3A. In periods 2 and 3, there is merely an increase of one in atomic number in going from family 2A to family 3A. In this case the *np* level clearly is above the *ns* level in potential energy, and the ionization potential for the element in family 3A is lower than the ionization potential for the element in family 2A. For example, the ionization potential for aluminum at 2.29×10^{-22} kcal/atom is less than the ionization potential for magnesium at 2.92×10^{-22} kcal/atom.

However, in later periods there is a bigger jump in atomic number from family 2A to family 3A. This allows the *np* level to decrease substantially in potential energy from family 2A to family 3A, so that the ionization potential for the element in family 3A is nearly the same or even greater than the ionization potential for the element in family 2A. For example, in period 6, barium, in family 2A, has an ionization potential of 1.99×10^{-22} kcal/atom, whereas thallium, in family 3A, has an ionization potential of 2.34×10^{-22} kcal/atom. Of course the nuclear charge on the barium atom is only 56 quanta compared to 81 quanta for the thallium atom.

TABLE 5.5

Periodic Table with First Ionization Potentials in kcal x 10^{22}/atom

	1A	2A	3B	4B	5B	6B	7B	8	8	8	1B	2B	3A	4A	5A	6A	7A	0
1	1 **H** 5.20																1 **H** 5.20	2 **He** 9.41
2	3 **Li** 2.06	4 **Be** 3.57											5 **B** 3.18	6 **C** 4.31	7 **N** 5.56	8 **O** 5.21	9 **F** 6.67	10 **Ne** 8.25
3	11 **Na** 1.97	12 **Mg** 2.92											13 **Al** 2.29	14 **Si** 3.12	15 **P** 4.01	16 **S** 3.97	17 **Cl** 4.98	18 **Ar** 6.03
4	19 **K** 1.66	20 **Ca** 2.34	21 **Sc** 2.51	22 **Ti** 2.61	23 **V** 2.58	24 **Cr** 2.59	25 **Mn** 2.84	26 **Fe** 3.02	27 **Co** 3.01	28 **Ni** 2.92	29 **Cu** 2.96	30 **Zn** 3.59	31 **Ga** 2.30	32 **Ge** 3.02	33 **As** 3.75	34 **Se** 3.73	35 **Br** 4.53	36 **Kr** 5.36
5	37 **Rb** 1.60	38 **Sr** 2.18	39 **Y** 2.44	40 **Zr** 2.62	41 **Nb** 2.63	42 **Mo** 2.73	43 **Tc** 2.77	44 **Ru** 2.82	45 **Rh** 2.86	46 **Pd** 3.19	47 **Ag** 2.90	48 **Cd** 3.44	49 **In** 2.21	50 **Sn** 2.81	51 **Sb** 3.31	52 **Te** 3.45	53 **I** 4.00	54 **Xe** 4.64
6	55 **Cs** 1.49	56 **Ba** 1.99	57 **La** 2.15	72 **Hf** 2.1	73 **Ta** 2.9	74 **W** 3.05	75 **Re** 3.01	76 **Os** 3.3	77 **Ir** 3.5	78 **Pt** 3.4	79 **Au** 3.53	80 **Hg** 3.99	81 **Tl** 2.34	82 **Pb** 2.84	83 **Bi** 2.79	84 **Po** 3.23	85 **At**	86 **Rn** 4.11
7	87 **Fr**	88 **Ra** 2.02	89 **Ac**															
					58 **Ce** 2.64	59 **Pr** 2.20	60 **Nd** 2.4	61 **Pm**	62 **Sm** 2.1	63 **Eu** 2.17	64 **Gd** 2.36	65 **Tb** 2.58	66 **Dy** 2.61	67 **Ho**	68 **Er**	69 **Tm**	70 **Yb** 2.38	71 **Lu** 2.35
					90 **Th**	91 **Pa**	92 **U** 1.5	93 **Np**	94 **Pu**	95 **Am**	96 **Cm**	97 **Bk**	98 **Cf**	99 **Es**	100 **Fm**	101 **Md**	102 **No**	103 **Lw**

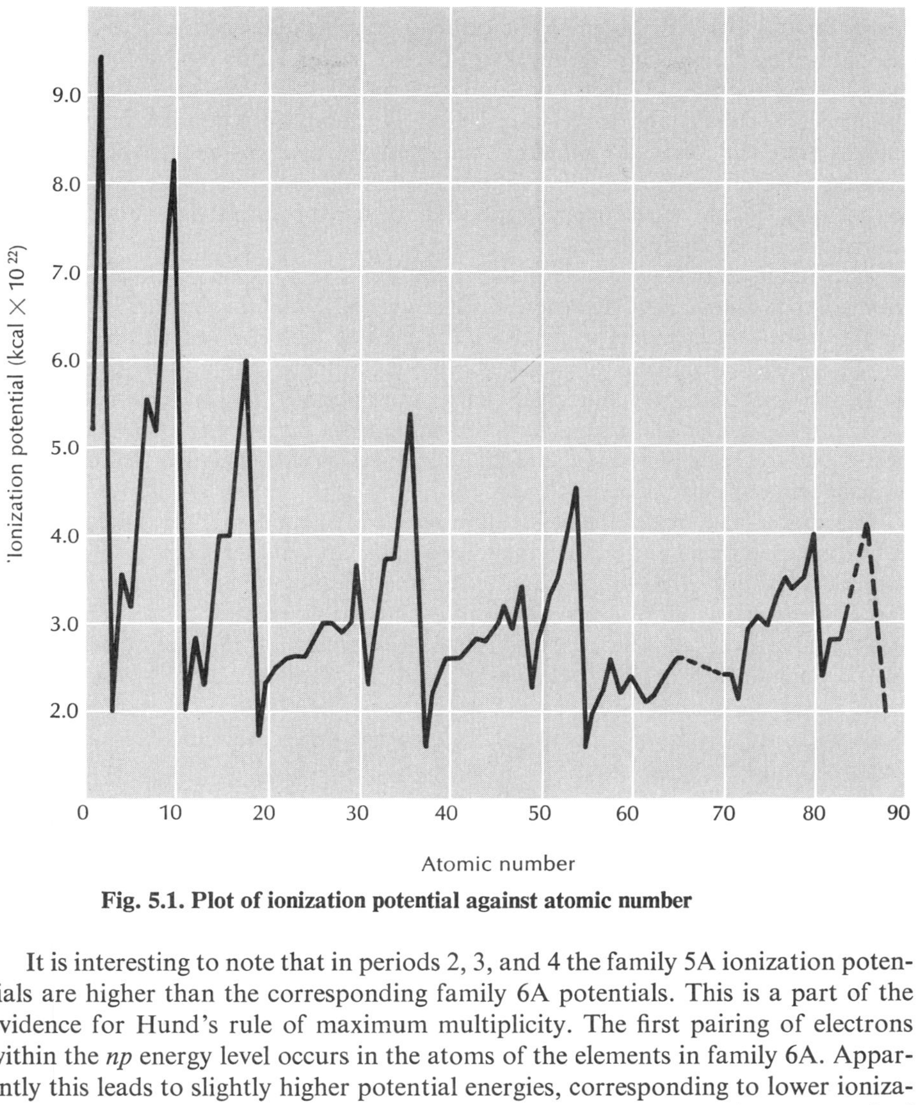

Fig. 5.1. Plot of ionization potential against atomic number

It is interesting to note that in periods 2, 3, and 4 the family 5A ionization potentials are higher than the corresponding family 6A potentials. This is a part of the evidence for Hund's rule of maximum multiplicity. The first pairing of electrons within the *np* energy level occurs in the atoms of the elements in family 6A. Apparently this leads to slightly higher potential energies, corresponding to lower ionization potentials. Thus such a paired electron is more easily removed than an unpaired electron.

Within the families of the representative elements, there is a general decrease in ionization potential down the family. Although this often is attributed to the fact that the electrons removed are in successively higher principal energy levels, the regularity of this trend is somewhat more fortuitous than might be expected otherwise. Considering the diversity of paths through which secondary energy levels change with atomic number, as shown in Figs. 4.11 and 4.12, it is somewhat amazing that there are as few exceptions to this general trend as there are.

As seen in Table 5.5 and Fig. 5.1, among the transition elements, the ionization potential generally increases across a period of the periodic table from left to right. Again this trend is attributed to the decrease in potential energy for electrons in the *ns* level as atomic number increases. Breaks in the trends tend to occur where there are peculiarities of electronic configuration involved. For example, in period 4, vanadium, cobalt, and nickel may have low ionization potentials for the same reasons that they tend to undergo configuration rearrangement when the first electron is removed. The low ionization potentials recorded for chromium and copper may be due to electronic configurations for the atoms that show only one electron in the *ns* energy level.

The trends in ionization potentials within families of transition elements appear anomalous at first glance. However, it is true generally that the ionization potentials for the transition elements of period 6 are greater than the ionization potentials for the transition elements of period 5, within the same family. This is attributed to the very large increase in nuclear charge from period 5 to period 6 atoms, compared to the smaller increase in nuclear charge from period 4 to period 5. This in turn is due to the presence of the lanthanide series of elements ahead of the transition elements of period 6.

For example, consider the elements of family 1B, copper, silver, and gold. The ionization potentials for both copper and gold are greater than the ionization potential for silver. That the high ionization potential for gold is due to the presence of the lanthanide series of elements ahead of the period 6 transition elements can be seen in Fig. 5.2, which is a portion of Fig. 4.12, showing the effect of nuclear charge on the potential energy of electrons in the 4*s*, 5*s*, and 6*s* energy levels. In these three cases, the ionization potentials are the energies required to remove the 4*s* electron from an atom of copper, the 5*s* electron from an atom of silver, and the 6*s* electron from an atom of gold. The horizontal dashed lines leading from the circles on the curves in Fig. 5.2 correspond to the $-\log(-E_p)$ values for the electrons to

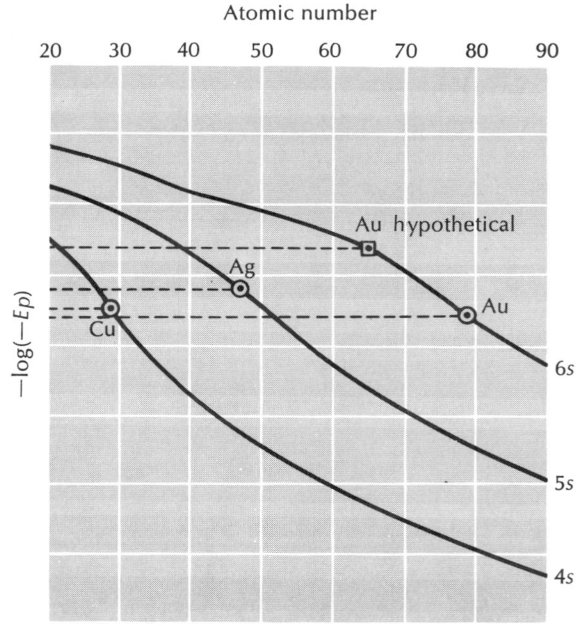

Fig. 5.2. Plot of $-\log(-E_p)$ against atomic number for the 4*s*, 5*s*, and 6*s* levels

be removed from the respective atoms when the energy corresponding to the ionization potential is supplied. Clearly the heights of these three lines are in the proper order to explain the ionization potential order. Thus the 6*s* electron in the gold atom is at the lowest potential energy corresponding to the highest ionization potential and the 5*s* electron in the silver atom is at the highest potential energy corresponding to the lowest ionization potential.

However, if we imagine that the fourteen lanthanide series elements do not exist, then the gold atom would have an atomic number fourteen units less than its real value of 79. At this atomic number of 65, the potential energy of the 6*s* electron is shown by the horizontal line from the square in Fig. 5.2. Clearly this potential energy is higher than the potential energy of the 5*s* electron in the silver atom. Under these hypothetical circumstances, gold would have a lower ionization potential than silver.

Thus the fact that the ionization potentials for period 6 transition elements are greater than the ionization potentials for period 5 transition elements within the same family can be attributed to the large increase in nuclear charge in going from period 5 to period 6. Actually this effect appears to carry over into the representative elements of families 3A and 4A, with thallium and lead exhibiting higher ionization potentials than indium and tin respectively.

Higher Ionization Potentials

It was suggested earlier that values of successively higher ionization potentials for a given element increase regularly. This has been attributed to the decreasing potential energy of the electrons within the ion as the electron repulsions are reduced with removal of electrons. From another point of view, it is increasingly difficult to remove negatively charged electrons from cations of increasingly high positive charge. In other words, the attraction for electrons is greater for a cation than for a similar neutral atom. A few successive ionization potentials are shown in Table 5.6.

TABLE 5.6
The First Eight Ionization Potentials for the Elements of Period 3

		IONIZATION POTENTIALS (kcal × 10^{22}/atom)							
ELEMENT	AT. NO.	1ST	2ND	3RD	4TH	5TH	6TH	7TH	8TH
Na	11	1.97	18.1	27.4	37.8	52.9	65.9	79.8	101
Mg	12	2.92	5.75	30.7	41.8	54.1	71.4	86.1	102
Al	13	2.29	7.20	10.9	45.9	58.8	72.9	92.4	109
Si	14	3.12	6.25	12.8	17.3	63.8	78.5	94.3	116
P	15	4.01	7.55	11.5	19.6	24.9	84.4	101	118
S	16	3.97	8.96	13.4	18.1	27.7	33.7	108	126
Cl	17	4.98	9.11	15.3	20.5	26.0	37.0	43.7	133
Ar	18	6.03	10.6	15.6	22.9	28.7	34.9	47.4	54.9

Summary

Although ionization potentials are measured on large quantities of the elements, the conditions are chosen so that interatomic interactions are minimized. Thus the ionization potential is the energy required to remove electrons from essentially isolated atoms.

The ionization potential is a measure of the attraction which an atomic nucleus has for its most loosely held electron. This suggests that atomic nuclei can attract additional electrons, and in some measure the ionization potential is a measure of this ability. However, additional properties can be defined to provide somewhat more useful measures of the tendency for neutral atoms to attract additional electrons.

ELECTRON AFFINITY

Since neutral atoms may attract additional electrons, the addition of such electrons should be accompanied by a decrease in potential energy for the added electron. Under the same conditions used for measuring ionization potentials, the potential energy decrease may be observed when an isolated electron is added to a neutral atom. The decrease is called the *electron affinity* of that atom. Thus the electron affinity for the hydrogen atom is the energy shown in Equation (5.4).

$$H + e^- = H^- + \text{Energy} \tag{5.4}$$

The product of the addition of an electron to a neutral atom is an ion with negative charge, that is, a particle with one more electron outside the nucleus than there are protons in the nucleus. Such negatively charged ions are called *anions*. The electrostatic and magnetic fields associated with anions are more like those associated with electrons than with protons, since the total fields have a greater contribution from the electron cloud than from the nucleus. This also means that the addition of an electron to an anion requires that work be done; it proceeds with the absorption of energy, rather than the liberation of energy, since the negatively charged anion repels an approaching electron.

The actual measurement of electron affinities is very difficult. As a result only spotty data of questionable accuracy are available. Generally the values are computed indirectly. With very little data available, trends in electron affinities are difficult to discern, although it is assumed generally that they are the same as those found for ionization potentials.

ATOMIC AND IONIC SIZES

In principle, electron affinities are a direct measure of the ability of neutral atoms to attract electrons. However, since electron affinities are not available, it is desirable to have an experimentally determined property that is related rather directly to this ability.

The attraction for electrons exhibited by neutral atoms is due to the positive charge on the nucleus. The electrostatic force of attraction between a nucleus and an electron is measured in terms of the sizes of the charges on the nucleus and the electron and the distance between the nucleus and the electron. Since the force of attraction is related directly to the size of charge, but inversely to the *square* of the distance, the distance tends to play a more important role than the charge. Thus the distance from the nucleus to the most remote electron in the atom can be a measure of electron-attracting ability. The shorter this distance is, the greater is the general tendency to attract electrons, if the importance of the size of the nuclear charge is ignored. If this distance is short, an external electron can be fairly close to the nucleus before it is repelled appreciably by the electron cloud of the atom.

In measurement of atomic sizes, it is assumed that atoms are spherical. The size is recorded as the radius of the sphere, with the nucleus at the center. Since it is impossible to observe atoms visually, various techniques have been used to try to determine atom sizes. Whatever technique is used, some observed property is related to atom size and the size is inferred from measurements of the property. For example, it is possible to use very short wavelength electromagnetic radiation to "x-ray" bulk quantities of the elements. The diffraction patterns obtained can be interpreted to provide information on the distance between two nuclei. Then, on the assumption that the atoms are tightly packed and touching, it is possible to determine what is called the atomic radius. However, since different techniques measure different properties, it is not surprising that there are many so-called atomic radii recorded in the literature. Nevertheless, significant trends in atomic radii can be discerned by studying a self-consistent set of values. Such a set of atomic radii is shown in Table 5.7, and these values are plotted against atomic number in Fig. 5.3.

The electrons in the highest principal energy level of an atom have the greatest probability of location at the greatest distance from the nucleus. Thus the distance of these electrons from the nucleus determines the atomic radius. Within any period of the periodic table, the electrons of highest potential energy are in the same principal energy level. For example, for all of the atoms of the elements in period 3, the electrons of highest potential energy are in principal energy level 3. As the nuclear charge increases, the attraction between the nucleus and these electrons increases. Consequently, the probable distance from the nucleus of the electrons in principal energy level 3 decreases. This accounts for the observation that atomic radii tend to decrease with increasing atomic number within any period of the periodic table.

The trends within the periods are somewhat irregular among the transition elements. At least a part of the reason for this appears to be the increase in the number of $(n - 1)d$ electrons in the atoms as the atomic number increases. The probable distance of the $(n - 1)d$ electrons from the nucleus is less than the probable distance of the ns electrons from the nucleus. This means that the $(n - 1)d$ electrons tend to be located between the ns electrons and the nucleus. As the number of $(n - 1)d$ electrons increases, their repulsions of the ns electrons tend to counteract somewhat the increase in nuclear charge tending to attract the ns electrons toward the nucleus. This is called the *shielding effect.* Due to the shielding

TABLE 5.7

Periodic Table Showing Atomic Radii in Angstroms

	1A	2A	3B	4B	5B	6B	7B	8	8	8	1B	2B	3A	4A	5A	6A	7A	0
1	1 **H** 0.371																1 **H** 0.371	2 **He**
2	3 **Li** 1.520	4 **Be** 1.113											5 **B** 0.794	6 **C** 0.772	7 **N** 0.74	8 **O** 0.74	9 **F** 0.709	10 **Ne**
3	11 **Na** 1.858	12 **Mg** 1.598											13 **Al** 1.432	14 **Si** 1.176	15 **P** 1.10	16 **S** 1.02	17 **Cl** 1.00	18 **Ar**
4	19 **K** 2.272	20 **Ca** 1.974	21 **Sc** 1.606	22 **Ti** 1.448	23 **V** 1.311	24 **Cr** 1.249	25 **Mn** 1.366	26 **Fe** 1.241	27 **Co** 1.253	28 **Ni** 1.246	29 **Cu** 1.278	30 **Zn** 1.332	31 **Ga** 1.221	32 **Ge** 1.225	33 **As** 1.24	34 **Se** 1.16	35 **Br** 1.142	36 **Kr**
5	37 **Rb** 2.48	38 **Sr** 2.152	39 **Y** 1.776	40 **Zr** 1.590	41 **Nb** 1.429	42 **Mo** 1.362	43 **Tc** 1.325	44 **Ru** 1.325	45 **Rh** 1.345	46 **Pd** 1.376	47 **Ag** 1.444	48 **Cd** 1.490	49 **In** 1.626	50 **Sn** 1.405	51 **Sb** 1.45	52 **Te** 1.432	53 **I** 1.333	54 **Xe**
6	55 **Cs** 2.654	56 **Ba** 2.174	57 **La** 1.870	72 **Hf** 1.563	73 **Ta** 1.43	74 **W** 1.370	75 **Re** 1.370	76 **Os** 1.338	77 **Ir** 1.358	78 **Pt** 1.388	79 **Au** 1.442	80 **Hg** 1.502	81 **Tl** 1.704	82 **Pb** 1.750	83 **Bi** 1.548	84 **Po** 1.67	85 **At**	86 **Rn**
7	87 **Fr**	88 **Ra**	89 **Ac** 1.878															
					58 **Ce** 1.825	59 **Pr** 1.820	60 **Nd** 1.814	61 **Pm**	62 **Sm**	63 **Eu** 1.994	64 **Gd** 1.786	65 **Tb** 1.762	66 **Dy** 1.752	67 **Ho** 1.743	68 **Er** 1.734	69 **Tm** 1.724	70 **Yb** 1.940	71 **Lu** 1.718
					90 **Th** 1.798	91 **Pa** 1.606	92 **U** 1.38	93 **Np** 1.3	94 **Pu** 1.513	95 **Am**	96 **Cm**	97 **Bk**	98 **Cf**	99 **Es**	100 **Fm**	101 **Md**	102 **No**	103 **Lw**

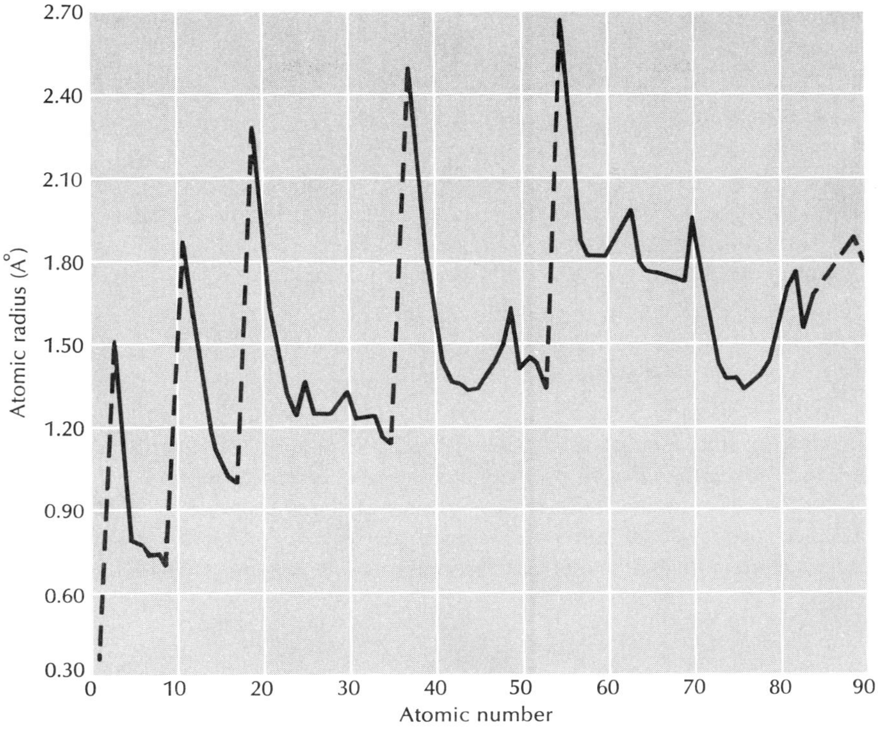

Fig. 5.3. Plot of atomic radius against atomic number

effect, the atomic radii of the transition elements within a given period tend to fall within a very narrow range.

Within the families of the representative elements, atomic radii tend to increase with increasing atomic number. For example, in family 6A, the order of increasing atomic radius is oxygen, sulfur, selenium, tellurium, and polonium. Here the electrons of highest potential energy lie in higher principal energy levels as the atomic number increases. For example, the electrons of highest potential energy lie in principal energy level 2 for oxygen and in principal energy level 3 for sulfur. The relative sizes of the O and S atoms depend upon the size of the nuclear charge difference. Is the higher nuclear charge of the S atom sufficient to make the probable distance from the nucleus of the electrons in principal energy level 3 of the S atom less than the probable distance from the nucleus of the electrons in principal energy level 2 of the O atom? No, it is not, as revealed in Table 5.7. What is more, this is true in most cases. In scanning down a periodic family, we see that the successive increases in nuclear charge are insufficient to cause a decrease in atomic radius.

Some interesting effects can be seen, however. Whereas the S atom is much larger than the O atom, the Se atom is only slightly larger than the S atom. In some measure we may account for this by noting that the difference in nuclear charge between oxygen and sulfur is 8 charge quanta, whereas the difference between sulfur and selenium is 18 charge quanta.

This same effect is seen when the radii of the atoms of the transition elements of period 6 are compared with the radii of their family members in period 5. In this case there is very little difference in size. For example, the silver and gold atoms have nearly the same radius, although the Ag atom is larger than the Cu atom. Again this can be attributed to the relative differences in nuclear charge. Between the Cu atom and the Ag atom, the difference is 18 charge quanta, whereas between the Ag atom and the Au atom, the difference is 32 charge quanta.

Among the elements in the lanthanide series, the trend is toward smaller radii with increasing atomic number, which is consistent with other trends within given periods. However, the differences are small, due to the shielding effect of the $(n - 2)f$ electrons.

In viewing the periodic table as a whole, the smallest atoms are in the upper right-hand corner and the largest atoms are in the lower left-hand corner. If there is a trend in electron attraction which follows the size trend, then the atoms that attract electrons most strongly are in the upper right-hand corner of the table and the atoms that attract electrons least are in the lower left-hand corner of the table.

Ionic Radii

Ions differ in size from their parent atoms. Cations have fewer orbital electrons. As a result the net repulsions among the electrons are less and the electron cloud shrinks in size. Consequently, all cations are smaller than the atoms from which they are formed.

Similarly, all anions are larger than the atoms from which they are formed. In this case the addition of electrons increases the net repulsions among the electrons and the electron cloud expands.

Thus cations are good electron attractors, due to their small size and over-all positive electrostatic charge. Similarly, anions are poor electron attractors, since they are large and have over-all negative electrostatic charges.

Some typical ionic radii are listed in Table 5.8 along with the atomic radii of the parent atoms.

TABLE 5.8 Comparisons of Some Atomic and Ionic Radii in Angstroms

	ATOM	1+ ION		ATOM	2+ ION		ATOM	2− ION		ATOM	1− ION
Li	1.52	0.60	Be	1.11	0.31	O	0.74	1.40	F	0.71	1.36
Na	1.86	0.95	Mg	1.60	0.65	S	1.02	1.84	Cl	1.00	1.81
K	2.27	1.33	Ca	1.97	0.99	Se	1.16	1.98	Br	1.14	1.95
Rb	2.48	1.48	Sr	2.15	1.13	Te	1.43	2.21	I	1.33	2.16

ELECTRONEGATIVITY

Because of differences in nuclear charge as well as atomic sizes, atomic radii are not accurate measures of the electron-attracting ability of atoms. It has become popular in the past few years to try to devise a scale of numbers for the elements that will provide a good idea of the relative degree to which atoms will attract electrons when the atoms are in close contact with many other atoms. This electron-attracting property has been called the *electronegativity* of the atom.

Electronegativity is an atomic property associated with the atom in contact with many other atoms; it measures the tendency of the atom to attract electrons. The definition of this property does not suggest any specific way to measure electronegativity. Indeed a number of different approaches have been used. The ionization potential, the relationship between the density of charge in the electron cloud and the atomic number, and the measurable potential energy changes accompanying chemical reactions have all been taken as starting points in devising sets of electronegativity values. However, regardless of the approach used and regardless of the numerical values listed, the trends in electronegativity among the elements are similar. Consequently, it is most important to choose a consistent set of values from which to observe the trends.

One consistent set of values for electronegativities, as proposed by Linus Pauling is given in Table 5.9; these are plotted against atomic number in Fig. 5.4.

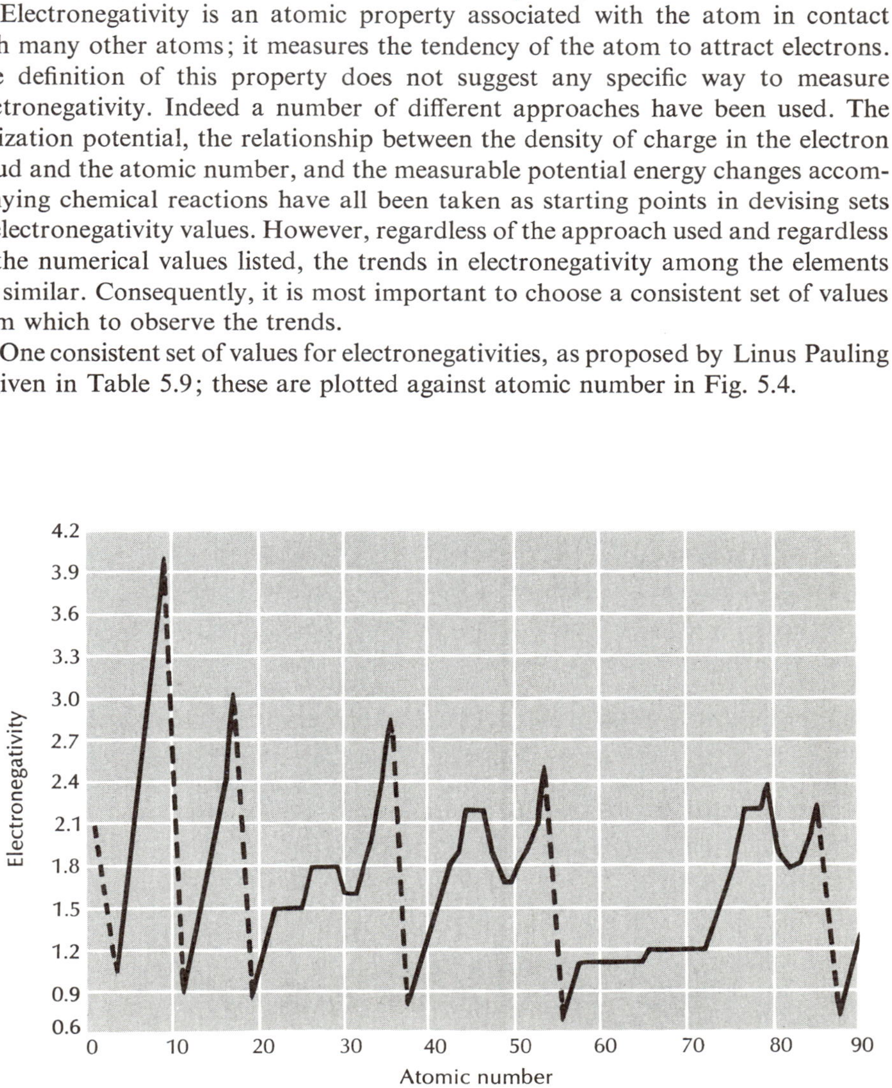

Fig. 5.4. Plot of electronegativity against atomic number

TABLE 5.9

Periodic Table Showing Electronegativities

	1A	2A	3B	4B	5B	6B	7B	8			1B	2B	3A	4A	5A	6A	7A	0
1	1 **H** 2.1																1 **H** 2.1	2 **He**
2	3 **Li** 1.0	4 **Be** 1.5											5 **B** 2.0	6 **C** 2.5	7 **N** 3.0	8 **O** 3.5	9 **F** 4.0	10 **Ne**
3	11 **Na** 0.9	12 **Mg** 1.2											13 **Al** 1.5	14 **Si** 1.8	15 **P** 2.1	16 **S** 2.5	17 **Cl** 3.0	18 **Ar**
4	19 **K** 0.8	20 **Ca** 1.0	21 **Sc** 1.3	22 **Ti** 1.5	23 **V** 1.6	24 **Cr** 1.6	25 **Mn** 1.5	26 **Fe** 1.8	27 **Co** 1.8	28 **Ni** 1.8	29 **Cu** 1.9	30 **Zn** 1.6	31 **Ga** 1.6	32 **Ge** 1.8	33 **As** 2.0	34 **Se** 2.4	35 **Br** 2.8	36 **Kr**
5	37 **Rb** 0.8	38 **Sr** 1.0	39 **Y** 1.2	40 **Zr** 1.4	41 **Nb** 1.6	42 **Mo** 1.8	43 **Tc** 1.9	44 **Ru** 2.2	45 **Rh** 2.2	46 **Pd** 2.2	47 **Ag** 1.9	48 **Cd** 1.7	49 **In** 1.7	50 **Sn** 1.8	51 **Sb** 1.9	52 **Te** 2.1	53 **I** 2.5	54 **Xe**
6	55 **Cs** 0.7	56 **Ba** 0.9	57 **La** 1.1	72 **Hf** 1.3	73 **Ta** 1.5	74 **W** 1.7	75 **Re** 1.9	76 **Os** 2.2	77 **Ir** 2.2	78 **Pt** 2.2	79 **Au** 2.4	80 **Hg** 1.9	81 **Tl** 1.8	82 **Pb** 1.9	83 **Bi** 1.9	84 **Po** 2.0	85 **At** 2.2	86 **Rn**
7	87 **Fr** 0.7	88 **Ra** 0.9	89 **Ac** 1.1															
					58 **Ce** 1.1	59 **Pr** 1.1	60 **Nd** 1.1	61 **Pm** 1.1	62 **Sm** 1.1	63 **Eu** 1.1	64 **Gd** 1.1	65 **Tb** 1.2	66 **Dy** 1.2	67 **Ho** 1.2	68 **Er** 1.2	69 **Tm** 1.2	70 **Yb** 1.2	71 **Lu** 1.2
					90 **Th** 1.3	91 **Pa** 1.5	92 **U** 1.7	93 **Np** 1.3	94 **Pu** 1.3	95 **Am** 1.3	96 **Cm** 1.3	97 **Bk** 1.3	98 **Cf** 1.3	99 **Es** 1.3	100 **Fm** 1.3	101 **Md** 1.3	102 **No** 1.3	103 **Lw** 1.3

Within a period of the periodic table, there is a general increase in electronegativity as atomic number increases. This is consistent with the observed decrease in atomic radius with increasing atomic number within a period. The shielding effect of the $(n - 1)d$ electrons is apparent among the transition elements, where electronegativities vary only slightly and then somewhat erratically.

Within families of the periodic table, there tend to be decreases in electronegativity as the atomic number increases. Again this agrees with the trend toward larger atomic radii as atomic number increases. Exceptions tend to occur where there are exceptionally large differences in atomic number. Thus the electronegativities of the period 6 transition elements are nearly the same as the electronegativities of the period 5 transition elements.

In Table 5.10, the electronegativities of the atoms of some elements are compared to the electronegativities of cations and/or anions derived from these atoms. As

TABLE 5.10

Some Atomic and Ionic Electronegativities

ATOM		ION		ATOM		ION		ATOM		ION	
H	2.1	H^+	7.5					H	2.1	H^-	Neg.
Li	1.0	Li^+	2.5	Be	1.5	Be^{++}	7.7	F	4.0	F^-	0.8
Na	0.9	Na^+	2.4	Mg	1.2	Mg^{++}	6.8	Cl	3.0	Cl^-	0.3
K	0.8	K^+	2.1	Ca	1.0	Ca^{++}	5.8	Br	2.8	Br^-	0.1
Rb	0.8	Rb^+	2.0	Sr	1.0	Sr^{++}	5.5	I	2.5	I^-	Neg.
Cs	0.7	Cs^+	1.9	Ba	0.9	Ba^{++}	5.2				

expected from the atomic and ionic sizes, the electronegativities of cations are higher than the electronegativities of the atoms from which they are derived. Similarly, the electronegativities of anions are lower than the electronegativities of the atoms from which they are derived. In fact, the electronegativities of some of the anions are negative on the arbitrary scale adopted for recording electronegativities.

ELECTROSTATIC POLARIZATION

If an atom is placed between two oppositely charged plates, the somewhat mobile and diffuse electron cloud may be distorted as shown in Fig. 5.5. The original atom is spherically symmetrical, which means that over-all, the electrons are equally distributed in all directions.

It is possible to define a center of negative charge for the diffuse cloud in much the same way that a center of gravity may be defined for a material body. If the body is suspended from the center of gravity, it is balanced, and in many ways resembles a small point object of the same mass. In other words the center of gravity is a point at which all of the mass of the body may be considered to be located, although it is more diffusely distributed than that in reality.

In the same way it is possible to define a center of negative charge for the diffusely distributed electron cloud. If the cloud is spherically symmetrical, then the

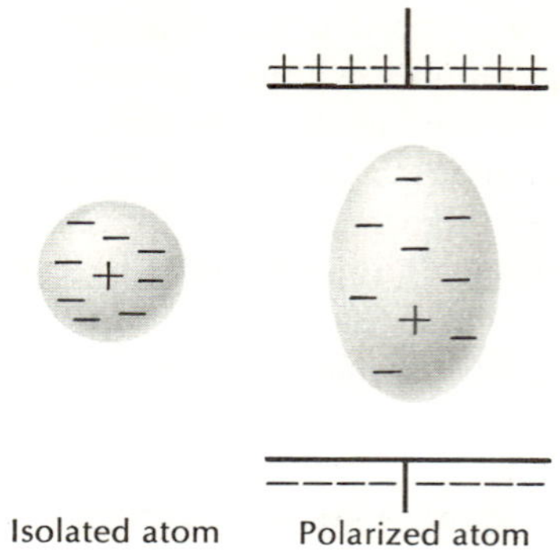

Fig. 5.5. Electrostatic polarization of an atom

electrons are equally distributed around the center of the sphere, and the net result is that electrically the atom behaves as though all of the negative charge is centered in the nucleus. Since this also is the center of positive charge, the atom as a whole acts like an uncharged particle, despite the fact that it is composed of separate positive and negative components.

From this point of view, the electrical properties of ions can be explained as well. Within a cation, for example, the electrons are still spherically symmetrical around the central nucleus. Thus the total negative charge associated with the electron cloud may be imagined at the center of the sphere, that is, at the nucleus. However, there are fewer negative charges than positive charges centered at the nucleus and the cation acts like a simple positively charged particle. Similarly, in the anion, the electrons are spherically symmetrical around the nucleus, and the negative electron charge may be imagined to be centered at the nucleus. However, in this case, there is more negative charge than positive charge centered at the nucleus and the anion acts like a simple negative particle.

However, when an atom, or an ion, is placed between two electrostatically charged plates, as shown in Fig. 5.5, the electron cloud is attracted to the positive plate and repelled from the negative plate. At the same time, the nucleus is attracted to the negative plate and repelled from the positive plate. The net result is distortion of the atom, the electron cloud being displaced toward the positive plate and the nucleus displaced toward the negative plate. Here the centers of negative and positive charge do not coincide. Such a particle, in which the two centers of charge do not coincide, is said to be *electrostatically polarized,* since it may be visualized in terms of positive and negative poles. The particle is then called an *electrostatic dipole.* An isolated atom is not electrostatically polarized, and is not an electrostatic dipole. However, when it is placed between the two charged plates, it becomes electrostatically polarized due to the action of the electrostatic field. Thus it is said to have an *induced electrostatic dipole,* as opposed to an intrinsic or *permanent electrostatic dipole.*

Although we have not yet seen how it is possible for a particle to be a permanent electrostatic dipole, suppose that this is possible. Then imagine this particle introduced between two electrostatically charged plates as shown in Fig. 5.6. The particle

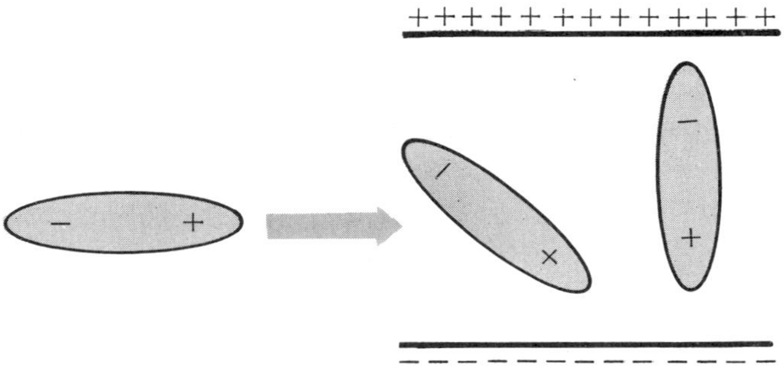

Fig. 5.6. Orientation of an electrostatic dipole in an electrostatic field

undergoes a twisting motion while orienting its positive pole toward the negative plate and its negative pole toward the positive plate. In other words, a torque is exerted on the particle. This torque depends upon the size charges and the distance between them. The torque is greater for larger charges and for greater distances. A useful property of electrostatic dipoles is the *electrostatic dipole moment*, μ, defined as the product of one of the charges, q, and the distance d between them, as shown in Equation (5.5).

$$\mu = qd \tag{5.5}$$

In a similar way, the induced electrostatic dipole of the atom between the two plates has an induced electrostatic dipole moment.

The units in which the electrostatic dipole moment is expressed consist of the product of the charge and distance units. Thus a common unit for the electrostatic dipole moment is the coulomb × angstrom.

For a particular atom, the charge q is fixed, but the distance of charge separation or the degree of atomic distortion depends upon the magnitude of the electrostatic field associated with the charged plates. Consequently the magnitude of the electrostatic dipole moment depends upon the electrostatic field in such a way that the electrostatic dipole moment is directly proportional to the electrostatic field strength E. This is shown by Equation (5.6).

$$\mu = \alpha E \tag{5.6}$$

The proportionality constant α in Equation (5.6) is a property, characteristic of each atom. Suppose that many different atoms are examined with a constant electrostatic field strength E. The magnitude of the electrostatic dipole moment μ will vary from atom to atom depending upon the degree to which the atom may be polarized. If α is large, then the electrostatic dipole moment is large, and correspondingly, the electrostatic polarization or distortion of the atom is large. Likewise, if α is small, then the electrostatic dipole moment is small, and correspondingly, the electrostatic polarization or distortion of the atom is small. The proportionality constant α is called the *electrostatic polarizability*.

The units in which the electrostatic polarizability is expressed are derived from Equation (5.6). The electrostatic field strength E is measured as charge per unit area of the plates; that is, E is the charge density on the surface of the plates. A convenient unit for E is coulomb/(angstrom)^2. If the electrostatic dipole moment is expressed in coulomb × angstrom, then the units for the electrostatic polarizability α are determined as follows:

$$\alpha = \frac{\mu}{E} = \frac{\text{coulomb} \times \text{angstrom} \times (\text{angstrom})^2}{\text{coulomb}} = (\text{angstrom})^3$$

Clearly the electrostatic polarizability is expressed in volume units.

The electrostatic polarizability of an atom depends, in some measure, on the radius of the atom, suggesting that the electron cloud is rather diffuse. Data for the electrostatic polarizabilities of atoms are sparse, although there is somewhat more data available for ions. Some of the experimentally determined values for electrostatic polarizability are recorded in Table 5.11.

TABLE 5.11 Electrostatic Polarizabilities in Å^3

1A		2A	3B	4B	3A	4A	6A		7A	0
										He
										0.20
Li	Li^{1+}	Be^{2+}			B^{3+}	C^{4+}	O	O^{2-}	F^{1-}	Ne
12	0.03	0.008			0.003	0.002	0.15	3.9	1.0	0.39
Na	Na^{1+}	Mg^{2+}			Al^{3+}	Si^{4+}		S^{2-}	Cl^{1-}	Ar
27	0.18	0.009			0.05	0.016		10.2	3.66	1.62
K	K^{1+}	Ca^{2+}	Sc^{3+}	Ti^{4+}				Se^{2-}	Br^{1-}	Kr
34	0.83	0.47	0.29	0.19				10.5	4.77	2.46
Rb	Rb^{1+}	Sr^{2+}	Y^{3+}	Zr^{4+}				Te^{2-}	I^{1-}	Xe
50	1.40	0.86	0.55	0.37				14.0	7.10	3.99
Cs	Cs^{1+}	Ba^{2+}	La^{3+}							
61	2.42	1.55	1.04							

In this abbreviated form of the periodic table, some approximate trends can be discerned. Within the 1A and 0 families, the electrostatic polarizability of the atoms increases as atomic number increases. Thus high electrostatic polarizability corresponds to large atomic radius. This same trend is observed among similar ions within families 1A, 2A, 3B, 4B, 3A, 4A, 6A, and 7A.

In comparing the ions and atoms in family 1A, it is clear that the cations are polarized with much more difficulty than the neutral atoms. This corresponds to the previously mentioned observation that cations are more compact than their parent atoms. The correspondingly higher ionization potentials for the cations also suggest that the electrons are held much more firmly in the cation and that the cation, therefore, should be less easily deformed.

Similarly, the one comparison of an atom with an anion, which is made in family 6A, suggests that anions are more easily polarized electrostatically than the parent atoms. The high values for the anions of the chalcogens (family 6A) and the halogens (family 7A) also suggest this. Of course, this is in agreement with the previous observation that the anions are always larger than the parent atoms. As a result, the electron cloud is more diffuse and more easily polarized.

MAGNETIC POLARIZATION

As indicated in Chapter 1, matter behaves in either of two ways in a magnetic field. Paramagnetic materials enhance the magnetic field, whereas diamagnetic materials diminish it. Although some materials exhibit an extremely strong paramagnetism (ferromagnetism), this appears to be an aggregate property that is not due entirely to the individual atoms.

The magnetic properties of atoms are attributed to electron motions inside the atoms. These motions are of two types: translational and spin. The total magnetic field associated with any atom is the sum of the translational and spin contributions of all electrons in the atom.

When two electrons are paired in an atomic orbital, both the translational and spin motions of the two electrons are opposite to each other, and the contributions from one electron cancel the contributions from the other. This can be envisioned by reference to Fig. 1.11 (Chapter 1) where the lines of force associated with a north and south pole are shown. Imagine that a new north pole is superimposed on the old south pole and a new south pole is superimposed on the old north pole. The new lines of force are exactly opposite to those of the original field and the two effects cancel each other. The net effect is no magnetic field at all.

However, it is known that if a permanent bar magnet is brought close to a bar of unmagnetized iron, a magnetic field is *induced* in the iron bar as shown in Fig. 5.7. The induced magnetic field opposes the permanent magnetic field, since the induced south pole is adjacent to the permanent north pole and the induced north pole is adjacent to the permanent south pole. The net effect of this induced magnetic

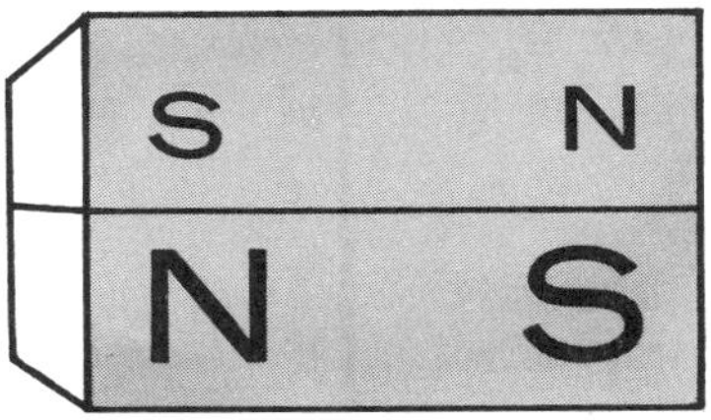

Fig. 5.7. Induced magnet

field is to reduce the apparent strength of the permanent magnetic field. In other words, this is a diamagnetic effect.

In a similar way, when atoms are introduced into an external magnetic field, the paired electrons are uncoupled, so to speak, and a magnetic field is induced in the atom. The induced field opposes the external field, thus reducing the apparent strength of the external field. In this way, all atoms exhibit diamagnetism, an effect induced in the atom by an external magnetic field. Thus all atoms are said to possess *induced magnetic dipoles* when they are placed in a magnetic field.

On the other hand, atoms with unpaired electrons have permanent magnetic fields associated with them, since for the unpaired electrons there are no electrons with opposing spin and translational motion. These atoms are said to possess *permanent magnetic dipoles*. When these atoms are placed in an external magnetic field, they become aligned with the external field and enhance its field strength. Thus atoms with unpaired electrons exhibit paramagnetic properties.

It should be noted that atoms with unpaired electrons exhibit both diamagnetism and paramagnetism. However, the paramagnetic effects due to the permanent dipole always exceed the induced diamagnetic effects.

Just as permanent electrostatic dipoles exhibit an aligning twisting motion when introduced into an external electrostatic field, permanent magnetic dipoles exhibit the same kind of motion when introduced into an external magnetic field. This twisting tendency is measured by the *magnetic dipole moment*. Magnetic dipole moments often are measured in *Bohr magnetons*. The Bohr magneton is the magnetic dipole moment associated with the spin contribution of a single electron.

Atomic magnetic moments are difficult to measure, since it is necessary to make measurements on individual atoms rather than on aggregates of atoms. However, the atoms of the elements in periodic families 2A, 2B, and 0, probably have zero magnetic dipole moments, consistent with the complete pairing of electrons in these atoms.

Magnetic dipole moments have been determined for more ions than atoms. If the electrons are all paired in the ion, the ion exhibits only diamagnetic properties. Whereas the atoms of sodium, potassium, copper, silver and gold appear to exhibit paramagnetic properties, the Na^+, K^+, Cu^+, Ag^+, and Au^+ ions are only diamagnetic.

Good correlation can be made between magnetic dipole moments and the number of unpaired $(n - 1)d$ electrons in the ions of period-4 transition elements. Data are recorded in Table 5.12. In some instances, ranges are given for the magnetic dipole moments corresponding to minimum and maximum values found among the ions listed in that group.

TABLE 5.12

Magnetic Dipole Moments of Some Fourth Period Transition Element Ions

IONS	UNPAIRED 3*d* ELECTRONS	MAGNETIC MOMENT (BOHR MAGNETONS)
Sc^{3+}, Ti^{4+}	0	0
Sc^{2+}, Ti^{3+}, V^{4+}, Cr^{5+}	1	1.73
Ti^{2+}, V^{3+}, Cr^{4+}	2	2.8
V^{2+}, Cr^{3+}, Mn^{4+}	3	3.8–3.9
Cr^{2+}, Mn^{3+}	4	4.8–4.9
Mn^{2+}, Fe^{3+}	5	5.7–5.9
Fe^{2+}, Co^{3+}	4	5.2–5.5
Co^{2+}, Ni^{3+}	3	4.4–5.2
Ni^{2+}, Cu^{3+}	2	3.2–3.4
Cu^{2+}	1	1.9
Cu^{1+}, Zn^{2+}	0	0

Clearly, the magnetic dipole moment increases as the number of unpaired $(n - 1)d$ electrons increases. This is shown more clearly in Table 5.13 where the appropriate values from Table 5.12 have been collected to show ranges of values for the magnetic dipole moment that correspond to the numbers of unpaired $3d$ electrons. It appears that the same brackets of values may be used to describe the numbers of unpaired $4d$ and $5d$ electrons among species of the transition elements in periods 5 and 6.

TABLE 5.13

Magnetic Dipole Moments Related to the Number of Unpaired 3d Electrons

NUMBER OF UNPAIRED $3d$ ELECTRONS	MAGNETIC DIPOLE MOMENT RANGE (BOHR MAGNETONS)
0	0
1	1.7–2.0
2	2.8–3.4
3	3.8–5.2
4	4.8–5.5
5	5.5–6.0

SUMMARY

The ability of atoms to attract electrons can be related to the properties of ionization potential, electron affinity, atomic and ionic radii, electronegativity, and electrostatic polarizability. Atoms with high ionization potentials, high electron affinities, small atomic radii, high electronegativity and small electrostatic polarizability all have strong attractions for electrons. These atoms tend to be clustered toward the upper right-hand corner of the periodic table.

Magnetic dipole moments are related to the number of unpaired electrons in atoms and ions. Paramagnetic species have unpaired electrons whereas in nonparamagnetic species all electrons are paired.

QUESTIONS AND PROBLEMS

5.1 From the data in Table 5.5, compute the energy in kcal required for each of the following:

a) The production of 1.0 mole of free electrons from lithium.
b) The production of 3.1 moles of sodium ions from sodium.
c) The production of 1.4 g of Xe^{+} from Xe.

5.2 Using a periodic table that does *not* list electronic configurations, write the electronic configuration for each of the following:

a) F	e) Cr^{+}	i) O
b) F^{+}	f) Cr^{++}	j) O^{--}
c) F^{-}	g) Y	k) Co
d) Cr	h) Y^{+}	l) Co^{+}

5.3 Write equations like (5.2) and (5.3) for each of the following:
- a) The first ionization of carbon.
- b) The first ionization of copper.
- c) The fourth ionization of aluminum.
- d) The eighth ionization of chlorine.
- e) The sum of the first four ionizations of silicon.

5.4 In principle, ionization potentials have been compared to nuclear binding energies.
- a) What are the similarities and differences in ionization potentials and nuclear binding energies?
- b) In general, which is larger, ionization potentials or nuclear binding energies? By how much?

5.5 Using a periodic table that does *not* contain ionization potentials, predict which element in each of the following pairs has the higher ionization potential, and why:

a) S, P	d) S, F	g) I, Br
b) Cs, Au	e) Na, O	h) Cr, Mn
c) Rn, At	f) Ne, Cl	i) Ru, Os

5.6 Give an explanation for each of the following observations:
- a) The first ionization potential for indium is less than the first ionization potential for cadmium.
- b) The first ionization potential for gallium is about the same as the first ionization potential for aluminum.
- c) There is a sharp rise between the first and second ionization potentials of aluminum but a still sharper rise between the third and four ionization potentials of aluminum.
- d) The first ionization potential for magnesium is higher than the first ionization potential for sodium, but the second ionization potential for sodium is higher than the second ionization potential for magnesium.
- e) The first ionization potential for nitrogen is higher than the first ionization potential for oxygen, but the second ionization potential for oxygen is higher than the second ionization potential for nitrogen.
- f) The first ionization potential for argon is greater than the first ionization potential for chlorine, but the eighth ionization potential for chlorine is *much* greater than the eighth ionization potential for argon.
- g) The fifth ionization potential for silicon is greater than the fourth ionization potential for aluminum.
- h) In period 3, the third ionization potential does not increase very regularly with increasing atomic number.

5.7 Does an atom with high electron affinity probably have a high or low ionization potential? Why?

5.8 All second-electron affinities probably are negative; that is, energy is absorbed instead of released. Why?

5.9 Using a periodic table that does *not* list atomic radii, predict which atom in each of the following pairs has the larger atomic radius, and indicate your reasons for making the choice:

a) H, Li	d) Cd, In	g) As, Sb
b) Nb, Ta	e) P, S	h) Ca, Ba
c) Na, Be	f) Th, U	

5.10 In Fig. 5.3, what elements are represented by the highest peaks? Why?

5.11 Using a periodic table only, predict which species in each of the following pairs is the larger, and give reasons for your choice:

a) Li^+, Be^{++} d) O^{--}, F^- g) Na^+, F^-
b) Se^{--}, Cl^- e) S^{--}, S^{++}
c) Cu^{++}, Zn^{++} f) Fe^{++}, Fe^{3+}

5.12 Using a periodic table that does *not* list electronegativities, predict which species in each of the following pairs should have the higher electronegativity, and give reasons for your choice:

a) Al, C d) Na^+, Mg^{++} g) Au^+, Au^{3+}
b) O, S e) S^{--}, F^- h) K^+, Cl^-
c) Na, Mg f) Cu^{++}, Zn^{++} i) P^{3+}, P^{3-}

5.13 In Fig. 5.4, what elements are represented by the highest peaks? By the lowest minima? Why?

5.14 Using Table 5.10, what trends can you discern for the electronegativities of ions?

5.15 Electrostatic dipole moments can be expressed in coulomb $\times$ angstrom. Show how an electrostatic dipole moment of 6.40×10^{-19} coulomb $\times$ Å might be expressed in the unit: electron $\times$ angstrom. What interpretation can be given to the result?

5.16 If the electrostatic dipole moment is expressed in electron $\times$ angstrom and the electrostatic polarizability is in cubic angstroms, in what units must E be expressed in Equation (5.6)?

5.17 If each of the following atoms or ions is placed in an electrostatic field of strength 0.25 electrons/Å^2, what is the induced electrostatic dipole moment in electron $\times$ angstrom for each of the following? (See Table 5.11.)

a) Li b) Li^+ c) F^- d) O^{--}

5.18 Using a periodic table that does *not* list electrostatic polarizabilities, predict which species in each of the following pairs should have the lower electrostatic polarizability, and give reasons for your choice:

a) O^{--}, F^- b) Na, K c) Na^+, Mg^{++} d) S, S^{--} e) Cl^-, Ar

5.19 Which of the following species are paramagnetic?

a) Li f) Sc k) Sc^{3+}
b) Be g) Tl l) H^-
c) B h) Pt m) C^{++}
d) F i) Ir n) W^{4+}
e) Ne j) Fe^{++} o) Pr^{3+}

5.20 What are the approximate magnetic dipole moments to be expected for each of the following species?

a) Zr^{4+} b) Ag^{++} c) Au^{3+} d) Mo^{++} e) Hg^{++} f) Ru^{3+}

5.21 In Tables 5.12 and 5.13 why are there no data for more than five unpaired $3d$ electrons?

5.22 What experimental data in this chapter substantiates Hund's rule of maximum multiplicity?

5.23 In Table 4.11 some of the electronic configurations appear to be out of the regular progression: for example, Cr, Cu, and Rh. What experimental evidence might be used to justify the proposed configurations as opposed to the "expected" ones?

6

Molecules and Chemical Bonds

INTRODUCTION Experience has shown that many discrete substances are combinations of two or more elements. In any specific substance, the relative amounts of the elements present are always the same. These substances are called *compounds*. One convenient system of classification of compounds depends on the number of different elements present. Compounds containing two different elements are called *binary compounds*, whereas those containing three and four different elements are called *ternary compounds* and *quaternary compounds*, respectively. For example, water, composed of hydrogen and oxygen, is a binary compound; common table sugar, composed of carbon, hydrogen, and oxygen, is a ternary compound; nitroglycerin, composed of carbon, hydrogen, oxygen, and nitrogen, is a quaternary compound.

The formation of compounds can be explained in terms of combinations of atoms of the elements to form larger, composite particles, called *molecules*.

Formulas are the shorthand abbreviations that show the over-all composition of compounds. They consist of symbols for the component elements with subscript numbers following each symbol to indicate the number of each kind of atom in the molecule. (Usually the number "1" is omitted.) Thus the formula for water is H_2O, indicating that the water molecule consists of 2 atoms of hydrogen and 1 atom of oxygen. Similarly, the formula for common table sugar is $C_{12}H_{22}O_{11}$

indicating that the molecule of sugar consists of 12 atoms of carbon, 22 atoms of hydrogen, and 11 atoms of oxygen.

The relative mass of a molecule may be determined on the atomic weight scale by adding the atomic weights of the atoms concerned. The result of this addition is called the *molecular weight*. Thus for water, H_2O, the molecular weight may be found as follows:

$$2 \text{ atoms H} \times \frac{1.01 \text{ amu}}{1 \text{ atom H}} + 1 \text{ atom O} \times \frac{16.00 \text{ amu}}{1 \text{ atom O}} = 18.02 \text{ amu}$$

In a similar way, the molecular weight for sugar is found to be 342.3 amu:

$$12 \text{ atoms C} \times \frac{12.01 \text{ amu}}{1 \text{ atom C}} + 22 \text{ atoms H} \times \frac{1.01 \text{ amu}}{1 \text{ atom H}} + 11 \text{ atoms O} \times \frac{16.00 \text{ amu}}{1 \text{ atom O}} = 342.3 \text{ amu}$$

Clearly, the formula may represent a single molecule. Thus the formula H_2O may represent one molecule of water and $4H_2O$ may represent four molecules of water. However, for most chemically important systems, very large numbers of molecules are involved. Therefore it is convenient to allow the formula to represent 1 mole of molecules, that is 6.0229×10^{23} molecules. From this point of view, the formula H_2O may represent 1 mole of water molecules and $4H_2O$ may represent 4 moles of water molecules.

This is consistent with the interpretation of the symbol as a mole of atoms. The formula H_2O may represent 1 mole of water molecules composed of 2 moles of H atoms and 1 mole of O atoms. When the formula is interpreted as a mole of molecules, then the subscripts represent moles of atoms. As another illustration, the formula, $C_{12}H_{22}O_{11}$, represents 1 mole of sugar molecules, composed of 12 moles of C atoms, 22 moles of H atoms, and 11 moles of O atoms.

It should be clear that the mass of a mole of molecules of any substance is numerically equal to the molecular weight expressed in grams. Thus a mole of water molecules has a mass of 18.02 g and a mole of sugar molecules has a mass of 342.3 g.

The percentage composition of compounds can be determined directly from the formulas for those compounds. For example, benzene has the formula C_6H_6. From the formula it is clear that 1 mole of benzene molecules consists of 6 moles of C atoms and 6 moles of H atoms. The 6 moles of C atoms have a mass of 72.06 g and the 6 moles of H atoms have a mass of 6.06 g. Thus the total mass of 1 mole of benzene molecules is 78.12 g, which is the molecular weight of benzene expressed in grams.

Since percentage composition always is expressed in mass terms unless specified to the contrary, the percent of carbon in benzene is the number of grams of carbon in 100 g of benzene, and the percent of hydrogen in benzene is the number of grams of hydrogen in 100 g of benzene. Thus the percentage composition of benzene may be computed as follows:

$$100 \text{ g } C_6H_6 \times \frac{72.06 \text{ g C}}{78.12 \text{ g } C_6H_6} = 92.24 \text{ g C}$$

$$100 \text{ g } C_6H_6 \times \frac{6.06 \text{ g H}}{78.12 \text{ g } C_6H_6} = 7.76 \text{ g H}$$

Thus benzene is said to be 92.24 percent carbon and 7.76 percent hydrogen.

Experimentally, the percentage composition of new compounds can be determined. From this kind of data, it is desirable to be able to write the formula for the compound. However, some limitations must be noted. More than one compound may have the same percentage composition. For example, the compound acetylene has the formula C_2H_2. Thus a mole of acetylene molecules has a mass of 26.02 g and contains 24.02 g of carbon and 2.02 g of hydrogen. The acetylene is 92.24 percent carbon and 7.76 percent hydrogen, which can be shown as follows:

$$100 \text{ g } C_2H_2 \times \frac{24.02 \text{ g C}}{26.04 \text{ g } C_2H_2} = 92.24 \text{ g C}$$

$$100 \text{ g } C_2H_2 \times \frac{2.02 \text{ g H}}{26.04 \text{ g } C_2H_2} = 7.76 \text{ g H}$$

It is clear that acetylene and benzene have the same percentage composition although they have different formulas and different molecular weights.

The similarity between acetylene and benzene lies in the fact that the elements occur in the same atomic ratio in both compounds. In acetylene, the elements carbon and hydrogen occur in the ratio of 2 moles of C atoms to 2 moles of H atoms, which is the same as 1 mole of C atoms to 1 mole of H atoms. In benzene, carbon and hydrogen occur in the ratio of 6 moles of C atoms to 6 moles of H atoms, which is the same as 1 mole of C atoms to 1 mole of H atoms. This simplest atomic ratio may be expressed by an *empirical formula*, which in this case is CH. Thus compounds with the same empirical formula have the same percentage composition. By the same token, only the empirical formula may be determined from percentage composition data alone.

The compound butyric acid is 54.5 percent carbon, 9.2 percent hydrogen and 36.3 percent oxygen. This means that 100 g of butyric acid contains 54.5 g of carbon, 9.2 g of hydrogen, and 36.3 g of oxygen. Since the subscripts in an empirical formula represent the relative numbers of moles of atoms of each element, the first step in writing the empirical formula is to find the numbers of moles of atoms in some specified quantity of butyric acid. Since the data are already given for 100 g of butyric acid, this is a convenient quantity to use, although any quantity could be used just as well. The calculation for 100 g of butyric acid is carried out as follows:

$$54.5 \text{ g C} \times \frac{1 \text{ mole C atoms}}{12.0 \text{ g C}} = 4.54 \text{ moles C atoms}$$

$$9.2 \text{ g H} \times \frac{1 \text{ mole H atoms}}{1.01 \text{ g H}} = 9.1 \text{ moles H atoms}$$

$$36.3 \text{ g O} \times \frac{1 \text{ mole O atoms}}{16.0 \text{ g O}} = 2.27 \text{ moles O atoms}$$

Once the number of moles of atoms of each element has been determined, it is possible to write the tentative empirical formula as $C_{4.54}H_{9.1}O_{2.27}$. Although the atomic ratios are correct, such a formula cannot be interpreted readily on an atomic basis, since the "molecules" would contain fractions of atoms. To reduce the formula to its lowest integral terms, each of the subscripts is divided by the smallest. This does not alter the *relative* numbers of moles of atoms. The empirical formula is then determined as follows:

$$C_{\frac{4.54}{2.27}}H_{\frac{9.1}{2.27}}O_{\frac{2.27}{2.27}} = C_2H_4O$$

Although the empirical formula for butyric acid is C_2H_4O, there is no guarantee that this is the molecular formula, that is, the actual formula representing the molecule. However, the molecular formula of butyric acid must be some integral multiple of this empirical formula, such as C_2H_4O, $C_4H_8O_2$, and $C_6H_{12}O_3$.

Several other compounds have the same percentage composition as butyric acid and therefore the same empirical formula. Two of these are vinyl alcohol and paraldehyde. Experimentally, the molecular weights of butyric acid, vinyl alcohol, and paraldehyde have been shown to be 88.10 amu, 44.05 amu, and 132.2 amu respectively. The molecular weight derived from the empirical formula is 44.06 amu. Thus the empirical formula is the correct molecular formula for vinyl alcohol. Since the molecular weight of butyric acid is twice that derived from the empirical formula, its molecular formula must be twice the empirical formula, or $C_4H_8O_2$. Similarly, the formula for paraldehyde must be $C_6H_{12}O_3$, which is three times the empirical formula, accounting for the fact that the molecular weight of paraldehyde is three times the molecular weight derived from the empirical formula.

Molecular and empirical formulas and the mass relationships associated with them constitute the gross features of compound and molecule formation. In the remainder of this chapter, we shall look at some of the finer features of compound and molecule formation, in an attempt to find some answers to the question: What are the factors responsible for the bonds which exist in these composite particles which we call molecules?

COMBINATIONS OF ATOMS WITH WIDELY DIFFERENT ELECTRONEGATIVITIES: IONIC BONDS

If an atom of very high electronegativity approaches an atom of low electronegativity, the very strong attraction for electrons exerted by the atom of high electronegativity conceivably may be sufficient to remove the electron from the atom of low electronegativity. For example, if an atom of fluorine approaches an atom of cesium, the small, highly electronegative atom of fluorine may polarize the atom of cesium and then actually remove an electron from the Cs atom, as shown schematically in Fig. 6.1.

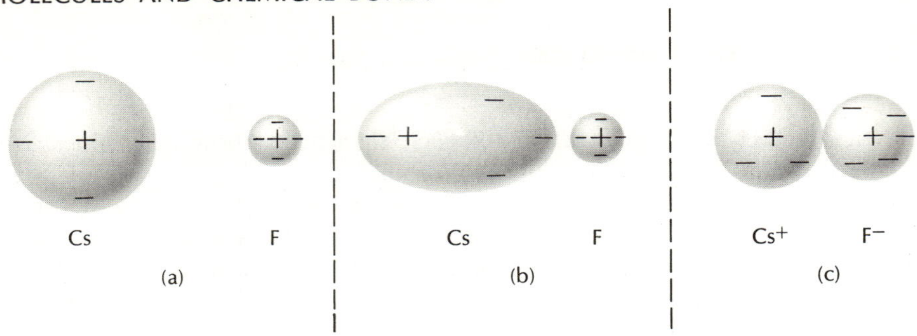

Fig. 6.1. Schematic representation of the formation of cesium and fluoride ions from the atoms

The energetics of this process may be viewed in two steps. One step is the absorption of energy by the Cs atom with removal of an electron. The energy involved is the ionization potential. The process is represented by Equation (6.1). The other step is the addition of the electron to the F atom with a decrease in potential energy equivalent to the electron affinity. This is shown in Equation (6.2). The over-all process is the sum of these two steps, shown in Equation (6.3).

$$\mathrm{Cs} + 1.50 \times 10^{-22}\ \mathrm{kcal/atom} = \mathrm{Cs}^+ + \mathrm{e}^- \tag{6.1}$$

$$\mathrm{F} + \mathrm{e}^- = \mathrm{F}^- + 1.40 \times 10^{-22}\ \mathrm{kcal/atom} \tag{6.2}$$

$$\mathrm{Cs} + \mathrm{F} + 0.10 \times 10^{-22}\ \mathrm{kcal} = \mathrm{Cs}^+ + \mathrm{F}^- \tag{6.3}$$

Obviously the production of two simple ions by this process requires an energy input, and the two ions produced are at a higher state of potential energy than the two original atoms. If this process is to occur spontaneously, another potential energy decrease must be possible.

The summary given by Equations (6.1), (6.2), and (6.3) assumes that the ions produced are isolated and not interacting. However, it has been pointed out that cations and anions tend to act like simple point charges. Thus there is an additional decrease in potential energy associated with the coming together of the two ions. Considering the two ions as point quantum charges, and using the ionic radii from Table 5.8, we find that the decrease in potential energy is about 1.69×10^{-22} kcal. This is shown in Equation (6.4), which added to Equation (6.3) produces Equation (6.5), summarizing the over-all process.

$$\mathrm{Cs} + \mathrm{F} + 0.10 \times 10^{-22}\ \mathrm{kcal} = \mathrm{Cs}^+ + \mathrm{F}^- \tag{6.3}$$

$$\mathrm{Cs}^+ + \mathrm{F}^- = \mathrm{Cs}^+\mathrm{F}^- + 1.69 \times 10^{-22}\ \mathrm{kcal} \tag{6.4}$$

$$\mathrm{Cs} + \mathrm{F} = \mathrm{Cs}^+\mathrm{F}^- + 1.59 \times 10^{-22}\ \mathrm{kcal} \tag{6.5}$$

Thus the interaction of the ions is necessary to provide sufficient potential energy decrease to make ion formation feasible. However, if large numbers of Cs atoms and F atoms are permitted to react, and many Cs and F ions are formed, even greater potential energy decrease is possible through the formation of large aggregates of cations and anions. Without trying to be very accurate at this time about the exact nature of the matrix, we can visualize a cross section of the network of Cs and F ions schematically represented in Fig. 6.2. In this aggregate each positive Cs ion is surrounded by several negative F ions and each negative F ion is surrounded by several positive Cs ions. As a result each ion experiences electrostatic attractions to several other ions, leading to large potential energy decreases.

Substances like the one described here consist of ions and are called *ionic substances*, or *ionic compounds*. The attractions among the ions that lead to the significant potential energy decrease constitute the *ionic bond*.

Fig. 6.2. Schematic cross section of a typical ionic lattice

$F^- \; Cs^+ \; F^- \; Cs^+ \; F^- \; Cs^+ \; F^- \; Cs^+ \; F^- \; Cs^+ \; F^- \; Cs^+$
$Cs^+ \; F^- \; Cs^+ \; F^- \; Cs^+ \; F^- \; Cs^+ \; F^- \; Cs^+ \; F^- \; Cs^+ \; F^-$
$F^- \; Cs^+ \; F^- \; Cs^+ \; F^- \; Cs^+ \; F^- \; Cs^+ \; F^- \; Cs^+ \; F^- \; Cs^+$
$Cs^+ \; F^- \; Cs^+ \; F^- \; Cs^+ \; F^- \; Cs^+ \; F^- \; Cs^+ \; F^- \; Cs^+ \; F^-$
$F^- \; Cs^+ \; F^- \; Cs^+ \; F^- \; Cs^+ \; F^- \; Cs^+ \; F^- \; Cs^+ \; F^- \; Cs^+$
$Cs^+ \; F^- \; Cs^+ \; F^- \; Cs^+ \; F^- \; Cs^+ \; F^- \; Cs^+ \; F^- \; Cs^+ \; F^-$
$F^- \; Cs^+ \; F^- \; Cs^+ \; F^- \; Cs^+ \; F^- \; Cs^+ \; F^- \; Cs^+ \; F^- \; Cs^+$
$Cs^+ \; F^- \; Cs^+ \; F^- \; Cs^+ \; F^- \; Cs^+ \; F^- \; Cs^+ \; F^- \; Cs^+ \; F^-$

From Fig. 6.2, it should be clear that it is impossible to pick out specific molecules of ionic compounds. Actually each positive Cs ion is bonded to several negative F ions, and each negative F ion is bonded to several positive Cs ions. However, from the electron process described, it is apparent that there must be equal numbers of positive Cs ions and negative F ions. As a result the empirical formula for this compound is CsF. It is customary to represent ionic compounds by their empirical formulas. However, it is not correct to interpret the formula CsF as 1 mole of molecules of cesium fluoride, as the compound is named; rather the formula represents a mole of cesium ions and a mole of fluoride ions, as the negative ions are called.

Ion Formation

Since the important contribution to the energetics of the ionic bond is made by the mutual attraction of oppositely charged ions, it might appear energetically advantageous to increase the charges on either or both of the ions. Thus the following question might be raised: Why doesn't cesium form the Cs^{++} ion and fluorine the F^{--} ion, since the electrostatic attractions between these ions should be much greater in the ionic aggregate, thus leading to still lower potential energies?

However, in considering cation formation, it must be remembered that the removal of a second electron is always more difficult than the removal of the first. It requires the expenditure of considerably more energy in the formation of the cation. In order to have a favorable energy decrease over all, the decrease in potential energy due to the interionic attractions must be greater than the increase in ionization potential for the removal of additional electrons. In the case of cesium, the first electron removed is a 6*s* electron, whereas the second one to be removed is a 5*p* electron. The 5*p* electron is in a much lower energy state than the 6*s* electron. In fact, the second ionization potential for cesium is more than six times the first ionization potential. Thus the energy expended in the ionization step is much greater for the formation of Cs^{++} than it is for the formation of Cs^{+}. Actually it is so large that it exceeds the decrease in potential energy due to interionic attractions among ions of higher charge, and CsF_2 is not formed.

Similarly it might seem feasible to add more than one electron to an atom of fluorine. Several factors must be considered. First, second-electron affinities are negative, which means that due to the negative charge already on the ion, work must be done to add an electron to the ion. In addition, the relative size of the electron affinity depends upon the available energy levels in the atom or ion. The lower the energy level available, the more readily the addition will occur. Thus a second electron will be added to an ion only if the additional expenditure of energy is less than the decrease in potential energy due to increased interionic attractions. In the case of fluorine, the first electron added enters the 2*p* level, whereas the second must enter the much higher 3*s* level. Thus the over-all expenditure of energy to add the second electron is more than the decrease in potential energy to be gained due to increased interionic attractions. As a result, Cs_2F is not formed.

Since ionization potentials increase with the removal of electrons, it appears likely that there is some practical upper limit to the number of electrons that may be removed from simple atoms by interaction with other atoms. In practice, it is doubtful that simple cations with charges in excess of 3+ quanta actually exist in compounds under usual conditions. Similarly, because of the increased difficulty of adding successive electrons to atoms, it is doubtful that simple anions with charges in excess of 3− quanta actually exist in compounds under usual conditions.

Since cation formation is enhanced by low ionization potentials, the elements on the left side and toward the bottom of the periodic table are the best cation formers. The alkali metals (family 1A) all tend to form 1+ ions corresponding to the removal of the *ns* electron. The removal of additional electrons from the $(n - 1)p$ level requires too much energy to be feasible. Among the members of the family, lithium is the least likely to form ions, and cesium (ignoring the uncommon francium) is the most likely to form ions.

Among the alkaline earths (family 2A), the 2+ ions are common. Although it takes somewhat more energy to remove the two *ns* electrons than to remove just one, the second ionization potential for the alkaline earths is much less than the second ionization potential for the alkali metals. Therefore the decrease in potential energy due to the increase in interionic attraction more than compensates for the increase in ionization potential. Since the third electron to be removed must come from the $(n - 1)p$ level, the third ionization potential is quite high and there is no tendency for the alkaline earths to form 3+ ions.

Among the transition elements, the 2+ and 3+ ions are most common. Roughly, these correspond to the removal of the two *ns* electrons in the one case and the removal of the two *ns* electrons and one $(n - 1)d$ electron in the other case. Since the energy difference between the *ns* and $(n - 1)d$ electrons is not very great among the transition elements, it is possible to remove three electrons in some cases. Within each period the more highly charged ions tend to be formed by the elements to the left (lower atomic number), which have the lower ionization potentials. The 2+ ions are formed more readily by those transition elements which have the higher ionization potentials, that is, those to the right. In addition, those elements in family 1B (the coinage metals) also tend to form 1+ ions corresponding to the removal of the single *ns* electron. However, they do form ions with higher charges under appropriate conditions.

Since anion formation is enhanced by high electron affinity, and consequently high electronegativity, the elements to the right and to the top of the periodic table are the best anion formers. The halogens (family 7A) all form 1− ions corresponding to the addition of one electron to the *np* level. Since any additional electrons must be placed in higher levels, such as the $(n + 1)s$ level, the 2− ions of these elements are not found.

Among the chalcogens (family 6A), the 2− ions are common. Despite the fact that the second electron affinity is negative, so that the energy change accompanying electron addition might appear unfavorable, there is still sufficient increase in interionic attractions to lead to a favorable net decrease in energy. However, since a third electron must be placed in the much higher $(n + 1)s$ energy level, the 3− ions are not formed. Within this family it becomes increasingly difficult for the 2− ions to be formed as atomic number increases.

In the nitrogen family (family 5A), simple anion formation occurs with the addition of three electrons. Probably only nitrogen forms the 3− ion. In this case, three electrons may be added to the *np* level with resultant decrease in potential energy due to favorable interionic attractions. Any additional electrons must enter the $(n + 1)s$ level.

Formulas and Nomenclature

The process of electron removal is called *oxidation*. Thus the formation of cations from atoms is a process of oxidation. As a result the quantum charge on ions is called the *oxidation number* or *oxidation state*. This name applies to anions as well as cations, and the number when stated must be accompanied by the correct positive or negative sign. Thus a positive cesium ion has an oxidation number of 1+ and a positive barium ion has an oxidation number of 2+. Similarly the negative fluorine ion has an oxidation number of 1− and the negative oxygen ion has an oxidation number of 2−.

In naming the ions, the simple cations retain the name of the element from which they are formed. Thus Cs^+ is the cesium ion. However, the negative ions have names composed of a root from the name of the element to which the suffix *-ide* is added. Thus the negative fluorine ion, F^-, is called the fluoride ion. Similarly O^{--} is the oxide ion and N^{3-} is the nitride ion.

Ionic compounds are named from the ions. Thus the compound composed of cesium ions and fluoride ions is called cesium fluoride. Similarly the compound formed of sodium ions and oxide ions is sodium oxide.

Formulas may be written for ionic compounds in such a way that the sum of all of the oxidation numbers of the ions represented is zero. This is equivalent to saying that within any sample of the compound there must be equal numbers of positive and negative charges, or that the matter is electrostatically neutral. Thus sodium oxide must have the formula Na_2O since the sodium ion is Na^+ and the oxide ion is O^{--}. It should be remembered, however, that this is an empirical formula and does not represent molecules as such. This formula represents either two sodium ions and one oxide ion or it represents one mole of oxide ions and two moles of sodium ions.

COMBINATIONS OF ATOMS WITH IDENTICAL BUT HIGH ELECTRONEGATIVITIES: NONPOLAR COVALENT BONDS

Ionic bonds are formed between atoms of high and low electronegativity. However, electron transfer appears to be out of the question if the two atoms involved are identical and if they have high electronegativities. One atom does not have greater attraction for electrons than the other, but both atoms have a high attraction for electrons.

The simplest illustration of such a system is two approaching hydrogen atoms. As the two atoms of hydrogen approach each other, there are electrostatic repulsions between the electrons in the two atoms and also between the two nuclei of the atoms. At the same time there is attraction between the nucleus of one atom and the electrons of the other atom. This is summarized in Fig. 6.3. The top curve shows the increase in potential energy due to repulsions, ignoring attractions. The bottom curve shows the decrease in potential energy due to attractions, ignoring repulsions. The over-all effect is obtained by adding the two curves. This is shown by the solid line curve in the middle.

From Fig. 6.3 it can be seen that the effects of repulsion and attraction almost balance each other over large distances. However, as the two atoms approach, the attraction becomes a little greater than the repulsion, resulting in a net decrease in

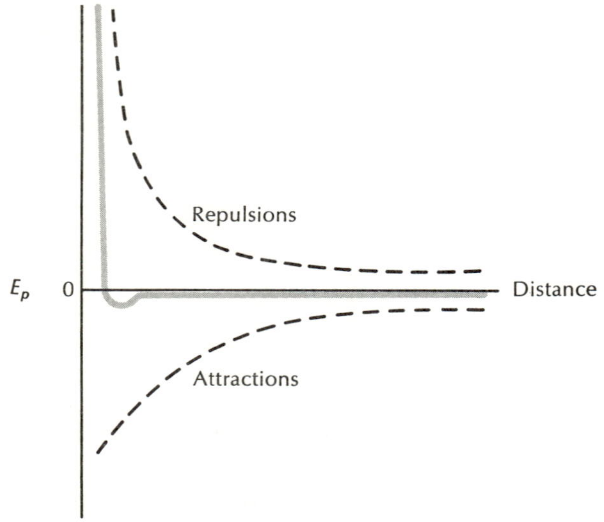

Fig. 6.3. Potential energy change as two hydrogen atoms approach and form a covalent bond

potential energy. The potential energy reaches a minimum when the nuclei of the two atoms are 0.74 Å apart. After this, the potential energy rises sharply due to the strong nuclear repulsions at very short distances.

Thus the system of two hydrogen atoms is at minimum potential energy when the nuclei are 0.74 Å apart, and, in general, the two atoms are held in this position. In energy terms, the system composed of the two atoms is at the bottom of a potential energy well similar to the one described earlier for nucleons in a nucleus. The bottom of this potential energy well lies at -1.73×10^{-22} kcal. Thus, to separate the two atoms, 1.73×10^{-22} kcal of energy must be supplied. This energy is called the *bond energy*. It is a measure of the strength of the bond holding the two atoms together. The bond energy is to atoms in molecules as the binding energy is to nucleons in nuclei.

The formation of the bond between two H atoms may be depicted by an equation, as follows:

$$2H = H_2 + 1.73 \times 10^{-22} \text{ kcal/molecule} \qquad (6.6)$$

This equation indicates that two H atoms combine to form a hydrogen molecule, with a decrease in potential energy of 1.73×10^{-22} kcal. Equation (6.6) may also be written in mole quantities as well. The decrease in potential energy associated with the formation of a mole of hydrogen molecules may be computed as follows:

$$1 \text{ mole } H_2 \text{ molecules} \times \frac{6.02 \times 10^{23} \, H_2 \text{ molecules}}{1 \text{ mole } H_2 \text{ molecules}} \times \frac{1.73 \times 10^{-22} \text{ kcal}}{1 \, H_2 \text{ molecule}} = 104 \text{ kcal}$$

Using this value, Equation (6.6) may be rewritten as Equation (6.7):

$$2H = H_2 + 104 \text{ kcal/mole} \qquad (6.7)$$

That is, 2 moles of H atoms combine to form one mole of hydrogen molecules, with a decrease in potential energy of 104 kcal.

The Covalent Bond

A more pictorial representation of the approach of two H atoms to form a hydrogen molecule is given in Fig. 6.4. As the two atoms approach they are polarized, the electrons being attracted more and more toward the space between the nuclei. Whereas the probable space occupied by each electron at the outset was spherical, corresponding to the probable charge distribution for the 1*s* electron, now the probable location for the two electrons is much more localized between the two nuclei.

Of course, this localization within a somewhat smaller space increases electrostatic repulsions between the two electrons. However, a very important property of electrons allows them to be packed into small spaces. This property is associated

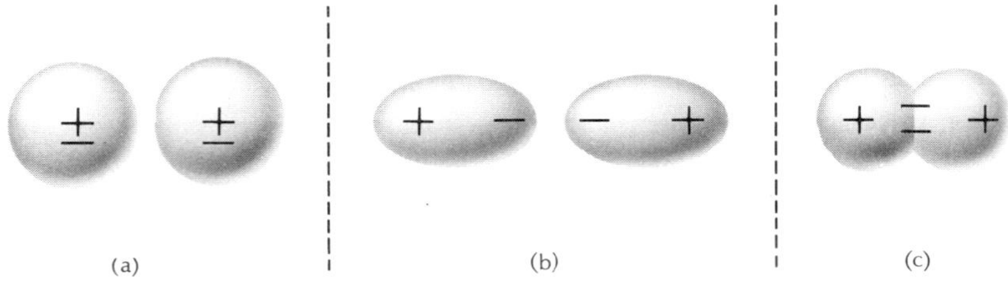

Fig. 6.4. Representation of the electron charge distribution in the covalent bond of the hydrogen molecule

with electron spin. It has already been shown that two electrons may occupy the same space if they have opposite spins. This is demonstrated by the fact that if two electrons are in the same atomic secondary energy level, they have the greatest probability of being in two different orbitals. Thus they have different spatial orientations, permitting the greatest separation of electrons, a situation which is consistent with electrostatic repulsion. However, when each atomic orbital within a given secondary energy level is occupied by one electron, it is energetically preferable to have two electrons of opposite spin in one orbital of that energy level, rather than for the second electron to be placed in a level of higher potential energy. Thus the spin interaction of electrons, whatever its nature may be, permits two electrons to occupy the same space despite electrostatic repulsions.

Thus, if the spins of the electrons in the two hydrogen atoms are opposed, the two electrons may be localized between the two nuclei, with resultant potential energy decrease, despite the increased electrostatic repulsions.

Thus the chemical bond between two highly electronegative atoms is due largely to the increased electrostatic attraction between the atomic nuclei and the electrons packed between them. In multiple electron systems, the most energetically economical way for the increase to occur is for electrons to be paired so that the spin interactions permit two electrons to occupy the same space. In this way it is possible for the maximum number of electrons to be packed between the nuclei, thus maximizing electrostatic attractions between the nuclei and the electrons.

From one point of view, the electron pair localized between the two nuclei occupies a distorted 1*s* orbital in each of the two atoms. However, as an alternative, it is possible to think of a new pair of orbitals related to the two atomic orbitals but now associated with the molecule as a whole. These new orbitals are called *molecular orbitals.* A pictorial representation of the changing probable charge distribution as two hydrogen atoms approach is given in Fig. 6.5. From the charge distributions associated with the two 1*s* orbitals of the atoms, two new charge distributions are formed, associated with the molecular orbitals. One of these charge distributions lies mainly between the two nuclei. The other consists of two parts on opposite sides of the two nuclei, away from the space between the two nuclei. The molecular orbital associated with the charge distribution concentrated between the two nuclei is called a *bonding orbital.* The molecular orbital associated with the charge distribution not lying between the two nuclei is called an *anti-*

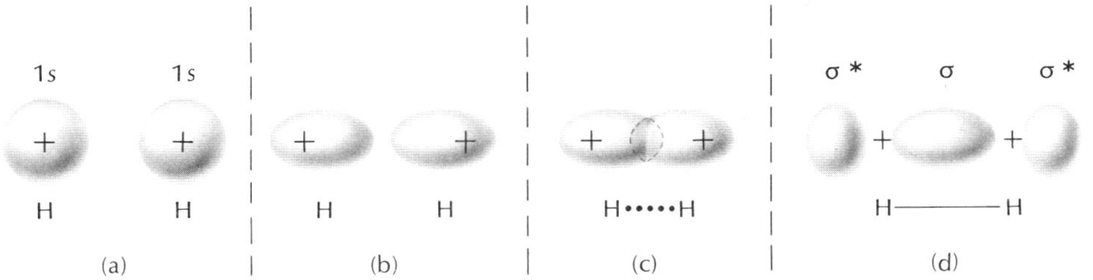

Fig. 6.5. Schematic representation of the formation of the sigma orbitals from the 1*s* orbitals of two hydrogen atoms

bonding orbital. These orbitals which have been pictured by combining the charge distributions of atomic *s* orbitals are called sigma (σ) orbitals.

Thus in the hydrogen molecule, the two electrons are in the sigma bonding orbital, since this orbital represents a lower potential energy than the sigma antibonding orbital.

It is somewhat unfortunate that in common chemical parlance so much emphasis is placed upon electron pairing as a source of bond energy. Certainly the pairing of spins plays an important role in allowing the close packing of the maximum number of electrons between the nuclei, but the bond itself probably is more electrostatic in nature than might be suspected otherwise. With this in mind, the *covalent bond* may be defined as a pair of electrons shared between two atoms.

From the point of view developed here, it is possible to see why two helium atoms do not form a covalent bond. Within the individual, isolated atoms, the 1*s* level is occupied completely, and the electrons are paired. Any attempt to localize the two pairs of electrons between the two nuclei results in electrostatic repulsion without the benefit of additional spin pairing. From the point of view of molecular orbitals, the four electrons must occupy both the bonding and antibonding sigma orbitals. The electrons in the bonding sigma orbital would be at a low potential energy and would contribute, therefore, to the formation of a covalent bond. However, the electrons in the antibonding sigma orbital would be at a high potential energy and would oppose the formation of a covalent bond. A part of the reason for the high potential energy of the electrons in the antibonding orbital is the repulsion of these electrons by the localized pair of electrons in the bonding orbital. This repulsion tends to force the electrons in the antibonding orbital farther away from the nuclei, and consequently to a higher potential energy state. The result is that no covalent bond is observed between neutral helium atoms. Thus elemental helium exists in the form of atoms, whereas elemental hydrogen exists as diatomic molecules, that is, as molecules composed of two atoms.

Thus it appears that the only fruitful interactions between two atomic orbitals occur when the sum of the electrons in the two orbitals is less than four. If there is a total of four electrons in the two atomic orbitals, then in the molecules they must fill completely both the bonding and antibonding molecular orbitals, the increased potential energy of the antibonding electrons canceling the favorable effect of the decreased potential energy of the bonding electrons. If there is a total of three electrons in the two atomic orbitals, then two of these may appear in the bonding molecular orbital, the other one being in the antibonding molecular orbital. If other possibilities are not available, this condition should lead to a bond between the two atoms. If there is a total of two electrons in the two atomic orbitals, both will appear in the bonding molecular orbital and the antibonding molecular orbital will be vacant. This is the most favorable situation and the one most frequently encountered. If there is a total of one electron in the two atomic orbitals, this one electron will appear in the bonding molecular orbital, and this leads to a bond between the two atoms.

Since the covalent bond energy can be attributed to the attraction of two atoms for electrons localized between them, it appears that covalent bonds should be found predominantly among atoms of high electronegativity. For this reason, we shall look at other highly electronegative atoms to see what factors affect covalent bond formation.

FLUORINE. Fluorine is the most electronegative element known. Although some ions have higher electronegativities, no other neutral atoms attract electrons so strongly.

The electronic configuration of the fluorine atom is $1s^2 2s^2 2p^5$. Consequently all orbitals in the second principal energy level are filled, with the exception of one of the $2p$ orbitals. From the previous discussion, it appears that if a bond is to form between two atoms of fluorine, the significant atomic orbital interaction will occur between two of these half-filled, $2p$ orbitals, one in each atom. A pictorial representation of the interaction of two atomic $2p$ orbital charge distributions to form two molecular orbital charge distributions is given in Fig. 6.6.

The over-all result resembles the formation of the bonding and antibonding molecular orbitals from the interaction of the s orbitals of two hydrogen atoms. In fact these orbitals also are called sigma orbitals. They differ in having been originated through the interaction of atomic p orbitals instead of atomic s orbitals, and they differ in potential energy. However, in terms of the type of charge distribution they are similar.

Clearly, the two electrons that are unpaired in the single F atoms may be paired in a diatomic fluorine molecule with a pair of electrons occupying the bonding molec-

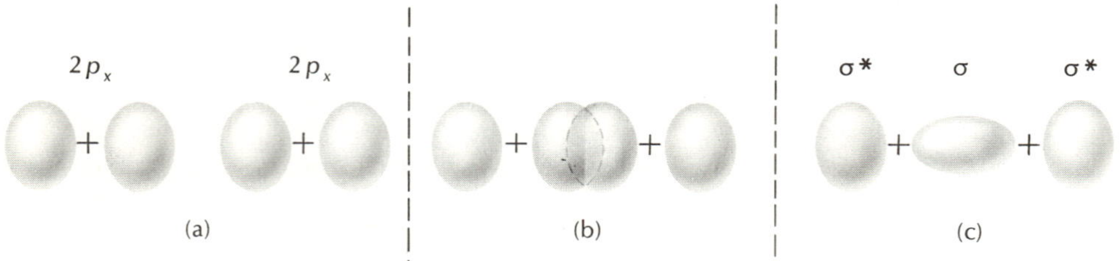

Fig. 6.6. Schematic representation of the formation of the sigma orbitals from two $2p_x$ orbitals

ular orbital. Thus the fluorine molecule consists of two atoms of F held together by a covalent bond. Elemental fluorine may be represented by the formula F_2.

Sometimes more descriptive formulas for molecules are desirable. For this purpose, the individual atoms may be represented by the atomic symbols, with a line between them to represent the pair of shared electrons. In this way, the fluorine molecule may be shown as F—F. In some contexts, it is important to remember the *ns* and *np* electrons, which are not directly involved in the bonding and which we shall call the *nonbonding electrons*. These may be represented individually by dots or in pairs by straight lines distributed around the atomic symbols. Thus the fluorine molecule may be represented as $:\ddot{\underset{..}{F}}—\ddot{\underset{..}{F}}:$ or $|\overline{\underline{F}}—\overline{\underline{F}}|$. These formulas are called *electronic formulas* or *structural formulas*.

OXYGEN AND SULFUR. Oxygen is the second most electronegative element. The oxygen atom has the electronic configuration $1s^22s^22p^4$. Consequently, there are two $2p$ orbitals with one electron each. The charge distributions for electrons in these two orbitals are perpendicular to each other. Thus as two atoms of oxygen approach, two of the $2p$ charge distributions may meet head-on, so to speak, as represented in Fig. 6.6. The charge distributions of the other half-filled $2p$ orbitals approach each other sideways as depicted in Fig. 6.7. In this way a bonding orbital

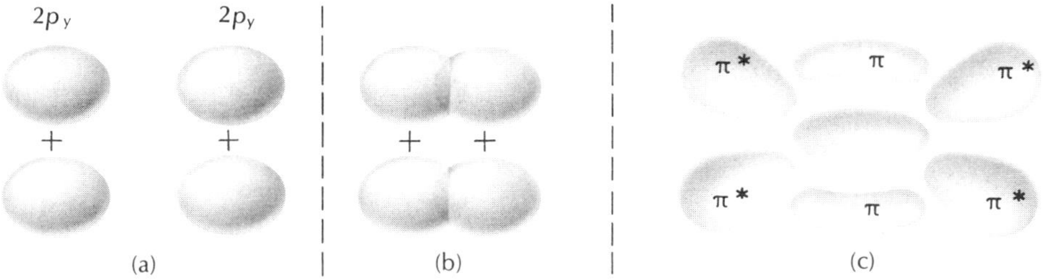

Fig. 6.7. Schematic representation of the formation of the pi orbitals from two $2p_y$ orbitals

is formed, with a charge distribution consisting of two lobes that lie somewhat between the two nuclei but not on the line between them. In a similar fashion, an antibonding orbital is formed, with a charge distribution consisting of four lobes that lie away from the region between the nuclei. This set of orbitals is called the pi (π) set of orbitals.

Thus when two O atoms approach each other, two bonding orbitals (one σ and one π) and two antibonding orbitals (one σ and one π) are formed by the interaction of the two half-filled $2p$ orbitals. The resultant charge distributions are shown in Fig. 6.8. The spatial orientations of the charge distributions for the two bonding orbitals might be compared with a frankfurter sandwich, the sigma charge distribution being the frankfurter and the pi charge distribution the two halves of the bun.

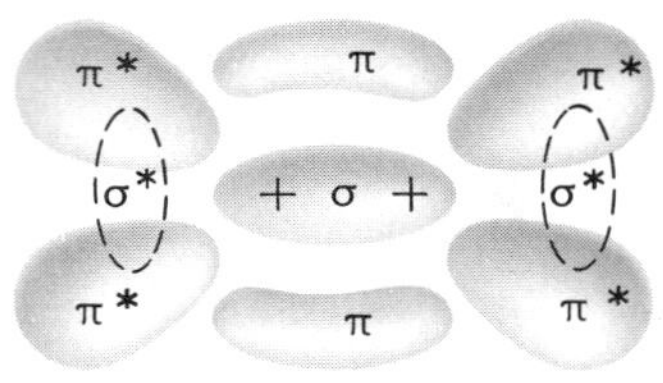

Fig. 6.8. Schematic representation of a double bond showing both the sigma and pi orbitals

Of course, the four unpaired electrons in the two O atoms can be paired and localized between the two nuclei in the two bonding orbitals. This type of covalent bond, in which two pairs of electrons are shared between two atoms, is called a *double covalent bond*, or more commonly, just a *double bond.* Note that in this double bond, one pair of bonding electrons is in the sigma bonding orbital and the other pair of bonding electrons is in the pi bonding orbital.

Thus elemental oxygen is diatomic, having the formula O_2, which may also be represented as O—O.

Sulfur is in the same periodic family as oxygen, and the electronic configuration of the S atom is $1s^22s^22p^63s^23p^4$. The similarity to the oxygen configuration suggests that sulfur should form diatomic molecules, S_2, in which there is a double bond with one shared pair of electrons in a sigma bonding orbital and a second shared pair of electrons in a pi bonding orbital.

However, experimental determinations of molecular weight indicate a value of about 256 amu for the most common form of elemental sulfur. This corresponds to a formula of S_8 for the molecule. Clearly, the sulfur molecule is quite different from the oxygen molecule. The results of x-ray and electron diffraction studies have been interpreted to indicate that the eight atoms in the sulfur molecule lie at the vertices of a puckered octagon, as shown in two views in Fig. 6.9.

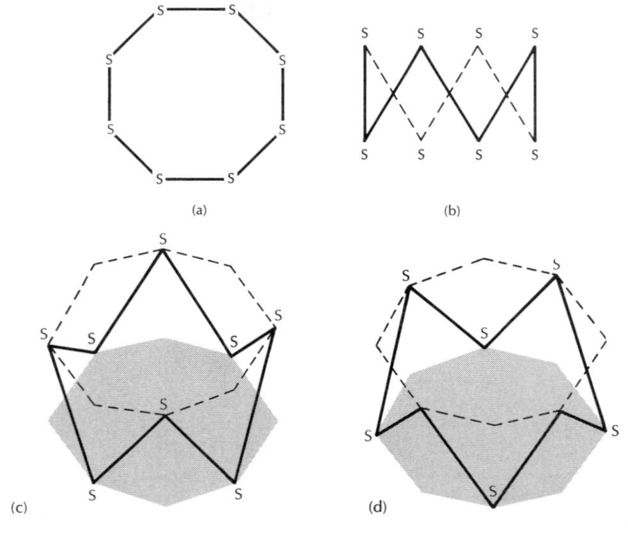

Fig. 6.9. Molecular structure of S_8

The bonding suggested in Fig. 6.9 is a reasonable alternative to double-bond formation. In this case each atom of sulfur forms two single bonds, one with each of the adjacent atoms. Each of these bonds is a pair of electrons shared between two atoms in a sigma bonding orbital formed by the interaction of two $3p$ atomic orbitals.

We are faced now with the problem of explaining why S and O atoms behave differently, despite the similarities in the electronic configurations of their atoms.

First, we recognize that oxygen is more electronegative than sulfur. This means that O atoms attract electrons more strongly than S atoms. Consequently, the decrease in potential energy brought about by localizing a pair of electrons between two O atoms should be greater than the potential energy decrease brought about by localizing a pair of electrons between two S atoms. This should be true because the localized pair of electrons will experience stronger attractions to the O atoms than to the S atoms. Since the packing of two pairs of electrons between two atoms involves an increase in electrostatic repulsions between these two pairs of electrons, the larger electronegativity for the atoms helps to offset this possible increase in potential energy.

From this point of view, double-bond formation appears to be most likely between atoms of high electronegativity.

We also recognize that the O atom is much smaller than the S atom. In viewing the sulfur molecule as it has been proposed in Fig. 6.9, we see that there are four different groups of electrons to be accommodated on the surface of each atom, namely the two pairs of shared electrons in the sigma bonding orbitals and the two pairs of nonbonding electrons resulting from the one ns orbital and the one np orbital which were full on the individual atoms prior to bonding. In a sense, these four pairs of electrons are competing for space on the surface of each atom. Since they repel each other, the smaller the surface available, the greater will be the repulsions, and consequently the higher the potential energy will be.

If the structure shown in Fig. 6.9 were composed of smaller O atoms, the repulsions among electron pairs on the surface of each atom would be larger than the repulsions among the electron pairs on the surface of the larger S atoms. When two pairs of electrons are packed between two atoms in double-bond formation, more surface area is provided for the remaining four electrons than would be provided otherwise. This allows for less repulsion and a lower potential energy.

In addition, the orbital interactions required for pi molecular orbital formation are less effective among larger atoms. The reason for this can be seen in the schematic diagram in Fig. 6.10. When small atoms combine to form sigma orbitals,

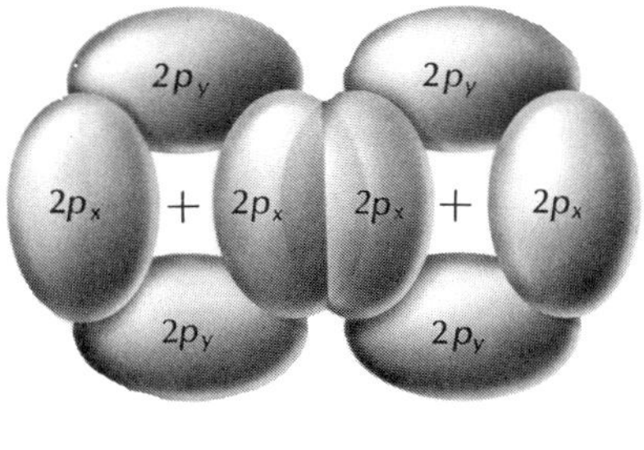

Fig. 6.10. Representation of the simultaneous interaction of the $2p_x$ and the $2p_y$ orbitals as two atoms approach

the perpendicular *p* charge distributions that must interact in a sideways fashion are closer together than they are when large atoms combine.

From this point of view, double-bond formation appears to be most likely between atoms with small radii.

Over all, this point of view suggests that double-bond formation should be most prevalent between small, highly electronegative atoms. Of course this corresponds to what appears to be true for oxygen and sulfur, where the small, highly electronegative O atoms form diatomic molecules through double-bond formation and the larger, less electronegative S atoms form octatomic molecules through single bond formation.

The over-all picture for oxygen and sulfur may be summarized with experimental data. Bond energy has been defined as the decrease in potential energy when two atoms are brought together from an infinite distance to form a bond. Numerically, this is the same as the energy required to break the bond and separate the resultant atoms. This dissociation energy can be measured and expressed as the energy required to break a mole of bonds, or the energy released when a mole of bonds is formed. The values for the oxygen-oxygen and sulfur-sulfur bonds are given in Table 6.1.

TABLE 6.1

Oxygen-Oxygen and Sulfur-Sulfur Bond Energies in kcal/Mole of Bonds

O—O	35
O═O	119
S—S	54
S═S	84

These values may be used to compare the decrease in potential energy for the formation of O_2, O_8, S_2, and S_8 as follows:

$$8\,O = 4\,O_2 + 476\text{ kcal} \quad (6.8)$$

$$8\,O = O_8 + 280\text{ kcal} \quad (6.9)$$

$$8\,S = 4\,S_2 + 336\text{ kcal} \quad (6.10)$$

$$8\,S = S_8 + 432\text{ kcal} \quad (6.11)$$

It is clear that given 8 moles of O atoms, the formation of 4 moles of diatomic oxygen molecules is highly favored over the formation of 1 mole of octatomic oxygen molecules. However, given 8 moles of S atoms, the formation of 1 mole of octatomic sulfur molecules is favored over the formation of 4 moles of diatomic sulfur molecules. From the list of bond energies in Table 6.1, it is evident that the reason for this difference between oxygen and sulfur is the very high double-bond energy and low single-bond energy for oxygen.

The viewpoint developed here may also be correlated with one other piece of experimental data for the sulfur molecule. The determinations of the positions occupied by the S atoms in the molecule suggest that the angle between the two bonds to a single S atom is about 107°. This is called the *bond angle.*

If we focus our attention on any single S atom in the sulfur molecule represented in Fig. 6.9, it is clear that there are four groups of electrons on the surface of the atom. There are two pairs of bonding electrons and two pairs of nonbonding electrons. These four groups of electrons repel each other and thus compete for space on the atomic surface. If the four groups of electrons are alike, they should assume positions on the spherical atomic surface such that they all are equidistant from each other. This represents the arrangement of lowest potential energy. These positions on the sphere correspond to the vertices of a tetrahedron inscribed in the sphere, as shown in Fig. 6.11. If lines are drawn from the center of the sphere to the electron pairs, as shown in Fig. 6.11, then the angle between any two lines is the *tetrahedral angle* of 109°28′.

The bond angle of about 107° in the S molecule is very nearly the tetrahedral angle. It is not exactly the tetrahedral angle because the four pairs of electrons on

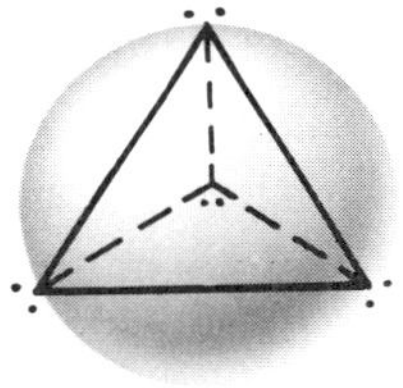

Fig. 6.11. Tetrahedral distribution of four electron pairs on the surface of a spherical atom

the surface of the S atom are not equivalent. The two shared pairs of electrons are localized between the two atoms and therefore do not tend to occupy as much atomic surface space as the two pairs of nonbonding electrons. Therefore, they may be somewhat closer together than the less restricted nonbonding pairs of electrons.

NITROGEN AND PHOSPHORUS. The electronic configuration for the nitrogen atom is $1s^2 2s^2 2p^3$. There is one electron in each of the three $2p$ orbitals, so it is possible for all three of these orbitals to interact with the three $2p$ orbitals of another N atom. This permits the formation of a second set of pi molecular orbitals. The charge distribution of this second set of pi orbitals lies in a plane perpendicular to the plane of the charge distribution of the first set. Figure 6.12 shows the approximate orientation of the charge distributions of the two bonding pi orbitals and the one bonding sigma orbital. Spatially these may be compared with a frankfurter (σ charge distribution) surrounded on four sides by halves of buns (two π charge distributions). When two N atoms combine in this way, the result is the diatomic nitrogen molecule N_2, which has a *triple bond;* that is, it has three pairs of electrons shared between the two atoms. The molecule also may be written as $N{\equiv}N$, $:N{\equiv}N:$, and $|N{\equiv}N|$.

For the small, electronegative N atom, this arrangement is the preferred one. The three pairs of electrons localized between the two atoms are highly attracted

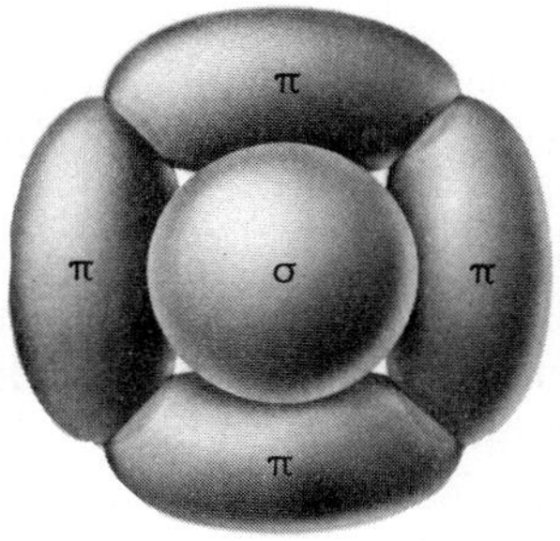

Fig. 6.12. End view, cross section, of a triple bond showing two pi and one sigma bonding orbitals

to the nuclei of both atoms and are in a low state of potential energy. When six of the peripheral electrons associated with each atom are localized between the atoms, the repulsions among the nonbonding electron pairs are minimized.

By analogy with oxygen and sulfur, it might be expected that phosphorus does not form the triply bonded P_2 molecule as readily as a singly-bonded molecule. This appears to be correct. Molecular weight determinations suggest the formula P_4 for elemental phosphorus.

If P atoms form only single bonds, each atom may form single bonds with three other atoms. Electron diffraction data have been interpreted to indicate that the four P atoms lie at the vertices of a regular tetrahedron. This allows each atom to be bonded to three other atoms, with a total of six bonds among the four atoms. These bonds lie along the edges of the tetrahedron as shown in Fig. 6.13.

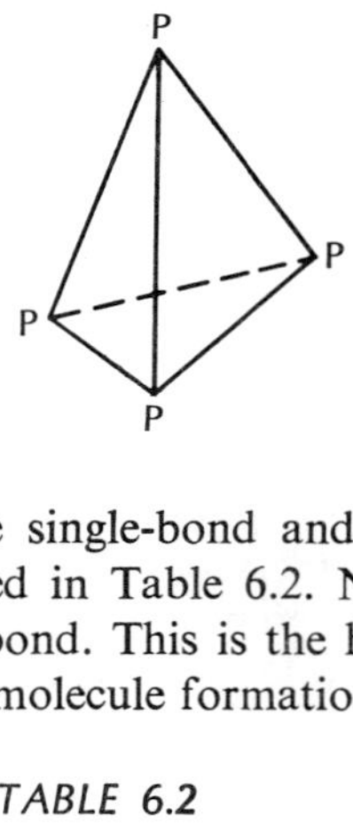

Fig. 6.13. Molecular structure of P_4

The single-bond and triple-bond energies for nitrogen and phosphorus are recorded in Table 6.2. Note the extremely high value for the nitrogen-nitrogen triple bond. This is the highest bond energy known. The comparison between N and P molecule formation is given by Equations (6.12) through (6.15).

TABLE 6.2

Nitrogen-Nitrogen and Phosphorus-Phosphorus Bond Energies in kcal/Mole of Bonds

N—N	39
N≡N	226
P—P	48
P≡P	117

$$4N = 2N_2 + 452 \text{ kcal} \tag{6.12}$$

$$4N = N_4 + 234 \text{ kcal} \tag{6.13}$$

$$4P = 2P_2 + 234 \text{ kcal} \tag{6.14}$$

$$4P = P_4 + 288 \text{ kcal} \tag{6.15}$$

Clearly for nitrogen, the triply-bonded diatomic molecule is very highly favored over a hypothetical tetratomic molecule with the four nitrogen atoms at the vertices of a tetrahedron. Clearly the tetratomic phosphorus molecule is favored over the diatomic molecule, although these two possible forms of phosphorus are close in potential energy. Obviously, it is by virtue of the tetrahedral arrangement permitting six bonds among four atoms that the tetratomic phosphorus molecule is favored. It should not be surprising that P_2 is found under some conditions.

COMBINATIONS OF ATOMS WITH HIGH BUT DIFFERENT ELECTRONEGATIVITIES: POLAR COVALENT BONDS

The previous discussion of the covalent bond has been limited to bonds between identical atoms. Under these circumstances, the formation of molecular orbitals can be envisioned from the interactions of identical orbitals on the atoms involved. In this section, the concept of the covalent bond is extended to include bonding between unlike atoms. Here it is necessary to envision the formation of molecular orbitals through the interaction of unlike orbitals.

However, the situation is not as complex as it might appear at first sight. In Chapter 5 it was pointed out that the electrons of highest potential energy in nearly all atoms are the *ns* and *np* electrons. By the same token, these are the electrons that are normally on the periphery of the atoms and are involved in bonding. In fact, the *ns* and *np* electrons of the atoms often are called the *valence electrons.*

Thus the main interactions involved in bonding are between two *ns* orbitals, between two *np* orbitals, or between an *ns* and an *np* orbital. Interactions between two *ns* orbitals and between two *np* orbitals have been discussed in situations where the orbitals are identical. The formation of molecular orbitals from different *ns* and *np* orbitals is considered to be the same in principle with some differences in details. Thus the interaction between one *ns* orbital on one atom with another *ns* orbital on a different kind of atom leads to a pair of sigma orbitals, bonding and antibonding. The interaction between one *np* orbital on one atom with another *np* orbital on another type of atom leads to either a pair of sigma molecular orbitals or a pair of pi molecular orbitals, depending upon the spatial orientations of their charge distributions.

Due to the nondirectional character of the spherical *s* type charge distribution, the interaction of an *ns* orbital on one atom with an *np* orbital on another atom also leads to the formation of a pair of sigma molecular orbitals. This is shown in Fig. 6.14.

To illustrate some of the characteristics of bonding between unlike atoms of

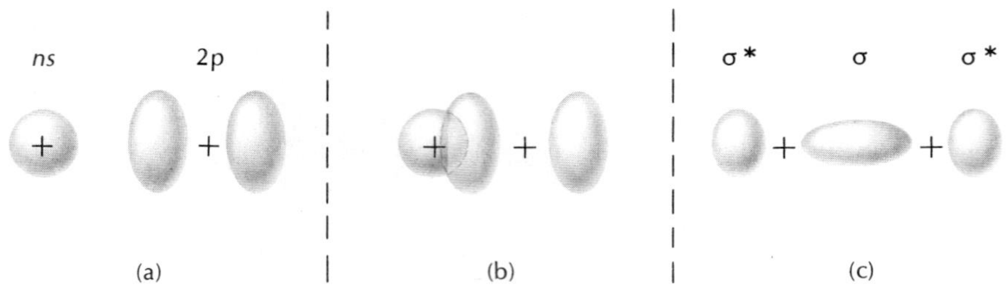

Fig. 6.14. Representation of the formation of sigma orbitals from the overlap of *ns* and 2*p* orbitals

relatively high electronegativity, we shall look at some of the binary compounds of hydrogen with some of the highly electronegative elements.

The Hydrogen Halides

The electronic configuration for the fluorine atom is $1s^22s^22p^5$. This permits one atom of hydrogen to combine with one atom of fluorine through the interaction of the half-filled $1s$ orbital of hydrogen and the one half-filled $2p$ orbital of fluorine. Thus one bonding sigma molecular orbital permits the localization of one pair of electrons between the nuclei of the two atoms. The molecule formed has the formula HF, which may also be represented as H—F, H—$\ddot{\underset{..}{F}}$:, or H—$\overline{\underline{F}}$|. This compound is called hydrogen fluoride. In a similar way, hydrogen forms compounds with the other halogens with the formulas HCl, HBr, and HI.

Bond polarity. Since the bond in the HF molecule consists of a pair of electrons shared between two atoms, it is a covalent bond. However, since the F atom is much more electronegative than the H atom, the shared electrons are attracted more strongly to the F atom than they are to the H atom. Thus the probability of finding this pair of electrons close to the F nucleus is greater than the probability of finding this pair of electrons close to the H nucleus. This suggests that the charge distribution of the sigma bonding orbital resembles that shown in Fig. 6.15.

Fig. 6.15. Representation of the electron charge distribution in HF

If the F atom attracted electrons strongly enough, it would remove the electron entirely from the hydrogen, and the compound would be ionic and would consist of positively charged hydrogen ions and negatively charged fluoride ions. This is not so, but we can imagine that the transfer of electrons has progressed part way, so that the fluorine has partially removed the electron from the hydrogen. In this way it is possible to think of the covalent bond between hydrogen and fluorine possessing *partial ionic character*.

The partial ionic character of this covalent bond can be described in terms of the separation of the centers of positive and negative charge associated with the bonded atoms, due to the unequal attraction for electrons exhibited by the bonded atoms. In Chapter 5 we discussed induced electrostatic dipoles, where the centers of positive and negative charge were separated due to an external electrostatic field. However, this is an example of a *permanent dipole* which results from the electronegativity differences in the covalently bonded atoms. Since this separation of the charge centers leads to the formation of two separate electrostatic poles, this type of covalent bond is called a *polar covalent bond.*

Often the separation of charge centers is indicated in structural formulas by the symbols $\delta+$ and $\delta-$ used as superscripts to the appropriate atomic symbols. Thus hydrogen fluoride may be represented as $H^{\delta+}—F^{\delta-}$. The deltas suggest a *partial* ionic charge on the H and F atoms, but more accurately, they suggest that the center of negative charge has been displaced toward the F atom.

It is a general rule that bonds between atoms of unequal electronegativity are polar; that is, the centers of positive and negative charge do not coincide. The greater the electronegativity difference, the greater the charge separation; that is, the greater the polarity of the bond.

Since the electronegativity of hydrogen is constant and the electronegativities of the halogens decrease with increasing atomic number, there is a decrease in bond polarity from HF to HCl to HBr to HI. This is the same as saying that the effective separation of charge centers is less in HI than in HF, for example. Or, from a different point of view, the sizes of the partial ionic charges decrease from HF to HI. The bond in the HF molecule is quite polar, whereas the bond in the HI molecule is hardly polar at all.

Since the separation of charge centers in the hydrogen fluoride molecule means that the molecule has a permanent electrostatic dipole, it should exhibit a measurable electrostatic dipole moment. In fact, if the bonds decrease in polarity from HF to HCl to HBr to HI, there should be a noticeable decrease in measured electrostatic dipole moments in that order. That this is so is shown very clearly in Table 6.3.

TABLE 6.3

Electrostatic Dipole Moments for the Hydrogen Halides

HF	6.36×10^{-30} coulomb × angstrom
HCl	3.43×10^{-30} coulomb × angstrom
HBr	2.60×10^{-30} coulomb × angstrom
HI	1.27×10^{-30} coulomb × angstrom

If the bonds were not polar, the electrostatic dipole moments would be zero. Those covalent bonds in which there is no separation of charge centers are called *nonpolar covalent bonds.*

OXIDATION NUMBERS. The partial ionic character of polar covalent bonds suggests an extension of the oxidation number concept. Recall that the oxidation number of an element in an ionic compound is the ionic charge on the ion of that element. Thus in cesium fluoride, the oxidation number for cesium is 1+ and the oxidation number for fluorine is 1−.

In extending the concept of oxidation number to atoms joined by polar covalent bonds, all shared electrons are considered to be associated with the more electronegative atom. For example, in hydrogen fluoride, since the fluorine is the more electronegative, the bonding electrons are displaced toward the fluorine, and consequently they are assigned to the fluorine for the purpose of oxidation number assignment. Since this requires ten electrons to be associated with the fluorine atom, which has an atomic nucleus containing only nine protons, the oxidation number assigned to fluorine is 1−. However, by this convention, hydrogen has no electrons assigned to it. This is one less than the number of electrons in the neutral atom and the assigned oxidation number for hydrogen is 1+. This is *not* to suggest that hydrogen fluoride consists of H^+ and F^- ions, but only that it is convenient to assign these oxidation numbers to the atoms in the molecule. Since molecules are electrostatically neutral, it should be obvious that the sum of the oxidation numbers of all atoms in the molecule must be zero.

By considering atomic electronic structures, often it is possible to determine, in simple cases, the formulas of compounds formed between different elements. For example, it is clear that the localization of one pair of electrons between a H atom and a F atom exhausts the bonding capabilities of the *ns* and *np* orbitals of hydrogen and fluorine. A similar argument may be given for hydrogen with any other halogen. Thus if X represents any of the halogens, it is expected that the generalized formula for the compounds formed between hydrogen and the halogens can be written as HX, which stands for HF, HCl, HBr, HI, and HAt.

Since all of the halogens are more electronegative than hydrogen, it seems reasonable to assign the halogens oxidation states of 1− in combination with hydrogen, which has an oxidation state of 1+. Of course, the oxidation numbers to be expected can be surmised as follows: The electronegative halogen atoms have just one unpaired electron each and so can accept only one more to form a localized pair. Thus they should have an oxidation number of 1− each. The hydrogen, which is less electronegative, has only one electron per atom to lose in the formation of a localized pair. Therefore the oxidation number for each hydrogen should be 1+. Thus to balance the oxidation states of hydrogen and a halogen in a compound requires equal numbers of atoms of each and the formula should be of the type, HX. This is an alternative way to look at the atomic configurations to decide what the formulas of simple compounds should be.

With the assignment of oxidation numbers, the binary nomenclature system described for the binary ionic compounds may be extended to the binary covalent compounds. The element with the positive oxidation number is named first, followed by the element with the negative oxidation number. The element with the negative oxidation number is named by adding the suffix *-ide* to the root of the name of the element. Thus HF is hydrogen fluoride; HCl is hydrogen chloride; HBr is hydrogen bromide; and HI is hydrogen iodide.

For some of the simple binary compounds of hydrogen, an alternative nomenclature system often is used. This is related to a property of these compounds that will be discussed in detail later. These compounds are called acids. To name them as acids, the hydrogen is designated by the prefix *hydro-*, which is added to the root of the name of the element with the negative oxidation number, which, in turn, has the suffix *-ic* added to it. Thus HF is hydrofluoric acid; HCl is hydrochloric acid; HBr is hydrobromic acid; and HI is hydroiodic acid which often is shortened to hydriodic acid.

The Binary Compounds of Hydrogen with Oxygen and Sulfur

WATER AND HYDROGEN SULFIDE. The O atom has the electronic configuration $1s^22s^22p^4$ and the S atom has the electronic configuration $1s^22s^22p^63s^23p^4$. Each of the atoms has a higher electronegativity than hydrogen. Each of the two half-filled *np* orbitals of the O or S atom may interact with the *ns* orbitals of separate H atoms to form two sets of sigma orbitals. Consequently, both the O and S atoms may accept a total of two electrons when a pair of electrons is localized between itself and two H atoms.

Thus the formulas of the two compounds are H_2O and H_2S. Systematically they are named as hydrogen oxide and hydrogen sulfide, or as hydroxic acid and hydrosulfuric acid. However, the common name for H_2O, water, is used almost exclusively.

The structural formula for water usually is written as

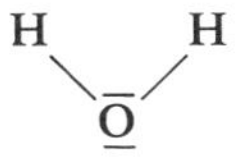

This suggests that the molecule is not linear, that is, the three atoms do not lie in a straight line. Considering the four groups of electrons on the surface of the central oxygen atom, it is to be expected that the bond angle should be near the tetrahedral angle, which is about 109°. Although the actual angle is difficult to measure, the best experimental evidence suggests that the angle is about 104.5°. That it is less than the tetrahedral angle is explained on the assumption that the restricted, localized pairs of electrons in the bonds require less surface area on the oxygen atom than the two nonbonding electron pairs. Consequently, the bonding pairs of electrons may be closer together than otherwise expected, and the bond angle is less than the tetrahedral angle.

The covalent bonds in the water molecule are polar, since oxygen is more electronegative than hydrogen. This suggests that the water molecule has an electrostatic dipole moment. In fact, it does, and the value is 6.13×10^{-30} coulomb $\times$ angstrom. However, the fact that water has an electrostatic dipole moment is good evidence that the molecule is not linear.

To see why this is so, consider a hypothetical linear molecule, H—O—H. Although the bonds are polar, the symmetry of the molecule indicates that the centers of positive and negative charge coincide at the nucleus of the O atom. Such a molecule could not have an electrostatic dipole moment. A molecule of this nature is called a *nonpolar molecule*. Obviously, a distinction must be made between *polar bonds* and *polar molecules*, since not all molecules containing polar bonds are polar molecules.

However, if the water molecule is angular, as indicated by the following diagram,

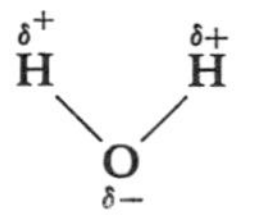

then the center of positive charge is not at the center of the O atom, but is displaced to a point slightly above the center of the O atom as represented in the diagram. Thus the water molecule is polar.

HYDROGEN PEROXIDE AND THE HYDROGEN POLYSULFIDES. Hydrogen and oxygen may combine in an alternative way. Each O atom still forms a covalent bond with each of two other atoms. However, one of the other atoms is another O atom. This permits the kind of bonding represented by the following structural formula:

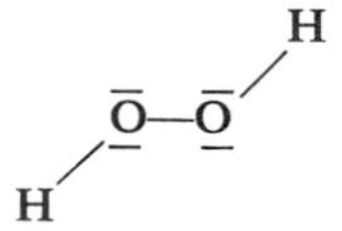

This compound, with the molecular formula H_2O_2, is called hydrogen peroxide.

There are two kinds of bonds in H_2O_2. The O—O bond is a nonpolar covalent bond, whereas the two H—O bonds are polar covalent bonds.

Considering the charge centers associated with individual polar bonds, partial ionic charges may be assigned as follows:

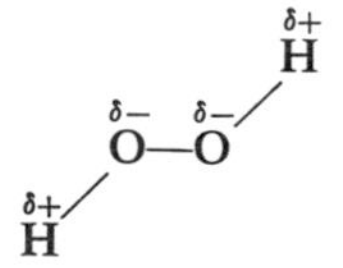

The adjacent partial negative charges close to the O atoms suggest a degree of instability associated with the O—O bond in H_2O_2. In addition, recall the low bond energy for the O—O single bond as compared to the high bond energy for the O=O double bond in O_2. Therefore it is not surprising that H_2O_2 is decomposed easily. If you have ever purchased a hydrogen peroxide solution, you have seen that it comes in a brown bottle. The brown glass filters out wavelengths of light that are energetic enough to break the O—O bond.

Self-bonding is called *catenation*. An earlier discussion of elemental oxygen and sulfur indicated that oxygen tends to form a double bond with itself, whereas sulfur has a greater tendency to form single bonds with itself. This tendency continues among the compounds, where S atoms tend to form long chains bonded together by single nonpolar covalent bonds. Thus the sulfur analog to hydrogen peroxide is H_2S_2, or hydrogen persulfide, more commonly called hydrogen disulfide. However, related compounds containing as many as five S atoms are known. The structural and molecular formulas of these compounds are shown in Table 6.4.

TABLE 6.4
The Hydrogen Polysulfides

H_2S_2	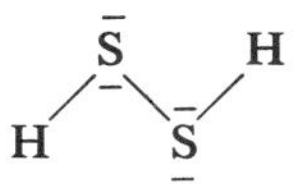	hydrogen disulfide
H_2S_3	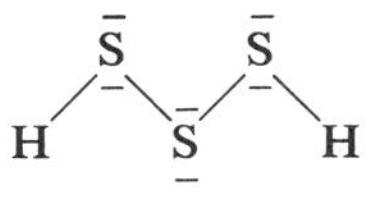	hydrogen trisulfide
H_2S_4	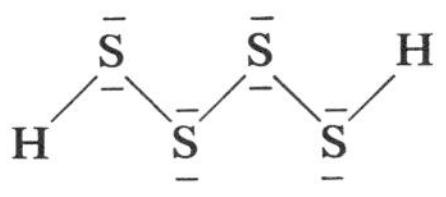	hydrogen tetrasulfide
H_2S_5	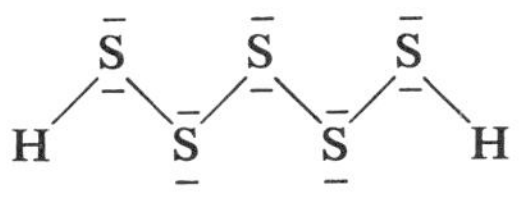	hydrogen pentasulfide

Their names are derived from *hydrogen sulfide* using prefixes to indicate the number of sulfur atoms per molecule. Thus the prefixes *di-*, *tri-*, *tetra-*, and *penta-* denote 2, 3, 4, and 5 atoms respectively. As a group these catenated hydrogen sulfides often are called the *hydrogen polysulfides*.

In order to assign oxidation numbers to the atoms in catenated molecules, several observations can be made. In the molecules just described, hydrogen is bonded to more electronegative elements and thus is assigned an oxidation number of 1+. However, the O—O and S—S bonds are nonpolar, since identical atoms are bonded. Consequently, in hydrogen peroxide, H_2O_2, each of the O atoms is assigned an oxidation number of 1−, indicated as follows:

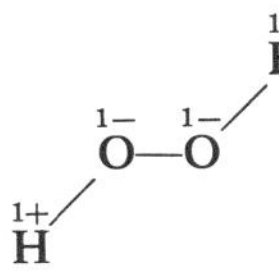

Similarly, in hydrogen disulfide, H_2S_2, each H atom is assigned an oxidation number of 1+ and each S atom is assigned an oxidation number of 1−.

In the molecules of the hydrogen polysulfides, the H atoms are assigned oxidation numbers of 1+ each. Then, each of the S atoms bonded to an H atom is assigned an oxidation number of 1−. Since all of the other S atoms in the molecule are bonded to S atoms only, and the bonds are nonpolar, these S atoms are assigned oxidation numbers of zero. For example, the assignments in hydrogen tetrasulfide, H_2S_4, are

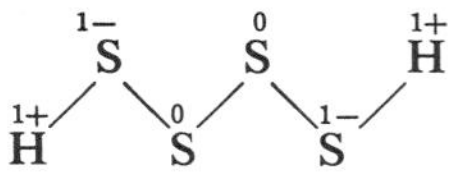

Oxidation number assignments often are made, however, on the basis of the simple molecular formula instead of the structural formula. For example, with H_2O_2 one begins by assigning an oxidation number of 1+ per atom to hydrogen. Then an oxidation number of 1− per atom is assigned to oxygen in order that the sum of the oxidation numbers may equal zero. If this same procedure is followed with H_2S_4, hydrogen is assigned an oxidation number of 1+ per atom as before. However, sulfur is assigned an oxidation number of $\frac{1}{2}-$ per atom so that the sum of the oxidation states equals zero. This is a perfectly good assignment. A fractional oxidation number simply means that the oxidation numbers of several atoms with different oxidation numbers have been averaged.

Two Binary Compounds of Hydrogen with Nitrogen

The electronic configuration of the N atom is $1s^22s^22p^3$. Nitrogen is more electronegative than hydrogen. Each of the three half-filled *np* orbitals of the N atom may interact with the *ns* orbitals of three separate H atoms to form three sets of sigma orbitals. Consequently, the N atom may accept a total of three electrons in localizing three pairs of electrons between itself and three H atoms.

As a result, the formula for one simple compound of nitrogen and hydrogen is H_3N. It is usual practice in writing molecular formulas for binary compounds to write the symbol of the element with the positive oxidation state first. This puts the elemental symbols in the same order in which they are given in the name of the compound. From this point of view, the formula H_3N is correct. However, by custom, it is almost always written as NH_3.

Systematically, the name for this compound is hydrogen nitride or hydronitric acid, but these names are rarely if ever used. The common name, *ammonia*, is used almost exclusively.

The structural formula for ammonia usually is written as

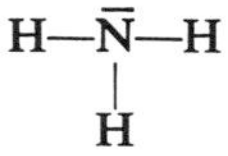

or

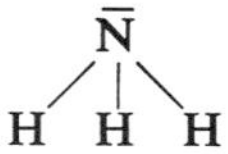

As expected, the four pairs of electrons on the surface of the nitrogen atom are pictured in the tetrahedral arrangement. Each of the bond angles in the ammonia molecule appears to be about 107°, in good accord with the tetrahedral arrangement of electron groups on the surface of the N atom. As depicted, the ammonia molecule has the shape of a trigonal pyramid in which the N atom is at the apex of the pyramid and the three H atoms are at the vertices of the triangular base.

The symmetry of the ammonia molecule and the polarity of the N—H bonds, like those of water, suggest that the molecule is polar. This prediction is substantiated by the experimentally determined value of 4.82×10^{-30} coulomb $\times$ angstrom for the electrostatic dipole moment.

Hydrogen and nitrogen atoms combine in an alternative way to form a binary compound. Each N atom still forms a covalent bond with each of three other atoms. However, one of these other atoms is another N atom. This permits the kind of bonding represented by the following structural formula:

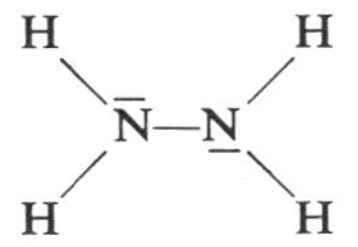

This compound, N_2H_4, is called hydrazine.

Like the hydrogen peroxide molecule, the hydrazine molecule contains two different kinds of bonds. The N—N bond is a nonpolar covalent bond, whereas the four N—H bonds are polar covalent bonds. Again the assignment of partial ionic charges implies adjacent negative charge centers:

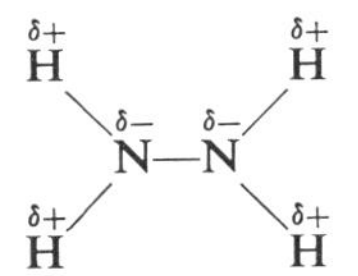

In addition, the very high bond energy for the N≡N triple bond suggests that the triple bond is preferred over the N—N single bond. All of this suggests that hydrazine is a somewhat unstable compound.

In assigning oxidation numbers in NH_3 and N_2H_4, hydrogen is assigned a value of 1+ per atom since it is less electronegative than nitrogen. Then in ammonia, nitrogen has an oxidation number of 3− per atom, and in hydrazine an oxidation number of 2− per atom.

The Hydrocarbons

The electronic configuration of the C atom is $1s^22s^22p^2$. At first glance it would appear that each of the two half-filled $2p$ orbitals could interact with the $1s$ orbitals of separate H atoms to form a simple compound with the formula CH_2. However, this compound does not form.

In fact, one of the simplest binary compounds of carbon and hydrogen has the formula CH_4. This suggests that all four valence electrons of the C atom are involved in bonding. To explain this fact, it has been proposed that in the bonding process an electron is promoted from the 2*s* level to the 2*p* level. Thus in preparation for bonding, the atom assumes an excited state with the electronic configuration $1s^22s^12p^3$. The energy required to promote the electron from the 2*s* to the 2*p* level is more than regained through the bond energy of the two additional bonds that may form as a result.

However, the fact that three of the valence electrons are 2*p* electrons whereas the fourth is a 2*s* electron suggests that three of the four C—H bonds differ from the other two. In fact, no experimental evidence is available to support this suggestion. All four bonds in the CH_4 molecule appear to be identical.

In order to explain this apparent equality among the bonding orbitals, an alternative view of the carbon atom is taken. According to this view, the *ns* orbital and one, two, or three of the *np* orbitals may be *hybridized.* That is, the electrons in these orbitals become equivalent in potential energy and in directional characteristics.

METHANE AND ETHANE. For example, when the *ns* and all three of the *np* orbitals hybridize, the four electrons appear in four sp^3 hybrid orbitals. The notation indicates that the hybrid orbitals have been formed from one *ns* and three *np* atomic orbitals. Each of the four sp^3 hybrid orbitals has a charge distribution similar to that shown in Fig. 6.16. These hybrid orbitals have charge distributions that are directed in space so as to minimize electron repulsions. Since there are four of them, they are directed toward the vertices of a tetrahedron.

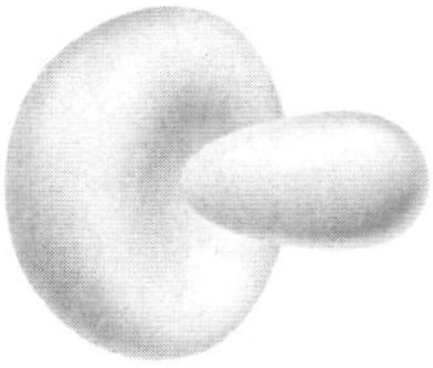

Fig. 6.16. Charge distribution for an sp^3 hybrid orbital

Each of these hybrid orbitals may interact with the *ns* orbital of an H atom to form four sets of sigma orbitals, thus accounting for four H atoms bonded by single bonds to one C atom. The formula for this compound, *methane*, is CH_4.

The molecule of methane is tetrahedral, with each of the four H atoms at one of the vertices of the tetrahedron, and the C atom at the center as shown in Fig. 6.17. Very commonly, the structural formula for methane is written as

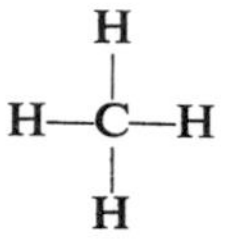

where the tetrahedral structure is understood. The bond angles in methane are all tetrahedral angles of 109°28′.

Because the electronegativities of carbon and hydrogen do not differ by very much, the bonds in methane are not very polar. However, despite any polarity, there is no separation of charge centers, since both the center for positive charge

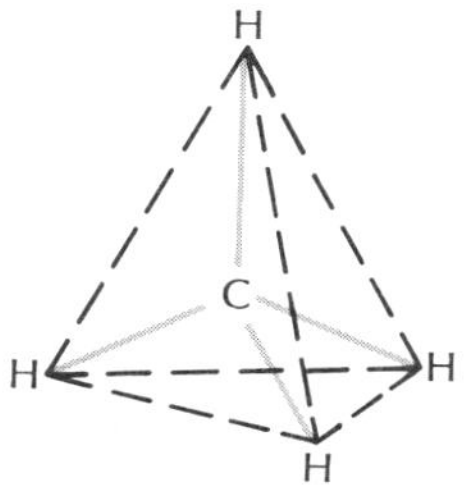

Fig. 6.17. Tetrahedral structure of the methane molecule

and the center for negative charge are in the nucleus of the central C atom. This is in accord with the experimental value of zero for the electrostatic dipole moment of methane.

Oxidation numbers are not often assigned to the atoms in the binary compounds of carbon and hydrogen. However, such assignments may be made as before. Since carbon is slightly more electronegative than hydrogen, each H atom is assigned an oxidation number of 1+. Thus in CH_4, the carbon has an oxidation number of 4−.

Carbon and hydrogen may combine in an alternative way. Each C atom still forms four covalent bonds, but one of these bonds is to another C atom, leading to a structure that may be represented as

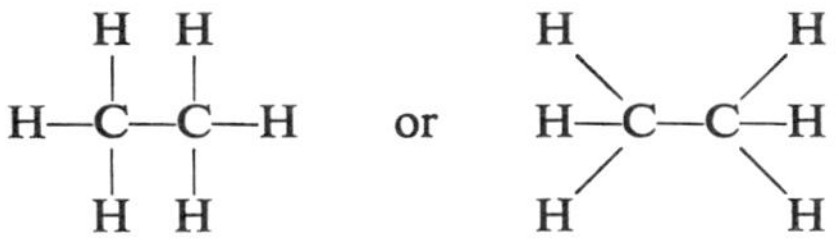

This compound has the formula C_2H_6 and is called *ethane*.

In the ethane molecule, the four electron pairs on the surface of each C atom are arranged tetrahedrally so that all bond angles are very close to the tetrahedral angle. Certainly the C—C single bond is nonpolar and the six C—H bonds are not very polar. The electrostatic dipole moment for ethane is zero.

If oxidation numbers are assigned to the atoms in the ethane molecule, hydrogen is assigned a value of 1+ per atom, which requires the assignment of a value of 3− per atom for carbon.

Of course, in theory, there is nothing to prevent further catenation among C atoms to form long chains. Likewise, in practice, the number of binary compounds of carbon and hydrogen is extremely large. As a group, the binary compounds of carbon and hydrogen are called the *hydrocarbons*. Hydrocarbons involving only single bonds between C atoms are called *alkanes*. Ethane is an example of an alkane.

Several factors contribute to the ability of carbon to catenate extensively. Among the alkanes, for example, the low polarity of the C—H bonds and the symmetrical geometry of the molecules, which accounts for the zero electrostatic dipole moments, also means that there are no adjacent negative charge centers like those in hydrogen peroxide and hydrazine. Thus the bond energy for the C—C bond is much higher than the bond energy for either the N—N or O—O bond.

In addition, with four orbitals available for bonding, it is possible to have double and triple bonds between C atoms without exhausting the bonding capacities of the atoms. The following are some examples of compounds in which double and triple bonds between C atoms are found.

ETHYLENE. When only two of the *np* orbitals of carbon hybridize with the *ns* orbitals, the three resulting sp^2 hybrid orbitals are directed toward the vertices of an equilateral triangle consistent with maximum separation of the electrons in these orbitals. The unhybridized *np* orbital lies perpendicular to the plane of the three sp^2 hybrid orbitals as shown in Fig. 6.18a. As a result, when one of the sp^2

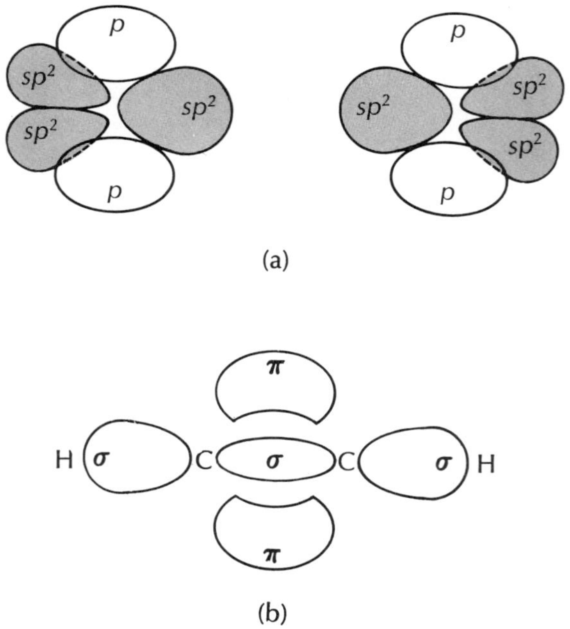

Fig. 6.18. Representation of sp^2 hybrid orbitals and the molecular structure for ethene

hybrid orbitals of one C atom interacts with one of the sp^2 hybrid orbitals of another C atom, the two unhybridized *np* orbitals of the two atoms are in position to interact to form a pair of pi orbitals. The bonding orbitals are shown in Fig. 6.18. The remaining two sp^2 hybrid orbitals on each carbon interact with the *ns* orbitals of H atoms to form pairs of sigma orbitals.

The compound thus formed is called *ethylene* and has the molecular formula C_2H_4. Corresponding to this is the following structural formula:

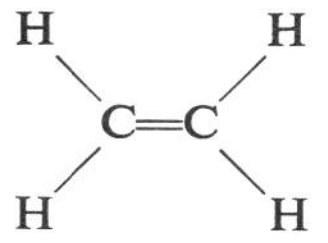

All six atoms in the ethylene molecule lie in the same plane. The H—C—H bond angles are 120°, which is consistent with the maximum separation of three electron groups on the surface of each C atom. One of these groups is the two pairs of electrons localized between the two C atoms, and the other two are pairs of electrons localized between the C atom and the two H atoms. Three groups of electrons on the surface of a sphere are at maximum separation when they are in the plane of a great circle on the sphere and at the vertices of an equilateral triangle inscribed in that circle, as shown in Fig. 6.19.

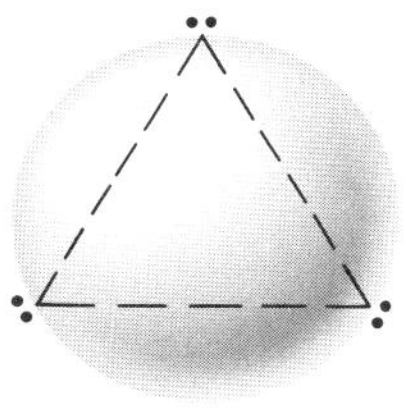

Fig. 6.19. Trigonal arrangement of three electron groups on the surface of a spherical atom

Of course, additional catenation is possible, with C atoms occupying the positions of the H atoms in the ethylene molecule. Ethylene is the first member of a large family of hydrocarbons in which there are carbon-to-carbon double bonds. This family is called the *alkenes*.

BENZENE. An interesting compound, called *benzene*, can be visualized as the result of the catenation of ethylene-type groups of atoms. The 120° bond angles in ethylene permit the combination of three such groups into a hexagonal ring of six C atoms, with all of the atoms in the same plane. Looking down on the plane, the structure appears somewhat as follows:

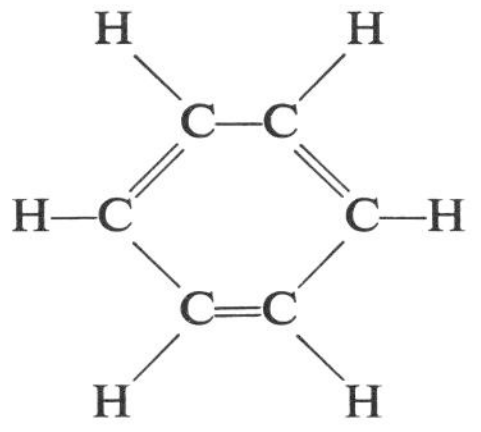

In this structure, as it is represented, there are three pairs of doubly bonded C atoms bonded to each other by single bonds. This results in alternating double and single bonds around this ring of six C atoms. As represented, each C atom has three groups of electrons on its surface, and all bond angles are 120°

However, if we change our perspective on the molecule and look at it more from the side than the top, we can begin to appreciate some of the possibilities inherent in this structure. In Fig. 6.20a, the benzene ring is shown with the carbon skeleton and the sigma bonding orbitals represented as lines between carbons. To avoid clutter in the diagram, the hydrogens and the covalent bonds between carbons and hydrogens have been omitted. In addition, the unhybridized $2p$ orbital charge distributions are shown prior to pi bonding orbital formation.

In Fig. 6.20b, the result of pi bonding orbital formation is shown between adjacent C atoms. The C atoms are numbered for convenience. The pi bonding orbital charge distributions are shown between atoms 1 and 2, 3 and 4, and 5 and 6. However, looking at Fig. 6.20a, one might wonder why this choice was made. As an alternative, the pi bonding orbital charge distributions might have been located between C atoms 2 and 3, 4 and 5, and 6 and 1, as shown in Fig. 6.20c. In fact, there is no way to distinguish between these two possibilities. The electron in the unhybridized *np* orbital of C atom 1 has as much chance of being localized between atoms 1 and 2 as it does of being localized between atoms 1 and 6. Similarly, the electron in the unhybridized *np* orbital of C atom 2 has equal probability of being localized between atoms 2 and 1 as it does between atoms 2 and 3. This argument may be continued for the other atoms in the ring.

As a result, it is not possible to view these electrons from the original unhybridized $2p$ orbitals as localized between any two C atoms. Rather they are *delocalized*, or in other terms, they are shared by more than two atoms. This delocalization is somewhat more apparent if an attempt is made to show the charge

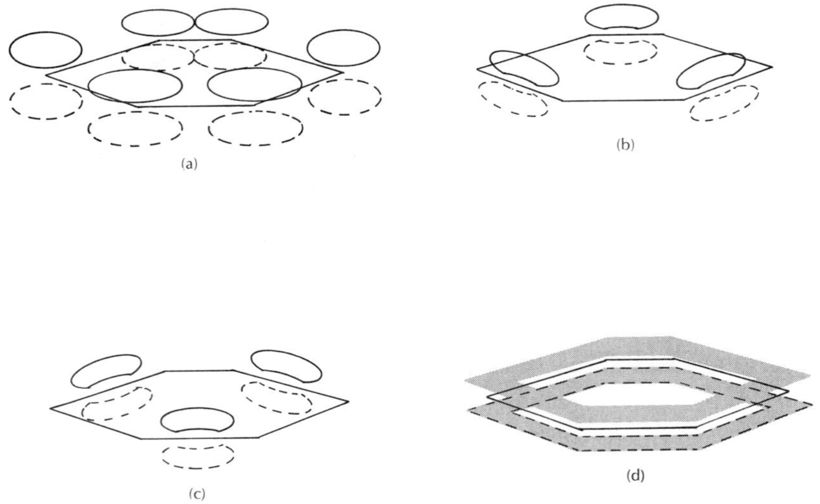

Fig. 6.20. Representation of orbital distribution in the benzene molecule

distribution associated with the pi bonding orbitals in the benzene ring. The result is two continuous hexagonal bands, on either side of the ring, as shown in Fig. 6.20d. This represents the probable charge distribution associated with the six electrons that occupy the pi bonding orbitals in benzene.

This delocalized system of pi electrons is characteristic of another large family of hydrocarbons usually called the *aromatic hydrocarbons*.

The structural formula for benzene often is written as

 or 

These representations omit all elemental symbols and do not show the hydrogens. They correspond to the alternating double and single bond picture of Figs. 6-20b and 6.20c. When these structural formulas are used, it must be remembered that the double bonds do not represent double bonds in the usual localized sense, but that they represent a delocalized system of six pi electrons.

An alternative structural formula for benzene is

In this formula the delocalized pi electrons are represented by a dotted circle inside the hexagon. Of course, the sides of the hexagon do represent localized pairs of sigma electrons. Again, the hydrogens and their bonds to carbon are omitted in this representation.

ACETYLENE. When only one of the *np* orbitals of carbon hybridizes with the *ns* orbital, a hybrid *sp* orbital of one atom may interact with a hybrid *sp* orbital of another C atom to form a pair of sigma orbitals. If the electrons in the *sp* hybrid orbitals minimize their repulsions, the two *sp* orbital charge distributions lie in a line with the atomic nucleus. This line is mutually perpendicular to both of the unhybridized *np* orbitals, as shown in Fig. 6.21. As a result, when one hybrid *sp* orbital of one atom interacts with one hybrid *sp* orbital of another atom, the four unhybridized *np* orbitals of the two C atoms are in position to form two sets of pi orbitals. Thus a carbon-to-carbon triple bond results, as suggested by Fig. 6.21. The remaining *sp* hybrid orbital of each carbon interacts with the *ns* orbital of a hydrogen atom to form a pair of sigma orbitals.

The compound in this case has the molecular formula C_2H_2, which is represented by the following structural formula:

$$H{-}C{\equiv}C{-}H$$

This compound is called *acetylene* and is the first member of a family of catenated hydrocarbons that have triple carbon-to-carbon bonds in their molecules. The members of this family are called the *alkynes*.

All four atoms in the acetylene molecule lie in the same straight line, which is consistent with the maximum separation of two groups of electrons on the surface

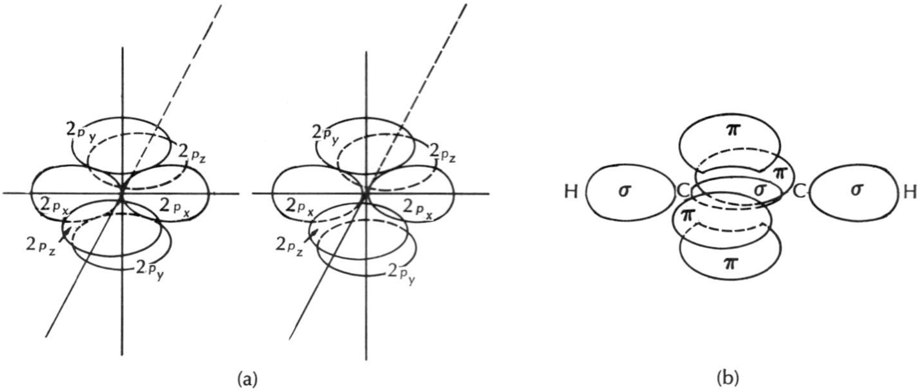

Fig. 6.21. Representation of *sp* hybrid orbitals and the molecular structure of acetylene

of each carbon atom. One of these groups consists of the six pairs of electrons localized between the two C atoms and the other is the pair of electrons localized between the C and H atoms.

Oxidation Numbers

So far, oxidation numbers have been assigned on an electron-transfer rationale. For ionic compounds, the electron transfer is complete and the oxidation number clearly is identical to the ionic charge. In molecules composed of identical atoms, there is no electron transfer and all atoms are assigned oxidation numbers of zero. However, where polar covalent bonds are involved, all bonding electrons are assigned to the more electronegative atom in assigning oxidation numbers.

Actually, oxidation-number assignment is a convenient bookkeeping system for electrons in molecules. When dealing with complex compounds, it is more convenient to adopt some reasonable set of bookkeeping rules, avoiding the necessity to determine relative electronegativities. Some of the fundamental rules follow, with parenthetical comments on the rationale behind them.

1. The sum of the oxidation numbers of the atoms in a molecule or ion equals the net quantum electrostatic charge on the molecule or ion. Thus for a neutral molecule the sum of the oxidation numbers of all of the atoms is zero.
2. All elements in the elementary state are assigned oxidation number zero for each atom. (There is no electron transfer between identical atoms.)
3. In compounds, alkali metal atoms (family 1A) always are assigned oxidation

number 1+. (These elements with low electronegativity have only one electron in the *ns* level that may be involved in bonding.)

4. In compounds, alkaline earth atoms (family 2A) always are assigned an oxidation number of 2+. (These elements with low electronegativity have only two *ns* electrons that may be involved in bonding.)
5. In compounds, fluorine always is assigned an oxidation number of 1−. (Since this is the most electronegative atom it should have a negative oxidation number. There is only one vacancy in the *np* level.)
6. In *binary* compounds, chlorine, bromine, and iodine normally are assigned oxidation numbers of 1− per atom, except when they are combined with other halogen atoms. (These elements are highly electronegative as a family and each atom is able to accept one electron in the *np* energy level.)
7. In compounds, oxygen is assigned an oxidation number of 2− per atom except when it is bonded to itself or to fluorine. (Oxygen is the second most electronegative element and each atom has two half-filled *np* orbitals to use in bonding.)
8. In compounds, hydrogen is assigned an oxidation number of 1+ per atom except when it is combined with the alkali metals or the alkaline earths. (Except for the elements mentioned, hydrogen usually bonds to elements of higher electronegativity, and each atom has only one electron that may be involved in bonding.)

With these few rules it is possible to assign oxidation numbers to most elements in most compounds. (Unless a statement is made to the contrary, it is the convention to indicate the oxidation state per atom without expressing the "per atom.") For example, consider the compound with the formula $K_2Cr_2O_7$. From rule 3, potassium is assigned the oxidation number of 1+; from rule 7, oxygen is assigned the oxidation number of 2−. From rule 1, it is possible to calculate the oxidation number for chromium. The sum of the oxidation numbers of potassium and oxygen is $(2 \times 1+) + (7 \times 2-) = 12-$. Therefore, the sum of the oxidation numbers of the two chromiums must be 12+ and the oxidation number of chromium is 6+.

THE COMMON OXIDATION NUMBERS. Some of the more commonly encountered oxidation numbers of the elements are listed in Table 6.5. It should be noted that not all known oxidation states of the elements are listed here, but only the more common ones. Among these, the less common ones are given in parentheses. Scrutiny of this table reveals several notable trends.

First, it is interesting that of the 103 elements listed, only 19 appear to exhibit negative oxidation numbers. Of these only 12 exhibit negative oxidation numbers commonly. Since the sum of the oxidation numbers of all atoms in a molecule must be zero, it follows that some atom must be assigned an oxidation number that is negative in nearly all compounds. For this reason, it follows that there is a very high probability that at least one of the following elements will be found in practically any chosen compound: H, F, O, N, C, Si, P, S, Cl, Br, Se, or I.

Among the transition elements there is a wide variety of oxidation states, all of which are positive. For these elements, the variety of oxidation numbers is due to the inclusion of $(n - 1)d$ electrons among the valence electrons, where they do not differ markedly in energy from the *ns* and *np* electrons. Thus the greatest

TABLE 6.5

The Common Oxidation Numbers of the Elements

H 1− 1+																**H** 1− 1+	**He**
Li 1+	**Be** 2+											**B** (3−) 3+	**C** 4− 4+	**N** 3− 1+ 2+ 3+ 4+ 5+	**O** 2− 1−	**F** 1−	**Ne**
Na 1+	**Mg** 2+											**Al** 3+	**Si** 4− 4+	**P** 3− 3+ 5+	**S** 2− 2+ 4+ 6+	**Cl** 1− 1+ 3+ 4+ 5+ 7+	**Ar**
K 1+	**Ca** 2+	**Sc** 3+	**Ti** 2+ 3+ 4+	**V** 2+ 3+ 4+ 5+	**Cr** 2+ 3+ (5+) 6+	**Mn** 2+ 3+ 4+ (6+) 7+	**Fe** 2+ 3+ (6+)	**Co** 2+ 3+	**Ni** 2+	**Cu** 1+ 2+	**Zn** 2+	**Ga** (1+) 3+	**Ge** (4−) (2+) 4+	**As** (3−) 3+ 5+	**Se** 2− 2+ 4+ 6+	**Br** 1− 1+ 5+	**Kr** (1+)

Rb 1+	Sr 2+	Y 3+	Zr 4+	Nb (3+) 5+	Mo 3+ 4+ 5+ 6+	Tc 7+	Ru 2+ 3+ 4+ 6+ (8+)	Rh 3+	Pd 2+ (3+) (4+)	Ag 1+ (2+)	Cd 2+	In (1+) 3+	Sn 2+ 4+	Sb (3−) 3+ 5+	Te (2−) 2+ 4+ 6+	I 1− 1+ 4+ 5+ 7+	Xe (2+) (4+) (6+)
Cs 1+	Ba 2+	La 3+	Hf 4+	Ta (3+) 5+	W (4+) (5+) 6+	Re (3+) 4+ (6+)	Os (2+) (3+) 4+	Ir 3+ (4+) (6+)	Pt 2+ 4+ (6+)	Au 1+ 3+	Hg 1+ 2+	Tl 1+ 3+	Pb 2+ 4+	Bi 3+ (5+)	Po (2−) 2+ 4+	At (1−)	Rn
Fr 1+	Ra 2+	Ac 3+															

Ce 3+ 4+	Pr 3+ 4+	Nd 3+	Pm 3+	Sm 2+ 3+	Eu 2+ 3+	Gd 3+	Tb 3+	Dy 3+	Ho 3+	Er 3+	Tm 3+	Yb 2+ 3+	Lu 3+
Th 4+	Pa 4+ 5+	U 3+ 4+ 5+ 6+	Np 3+ 4+ 5+ 6+	Pu 3+ 4+ 5+ 6+	Am 3+ (4+) 5+ 6+	Cm 3+ (4+)	Bk 3+ 4+	Cf 3+	Es	Fm	Md	No	Lw

variety in oxidation numbers displayed by individual elements occurs near the middle of the transition series, where the atoms have maximum numbers of unpaired $(n-1)d$ electrons. Among the transition elements, oxidation numbers of 2+ and 3+ are very common and the maximum positive oxidation state often reaches, but does not exceed, the family number in which the element appears in the periodic table. Within any family of transition elements, there is a tendency toward increasing stabilization of higher oxidation numbers as atomic number increases.

Among the representative elements of high electronegativity, there is great variety in oxidation numbers due to the rather large number of *ns* and *np* electrons available for bonding. Again it is true that the positive oxidation numbers normally do not exceed the family number, although the rare gases (family 0) are an exception.

Nomenclature

Compounds are named systematically by the application of a number of rules related to assigned oxidation numbers. It has been indicated already that simple binary compounds are named by the element with the positive oxidation state, followed by the root of the name of the element with the negative oxidation state, to which the suffix *-ide* has been added.

From Table 6.5 it is obvious that many elements have more than one oxidation number, and therefore may form several compounds with another element. We have already encountered H_2O and H_2O_2, both of which are binary compounds of hydrogen and oxygen. Obviously, both of them should not have the same name. From scrutiny of Table 6.5, it is clear that there are relatively few instances in which a single element exhibits two different negative oxidation numbers. Oxygen is one and H_2O and H_2O_2 are examples. In these few cases it is not difficult to remember some variant of the name that will distinguish the two, such as hydrogen oxide and hydrogen peroxide.

However, there are many examples of elements that exhibit more than one positive oxidation number. For example, iron and oxygen form two different compounds with the formulas FeO and Fe_2O_3. Both of these are iron oxides. Nitrogen and oxygen form six different compounds with the formulas N_2O, NO, N_2O_3, NO_2, N_2O_4, and N_2O_5. All of these compounds are nitrogen oxides.

Historically, three different nomenclature systems have evolved for use in distinguishing compounds of these types. As a result, many compounds have at least three different names, depending upon the system employed. We shall look at these three systems as they apply to the nomenclature of binary compounds. However, it should be kept in mind that these rules may apply also to more complex molecules.

SUFFIX SYSTEM. The earliest system used to distinguish positive oxidation states employs the suffixes *-ous* for the lower of two positive oxidation numbers and *-ic* for the higher of two positive oxidation numbers. These suffixes are added to the roots of the names of the elements exhibiting positive oxidation numbers. The Latin roots are used for those elements whose symbols are derived from the Latin names, except for mercury, where we are spared the tongue-twisters hydrargyrous and hydrargyric.

Thus FeO and Fe_2O_3 are named ferrous oxide and ferric oxide respectively. (Failure to use the Latin root in this case leads to some "ironic" compounds.) Generally, it is agreed that these suffixes need not be applied to elements that do not show multiple positive oxidation numbers, such as the elements in periodic families 1A, 2A, and 3A, and individual elements such as hydrogen, aluminum, zinc, and cadmium.

Table 6.6 lists some of the elements commonly named by this system, with the oxidation states and their corresponding names.

TABLE 6.6 Elements Commonly Named by the Suffix System

ELEMENT	SYMBOL	OX. NO.	NAME
Nitrogen	N	1+	nitrous
		2+	nitric
Phosphorus	P	3+	phosphorous
		5+	phosphoric
Sulfur	S	4+	sulfurous
		6+	sulfuric
Titanium	Ti	3+	titanous
		4+	titanic
Chromium	Cr	2+	chromous
		3+	chromic
Manganese	Mn	2+	manganous
		3+	manganic
Iron	Fe	2+	ferrous
		3+	ferric
Cobalt	Co	2+	cobaltous
		3+	cobaltic
Copper	Cu	1+	cuprous
		2+	cupric
Arsenic	As	3+	arsenous
		5+	arsenic
Selenium	Se	4+	selenous
		6+	selenic
Tin	Sn	2+	stannous
		4+	stannic
Gold	Au	1+	aurous
		3+	auric
Mercury	Hg	1+	mercurous
		2+	mercuric
Thallium	Tl	1+	thallous
		3+	thallic
Lead	Pb	2+	plumbous
		4+	plumbic
Cerium	Ce	3+	cerous
		4+	ceric

This suffix system has several weaknesses. First, it is adequate only when the positive element exhibits no more than two different oxidation numbers. Many elements exhibit more than this. Second, it requires memorization of the oxidation states of all of the elements in order that one may know that plumbous oxide is PbO, and not Pb_2O, Pb_2O_3, or PbO_2, for example.

NUMERICAL PREFIX SYSTEM. In this system, a prefix derived from the Latin or Greek, designating numbers of atoms of each element per molecule, is used. The prefixes through numerical value twelve are given in Table 6.7. For most of the

TABLE 6.7
Commonly Used Numerical Prefixes

NUMBER	GREEK	LATIN
$\frac{1}{2}$	hemi-	semi-
1	[mon(o)-]†	uni-
$1\frac{1}{2}$		sesqui-
2	[di-]	bi-
3	[tri-]	ter-
4	[tetr(a)-]	quadr(i)-, quater-
5	[pent(a)-]	quinqu(e)-
6	[hex(a)-]	sex(i)-
7	[hept(a)-]	sept(i)-
8	[oct(a)-], octo-, octi-	
9	enne(a)-	[non(a)-], novi-
10	[dec(a)-], deci-	
11	hendec(a)-	[undec(a)-]
12	[dodec(a)-]	

† Prefixes in brackets are those most often used in naming compounds.

nomenclature with which we are concerned here, the Greek prefixes are used. For some other purposes the Latin prefixes are preferred. Of course there are exceptions; and the prefixes in square brackets in Table 6.7 are those most often used in the naming of simple binary compounds.

Often the prefix for one, mon(o)- is omitted in naming compounds. It is used only when it clarifies a name. Also it has become customary in many cases to use the prefix only with the element of negative oxidation number.

For example, MnO_2 usually is called manganese dioxide instead of monomanganese dioxide. Similarly B_2O_3 is called boron trioxide instead of diboron trioxide. However, CO is named carbon monoxide to distinguish it from CO_2 which is named carbon dioxide. Some additional examples of compounds commonly named by this system are given in Table 6.8.

Certain difficulties have arisen with this system. First, some of the common usages have become confusing. For example, two of the oxides of chromium, Cr_2O_3 and CrO_3, commonly have been called chromium trioxide. In other cases this nomenclature system is too specific about the number of atoms per molecule. For example, nitrogen in oxidation state +4 forms two closely related oxides, NO_2 and N_2O_4. When it is important to distinguish between these two this system

TABLE 6.8
Some Examples of the Use of Numerical Prefixes

FORMULA	NAME
BrF_3	bromine trifluoride
IF_7	iodine heptafluoride
SO_2	sulfur dioxide
SO_3	sulfur trioxide
SF_6	sulfur hexafluoride
S_2Cl_2	disulfur dichloride
PCl_3	phosphorus trichloride
PCl_5	phosphorus pentachloride
As_2S_5	arsenic pentasulfide
CO	carbon monoxide
CO_2	carbon dioxide
CS_2	carbon disulfide
CCl_4	carbon tetrachloride
BF_3	boron trifluoride
B_2O_3	boron trioxide
Al_2O_3	aluminum trioxide
MnO_2	manganese dioxide
N_2O_4	dinitrogen tetroxide
NO_2	nitrogen dioxide

is used. However, when it is not important to distinguish between them, the system is too specific.

THE STOCK OXIDATION NUMBER SYSTEM. The most modern and generally preferred nomenclature system used for naming a wide variety of compounds is that devised by Alfred Stock. In the Stock system, the positive oxidation number is represented as a roman numeral in parentheses immediately after the name of the element having that oxidation number. Thus the name for FeS is iron(II) sulfide and Fe_2S_3 is iron(III) sulfide. Similarly, both NO_2 and N_2O_4 are named nitrogen(IV) oxide. Likewise, the name of S_2Cl_2 is sulfur(I) chloride.

It should be pointed out again that in this system as well as the others discussed so far, the oxidation state designation is not needed for elements that do not exhibit multiple oxidation states. Thus $CaCl_2$ may be called calcium chloride and need not be specified as calcous chloride, calcic chloride, calcium dichloride, or calcium(II) chloride.

Bond Energy

Bond energy has been introduced as the decrease in potential energy when two atoms come together from infinite separation to form a bond. The concept of bond energy is clear for diatomic molecules, but further explanation is needed for more complex molecules.

Experimentally, the bond energy is determined by measuring the *bond dissociation energy*, which is the energy required to break one specific bond within a molecule. Equations (6.16) and (6.17) show the bond dissociation energies for successive cleavage of the bonds in a mole of water molecules.

$$H_2O + 117.5 \text{ kcal} = H + O\text{—}H \tag{6.16}$$

$$O\text{—}H + 101.5 \text{ kcal} = H + O \tag{6.17}$$

This indicates that 117.5 kcal is required to break the first O—H bond in a mole of water molecules, whereas only 101.5 kcal is required to break the second O—H bond in a mole of water molecules after the first bond has been broken. Although both of these bonds are O—H bonds, they differ in strength when they are broken in succession. This suggests that it is not possible to specify an exact dissociation energy for a particular type of bond. The exact value depends upon the rest of the bonds in the molecule.

However, if both bonds are broken at the same time in a mole of water molecules, 219.0 kcal of energy are required, and on the average, 109.5 kcal of energy are required to break an O—H bond in a mole of water molecules. Many molecules contain O—H bonds, and the bond dissociation energy varies somewhat from molecule to molecule. However, it is found that the variation is relatively small, so the average bond dissociation energy is a good indicator of the approximate strength of the bond. This average bond dissociation energy value is called the *bond energy*, which is an approximate measure of the strength of the bond regardless of its surroundings.

Bond energies are listed for a number of bonds in Tables 6.9 and 6.10. In Table 6.9, the values are arranged by the nature of the atoms involved in the bond, and in Table 6.10 they are arranged in order of decreasing values. Double and triple bond values are included. In both tables the more approximate values are preceded by *ca* and questionable values are followed by (?).

Many factors affect bond energies. Several of these are discussed in the following sections and then a few trends are summarized.

BONDING ORBITALS. To insure maximum orbital interaction and consequently the formation of good bonding molecular orbitals, it is important that the bonding electrons have approximately the same potential energy within their respective atoms, prior to bonding. Figure 4.12 can be used as an approximate guide to the relative electron energies. From this diagram, it is not surprising to see, for example, that the 1*s* electron in a hydrogen atom is at about the same energy as an electron in the following atomic orbitals: 2*s* orbital of carbon, 2*p* orbital of oxygen, 3*s* orbital of phosphorus, 3*p* orbital of chlorine, 4*s* orbital of copper, and 4*p* orbital of selenium. It is a common fallacy to assume that a 5*s* electron, for example, must always be of higher energy than a 1*s* electron. This is true in any individual atom, but it is not true when different atoms are compared.

The fact is that the valence electrons of all atoms do not differ much in energy. Perhaps the best clue to this is the ionization potential data shown in Table 5.5. The first electron normally removed when an atom is ionized is one of the valence electrons, and the ionization potential is a direct measure of the energy of that electron. Therefore, the ionization potential often is a quick and convenient prop-

TABLE 6.9
Bond Energies in kcal/Mole of Bonds

Bonds are arranged in order of increasing atomic number of the atoms bonded. Thus all bonds of hydrogen are listed first. Then the bonds of lithium are listed, except the H—Li bond, which is listed in the first section under At. No. = 1.

At. No. = 1	
H—H	104
H—Li	58
H—B	93?
H—C	99
H—N	93
H—O	111
H—F	135
H—Na	48
H—Si	76
H—P	ca 77
H—S	83
H—Cl	103
H—As	ca 59
H—Se	66?
H—Br	87
H—Te	57?
H—I	71
H—Cs	42
At. No. = 3	
Li—Li	25
At. No. = 4	
Be—Cl	109
Be—Br	89
Be—I	69
At. No. = 5	
B—C	89
B—N	106
B—O	128
B—F	154
B—Cl	109
B—Br	90
At. No. = 6	
C—C	83
C=C	146
C≡C	200
C—N	73
C=N	147
C≡N	213
C—O	86
C=O	177
At. No. = 6	
C—F	116
C—Al	61
C—Si	72
C—P	63
C—S	65
C=S	128
C—Cl	81
C—Zn	40
C—As	48
C=Se	112
C—Se	58
C—Br	68
C—Cd	30
C—Sn	54
C—Sb	47
C—I	51
C—Hg	27
C—Pb	31
C—Bi	31
At. No. = 7	
N—N	39
N=N	100
N≡N	226
N—O	50
N=O	145
N—F	65
N—Cl	46
At. No. = 8	
O—O	35
O=O	119
O—F	45
O—Si	107
O—P	80?
O=S	ca 110
O—Cl	60
O—As	ca 80
O—Br	48
O—Zr	183
O—Sb	ca 80
O—I	48
O=Os	123
At. No. = 9	
F—F	37
F=Si	135
F=P	117
F—S	68
F=As	111
F—Se	68
F—Te	80
F—I	63
F—Ir	67?
F—Hg	66
F—U	123
At. No. = 14	
Si—Si	53
Si—S	70?
Si—Cl	91
Si—Br	74
Si—I	56
At. No. = 15	
P—P	48
P≡P	117
P—Cl	78
P—Br	63
P—I	44
At. No. = 16	
S—S	54
S=S	84
S—Cl	61
S—Br	52
At. No. = 17	
Cl—Cl	58
Cl—Ti	ca 105
Cl—V	ca 92
Cl—Cr	85
Cl—Te	81
Cl—Co	86
Cl—Ni	87
Cl—Ge	81
Cl—As	70
Cl—Se	58
Cl—Zr	116
At. No. = 17	
Cl—In	75
Cl—Sn	76
Cl—Sb	74
Cl—W	74
Cl—Hg	54
Cl—Bi	67
At. No. = 24	
Cr—I	56
At. No. = 26	
Fe—Br	69
Fe—I	54
At. No. = 32	
Ge—Ge	45
Ge—Br	66
Ge—I	51
At. No. = 33	
As—As	35
As—Br	58
As—I	43
At. No. = 34	
Se—Se	41
At. No. = 35	
Br—Br	46
Br—In	63
At. No. = 35	
Br—Sn	65
Br—Hg	44
At. No. = 42	
Mo—I	89
At. No. = 49	
In—I	55
At. No. = 50	
Sn—Sn	39
At. No. = 51	
Sb—Sb	29?
At. No. = 53	
I—I	36
I—Hg	35

TABLE 6.10 *Bond Energies in kcal/Mole of Bonds*

The list is arranged in order of decreasing values.

N≡N	226	V—Cl	ca 92	Be—I	69	Si—Si	53
C≡N	213	Si—Cl	91	Fe—Br	69	S—Br	52
C≡C	200	B—Br	90	C—Br	68	C—I	51
Zr—O	183	Be—Br	89	S—F	68	Ge—I	51
C=O	177	B—C	89	Se—F	68	N—O	50
B—F	154	Mo—I	89	Bi—Cl	67	Na—H	48
C=N	147	H—Br	87	Ir—F	67?	C—As	48
C=C	146	Ni—Cl	87	Hg—F	66	I—O	48
N=O	145	C—O	86	Ge—Br	66	P—P	48
H—F	135	Co—Cl	86	H—Se	66?	Br—O	48
Si=F	135	Cr—Cl	85	C—S	65	C—Sb	47
B—O	128	S=S	84	N—F	65	N—Cl	46
C=S	128	H—S	83	Sn—Br	65	Br—Br	46
Os=O	123	C—C	83	C—P	63	O—F	45
U—F	123	C—Cl	81	I—F	63	Ge—Ge	45
O=O	119	Fe—Cl	81	P—Br	63	P—I	44
P=F	117	Ge—Cl	81	In—Br	63	Hg—Br	44
P≡P	117	Te—F	80	C—Al	61	As—I	43
C—F	116	As—O	ca 80	S—Cl	61	Cs—H	42
Zr—Cl	116	Sb—O	ca 80	Cl—O	60	Se—Se	41
C=Se	112	P—O	80?	H—As	59?	C—Zn	40
H—O	111	P—Cl	78	Li—H	58	N—N	39
As—F	111	H—P	ca 77	C—Se	58	Sn—Sn	39
S=O	ca 110	Si—H	76	Cl—Cl	58	F—F	37
Be—Cl	109	Sn—Cl	76	Se—Cl	58	I—I	36
B—Cl	109	In—Cl	75	As—Br	58	O—O	35
Si—O	107	Si—Br	74	H—Te	57?	As—As	35
B—N	106	Sb—Cl	74	Si—I	56	Hg—I	35
Ti—Cl	ca 105	W—Cl	74	Cr—I	56	C—Pb	31
H—Cl	103	C—N	73	In—I	55	C—Be	31
N=N	100	C—Si	72	C—Sn	54	C—Cd	30
C—H	99	H—I	71	S—S	54	Sb—Sb	29?
N—H	93	As—Cl	70	Hg—Cl	54	C—Hg	27
B—H	93?	Si—S	70?	Fe—I	54	Li—Li	25

erty to note in this matter. Ignoring the rare gases of low atomic number, we find that the ionization potentials for all elements fall into a rather narrow range.

On this basis, bonds between identical atoms should be favored and should have larger bond energies than bonds between atoms in which bonding electrons are of different energies. It is to be expected that bond energies are smaller as the differences in valence-electron energies increase.

At the same time, it is to be expected that even in bonds between identical atoms, bond energies will be larger when the electrons involved in the bonding are in lower energy states. For example, since the $3p$ electrons are of lower energy in chlorine than the $4p$ electrons in bromine, as suggested by the ionization potential in Table 5.5, it is to be expected that the Cl—Cl bond energy is larger than the Br—Br bond energy.

ATOMIC SIZES. The spherical surfaces of the tiny atoms are quite small, and any electrons on these surfaces must be quite close to each other. Thus electrostatic repulsions among electrons on the atomic surfaces tend to be quite high. This is true not only among electrons on the surfaces of individual atoms, but also among electrons on the surfaces of two atoms bonded to each other, since these two atoms are close to each other. Consequently, very small atoms tend to form double and triple bonds to relieve the electronic congestion on the atomic surfaces. Nearly all multiple bonding involves one of the small atoms of carbon, nitrogen, oxygen, or fluorine.

NONBONDING ELECTRON PAIRS. The effect of the number of nonbonding electron pairs left on the surface of the atoms is closely related to atomic sizes. The electron pairs repel each other, and this repulsion becomes significant for the very small atoms. Many molecular phenomena can be explained by assuming that nonbonding electron pairs tend to spread out over the atomic surface more than bonding electron pairs. This is reasonable since the bonding electron pairs are restricted by the mutual attractions of two nuclei so that they are strongly localized. However, the nonbonding electrons are more like electrons in individual atoms, which tend to separate as much as possible to relieve the interelectronic electrostatic repulsions.

This leads to the conclusion that repulsions will be least among bonding electron pairs and greatest among nonbonding electron pairs. In the case of multiple bonds, the entire multiple bond system consisting of two or three pairs of electrons is incorporated into the term *bonding pair*. This helps to explain the tendency of carbon to bond through an excited state. The C atom is able to place all of its valence electrons into bonding pairs through a low-energy excitation. If one of the $2s$ electrons of the C atom was not promoted, we might expect that carbon could form a maximum of two covalent bonds and would have a pair of nonbonding electrons remaining.

The repulsions of nonbonding electron pairs should be smaller on larger atoms, and this effect should become increasingly insignificant as atomic sizes increase.

BOND POLARITY. Covalent bonds are polar when they lead to a separation of the positive and negative charge centers. There is increased attraction between atoms due to the separation of charge centers. This effect is similar to the effect of aggregation in ionic bonding. With this additional decrease in potential energy, polar covalent bonds should have higher bond energies than otherwise equivalent nonpolar covalent bonds.

SOME TRENDS. In bonds between identical atoms (nonpolar bonds), multiple-bond formation is favored for oxygen and nitrogen due to their small size and the electronic requirement that highly repelling nonbonding electron pairs are left on these atoms. Carbon also forms multiple bonds. However, the single-bond energy for the C—C bond is higher than the N—N and O—O single-bond energies, since all valence electrons of the C atoms are used as bonding electrons, which are less repelling than nonbonding electrons. Although the very small fluorine atom has a tendency to form double bonds with other atoms, it has no tendency to do so when catenated, since insufficient low-lying energy levels are available for the multiple bonding. Hydrogen, although small, has no nonbonding electrons and has no capacity for double-bond formation.

The catenated single-bond energies for families 4A, 5A, 6A, and 7A are shown in Table 6.11. The high single-bond energy for the C—C bond compared to the

TABLE 6.11

Some Single Bond Energies in kcal/mole of Bonds

C—C 83	N—N 39	O—O 35	F—F 37
Si—Si 53	P—P 48	S—S 54	Cl—Cl 58
Ge—Ge 45	As—As 35	Se—Se 41	Br—Br 46
Sn—Sn 39	Sb—Sb 29?		I—I 36

single-bond energies for N—N, O—O, and F—F bonds has been mentioned before. However, the N—N, O—O, and F—F bonds have about the same energies. Probably this is a consequence of a balance between opposing factors. In progressing from nitrogen to fluorine, the electrons used for bonding appear in lower and lower energy states as the nuclear charge increases. However, atomic size decreases in the same order and electron repulsions increase as a result. In addition, the electronic structures dictate an increasing number of nonbonding electrons as atomic size decreases.

On the other hand the Cl—Cl bond energy exceeds the S—S bond energy, which in turn exceeds the P—P bond energy. In these cases the atoms are larger and the electron repulsion problem is less acute, so the decrease in potential energy of *np* electrons with increasing nuclear charge is the predominating factor.

Within the halogens, for example, there is a general decrease in catenated single-bond energies from chlorine to bromine to iodine. This decrease in bond energy is in accord with the increasing potential energy of the electrons within the atomic orbitals, as indicated by the decreasing ionization potentials. (*See* Table 5.5.) The low value of the bond energy for the F—F bond can be attributed to the small size of the F atom and the electrostatic repulsions among nonbonding electron pairs.

Table 6.12 is a listing of the elements with which each electronegative atom forms covalent bonds; it includes all those for which bond energy data is available. Under each element, the arrangement is from highest to lowest bond energy. Thus the H—F bond has the highest bond energy of any involving the H atom. It also has the highest bond energy of any involving the F atom. Several important observations can be made from this tabulation.

For example, hydrogen forms strong bonds with many of these elements. Hydrogen has an intermediate ionization potential, suggesting that the 1*s* electron in hydrogen does not differ greatly in energy from the valence electrons of many other atoms. Also, hydrogen has an intermediate electronegativity that favors polar covalent bonds with the very highly electronegative atoms. Finally, the H atom never has any nonbonding electron pairs, so electron repulsion is at a minimum.

TABLE 6.12

Elements with Which Various Elements Form Covalent Bonds

The elements are arranged in the order of decreasing bond energy. All bonds are single bonds except where indicated by = in front of the atomic symbol.

				H
				F, O, Cl, C, N, B, Br, S, P, Si, I, Se, N, As, Li, Te, Na, Cs
B	**C**	**N**	**O**	**F**
F, O, Cl, N, H, Br, C	F ,H ,B ,O, Cl, N, Si, Br, S, P, Al, Se, Sn, I, As, Sb, Zn, Pb, Bi, Cd, Hg	B, H, C, F, O, Cl	H, C, As, Sb, P, Cl, N, Br, I, F	H, U, C, As, Te, S, Se, Ir, Hg, I, O
	Si	**P**	**S**	**Cl**
	=F, O, Cl, H, Br, C, S, I	=F, O, Cl, H, C, Br, I	H, Si, F, C, Cl, Br	Zr, Be, Ti, H, V, Si, Ni, Co, Cr, C, Fe, Ge, P, Sn, In, Sb, W, As, Bi, S, O, Se, Hg, N
		As	**Se**	**Br**
		F, O, Cl, H, Br, C, I	F, H, C, Cl	Be, H, Si, Fe, C, Ge, Sn, P, In, As, S, O, Hg
		Sb	**Te**	**I**
		O, Cl, C	F, H	Mo, H, Be, F, Si, Cr, In, Fe, C, Ge, O, P, As, Hg

The three most electronegative elements, fluorine, oxygen, and chlorine, tend to form strong bonds with many of these elements. In particular, elements with low electronegativity (2.1 or less) form their strongest bonds with these three elements, and this is consistent with maximum bond polarity. This is true for hydrogen, boron, silicon, phosphorus, arsenic, and antimony. Elements with an intermediate electronegativity (approximately 2.4 to 3.0) tend to form moderately strong bonds with these three elements. Elements in this category include carbon, sulfur, selenium, and nitrogen. The elements of high electronegativity (greater than 2.4) tend to form rather weak bonds with fluorine, oxygen, and chlorine, consistent with low bond polarity. Elements in this group include oxygen, fluorine, chlorine, bromine, and iodine.

The bond energies for the hydrogen halides are given in Table 6.13. Clearly the bond energies decrease as the atomic number of the halogen increases. This is consistent with the increasing difference in energies of the bonding electrons and the decrease in bond polarity.

TABLE 6.13

Hydrogen Halide Bond Energies in kcal/mole of Bonds

H—F	135
H—Cl	103
H—Br	87
H—I	71

Multiple bonds occur between different atoms as well as between identical atoms. Some of the commonly occurring multiple bonds are C≡N, C═O, C═N, N═O, Si═F, C═S, P═F, and S═O. It is significant that these multiple bonds almost always involve one of the very small atoms of carbon, nitrogen, oxygen, or fluorine. This is consistent with the tendency to pack more bonding electrons between the atoms to reduce electron repulsions on the surfaces of these small atoms.

In fact this tendency of very small atoms to reduce electron repulsions appears to be so great that they will force multiple bonding upon other atoms where it normally is not expected. For example, the F atom has a strong tendency to form double bonds. How this is possible can be demonstrated with phosphorus trifluoride. Since the P atom has three unpaired *np* electrons we expect that one P atom will bond to three F atoms with the formation of three single bonds, as follows:

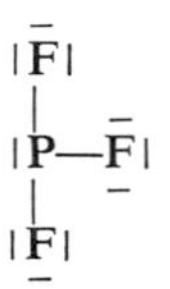

However, if the P atom has suitable vacant orbitals available, one of the otherwise nonbonding pairs of electrons on the surface of the F atom can become a bonding pair through pi orbital formation with one of the available vacant orbitals of phosphorus. In phosphorus, the most favorable orbitals from the standpoint of energy *and* geometrical arrangement of probable charge distributions are the 3*d* orbitals. Thus the structural formula may be written as follows, where the πd stands for the pair of electrons in the pi orbital:

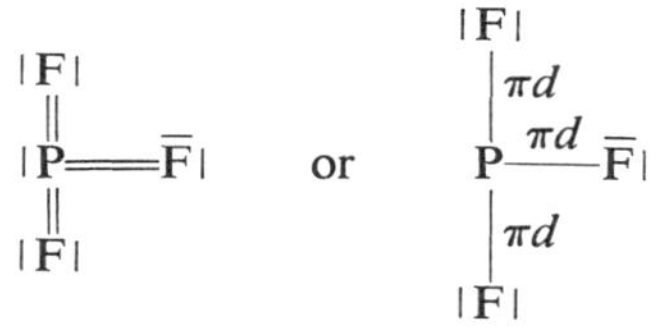

As a general rule, the small atoms of nitrogen, oxygen, and fluorine tend to force πd bonding, because of the electrostatic repulsions among otherwise nonbonding electron pairs. Not only does this decrease the number of competing electron groups on the surfaces of the small atoms, but it also localizes another pair of electrons between the two atoms. The atoms that have the most suitable *nd* orbitals avilable for πd bonding appear to be silicon, phosphorus, sulfur, and chlorine, all in period 3 of the periodic table.

This representation of the bonding in PF_3 introduces a slightly different point of view towards possible covalent bond formation. In all previous examples, covalent bond formation was pictured as the pairing of unpaired electrons on the two atoms involved. In the present view, a pair of electrons on one atom is shared with a vacant orbital of another atom. As a matter of convenience we sometimes distinguish between these two kinds of bonds. The *normal covalent bond* (polar and nonpolar) is a shared pair of electrons to which each atom has contributed one electron. The *coordinate covalent bond* is a shared pair of electrons that has been donated by one atom to another.

It must be emphasized that the definition of the coordinate covalent bond is a matter of convenience, and that a covalent bond is a covalent bond, no matter what the source of electrons is. Atoms really cannot distinguish between electrons, nor can we, except in the mind's eye.

COMBINATIONS OF ATOMS WITH SIMILAR BUT LOW ELECTRONEGATIVITIES: METALLIC BONDS

The metals have been defined in various ways. Perhaps the most satisfactory definition can be given in terms of some common physical properties. As a first definition, metals may be defined as those elements that have the following properties:

1. High electrical and thermal conductivity; electrical conductivity improves as temperature decreases.

2. Metallic luster: they take a polish and serve as reflectors of light.
3. Ductility: they can be drawn into thin wires.
4. Malleability: they can be pounded into various shapes.
5. High tensile strength: they resist fracture when a stress is applied.

Perhaps the best word to describe metals is *tough.* They can be deformed and tend to be somewhat elastic, but they are fractured only with difficulty.

Nearly all of the elements are metals. Although it is difficult to draw a sharp line of distinction between the metals and the nonmetals, the separation can be made in an approximate way on the periodic table by drawing a stair-step line above and to the right of the following elements: Be, Al, Ge, Sb, and Po. The metals are to the left and below the line, and the nonmetals are to the right and above the line. This suggests that the elements of high electronegativity, which tend to form the negative oxidation states, are the elements we classify as nonmetals, and all of the rest of the elements are metals. Of course, there are elements on the borderline; these are sometimes called *metalloids.*

The toughness of metals indicates some kind of cohesiveness that holds the atoms together. This we call the *metallic bond.* However, the metallic bond appears to be an aggregate property depending upon large numbers of atoms. Some of the features of a model of the metallic bond are as follows:

1. The atoms of the metals are closely packed into a network somewhat similar to that suggested for aggregated ionic compounds.
2. This network consists of positive ions held together by a sea of delocalized electrons that are not held rigidly between specific atoms, but are associated with the entire network.
3. The sea of electrons consists primarily of the valence electrons of the metal atoms.
4. The metallic bond is due to the interaction of the positive ions with the negative sea of delocalized electrons, and in a sense, the electron sea insulates the positive ions from each other.

From one point of view, atoms of metals are electron deficient with respect to the atoms of the nonmetals. In general they have too many empty orbitals to allow all to be occupied through covalent electron pair sharing between two atoms. In addition, the atoms of the metals have low electronegativities, and thus do not tend to attract electrons strongly enough to localize pairs between atoms. The close packing of the atoms in a network or lattice permits interactions among many atomic orbitals, with the formation of delocalized bands somewhat similar in principle to the delocalized electron system mentioned for benzene.

Thus, a few electrons may be shared with many different ions in the lattice. At the same time, the lattice provides a structure that is not rigid. In a covalent structure, if two atoms are moved with respect to each other, the bond is disturbed because electron pairs are localized between the two atoms. However, in the metallic structure, the delocalized electrons provide a freedom of movement not permitted in the more rigid covalent structures. This helps to account for the electrical conductivity, since the more mobile electrons may be transferred from one location to another without disturbing the over-all bonding very much. At the same time, it provides an explanation for metallic deformability, since the ions may be moved with respect to each other without disturbing the bonding very much.

SUMMARY

Bonding between atoms has been described from a highly electrostatic point of view. The ionic bond is formed between atoms with large differences in electronegativity. A major portion of the compound stability results from electrostatic attractions between cations and anions in large aggregates.

The covalent bond is viewed as a pair of electrons shared between two atoms. The electron spin permits the packing of electrons in pairs, thus maximizing the number of electrons between atoms. The electrons shared between atoms are attracted electrostatically by the nuclei of both atoms and thereby are in a state of low potential energy.

Other facets of covalent bonds may be understood in electrostatic terms as well. Both multiple bond formation and molecular geometry may be understood in terms of the repulsions of electron pairs on the surfaces of atoms. Polar covalent bonds involve the separation of the positive and negative charge centers due to electronegativity differences between atoms. The electrostatic dipole moments of molecules can be explained in terms of molecular geometry and bond polarity.

Electron delocalization permits electrons to be shared with more than two atoms. This has been illustrated with the delocalized pi electron system of benzene. However, delocalized electron systems appear to play an important role also in the bonding among the atoms of the metals. The metallic bond couples some features of the ionic bond with the delocalized electron system.

QUESTIONS AND PROBLEMS

6.1 Consider the compound with the molecular formula $H_4P_2O_7$.

a) How many atoms are there in a molecule of $H_4P_2O_7$?

b) How many atoms are there in a mole of $H_4P_2O_7$ molecules?

c) How many moles of hydrogen atoms are there in four moles of $H_4P_2O_7$ molecules?

d) How many moles of phosphorus atoms are there in 50 g of $H_4P_2O_7$?

e) How many molecules of $H_4P_2O_7$ are there in 145 g of $H_4P_2O_7$?

f) What is the percent composition of $H_4P_2O_7$?

g) If this compound were to be made from other materials, how many grams of PH_3 would be required to provide enough phosphorus to make 85 g of $H_4P_2O_7$?

6.2 Consider the compound with the molecular formula $C_8H_{16}O_2$.

a) What is the empirical formula for this compound?

b) What is the percent composition of this compound?

c) How many milligrams of carbon are there in 45 mg of $C_8H_{16}O_2$?

d) How many atoms of carbon, hydrogen, and oxygen are there in 100 mg of $C_8H_{16}O_2$?

e) Suppose that 100 g of this compound is to be prepared from other materials with the necessary carbon obtained from coal which is 78 percent carbon, the necessary hydrogen obtained from water, and the necessary oxygen obtained from air which is 23 percent oxygen. How much coal, water, and air should be needed theoretically?

6.3 What is the empirical formula for each of the following compounds having the indicated molecular formulas?
a) hydrazine, N_2H_4
b) hydrogen peroxide, H_2O_2
c) octane, C_8H_{18}
d) peroxydisulfuric acid, $H_2S_2O_8$
e) sucrose, $C_{12}H_{22}O_{11}$
f) sodium trimetaphosphate, $Na_3P_3O_9$

6.4 Write the empirical formula for each of the compounds represented by the following percent compositions:
a) K = 9.29%; I = 90.71%
b) C = 53.0%; O = 47.0%
c) Ca = 25.3%; C = 30.4%; H = 3.80%; O = 40.5%
d) H = 3.69%; Cl = 26.0%; O = 70.3%
e) C = 39.1%; H = 8.77%; O = 52.1%

6.5 If 4.00 g of a binary compound of phosphorus and sulfur contains 1.30 g of phosphorus, what is the empirical formula for the compound? If the molecular weight is about 286, what is the correct molecular formula?

6.6 Carbohydrates are compounds containing only carbon, hydrogen and oxygen and having the atomic ratio H:O = 2:1, the same as water, H_2O. When heated in the absence of air, these compounds decompose to form carbon and water. If 310 g of a carbohydrate leaves a residue of 124 g of carbon after heating, what is the empirical formula for the carbohydrate? If 0.165 mole of the carbohydrate molecules contains 1.00 g of hydrogen, what is the molecular formula for the carbohydrate?

6.7 Platinum forms two compounds with chlorine, $PtCl_2$ and $PtCl_4$. Both of these compounds are thermally unstable, and when they are heated, they decompose to elemental metallic platinum and chlorine, which is a gas and escapes. After heating 6.40 g of a mixture of these two compounds, 4.48 g of platinum residue remains. What is the percent of $PtCl_2$ in the original mixture?

6.8 A mixture of NaBr and CsBr has a combined percentage of bromine equal to 60.0. What is the percent of NaBr in the mixture?

6.9 The nitrogen in a 4.75-g sample of a mixture of Li_3N and Mg_3N_2 is just sufficient to make 2.01 g of NH_3. What is the percent of Li_3N in the mixture?

6.10 Why do ionic compounds tend to aggregate?

6.11 Write equations similar to Equations (6.1), (6.2), and (6.3) for the formation of the isolated ions for each of the following compounds. Which of these appears to be the most favorable?
a) KF b) NaF c) LiF d) MgF_2 e) AlF_3

6.12 Indicate, giving your reasons, what ions you expect each of the following elements to form in ionic compound formation:
a) K b) Sc c) N d) S e) Br f) He

6.13 Write the formulas and names for the simple ionic compounds that you would expect to form between each of the following pairs of elements:
a) Na, S b) Mg, P c) La, F d) Ra, O e) Fr, I

6.14 Write a formula for each of the following:
a) potassium oxide
b) magnesium chloride
c) lithium nitride
d) barium iodide
e) sodium fluoride

6.15 Explain why hydrogen forms H_2, but helium does not form He_2, whereas lithium forms Li_2 only under unusual circumstances.

6.16 Describe the bonding in each of the following molecules or ions:
a) Li_2 b) H_2^+ c) He_2^+

6.17 Describe the types of molecules which the following elements probably form in the elementary state:
a) chlorine b) selenium c) arsenic

6.18 Using lines for electron pairs, write the structural formula for each of the following:
a) Cl_2 b) N_2 c) P_4 d) O_2 e) S_8

6.19 Describe the electron distribution on the surface of an iodine atom in the I_2 molecule.

6.20 Why does fluorine not form a double bond in F_2, whereas oxygen does in O_2?

6.21 Distinguish between sigma and pi orbitals.

6.22 Why does sulfur catenate into long chains more readily than oxygen?

6.23 Show how to compute the energy terms in Equations (6.12), (6.13), (6.14), and (6.15) from the data in Table 6.2.

6.24 Why is it to be expected that atoms will form the maximum number of bonds in molecule formation?

6.25 How can you account for the fact that the bond angles in elemental phosphorus are much smaller than the tetrahedral angle, although the molecule has the shape of a tetrahedron?

6.26 Why are the F—F, O—O, and N—N bond energies apparently anomalously low when compared to other elements in their respective families?

6.27 Is it possible for a molecule containing only nonpolar bonds to be polar? For example, would the water molecule be polar even if the bonds were nonpolar?

6.28 Since the H—F bond has partial ionic character is there any potential energy advantage associated with the aggregation of hydrogen fluoride molecules, as there is in the aggregation of ions in ionic compounds? If not, why not? If there is, suggest how the hydrogen fluoride molecules might tend to aggregate.

6.29 How would you expect the bond angle in the hydrogen sulfide molecule to compare to the bond angle in the water molecule? Why?

6.30 Suggest the types of binary compounds which selenium forms with hydrogen.

6.31 Approximately what is the H—O—O bond angle in hydrogen peroxide?

6.32 Hydrogen peroxide has an electrostatic dipole moment of 7.09×10^{-30} coulomb $\times$ angstrom. What does this imply about the molecular structure of hydrogen peroxide?

6.33 Suggest what molecules may result from the decomposition of hydrogen peroxide. Write an equation for this decomposition and include the correct energy term.

6.34 Account for the fact that the bond angle in ammonia is slightly less than the tetrahedral angle. Should the bond angle in PH_3 be the same, smaller, or greater than the bond angle in ammonia? Why?

6.35 Nitrogen and hydrogen form binary compounds with the following molecular formulas. Suggest possible structural formulas and molecular geometries for them: a) N_3H_5; b) HN_3.

6.36 Why doesn't nitrogen or oxygen benefit from bonding through an excited state the way carbon does?

6.37 What factors account for the existence of a very large number of hydrocarbons?

6.38 Write the structural and simple molecular formulas for the uncatenated binary compounds formed by each of the following pairs of elements:

a) H, I	d) O, F	g) C, I
b) H, Se	e) S, Cl	h) C, O
c) N, Cl	f) As, Cl	i) Cl, O

6.39 Give the simple molecular formulas and the structural formulas for the following:
a) hydrogen peroxide
b) sodium trisulfide
c) carbon tetrachloride

6.40 Describe the geometry of each of the following molecules:

a) water	e) elemental arsenic	i) acetylene
b) ammonia	f) hydrazine	j) H_2Se
c) AsH_3	g) benzene	k) Si_2H_6
d) methane	h) ethylene	

6.41 In each of the following, assign the proper oxidation number to each element:

a) $KMnO_4$	g) $H_2S_2O_7$	m) HN_3
b) $POCl_3$	h) C_2H_4	n) FeS_2
c) $CHCl_3$	i) C_2H_2	o) Fe_3O_4
d) MnO_2	j) C_6H_6	p) K_2MoOCl_5
e) Cl_2O	k) KNO_2	q) $K_3W_2Cl_9$
f) SO_2Cl_2	l) H_5IO_6	

6.42 Criticize the following statement: In different molecules with the same empirical formula, the elements will have the same oxidation numbers.

6.43 Suggest possible structural formulas for the following binary compounds:

a) N_2O	e) NO_2	h) C_3H_8
b) Cl_2O	f) CO_2	i) C_3H_6
c) NO	g) SO_2	j) C_3H_4
d) CO		

6.44 Name each of the following compounds in as many ways as possible:

a) Mg_3N_2	e) CuCl	h) H_2O
b) PF_5	f) Mn_2O_3	i) Na_2Se
c) XeF_4	g) HI	j) N_2H_4
d) $FeCl_2$		

6.45 Write the molecular formula for each of the following:

a) antimony pentafluoride	g) hydrobromic acid
b) chromic oxide	h) stannous fluoride
c) cupric iodide	i) boron trichloride
d) carbon tetrachloride	j) arsenic(III) oxide
e) mercury(II) sulfide	k) cobaltous chloride
f) selenium(IV) oxide	

6.46 Suggest appropriate names for each of the following compounds:

a) Mn_3O_4	d) H_2Se_2	f) OF_2
b) Na_2O_2	e) PH_3	g) ICl
c) K_2S_4		

6.47 Name the six oxides of nitrogen by the numerical prefix system.

6.48 Describe the bonding in SiF_4.

6.49 Explain each of the following observations:

a) The bond energy for the C—F bond is higher than the bond energy for the C—Cl bond.

b) The Br—Br bond energy is higher than the I—I bond energy.

c) The C═O bond energy is higher than the bond energy for the C—O, C—S, or C═S bonds.

d) The strongest bonds formed by boron and carbon are with fluorine, whereas the bonds between fluorine and nitrogen and fluorine and oxygen are not the strongest bonds formed by nitrogen or oxygen.

6.50 The following are structural formulas for some ternary compounds. In each case describe the bonding and geometry represented by these formulas.

a) 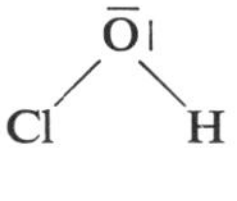

b) 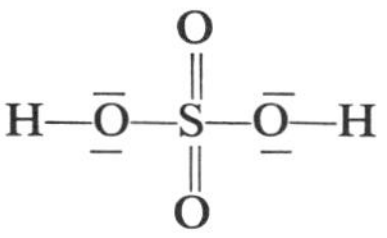

c) 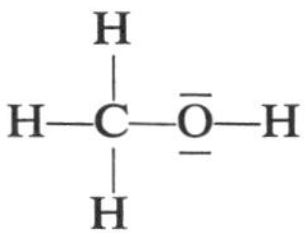

d) 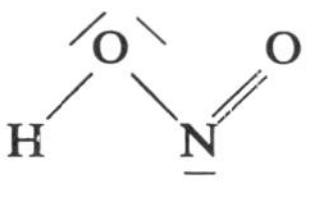

e) 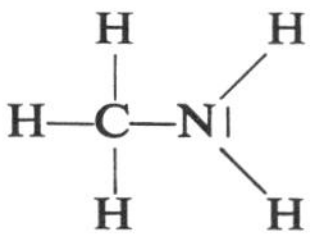

f) 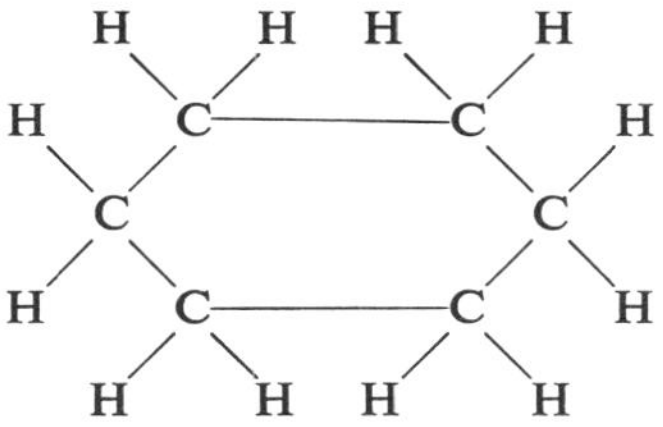

g) 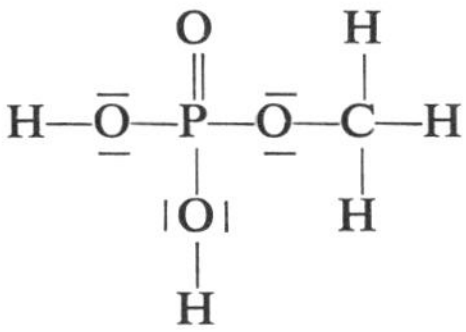

h) 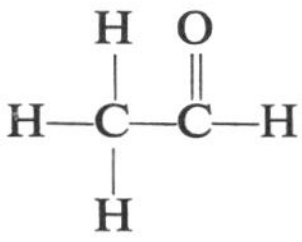

6.51 Describe the model of the metallic bond and indicate how it can be used to explain some of the physical properties of the metals.

6.52 Benzene does not exhibit the properties of electrical and thermal conductivity, luster, or malleability, ductility, and tensile strength exhibited by the metals. Yet benzene has a delocalized electron system. Explain this apparent anomaly.

7

Three States of Matter

INTRODUCTION Chapter 6 has dealt with several types of chemical bonds that account for the relative proportions of elements present in compounds. It has been noted that the ionic and metallic bonds account for high degrees of aggregation found among the ionic compounds and the metals. It is the purpose of this chapter to discuss the aggregation of atoms, ions, and molecules.

The familiar manifestations of aggregation are summarized in the three common states of matter: gas, liquid, and solid. The tendency of gases to diffuse and fill any container they occupy is familiar. Although liquids do not tend to fill any container they occupy, they do flow and do assume the shape of their containers. In contrast, the notable characteristics of solids are their rigidity and their ability to hold their own shape independent of any containers.

GASES

A distinguishing property of gases is the great deformability of the gaseous state. Gases may be compressed or expanded. The fact that gases spontaneously expand into a larger volume suggests that gases exert pressure, another familiar fact.

Recall that pressure is defined as force per unit area. Thus some typical units of pressure might be pounds per square inch and dynes per square centimeter. Sometimes pressure is defined in terms of the force per unit area exerted by a column of liquid on the base of the column. Consider such a column as shown in Fig. 7.1. The pressure is defined by Equation (7.1) where P, F, and A are pressure, force, and area respectively.

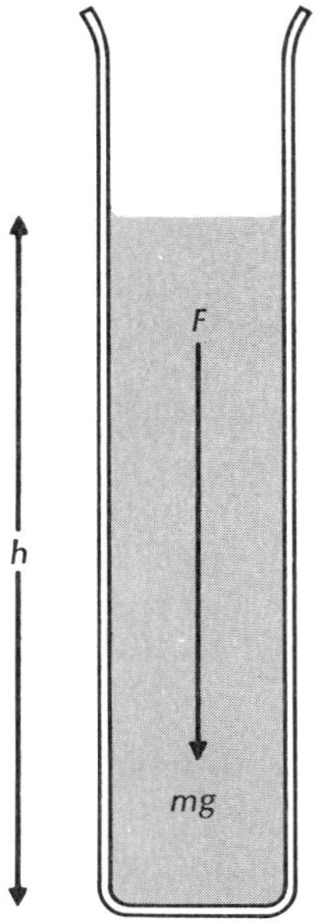

Fig. 7.1. Pressure exerted by a column of liquid

$$P = \frac{F}{A} \tag{7.1}$$

However, force is the product of mass and acceleration as shown by Equation (1.1), and under the acceleration of gravity this may be expressed by Equation (7.2) where g is the acceleration of gravity and m is mass.

$$F = mg \tag{7.2}$$

If the expression for F in Equation (7.2) is substituted into Equation (7.1), then Equation (7.3) results.

$$P = \frac{mg}{A} \tag{7.3}$$

By definition, the density of a substance, such as the liquid in the column of Fig. 7.1, is its mass per unit volume as expressed in Equation (7.4), where d is density and V is volume.

$$d = \frac{m}{V} \tag{7.4}$$

If Equation (7.4) is solved for *m*, as shown by Equation (7.5), this expression may be substituted into Equation (7.3) to obtain Equation (7.6).

$$m = Vd \tag{7.5}$$

$$P = \frac{Vdg}{A} \tag{7.6}$$

The volume of a cylindrical column as shown in Figure 7.1 is the product of the area *A* of the base and the height *h* as shown in Equation (7.7).

$$V = Ah \tag{7.7}$$

This expression for *V* may be substituted into Equation (7.6) to obtain Equation (7.8), which shows that the pressure exerted by a column of a liquid upon its base is the product of the density of the liquid, the height of the column, and the standard acceleration of gravity.

$$P = \frac{dAhg}{A} = dhg \tag{7.8}$$

It is clear from Equation (7.8) that the pressure exerted by the column of liquid is independent of the diameter of the column.

If a specific liquid is chosen for the column, the density at a specified temperature is constant. We have already indicated that the standard acceleration of gravity, *g*, is approximately constant. Thus in Equation (7.8) the product *dg* may be considered constant, so that the pressure is directly proportional to the height of the liquid column.

Because of its high density, 13.6 g/cm³ at 0°C, mercury often is chosen as the suitable liquid and pressures are expressed as cm Hg or mm Hg, or sometimes as inches of mercury.

To see how Equation (7.8) may be used to show the interrelationship among pressure units, consider the problem of expressing a pressure of 1.0 cm Hg in dyne/cm². Equation (7.8) may be used as follows:

$$P = dhg = \frac{13.6\ \text{g}}{\text{cm}^3} \times 1.00\ \text{cm} \times \frac{981\ \text{cm}}{\text{sec}^2} = 1.33 \times 10^4\ \text{dynes/cm}^2$$

Convince yourself that g/(cm × sec²) is equivalent to dynes/cm².

A part of the reason for the popularity of the height of a column of mercury as a pressure unit is the way in which pressures often are measured. The device for measuring gas pressures is called a *manometer* and a manometer used for measuring the pressure exerted by our gaseous atmosphere is called a *barometer*. A common type of barometer is the Torricelli barometer schematically represented in Fig. 7.2. The atmosphere exerts a pressure on the pool of liquid mercury, forcing it into the tube. The space above the column of mercury in the tube is a vacuum, so that the only downward pressure in the tube is that exerted by the weight of the liquid column of mercury. Thus, the atmospheric pressure may be measured directly by the height of the column of mercury.

The pressure exerted by the earth's atmosphere is used to define another pressure unit. Although atmospheric pressure varies from altitude to altitude, and even

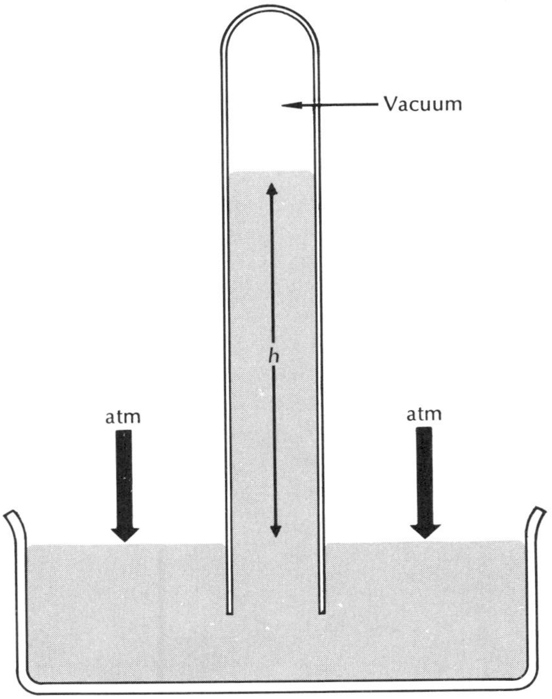

Fig. 7.2. Torricelli barometer

from day to day in the same location, it is found that at sea level the atmosphere exerts a pressure of approximately 76 cm Hg. Thus the atmosphere of pressure (atm) is defined as the pressure equivalent to that exerted by a column of mercury 76.00 cm high, at 0°C under standard gravity, for which the acceleration is 981 cm/sec^2 or 32.2 ft/sec^2.

Experiments with Gases

VOLUME-PRESSURE DEPENDENCE. If a gas is placed in a container, the walls of the container must exert a pressure equal to the pressure exerted by the gas. If this were not so, then, if the pressure exerted by the gas were greater than that exerted by the walls, the walls would be pushed out. If the walls exerted the greater pressure, the gas would be compressed. This may be demonstrated with a movable piston confining a gas in a container as shown in Fig. 7.3. If additional force is

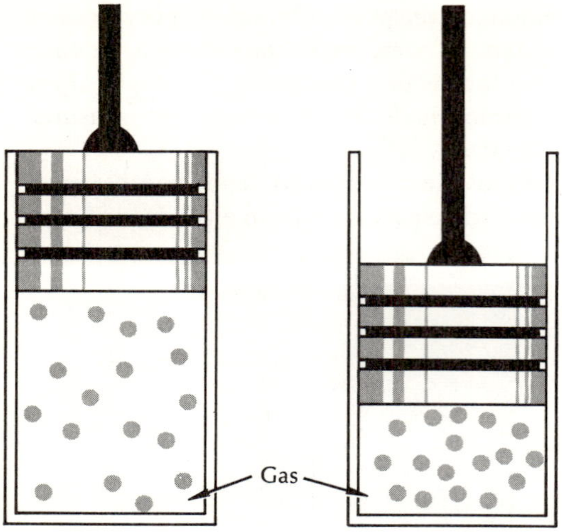

Fig. 7.3. Representation of a gas in a piston

applied to the piston, the volume of the gas decreases until the pressure exerted by the gas equals that exerted on the piston.

Typical results of the measurement of the effect of pressure upon gas volume at constant temperature are summarized graphically in Fig. 7.4. Although these relationships are strictly valid only when the measurements are made to a few significant figures, they serve as a basis for understanding the nature of gases.

One important observation is that each curve in Fig. 7.4 represents a different size of gas sample, but each curve is approximately independent of the nature of the gas. Thus the top curve may represent the volume-pressure relationship for 0.1 mole of neon atoms, 0.1 mole of nitrogen molecules, or 0.1 mole of hydrogen chloride molecules. In fact, it may represent a mixture of two or more gases with a combined total of 0.1 mole of particles, 0.05 mole of neon atoms and 0.05 mole of nitrogen molecules, for example.

From this observation, it appears that the behavior of gases depends mostly upon the number of particles (atoms or molecules) present and not very much upon the nature of those particles. Two very important laws are related to this observation.

The first of these laws has been indicated already. It states that *the total pressure exerted by a mixture of two or more gases equals the sum of the pressures exerted by the individual gases alone*. This is known as *Dalton's law*, which is expressed mathematically as Equation (7.9), where P_T is the total pressure exerted by a mixture of gases, and P_A, P_B, P_C, and so on, represent the independent pressures exerted by separate components of the mixture.

$$P_T = P_A + P_B + P_C + \dots \tag{7.9}$$

The second of these laws is extremely important to chemists. Known as *Avogadro's law*, it states that *equal volumes of gases at the same pressure and tem-*

perature contain equal numbers of particles. This fact makes it possible to compare the numbers of molecules or atoms in two gas samples by just comparing their volumes.

The shapes of the curves in Fig. 7.4 suggest an inverse relationship between the volume and pressure for gases, that is, the volume decreases as the pressure increases. In fact, if the volume is plotted against the reciprocal pressure as shown in Fig. 7.5, a family of approximately straight lines results. The extrapolation of

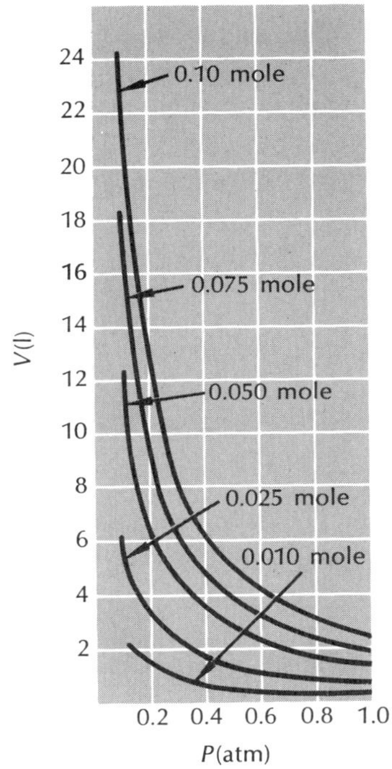

Fig. 7.4. Plot of volume against pressure at 25°C for various samples of a gas

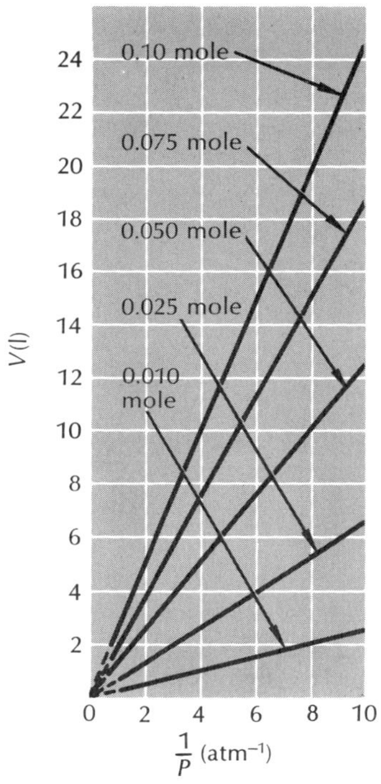

Fig. 7.5. Plot of volume against reciprocal pressure at 25°C for various samples of a gas

each of these lines passes through the origin. This suggests that at constant temperature and sample size, *the volume of a gas is inversely proportional to the pressure.* This is known as *Boyle's law*, which is expressed mathematically by Equation (7.10), where V is volume, P is pressure, and k_1 is the proportionality constant.

$$V = \frac{k_1}{P} \tag{7.10}$$

VOLUME-TEMPERATURE DEPENDENCE. Typical results of the measurement of the effect of temperature upon gas volume at constant pressure are summarized graphically in Fig. 7.6. Again these results are valid only when measurements are made

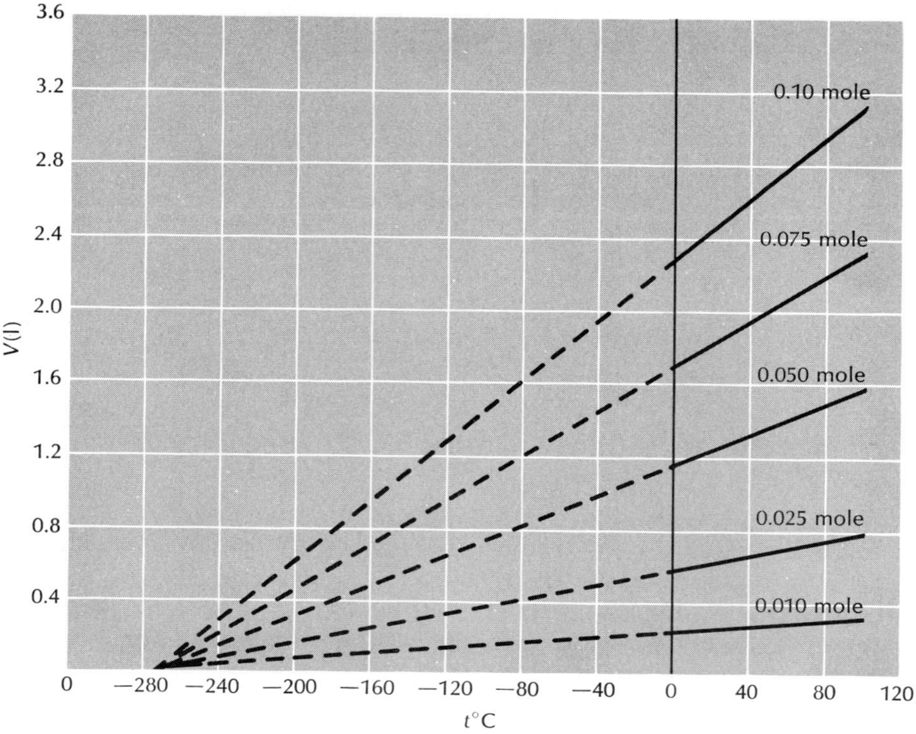

Fig. 7.6. Plot of volume against Celsius temperature for various samples of a gas at 1 atm pressure

to a few significant figures. Each solid curve in Fig. 7.6 represents a different size of gas sample, but again the curve is nearly independent of the gas used.

The solid straight lines in Fig. 7.6 indicate that the volume of a gas at constant sample size and constant pressure is pseudoproportional to the Celsius temperature. This is expressed by Equation (7.11), where V is volume, t is Celsius temperature, and k_2 and k_3 are the two proportionality constants.

$$V = k_2 t + k_3 \tag{7.11}$$

However, it is interesting to note that the extrapolations of the experimental lines, shown by the dashed lines, all intersect the temperature axis at the same point, approximately −273°C. If the extrapolated lines represented experimentally realizable conditions, then the volume of any gas sample should be zero at approximately −273°C. This temperature is called *absolute zero.*

If the volume axis of the graph in Fig. 7.6 could be picked up and moved to the left 273 temperature degrees, it would be located in such a way that the extrapolated lines would pass through the origin, and in this way the volume-temperature relationship could be stated in terms of a direct proportionality. This has been done in Fig. 7.7. Thus, a new temperature scale is defined with its zero value at −273°C. This scale is called the *absolute* or *Kelvin temperature scale* and is usually represented by °K. Since the Kelvin degree has the same magnitude as the Celsius degree, the relationship between the Kelvin and Celsius temperatures may be expressed by Equation (7.12).

$$°K = °C + 273 \tag{7.12}$$

The graph in Fig. 7.7 suggests that at constant pressure and sample size, *the volume of a gas is directly proportional to the Kelvin temperature*. This is known as Charles' law, expressed mathematically as Equation (7.13), where V is volume, T is Kelvin temperature, and k_4 is the proportionality constant.

$$V = k_4T \tag{7.13}$$

GASEOUS DIFFUSION. It is a well-known fact that gases diffuse. One need not be on top of a skunk to smell him. Also, as indicated above, gases tend to fill any space into which they are released. The results of the work of Thomas Graham

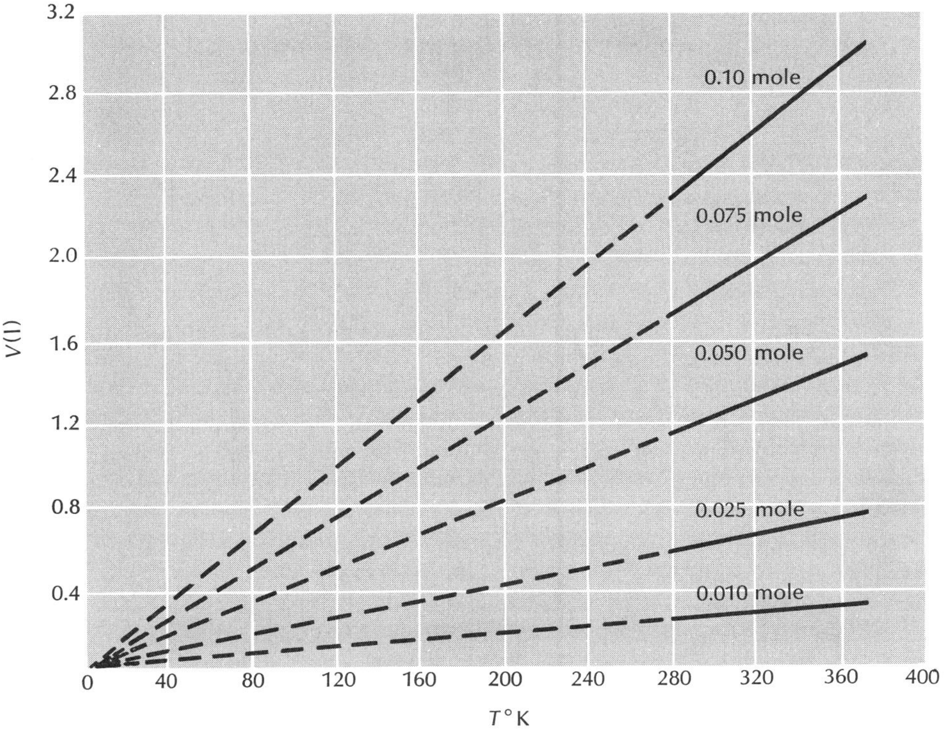

Fig. 7.7. Plot of volume against absolute temperature for various samples of a gas at 1 atm pressure

have shown that *the rate of diffusion is inversely proportional to the square root of the molecular weight of the gas.* This is expressed by Equation (7.14), where r is the rate of diffusion and M is the molecular weight of the gas.

$$r = \frac{k_5}{\sqrt{M}} \tag{7.14}$$

Often this relationship is used to compare the rates of diffusion of two gases A and B. Then Equation (7.14) may be rewritten as Equations (7.15) and (7.16) where the subscripts denote the individual gases.

$$r_A = \frac{k_5}{\sqrt{M_A}} \tag{7.15}$$

$$r_B = \frac{k_5}{\sqrt{M_B}} \tag{7.16}$$

Then Equation (7.15) may be divided by Equation (7.16) to derive Equation (7.17) in which the proportionality constant has been cancelled out.

$$\frac{r_A}{r_B} = \sqrt{\frac{M_B}{M_A}} \tag{7.17}$$

Kinetic Molecular Theory

These experimental observations have led to a model of gases which has been called the *kinetic molecular theory.* The following four postulates are the keys to this theory.

1. Gases consist of small atoms or molecules whose volume is insignificant with respect to the total volume of the gas.
2. There are negligible attractive forces between the particles.
3. The particles are in rapid, random, straight-line motion, colliding with each other and with the walls of the container, and for all practical purposes all collisions are perfectly elastic; that is, they result in no over-all change in kinetic energy.
4. In any sample of gas, particles have different kinetic energies, but the average kinetic energy is directly proportional to the Kelvin temperature.

Postulate 1 is equivalent to a statement that a gas is mostly empty space. This helps to account for the compressibility of gases and the ability of gases to mix homogeneously. Postulate 3 accounts for the pressure exerted by gases, the ability of gases to expand into any space in which they are placed, and the diffusion of gases. Postulate 4 helps to account for the increase in volume as temperature increases at constant pressure or for the increase in pressure at constant volume as temperature increases.

THE IDEAL GAS EQUATION. Something of the picture of gases is shown in Fig. 7.8, where the random motion of molecules in a gas is shown. With this and the four postulates, the relationship of pressure, volume, sample size, and temperature may be derived. A rigorous derivation is beyond our needs here, but a simplified derivation will serve the purpose to show how these postulates can be put to use.

According to the kinetic molecular theory, the pressure exerted by a gas is due to the collisions of the gas molecules with the walls of the container. The pressure exerted on the walls of the container is proportional to the number of impacts per unit area per unit time, and also to the momentum of the particles striking the walls of the container.

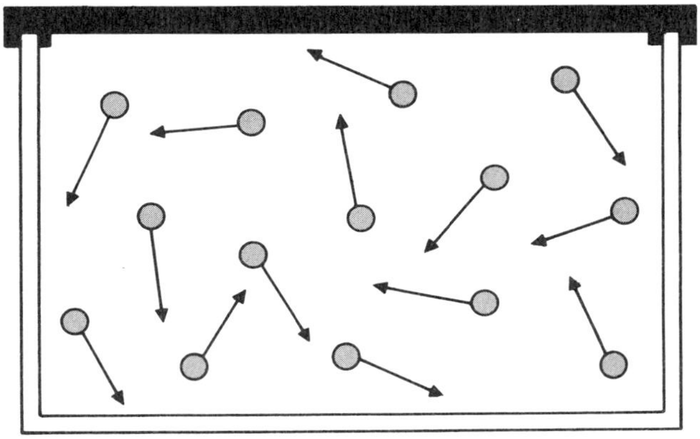

Fig. 7.8. Representation of a gas confined in a box

The number of impacts per unit area per unit time, I, is proportional to the number of particles in the box, N, inversely proportional to the volume of the box, V, and directly proportional to the average speed of the particles, s, as shown in Equation (7.18) where K_1 is the proportionality constant.

$$I = K_1 N\left(\frac{1}{V}\right)s \tag{7.18}$$

This equation makes intuitive sense, since an increase in the number of particles should increase the number of impacts per unit area per unit time, an increase in the volume of the box should decrease the number of impacts per unit area per unit time, and an increase in the average speed should increase the number of impacts per unit area per unit time.

The momentum of a particle is defined as the product of the mass and speed, as shown by Equation (7.19) where m is particle mass and s is the speed.

$$\text{momentum} = ms \tag{7.19}$$

It is reasonable that the pressure should depend upon momentum, since the force with which a particle collides with the container wall should be great for large masses as well as high speeds.

Since the pressure is proportional to both the number of impacts per unit area per unit time and the momentum, Equations (7.18) and (7.19) may be used to write Equation (7.20) where P is pressure and K_2 is the proportionality constant.

$$P = K_2K_1N\left(\frac{1}{V}\right)sms \tag{7.20}$$

However, the number of particles, N, is directly proportional to the number of moles of particles, n, as shown by Equation (7.21), in which K_3 is the proportionality constant.

$$N = K_3n \tag{7.21}$$

The expression for N given by Equation (7.21) may be substituted into Equation (7.20) to provide Equation (7.22).

$$P = K_3K_2K_1n\left(\frac{1}{V}\right)ms^2 \tag{7.22}$$

However, in Chapter 1 it was shown that the kinetic energy of a particle is given by Equation (7.23), in which E_k is the kinetic energy.

$$E_k = \tfrac{1}{2}ms^2 \tag{7.23}$$

Equation (7.23) may be rearranged as Equation (7.24) and the expression for ms^2 substituted into Equation (7.22).

$$ms^2 = 2E_k \tag{7.24}$$

$$P = 2K_3K_2K_1n\left(\frac{1}{V}\right)E_k \tag{7.25}$$

One of the postulates of the kinetic molecular theory is that the average kinetic energy is proportional to the Kelvin temperature, as expressed by Equation (7.26) where K_4 is the proportionality constant, and T is the Kelvin temperature.

$$E_k = K_4T \tag{7.26}$$

The expression for the kinetic energy given in Equation (7.26) may be substituted into Equation (7.25).

$$P = 2K_4K_3K_2K_1n\left(\frac{1}{V}\right)T \tag{7.27}$$

Since K_4, K_3, K_2, and K_1 are all constants, their product is constant. Thus a new composite proportionality constant R may be defined.

$$R = 2K_4K_3K_2K_1 \tag{7.28}$$

Then R may be substituted into Equation (7.26), which may then be rearranged to produce Equation (7.29).

$$PV = nRT \tag{7.29}$$

Equation (7.29) is called the *ideal gas equation.* It expresses the combination of Boyle's and Charles' laws. Thus when both n and T are constant, the product PV is a constant, and the volume is inversely proportional to the pressure. This is Boyle's law. Similarly, when both n and P are constant, the volume is directly proportional to the Kelvin temperature; and this is Charles' law.

In the derivation of Equation (7.29), the postulate that the average kinetic energy is proportional to the Kelvin temperature is an important one. Although this postulate is made to provide a way for the model to lead to the derivation of the experimentally verifiable relationship expressed by Equation (7.29), it also provides a definition of temperature as a measure of kinetic energy, or more fundamentally as a measure of molecular motion.

It must be remembered that temperature is a measure of the *average* kinetic energy. It is not true that all molecules within a given sample of gas have the same kinetic energy. As the molecules move about, they collide with each other. Thus kinetic energy may be transferred from one molecule to another. At any one time within a sample there may be a molecule that is motionless, but correspondingly there is another molecule with very high kinetic energy.

Throughout a sample at constant temperature, the total kinetic energy is constant. Figure 7.9 shows a typical kinetic energy distribution as a plot of fraction of molecules against kinetic energy. Note that a very small fraction consists of molecules having very low kinetic energy and another small fraction consists of molecules having very high kinetic energy. Most of the molecules tend to cluster into a narrow region near the average kinetic energy.

Figure 7.10 shows the effect of temperature changes upon the kinetic energy distribution. Note that at low temperature there is a greater tendency for the molecules to cluster around the average kinetic energy than there is at high temperature.

The ideal gas Equation (7.29) is a very useful relationship. Experimental techniques for measuring pressures and volumes of gases have been perfected so that measurements of these properties can be made with a high degree of accuracy. The proportionality constant R, often called the *ideal gas constant*, has been evaluated experimentally, using samples of gas for which n, P, and V are known. The units in which R is expressed depends upon the units in which P, V, and T are expressed. Equation (7.30) shows Equation (7.29) solved for R:

$$R = \frac{PV}{nT} \tag{7.30}$$

Obviously R must be expressed in units corresponding to the product of pressure and volume units, divided by the product of moles of particles and Kelvin temperature. When the pressure is expressed in atmospheres and the volume in liters, then $R = 0.082055$ l $\times$ atm/(mole of particles $\times$ °K).

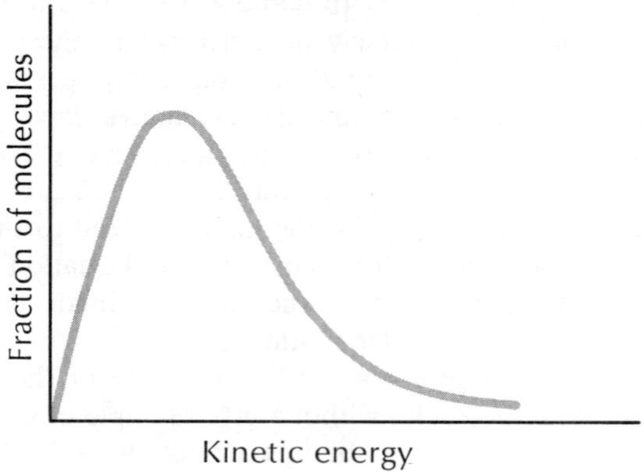

Fig. 7.9. Distribution of molecular kinetic energies in a gas sample

When P, V, and R are known, Equation (7.29) may be solved for n, the number of moles of particles. This provides an excellent experimental way to determine the number of moles of particles in a sample of gas.

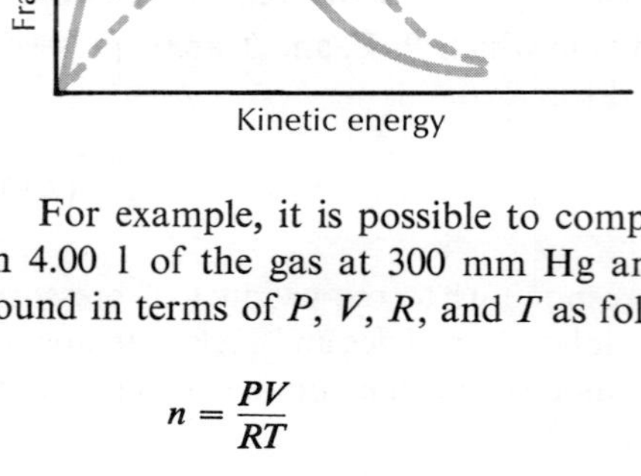

Fig. 7.10. Effect of temperature change on the kinetic energy distribution among molecules in a sample of gas

For example, it is possible to compute the number of moles of helium atoms in 4.00 l of the gas at 300 mm Hg and 15°C. From Equation (7.29), n may be found in terms of P, V, R, and T as follows:

$$n = \frac{PV}{RT}$$

The values for P, V, R, and T are as follows:

$$P = 300 \text{ mm Hg} \times \frac{1 \text{ atm}}{760 \text{ mm Hg}}$$

$$V = 4.00 \text{ l}$$

$$R = 0.0821 \text{ l} \times \text{atm}/(\text{mole of atoms} \times {}^\circ\text{K})$$

$$T = 15^\circ\text{C} = (15 + 273)^\circ\text{K} = 288^\circ\text{K}$$

Then n may be found as follows:

$$n = 300 \text{ mm Hg} \times \frac{1 \text{ atm}}{760 \text{ mm Hg}} \times 4.00 \text{ l} \times \frac{\text{mole He atoms} \times {}^\circ\text{K}}{0.0821 \text{ l} \times \text{atm}} \times \frac{1}{273^\circ\text{K}}$$

$$= 0.067 \text{ moles He atoms}$$

A similar procedure is used to determine molecular weights. Suppose that a 6.00-g sample of a gas occupies 806 ml at 25°C and 110 cm Hg. What is the molecular weight of the gas? Since the molecular weight numerically equals the number of grams per mole of particles, the ideal gas equation can be used to compute the number of moles of gas present, and this can be divided into the mass to provide the molecular weight.

$$n = \frac{PV}{RT}$$

$$= 110 \text{ cm Hg} \times \frac{1 \text{ atm}}{76 \text{ cm Hg}} \times 806 \text{ ml} \times \frac{1 \text{ l}}{10^3 \text{ ml}} \times \frac{1 \text{ mole molecules} \times {}^\circ\text{K}}{0.0821 \text{ l} \times \text{atm}} \times \frac{1}{298^\circ\text{K}}$$

$$= 0.0477 \text{ mole molecules}$$

$$\text{Molecular weight} = \frac{6.00 \text{ g}}{0.0477 \text{ mole molecules}} = 126 \text{ g/mole molecules}$$

Sometimes it is useful to be able to know how a change of conditions may affect one of the properties of a gas. Then Equation (7.29) may be rewritten as Equations (7.31) and (7.32), where the subscripts denote two sets of conditions.

$$P_1V_1 = n_1RT_1 \tag{7.31}$$

$$P_2V_2 = n_2RT_2 \tag{7.32}$$

These two equations may be divided to cancel out R.

$$\frac{P_1V_1}{P_2V_2} = \frac{n_1T_1}{n_2T_2} \tag{7.33}$$

Suppose that a sample of gas occupies 230 ml at 25°C and 720 mm Hg. The conditions are to be changed to 50°C at constant volume. What will the pressure be? In this case $V_1 = V_2$ and $n_1 = n_2$, so that Equation (7.33) reduces to

$$\frac{P_1}{P_2} = \frac{T_1}{T_2}$$

This may be solved for P_2 as follows:

$$P_2 = \frac{P_1T_2}{T_1} = \frac{620 \text{ mm Hg}}{} \times \frac{323\ °\text{K}}{298\ °\text{K}} = 672 \text{ mm Hg}$$

DEVIATIONS FROM IDEAL GAS BEHAVIOR. In discussing Boyle's and Charles' laws earlier, it was indicated that these laws are approximate. Since the ideal gas equation contains these two laws within it, this equation is approximate also. Thus the ideal gas equation and the kinetic molecular theory describe a hypothetical *ideal gas*. Real gases tend to deviate more or less from this model of an ideal gas. Why this is true can be seen by reviewing some of the postulates of the kinetic molecular theory.

According to postulate 1, the volumes of the molecules are negligible with respect to the total volume occupied by the gas. In other words, the molecules are so far apart, on the average, that they may be compressed into a small space without coming sufficiently close together to hinder the compression. In general this is true at low pressures, but at high pressures, the molecules are much closer and postulate 1 becomes less valid.

In addition, postulate 2 depends upon the nature of the gas particles. It is possible that some molecules attract each other rather strongly. If such attractions exist, they should be most effective at high pressures, where molecules are close together on the average, and at low temperatures where the kinetic energy is small and thus less effective in overbalancing any decrease in potential energy due to attractive forces.

In fact, gases do tend to behave like ideal gases at very low pressures, and as the pressure increases they tend to deviate from ideal gas behavior. Most substances may act somewhat ideally over certain ranges of temperature and pressure. However, at high pressures, gases tend to be less compressible than predicted by the ideal gas equation. Also at sufficiently low temperatures and high pressures, most gases can be liquefied; this is due to attractive force effects.

Thus a study of liquids can throw some light on the reasons for the deviation of real gas behavior from ideal gas behavior.

LIQUIDS

It is possible to consider an elementary model of the liquid state as an extension of the model for the gaseous state. The difference is that in the liquid state, the molecules are much closer and are held together by attractive forces. These attractions are not sufficiently great to maintain the molecules in a rigid network, but there is a degree of aggregation in the liquid state, as demonstrated by the failure of liquids to fill any container in which they are placed. The molecules of the liquid are in random motion, although it is a more restricted motion than that of the molecules in a gas. With less empty space around them, molecules are in more intimate contact; they hit each other more frequently than they do in the gaseous state.

Like molecules in gases, molecules in liquids have various kinetic energies related to the temperature. If a molecule has a high kinetic energy and is near the surface of the liquid, it may escape from the liquid into the surrounding atmosphere. As a result of this loss, the average kinetic energy of the liquid decreases, and consequently the temperature decreases. The process by which molecules escape from the liquid to the gaseous state is called *evaporation;* it is always accompanied by a cooling of the residual liquid.

Vapor Pressure

If an open container of liquid is placed in an enclosure, as shown in Fig. 7.11, high kinetic energy molecules reaching the surface will escape into the surrounding atmosphere, diffusing through any other gas present there. As evaporation continues, the concentration of these molecules in the gaseous phase increases, and the probability increases that some of these molecules will strike the liquid surface and be recaptured into the liquid phase. This process, whereby molecules leave the gaseous phase and enter the liquid phase, is called *condensation.* Thus a dynamic two-way process is established. Some molecules are evaporating from the liquid while others are condensing to the liquid. As the concentration of molecules increases in the gaseous phase, the rate of condensation increases. Eventually, the rate of condensation equals the rate of evaporation. At this point the system is said to be in *dynamic equilibrium,* since there is much molecular activity with no apparent change in either the gaseous or the liquid state. When the system is in equilibrium there is no further change in the *numbers* of molecules in either the gaseous or liquid state, although the identity of the molecules in each state does change.

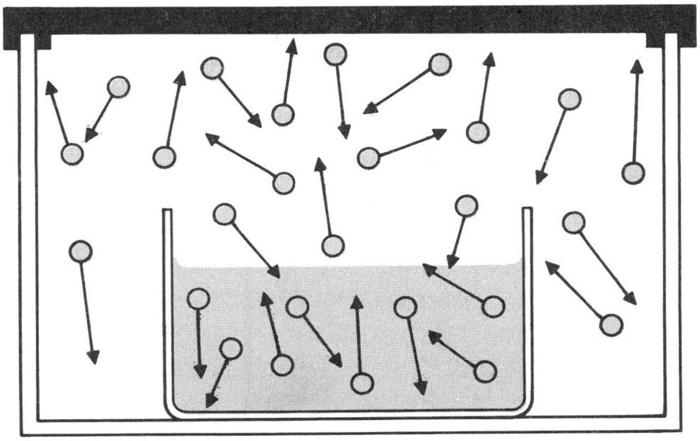

Fig. 7.11. Representation of a liquid-vapor equilibrium

At equilibrium, the vapor in the enclosure, like any other gas, exerts a pressure on the walls of the enclosure. This pressure is called the *equilibrium vapor pressure*, or more frequently just the *vapor pressure* of the liquid. At a given temperature, the vapor pressure is constant for any specified liquid.

At a given temperature, the vapor pressure of a liquid is a measure of the tenacity with which the molecules of the liquid are attracted to each other. At this given temperature, the average kinetic energy is the same, regardless of the substance involved. However, if one substance has high intermolecular attractions, then only the very few molecules that have very high kinetic energy and happen to be at the surface of the liquid will escape into the gaseous phase. Thus the rate of evaporation will be low and will be equaled by a low rate of condensation when relatively few molecules are in the gaseous phase. Thus a low vapor pressure for a liquid suggests high intermolecular attractions.

If heat is applied to a liquid, the average kinetic energy of the molecules in the liquid increases, and the rate of evaporation increases. Consequently the concentration of molecules in the vapor must be high to produce a high rate of condensation equal to the high rate of evaporation. Thus when a new equilibrium is established, there is a greater concentration of molecules in the gaseous state than at a lower temperature, and the vapor pressure is higher at the higher temperature. An increase in temperature leads to an increase in vapor pressure.

When a beaker of a liquid is heated, eventually small bubbles of vapor begin to rise within the liquid and burst upon hitting the surface. This process is called *boiling*. In some respects boiling is like evaporation, but in other respects it is different. When molecules evaporate from a liquid they diffuse from the liquid into the atmosphere around the liquid. The pressure of the atmosphere around the liquid does not affect the evaporation process. However, when a liquid boils, the vapor escapes from the liquid with sufficient pressure to displace the atmosphere around the liquid. Thus the temperature at which the vapor pressure of the liquid equals the pressure of the atmosphere around the liquid is called the *boiling point* of the liquid. The *normal boiling point* is defined as the temperature at which the vapor pressure of the liquid equals 1 atm.

The boiling point is a guide to the strength of the intermolecular attractions. If the boiling point of a liquid is high, the kinetic energy must be raised to a high level to overcome the intermolecular attractions, so as to allow the vapor pressure to be as high as 1 atm. Correspondingly, a low boiling point indicates small intermolecular attractions.

Heat of Vaporization

While a liquid is boiling, the temperature of the liquid remains rather constant, even as heat is being supplied. As the liquid boils, the kinetic energy of the molecules in the liquid decreases, since the molecules with high kinetic energy escape from the liquid. Consequently a heat source is necessary to maintain the average kinetic energy so that boiling can continue. Clearly then, heat must be absorbed to convert the liquid to a gas even without a temperature change. This heat is called the *heat of vaporization.*

Usually the heat of vaporization is expressed as the amount of heat required to convert a mole of molecules from the liquid to the gaseous state at the same temperature. This is called the *molar heat of vaporization.* The values for the molar heat of vaporization often are expressed in kcal/mole, where *mole* is understood to be mole of molecules or mole of atoms. Unless specified to the contrary, the molar heat of vaporization is measured at the normal boiling point.

The molar heat of vaporization also is a measure of the attractive forces among the molecules in the liquid state. A high heat of vaporization suggests that much energy must be absorbed to overcome these forces. Thus a high heat of vaporization indicates high intermolecular attractions.

Critical Temperature

Normally gases may be liquefied by cooling and/or by applying pressure. However, in practice, it is found that there is a characteristic temperature above which gases cannot be liquefied regardless of the pressure applied. In other words, when the kinetic energy of the molecules in the gas reaches a certain point, it is too great to be surpassed by potential energy decreases that can possibly result from increased attractions due to pressure applied. This temperature is called the *critical temperature.*

Thus a high critical temperature means that the gas can be liquefied even at high kinetic energies. That is, strong intermolecular attractions provide large decreases in potential energy, allowing these molecules to stay close together. Thus a high critical temperature is associated with high intermolecular attractions.

Similarly, a low critical temperature means that the gas can be liquefied by increasing the pressure only when the kinetic energy of the molecules is low. Therefore the intermolecular forces must be weak, since they are easily overcome by the kinetic energy. Thus a low critical temperature is associated with weak intermolecular attractions.

Intermolecular Forces of Attraction

Using the criteria of boiling points, molar heats of vaporization, and critical temperature, it is possible to study some trends among various liquids to examine the nature of the forces acting between molecules.

VAN DER WAALS FORCES. The rare gases (family 0) exist as monatomic species and do not tend to form molecules containing two or more atoms of the rare gas. Thus the usual ionic, covalent, or metallic bonding pictures appear not to pertain to these elements. However, the rare gases can be liquefied at reduced temperatures and at high pressures.

Some data for the rare gases are given in Table 7.1. The trends are very clear. The boiling points, molar heats of vaporization, and critical temperatures all increase as atomic number increases. This suggests that interatomic forces, whatever their nature, become stronger as atomic number increases. Also, the low boiling points and critical temperatures suggest that the forces are moderately weak.

These relatively weak forces existing between otherwise nonattracting atoms

TABLE 7.1

Some Data for the Rare Gases

	ATOMIC NUMBER	BOILING POINT °K	MOLAR HEAT OF VAPORIZATION kcal/MOLE	CRITICAL TEMPERATURE °K
He	2	4	0.020	5
Ne	10	27	0.421	44
Ar	18	87	1.557	151
Kr	36	120	2.157	210
Xe	54	165	3.019	290
Rn	86	211	3.918	378

and molecules are called van der Waals forces. They are attributed to electrostatic polarizations of the molecules as suggested by Fig. 7.12. Thus the rare gas atoms become slightly polarized, inducing dipoles in each other, and since the charge centers are separated, opposite charge centers attract each other.

If this is true, van der Waals forces should be strongest among the largest atoms and molecules where electrostatic polarization is most possible. This appears to

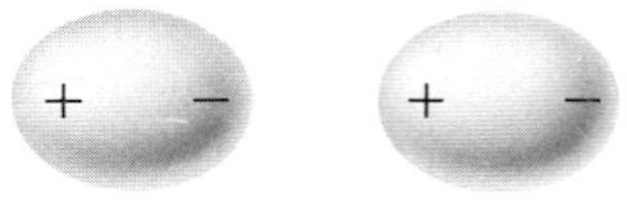

Fig. 7.12. Dipoles induced in rare-gas atoms

be true here, since the largest rare gas atoms, those of radon, have the strongest van der Waals forces.

To observe this same effect in molecules instead of atoms, consider the data for hydrogen and the halogens in Table 7.2. In this case, all of the molecules are diatomic, and the boiling points, molar heats of vaporization, and critical temperatures all increase as the size of the molecules increases.

DIPOLE-DIPOLE INTERACTIONS AND THE HYDROGEN BOND. Van der Waals forces have been explained above in terms of the interactions of induced electrostatic dipoles. This suggests that molecules having permanent dipoles might interact

TABLE 7.2

Some Data for the Halogens

	BOILING POINT °K	MOLAR HEAT OF VAPORIZATION kcal/MOLE	CRITICAL TEMPERATURE °K
H_2	20	0.216	33
F_2	86	1.56	144
Cl_2	238	4.876	417
Br_2	332	7.179	584
I_2	457	10.383	826

similarly. Uncomplicated examples of this effect are difficult to find, since permanent dipoles tend to cause induced dipoles as well. However, the data for the hydrogen halides given in Table 7.3 provide some interesting observations.

Within this group of compounds, the boiling point, molar heat of vaporization, and critical temperature increase from HCl to HBr to HI, which is the same order predicted on the basis of van der Waals forces. This is the order of increasing size of the halogen atom.

However, the HF appears to be anomalous in this series, with a very high boiling point, molar heat of vaporization, and critical temperature. It is true that HF contains the most polar bond of all of these molecules, and thus might be expected to provide the strongest interaction between permanent dipoles. However, the intermolecular attractions appear to be greater than expected.

TABLE 7.3

Some Data for the Hydrogen Halides

	BOILING POINT °K	MOLAR HEAT OF VAPORIZATION kcal/MOLE	CRITICAL TEMPERATURE °K
HF	292	7.20	503
HCl	188	3.86	324
HBr	207	4.21	363
HI	238	4.72	423

Some notion of the situation existing in the HF molecule can be seen with the aid of Fig. 7.13. When hydrogen bonds to the very highly electronegative fluorine atom, the only electrons associated with the H atom are highly localized between the two nuclei and are strongly attracted toward the fluorine. This means that one end of the HF molecule contains a highly exposed proton that is not shielded by electrons.

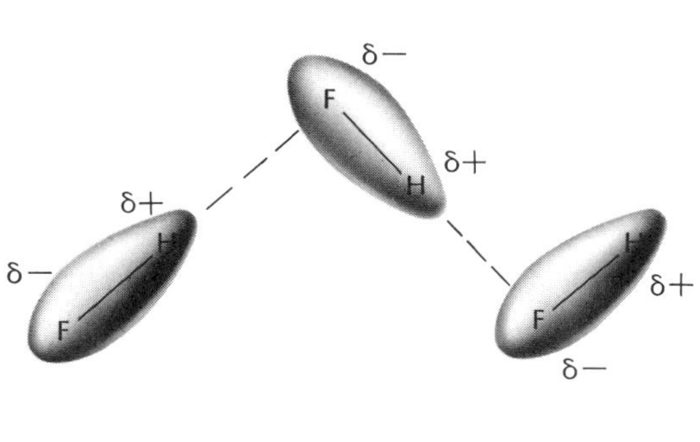

Fig. 7.13. Representation of hydrogen bonding in HF

At the same time, the electrons in the molecule are attracted toward the fluorine, where the nonbonding electrons are repelling each other rather strongly on the tiny surface of the F atom. As a result it is energetically feasible to reduce these repulsions by partially localizing a pair of these nonbonding electrons between the F atom and the H atom of an adjacent molecule toward which these electrons are attracted by the highly exposed proton.

This attraction between a hydrogen and a tiny, very electronegative atom is called the *hydrogen bond.* The hydrogen bond is a moderately strong bond and in the case of HF it has a profound effect upon the properties of the compound. This can be seen in Table 7.3. The effect upon the boiling point, for example, can be seen more clearly in Fig. 7.14, where the boiling points of the hydrogen halides are plotted against the atomic number of the halogen.

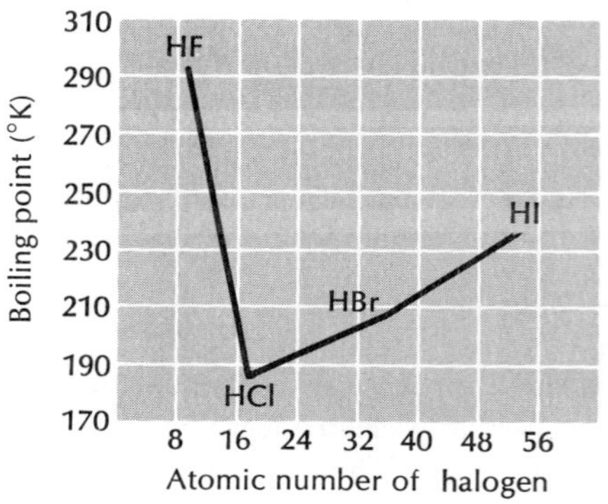

Fig. 7.14. Trends among the boiling points of the hydrogen halides

The tenacity of the hydrogen bonds is illustrated by the fact that at temperatures only slightly above the boiling point, gaseous HF exists in molecular aggregates. For example, it is found that at 292°K (the boiling point of HF), and at 1 atm of pressure, 0.276 g of gaseous HF occupy 96.0 ml. Using the ideal gas equation, the molecular weight may be found as follows:

$$n = \frac{PV}{RT} = \frac{1 \text{ atm} \times 0.096 \text{ l} \times \text{mole} \times {}^\circ\text{K}}{0.082 \text{ l} \times \text{atm} \times 292\ {}^\circ\text{K}} = 0.00401 \text{ mole}$$

$$\text{molecular weight} = \frac{0.276 \text{ g}}{0.00401 \text{ mole}} = 68.8 \text{ g/mole}$$

Thus the molecular weight of HF vapor under these conditions is 68.8, whereas if the formula is simply HF, the molecular weight should be 20.0. The experimental molecular weight is 3.44 times the simple molecular weight based upon the formula HF. This has been attributed to the aggregation of simple HF molecules through hydrogen bonding. From the data it would appear that these aggregates consist of about three or four simple HF molecules. At higher temperatures or very low pressures, this aggregation does not tend to occur.

Heat of Fusion

For the most part, pressure changes do not affect liquids significantly. However, upon cooling, liquids condense further to solids. Under these conditions, the kinetic energy of the molecules is decreased sufficiently to allow the intermolecular attractions to predominate, so that the molecules are attracted into a tightly packed framework in which they are free only to vibrate around some average fixed position within the network. The temperature at which this occurs is called the *freezing point*.

The reverse process occurs if a solid substance is heated. As heat is added, the kinetic energy increases so that some molecules are able to break away from the lattice of the solid and the solid begins to melt. During the melting process, the temperature remains constant. If the supply of heat is withdrawn, the solid and the liquid are in equilibrium, the solid melting at the same rate that the liquid is solidifying. This temperature at which the solid and liquid are in equilibrium is called the *melting point* of the solid. Normally the melting point of the solid is the same as the freezing point of the liquid.

Thus, during the melting process, energy must be supplied to overcome the attractive forces within the solid. This energy is called the *heat of fusion*, which often is given as the *molar heat of fusion*. The molar heat of fusion is the amount of energy required to melt one mole of solid at constant temperature.

Like the molar heat of vaporization, the molar heat of fusion can be used as a measure of the intermolecular forces of attraction. The molar heats of vaporization

TABLE 7.4
Some Molar Heats of Vaporization and Fusion

	MOLAR HEAT OF VAPORIZATION kcal/MOLE	MOLAR HEAT OF FUSION kcal/MOLE
Rare gases		
He	0.020	0.0050
Ne	0.421	0.080
Ar	1.557	0.281
Kr	2.157	0.391
Xe	3.019	0.548
Rn	3.918	0.638
Halogens		
H_2	0.216	0.028
F_2	1.56	0.372
Cl_2	4.876	1.530
Br_2	7.179	2.528
I_2	10.383	3.736
Hydrogen halides		
HF	7.20	0.938
HCl	3.86	0.476
HBr	4.21	0.575
HI	4.72	0.686

and the molar heats of fusion are compared in Table 7.4 for the rare gases, the halogens, and the hydrogen halides. It is clear that the molar heats of fusion show the same trends as the molar heats of vaporization. However, it is clear also in Table 7.4 that molar heats of vaporization tend to be considerably higher than molar heats of fusion. This is a general observation, indicating that it takes less energy to break down the framework of the solid to the point where the molecules have random, if limited, motion, than it takes to overcome these intermolecular attractions sufficiently to transform the liquid into a gas.

SOLIDS

The molecules in the solid state may be arranged in a random way; then the solid is said to be *amorphous*. However, more commonly, the molecules are arranged in rather regular geometrical patterns. Such solids are said to be *crystalline*.

Techniques of x-ray diffraction, and to some extent neutron and electron diffraction, have permitted the determination of the arrangements of atoms in solids. These patterns shed some light on the nature of solids and the nature of the particles of which the solids are composed.

Crystals Composed of Identical Atoms

Metals and rare gases are composed of atoms rather than polyatomic molecules. It is assumed that these atoms are spherical. If the reason for the atom aggregation is decreased potential energy through interatomic attractions that are largely electrostatic, the arrangements of the atoms will be such that they are as closely packed as possible, each atom touching as many other atoms as is possible.

Thus an important factor in the packing of atoms into a lattice is the number of *nearest neighbors* each atom has. This is called the *coordination number* for the surrounded atom and usually refers to the number of other atoms it touches. In general, arrangements that provide the highest coordination numbers are the closest packed arrangements and thus the preferred ones.

HEXAGONAL CLOSE PACKING. If a group of spheres or balls is placed on a flat surface, so they are packed into a planar structure, the closest packing is that shown in Fig. 7.15. This is called hexagonal packing since a group of six spheres surrounds a seventh sphere at the corners of a regular hexagon, as outlined in Fig. 7.15. In this planar arrangement, each atom has six nearest neighbors.

A second layer of hexagonally packed spheres may be placed on top of this layer in the manner depicted in Fig. 7.16.

The spheres in the top layer fit into the depressions formed by triangular groups of spheres in the bottom layer.

Since the spheres do not fit together so that they are everywhere in contact with another sphere, there are empty spaces between these two layers. These are marked in Fig. 7.16 as T_1, T_2, and O.

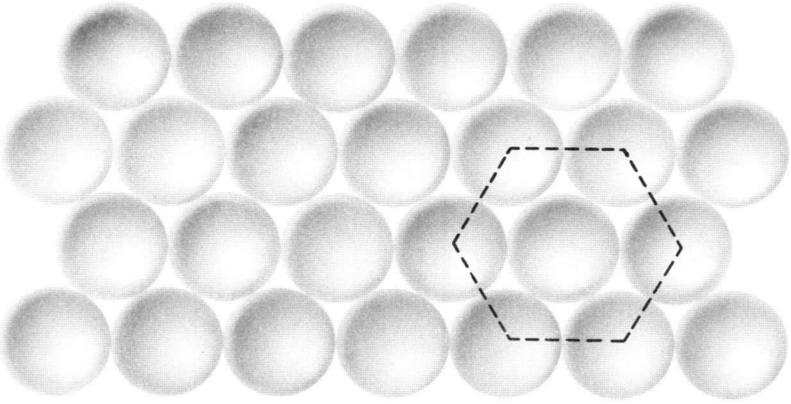

Fig. 7.15. Hexagonal close packing in one layer

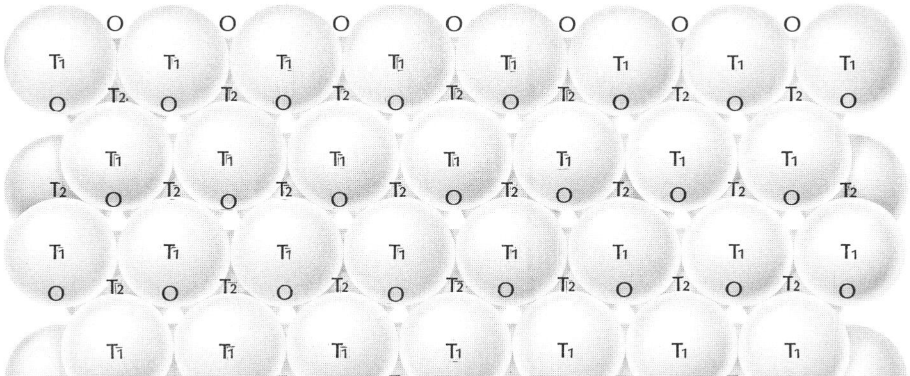

Fig. 7.16. Representation of two hexagonally close-packed layers

A typical T_1 hole is shown in Fig. 7.17. It consists of the space formed by a triangular group of spheres in the bottom layer and capped by a single sphere in the top layer. These four spheres lie at the vertices of a regular tetrahedron as shown in Fig. 7.17b. Thus the space marked by T_1 is called a *tetrahedral hole*.

A typical T_2 hole is shown in Fig. 7.18. It has the appearance of an inverted T_1

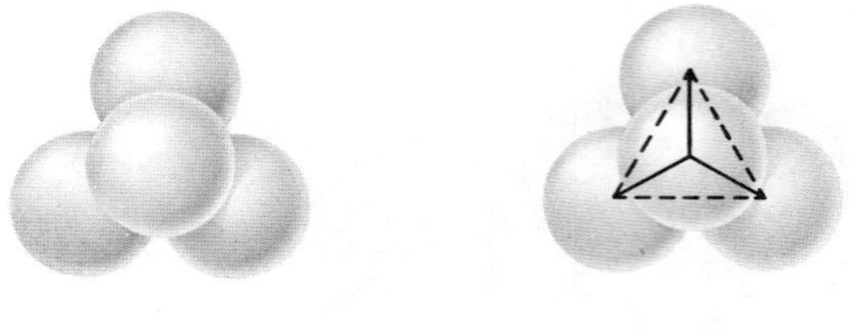

(a) (b)

Fig. 7.17. Representation of a tetrahedral hole, T_1

hole with the triangular group of spheres in the top layer and the cap in the bottom layer. As before, these four spheres lie at the vertices of a regular tetrahedron. Thus the space marked by T_2 also is called a tetrahedral hole.

A typical O hole is shown in Fig. 7.19. This hole is not capped and serves as a tunnel through the two layers. It consists of the space between a triangular group of spheres in the bottom layer and a triangular group of spheres in the top layer. These six spheres lie at the vertices of a regular octahedron, as shown in Fig. 7.19b. Thus the space marked by O is called an *octahedral hole*.

From the arrangement shown in Fig. 7.16, it should be clear that there are equal numbers of holes labeled T_1, T_2, and O in a large matrix composed of two hexagonally packed layers. This means that there are twice as many tetrahedral holes as octahedral holes.

The matrix may be extended by adding a third layer of hexagonally packed spheres to the two layers shown in Fig. 7.16. This may be done in either of two ways. First, imagine that the third layer is added so that the spheres are over T_2 holes. This means that the spheres in layer 3 are directly over the spheres in layer 1.

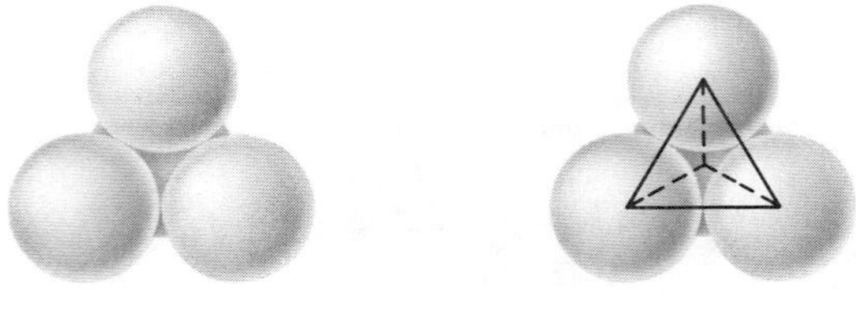

(a) (b)

Fig. 7.18. Representation of a tetrahedral hole, T_2

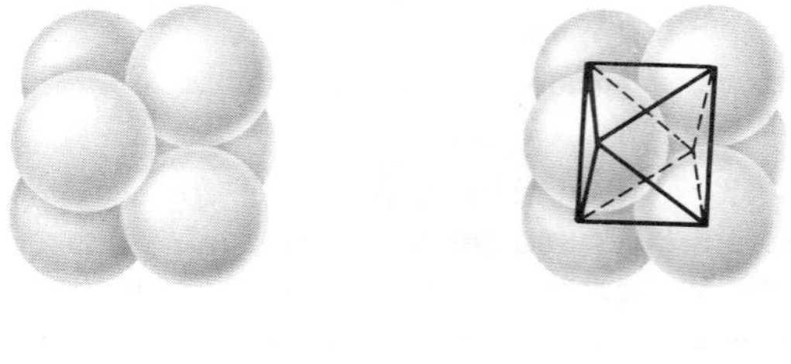

(a) (b)

Fig. 7.19. Representation of an octahedral hole

Then a fourth layer of spheres could be placed directly over layer 2. This often is called the ABABAB . . . arrangement to indicate that the odd numbered layers are alike and the even numbered layers are alike. Specifically it is called the *hexagonally close-packed* arrangement.

In the hexagonally close-packed structure, the octahedral holes are not capped and thus serve as channels or tunnels through the entire structure.

It should be noted also that each atom within this structure has a coordination number of 12. This may be seen by looking at one of the solid circles in the middle of Figure 7.16. It is surrounded by six spheres in the same layer; it touches three spheres in the layer below it and touches three more in the layer above it when that layer is added.

Within the hexagonally close-packed arrangement of spheres, there are as many octahedral holes as there are spheres. This may be seen if a solid sphere is located in the middle of Fig. 7.16. This sphere is in contact with six octahedral holes, three below it and three above it. In addition, each octahedral hole is bounded by six spheres, which means that one-sixth of the hole is associated with each sphere. Since each sphere is in contact with six octahedral holes, and one-sixth of each of these holes is associated with the sphere, there is one octahedral hole per sphere. Thus the number of octahedral holes in the lattice equals the number of spheres in the lattice. Also, since there are twice as many tetrahedral holes as octahedral holes in the lattice, there are twice as many tetrahedral holes as atoms.

CUBIC CLOSE PACKING. The second way in which a third layer of hexagonally packed spheres may be placed on top of the two-layer structure shown in Fig. 7.16 may be imagined if spheres are placed over the octahedral holes. In this case, the third layer does not coincide in position with either of the previous two layers of spheres. When this sequence is repeated with the addition of more layers, the arrangement is represented as *ABCABCABC* . . . to indicate that the first, fourth, and seventh layers, and so on, are in the same positions. This arrangement is called the *cubic close-packed* arrangement or the *face-centered cubic* arrangement.

The name for this arrangement is derived from the concept of the *unit cell.* It is possible to describe crystals in terms of some simple repeating geometrical unit, called the unit cell, from which the total crystal may be imagined to have been constructed. In this case the unit cell is composed of 14 spheres, one at each of the 8 corners of a cube and one in each of the 6 faces of the cube. Thus this unit cell is called the face-centered cube.

To see how the cube is derived from the close packed layers, see Fig. 7.20. Parts of four hexagonally packed layers are shown as follows: Layer 1, sphere *a*; Layer 2, spheres *b*, *c*, *d*, *e*, *f*, and *g* with sphere *f* hidden behind spheres *b*, *c*, and *d*; Layer 3, spheres *h*, *i*, *j*, *k*, *l*, and *m* with sphere *m* hidden behind spheres *i*, *k*, and *l*; Layer 4, sphere *n*. The face centered cube has six faces with each face having a sphere at the corner and one in the center as follows: Front face *abhe*, center *c*; Back face *nmgj*, center *l*; Top face *bjnh*, center *i*; Bottom face *aemg*, center *f*; Right face *abjg*, center *d*; Left face *ehmn*, center *k*.

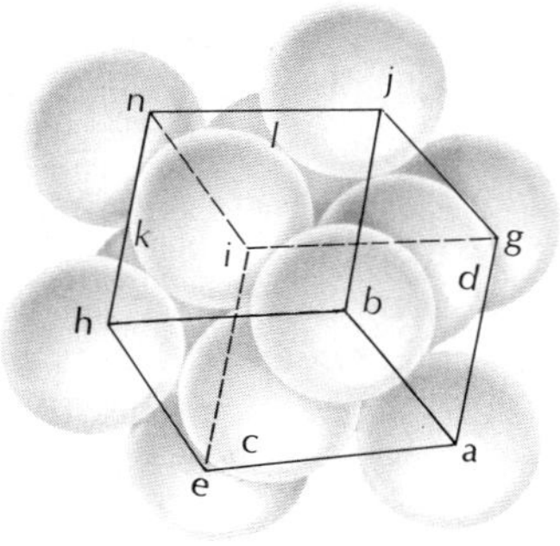

Fig. 7.20. Representation of the face-centered cubic unit cell and the packing of hexagonally close-packed layers

Like the hexagonal close packed arrangement, the coordination number for each atom is 12. Similarly, there are as many octahedral holes as atoms in the matrix and there are twice as many tetrahedral holes as atoms in the lattice.

BODY-CENTERED CUBIC PACKING. There is a third but spatially less economical way to pack spheres together. This is shown in Fig. 7.21. The spheres in any given layer are not packed as closely as in the hexagonal packing. Within the layer the spheres do not fit in the depressions formed by other spheres. The spheres in a layer are arranged in parallel rows but not touching. Then the second similar layer is placed on top, so that the spheres in this layer fit into the depressions of the first layer. The third layer then is located directly over the first layer and the over-all arrangement can be represented as ABABAB . . . .

This is called the *body-centered cubic* arrangement. The base of a unit cell is

Fig. 7.21. Representation of body-centered cubic packing

outlined in Fig. 7.21. The four atoms represent the base of the cube, and four other atoms directly over these four in the third layer represent the top of the cube. There is just one other sphere in the unit cell, at the center of the cube.

In the body-centered cubic arrangement, each sphere has a coordination number of eight. Thus the sphere at the center of the unit cell cube touches the spheres on the eight corners of the cube. However, each sphere has six *next* nearest neighbors at the centers of the six adjacent unit cell cubes. These six next nearest neighbors are only about 15 percent more distant than the nearest neighbors. Thus the *effective* coordination number in the body-centered cube is more than 8.

SUMMARY. There are three common ways in which atoms of the same size are packed into crystals. The hexagonal and cubic close packings are the favored arrangements, providing a coordination number of 12. This maximizes attractions between atoms by allowing the atoms to be as close as possible. A second-best arrangement is the body-centered cubic structure with a coordination number of eight or slightly more. However, it does not permit the atoms to be as close as they are in the hexagonal and cubic close packings.

Although the basis for crystal classification has been close packing of spheres without regard to any directional effects caused by any bonding that may occur, most of the rare gases and the metals can be described by these arrangements. Among the rare gases, neon, argon, krypton, and xenon all form face-centered cubic crystals. Approximately 80 percent of the metals crystallize in either the cubic close-packing or the hexagonal close-packing arrangements. In addition, molecular hydrogen, the elemental halogens, and the hydrogen halides form crystals in which the diatomic molecules are in close-packed structures as described here.

Ionic Crystals

Crystals of ionic compounds are composed of ions. In order for ions to be packed into crystals, it is necessary that the cation be surrounded by anions and the anion be surrounded by cations to allow for maximum potential energy decrease through interionic attractions. In addition, the cations and anions are of different sizes, so that the manner in which they are packed together depends upon the relative sizes of the two. Furthermore, the packing arrangement in the crystal is affected by the relative numbers of each kind of ion necessary to maintain electrostatic neutrality. For example, in sodium chloride, NaCl, there must be equal numbers of sodium ions, Na^+, and chloride ions, Cl^-. However, in calcium chloride, $CaCl_2$, there must be twice as many chloride ions, Cl^-, as there are calcium ions, Ca^{++}.

In ionic crystals, the coordination number for the cation is the number of *anionic* nearest neighbors, whereas the coordination number for the anion is the number of *cationic* nearest neighbors. If there is just one kind of cation and one kind of anion, and if the cation has a coordination number twice that of the anion, there must be twice as many anions as there are cations, as found in compounds such as $CaCl_2$ and TiO_2. On the other hand, if the anion has a coordination number twice that of the cation, there must be twice as many cations as anions, as found in Na_2O, for example.

As a matter of convenience we speak of 8:8 coordination, meaning that both the cation and the anion have coordination numbers of 8. Similarly we speak of 4:8 coordination, meaning that the cation has a coordination number of 4 and the anion has a coordination number of 8.

Many of the simple ionic crystals can be viewed as matrices of close-packed anions or cations, with the oppositely charged ions inserted in the tetrahedral and octahedral holes. Clearly the holes are smaller than atoms in the network. Thus the lattice may be viewed as a close-packed arrangement of the larger ions with the smaller ions in the holes. Since anions tend to be large, and cations small, often the close-packed lattice is composed of anions with cations in the holes, but this is not always true.

For the same sizes of ions in the close-packed lattice, octahedral holes are larger than tetrahedral holes. At the same time, the actual sizes of the holes depend upon the sizes of the ions in the close-packed lattice. For this reason, it is convenient to define a radius ratio as follows:

$$\text{radius ratio} = \frac{\text{radius of ion in hole}}{\text{radius of lattice ion}}$$

Clearly the value of this ratio is less than one. However, when the value is large, that is, near one, the two kinds of ions are nearly the same size and the chances of fitting the smaller one into one of the holes in the lattice is small. However, as the radius ratio value becomes smaller, the radius of the ion in the hole becomes smaller with respect to the radius of the lattice ion. Then the chances improve for fitting the smaller ion into an octahedral hole, and when the value becomes small enough, the smaller ion will fit into a tetrahedral hole. Determining the actual values of the radius ratio at which it becomes possible to fit the smaller ion into one of the holes is a problem in solid geometry. The solution to the problem is shown in Table 7.5.

TABLE 7.5
Effect of Radius Ratio on Fitting Ions in Crystals

>0.732	Too large for close packing
<0.732	Suitable for fitting in octahedral hole
<0.414	Suitable for fitting in tetrahedral hole

According to this table, only when the radius ratio is less than 0.732 is it possible to fit the smaller ion into one of the holes. Below this value the smaller ion will fit into an octahedral hole and below 0.414 it will fit into either an octahedral or a tetrahedral hole. However, when the radius ratio is below 0.414, the smaller ion will "rattle around," so to speak, in an octahedral hole, and it is more advantageous to place it in a tetrahedral hole where it can be closer to the lattice ions.

Table 7.6 shows some ionic radii that may be used in evaluating radius ratios. This table is arranged alphabetically by element symbol and includes both cations and anions.

TABLE 7.6

Some Ionic Radii for Radius Ratio Computations (Å)

Ag^{+}	1.26	Cu^{++}	0.72	Pt^{++}	0.96
Al^{3+}	0.50	F^{-}	1.36	Rb^{+}	1.48
Au^{+}	1.37	Fe^{++}	0.76	S^{--}	1.84
Au^{3+}	0.87	Fe^{3+}	0.64	Sb^{3+}	0.92
Ba^{++}	1.35	Hg^{++}	1.10	Sc^{3+}	0.81
Be^{++}	0.31	I^{-}	2.16	Se^{--}	1.98
Bi^{3+}	1.08	K^{+}	1.33	Sn^{++}	1.10
Br^{-}	1.95	La^{3+}	1.15	Sr^{++}	1.13
Ca^{++}	0.99	Li^{+}	0.60	Te^{--}	2.21
Cd^{++}	0.97	Mg^{++}	0.65	Ti^{++}	0.85
Cl^{-}	1.81	Mn^{++}	0.78	Tl^{+}	1.59
Co^{++}	0.74	N^{3-}	1.71	Tl^{3+}	0.95
Co^{3+}	0.63	Na^{+}	0.95	V^{++}	0.85
Cr^{++}	0.80	Ni^{++}	0.73	V^{3+}	0.69
Cr^{3+}	0.62	O^{--}	1.40	Y^{3+}	0.93
Cs^{+}	1.69	Pb^{++}	1.27	Zn^{++}	0.74
Cu^{+}	0.96				

With this background, we shall look at a few typical crystal types for some of the simple ionic substances. In the following summary, the common practice of naming crystal types by the name of a typical example is followed.

Two Ions with the Same Coordination Number

1. *Cesium Chloride Structure, CsCl,* 8:8 *Coordination, Radius Ratio* >0.732. In cesium chloride, the two ions are close to the same size and the radius ratio exceeds 0.732. Therefore, it is not possible to consider a close-packed lattice of chloride ions with cesium ions in the holes. Cesium chloride adopts a body-centered cubic arrangement, in which there are alternating layers of cesium and chloride ions. Thus both the cation and the anion have coordination numbers of eight.
2. *Rock Salt Structure, NaCl,* 6:6 *Coordination, Radius Ratio* <0.732. In this structure the chloride ions form a cubic close-packed lattice in which the sodium ions fit into the octahedral holes. Since there are as many octahedral holes as there are ions in the lattice, this allows equal numbers of cations and anions. Since each chloride ion in the lattice is surrounded by six octahedral holes and since each octahedral hole is surrounded by six chloride ions, both the cation and the anion have coordination numbers of six.
3. *Zinc Blende Structure, ZnS,* 4:4 *Coordination, Radius Ratio* <0.414. The larger sulfide ions form a cubic close-packed lattice. With the small radius ratio, the zinc ions can fit into the tetrahedral holes. Thus the zinc ions occupy half of the tetrahedral holes, since there are twice as many tetrahedral holes in the lattice as there are sulfide ions.
4. *Wurtzite Structure, ZnS,* 4:4 *Coordination, Radius* <0.414. Zinc sulfide has the ability to crystallize in two different ways and for this reason it is said to be

dimorphic. The only difference between zinc blende and wurtzite is the way the sulfide ions are packed. In the wurtzite structure, the sulfide ions form a hexagonal close-packed lattice with the zinc ions filling half of the tetrahedral holes.

Two Ions with Different Coordination Numbers

1. *Fluorite Structure*, CaF_2, 8:4 *Coordination, Radius Ratio Near* 0.732. Although the radius ratio appears to be just under 0.732 in this case, calcium fluoride behaves as though the radius ratio is too large to permit close packing. It crystallizes in a body-centered lattice with the larger fluoride ions at the corners of the cubes and alternately half of the body-center positions vacant, the other half of the body-center positions being occupied by calcium ions.
2. *Antifluorite Structure*, Na_2O, 4:8 *Coordination, Radius Ratio Near* 0.732. Again the radius ratio appears to be small enough to allow close packing. However, sodium oxide behaves as though the radius ratio is too large to allow close packing. The antifluorite structure differs from the fluorite structure in having the cations at the corners and the anions in half of the cube centers in a body-centered cubic structure.
3. *Rutile Structure*, TiO_2, 6:3 *Coordination, Radius Ratio* <0.732. In this structure, the larger anions are in a hexagonal, close-packed formation, the cations filling half of the octahedral holes.

SUMMARY

The three states of matter represent three degrees of aggregation of molecules (or atoms). Ideally gases exhibit no aggregation, each particle behaving independently of all other particles. Thus gases diffuse, exert pressure, and are compressible. However, in practice, the particles of real gases do exhibit attractions for one another, and deviations from ideal behavior are observed.

In view of these forces of attraction, gases may be converted to liquids by a decrease in kinetic energy (cooling) or by a decrease in average distance between molecules (increase in pressure). The liquid state is a state of intermediate aggregation where the forces of attraction are sufficient to cause the molecules to tend to stay together, but are insufficient to prevent them from moving past each other in a somewhat random way. Liquids may be converted to gases by the addition of energy, or to solids by the removal of energy.

The solid state represents the highest state of aggregation, in which the kinetic energy is low enough and the molecules are close enough so that the forces of attraction are strong enough to keep molecules (or atoms or ions) in a relatively rigid structure, where at best they can vibrate around certain fixed positions. Because of the tendency of the molecules to attain the lowest potential energy state, that is, the state of maximum intermolecular attractions, the molecules tend to pack as losely as possible. This dictates the geometrical patterns they assume in the crystal lattice.

QUESTIONS AND PROBLEMS

7.1 Convert a pressure of 1 atm to dyne/cm^2.

7.2 The bar is a unit of pressure often used by meteorologists. It is 10^6 dyne/cm^2. Express a pressure of 500 mm Hg in bars.

7.3 In making a Torricelli barometer, what is the advantage of using a liquid with a high density?

7.4 At room temperature, water has a density of 1.00 g/ml. Express a pressure of 1 atm as the height of a column of water at room temperature.

7.5 In accurate work, pressures measured with a Torricelli barometer require corrections. Why?

7.6 Use the data in Table P.4 to answer the following:
a) What is the height of a column of mercury in a Torricelli barometer when the pressure is 1 atm and the temperature is 100°C?
b) Using Equation (P.9), derive an equation to express the pressure in terms of the height of a column of mercury and the Celsius temperature.

7.7 If d, h, and g in Equation (7.8) are expressed in mks units, what is the corresponding unit of pressure? Similarly, what is the pressure unit when d, h and g are expressed in British units?

7.8 A sample of gas is at a pressure of 850 mm Hg and is composed of equal numbers of moles of N_2 and H_2 molecules. What pressure is exerted by each of these gases?

7.9 At a given temperature, the average speed of argon atoms is 1000 mph. What is the average speed of radon atoms at the same temperature?

7.10 At 100°C, the rate of diffusion of an unknown gas is 22 percent of the rate of diffusion of hydrogen gas at the same temperature. What is the molecular weight of the unknown gas?

7.11 What is the value of the proportionality constant k_1 in Boyle's law Equation (7.10)?

7.12 Show how the kinetic molecular theory accounts for each of the following observations:
a) Gases are easily compressed.
b) Gases are miscible, that is, they may be mixed homogeneously in all proportions.
c) Gases exert pressure.
d) Pressure exerted by gases increases as temperature increases.
e) When observed with a microscope, smoke particles are in constant motion.

7.13 Helium in a 10.0-l steel cylinder is under a pressure of 4.00 atm at 22.0°C. If the temperature is raised to 600°C, what will the pressure be inside the steel cylinder?

7.14 The density of a gas is 2.07 g/l at 30°C and 2.00 atm pressure.
a) What is the molecular weight of the gas?
b) What is the density of the gas at 0°C and 760 mm Hg?

7.15 What is the density of radon at 100°C and 70 cm Hg?

7.16 A sample of argon has a volume of 200 liters at 12°C and 750 mm Hg. What volume will it occupy at 40°C and 72 cm Hg?

7.17 A mixture of gases is composed of 5.00 g of H_2, 1.00 g of He, and 25.0 g of O_2 in a 8.33-l container at 25°C. What is the pressure of the gas in the container?

7.18 What volume does a mole of molecules of gas occupy at 25°C and 610 mm Hg?

7.19 State postulate 4 of the kinetic molecular theory in terms of the Celsius temperature and rewrite the ideal gas equation in terms of t = °C.

7.20 If 2.00 g of a gas occupies 1.84 l at 20°C and 620 mm Hg, what is its molecular weight?

7.21 A sample of argon gas was collected in a bottle by bubbling the argon through water into the bottle. The 266 ml bottle was exactly half full of argon at 25°C, and the total gas pressure inside the bottle was 720 mm Hg. The vapor pressure of water at 25°C is 23.8 mm Hg.

a) What volume would the dry argon gas occupy at 0°C and 760 mm Hg?
b) What is the mass of argon in the half-filled bottle?

7.22 When solid magnesium nitride, Mg_3N_2, is treated with water, the nitrogen appears as gaseous ammonia, NH_3. A 0.150-g sample of a mixture of Mg_3N_2 and $MgCl_2$ was treated with water, and 3.10 ml of dry, gaseous NH_3 was liberated at 20°C and 730 mm Hg. What was the percent of Mg_3N_2 in the sample?

7.23 A high vacuum has a pressure of 10^{-8} mm Hg. In this vacuum, how many molecules are there per milliliter at 20°C?

7.24 The density of elemental sulfur vapor is given below at various conditions. For each set of conditions, determine the molecular weight of the sulfur and write a correct chemical formula for the sulfur vapor.

a) density = 3.29 g/l at 600°C and 700 mm Hg
b) density = 0.571 g/l at 900°C and 650 mm Hg

7.25 A compound is found to contain only carbon, hydrogen, and oxygen. At 400°C and 200 mm Hg it is a gas with a density of 0.420 g/l. In addition a 18.9-mg sample of the compound contains enough hydrogen to form 15.4 mg of water and enough carbon to form 22.8 ml of CO_2 at 25°C and 700 mm Hg. What is the molecular formula for the compound?

7.26 In practice, the boiling point of water is not exactly constant. Why?

7.27 Given a substance in which the liquid and gaseous states are in equilibrium, compare the following properties for the liquid and gas phases:

a) compressibility
b) rate of diffusion
c) average molecular kinetic energy

7.28 The experimentally determined vapor pressure for liquid boron tribromide at various temperatures is given in the following table:

TEMP. °C	VAPOR PRESSURE mm Hg	TEMP. °C	VAPOR PRESSURE mm Hg
−40	1.55	+10	32.2
−30	3.13	+20	52.2
−20	6.01	+30	81.8
−10	10.5	+40	127
0	19.2	+50	186

a) Why does the vapor pressure increase as temperature increases?
b) Make a plot of vapor pressure against temperature.
c) What function of the vapor pressure might be plotted against temperature to give a linear graph? Plot your suggestion.
d) When you have prepared a linear graph, use it to write the relationship between the vapor pressure of boron tribromide and temperature.

e) Assuming this relationship holds to the boiling point, what is the normal boiling point for boron tribromide?

f) How does a decrease of 50 mm Hg in the pressure affect the boiling point?

7.29 The vapor pressure for liquid sodium chloride at various temperatures is given in the following table:

TEMP. °C	VAPOR PRESSURE mm Hg	TEMP. °C	VAPOR PRESSURE mm Hg
1180	70.6	1280	184
1200	86.5	1300	220
1220	105	1320	262
1240	128	1340	310
1260	154	1360	366

a) Determine the mathematical relationship between the vapor pressure and the temperature.

b) Assuming that this relationship is valid to the normal boiling point, what is the normal boiling point for sodium chloride?

c) How does a decrease of 50 mm Hg in pressure affect the boiling point?

7.30 A container is half-filled with water at standard conditions and then sealed. Then it is heated. Will the water boil at 100°C? Explain.

7.31 The molar heat of vaporization of krypton is 2.240 kcal/mole. What is this in calories per gram?

7.32 The molar heat of fusion for xenon is 549 cal/mole. How much heat is required to melt 200 g of solid xenon at the melting point?

7.33 Predict the trends in the following properties for the binary hydrogen compounds of the elements of family 6A (H_2O, H_2S, H_2Se, H_2Te).

a) Boiling and melting points

b) Molar heats of vaporization and fusion

c) Critical temperature

7.34 In family 4A, the normal boiling point for SiH_4 is higher than the normal boiling point for CH_4, whereas in family 7A, the normal boiling point for HF is higher than the normal boiling point for HCl. Explain.

7.35 At any given temperature, which has the higher vapor pressure, NH_3 or PH_3? Explain.

7.36 In a sample of elemental hydrogen, H_2, are the molecules apt to be hydrogen bonded? Explain.

7.37 For each of the following pairs of substances, predict which should have the higher critical temperature:

a) NaCl, HCl   b) O_2, H_2O c) HF, He

7.38 In Table 7.4, why is the molar heat of fusion higher for HF than for the other hydrogen halides?

7.39 As a general rule, ionic compounds have much higher melting and boiling points than nonionic compounds. Suggest a reason for this.

7.40 Why is elemental iodine a solid at normal room conditions, whereas fluorine and chlorine are gases at these same conditions?

7.41 Normally, if the Kelvin temperature of a gas is doubled at constant pressure, the volume of the gas doubles. However, when the Kelvin temperature is doubled for a sample of gaseous hydrogen fluoride, the volume more than doubles. Why?

7.42 Criticize the following statement: Molecular motion ceases in a substance going from the liquid state to the solid state.

7.43 What trends in density would you expect in a substance that is going from the gaseous to the liquid to the solid state?

7.44 Distinguish between the body-centered and the face-centered cubic crystal arrangements. Also distinguish between hexagonal close packing and cubic close packing.

7.45 If a third layer of spheres is placed on top of the two layers shown in Fig. 7.16, to form a hexagonally close packed structure, where are the T_1, T_2, and O holes between the second and third layers?

7.46 If a third layer of spheres is placed on top of the two layers shown in Fig. 7.16 to form a cubic close-packed structure, where are the T_1, T_2, and O holes between the second and third layers?

7.47 What kinds of holes are found in the body-centered cubic structure?

7.48 It has been suggested that cesium chloride might have a close-packed structure with alternating positive and negative ions in each layer, giving each ion a coordination number of 12. What is wrong with this suggestion?

7.49 Can there be an anticesium-chloride structure the way there is an antifluorite structure?

7.50 In the description of the body-centered cubic structure for identical atoms, it was pointed out that each atom has six next nearest neighbors to increase the effective coordination number beyond eight. Is this true also in cesium chloride?

7.51 Use Fig. 7.21 to identify the locations of the Ca^{++} and F^- ions in the fluorite structure and the Na^+ and O^{--} ions in the antifluorite structure.

7.52 Justify the 8:4 coordination in fluorite and the 6:3 coordination in rutile.

7.53 Criticize the following statement: In both the fluorite and antifluorite structures, the larger ions are at the corners of the body-centered cube and the smaller ion in the center of the cube.

7.54 The following compounds crystallize in the rock salt structure: CsF, AgCl, NaI, MgO, RbCl, and LiCl. Which of these would be expected to crystallize this way and which are anomalies?

7.55 For each of the following compounds suggest the most likely crystal structure for the solids:

a) BaF_2
b) CuCl
c) TlCl
d) MnF_2
e) VCl_2
f) CsI
g) ZnO
h) CaS
i) Rb_2O
j) $SrCl_2$
k) AgBr
l) Li_2O
m) BaI_2

8

The Nonmetals and Some of Their Binary Compounds

INTRODUCTION In this chapter we shall look at some of the typical molecules and ions formed by nonmetals. The chapter will serve to review ionic and covalent bonds and to illustrate some related properties. At the same time it will introduce some typical chemical species.

Following a section on covalent bond representation, the chapter describes molecular types beginning with the elements and progressing to uncatenated binary compounds and catenated binary compounds. Within each of these sections, the organization is by family in the periodic table, beginning with family 0 and proceeding to families 7A, 6A, 5A, 4A, and 3A in that order.

COVALENT BOND REPRESENTATION

The molecules used for illustrative purposes in Chapter 6 were relatively simple molecules in which the formation of bonding orbitals and their corresponding charge distributions was comparatively easy to describe.

In general, it was possible to represent molecular structures by employing dots for individual electrons or lines for electron pairs. Thus water was represented as follows:

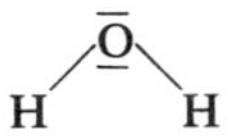

This shows two pairs of bonding electrons in sigma bonding orbitals and two pairs of nonbonding electrons. Similarly, elemental nitrogen was represented as

|N≡N|

This shows two pairs of nonbonding electrons and three pairs of bonding electrons localized between the two atoms in one sigma bonding orbital and two pi bonding orbitals.

However, there was some difficulty in representing benzene, which has a delocalized pi electron system. The framework of the benzene molecule can be represented adequately with six carbons in a planar hexagonal ring bonded together by pairs of electrons in sigma bonding orbitals. In addition, one hydrogen is bonded to each carbon by a pair of electrons in a sigma bonding orbital. Thus the framework of the molecule may be represented as follows:

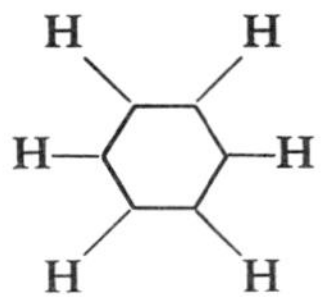

The remaining six electrons may be shown as three localized bonding pairs alternately around the ring, although we find it impossible to distinguish the two alternative formulations shown in Fig. 8.1. Although each of these representations

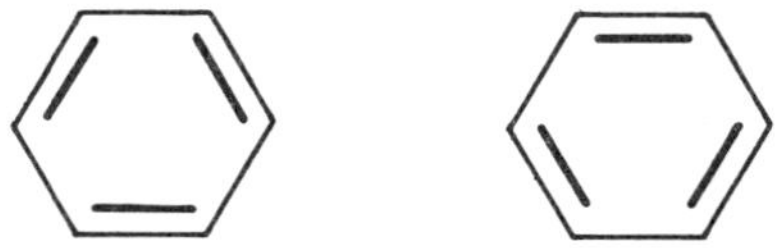

Fig. 8.1. Canonical forms for benzene

satisfies the stipulation that each atom has its *ns* and *np* electrons involved to the maximum, neither of them gives a good picture of the bonding in the benzene molecule. The best representation is one in which the six electrons here represented as three *localized* pairs in pi bonding orbitals are *delocalized* and shared with the entire ring.

In order to show this, it has become customary to use the two structural formulas in Fig. 8.1, which show the alternative ways in which *electrons* may be *localized.* These are called *canonical forms.* However, it is recognized that neither of the canonical forms adequately represents the molecule. Rather the molecule is a *hybrid* of the canonical forms. The hybrid may be conceived as an average of the canonical forms. In addition, the hybrid is the only form in which the molecule exists, since neither of the canonical forms has any existence of its own. The hybrid then represents *electron delocalization.*

It must be emphasized that canonical forms differ only in the arrangement of localized electrons. They cannot differ in atom arrangement. For example, consider the molecule of sulfur(IV) oxide, SO_2. If we desire to write a structural formula for this molecule, we must have some experimental evidence for the atom arrangement. For example, the two oxygens may be bonded to each other with a sulfur bonded to one of the oxygens, or the sulfur may lie between the two oxygens and be bonded to both.

Once it is known that the sulfur lies between the two oxygens, an attempt may be made to represent the electron distribution. The following is one stepwise approach that may be used:

1. Represent the atoms in correct sequence, ignoring details of geometry.

   O S O

2. Find the total valence electrons for all of the atoms, 18 in this case.
3. Distribute *pairs* of electrons, 9 in this case, among the three atoms, taking care first to use all *ns* and *np* orbitals of each atom.

   $|\overline{\underline{O}}-\overline{S}=\underline{O}|$

4. Consider alternative ways in which electrons might be distributed, using the *ns* and *np* orbitals of each atom.

   $|\underline{O}=\overline{S}-\overline{\underline{O}}|$

5. Think of other factors pertaining to the individual molecule that might account for other possible electron distributions. In this case the sulfur has available *nd* orbitals and the electron repulsions on the very tiny oxygen atoms tend to force πd bonding. This may be represented as follows:

   $\overline{\underline{O}}\overset{\pi d}{-\!-\!-}\overline{S}=\underline{O}|$ $\qquad$ $|\underline{O}=\overline{S}\overset{\pi d}{-\!-\!-}\underline{O}|$

6. Describe the geometry of the molecule in terms of the canonical forms. In this case all four canonical forms show three groups of electrons on the surface of the central sulfur. These three groups of electrons (two bonding groups and one nonbonding group) should be directed toward the corners of an equilateral triangle, and the sulfur dioxide molecule should be angular, with an angle of about 120°.
7. Describe the molecule as a whole. The canonical forms of sulfur dioxide are shown in Table 8.1. Sulfur(IV) oxide is an angular molecule with a central sulfur bonded to two oxygens. There are two localized pairs of electrons in sigma bonding orbitals, with one pair between sulfur and each oxygen. In addi-

TABLE 8.1

Canonical Forms for Sulfur(IV) Oxide

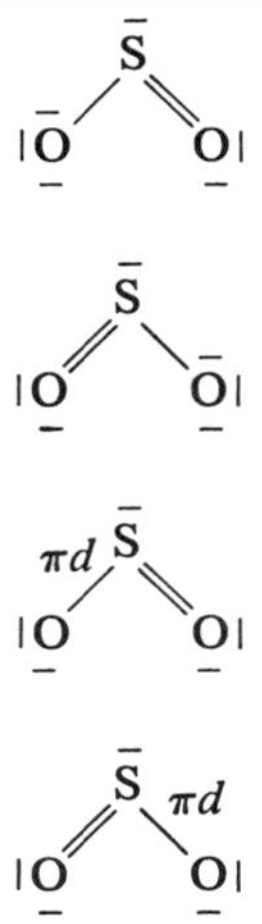

tion there are two delocalized pairs of electrons including one pair of πd electrons.

For review and reference, Table 8.2 indicates the common geometrical distributions of electron groups on atomic surfaces. These are consistent with maximum separation of charge. In the figures in Table 8.2, positions are labeled. In all cases the position labeled by *a* is the center of the atom, and all other positions are locations for electron groups. Potential bonds could be represented by straight lines from position *a* to the other positions, and the angles which these lines make are potential bond angles. These are listed in the last column of the table.

The disposition of two or three electron groups is fairly easy to see, consisting of two groups lying in a straight line including the atomic center, and three groups lying at the corners of an equilateral triangle, with the atomic center at the center of the triangle. Four groups attain maximum separation at the vertices of a regular tetrahedron with the center atom at the center of the tetrahedron. In this case all internal angles are the tetrahedral angles.

When there are five electron groups, they appear to be disposed toward the vertices of a trigonal bipyramid with the center atom at the center of the trigonal bipyramid. The trigonal bipyramid may be pictured as two tetrahedra located base to base, with the center atom at the center of the equilateral triangle formed by the common tetrahedron bases. Thus the triangle in the center forms an equatorial plane with three of the five groups of electrons at the triangle corners. Internal angles within this plane are 120°. The other two electron groups lie above and below the center of the equatorial triangle respectively. The internal angles formed by either of these groups with one of the equatorial groups is 90°. Finally the two groups above and below the equatorial plane lie on an axis perpendicular to the equatorial plane, and they pass through the center of the triangle. The internal angle between these two groups is 180°.

Where there are six electron groups, they appear to be disposed toward the vertices of an octahedron, with the center atom at the center of the octahedron.

TABLE 8.2

Geometrical Distributions for Electron Groups on Atomic Surfaces

NUMBER OF ELECTRON GROUPS	GEOMETRY		
2	Linear	*b—a—b*	$\angle bab = 180^\circ$
3	Trigonal	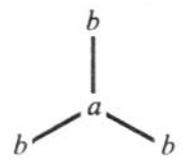	$\angle bab = 120^\circ$
4	Tetrahedral	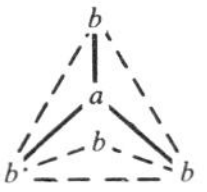	$\angle bab = 109^\circ 28'$
5	Trigonal bipyramidal	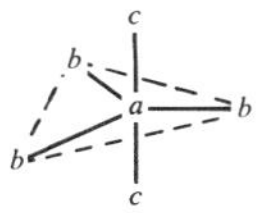	$\angle bab = 120^\circ$ $\angle bac = 90^\circ$ $\angle cac = 180^\circ$
6	Octahedral	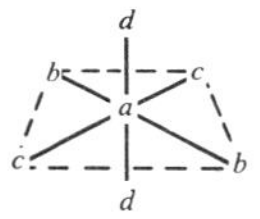	$\angle bab = \angle cac$ $= \angle dad = 180^\circ$ $\angle bac = \angle bad$ $= \angle cad = 90^\circ$
7	Pentagonal bipyramidal	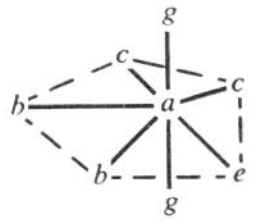	$\angle bac = \angle cad$ $= \angle dae = \angle eaf$ $= \angle baf = 72^\circ$ $\angle bad = \angle cae$ $= \angle daf = \angle bae$ $= \angle caf = 144^\circ$ $\angle bag = \angle cag$ $= \angle dag = \angle eag$ $= \angle fag = 90^\circ$

The octahedron may be pictured as two square pyramids located base to base, with the center atom at the center of the square formed by the common pyramidal bases. Thus the square in the center forms an equatorial plane with four of the six electron groups at the corners of the square. Internal angles within this plane are 90° between adjacent electron groups and 180° between opposite electron groups. The remaining two electron groups lie above and below the equatorial plane on an axis perpendicular to the plane and through the center of the square. The internal angle between either of these groups and any group in the equatorial plane is 90°, and the internal angle between these two groups themselves is 180°.

When there are seven electron groups, they appear to be disposed toward the vertices of a pentagonal bipyramid with the center atom at the center of the structure. The pentagonal bipyramid may be pictured as two pentagonal pyramids located base to base with the center atom at the center of the common pentagonal bases. Thus the pentagon in the center forms an equatorial plane with five of the seven electron groups lying at the corners of the pentagon. Internal angles within this plane are 72° between adjacent groups and 144° between nonadjacent groups. The remaining two electron groups lie above and below the equatorial plane on an axis perpendicular to the plane and passing through the center of the pentagon. The internal angle between either of these groups and any group in the equatorial plane is 90° and the internal angle between these two groups themselves is 180°.

THE ELEMENTS

Rare Gases

Some of the properties of the rare gases are summarized in Table 8.3. The elemental rare gases do not tend to form covalent bonds among themselves, primarily

TABLE 8.3
Some Properties of the Rare Gases

	HELIUM	NEON	ARGON	KRYPTON	XENON	RADON
Atomic number	2	10	18	36	54	86
Atomic weight	4.00	20.18	39.95	83.80	131.3	(222)
Atomic radius (Å)	0.93	1.31	1.74	1.89	2.09	2.14
Melting point (°K)	2(40 atm)	24	84	116	161	202
Boiling point (°K)	4	27	87	121	165	211
Heat of fusion (kcal/mole)	0.0050	0.080	0.281	0.391	0.548	0.638
Heat of vaporization (kcal/mole)	0.020	0.421	1.557	2.157	3.019	3.918
Critical temperature °K	5	44	151	210	290	378

because suitable low energy levels are not available. Interatomic forces appear to be limited to van der Waals forces, which increase as atomic sizes increase. This accounts for the increase in melting point, boiling point, molar heat of fusion, molar heat of vaporization, and critical temperature with increasing atomic number. Atomic radii have been included in Table 8.3, whereas they were not included in Table 5.7. The values given here appear to be consistent within the family, but may not be consistent with the values for the other atoms shown in Table 5.7.

Since the elements consist of individual atoms, the rare gases crystallize in close-packed lattices of identical atoms. Data are inconclusive for helium and not available for radon, but the rest of the rare gases crystallize in the cubic close-packed structure.

The Halogens

The halogens all form diatomic molecules, one pair of electrons being shared between two halogens in a sigma bonding orbital. This may be represented in general terms using X as a symbol for a halogen atom.

$$|\overline{\underline{X}}—\overline{\underline{X}}|$$

Some of the properties of hydrogen and the elemental halogens are summarized in Table 8.4. The increase in melting point, boiling point, molar heats of fusion and

TABLE 8.4

Some Properties of Hydrogen and the Halogens

	HYDROGEN	FLUORINE	CHLORINE	BROMINE	IODINE
Atomic number	1	9	17	35	53
Atomic weight	1.008	19.00	35.45	79.91	126.9
Atomic radius (Å)	0.371	0.709	1.00	1.14	1.33
Molecular form	H_2	F_2	Cl_2	Br_2	I_2
Molecular weight	2.016	38.00	70.90	159.82	253.8
Melting point °K	14	53	172	266	387
Boiling point °K	20	86	238	332	457
Heat of fusion (kcal/mole)	0.028	0.372	1.530	2.528	3.736
Heat of vaporization (kcal/mole)	0.216	1.56	4.876	7.179	10.383
Critical temperature °K	33	144	417	584	826

vaporization, and critical temperature with increasing atomic number are consistent with the increases in atomic radii and increases in van der Waals forces. The same effects can be seen in the states of the elements at room conditions. Hydrogen, fluorine, and chlorine are gases, bromine is a liquid, and iodine is a solid.

Crystals of the halogens have diatomic molecules in close-packed structures. Data are not available for fluorine, but in the crystals of chlorine, bromine, and iodine, three different distances have been measured between adjacent atoms. The shortest of these three distances is attributed to the distance between two atoms in a molecule. The next shortest distance is between atoms in adjacent molecules in a close packed layer. The longest distance is between atoms in molecules in adjacent layers.

The Chalcogens

The common molecular forms of elemental oxygen and sulfur have been compared before. The differences have been attributed to the strong tendency toward double bond formation between the very small oxygen atoms.

Thus the common molecular form of oxygen is the diatomic O_2, containing a double covalent bond between the two atoms. However, oxygen is capable of forming a second molecular form, called ozone, which exists as a triatomic molecule, O_3. Such different forms of a single *element* are called *allotropes*. Specifically, allotropes are different *molecular* or *crystalline* forms of the same *element*. Note that allotropes are elements and not compounds and they may be either different molecular or crystal forms. Sometimes it is convenient to distinguish between *molecular allotropes* and *crystal allotropes*.

Crystal allotropes should not be confused with *polymorphs*. Polymorphs are different crystal forms of the same *compound*, such as zinc blende and wurtzite, which are both zinc sulfide. Crystal allotropes are different crystal forms of the same *element*.

The two canonical forms of ozone are as follows:

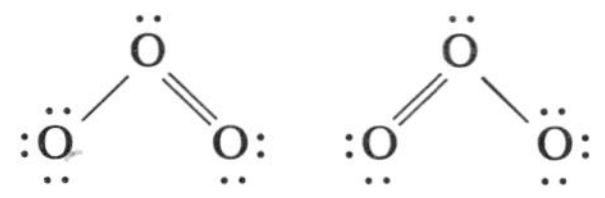

The molecule may be described as angular with three groups of electrons on the central oxygen. The expected bond angle is about 120° and the experimentally determined angle is 117°. The slightly smaller angle is attributed to the strong repulsions of the nonbonding electrons as compared to the repulsions of the more localized bonding electrons. The molecule also is a hybrid of the two canonical forms and may be described as three oxygen atoms bonded together with two pairs of electrons in sigma bonding orbitals and one delocalized pair of electrons.

In looking at the two canonical forms of ozone, it appears that due to the bonding there has been a shifting of electron density away from the central atom. As the forms have been written, the electrons tend to be toward the bottom of the molecule and away from the top. The result is that there is a separation of charge centers and ozone has an electrostatic dipole moment of 0.46 debye where the debye is defined as 3.33×10^{-30} coulomb $\times$ angstrom.

To represent such an electron density shift, we use a new convention called the *formal charge*. To assign formal charges to the atoms in a molecule, all nonbonding electrons are assigned to their respective atoms, and all bonding electrons are divided equally between the bonded atoms. The difference between the number of electrons assigned to each atom in this way and the number of valence electrons in the neutral unbonded atom is the *formal charge*. In this way formal charges may be assigned to each of the atoms in the canonical forms of ozone as follows:

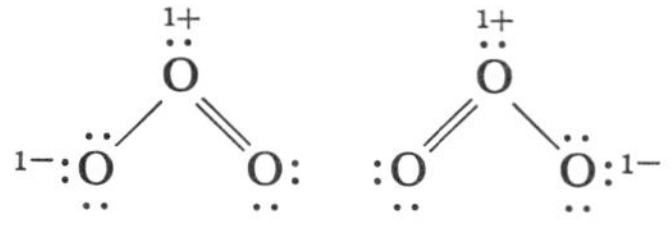

In each canonical form, the two nonbonding electrons shown on the central oxygen are assigned to that oxygen along with three of the six bonding electrons. This is a total of five electrons, which is one less than the number of electrons in the *ns* and *np* orbitals of the unbonded oxygen atom. Thus the formal charge on the central oxygen is 1+. The doubly bonded terminal atom is assigned two of the four bonding electrons and the four nonbonding electrons for a total of six, which is the same as the number of valence electrons for the unbonded oxygen atom. Therefore this atom is assigned a formal charge of zero. The singly bonded terminal atom is assigned one of the two bonding electrons and the six nonbonding electrons for a total of seven, which is one more than the valence electrons in an unbonded oxygen atom. This oxygen is assigned a formal charge of 1−.

In averaging the two canonical forms to represent the hybrid, we assign the central oxygen atom a formal charge of 1+ and each of the terminal oxygens a formal charge of $\frac{1}{2}-$. This represents a shift of electron density toward the bottom portion of the molecule as it has been represented, and suggests a separation of charge centers.

The formal charge represents a shift of electron density due to bonding and molecular geometry, without regard for electronegativity differences. Formal charges should not be confused with oxidation numbers or partial ionic charges. The representation of formal charges in the two canonical forms of ozone does not indicate polar covalent bonds. Thus ozone is an example of a substance that has nonpolar covalent bonds but has polar molecules.

Since ozone has a permanent dipole moment, ozone molecules tend to attract each other through dipole-dipole interactions, and ozone has a higher melting point, boiling point, molar heat of vaporization, and critical temperature than diatomic oxygen.

The structural formula for oxygen has not been represented because oxygen is paramagnetic and must therefore have unpaired electrons. However, ozone is diamagnetic with all electrons paired as indicated in the structures on the previous page. The paramagnetism of oxygen may be explained in terms of the molecular orbital approach used in Chapter 6. When the three $2p$ orbitals of the oxygen atoms interact, six molecular orbitals result: three bonding orbitals (one and two) and three antibonding orbitals (one and two). Between the two atoms there are eight $2p$ electrons to occupy these orbitals. Presumably six of these are in the bonding orbitals, and the other two serve as antibonding electrons to cancel the bonding effect of one of the bonding pairs. This leads to a net effect of a double bond in O_2, which is what we usually represent. However, it appears that these two antibonding electrons occupy the two equivalent antibonding pi orbitals where their spins are unpaired, thus producing the paramagnetism characteristic of oxygen.

The common molecular form of sulfur was discussed in Chapter 6, where the puckered octagonal ring was pictured in Fig. 6.9. Although it is not clear why sulfur tends to have eight atoms per ring unit instead of five, six, seven or some other number, the higher degree of catenation appears to be due to less tendency in the larger sulfur atoms to form double bonds. In the puckered octagonal ring, the bond angles are 107°, close to the anticipated tetrahedral angle.

Sulfur has two crystal allotropes, both of which appear to be composed of close packed S_8 molecules. When heated, solid sulfur melts at about 120°C to a free flowing liquid which is thought to consist of S_8 molecules. At about 160°C the liquid becomes very thick and viscous and probably consists of long chains of sulfur atoms formed by the breaking of the S_8 rings followed by the joining of the short chains to form long chains. If this viscous liquid is quenched, that is, cooled very quickly, the solid formed is quite rubbery and probably consists of long chains of sulfur atoms. Liquid sulfur boils near 445°C, and the gaseous sulfur at this temperature consists of S_8 molecules. Above 800°C the vapor consists mostly of S_2 molecules, which must be bonded in a manner similar to O_2. At temperatures in excess of 2000°C the vapor consists mostly of atomic sulfur.

Below 60°C two crystal allotropes of selenium are known, both probably composed of Se_8 rings analogous to the S_8 rings. However, above 60°C, the stable allotrope is the so-called metallic selenium consisting of parallel, close-packed infinite chains of atoms, that is, the chains are of indefinite length. Selenium melts at about 220°C and boils at about 685°C. Relatively little is known of the molecular forms in the liquid, but the vapor near the *melting point* consists of Se_6 molecules, whereas the vapor at the *boiling point* consists of highly catenated molecules. Above 900°C the molecular form is mostly Se_2 and above 2000°C it is mostly atomic selenium.

Tellurium apparently does not form the Te_8 ring structure, but it exists only in a metallic form similar to the infinite chain form of selenium. The boiling point of tellurium is about 1390°C and the vapor at this temperature consists of Te_2 molecules.

The family trends are rather clear. Oxygen tends to form diatomic molecules. Sulfur and selenium catenate to a higher degree, forming the puckered octagonal ring as well as infinite chains. Crystals consist of close-packed rings or parallel packing of chains. Chain formation reaches its maximum in tellurium, which forms only this kind of structure.

A few properties of the elements of Family 6A are recorded in Table 8.5. It is difficult to make direct comparisons among all of the elements because of the wide

TABLE 8.5

Some Properties of the Chalcogens

	OXYGEN	SULFUR	SELENIUM	TELLURIUM	POLONIUM
Atomic number	8	16	34	52	84
Atomic weight	16.00	32.06	78.96	127.6	(210)
Atomic radius (Å)	0.74	1.02	1.16	1.432	1.67
Melting point °K	54	386-392	417-493	723	527
Boiling point °K	90	718	958	1260	1235
Heat of fusion (kcal/mole of atoms)	0.053	0.29	1.6	4.28	
Heat of vaporization (kcal/mole of atoms)	0.815	3	3.4	11.9	29
Critical temperature °K	155	1313			

variety of allotropic forms in which they exist. The properties listed in Table 8.5 vary according to the molecular form of the element. In some cases ranges are given. Melting points vary, especially with crystalline form since there may be a variety of coordination numbers that will determine the number of bonds that must be broken in the melting process. Roughly, the trends are as expected if interatomic forces increase as atomic size increases.

The Nitrogen Family

The nitrogen and phosphorus molecules were discussed in Chapter 6. Nitrogen forms a diatomic molecule with three pairs of electrons localized between the two atoms. The tetrahedral P_4 molecule was pictured in Fig. 6.13, and the point was made that this type of molecule is the result of lesser tendency to form multiple bonds.

However, the P_4 molecule is the common form of phosphorus only in the vapor state. In the solid state, phosphorus exhibits several crystal allotropes, usually specified by color. The stable form at room conditions is black phosphorus, which forms a so-called *giant* molecule in which very large numbers of atoms are bonded into a giant three-dimensional matrix with covalent bonds. Thus it is not possible to distinguish any individual molecules. Figure 8.2 shows a schematic representation of the black phosphorus structure. The solid spheres are in a top layer and the dashed spheres in a layer below. Each atom forms three covalent bonds, one with each of three other atoms. Those atoms labeled *a* have all of their bonds between atoms in the two layers represented in the diagram. Those atoms in the top layer, labeled *b* have bonds to two atoms in the top layer, and a third bond to the layer that lies on top of it. Those atoms in the bottom layer that are labeled *c* have

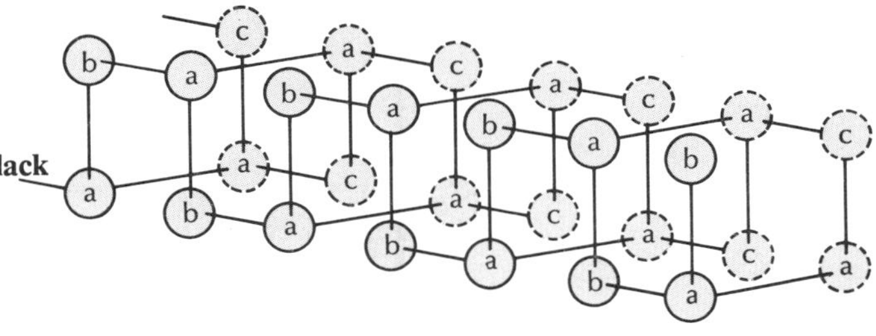

Fig. 8.2. Structure for black phosphorus

bonds to two atoms in the bottom layer with the third bond to the layer that lies below it. Bond angles in this structure are approximately 100°.

Giant molecule formation is rather characteristic of those nonmetals that tend to catenate to a high degree and have high bond energies.

White and red phosphorus are common allotropes but their structures are unknown.

In the vapor state arsenic exists mainly as the As_4 molecule. However, antimony appears to consist of a mixture of Sb_4 and Sb_2 in the vapor, and bismuth appears as a mixture of Bi_2 molecules and Bi atoms.

Both arsenic and antimony form allotropes for which the structures are unknown. However the stable modification for both arsenic and antimony in the solid state is the so-called gray or metallic form. Bismuth forms only the metallic form.

The structure of the metallic modification is represented in Fig. 8.3. The representation in Fig. 8.3 has been shown in a more open network than really exists in order to show the geometrical arrangement a little more clearly. The over-all structure is composed of molecular layers, one of which is shown in Fig. 8.3. The solid spheres form a top layer and the dashed spheres form a bottom layer. Each atom in the top layer appears to be at the apex of a trigonal pyramid, with three atoms in the lower layer lying at the corners of the triangular base. The over-all

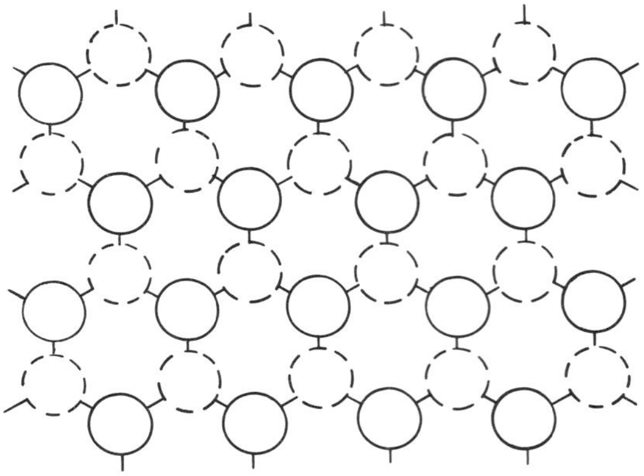

Fig. 8.3. Structure of the metallic forms of arsenic, antimony, and bismuth

effect is that of a series of fused, puckered, hexagonal rings. The total crystal is formed by stacking these layers on top of each other. Any one molecular layer consists of two layers of atoms bonded covalently. However, the molecular layers must be held together by van der Waals forces.

As a result, atoms in two adjacent molecular layers are farther apart than atoms in the two atomic layers within the same molecular layer. This means that each atom has three nearest neighbors and three more next nearest neighbors. However, in scanning down the family from arsenic to antimony to bismuth, the difference in distance between nearest and next nearest neighbors diminishes as shown in Table 8.6. This means that the atoms in bismuth are more tightly packed

TABLE 8.6 Spatial Relationships of Atoms in As, Sb, and Bi

ELEMENT, METALLIC FORM	DISTANCE TO NEAREST NEIGHBOR ATOM (A)	DISTANCE TO NEXT NEAREST NEIGHBOR ATOM (A)	DIFFERENCE BETWEEN TWO DISTANCES (A)
As	2.51	3.15	0.64
Sb	2.87	3.37	0.50
Bi	3.10	3.47	0.37

than the atoms in arsenic and it also means that the bismuth structure more nearly approximates a close-packed metallic structure. This is a part of the basis for thinking that metallic properties increase down the family. Arsenic and antimony often are considered to be metalloids, and bismuth a metal.

A few properties of the elements of family 5A are recorded in Table 8.7. It is difficult to make comparisons within the family, since there are tremendous differences in structure and aggregation from discrete molecular nitrogen to metallic bismuth. In general the trends are what might be expected. Melting points, boiling points, and heats of fusion and vaporization are low for nitrogen, as expected, and increase down the family, as there is an increase in aggregation and catenation and thus many more bonds to break in a change to a less condensed state.

TABLE 8.7

Some Properties of the Elements of the Nitrogen Family

	NITROGEN	PHOSPHORUS	ARSENIC	ANTIMONY	BISMUTH
Atomic number	7	15	33	51	83
Atomic weight	14.01	30.97	74.92	121.8	209.0
Atomic radius (Å)	0.74	1.10	1.24	1.45	1.548
Melting point °K	63	317		903	544
Boiling point °K	77	553		1653	1833
Heat of fusion (kcal/mole of atoms)	0.0861	0.15		4.88	2.61
Heat of vaporization (kcal/mole of atoms)	1.33	2.96	7.75	30.6	36.2

The Carbon Family

Carbon, silicon, and germanium appear to be typical nonmetals, forming large three-dimensional networks in the elemental state. Tin appears to be somewhat intermediate in nature, forming both a three-dimensional covalent network and a more close-packed metallic structure. Lead appears to be metallic, forming a slightly distorted cubic close-packed crystal.

Carbon forms two crystal allotropes as shown in Fig. 8.4. In diamond, each carbon atom is covalently bonded to four other carbon atoms arranged tetrahedrally around it. Thus diamond is a very strongly bonded three dimensional structure. This accounts for the hardness of diamond and the difficulty encountered in trying to melt or vaporize diamond.

Graphite is the second allotrope of carbon, and has molecular layers of fused

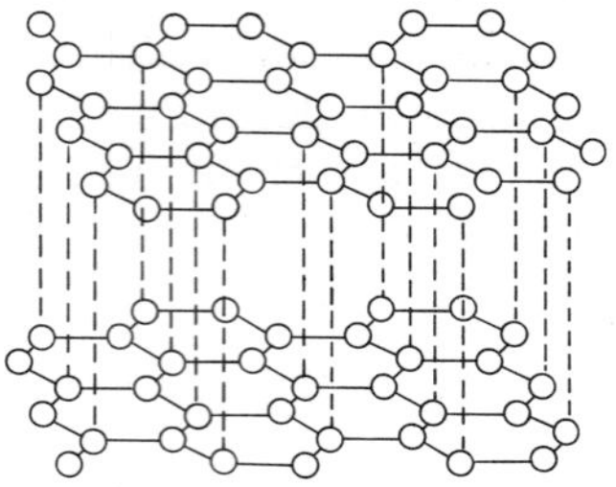

(a) Graphite

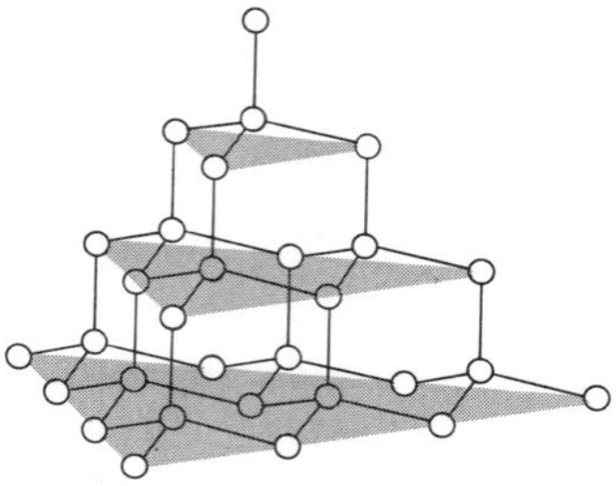

(b) Diamond

Fig. 8.4. Structures for graphite and diamond

benzene rings that are held together by van der Waals forces. The individual rings in the layers differ from benzene in that all bonds are to other carbon atoms.

Silicon and germanium have one solid modification each, namely the diamond-type structure. Tin exhibits three allotropes, one of which is the diamond-type structure, called gray tin. The other two forms appear to have metallic-type lattices of close-packed atoms. Lead exhibits only the metallic cubic close-packed structure.

Some of the properties of the elements of the carbon family are summarized in Table 8.8. In general, boiling points are universally high. In all cases many bonds must be broken. However, the metallic bond permits more mobility for atoms as indicated by the general metallic properties of malleability, ductility, etc. Therefore it is not surprising that the metallic tin and lead have rather lower melting points and heats of fusion, although their boiling points and heats of vaporization are high.

UNCATENATED BINARY COMPOUNDS

Rare Gases

In the ground state, rare gas atoms have configurations in which all electrons are paired and the usual valence orbitals are occupied. In order for these atoms to form covalent bonds, they must employ higher-energy orbitals. Thus to make

TABLE 8.8

Some Properties of the Elements of the Carbon Family

	CARBON	SILICON	GERMANIUM	TIN (WHITE)	LEAD
Atomic number	6	14	32	50	82
Atomic weight	12.01	28.09	72.59	118.7	207.2
Atomic radius (Å)	0.772	1.176	1.225	1.405	1.750
Melting point °K	3823	1680	1231	505	600
Boiling point °K	5100	2628	3100	2533	2017
Heat of fusion (kcal/mole of atoms)		11.1	7.12	1.69	1.14
Heat of vaporization (kcal/mole of atoms)		94.2	79.8	69.3	42.4

bonding energetically feasible, the other element should be highly electronegative to decrease electron potential energy. In addition, bonding among the rare gases should be most feasible for the elements with highest atomic number, where relatively little energy is required to promote electrons to the next higher energy level.

Although there is some evidence that radon may have a rich chemistry, most of the compounds of the rare gases studied so far have been fluorides and oxides of xenon and krypton.

The main oxidation states displayed by the rare gases are even, which is consistent with the notion that for bonding to occur one or more pairs of electrons in the atom at its ground state need to be promoted and unpaired, so that orbitals become available for bonding in groups of two. Thus the common oxidation states are 2+, 4+, and 6+.

Xenon(II) fluoride may be represented as follows:

The five groups of electrons on the surface of the xenon atom should be directed toward the vertices of a trigonal bipyramid. Since the nonbonding electrons are less localized and tend to be more strongly repelling, they are located in the equatorial plane where they have maximum separation. Thus the two bonding electron pairs are in the axial positions, and the molecule is linear.

Xenon(IV) fluoride may be represented as follows:

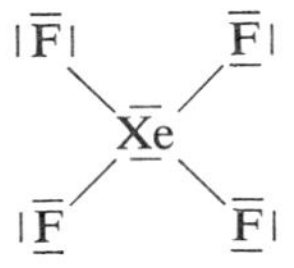

The six electron groups on the surface of the xenon atom should be directed toward the vertices of a regular octahedron. The two nonbonding pairs have maximum separation when they are in the axial positions. Thus the bonding pairs are in the equatorial positions and XeF_4 is a square planar molecule.

Xenon(VI) fluoride has been prepared, but its structure still is in doubt. The only oxide of xenon prepared so far is the highly explosive xenon(VI) oxide, XeO_3, and its structure is not known. Both krypton(II) fluoride and krypton(IV) fluoride have been prepared.

The Halogens

THE HYDROGEN HALIDES. Some of the properties of the hydrogen halides were discussed in Chapter 6. These and other properties are summarized in Table 8.9. The relatively high melting and boiling points, molar heats of fusion and vaporization, and critical temperature for hydrogen fluoride have already been attributed to hydrogen bonding. This also accounts for the fact that HF is a liquid at standard conditions, where standard conditions are 0°C and a pressure of 1 atm, whereas the other hydrogen halides are gases.

The electrostatic dipole moments are measures of molecule polarity and the fact that they decrease from HF to HCl to HBr to HI indicates decreasing molecule polarity in that same order. This coincides with decreasing bond polarity as the difference in electronegativity between hydrogen and the halogen decreases.

Dielectric constants have been included in Table 8.9. Recall from Chapter 1 that the dielectric constant is a measure of the ability of a medium to insulate electrostatic charges from each other compared to the same ability of a vacuum.

TABLE 8.9

Some Properties of the Hydrogen Halides

	HYDROGEN FLUORIDE	HYDROGEN CHLORIDE	HYDROGEN BROMIDE	HYDROGEN IODIDE
Molecular weight	20.01	36.46	80.93	127.9
State at S.C.	liquid	gas	gas	gas
Melting point, °K	190	162	187	222
Boiling point, °K	292	188	207	238
Heat of fusion, kcal/mole	0.938	0.476	0.575	0.686
Heat of vaporization, kcal/mole	7.20	3.86	4.21	4.72
Critical temperature, °K	503	324	363	423
Dipole moment, debyes	1.91	1.03	0.78	0.38
Dielectric constant at (°K)	84 (273°)	12 (165°)	7.00 (188°)	3.39 (223°)

Thus, to say that HF has a dielectric constant of 84 is to say that HF insulates electrostatic charges 84 times as well as a vacuum.

To see how hydrogen halide molecules can have an insulating effect on electrostatic charges, consider two oppositely charged plates at a fixed distance from each other, as indicated in Fig. 8.5. The space between the plates is a vacuum in Fig. 8.5a, and the two plates exert a certain attraction for each other. Then in Fig. 8.5b, imagine the charges removed and the space filled with polar hydrogen halide molecules. Then in Fig. 8.5c, the charges have been restored to the plates, causing the polar molecules to orient themselves in the electrostatic field with their negative charge centers toward the positive plate and their positive charge centers toward the negative plate. The result of this orientation is that the bulk sample between the plates takes on the appearance of an electrostatic dipole, with the

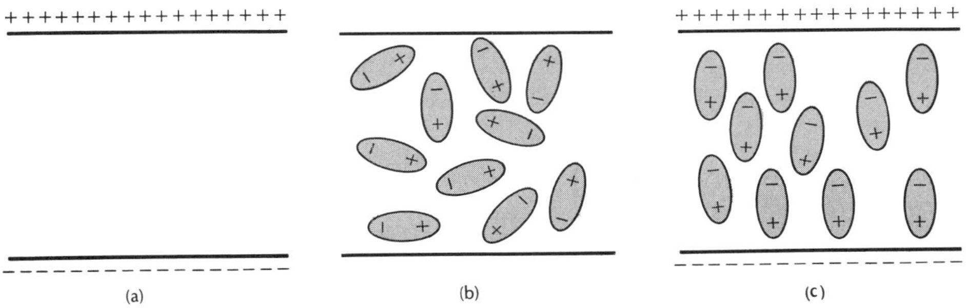

Fig. 8.5. Orientation of polar molecules in an electrostatic field

negative center displaced toward the positive plate and the positive center displaced toward the negative plate.

This is more obvious in the extreme, idealized case illustrated in Fig. 8.6, where the oriented polar molecules are closely packed between the two plates. Here it is seen that there is a layer of negative charge, adjacent to the positive plate, which in effect partially neutralizes the effect of the positive charges on the negative plate. Similarly, there is a layer of positive charge, adjacent to the negative plate, partially neutralizing the effect of the negative charges on the positive plate. Consequently the two charged plates no longer attract each other as forcefully as they did when there was a vacuum between them.

From this model it appears that the dielectric constant of a substance is related directly to the molecular polarity. In general this is true, and more polar substances have higher dielectric constants. However, even nonpolar substances tend to have dielectric constants slightly greater than one, due to the induced electrostatic dipoles caused by the electrostatic field.

The dielectric constant does depend upon both temperature and state of aggregation. As temperature increases, the molecular kinetic energy increases. This

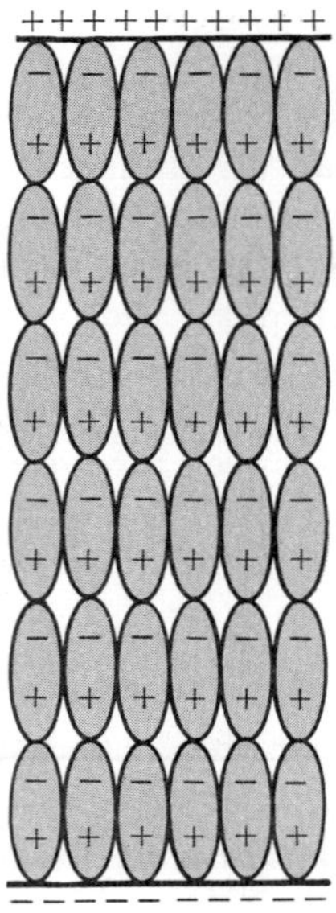

Fig. 8.6. Effect of the orientation of polar molecules upon the capacity of the condenser

kinetic energy increase opposes the orderly alignment of the molecules in the electrostatic field and the over-all insulating ability of the medium decreases. Thus dielectric constants decrease with increasing temperature. This effect for hydrogen fluoride is shown in Table 8.10.

Regardless of the molecular polarity of the substance, dielectric constants for gases tend to be near unity, since there are relatively few molecules per unit of space. Although the molecules may be oriented somewhat, the total effect is small and dielectric constants are small. Table 8.11 shows the dielectric constants for three gaseous hydrogen halides at 0°C.

In Table 8.9 the dielectric constants have been recorded for the liquids. Note that the values are recorded at different temperatures, and this must be taken into account in making comparisons. However, it is clear that the dielectric constant of hydrogen fluoride is extremely high. This is generally true of hydrogen bonded substances, since the hydrogen bonding tends to aid in the orientation of the di-

TABLE 8.10 Effect of Temperature on the Dielectric Constant of Hydrogen Fluoride

°K	κ
200	170
231	130
246	110
273	84

TABLE 8.11
Dielectric Constants for Gaseous Hydrogen Halides at 273°K

HCl	1.0046
HBr	1.0031
HI	1.0023

poles in the liquid. Due to the different temperatures at which the dielectric constants are recorded in Table 8.9, it is not possible to be sure of the comparable order of the dielectric constants for HCl, HBr, and HI in the liquid state, but from data in Table 8.11 for gaseous hydrogen halides, it appears that the dielectric constant decreases as bond polarity decreases.

METALLIC HALIDES. The hydrogen halides are all covalent in character. However, when the halogens are bonded to elements of low electronegativity, the bonds are much more ionic. The alkali metal halides are typical ionic compounds, and some of the properties of these compounds are shown in Table 8.12. The properties are in the expected order, the most ionic compound, sodium fluoride, displaying the highest melting and boiling point and the highest molar heats of fusion and

TABLE 8.12
Some Properties of the Sodium Halides

	NaF	NaCl	NaBr	NaI
Melting point, °K	1265	1073	1028	924
Boiling point, °K	1977	1686	1663	1573
Heat of fusion (kcal/mole of cations)	8.03	6.88	5.519	5.23
Heat of vaporization (kcal/mole of cations)	49.93	40.61	44.2	38.2

vaporization. The only apparent anomaly is the slightly high value for the heat of vaporization of sodium bromide.

The high melting points and high heats of fusion for the ionic halides can be understood in terms of the close packing of ions in the ionic solids. Thus these compounds tend to aggregate with high coordination numbers for the ions. The strong electrostatic forces among these ions must be overcome in converting a substance from a condensed to a less condensed state. However, in covalent compounds like the hydrogen halides, it is not necessary to break the covalent bonds within the molecules, but only the attractions between the less closely packed molecules. As a matter of interest it should be noted that elemental carbon, silicon, and germanium, which form crystals in which the atoms are all covalently bonded together, exhibit comparably high melting points and heats of fusion. See Table 8.8. In these cases covalent bonds must be broken in converting from the condensed to the less condensed state.

The Chalcogens

HYDROGEN CHALCOGENIDES. The water molecule was discussed in Chapter 6. Sulfur, selenium and tellurium form compounds with analogous formulas that may be represented as H_2Y where Y is one of the chalcogens. These compounds are called the hydrogen chalcogenides. Some properties of these compounds are summarized in Table 8.13.

TABLE 8.13

Some Properties of the Hydrogen Chalcogenides

	H_2O	H_2S	H_2Se	H_2Te
Bond angle	104.5°	92.2°	91.0°	89.5°
State at S.C.	liquid	gas	gas	gas
Melting point, °K	273	190	207	225
Boiling point, °K	373	213	232	271
Heat of fusion, kcal/mole	1.4	0.568	0.600	0.98
Heat of vaporization, kcal/mole	9.7	4.46	4.71	4.59
Critical temperature, °K	647	373	411	
Dipole moment, debyes	1.84	0.931		
Dielectric constant at (°K)	78.5 (298°)	8.0 (212°)		

The trends are analogous to those exhibited by the hydrogen halides. Like hydrogen fluoride, water has a very high melting point, boiling point, molar heat of fusion, molar heat of vaporization, critical temperature and dielectric constant. This is attributed to hydrogen bonding between the hydrogen of one molecule and the tiny, electronegative oxygen of another molecule.

The strong hydrogen bonding effects in water even dictate the crystal structure of ice. The four pairs of electrons associated with each O atom in water should be directed toward the vertices of a tetrahedron. Thus when hydrogen bonds form between water molecules, the positive hydrogens tend to be attracted toward the electron pairs on the O atom, and the O atom tends to be surrounded by four H atoms, two through regular covalent bonding and two by hydrogen bonding. This is shown in Fig. 8.7a, where the solid lines represent covalent bonds and the dotted lines represent hydrogen bonds. The over-all structure of ice may be viewed as a matrix of oxygens, each of which is surrounded tetrahedrally by four hydrogens as seen in Fig. 8.7b. The black circles represent O atoms in a top layer, and the gray circles are O atoms in a bottom layer. Each oxygen in the top layer is bonded to three hydrogens between the two layers and one hydrogen between the top represented layer and the layer above it. Two of these bonds are regular polar covalent bonds and the other two are hydrogen bonds.

The ice structure is a rather open one and not a close-packed structure at all. Hydrogen sulfide, on the other hand, apparently has a crystal structure in which the individual molecules are in a close-packed structure and each molecule has twelve nearest neighbors.

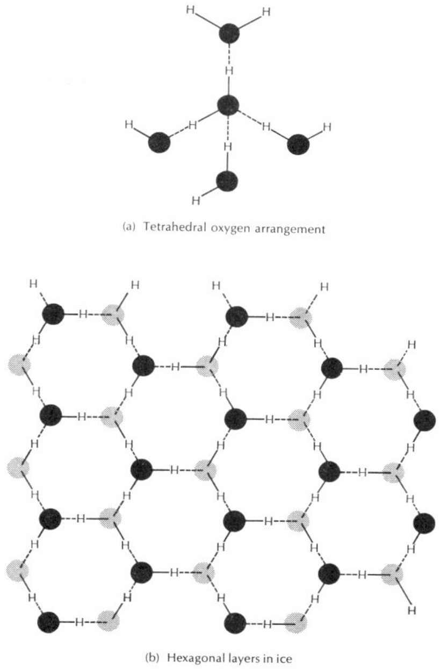

Fig. 8.7. Structure of ice, showing hydrogen bonds

Normally the melting of solids to liquids is accompanied by a decrease in density, as the increased kinetic energy tends to increase the average distance between molecules. However, when ice melts, the density of the liquid increases slightly as temperature increases, before decreasing again beyond about 4°C. This has been attributed to the breakdown of the hydrogen-bonded structure as temperature increases, with consequent closer packing of the molecules. Some densities of liquid water are shown in Table 8.14.

The hydrogen chalcogenide molecules are angular, as expected, since there are four groups of electrons on each central chalcogen. The bond angle in water is near the anticipated tetrahedral angle, and in ice the angle appears to be exactly the tetrahedral angle. The slight deviation from the tetrahedral angle among water

TABLE 8.14
Density of Liquid Water

TEMP. °C	DENSITY g/ml	TEMP. °C	DENSITY g/ml
0.0	0.999841	6.0	0.999941
1.0	0.999900	7.0	0.999902
2.0	0.999941	8.0	0.999849
3.0	0.999965	9.0	0.999781
4.0	0.999973	10.0	0.999700
5.0	0.999965		

molecules in the gaseous state is attributed to the repulsions of the nonbonding electron pairs, which force the bonding pairs together somewhat.

The larger deviations among the other similar compounds is attributed to the decreasing electronegativity of the chalcogens with increasing atomic number. As the electronegativity decreases, the bonding electrons are drawn away from the chalcogen toward the hydrogen so that the bonding electrons are further removed from the chalcogen atom surface. As a result, the bond angles may be somewhat smaller without making the bonding electrons much closer, and the repelling nonbonding electrons may occupy a greater portion of the surface of the central atom.

CHALCOGEN OXIDES. The characteristic positive oxidation states for the chalcogens are 4+ and 6+, corresponding to the involvement of four *np* electrons or both the *ns* and *np* electrons in the bonding. It might be expected that only the two unpaired *np* electrons would be involved in the bonding, as is the case among the hydrogen chalcogenides. On the other hand, it is possible for completed orbitals in one atom to interact with empty orbitals in another atom to form molecular orbitals. This can be accomplished with oxygen by having the two unpaired electrons paired in a single orbital, making an empty orbital available.

The chalcogen(IV) oxides have the type formula YO_2. Sulfur(IV) oxide was used as an example earlier in this chapter. The structure of the molecule is angular and the bonding involves a pair of bonding electrons in delocalized pi orbitals.

The larger selenium and tellurium have less tendency to form the double bonds, that is, they have less tendency to form pi bonding orbitals. As a result the selenium(IV) oxide exists in long chains of alternating oxygen and selenium atoms as follows:

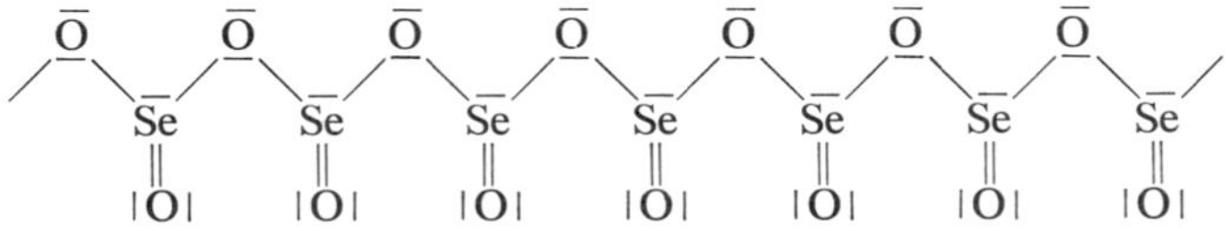

Thus there is no discrete molecule of selenium dioxide, although this chain is composed of repeating SeO_2 units. Compounds composed of large structural units consisting of smaller repeating units are called *polymers*. The SeO_2 polymeric chain is not planar. The oxygens lie in one plane and the seleniums in another, so that each selenium lies at the apex of a trigonal pyramid with the three oxygens at the corners of the triangular base. The structure of tellurium(IV) oxide is analogous to the structure for selenium(IV) oxide.

The vapor of sulfur(VI) oxide consists of discrete SO_3 molecules for which the following canonical forms may be written:

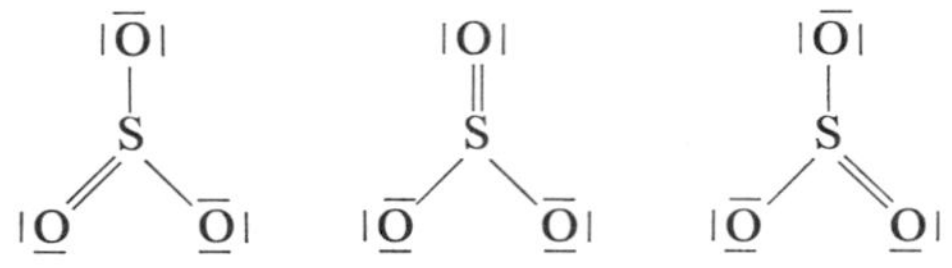

Thus in addition to any πd electrons, one pair of bonding electrons is delocalized. Since there are three groups of electrons on the surface of the central sulfur, the molecule should be planar with bond angles of 120°. This is in accord with experimental results.

The bonds in sulfur(VI) oxide are polar covalent, since oxygen is more electronegative than sulfur, but the molecule is nonpolar and has a zero electrostatic dipole moment. The symmetry of the molecule indicates that the center of the sulfur atom is the center of positive charge as well as the center of negative charge.

Solid sulfur(VI) oxide is trimorphic, that is, it exists in three different modifications, and none of these involves the discrete SO_3 molecule. Two of them consist of different packings of polymeric chains of repeating SO_3 units that may be depicted as follows:

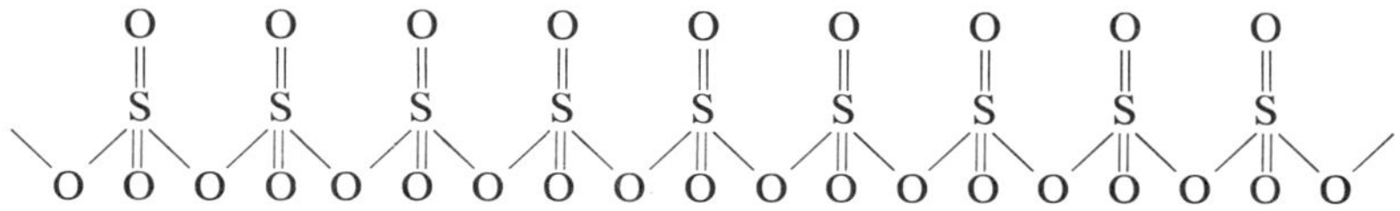

In this chain each sulfur is surrounded tetrahedrally by four oxygens, so that the chain is staggered rather than straight. The bond angles are approximately tetrahedral angles.

The third form of solid sulfur(VI) oxide consists of trimeric molecules, that is, the molecules consist of three repeating units. The structure is cyclic with three sulfurs and three oxygens alternated around a puckered hexagonal ring as shown in Fig. 8.8. In this structure, each sulfur is surrounded by four oxygens in tetrahedral positions. From one point of view, the ring may be considered to be a short chain which has closed to form the ring.

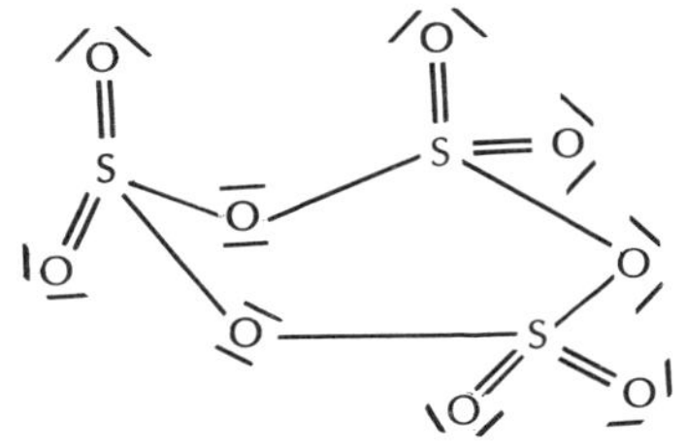

Fig. 8.8. Structure of the cyclic trimer of sulfur(VI) oxide

Selenium(VI) oxide is dimorphic with both modifications consisting of polymeric chains similar to those depicted for sulfur(VI) oxide. The molecular structure for tellurium(VI) oxide is unknown although it is dimorphic.

OXIDES OF OTHER ELEMENTS. Nearly all elements form oxides. Features of specific oxides are discussed under the headings of the elements that are combined with oxygen. However, in a general way, it should be noted that oxides of the metals tend to be ionic and thus form ionic-type crystals. However, among the metals, the covalent character of the bonds increases as the oxidation state of the metal increases. In an oxide like Mn_2O_7, for example, competition among the oxygens for electrons is great, and the bonds display a great deal of covalent character.

The oxides of the nonmetals are predominantly covalent, although within families of nonmetals the elements of high atomic number tend to be less covalent.

SULFIDES, SELENIDES, AND TELLURIDES. Sulfides are less numerous than oxides. Sulfides of elements with high oxidation numbers are rare because elements in high oxidation states tend to be highly electronegative, often sufficiently so to remove electrons from the sulfide leading to elemental sulfur.

However, many elements form sulfides in their lower oxidation states. In bonds with metals, sulfides tend to be less ionic than oxides. Many of the sulfides of the alkali metals, alkaline earths, and transition metals may be considered to be predominantly ionic.

Selenides and tellurides are much less common than sulfides and they usually occur only for elements in relatively low oxidation states.

CHALCOGEN HALIDES. Halides exist for the chalcogens in 2+, 4+, and 6+ oxidation states corresponding to the utilization of 2, 4, or 6 valence electrons. In addition, only the fluorides are known for the chalcogens in the 6+ oxidation state since only the small fluorines can fit around the central chalcogen in the required close-packed arrangement. Consistent with the electronegativity of fluorine, the fluorides are best known in the higher oxidation states (4+ and 6+), whereas the less electronegative chlorine, bromine, and iodine tend to form compounds in which the chalcogen oxidation number is 2+ or 4+. Sulfur forms no simple bromides or iodides, and selenium forms no iodides. This is consistent with the relative sizes of the atoms involved. The large tellurium atom is the only one to be surrounded by large iodine atoms, for example. The important simple halides of the chalcogens are listed in Table 8.15.

TABLE 8.15 The Principal Uncatenated Chalcogen Halides

	SCl_2			$SeBr_2$		$TeCl_2$	$TeBr_2$	
SF_4	SCl_4	SeF_4	$SeCl_4$	$SeBr_4$	TeF_4	$TeCl_4$	$TeBr_4$	TeI_4
SF_6		SeF_6			TeF_6			

The dihalides have the following type structure, where Y is a chalcogen and X is a halogen.

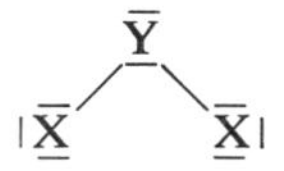

The molecules are angular with bond angles near 100°, which is slightly less than the tetrahedral angle.

In the vapor, the three tetrafluorides, SF_4, SeF_4, and TeF_4, along with $TeCl_4$, exist as discrete molecules that may be represented as follows:

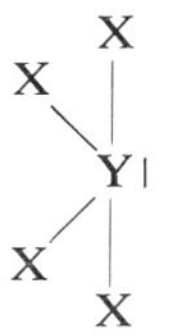

The five electron groups on the surface of the central chalcogen are directed toward the vertices of a trigonal pyramid. The highly repelling nonbonding electron pair is in an equatorial position, which places the axial halogens at a potential energy disadvantage, since they are at an angle of only 90° from the equatorial positions. Consequently, only small atoms, like fluorine atoms, can fit into these positions.

This helps to explain why both SCl_4 and $SeCl_4$ exist only in the solid state. They appear to be ionic compounds containing the ions SCl_3^+ and $SeCl_3^+$ respectively. These ions may be represented by the following type structural formula:

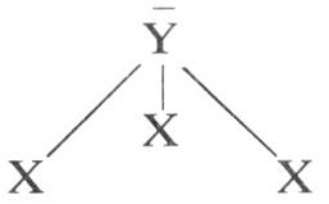

The ions are pyramidal, with the chalcogen at the apex and the three chlorines at the trigonal base corners. Energetically this provides a more advantageous arrangement in which the larger halogens are further removed from each other along the larger tetrahedral angle, which approximates 109°. The existence of these substances in the solid state also allows for a potential energy decrease in the close-packed crystals, due to electrostatic attractions. In fact, this decrease in potential energy is essential to provide the impetus for the ionization where the electronegativity differences certainly would not suggest ion formation.

The Nitrogen Family

HYDROGEN COMPOUNDS OF THE NITROGEN FAMILY. The structure of ammonia, NH_3, was discussed in Chapter 6. The other members of this family form analogous compounds with the common names phosphine (PH_3), arsine (AsH_3), stibine (SbH_3), and bismuthine (BiH_3). Each of these may be represented by the following type structural formula, where Z represents one of the nitrogen family members.

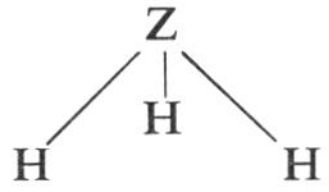

Each of these molecules has the shape of a trigonal pyramid with four groups of electrons on the surface of the atom of the nitrogen family element.

Some properties of these compounds are shown in Table 8.16. It is evident that the same trends are found here that were noted among the hydrogen halides

TABLE 8.16

Some Properties of Hydrogen Compounds of the Nitrogen Family

	NH_3	PH_3	AsH_3	SbH_3
Bond angle	107°	94°	92°	92°
Melting point, °K	195	139	159	185
Boiling point, °K	240	185	218	185
Heat of fusion, kcal/mole	1.35	0.270	0.285	
Heat of vaporization, kcal/mole	5.58	3.49	3.99	5.09
Dipole moment, debyes	1.44	0.55	0.15	
Critical temperature, °K	405	324		
Dielectric constant at (°K)	22.4 (240°)	2.6 (223°)	2.6 (223°)	2.9 (193°)

and hydrogen chalcogenides. The bond angles are all somewhat under the predicted tetrahedral angle, again decreasing as the electronegativity of the central atom decreases. The extraordinarily high melting and boiling points, heats of fusion and vaporization, and dielectric constant for ammonia must be the result of hydrogen bonding. Indeed the crystal structure of solid ammonia is described best as distorted cubic close packing of nitrogen atoms that has been expanded to permit hydrogen atoms between all adjacent nitrogens. In an undistorted structure, each nitrogen atom would have a coordination number of 12. In the ammonia structure, however, nitrogen has six nearest neighbors and six next nearest neighbors that are slightly more distant. This suggests that there may be both strong and weak hydrogen bonds.

TRIHALIDES OF THE NITROGEN FAMILY ELEMENTS. Except for nitrogen(III) bromide, all of the trihalides of the nitrogen family elements are known. Some of the properties of these compounds are summarized in Table 8.17. As in the simple binary compounds with hydrogen, the molecules of most of these compounds are pyramidal, with four groups of electrons arranged tetrahedrally around the central atom. One exception is bismuth(III) fluoride, which is ionic and exists in the solid state as an ionic crystal. Attempts to melt or vaporize it cause decomposition. Except for nitrogen(III) fluoride, the nitrogen trihalides tend to be quite explosive.

As seen in Table 8.17, for any given nitrogen family element the bond polarity, and consequently the molecular polarity, decrease with increasing atomic number of the halogen. This is shown by the electrostatic dipole moments. Similarly, in the trifluorides, for example, the dipole moments increase with increasing atomic number of nitrogen family element. This is due to the decreasing electronegativity in the nitrogen family and leads to greater electronegativity differences between the nitrogen family element and fluorine.

Although there are some fluctuations, in general the trends in melting and boiling points and molar heats of fusion and vaporization follow the trends in dipole moments. Thus high molecular polarity leads to dipole-dipole interactions and high melting and boiling points, for example.

PENTAHALIDES OF THE NITROGEN FAMILY. Phosphorus, arsenic, and antimony form pentafluorides and pentachlorides. Nitrogen probably is too small to permit

TABLE 8.17

Some Properties of the Trihalides of the Nitrogen Family

		NX_3	PX_3	AsX_3	SbX_3	BiX_3
F	Melting point, °K	65	121	265	564	
	Boiling point, °K	154	172	333	591	
	Heat of fusion, kcal/mole	0.095		2.48		
	Heat of vaporization, kcal/mole	2.77	3.94	7.10		
	Bond angle	102°	104°	102°		
	Dipole moment, debyes	0.24	1.02	2.65 in C_6H_6	3.8 in CS_2	
Cl	Melting point, °K	146	179	260	340	503
	Boiling point, °K	344	348	403	496	720
	Heat of fusion, kcal/mole		1.08	2.41	3.03	2.61
	Heat of vaporization, kcal/mole		7.29	8.36	10.8	17.4
	Bond angle		100°	98°	100°	100°
	Dipole moment, debyes		0.90 in CCl_4	2.15 in C_6H_6		
	Dielectric constant at (°K)		4.7 (295°)	12.6 (290°)	33 (348°)	
Br	Melting point, °K		233	334	369	490
	Boiling point, °K		448	494	561	726
	Heat of fusion, kcal/mole			2.82	3.51	
	Heat of vaporization, kcal/mole		9.27	9.99	14	18.0
	Bond angle		102°	101°	97°	100°
	Dipole moment, debyes		0.61 in CCl_4	1.60 in CCl_4	3 in C_6H_6	
	Dielectric constant at (°K)		3.9 (293°)	8.8 (308°)	20.9 (373°)	
I	Melting point, °K		334	414	443	683
	Boiling point, °K			673	990	
	Heat of fusion, kcal/mole			2.2	4.20	
	Heat of vaporization, kcal/mole		10.5	14.2	16.4	
	Bond angle		102°	102°	99°	
	Dipole moment, debyes		0 in CS_2	0.96 in CS_2	1.58 in CS_2	
	Dielectric constant at (°K)		4.1 (338°)	7.0 (423°)	13.9 (448°)	

the clustering of five atoms around it. In addition the large halides do not appear, since they are so large that five of them cannot cluster around a central atom.

The bonding in these molecules must involve the *nd* orbitals of nitrogen family element, which is another reason for the failure of nitrogen to form pentahalides. The five halogens are arranged around the central atom at the vertices of a trigonal bipyramid as expected. Exceptions to this occur for PCl_5 and PBr_5 in which there is a crowding problem for the halogens around the relatively small phosphorus.

In the solid state, PCl_5 exists as a matrix of the PCl_4^+ and PCl_6^- ions. In the PCl_4^+ ions the four halogens are arranged tetrahedrally around the central phosphorus. This permits less crowding than exists in the trigonal bipyramidal PCl_5 type structure. In PCl_6^- the six chlorines are arranged octahedrally around the central phosphorus. There is crowding here, but it is not much more severe than it would be in the trigonal bipyramid and the decrease in potential energy accompanying the packing of the ions into the crystal lattice compensates for this.

In solid PBr_5, the crystal is composed of PBr_4^+ and Br^- ions. In this case the bromines are too large to permit the octahedral packing around the phosphorus.

NITROGEN OXIDES. Nitrogen forms a large number of oxides corresponding to oxidation numbers of 1+, 2+, 3+, 4+, and 5+ for nitrogen. Of these the uncatenated oxides are nitrogen(II) oxide, nitrogen(III) oxide, nitrogen(IV) oxide and nitrogen(V) oxide.

In the vapor, nitrogen(II) oxide has the molecular formula NO. Since this molecule contains an odd number of electrons, it is impossible that all electrons are paired, and nitrogen(II) oxide gas is paramagnetic. The data suggest that there is one unpaired electron. Several canonical forms may be written for NO. Three of them are as follows:

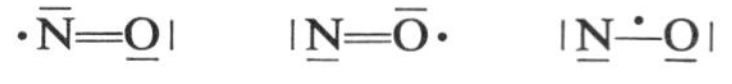

In the liquid and solid states, nitrogen(II) oxide is diamagnetic suggesting that two NO molecules have bonded together to form a dimeric molecule. Although the structure of the solid is not known, it is not unlikely that the molecular unit has a structure for which one of the canonical forms is

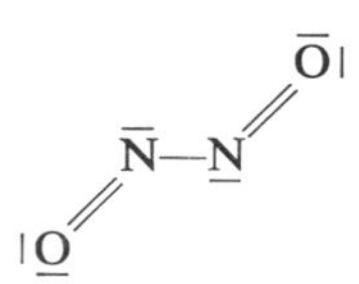

Nitrogen(III) oxide has the molecular formula N_2O_3. However, it is a fairly unstable species and its structure is not definitely established. There is some evidence that the molecule is a staggered chain of alternating oxygen and nitrogen atoms, as suggested by the following canonical form:

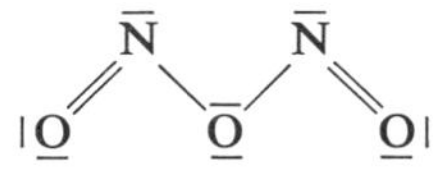

In the vapor, nitrogen(IV) oxide has the molecular formula NO_2. This is another example of a molecule containing an odd number of electrons. In the vapor, nitrogen(IV) oxide is a brownish-red, paramagnetic gas. The molecules may be described by a very large number of canonical forms. The following are three typical canonical forms:

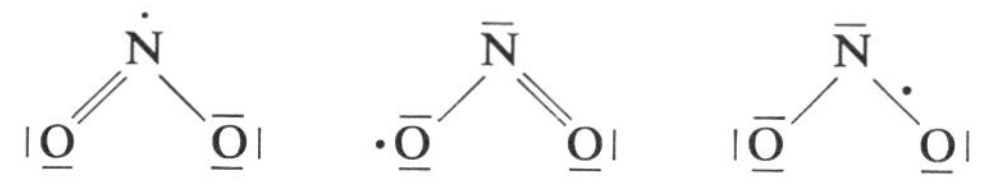

Like nitrogen(II) oxide, which also has an odd number of electrons in its simple molecule, nitrogen(IV) oxide tends to dimerize to N_2O_4 in the liquid and solid states. The experimental evidence suggests that this molecule contains a N—N bond, and that the molecule is planar. Many canonical forms may be written. Three of these are as follows:

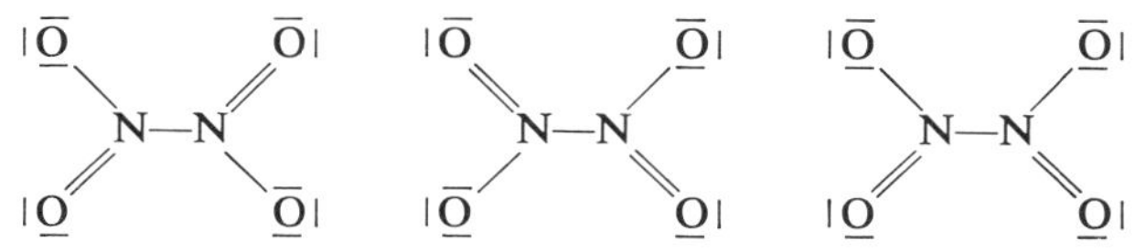

Nitrogen(V) oxide is not very stable but it has been prepared. Although the molecular structure has not been determined in the gas phase, the molecular weight corresponds to the molecular formula N_2O_5. In the solid state, the crystals appear to be composed of aggregates of nitryl ions, which have the formula NO_2^+, and nitrate ions, which have the formula NO_3^-.

PHOSPHORUS OXIDES. Phosphorus forms two oxides corresponding to the oxidation states 3+ and 5+ for phosphorus. This corresponds to the involvement of the three 3*p* electrons or the two 3*s* and three 3*p* electrons respectively.

The empirical formula for phosphorus(III) oxide is P_2O_3. However, in the vapor the compound exists as P_4O_6 molecules. The structure for P_4O_6 is given in Fig. 8.9. The four phosphorus atoms are at the vertices of a tetrahedron, and each of the oxygen atoms lies just off one of the edges of the tetrahedron. In this way each phosphorus is bonded to three oxygens.

Phosphorus(V) oxide has the empirical formula P_2O_5. However, the molecular form of the vapor is P_4O_{10}. The structure is shown in Fig. 8.10. It is similar to the P_4O_6 structure except that each phosphorus is bonded to an additional oxygen. Thus each phosphorus is surrounded tetrahedrally by oxygens.

The crystal structure for phosphorus(III) oxide is unknown, but phosphorus(V) oxide is trimorphic. One modification consists of close-packed P_4O_{10} molecular units. The other two are three-dimensional giant molecules of interconnected PO_4 tetrahedra, in which oxygens are shared between adjacent tetrahedra. In one modification, there are ten tetrahedra arranged in a cyclic formation, and in the other there are six tetrahedra arranged in a ring. Some feeling for this latter struc-

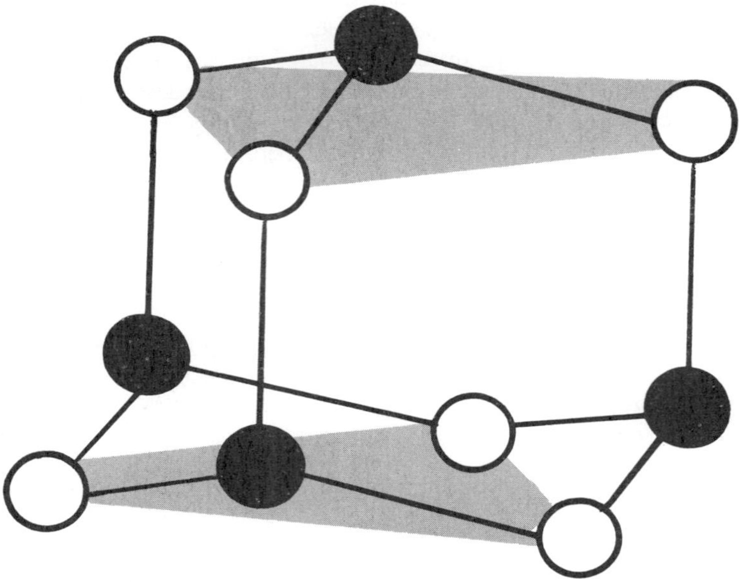

Fig. 8.9. Molecular structure for P_4O_6

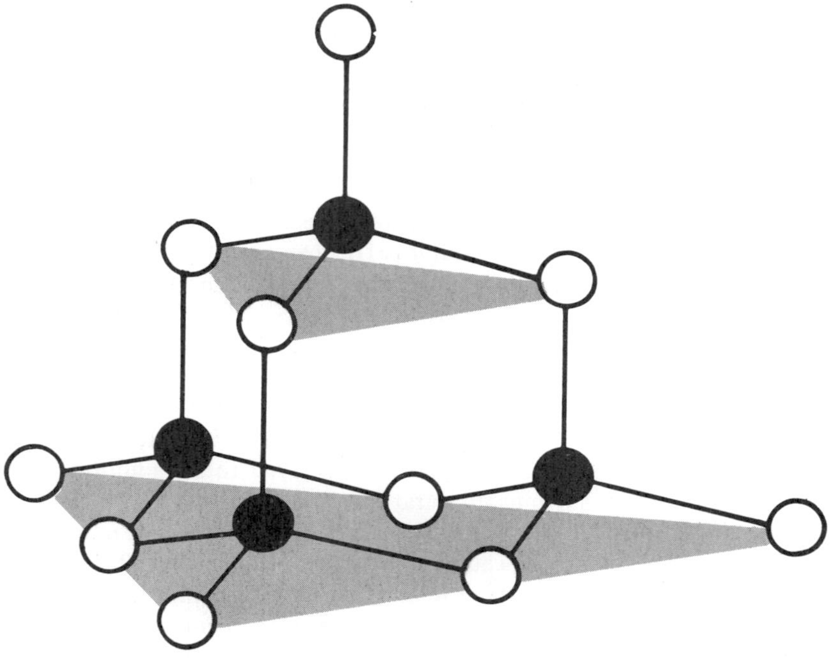

Fig. 8.10. Molecular structure for P_4O_{10}

ture can be obtained from Fig. 8.11 where the P and O atoms are in two layers. The P atoms are in the top layer and the O atoms are in the bottom layer. Each P atom is bonded to three O atoms in the bottom layer, and from this point of view is located at the apex of a trigonal pyramid with the three O atoms at the corners of the base. In addition, each P atom is bonded to another O atom in a layer above it and this O atom is bonded to another P atom in the layer above that.

The Carbon Family

In considering the compounds of the carbon family elements, it must be noted that the elements can display either of two oxidation states, namely 2+ and 4+. The 4+ state corresponds to bonding through the excited state of the atom in which there are four unpaired electrons. The 2+ state involves only the two unpaired *ns* electrons. The 4+ state is the most important state for carbon, silicon, and germanium. The 2+ state is more important for tin and lead, the more metallic elements.

HYDROGEN COMPOUNDS OF THE CARBON FAMILY. All elements of this family form compounds of the type QH_4, where Q is one of the family members. The stability of these compounds decreases as the atomic number of Q increases and PbH_4 is rare and relatively unstudied. The compounds and their names are CH_4 (methane), SiH_4 (silane), GeH_4 (germane), SnH_4 (stannane), and PbH_4 (plumbane).

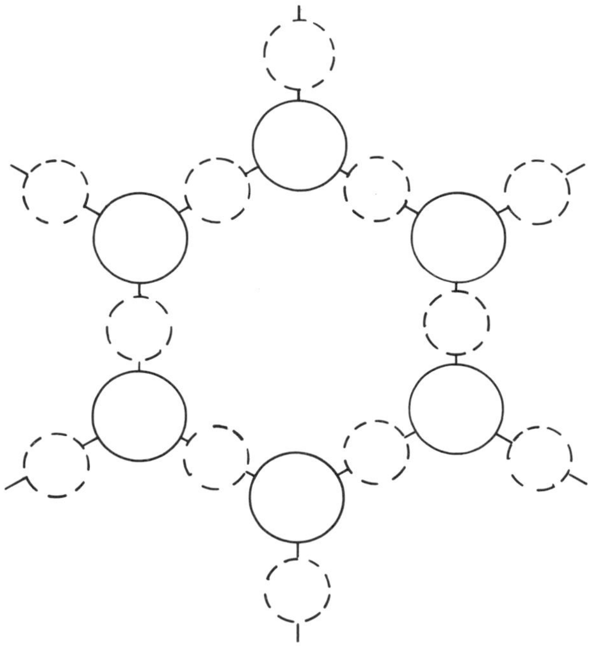

Fig. 8.11. Crystal structure for P_4O_{10}

All molecules are tetrahedral and as a result have zero electrostatic dipole moments.

Some of the properties of these compounds are compiled in Table 8.18. All of them have low melting and boiling points and heats of fusion and vaporization consistent with the lack of dipole-dipole interactions among these nonpolar molecules. Hydrogen bonding is not apparent in methane, due to the low polarity of the carbon-hydrogen bond. The compounds appear to exist as discrete molecules in all phases.

TABLE 8.18

Some Properties of Hydrogen Compounds of the Carbon Family

	CH_4	SiH_4	GeH_4	SnH_4	PbH_4
Melting point, °K	90	88	108		
Boiling point, °K	111	161	183	221	260
Heat of fusion, kcal/mole	0.224	0.159	0.200		
Heat of vaporization, kcal/mole	1.95	2.90	3.36	4.55	
Critical temperature, °K	191	269			
Dipole moment, debyes	0	0	0		

HALIDES OF THE CARBON FAMILY. All the elements in this family form tetrahalides of the type QX_4. The molecules are tetrahedral. Germanium forms the dihalides GeX_2, which are relatively unstable, but both tin and lead form stable and well-characterized dihalides.

Some properties of the carbon tetrahalides are shown in Table 8.19. Carbon tetraiodide has been omitted since it decomposes upon heating before melting.

TABLE 8.19

Some Properties of the Carbon Tetrahalides

	CF_4	CCl_4	CBr_4
Melting point, °K	89	250	321
Boiling point, °K	145	350	463
Heat of fusion, kcal/mole	0.147	0.598	0.944
Heat of vaporization, kcal/mole	2.94	7.17	10.4
Critical temperature, °K		556	
Dielectric constant at (°K)		2.24 (293°)	

The tetrahalides are nonpolar due to tetrahedral geometry; they appear to exist in all states as discrete molecules.

Some properties of the tetrachlorides of the carbon family are shown in Table 8.20.

In general the dihalides of tin and lead have considerable ionic character. In the vapor they appear as nonlinear molecules which are aggregated to some extent.

TABLE 8.20
Some Properties of the Tetrachlorides of the Carbon Family

	CCl_4	$SiCl_4$	$GeCl_4$	$SnCl_4$	$PbCl_4$
Melting point, °K	250	203	223	243	258
Boiling point, °K	350	331	359	387	
Heat of fusion, kcal/mole	0.598	1.84		2.19	
Heat of vaporization, kcal/mole	7.17	6.86	7.10	8.34	
Critical temperature, °K	556	506	550	592	
Dipole moment, debyes	0	0	0	0	
Dielectric constant at (°K)	2.24 (293°)	2.4 (289°)	2.43 (298°)	2.89 (293°)	2.78 (293°)

However, in the solid they take on the structure of ionic solids. The iodides are more covalent than the bromides, which in turn are more covalent than the chlorides and fluorides. Undoubtedly these are examples of substances in which the ionization is induced by the close-packing arrangements in the solid state.

OXIDES OF THE CARBON FAMILY. Carbon forms two oxides corresponding the 2+ and 4+ oxidation states for carbon. Carbon monoxide is best represented with a triple bond between the carbon and oxygen as follows:

$$|C\equiv O|$$

Carbon dioxide consists of linear molecules with the following structure:

$$|\underline{O}=C=\underline{O}|$$

The carbon monoxide molecule is polar but the carbon dioxide molecule is nonpolar despite rather highly polar bonds. This is due to the molecular geometry.

The common oxide of silicon has the empirical formula SiO_2. However, this compound does not exist as discrete molecules but rather as very high polymers, so that the molecular formula might better be written as $(SiO_2)_x$ to indicate that it is composed of repeating SiO_2 units.

The common name for silicon(IV) oxide is silica. Solid silica exists in three crystalline modifications as well as a glass. Glasses are solids in which the somewhat random arrangement of the liquid state is retained. Naturally occurring glassy silica is called *obsidian*.

The crystalline forms of silica are all giant three-dimensional molecules consisting of tetrahedral SiO_4 units in which each oxygen is shared by two tetrahedra.

Boron

Boron does not form the simple compound BH_3 with hydrogen, but it does form the trihalides which have the type formula BX_3. These molecules are planar as suggested by the following structural formula.

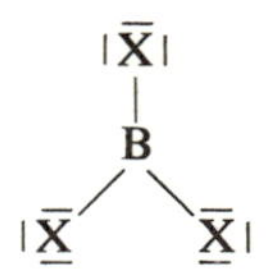

Boron also forms an oxide with the empirical formula, B_2O_3. However, in the crystalline state it consists of a giant three-dimensional molecular network somewhat resembling the silica structures. However, all four oxygen atoms in the BO_4 tetrahedra are not at the same distance from the boron.

CATENATED BINARY COMPOUNDS

The Chalcogens

The simple catenated species in this family were discussed somewhat in Chapter 6.

HYDROGEN PEROXIDE. Hydrogen peroxide, H_2O_2, is the single prominent example of oxygen catenation among the binary compounds. The structure of H_2O_2 may be represented as follows:

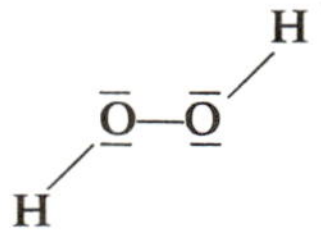

The bond angles in H_2O_2 are 101.5°, somewhat less than the tetrahedral angle due to the repulsions of the nonbonding electrons.

Hydrogen peroxide is unstable and explodes violently at 144°C. Its boiling point has been estimated at 150°C. However, its freezing point has been determined at −2°C. These high values, coupled with a high dielectric constant of 89.2 at 0°C, suggest that H_2O_2 is highly hydrogen bonded, which is not unexpected.

As a matter of organizational convenience, we often look upon compounds containing hydrogen as parents of families of compounds. The compounds in the family may be derived from the parent by replacing one or more of the hydrogens with some other atom or group of atoms. Note that this derivation is a formal one used for classification purposes and does not necessarily represent a way of preparing compounds in the laboratory.

From this point of view, for example, H_2O_2 may be considered as the parent of a compound with the formula Na_2O_2, in which the two hydrogens have been replaced by sodiums. Since sodium has a very low electronegativity and the oxygens have very high electronegativities. it is expected that any sodium-oxygen bonds are ionic and a type of structural formula may be written as follows:

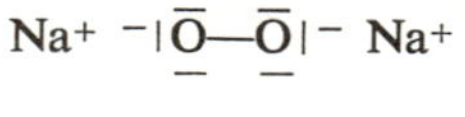

However, any ionic compound of this type is highly aggregated and the formula Na_2O_2 really is a modified empirical formula suggesting a structure composed of Na^+ ions and O_2^{--} ions. The O_2^{--} ion is called the peroxide ion. One can imagine other compounds composed of other cations with the peroxide ion. For example, barium peroxide is BaO_2.

Hydrogen peroxide, then, is the parent of a series of ionic peroxides. However, the hydrogens in H_2O_2 may be replaced by other atoms so that the bonds are covalent. Such compounds generally are called *peroxo* or *peroxy* compounds, and are characterized by an O—O single bond somewhere in the molecule.

It has been pointed out that H_2O_2 is unstable. Often by replacing hydrogens in a parent compound with something else the stability can be affected rather drastically. One common observation is that stability often is increased when the resulting derived compound is ionic. To a large degree this must be attributed to the decreased potential energy state due to aggregation of oppositely charged ions. Whereas H_2O_2 decomposes at 144°C, Na_2O_2 is a solid which may be heated to its melting point of 460°C before appreciable decomposition occurs. However, upon melting, it decomposes, since the close-packed ionic structure upon which its stability depends is partially destroyed.

THE POLYSULFIDES. The hydrogen polysulfides also were discussed in Chapter 6. Although sulfur tends to catenate to a larger degree than oxygen because of its lesser tendency to form double bonds, the hydrogen polysulfides are not very stable.

However, the hydrogen polysulfides may be considered as parents for a family of polysulfides in which the hydrogens have been replaced by less electronegative elements, for example, sodium. These more ionic polysulfides tend to be somewhat more stable. The polysulfide ions usually are represented collectively as S_x^{--} where x has values from 2 to 6.

The disulfide ion, the sulfur analog of the peroxide ion, occurs more readily than the peroxide with the transition elements. Some of the common disulfides are FeS_2, CoS_2, NiS_2, MnS_2, OsS_2, and RuS_2. The most familiar of these is FeS_2, called pyrite or fool's gold. All of the disulfides mentioned here form a class called *pyrites*. All of the pyrites have the same type crystal structure resembling the rock salt structure composed of a cubic close-packed lattice of disulfide ions, S_2^{--}, with the metallic cations in the octahedral holes. In general the pyrites have a metallic luster and resemble gold in appearance.

The Nitrogen Family

HYDRAZINE AND DIPHOSPHINE. The structure of hydrazine was discussed in Chapter 6. Phosphorus forms a similar compound called diphosphine, which may be represented by the following structural formula:

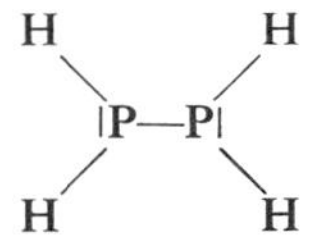

NITROGEN OXIDES. The dimerization of nitrogen(II) oxide and nitrogen(IV) oxide was mentioned earlier in this chapter. However, nitrogen(I) oxide, N_2O, involves the catenation of nitrogen. The molecule is linear and may be represented by the following two canonical forms:

The molecule is polar with an electrostatic dipole moment of 0.17 debyes.

HYDROGEN AZIDE. A higher degree of catenation is found in hydrogen azide, which has the molecular formula HN_3. The molecule may be represented by the following two canonical forms:

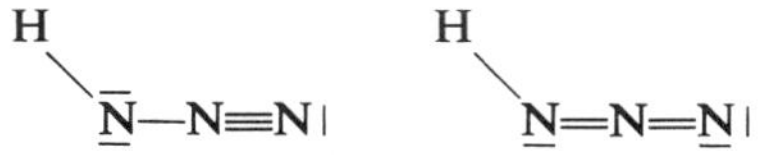

The three nitrogen atoms do lie in a straight line as suggested by these two canonical forms.

There are azides known in which the hydrogen has been replaced by other atoms. For example, the alkali metal azides are aggregates of alkali metal cations and azide ions which have the formula N_3^-, and which may be represented structurally by the following three canonical forms:

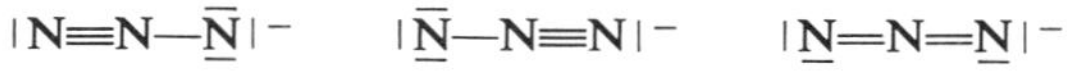

The third of these canonical forms is the average of the other two and by itself is probably a good representation of the structure of the azide ion. Hydrogen azide also is called hydrazoic acid.

The Hydrocarbons

An introduction to some parent hydrocarbons was made in Chapter 6. Here we shall expand upon that introduction to glimpse some of the multitude of different hydrocarbons known.

ALKANES. Methane, CH_4, is the simplest hydrocarbon. If we imagine a hydrogen removed from methane, the *methyl group* results. The methyl group consists of a carbon and three hydrogens and usually is written as CH_3. If two methyl groups are bonded together, another hydrocarbon results which may be represented as follows:

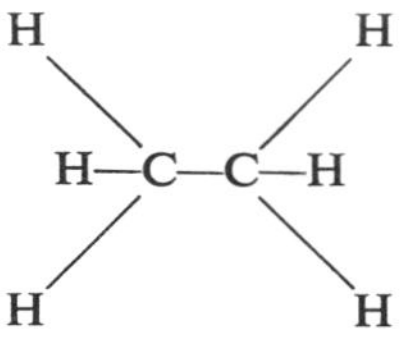

In this structure each carbon is surrounded tetrahedrally by four other atoms, namely three hydrogens and one carbon. This compound is called *ethane*, and has the molecular formula C_2H_6; it is often represented by the following structural formula which does not represent correct bond angles but is understood to represent the correct double tetrahedral structure.

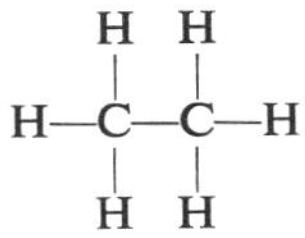

Ethane often is written in a more abbreviated form as $CH_3—CH_3$.

If a hydrogen in ethane is replaced by a methyl group, the compound *propane* results, which has a molecular formula of C_3H_8 and may be represented structurally as follows:

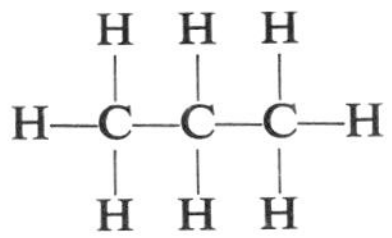

In more abbreviated form propane often is represented as $CH_3—CH_2—CH_3$.

In this way, long chains of carbons may be constructed. Each of the resulting hydrocarbons is named according to the number of carbons in the molecule or chain. The names for hydrocarbons containing up to 12 carbons are listed in Table 8.21.

All of the molecular formulas listed in Table 8.21 may be written as a single type formula, C_nH_{2n+2}, which indicates that the number of hydrogens is two more

TABLE 8.21 *Alkane Nomenclature Related to the Number of Carbons*

NUMBER CARBONS	NAME	MOLECULAR FORMULA	RELATED GROUP FORMULA	RELATED GROUP NAME
1	methane	CH_4	CH_3	methyl
2	ethane	C_2H_6	C_2H_5	ethyl
3	propane	C_3H_8	C_3H_7	propyl
4	butane	C_4H_{10}	C_4H_9	butyl
5	pentane	C_5H_{12}	C_5H_{11}	pentyl or amyl
6	hexane	C_6H_{14}	C_6H_{13}	hexyl
7	heptane	C_7H_{16}	C_7H_{15}	heptyl
8	octane	C_8H_{18}	C_8H_{17}	octyl
9	nonane	C_9H_{20}	C_9H_{19}	nonyl
10	decane	$C_{10}H_{22}$	$C_{10}H_{21}$	decyl
11	undecane	$C_{11}H_{24}$	$C_{11}H_{23}$	undecyl
12	dodecane	$C_{12}H_{26}$	$C_{12}H_{25}$	dodecyl

than twice the number of carbons. The hydrocarbons consisting of chains of carbon atoms having a molecular formula of this type are called the *alkanes*. Thus the alkanes consist of chains of carbon atoms in which there are only single carbon-carbon bonds.

The related groups listed in Table 8.21 consist of hydrocarbon molecules from which a hydrogen has been removed. These groups can be used to replace hydrogens in hydrocarbons to build up bigger and more complex hydrocarbon molecules. The names of these groups are formed from the hydrocarbon name by substituting the suffix *-yl* for the suffix *-ane*.

If the hydrocarbon containing four carbons, C_4H_{10}, is to be derived from propane by substituting a methyl group for a hydrogen, there are two possibilities. The methyl group may be substituted for one of the hydrogens on one of the end carbons or it may be substituted for one of the hydrogens on the middle carbon. In the first case the resulting molecule has a continuous chain of four carbon atoms as follows:

$$CH_3—CH_2—CH_2—CH_3$$

In the second case the hydrocarbon has a branched chain of carbon atoms with three atoms in a continuous chain and the fourth carbon bonded to the middle carbon.

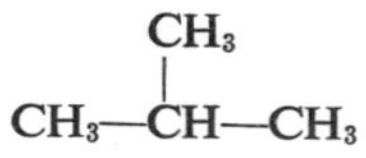

The hydrocarbon with the four carbons in a continuous chain is called a *straight chain hydrocarbon*, despite the fact that the chain of carbon atoms really is staggered and not straight. The other hydrocarbon is called a *branched chain hydrocarbon*. However, each of these two compounds has the same molecular formula, namely C_4H_{10}. These compounds are said to be *isomers*. *Isomers are different compounds with the same molecular formula.*

Since both hydrocarbons have the same formula, it is necessary to devise a system of nomenclature by which to distinguish the two compounds. The most commonly used system employs the following rules:

1. The hydrocarbon is named according to the number of carbons in the *longest continuous chain* of carbons in the molecule, using the names given in Table 8.21.
2. The carbons in the longest continuous chain are numbered consecutively from one end in such a way that any attached groups are bonded to carbons having the lowest numbers.
3. The parent name of the compound is prefixed by the names of the groups attached to the longest continuous chain, and these in turn are prefixed by the numbers of the carbons to which they are bonded.
4. If one group appears more than once on a chain, its name is prefixed with the suitable numerical prefix from Table 6.16, and this is preceded by the number

of each carbon atom to which the group is bonded in the longest continuous chain.

5. If several different groups appear on the same chain, they are named either in alphabetical order or in order of increasing size of the group.

For example, consider the following alkane:

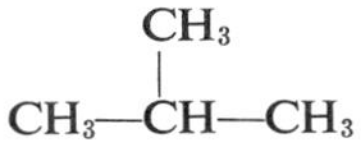

The longest continuous chain of carbons contains three carbons and this compound is named as a propane. There is a methyl group attached to the number 2 carbon, regardless of which end one uses to start the numbering. Thus the name of the compound is 2-methylpropane. Since the methyl group may be attached only to the number 2 carbon, sometimes the number is dropped and the compound is called methylpropane. If the methyl group is attached to the number 1 carbon, then the longest continuous chain contains four carbons and it is called butane.

For a second example, consider the following alkane:

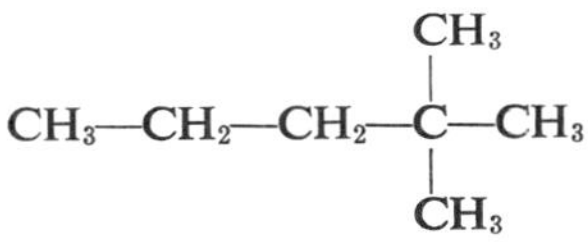

Here two methyl groups are attached to a chain of five carbons. Thus this is a dimethylpentane. The chain is numbered from right to left as written, so that the two methyl groups are attached to the number 2 carbon instead of the number 4 carbon. Thus the complete name for this hydrocarbon is 2,2-dimethylpentane.

For a third and final example, consider the following alkane:

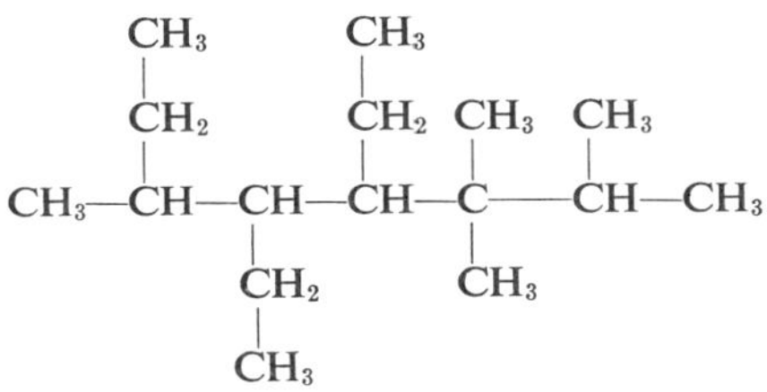

There is no reason to assume that the longest continuous chain of carbons will be written horizontally or even vertically on a page. Each structure must be inspected to find the longest continuous chain. In this case it begins in the upper left and proceeds down three carbons and then goes horizontally to the right for five carbons. Thus the longest continuous chain contains eight carbons and the compound is named as an octane. On this chain there are four methyl groups and two ethyl groups. If an alphabetical listing is chosen, the groups will be named in the order diethyl, tetramethyl, with ethyl before methyl. The longest continuous chain is numbered from the right as written to provide the lowest numbers for the carbons to which groups are attached. Thus the complete name for this compound is 4,5-diethyl-2,3,3,6-tetramethyloctane. If the longest continuous chain of carbons had been numbered in the reverse order, an *incorrect* name would have been given as 4,5-diethyl-3,6,6,7-tetramethyloctane. However, the groups might have been named in reverse order according to size to give another correct name: 2,3,3,6-tetramethyl-4,5-diethyloctane.

ALKENES. The hydrocarbons containing a carbon-carbon double bond are called the alkenes. The parent hydrocarbon for these compounds is ethylene, which has the molecular formula C_2H_4. Ethylene has the following structural formula, in which all of the atoms lie in the same plane.

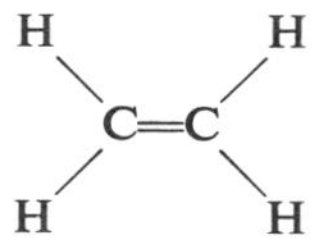

Ethylene is a common name for this compound. Systematically, it is named by the number of carbons, two in this case, using the suffix *-ene* in place of the suffix *-ane* used for the alkanes. Thus this compound is named ethene according to the systematic nomenclature.

If one of the hydrogens is replaced by a methyl group, the following compound results:

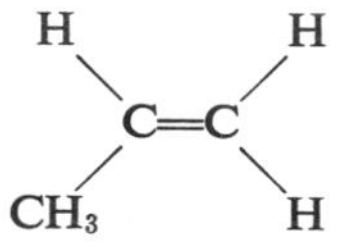

This compound has the molecular formula C_3H_6 and is called propene.

There are four different ways in which a hydrogen in propene may be replaced by a methyl group. The lone hydrogen on the carbon bearing the methyl group may be replaced.

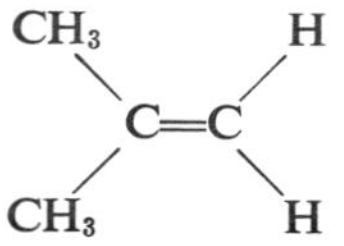

This compound is named from the longest continuous chain of carbon atoms *including both carbons that are doubly bonded.* Since this is three carbons in this compound, it is a propene. There is a methyl group attached to the chain, this is methyl propene.

The second and third ways to replace a hydrogen in propene with a methyl group are to replace either of the two hydrogens on the carbon not bonded to the methyl group. This produces the following two structures:

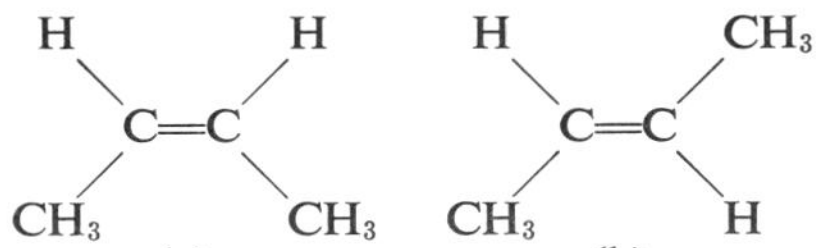

(a) (b)

These two compounds are alike in having four carbons in the longest continuous chain including the two double bonded carbons. In this respect both are butenes. In naming the alkenes, the longest continuous chain including the double bonded carbons is numbered so that *the first double bonded carbon will have the smallest number*. In this case it makes no difference and both of these compounds are 2-butenes. However compound (a) differs from compound (b). Since the molecules are planar, the two hydrogens and two methyl groups are located at the corners of a rectangle. In compound (a) the two methyl groups are on adjacent corners, whereas in compound (b) the two methyl groups are on opposite corners. Thus compounds (a) and (b) are isomers. Compound (a), in which the two methyl groups are *adjacent*, is called the *cis* isomer, whereas compound (b), in which the two methyl groups are *opposite*, is called the *trans* isomer. Thus to distinguish these two isomers, compound (a) is called *cis*-2-butene and compound (b) is called *trans*-2-butene.

The fourth way in which to replace a hydrogen in propene with a methyl group is to replace one of the hydrogens in the methyl group itself with a methyl group. This leads to the following structure:

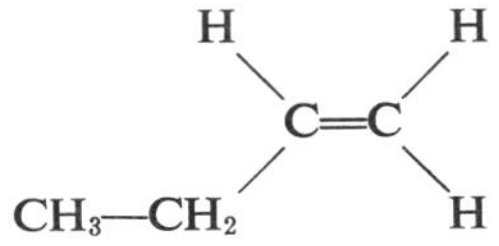

In this molecule the longest continuous chain of carbons including the two carbons that are doubly bonded contains four carbons, so the compound is a butene. If the carbons are numbered from right to left as the molecule is represented, then the correct name is 1-butene. If the carbons are numbered incorrectly, then the *incorrect* name is 3-butene.

Thus methylpropene, *cis*-2-butene, *trans*-2-butene, and 1-butene are all isomers with the common molecular formula of C_4H_8. This is in accord with the general type formula for the alkenes, C_nH_{2n}.

As a final example, consider the naming of the following compound:

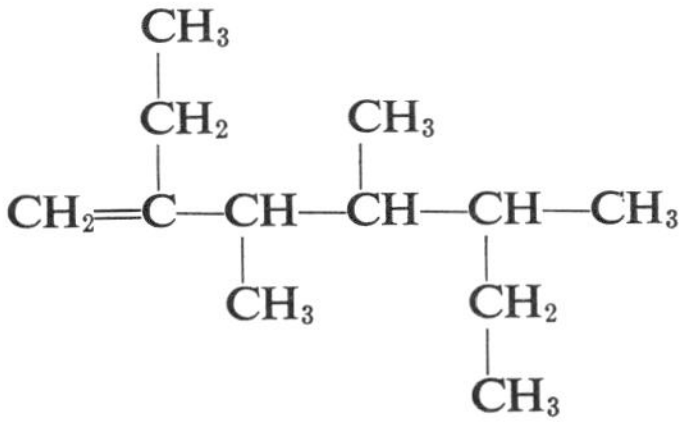

Note that there is an eight-carbon continuous chain starting in the upper left and ending in the lower right. However, this chain does not contain both doubly bonded carbons. Thus the compound is named according to the shorter seven-carbon chain horizontally from the left for five carbons and then down two carbons. The chain is numbered from the left so that the double bond is attached to carbon 1. Thus this compound is named as a 1-heptene. In addition there is an ethyl group on carbon 2 and single methyl groups on carbons 3, 4 and 5. Thus the complete name for this compound is 2-ethyl-3,4,5-trimethyl-1-heptene.

ALKYNES. The hydrocarbons whose molecules contain a carbon-carbon triple bond are called *alkynes*. The parent alkyne is acetylene, which has the molecular formula C_2H_2 and the following structural formula:

$$H{-}C{\equiv}C{-}H$$

Actually acetylene is a common name. The alkynes are named systematically according to the number of carbons in the longest continuous chain of carbons containing both triply bonded carbons. The suffix *-yne* replaces the suffix *-ane* in the similar alkane name. Thus acetylene also is called ethyne.

Thus if a methyl group replaces one of the hydrogens in ethyne, the resultant compound has the following structure and is called propyne:

$$CH_3{-}C{\equiv}C{-}H$$

If both hydrogens in ethyne are replaced by methyl groups, then the compound is a butyne and specifically it is 2-butyne with structure as follows:

$$CH_3{-}C{\equiv}C{-}CH_3$$

The other isomer of 2-butyne is 1-butyne, which has the following structural formula:

$$H{-}C{\equiv}C{-}CH_2{-}CH_3$$

AROMATIC HYDROCARBONS. Benzene is the parent compound of the so-called *aromatic hydrocarbons*. If one of the hydrogens in benzene is replaced by a methyl group, the compound is called methylbenzene and has the structural formula as follows:

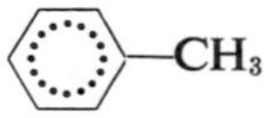

However, there are three ways in which a hydrogen in methyl benzene may be replaced by a methyl group. The replaced hydrogen may be adjacent to the methyl group, may be opposite the methyl group, or may be the second hydrogen from the methyl group. This leads to the following three isomers:

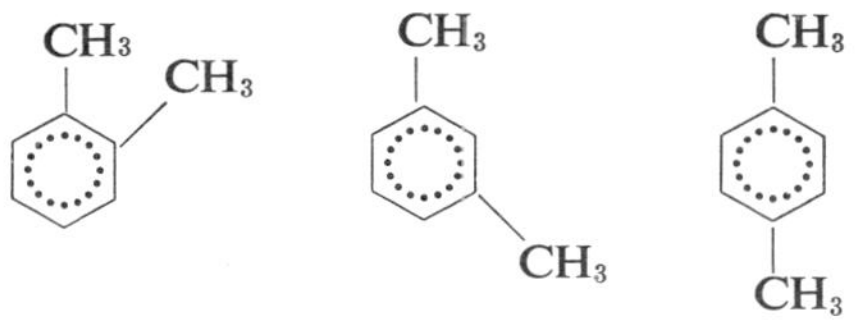

(a) (b) (c)

These three isomers are called *ortho*, *meta*, and *para* isomers. Compound (a), in which the two methyl groups are adjacent, is called *ortho*-dimethylbenzene. Compound (b), in which the two methyl groups are neither adjacent nor opposite, is called *meta*-dimethylbenzene. Compound (c), in which the two methyl groups are opposite, is called *para*-dimethylbenzene. Usually the prefixes are abbreviated by their first letters and the three names are written as *o*-dimethylbenzene, *m*-dimethylbenzene, and *p*-dimethylbenzene.

However, there is an alternative way to name these aromatic hydrocarbons. The six carbons in the benzene ring are numbered from one to six, so that any attached groups are bonded to carbons with the lowest numbers. Then the attached groups are identified by the number of the carbon to which they are bonded. For example, consider the following typical aromatic hydrocarbon:

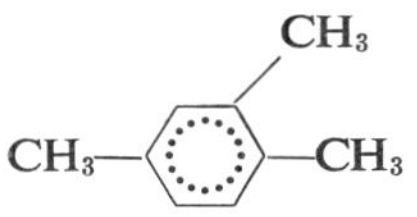

This compound has a methyl group attached to each of carbons 1, 2, and 4. Thus the correct name is 1,2,4-trimethylbenzene.

If a hydrogen is removed from a molecule of benzene, the resulting group, which is usable as a substitute for hydrogens in building other hydrocarbon molecules, is called the *phenyl* group and has the formula C_6H_5. Thus methylbenzene might be named alternately as phenyl methane. From this point of view the phenyl group has replaced a hydrogen in methane.

Some hydrocarbon properties. Most hydrocarbons have relatively small electrostatic dipole moments. Partly this is due to the relatively nonpolar carbon-carbon and carbon-hydrogen bonds. Partly this is due to the molecular geometries which often keep the centers of positive and negative charge together. For example, ethene, $CH_2{=}CH_2$, has zero electrostatic dipole moment. Similarly, *trans*-2-butene also has zero electrostatic dipole moment. However, *cis*-2-butene has an electrostatic dipole moment of 0.38 debye.

Another good example can be found among the aromatic hydrocarbons. Benzene has zero dipole moment, as expected from its symmetrical, planar structure. However, methyl benzene has an electrostatic dipole moment of 0.36 debye. If a second methyl group is placed in the para position, opposite the first methyl group, the electrostatic dipole moment is zero. However, if a second methyl group is placed in the ortho position, adjacent to the first methyl group, the electrostatic dipole moment is 0.62 debye.

Boron

Boron forms a number of binary compounds with hydrogen. The simplest of these is called diborane and has the molecular formula B_2H_6. Although the molecular formula is similar to that for ethane, C_2H_6, the structure is not the same, primarily because boron has too few electrons to permit the same kind of bonding.

Several attempts have been made to develop a picture of the bonding in diborane. The following is one of the more probable:

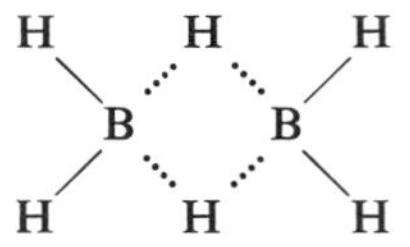

In this proposed structure two hydrogens are used as bridges between the two boron atoms. There are two hydrogens bonded to each boron by normal covalent bonds.

The other known boranes have the following molecular formulas: B_4H_{10}, B_5H_9, B_5H_{11}, B_6H_{10}, B_9H_{15}, and $B_{10}H_{14}$.

SUMMARY

The elements with small, highly electronegative atoms tend to form multiple bonds and thus discrete diatomic molecules. However, the other atoms tend to form single bonds and thus there is a tendency among atoms which can form more than one single bond to form giant molecules of atoms. For example, oxygen forms O_2, whereas sulfur forms the puckered octagonal ring, S_8, along with long chains of sulfurs. However, selenium tends to form the infinite chains. Among the elements with larger atoms, where covalent bond energies are low, metallic structures are preferred.

Catenation is characteristic of intermediate sizes of atoms which have sufficiently large radii to discourage multiple bond formation, but which still have relatively high single bond energies. Carbon is the most prolific in forming catenated compounds, and sulfur displays some tendency toward limited catenation.

Among the oxides, highly polymerized species tend to be formed by the elements with intermediate electronegativities where covalent bonds of high polarity can be formed. This tends to form molecules with adjacent positive and negative charge centers, and as electronegativity differences increase, the compounds tend to form ionic solids. For example, the oxides of sulfur are somewhat intermediate, forming both discrete molecules as well as partially polymerized oxides. However, selenium oxides are highly polymerized. Similarly, nitrogen forms discrete oxide molecules, whereas the phosphorus oxides are polymerized. In like manner, carbon forms discrete molecules of its oxides, but the oxide of silicon is very highly polymerized.

The halogens have little catenation capability due to their atomic electronic configurations. For the same reasons, the halides tend to form discrete molecules. However, since the halogens tend to have an oxidation state of 1−, relatively large numbers of halogens must cluster around central atoms with large oxidation numbers. As a result relative size plays an important role.

QUESTIONS AND PROBLEMS

8.1 Write feasible structural formulas for the following molecules, using canonical forms where appropriate, and describe the molecular geometry.
- a) S_2O : Two sulfurs are bonded together and oxygen is bonded to one sulfur.
- b) HCN : Atoms are bonded in the order indicated.
- c) $POCl_3$: Phosphorus is central with each of the other four atoms bonded to it.
- d) HNCS : Atoms are bonded in the order indicated, that is, H to N, N to C, and C to S.
- e) HNO_3 : Nitrogen is central with three oxygens bonded to it. The hydrogen is bonded to one of the oxygens.

8.2 What is meant by canonical forms and the hybrid of the canonical forms?

8.3 In crystalline bromine, three different distances have been measured between adjacent bromines: a) 2.27 Å; b) 3.31 Å; and c) 3.78 Å. How can this observation be interpreted?

8.4 Liquid fluorine, chlorine and bromine have the following dielectric constants at the temperatures indicated. Explain this trend.
a) F_2, 1.54 at −202°C b) Cl_2, 2.10 at −50°C c) Br_2, 3.09 at +20°C

8.5 What is the difference between formal charge and oxidation number?

8.6 What is the approximate density in grams per liter for ozone at 200°K and a pressure of 600 mm Hg?

8.7 Suppose that you had two unlabeled containers, one with gaseous ozone and the other with gaseous oxygen. What properties might you use to distinguish between the two?

8.8 The following alternate structure has been proposed for ozone.

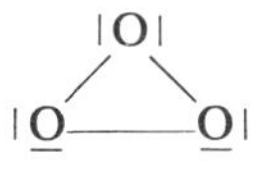

Which properties of ozone, which have been discussed here, are in agreement with this structure? Which are in disagreement?

8.9 If ozone had a linear molecule, would it have a permanent electrostatic dipole moment?

8.10 Suggest why the bond angles in S_8 are slightly less than the tetrahedral angle.

8.11 Are the chains in plastic sulfur straight?

8.12 Account for the fact that selenium has a higher boiling point than sulfur.

8.13 In Table 8.5, why is a range shown for the melting points of sulfur and selenium but not for oxygen and tellurium?

8.14 What is a moderately simple way to decide what the molecular form of the vapor of tellurium is at 1390°C?

8.15 Why is it difficult to give a molecular weight for arsenic but not for nitrogen?

8.16 In the metallic modifications of arsenic, antimony, and bismuth represented in Fig. 8.3, where are the three next nearest neighbors located?

8.17 Although many oxides are ionic, many are covalent. What factors tend to increase the covalent character of oxides?

8.18 Why do we expect the usual oxidation states of the chalcogens to be even?

8.19 How much heat is required to melt 150 g of silicon at the melting point of silicon?

8.20 Why does phosphorus show allotropes that are wider in range and more complex structurally than those of nitrogen?

8.21 The P_4 molecule appears in the vapor of phosphorus. Why do these molecules not appear as entities in the stable solid form?

8.22 How can bismuth be more close packed than arsenic when the interatomic distance shown in Table 8.6 is greater for bismuth than arsenic?

8.23 Use Fig. 4.12 to compare the energies required for promotion of valence electrons in the rare gas atoms.

8.24 Why are the nonbonding electrons in XeF_2 at maximum separation in the equatorial positions instead of the axial positions?

8.25 Speculate on the possible structures for XeO_3 and XeF_6 and give the reasons for your suggestions.

8.26 We have shown that in some cases if triatomic molecules were linear, they would have zero electrostatic dipole moments. Is it possible that a linear, triatomic molecule might have an electrostatic dipole moment?

8.27 Explain how polar substances can effectively decrease the force of attraction between two oppositely charged particles. Why should substances that are hydrogen bonded have high dielectric constants?

8.28 The following are dielectric constants for carbon disulfide at various temperatures.

40°C	2.588	−10°C	2.722
30°C	2.615	−20°C	2.749
20°C	2.641	−30°C	2.775
10°C	2.668	−40°C	2.812
0°C	2.695	−50°C	2.829

a) Why does the dielectric constant increase as the temperature decreases?
b) From the dielectric constant values, what would you say about the polarity of carbon disulfide molecules?
c) What is the mathematical relationship between the dielectric constant of carbon disulfide and temperature over this temperature range?

8.29 The following are dielectric constants for cyclohexanol at various temperatures:

65°C	10.02	40°C	12.88
60°C	10.54	35°C	13.55
55°C	11.09	30°C	14.26
50°C	11.67	25°C	15.00
45°C	12.25	20°C	15.78

a) From the dielectric constant values, what would you say about the polarity of the cyclohexanol molecules?
b) Over this temperature range, what is the mathematical relationship between the dielectric constant of cyclohexanol and temperature?

8.30 Fluorine forms an ion with the formula HF_2^-. Suggest a possible structural formula for this species.

8.31 In comparison to other uncatenated binary hydrogen compounds of the chalcogens, water has abnormally high melting and boiling points and molar heats of fusion and vaporization. Suggest a reason for this.

8.32 Suppose that you have four unlabeled containers, each with a sample of one of the four hydrogen chalcogenides, H_2S, H_2Se, H_2O, and H_2Te. Describe a process or techniques you might use to decide which one of these substances is in which container.

8.33 What are the formal charges on the atoms in water? What do these values imply?

8.34 Why is water more highly hydrogen bonded than hydrogen sulfide?

8.35 Criticize the following statement: The molecular weight of selenium(IV) oxide is 111.

8.36 Are there canonical forms to be written for TeO_2? If there are, write them. If there are not, why not?

8.37 The bond angle in SO_2 is 120°. What angle is expected in SeO_2?

8.38 The heat of vaporization of SO_2 is 5.96 kcal/mole of SO_2 units, and the heat of vaporization of TeO_2 is 51.6 kcal/mole of TeO_2 units. Account for this great difference.

8.39 The cyclic modification of sulfur(VI) oxide has a melting point of 290°K, whereas both chain modifications have melting points in excess of 300°K. Suggest a reason for this.

8.40 Sulfur(VI) oxide has three different melting points but only one boiling point. Why?

8.41 Describe the structure of the SF_6 molecule.

8.42 Why are the melting and boiling points of ammonia much higher than the melting and boiling points of phosphine?

8.43 It has been suggested that the structure of PF_5 might be described in terms of canonical forms involving phosphorus to fluorine double bonds. Draw several such canonical forms.

8.44 What differences in bonding probably occur in NF_3 and PF_3?

8.45 The stability of the phosphorus pentahalides decreases from phosphorus(V) fluoride to the phosphorus(V) iodide. Suggest reasons for this.

8.46 How can you account for the fact that NF_3 is stable when the other nitrogen trihalides tend to be unstable?

8.47 Account for the fact that the dipole moment of PF_3 is greater than the dipole moment of NF_3.

8.48 Phosphorus(V) fluoride has zero electrostatic dipole moment. Is this to be expected or is this an anomalous value? Explain.

8.49 Explain why PCl_5 has a higher melting point (420°K) than $SbCl_5$ (278°K).

8.50 Although some canonical forms have been written for the oxides of nitrogen, write all other possible canonical forms for NO, N_2O_3, NO_2, N_2O_4, and N_2O_5.

8.51 Some diffraction data suggest that the NON bond angle in N_2O_5 is 180°. What would this suggest about the canonical forms contributing to the delocalized hybrid?

8.52 Nitrogen(II) oxide has an electrostatic dipole moment. Is this consistent with the proposed electronic structure? Explain.

8.53 It has been suggested that solid nitrogen(V) oxide consists of nitryl ions, NO_2^+, and nitrate ions, NO_3^-. Suggest canonical forms for each of these two ions.

8.54 Nitrogen(V) oxide has a melting point of 314°K and a boiling point of 320°K. Suggest a reason why these two values are so close together.

8.55 The oxides of nitrogen tend to have highly delocalized electron systems. Is this true also for the phosphorus oxides? If yes, write canonical forms for phosphorus(III) oxide and phosphorus(V) oxide to show the delocalization. If not, why not?

8.56 Suggest formulas for arsenic oxides and sulfides.

8.57 In discussing the halogen, chalcogen, and nitrogen families, it was noted that the uncatenated binary compounds with hydrogen of the elements in period 2 had unusually high melting and boiling points and heats of fusion and vaporization in comparison to the analogous compounds of the other elements in the family. This is not true in the carbon family. Why?

8.58 The bonds in $SiCl_4$ are polar. However, this compound has zero dipole moment. Explain.

8.59 Explain the trends in properties shown in Tables 8.19 and 8.20.

8.60 Why is $|\mathrm{C{=}\underline{O}}|$ not considered to be a feasible canonical form for carbon monoxide?

8.61 The triple bond in CO should be highly polar due to the difference in electronegativity between carbon and oxygen. However, carbon monoxide has a very small electrostatic dipole moment of only 0.10 debye. Suggest an explanation for this.

8.62 Carbon dioxide tends to sublime, that is, it changes from a solid to a gas directly without passing through the liquid state. However, under a pressure of 5 atm, it will melt at 216°K. Silicon(IV) oxide is a solid at room conditions and has a melting point of 1983°K. Suggest a reason for this tremendous difference.

8.63 Sometimes the boron trihalides have been called electron deficient molecules. Suggest what this means.

8.64 Why do the halogens tend not to form catenated binary compounds?

8.65 Would you expect the H_2O_2 moleule to be planar? Why?

8.66 Suggest molecular-type formulas for the compounds with the names sodium hydrogen peroxide and sodium hydrogen trisulfide.

8.67 Suggest an experimental technique for estimating the normal boiling point of a substance like hydrogen peroxide which explodes before the normal boiling point is reached.

8.68 The following seven structural formulas represent relatively complex compounds. Find those in the family of peroxy compounds.

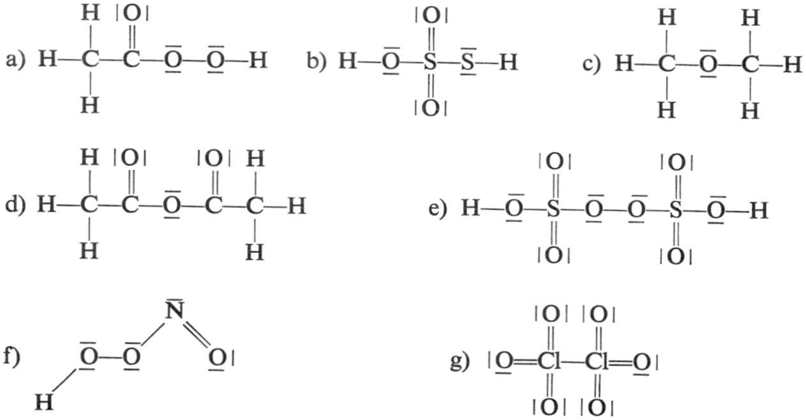

8.69 What is the oxidation number for iron in FeS_2? What systematic name can be given to this compound?

8.70 The melting point for hydrogen peroxide is $-2°C$ and the melting point for hydrogen disulfide is $-89°C$. What is the reason for this big difference?

8.71 Is it likely that selenium and tellurium form polyselenides and polytellurides? Explain.

8.72 The heat of vaporization of hydrazine is 9.99 kcal/mole and the heat of vaporization of diphosphine is 6.88 kcal/mole. Why is the value for hydrazine higher than the value for diphosphine?

8.73 Does the structural formula for dimeric nitrogen(II) oxide require that the molecule be planar?

8.74 What are the formal charges on each atom in N_2O? What does this imply about the contribution of bond distribution to the electrostatic dipole moment of the molecule?

8.75 Why does the azide ion have more canonical forms that the hydrogen azide molecule?

8.76 Distinguish among the following terms:
a) allotrope   b) polymorph   c) isomer.

8.77 Are wurtzite and zinc blende isomers? Explain.

8.78 Write structural formulas for five isomers with the formula C_6H_{14}; nine isomers with the formula C_7H_{16}; and eight isomers with the formula C_5H_{10}.

8.79 Evaluate the following statement for its validity: Since acetylene and benzene have the same empirical formula, they are isomers.

8.80 Write an appropriate name for each of the following hydrocarbons:

a) $CH_3-CH_2-CH_2-CH_2-CH_3$

b) 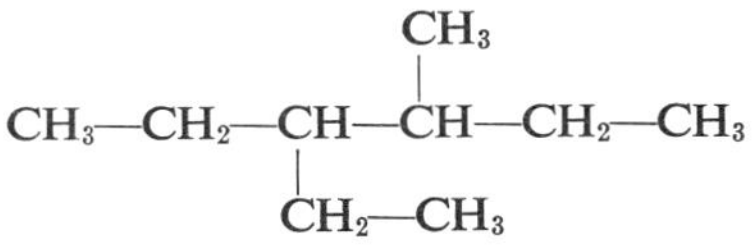

c) 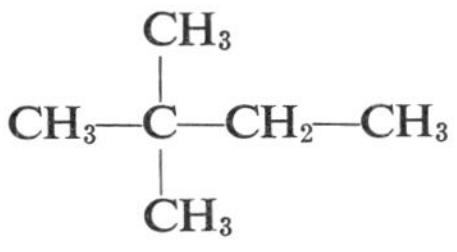

d) 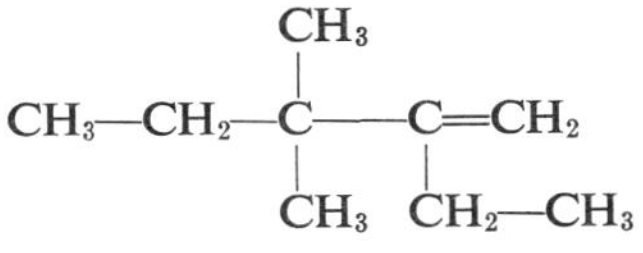

e) 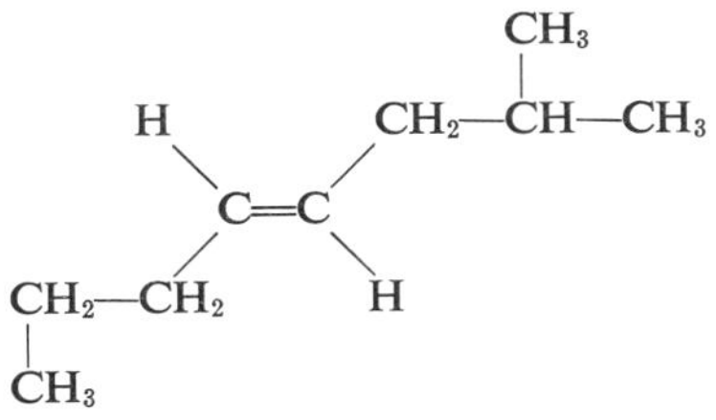

f) 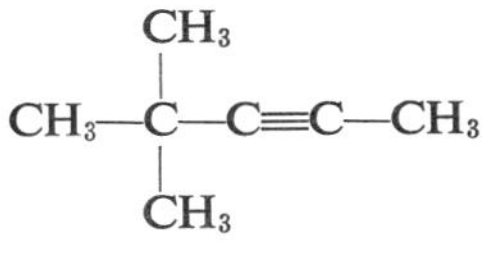

g) 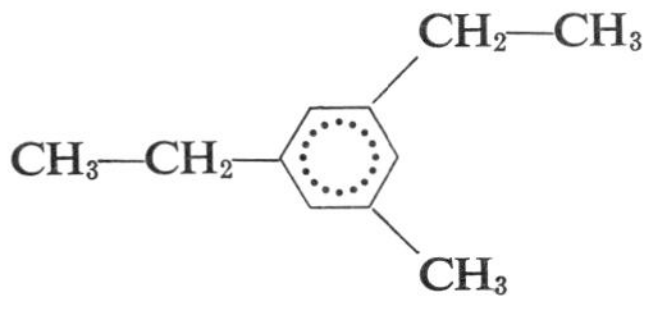

h) 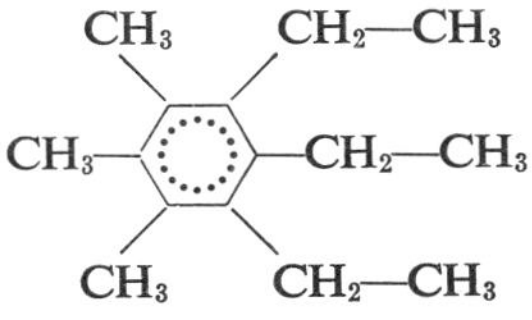

i) 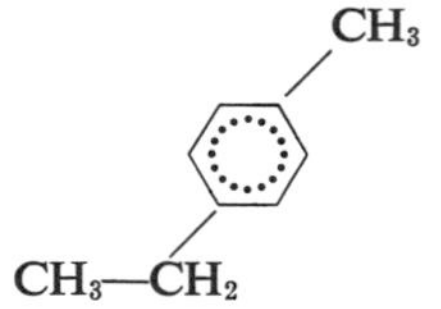

j) 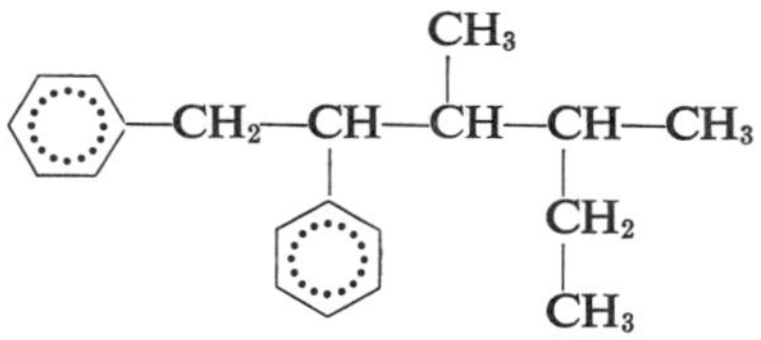

8.81 Write the correct structural formula for each of the following.
a) decane
b) 2,2,3,3-tetramethylbutane
c) 2-methylpropane
d) 3-methyl-3-ethylhexane
e) 2,3,5-trimethyl-4,4-diethylheptane
f) 2,3-dimethyl-2-butene
g) *trans*-2,3,7,8-tetramethyl-4-nonene
h) 4-ethyl-2-octene
i) 2-hexyne
j) 3-methyl-3-ethyl-1-pentyne
k) *o*-methylethylbenzene
l) 1,3-dimethyl-2-phenylbenzene
m) *p*-diphenylbenzene
n) 1,2,4-triphenyl-2,6-dimethyloctane
o) 1,2,4-triethylbenzene

8.82 There is something wrong with each of the following names. What is it?
a) 3,3-dimethyl-2-pentyne
b) 1,5-dimethylbenzene
c) 2,3,3,6,6-pentaethyl-5-methyloctane
d) *cis*-2-ethyl-1-butene
e) *p*-trimethyl benzene
f) 1,3,4-trimethylbenzene

8.83 Why is it that benzene has zero dipole moment, but methylbenzene has a dipole moment?

8.84 How is it possible to have trimethylbenzene with zero electrostatic dipole moment?

8.85 Why does *trans*-2-butene have zero electrostatic dipole moment, whereas *cis*-2-butene has an electrostatic dipole moment of 0.38 debye?

8.86 From the following list of hydrocarbons, choose those which should have zero electrostatic dipole moment:
a) ethyne
b) propene
c) phenylbenzene
d) *cis*-diphenylethene
e) 1,1-diphenylethene
f) triphenylmethane
g) 2,3-dimethyl-2-butene

8.87 *Cis*-2-butene has a melting point of −139°C whereas *trans*-2-butene has a melting point of −106°C. Suggest a reason for the difference.

9

Some Ternary and Higher Compounds of the Nonmetals

INTRODUCTION In the previous chapter we looked at some of the binary compounds of the nonmetals. Emerging from that survey was a way of looking at catenated compounds that is also useful for other complex compounds. There are certain characteristic groups of atoms that recur in many compounds and may be considered to be modules from which complex molecules are constructed.

Often hydrogen compounds are considered to be parents of families of compounds whose molecules may be derived by replacement of hydrogens in the parent with other atoms or groups of atoms. In addition, some of the simple atoms and groups of atoms may be derived by removal of a hydrogen atom from molecules of some of the simple binary compounds of hydrogen. The following five kinds of atoms and groups of atoms may serve as a beginning from which the structure of some more complex molecules may be developed.

1. The removal of a hydrogen from a hydrogen halide leaves a *halogen atom*, a very prominent type of atom in many complex molecules. For example, the replacement of three hydrogens in methane, CH_4, with three chlorines leads to chloroform, $CHCl_3$.
2. The removal of a hydrogen from water leaves the *hydroxy* or *hydroxo* group which has the formula OH. For example, the alcohols are a sub-family of the

hydroxy compound family in which the hydroxy group has replaced one or more hydrogens in an alkane. Thus in methyl alcohol, CH_3OH, one hydrogen of methane is replaced by the hydroxy group.

3. The removal of a hydrogen from ammonia leaves the *amino* or *amido* group, which has the formula NH_2. When this group replaces a hydrogen in methane, methyl amine, CH_3NH_2, is the result.
4. The removal of a hydrogen from an alkane leads to an *alkyl* group, of which the methyl and ethyl groups are examples. The removal of a hydrogen from benzene leads to the *phenyl* group, which has the formula C_6H_5.
5. The *metals* often may be considered as replacements for hydrogen. These have been illustrated earlier by the peroxides, in which sodium peroxide was considered to be derived from hydrogen peroxide by replacement of the hydrogens with sodiums.

It should be remembered that these are formal representations and do not necessarily represent practical laboratory preparations. This is a useful way of looking at the structures of complex molecules.

In this chapter we shall look first at a broad class of compounds called the *hydrates;* then we shall look more systematically at some of the compounds formed by the elements in periodic families 0, 7A, 6A, 5A, 4A and 3A, with particular reference to a very important group of compounds called the *hydroxy compounds*, which obviously contain the hydroxy group.

HYDRATES

Many compounds contain hydrogen and oxygen. The simplest compound of this type is water. Scrutiny of Table 6.9 reveals that the hydrogen-to-oxygen bond is the strongest single bond that oxygen forms and the second strongest bond that hydrogen forms. Thus the water molecule, which contains two hydrogen oxygen bonds, is at the bottom of a rather deep potential energy well, so to speak. The result is that many compounds containing hydrogen and oxygen are unstable with respect to the formation of water. In some cases they are so unstable that they can exist only under very special conditions, and in other cases they are unstable only at elevated temperatures. However, as a group, these compounds are called the *hydrates*.

Water of Constitution

In many cases the hydrogen and oxygen necessary to form water are not already bonded into H_2O molecules. The hydroxy compounds are examples in which one hydrogen of the water has been replaced by another atom or group of atoms. However, within the molecule or in an adjacent molecule, there may be a hydrogen which can combine with the hydroxy group to form water.

For example, the compound zinc hydroxide has the formula $Zn(OH)_2$, which indicates that there are two hydroxy groups combined with a single zinc. This might be viewed also as the replacement of a single hydrogen on two waters by a zinc atom. Upon heating, zinc hydroxide loses water and zinc oxide is left behind. This is represented by Equation (9.1).

$$Zn(OH)_2 = ZnO + H_2O \tag{9.1}$$

Since zinc hydroxide has the components of water in it, but not originally bonded in the form of water molecules, it is said to contain water of constitution, and is called a hydrate for this reason. The process of removing water is called *dehydration*, and the substrate remaining after the removal of the water is called the *anhydride*. In this case zinc oxide is the anhydride.

Hydrates Containing Molecular Water

In some cases hydrates contain molecules of water already bonded as H_2O entities. Molecular water may be incorporated into other structures in three common ways.

WATER OF COORDINATION. The water molecule contains two pairs of nonbonding electrons on the oxygen atom. These electrons may be used as bonding electrons with some species that has a completely vacant orbital suitable for bonding. Such typical species are the cations, in which the orbitals for the valence electrons are vacant. In addition, cations of transition elements appear to be able to use the $(n - 1)d$ orbitals as well as the ns and np orbitals for this purpose.

For example, the cobalt(II) ion appears to be able to form covalent bonds with water molecules in this way, and one of these bonds may be represented as follows:

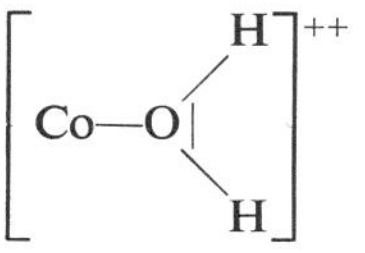

Since the electrons in the covalent bond between cobalt(II) ion and the oxygen in the water molecule are imagined to come from the water molecule, the bond is a *coordinate covalent bond* and the water is said to be coordinated to the cobalt(II) ion. In this way, molecules of water may be incorporated into larger structures. The cobalt(II) ion is capable of forming six coordinate covalent bonds, and thus six molecules of water may be bonded to a single cobalt(II) cation. This leads to a new complex ion often represented as $[Co(H_2O)_6]^{++}$.

Thus *anhydrous* cobalt(II) chloride has the formula $CoCl_2$ and *hydrated* cobalt (II) chloride may be represented as $[Co(H_2O)_6]Cl_2$. For some uses, it is common practice to represent the six waters of hydration with the cobalt(II) chloride as a whole: $CoCl_2 \cdot 6H_2O$. In this way hydrated cobalt(II) chloride is named cobalt(II) chloride-6-hydrate.

Upon heating, cobalt(II) chloride-6-hydrate forms anhydrous cobalt(II) chloride and water as shown by Equation (9.2).

$$CoCl_2 \cdot 6H_2O = CoCl_2 + 6H_2O \qquad (9.2)$$

Anion water. Water molecules also may be associated with anions through hydrogen bonding. One of the common anions that is often hydrogen bonded to water is the sulfate ion, SO_4^{--}. This ion has the following structure:

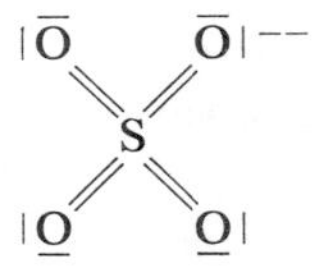

The hydration of the sulfate ion may be represented as follows where the dotted bonds are hydrogen bonds:

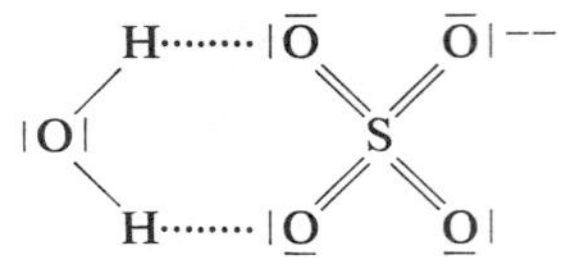

Copper(II) sulfate-5-hydrate contains four waters of coordination and one anion water, which may be represented as follows:

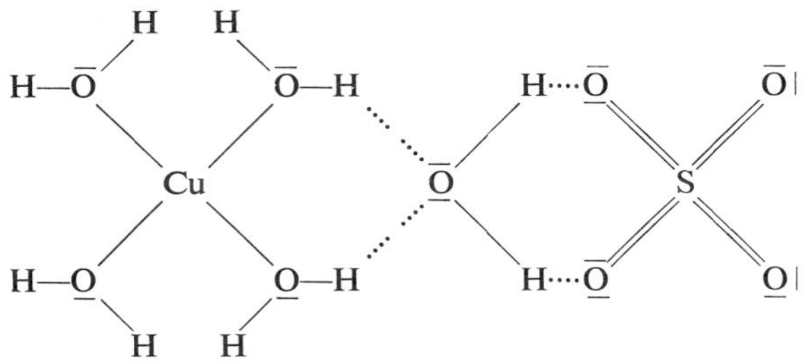

Thus four waters are coordinated to the copper(II) ion, and the fifth water is hydrogen bonded to the sulfate ion as well as to one of the coordinated waters. Without distinction between the two different kinds of water of hydration, hydrated copper(II) sulfate often is represented as $CuSO_4 \cdot 5H_2O$ and is called copper(II) sulfate-5-hydrate.

Crystal water. In many cases definite proportional amounts of water are incorporated into the holes in crystals. Many chlorides are of this type. Barium chloride-2-hydrate, $BaCl_2 \cdot 2H_2O$ is of this type.

In other cases, water appears in nonproportional amounts between layers in crystals and on the surfaces of crystals. This kind of hydrate cannot be represented adequately by a formula since different samples may have different percentages of water.

HYDROXY COMPOUNDS

Hydroxides

When the hydroxy group is combined with an atom of low electronegativity, that is, a metal atom, an electron is transferred from the metal atom to the hydroxy group to form the *hydroxide ion*, OH^-. These compounds are ionic and are composed of metal ions and hydroxide ions in a crystal matrix. Since the oxidation number of the hydroxide ion as a whole is 1−, sodium hydroxide has the formula NaOH, and barium hydroxide has the formula $Ba(OH)_2$, where the subscript indicates the number of hydroxide ions present.

Hydroxy Acids

When the hydroxy group is bonded to a more highly electronegative atom, such as chlorine, the result is a covalent bond. In this example the result may be shown by the following structural formula:

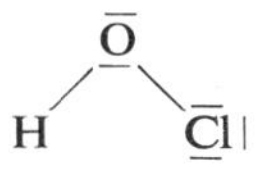

Compounds of this type are called *hydroxy acids*.

Since the nonmetal to which the hydroxy group is bonded is quite electronegative, the H—O bond is quite polar and the hydroxy acids tend to be highly hydrogen bonded. Of course, the hydrogen bonding enhances the stability of the hydroxy acid through aggregation, which represents a decrease in potential energy.

However, if the decrease in potential energy due to hydrogen bonding or other dipole-dipole interactions is less than the decrease in potential energy represented by dehydration and the formation of the very stable water molecule, then the hydroxy acid tends to decompose. This appears to be the case with HOCl. Although there is some evidence for the formation of small amounts of HOCl in water solutions, the compound has not been isolated.

Presumably, if the polarity of the H—O bonds could be increased, hydrogen bonding would be enhanced and the dehydration process would be less highly favored and the hydroxy acid would be more stable.

One common way in which this is accomplished is by bonding of another highly electronegative atom to the central nonmetal. This extra atom usually is oxygen, which can make available one empty orbital for bonding with an otherwise nonbonding pair on the chlorine. This new hydroxy acid has the following structure:

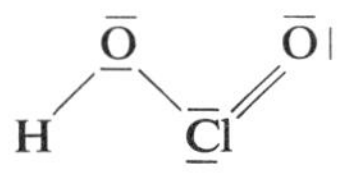

The group of two oxygens and one chlorine constitute a very electronegative group of atoms that tend to drain electrons away from hydrogen, making the H—O bond more polar than it is in HOCl. This should enhance the stability of HOClO over the stability of HOCl. If it does, it is not obvious, for the evidence for this compound is about the same as that for HOCl and it has not been isolated.

However, with two additional pairs of nonbonding electrons on the chlorine in HOClO, it is possible to bond one or two more oxygens to the central chlorine to form $HOClO_2$ and $HOClO_3$:

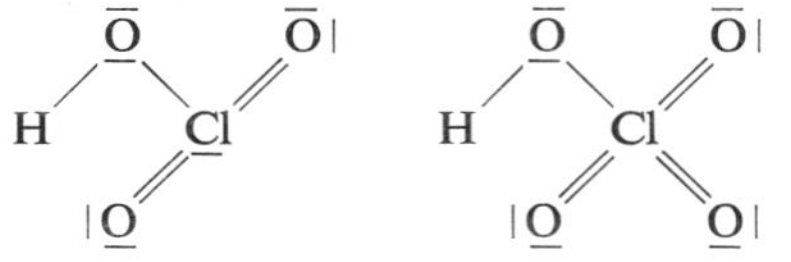

As the compounds become more highly oxygenated, the polarity of the H—O bond increases and molecular stability increases. Although $HOClO_2$ has not been isolated, there is good evidence for it in solution. However, pure $HOClO_3$ has been isolated.

This indicates the general pattern for the molecular structure of typical simple hydroxy acids. The hydroxy groups are bonded to a central atom to which additional oxygens may be bonded also. In general, the more highly oxygenated hydroxy acids tend to be the more stable with respect to dehydration.

It should be pointed out that although hydrogen bonding provides for some of the stability of hydroxy acids due to intermolecular attractions, it can also provide a path for dehydration, when dehydration is energetically feasible. Thus, if two molecules of HOCl are hydrogen bonded, the hydrogen involved in the bond lies close to an oxygen which already is bonded to another hydrogen. This can be seen schematically as follows:

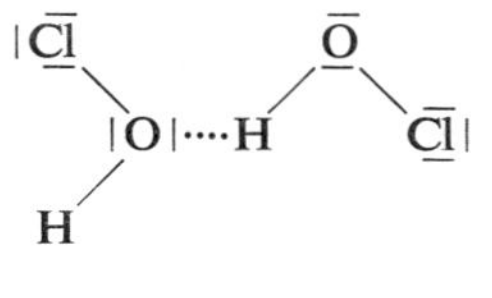

This makes it relatively simple for two hydrogens and one oxygen to come together to form H_2O. However, whether dehydration occurs or not depends upon the energetics of the situation.

Nomenclature. Formulas for the hydroxy acids commonly are written in three ways in addition to structural formulas. The first method does not indicate molecular structure and, for example, the four hydroxy acids of chlorine are written as HClO, $HClO_2$, $HClO_3$, and $HClO_4$. In the other two types of formulas, the hydroxy groups are indicated as such, either first or last. Thus the hydroxy acids of chlorine may be represented as HOCl, HOClO, $HOClO_2$ and $HOClO_3$ or as ClOH, ClO(OH), $ClO_2(OH)$ and $ClO_3(OH)$.

In a series of acids such as this, the oxidation number of the central atom varies with the compound. Thus chlorine has oxidation numbers of 1+, 3+, 5+, and 7+ in HClO, $HClO_2$, $HClO_3$, and $HClO_4$ respectively. This serves as a basis for naming the compounds. The compounds are named as acids, the name of the acid being derived from the root of the name of the central element, chlorine in this case. For the lower two oxidation states, the suffix *-ous* is added to the root and for the higher two oxidation states, the suffix *-ic* is added to the root. For the lowest of the four oxidation states, the prefix *hypo-* is added to the root and for the highest of the four oxidation states, the prefix *per-* is added to the root. Thus the four hydroxy acids of chlorine are named as follows.

HClO	hypochlorous acid
$HClO_2$	chlorous acid
$HClO_3$	chloric acid
$HClO_4$	perchloric acid

HYDROXY ACID DERIVATIVES. Several types of compounds are commonly considered to be derivatives of the hydroxy acids. Formally, these may be derived from the hydroxy acids by replacing the hydrogen in the hydroxy group with an atom or group of atoms or by replacing the entire hydroxy group with an atom or group of atoms.

To illustrate four common types of hydroxy acid derivatives, we shall use the most highly oxygenated simple hydroxy acid of sulfur, which is sulfuric acid. Sulfuric acid may be represented by the formulas H_2SO_4, $(HO)_2SO_2$, or $SO_2(OH)_2$, or structurally as follows:

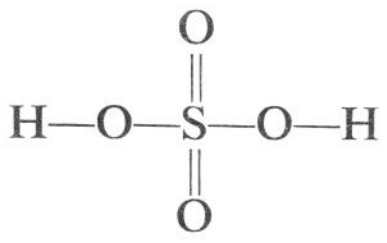

In this representation the sulfur is at the center of a tetrahedron, with the four oxygens at the vertices of the tetrahedron.

1. *Salts.* When a hydroxy hydrogen is replaced by an atom of low electronegativity, ionic compounds, called *salts*, are formed.

If the two hydrogens in sulfuric acid are replaced by sodiums, the resulting compound is composed of Na^+ and SO_4^{--} ions. The name of the SO_4^{--} ion is *sulfate.* In general, the names of the oxygenated anions are derived from the corresponding acid name by dropping the word *acid* and replacing the suffix *-ous* by *-ite* or the suffix *-ic* by *-ate.* Upon occasion, slight alterations are made to make the names more euphonious. Thus SO_4^{--} is the sulfate ion instead of the sulfurate ion.

To illustrate the nomenclature further, the anions derived from the hydroxy acids of chlorine have the following names:

ClO^-	hypochlorite
ClO_2^-	chlorite
ClO_3^-	chlorate
ClO_4^-	perchlorate

For salts, the metal is named first and then the anion. Thus Na_2SO_4 is sodium sulfate and $KClO_4$ is potassium perchlorate.

Sulfuric acid is an example of a dihydroxy acid; that is, it contains two hydroxy groups. In such compounds, one or both of the hydrogens may be replaced. In the case of sodium sulfate, Na_2SO_4, both have been replaced. However, if only one is replaced, the compound has the formula $NaHSO_4$. This compound is composed of the Na^+ and HSO_4^- ions. The HSO_4^- ion is called the hydrogen sulfate ion, and $NaHSO_4$ is sodium hydrogen sulfate. It should be noted that the hydrogen sulfate ion is a hydroxy acid also.

It is important to realize that salts are more stable than the acids from which they are derived. Certainly the removal of the hydrogens removes the source of any tendency toward dehydration. In addition, the ionic matrix is quite stable due to interionic attractions. For example, although chloric acid has not been prepared in pure form, potassium chlorate, $KClO_3$, is a well-known compound available in pure form.

Thus even the hydrogen salts, such as $NaHSO_4$, are more stable than the acids from which they are derived, due to interionic attractions.

2. *Esters.* When the hydrogen in a hydroxy acid is replaced by an alkyl or phenyl group, the resulting compound is called an *ester*. If the two hydrogens in sulfuric acid are replaced by methyl groups, the resulting compound may be represented by the formulas $(CH_3)SO_4$, $(CH_3O)_2SO_2$ or $SO_2(OCH_3)_2$ and structurally as follows:

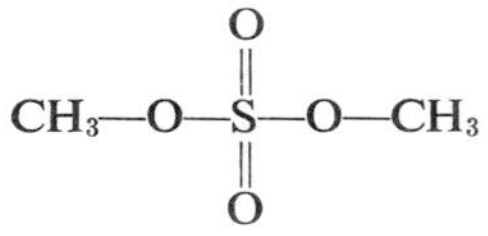

The esters are named like the salts, with the alkyl or phenyl group first. Thus $(CH_3)_2SO_4$ is dimethyl sulfate. Of course if only one hydrogen is replaced, the compound is methyl hydrogen sulfate, $(CH_3)HSO_4$.

Although there are hydrogens on alkyl and phenyl groups, these hydrogens are not involved in hydrogen bonding. As a result, hydrogens and oxygens tend not to be brought into as close proximity, and dehydration occurs less readily among the esters than among the hydroxy acids. Consequently, esters of unstable acids often can be isolated. For example, ethyl hypochlorite has been prepared.

3. *Acid Halides.* If the entire hydroxy group of a hydroxy acid is replaced by a halogen atom, the resulting compound is called an *acid halide*. For example, if the two hydroxy groups in sulfuric acid are replaced by chlorines, the compound has the formula SO_2Cl_2.

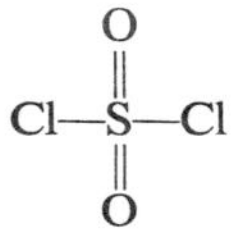

This compound is called sulfuryl chloride, and the central SO_2 group is called the sulfuryl group. Sometimes it is called the diacid chloride of sulfuric acid. Of course, if only one hydroxy group in sulfuric acid is replaced by chlorine, the resulting compound is a monohydroxy acid:

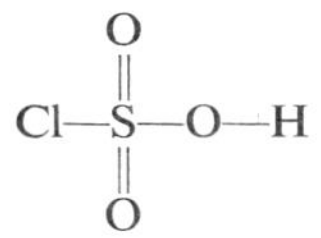

This is the monoacid chloride of sulfuric acid. Since the monohydroxy acids derived from sulfuric acid are called sulfonic acids, this has been called chlorosulfonic acid as well.

Of course chlorosulfonic acid is a hydroxy acid, but sulfuryl chloride is not. Thus sulfuryl chloride is not susceptible to dehydration or hydrogen bonding.

4. *Amides.* If the hydroxy group of the hydroxy acid is replaced by an amino group, the resulting compound is called an amide. For example, if the two hydroxy groups in sulfuric acid are replaced by amino groups, the resulting compound is the diamide of sulfuric acid which may be represented by the formulas $SO_2(NH_2)_2$, $(NH_2)_2SO_2$, or NH_2—SO_2—NH_2.

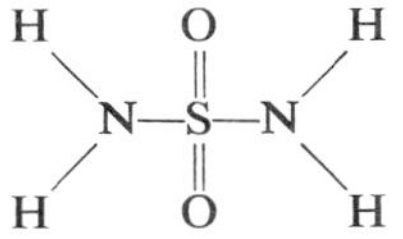

This diamide of sulfuric acid also is called sulfuric diamide. The monoamide is formed by replacing only one of the hydroxy groups in sulfuric acid by an amino group. This compound is a monohydroxy sulfuric acid, that is, it is a sulfonic acid. It has the formula NH_2—SO_2—OH and is called amino sulfonic acid, the monoamide of sulfuric acid, sulfuric monoamide, or most commonly, sulfamic acid.

The amides are susceptible to hydrogen bonding, although the bonds are between hydrogen and nitrogen atoms. Thus dehydration tends to occur less readily among the amides than among the hydroxy acids.

With these introductory remarks on the hydroxy acids, we now shall look at some of the common ternary and higher compounds of the nonmetals. Many, but not all, of these are related to the hydroxy acids.

RARE GASES

Salts of two hydroxy acids of xenon have been prepared. These correspond to xenon in oxidation states of 8+ and 6+. The oxyanions correspond to the formulas XeO_6^{4-} and XeO_6^{6-}. They have been named xenate and perxenate.

There have been suggestions that compounds corresponding to barium kryptate and kryptic acid may have been prepared.

HALOGENS

In addition to the hydroxy acids, the halogens form the alkyl halides in which halogen atoms have been substituted for some hydrogen atoms in hydrocarbons.

Alkyl Halides

If an alkyl group replaces a hydrogen in a hydrogen halide, a simple *alkyl halide* results. We may view these compounds as the result of substituting a halogen for a hydrogen in an alkane as well. In fact compounds of this general type result from the substitution of halogens for hydrogens in any hydrocarbons.

These compounds are named in the same way that the hydrocarbons are named, using the names *fluoro*, *chloro*, *bromo*, and *iodo* for the halogens. The following examples illustrate typical names:

$CH_3—Cl$ chloromethane or methyl chloride

$CH_3—CH_2—CH_2—Br$ 1-bromopropane

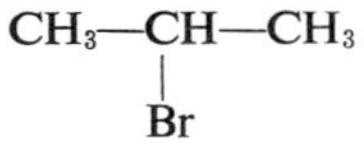

2-bromopropane

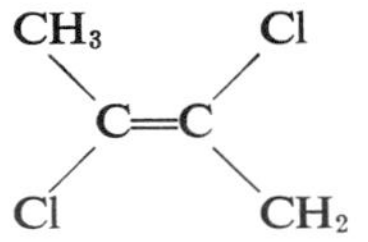

trans-2,3-dichloro-2-butene

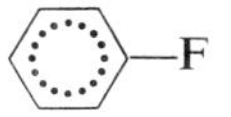

fluorobenzene or phenylfluoride

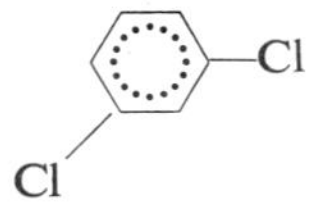

1,3-dichlorobenzene or
m-dichlorobenzene

A few electrostatic dipole moments are shown in Table 9.1 for some typical halides. Chloromethane has a moderately high dipole moment that is explained in terms of a highly polar carbon-to-chlorine bond. However, as the number of chlorines increases, and the polar bonds become more symmetrically arranged, the dipole moment decreases, finally reaching zero for tetrachloromethane.

Similarly, chlorobenzene has an electrostatic dipole moment, and the two unsymmetrical dichlorobenzenes have dipole moments, whereas the symmetrical 1,4-dichlorobenzene has zero dipole moment. In a similar way, *cis*-1,2-dichloroethene has an electrostatic dipole moment, whereas *trans*-1,2-dichloroethene does not.

Hydroxy Acids

Among the halogens, there are no known hydroxy acids or hydroxy acid derivatives for fluorine. Chlorine, bromine, and iodine form both hypohalous acids

TABLE 9.1
Some Electrostatic Dipole Moments (Debyes)

CH_3Cl	1.87
CH_2Cl_2	1.54
$CHCl_3$	1.02
CCl_4	0
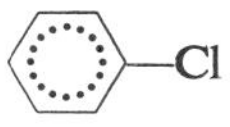	1.70
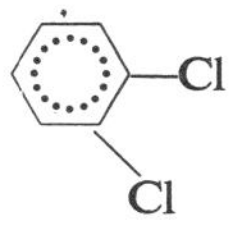	2.50
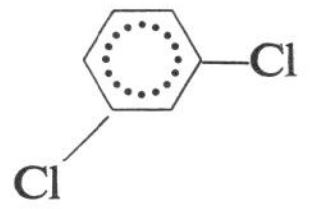	1.72
	0
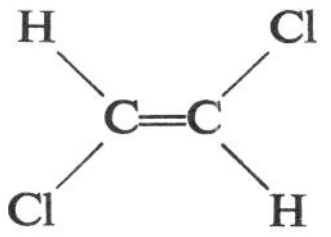	0
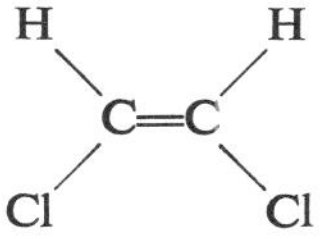	1.90

and halic acids and their derivatives. Only chlorine and iodine form perhalic acids and chlorine alone forms halous acid and its derivatives.

HYPOHALOUS ACIDS. Although none of the hypohalous acids has been isolated, there is good evidence for the existence of hypohalous acids of chlorine, bromine, and iodine in water solutions. Salts and esters of these acids have been prepared.

HALOUS ACIDS. It is not known why chlorine alone forms a halous acid. However, salts of chlorous acid have been prepared and the O—Cl—O bond angle has been measured at 110°, which is very close to the tetrahedral angle predicted by the following structural formula for chlorite ion:

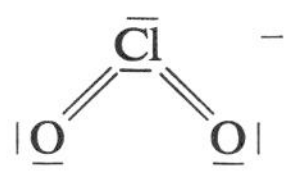

HALIC ACIDS. The halic acids are more stable than the hypohalous or the halous acids. However, only iodic acid, HIO_3, has been prepared and isolated, while chloric acid, $HClO_3$, and bromic acid $HBrO_3$, are known only in water solutions. However, many chlorate, bromate, and iodate salts are known in crystalline form in combination with the alkali metals, alkaline earth metals, and many of the transition metals. Most of these decompose, however, before they melt, although there are a few for which melting points have been determined.

The halate ions are pyramidal in shape as suggested by the four electron groups around the central halogen, three of these serving as bonding groups and the fourth acting as a nonbonding group. Thus the electron groups are arranged tetrahedrally and the ion is pyramidal, with the halogen at the apex and the three oxygens at the corners of the trigonal base. The O—X—O angle in all of these ions is nearly 110°, which is very close to the value for the tetrahedral angle. The typical structural formula is represented as follows, where the halogen atom is understood to be above the plane of the three oxygens:

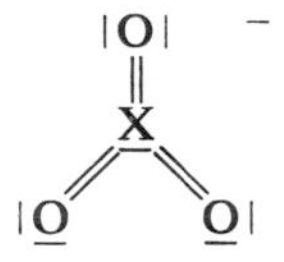

PERHALIC ACIDS. Pure perchloric acid, $HClO_4$, has been isolated as a liquid with a freezing point of −112°C. At normal pressure, it decomposes before it boils; however, it does boil at 19°C at a pressure of 11 mm Hg. Many perchlorate salts are known, including those of the alkali metals, the alkaline earth metals, and many transition metals. Many of these decompose before they melt.

The perchlorate ion is tetrahedral, with a central chlorine surrounded by four oxygens as suggested by the following structural formula:

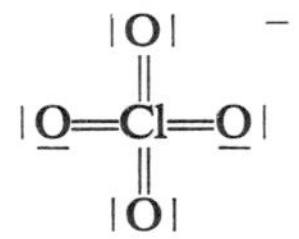

Iodine forms two periodic acids. The one analogous to perchloric acid has the formula HIO_4 and is called metaperiodic acid. The more common periodic acid for iodine has the formula H_5IO_6, and the following structure, in which the iodine is at the center and surrounded octahedrally by six oxygens:

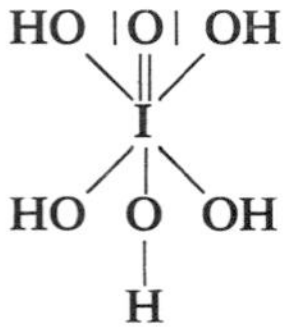

Apparently this kind of structure is permitted by iodine because of the large radius of the iodine atom. This permits six oxygens to cluster around the central iodine.

The nomenclature is based upon the degree of hydration of the two acids. In a formal way, orthoperiodic acid could be represented as the 2-hydrate of metaperiodic acid, thus $HIO_4 \cdot 2H_2O$. In this case the water is water of constitution. But from this point of view, the prefix *ortho-* is used with the most highly hydrated hydroxy acid and the prefix *meta-* with the least hydrated hydroxy acid.

CHALCOGENS

Alcohols and Ethers

If one alkyl group is substituted for a hydrogen in water, the resulting compound is an *alcohol.* From an alternate point of view, alcohols are alkanes in which hydroxy groups have replaced one or more hydrogens. Thus the simplest alcohol has the formula CH_3OH.

The alcohols are named by replacing the final *e* on the name of the alkane from which they are derived by the suffix *-ol.* Thus CH_3OH is methanol. The following examples illustrate some typical alcohols and their names.

$CH_3—CH_2—OH$	ethanol
$CH_3—CH_2—CH_2—OH$	1-propanol
$CH_3—CH(OH)—CH_3$	2-propanol
$CH_3—C(CH_3)(OH)—CH_3$	2-methyl-2-propanol

The longest continuous chain includes the carbon bonded to the hydroxy group, and the number in front of the name indicates the number of the carbon to which the hydroxy group is bonded.

When both hydrogens in water are replaced by alkyl or phenyl groups, the resulting compound is called an *ether.* The ethers usually are named by naming the two alkyl or phenyl groups. Thus $CH_3—O—CH_3$ is called dimethyl ether.

Some electrostatic dipole moments for some alcohols and ethers are shown in Table 9.2. The alcohols have higher dipole moments than the ethers. The alcohols are rather polar substances and like water, they exhibit a tendency to form hydrogen bonds.

TABLE 9.2
Electrostatic Dipole Moments for Some Alcohols and Ethers

COMPOUNDS	DIPOLE MOMENT (DEBYES)
CH_3OH	1.70
CH_3CH_2OH	1.69
$CH_3—O—CH_3$	1.30
$CH_3—O—CH_2CH_3$	1.23
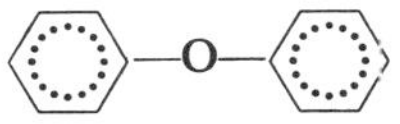	1.23

Uncatenated Hydroxy Acids of Sulfur

In the formation of the uncatenated hydroxy acids of sulfur, it is expected that two hydroxy groups will bond to sulfur, corresponding to the involvement of the two unpaired 3*p* electrons in sulfur. The two important hydroxy acids of this type are sulfurous acid and sulfuric acid, which have the formulas H_2SO_3 and H_2SO_4 respectively.

Actually, all attempts to prepare sulfurous acid have led to water and sulfur(IV) oxide, which is the formal anhydride of sulfurous acid. However, salts of sulfurous acid are known, consisting of both hydrogen sulfite ions, HSO_3^-, and sulfite ions, SO_3^{--}. The structures of these two ions are as follows:

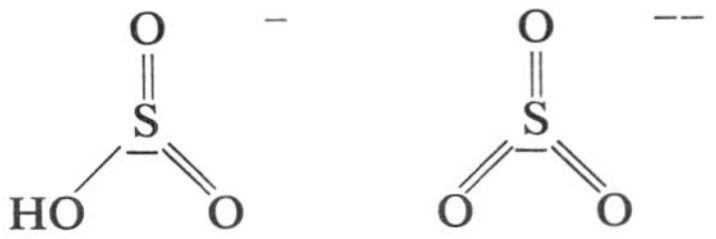

hydrogen sulfite ion sulfite ion

Both of these ions may be pictured as trigonal pyramids, with sulfur at the apex and three oxygens at the corners of the trigonal base. The O—S—O bond angles are close to the tetrahedral angle in both cases.

Not only the salts of sulfurous acid, but also the diesters, diacid halides, and the diamide have been prepared and they are more stable than the acid.

Sulfuric acid is more stable than sulfurous acid and can be isolated as a solid melting at 10°C. However, it decomposes before it boils. The molecular structure may be represented as follows:

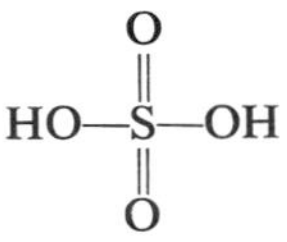

The sulfur is surrounded tetrahedrally by four oxygens. However the molecule is a distorted tetrahedron. The O—S—O bond angle formed by the two hydroxy oxygens is about 104° and the O—S—O bond angle formed by the other two oxygens is about 177°.

As indicated earlier, sulfuric acid forms two kinds of salts, those containing hydrogen sulfate ions, HSO_4^-, and those containing sulfate ions, SO_4^{--}. In addition, the acid halides, esters, and amides are known.

In a formal way, sulfuric acid may be considered to undergo partial dehydration, with the formation of polymeric hydroxy acids. This is represented schematically for the first step as follows:

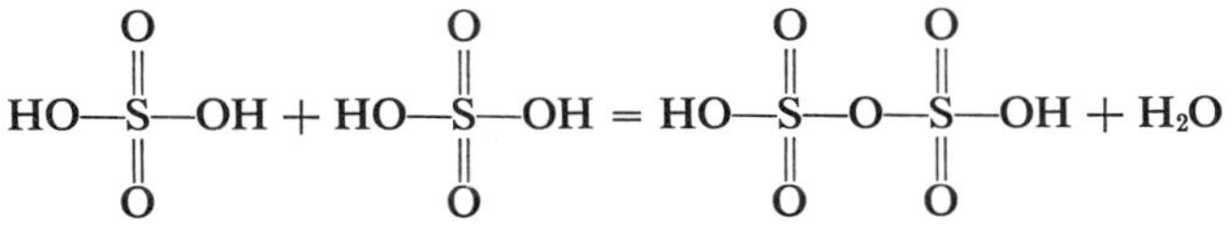

This partial dehydration amounts to the elimination of water between two molecules of sulfuric acid. The product is called *disulfuric acid* and is still a dihydroxy acid. Structurally it appears to consist of two SO_4 tetrahedra which share a common oxygen.

As a dihydroxy acid, disulfuric acid might undergo another partial dehydration with another molecule of sulfuric acid.

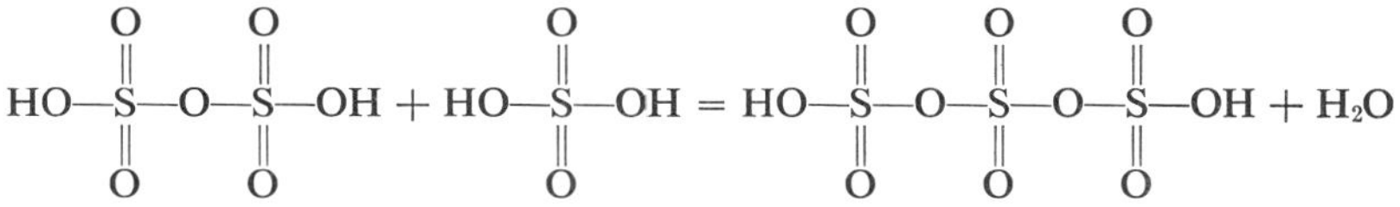

The product of this partial dehydration is another dihydroxy acid called *trisulfuric acid*, consisting of three SO_4 tetrahedra sharing oxygen corners.

Neither the disulfuric acid nor the trisulfuric acid have been isolated, but salts of both are known and the diacid chloride of disulfuric acid is known.

Peroxy Acids of Sulfur

Two common hydroxy acids of sulfur contain the peroxy grouping, O—O. These are called peroxomonosulfuric acid and peroxodisulfuric acid.

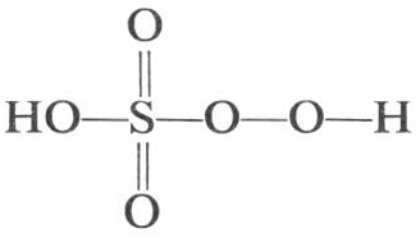

peroxomonosulfuric acid

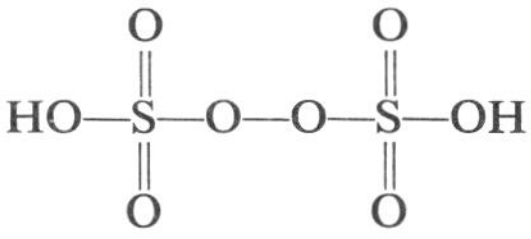

peroxodisulfuric acid

The peroxomonosulfuric acid may be considered to be tetrahedral with respect to the S atom. The peroxodisulfuric acid consists of two SO_4 tetrahedra joined through two oxygens but not sharing oxygens. Both of these acids are known along with their salts. The salts of peroxodisulfuric acid are especially well known laboratory reagents.

Catenated Hydroxy Acids of Sulfur

The hydroxy acids also are affected by the ability of sulfur to catenate. One of the catenated hydroxy acids of sulfur can be imagined if one of the hydroxy oxygens is replaced with a sulfur. This acid is called *thiosulfuric acid*. The prefix *thio-* is used to imply the replacement of oxygen by sulfur. Although thiosulfuric acid has not been isolated, the esters and salts of the acid are well known. The thiosulfate ion has the formula $S_2O_3^{--}$ and the following structure:

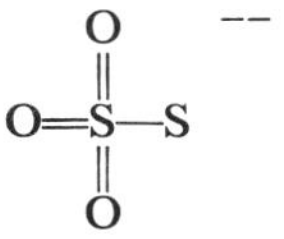

The *thionic acids* constitute a series in which the simplest member has two *sulfonic acid* groups bonded through the two S atoms. This provides a structure that can be represented briefly as HO_3S—SO_3H. The other members of the series have from one to four S atoms between the two sulfonic acid groups. Thus *dithionic acid* is HO_3S—SO_3H, ($H_2S_2O_6$); *trithionic acid* is HO_3S—S—SO_3H, ($H_2S_3O_6$); *tetrathionic acid* is HO_3S—S—S—SO_3H, ($H_2S_4O_6$); *pentathionic acid* is HO_3S—S—S—S—SO_3H, ($H_2S_5O_6$); *hexathionic acid* is HO_3S—S—S—S—S—SO_3H, ($H_2S_6O_6$). The thionic acids have not been isolated, but the salts are known and well characterized.

In the polythionate ions, the chain of S atoms is staggered, with the bond angles near the tetrahedral angle. The central S in each sulfonic acid group is surrounded tetrahedrally by three O atoms and one S atom. The structure of the dithionate ion may be represented as follows:

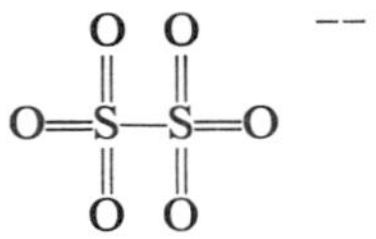

Hydroxy Acids of Selenium and Tellurium

Selenium hydroxy acids resemble the hydroxy acids of sulfur. Selenous acid, H_2SeO_3, has been isolated along with its salts containing both the hydrogen selenite ions, $HSeO_3^-$, and the selenite ions, SeO_3^{--}. In addition, esters and diacid halides are known for selenous acid.

Selenic acid, H_2SeO_4, is known along with its salts, which contain both the hydrogen selenate ion, $HSeO_4^-$, and the selenate ion, SeO_4^{--}. In addition esters of selenic acid are known.

Salts of polymers of selenic acid are known containing the diselenate ions, $Se_2O_7^{--}$ and the triselenate ions, $Se_3O_{10}^{--}$.

Salts of tellurous acid, H_2TeO_3, in which the tellurite ions, TeO_3^{--} are present, are known. The simple telluric acid, H_2TeO_4, has not been isolated, but salts containing the TeO_4^{--} ion have been prepared.

Like iodine, the larger tellurium is able to accommodate more hydroxy groups around it, and the common telluric acid is the orthotelluric acid, H_6TeO_6, which may be formulated also as $Te(OH)_6$, instead of the metatelluric acid, H_2TeO_4. Both the acid and its salts are known.

NITROGEN FAMILY

Amines

When an alkyl or phenyl group replaces one or more hydrogens in ammonia, the resulting compound is called an *amine*. From an alternate point of view, amines are formed by replacement of hydrocarbon hydrogens by amino groups. Thus the simplest amine is $CH_3—NH_2$, which is called methyl amine or aminomethane.

Ammonium Compounds

The ammonia molecule has a nonbonding pair of electrons on the nitrogen. This nonbonding pair may interact with species that have vacant orbitals suitable for bonding. The hydrogen ion, H^+, is such a species with a vacant 1*s* orbital. When the ammonia molecule bonds to a hydrogen ion, the resulting species is the *ammonium ion*, NH_4^+.

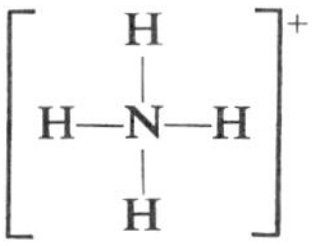

In many respects, the ammonium ion behaves like an alkali metal ion in the types of compounds which it forms. For example, it forms ionic ammonium chloride, NH_4Cl, and ionic ammonium sulfate, $(NH_4)_2SO_4$, among many others. The compounds are ionic and tend to exist primarily in the solid state. They tend to decompose either before melting or at temperatures slightly in excess of the melting point.

Hydroxy Acids of Nitrogen

Superficially, it would appear that hydroxy acids of nitrogen should have three hydroxy groups bonded to nitrogen, corresponding to the involvement of the three unpaired 2*p* electrons. However, the nitrogen atom appears to be too small to accommodate three hydroxy groups around it. Thus the commonly encountered hydroxy acids contain only one hydroxy group.

Nitrous acid, HNO_2, is the monohydroxy acid corresponding to nitrogen in the 3+ state of oxidation. The acid has not been isolated, but there is evidence for it in water solutions. The structure of the acid is not known, but from measurements on the salts of nitrous acid, it appears that the structure of the nitrite ion may be described by the following canonical forms:

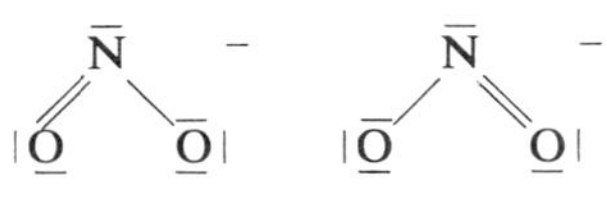

The ion is angular, and the bond angle is equal to 115°.

In addition to the salts, esters and acid halides of nitrous acid have been isolated.

Nitric acid, HNO_3, is the monohydroxy acid corresponding to nitrogen in the 5+ oxidation state. The acid has been isolated at low temperature; however, it tends to decompose at room temperatures. The structure of nitric acid can be represented by the following canonical forms:

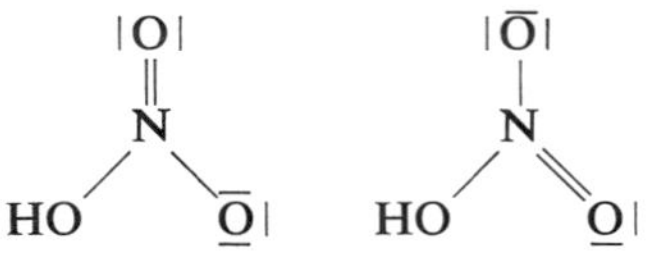

The three oxygens and the nitrogen are in the same plane, with the bonds directed toward the corners of a triangle.

The nitrate ion, NO_3^-, may be represented by the following canonical forms:

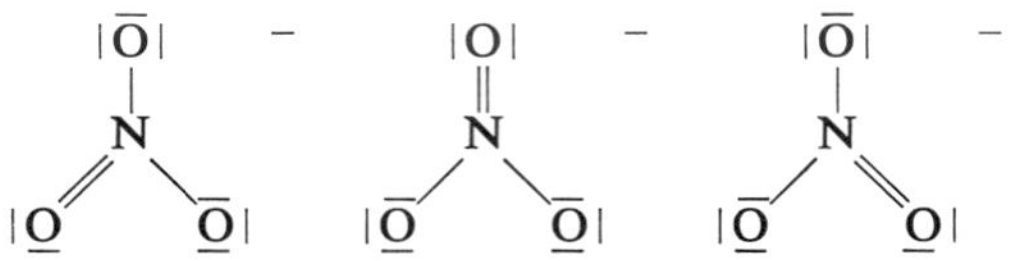

The nitrate ion is planar and the bond angles are 120°.

In addition to salts, esters and acid halides of nitric acid are known.

When the entire hydroxy group of nitric acid is replaced by an alkyl or phenyl group, the resulting compounds are called *nitro compounds*, and the covalently bonded NO_2 group is called the *nitro group*. The simplest nitro compound has the formula $CH_3—NO_2$ and is called nitromethane. Nitromethane may be represented by the following canonical forms:

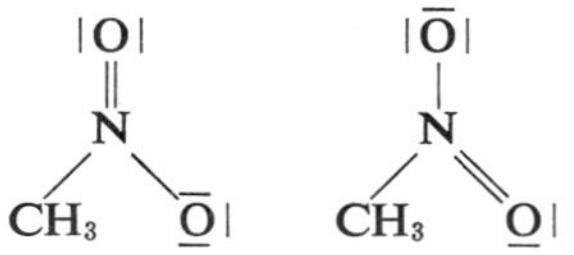

In general, the nitro compounds are highly polar. Several typical dipole moments are given in Table 9.3.

TABLE 9.3 Dipole Moments for Some Nitro Compounds (Debyes)

$CH_3—NO_2$	3.5
$CH_3—CH_2—NO_2$	3.5
$C_6H_5—NO_2$	4.3

Hydroxy Acids of Phosphorus

The larger phosphorus atom is able to accommodate three hydroxy groups around it. Thus phosphoric acid, H_3PO_4, is a trihydroxy acid that may be represented by the following structure:

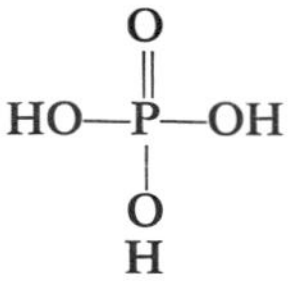

The phosphorus is at the center of a tetrahedron surrounded by four oxygens at the vertices.

Three different series of salts of phosphoric acid are known, those containing the dihydrogen phosphate ion, $H_2PO_4^-$, those containing the monohydrogen phosphate ion, HPO_4^{--}, and those containing the phosphate ion, PO_4^{3-}.

In addition to the salts, the esters, acid halides, and amides of phosphoric acid are known.

Partial dehydration between phosphoric acid molecules also occurs to form polymeric phosphates and phosphoric acids. *Diphosphoric acid* results from splitting out one water between two phosphoric acids. It has the following structure:

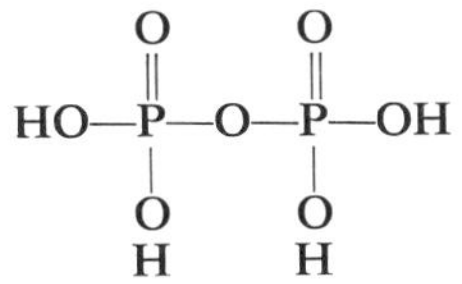

Diphosphoric acid is a tetrahydroxy acid with each phosphorus at the center of a tetrahedron and surrounded by four oxygens.

Triphosphoric acid results from splitting out two waters between three phosphoric acids. It has the following structure:

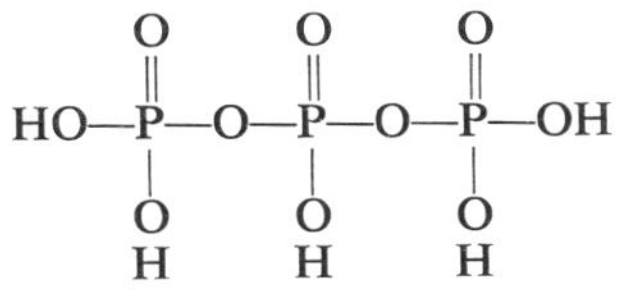

Triphosphoric acid is a pentahydroxy acid consisting of three PO_4 tetrahedra sharing oxygens.

More highly polymeric species tend to be cyclic in nature. It is possible to split out three molecules of water among three molecules of phosphoric acid to form *cyclotriphosphoric acid.*

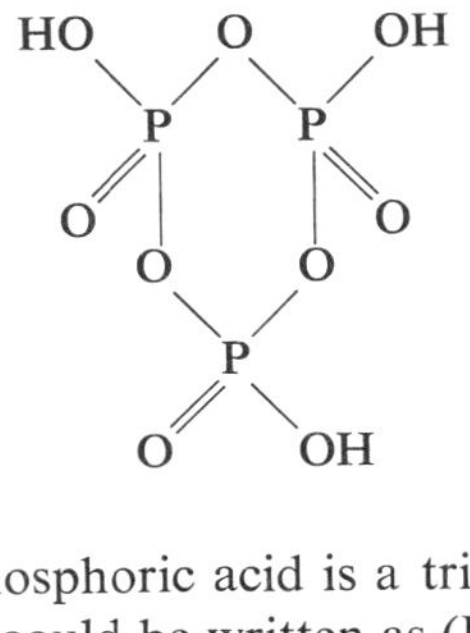

Cyclotriphosphoric acid is a trihydroxy acid which has the formula $H_3P_3O_9$. Formally this could be written as $(HPO_3)_3$, which indicates that it is a trimer consisting of three repeating HPO_3 units. Each of these units resembles HNO_3 in formula and also appears to be a less hydrated form of H_3PO_4. From this point of view, H_3PO_4

often is called orthophosphoric acid. No simple metaphosphoric acid is known, but the cyclotriphosphoric acid often is called trimetaphosphoric acid.

In a similar way, four molecules of water may be split out among four molecules of orthophosphoric acid to form *cyclotetraphosphoric acid.*

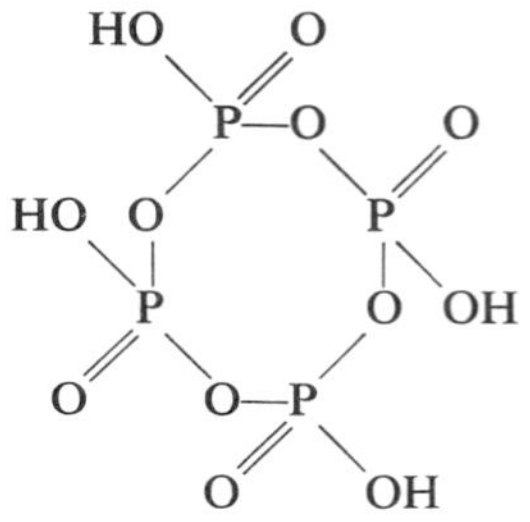

The molecular formula for cyclotetraphosphoric acid is $H_4P_4O_{12}$, which may be represented as $(HPO_3)_4$. Cyclotetraphosphoric acid also is called tetrametaphosphoric acid.

Phosphorous acid has the molecular formula H_3PO_3. However, its structure differs slightly from other hydroxy acids discussed thus far by having one of its hydrogens bonded directly to the phosphorus.

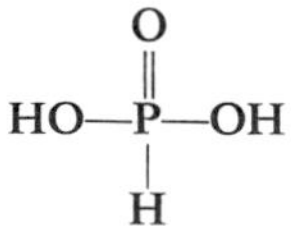

Phosphorous acid forms two sets of salts, those containing the hydrogen phosphite ion, $H_2PO_3^-$, and those containing the phosphite ion, HPO_3^{--}. The hydrogen bonded to the phosphorus is not replaced in salt formation.

Hypophosphorous acid has the molecular formula H_3PO_2 and the following structural formula:

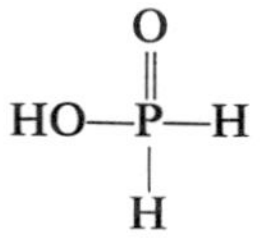

Thus hypophosphorous acid is a monohydroxy acid that forms salts containing the hypophosphite ion, $H_2PO_2^-$.

Hydroxy Acids of Arsenic

Arsenic acid has the molecular structure of phosphoric acid. Like phosphoric acid it forms three series of salts as well as polymeric hydroxy acid salts. Salts of diarsenic acid, $H_4As_2O_7$ are known along with salts of a polymetaarsenic acid which usually is represented as $(HAsO_3)_x$.

Arsenious acid differs from phosphorous acid although it has the analogous molecular formula, H_3AsO_3. First of all, arsenious acid is a trihydroxy acid with the following structural formula:

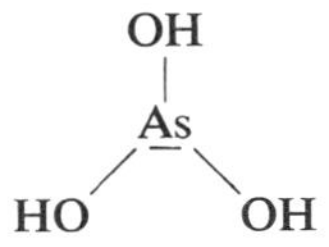

The four electron groups are arranged tetrahedrally around the central arsenic so that the molecule appears to be a trigonal pyramid with the arsenic at the apex and the three oxygens at the base.

Arsenious acid has not been isolated, but it does form three sets of salts, those containing the dihydrogen arsenite ion, $H_2AsO_3^-$, those containing the monohydrogen arsenite ion, $HAsO_3^{--}$, and those containing the arsenite ion, AsO_3^{3-}. In addition, a partially dehydrated form of arsenious acid is known in solution with the formula $HAsO_2$. This acid is called metaarsenious acid. As a result, sometimes arsenious acid, H_3AsO_3 is called orthoarsenious acid. Salts of metaarsenious acid are known.

CARBON FAMILY

Hydroxy Acids of Carbon

The simple dihydroxy acid of carbon, carbonic acid, with the formula, H_2CO_3, has not been isolated although it has been shown to exist in water and ether solutions. However, two series of salts of this acid are well known, one containing the hydrogen carbonate ion, HCO_3^-, and the other the carbonate ion, CO_3^{--}. The structures of these ions are represented by the following canonical forms:

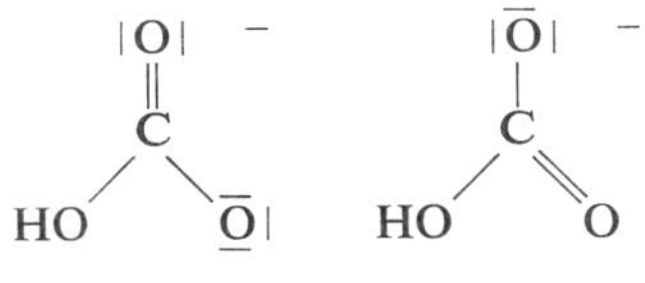

hydrogen carbonate ion

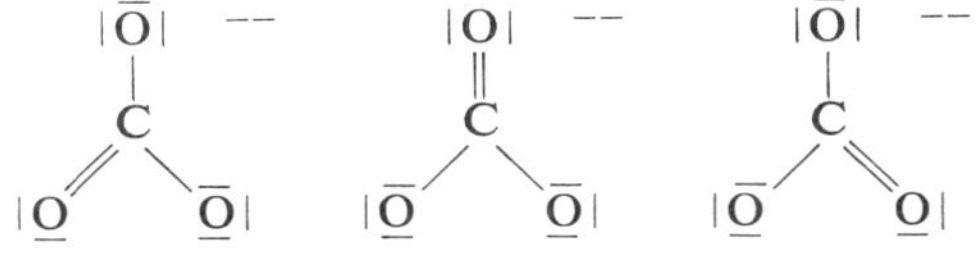

carbonate ion

In addition to the salts, esters, acid chlorides, and amides of carbonic acid are known.

A very important series of monohydroxy acids of carbon can be formed by replacement of one of the hydroxy groups in carbonic acid with either hydrogen, an alkyl group, or a phenyl group. These acids are called *carboxylic acids.* These acids appear to have stability and many of them have been isolated. One typical and common acid of this type is acetic acid, which has the formula CH_3COOH and may be represented structurally as follows.

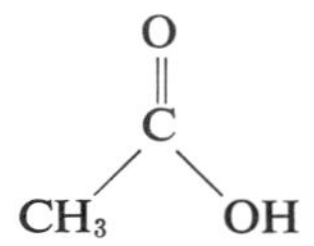

The methyl group, the hydroxyl group, and the oxygen surround the central carbon at the corners of a triangle. The central carbon, the oxygen, and the hydroxy group together are called the *carboxylic acid group.*

In a systematic way, these carboxylic acids are named according to the number of carbons in the chain, including the carbon in the carboxylic acid group, using the alkane names and replacing the final *e* in the name with the suffix *-oic*. Thus CH_3—COOH is named ethanoic acid. However, many of these acids have common names that are used more frequently than the systematic names. The following are a few examples.

H—COOH	methanoic acid formic acid
CH_3—COOH	ethanoic acid acetic acid
CH_3—CH_2—COOH	propanoic acid propionic acid
CH_3—CH_2—CH_2—COOH	butanoic acid butyric acid

When a phenyl group is attached to the carboxylic acid group, the compound is called *benzoic acid,* and has the formula C_6H_5—COOH.

Dicarboxylic acids contain two carboxylic acid groups. The simplest of these is called oxalic acid, which has the molecular formula $H_2C_2O_4$.

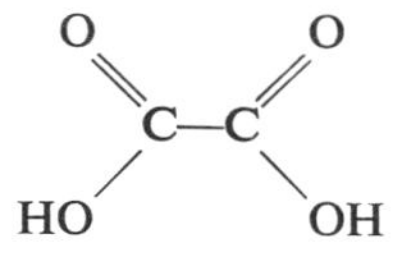

Many esters, acid halides, and amides of the carboxylic acids are known.

If both of the hydroxy groups of carbonic acid are replaced by alkyl or phenyl groups, the resulting compounds are known as *ketones*. One of the most commonly encountered ketones is *acetone*, known also as *dimethyl ketone*.

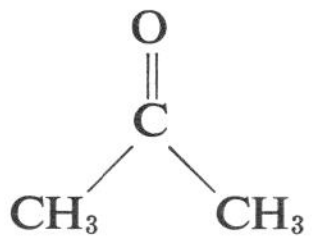

Hydroxy Acids of Silicon

Simple hydroxy acids of silicon apparently do not exist. Attempts to prepare them lead to silicon(IV) oxide. However, more complex hydroxy acids apparently do exist, at least as salts. These more complex hydroxy acids are derived from a hypothetical *orthosilicic acid* which consists of four hydroxy groups bonded to silicon, $Si(OH)_4$. Hypothetical though this acid may be, the tetramethylsilicate ester, $Si(OCH_3)_4$, has been prepared.

The partial dehydration of orthosilicic acid by splitting out one water between silicic acid molecules leads to disilicic acid, and a few salts containing the disilicate ion, $Si_2O_7^{6-}$ are known. The structure of disilicic acid may be represented as follows:

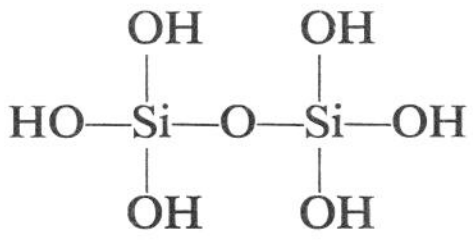

The dehydration of orthosilicic acid by splitting out six water moiecules among six acid molecules leads to the formation of cyclohexasilicic acid.

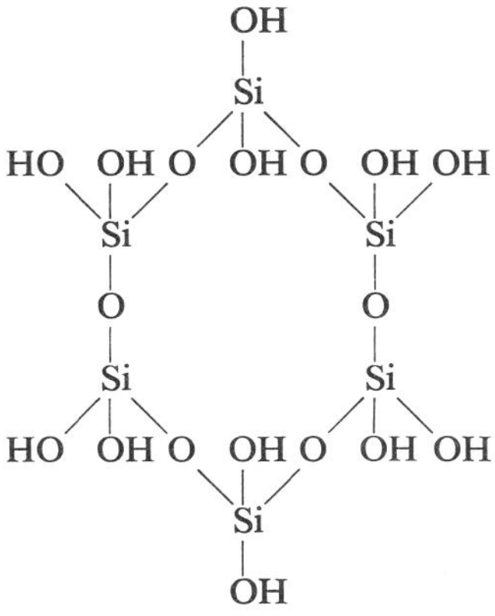

In this structure each silicon is surrounded tetrahedrally by four oxygens. Salts of cyclohexasilicic acid contain the $Si_6O_{18}^{12-}$ ion. Beryl and emerald contain this ion in combination with beryllium and aluminum ions.

Indefinitely long chains may be formed by the continuation of the kind of partial dehydration represented in the formation of disilicic acid. A portion of a typical chain may be represented as follows:

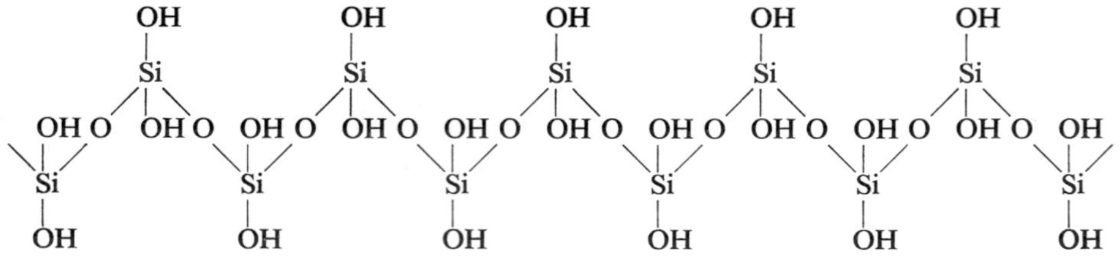

The repeating unit in this chain can be given by the formula H_2SiO_3 which resembles the formula proposed for carbonic acid, H_2CO_3. This chain-type polymer has been called metasilicic acid, but it certainly is not composed of discrete H_2SiO_3 molecules. The salts of this chain-type polymer have been called *pyroxenes*.

Another chain-type molecule of greater complexity can be visualized as the linking together of two parallel pyroxene chains to form an infinite chain of fused rings in which each ring is composed of six silicons and six oxygens, with each silicon at the center of a tetrahedron surrounded by four oxygens. Structurally this may be represented as follows:

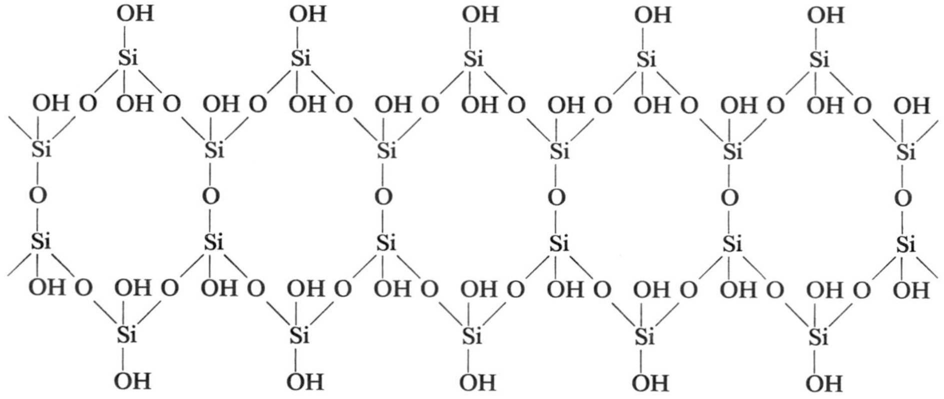

The repeating units in this polymer have the formula $H_6Si_4O_{11}$. The salts of this type of complex chain silicic acid are called *amphiboles*. Many amphiboles are fibrous, reflecting the chain-type structure. These fibrous amphiboles are called *asbestos*.

Continued partial dehydration leads to the bonding of parallel amphibole chains into infinite layer-type structures called *micas*, as shown on the next page. The repeating units in this polymer have the formula $H_2Si_2O_5$.

The final step in the polymerization of orthosilicic acid occurs by cross-linking the mica type layers into a three-dimensional framework. The most commonly known of these are the felspars, which constitute nearly two-thirds of the igneous rocks of the earth.

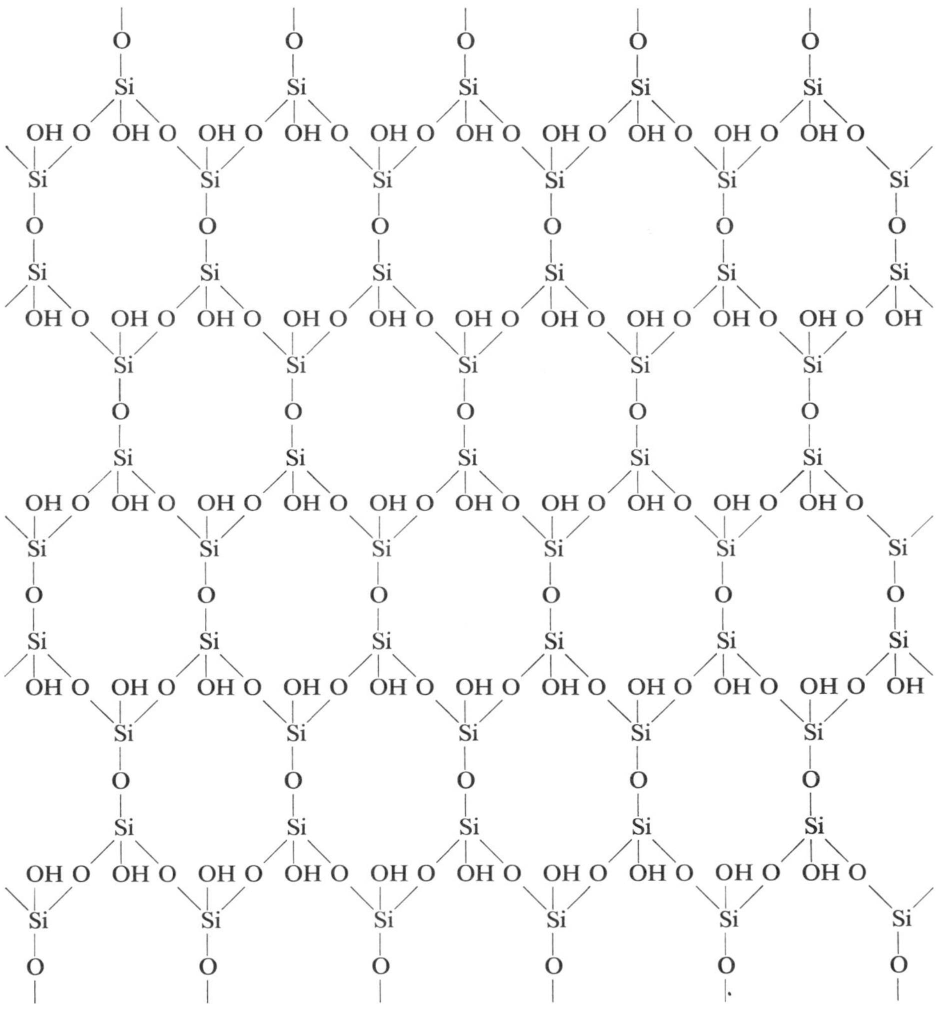

Miscellaneous Compounds

HYDROGEN CYANIDE, HCN, may be represented by the following structure:

$$H-C\equiv N|$$

The molecule is linear and very highly polar, and apparently highly hydrogen bonded. It has a melting point of 260°K, a boiling point of 299°K, and a critical temperature of 457°K. It has a dipole moment of 2.93 debyes and a dielectric constant of 115 at 20°C. It has one of the highest dielectric constants of the known liquids.

Hydrogen cyanide also is called *hydrocyanic acid*, and when the hydrogen is replaced by a metal, the salts are called *cyanides*. The cyanide ion has the following structure:

$$|C{\equiv}N|^-$$

Thiocyanic acid is a relatively unstable compound with the molecular formula HNCS. The structure may be represented by the following canonical forms:

$$H{-}N{\equiv}C{-}\overline{\underline{S}}| \qquad H{-}\overline{N}{=}C{=}\underline{S}|$$

Although thiocyanic acid can be prepared at low temperature, it is best known in the form of its salts, which contain the thiocyanate ion, NCS^-. The thiocyanate ion is linear and can be represented by the following canonical forms:

$$|N{\equiv}C{-}\overline{\underline{S}}|^- \qquad |\underline{N}{=}C{=}\underline{S}|^-$$

BORON

Three hydroxy groups may be bonded to boron to form orthoboric acid, H_3BO_3. The acid is stable and occurs as a solid at room conditions. The molecule consists of a central boron with three hydroxy groups arranged trigonally around it:

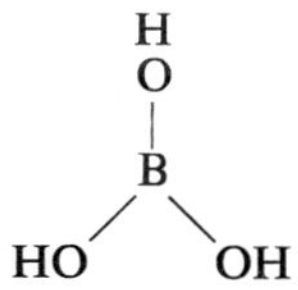

Esters and salts of this acid are known.

The polymerized acids are better known, however. Partial dehydration of orthoboric acid with the elimination of one water between two acid molecules leads to the diboric acid.

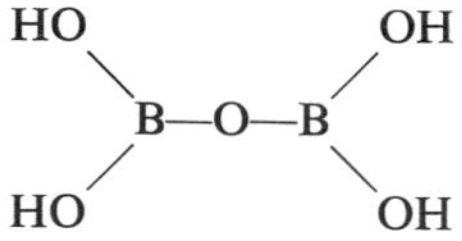

The salts of diboric acid contain the diborate ion, $B_2O_5^{4-}$.

Partial dehydration by elimination of three waters among three molecules of orthoboric acid lead to cyclotriboric acid.

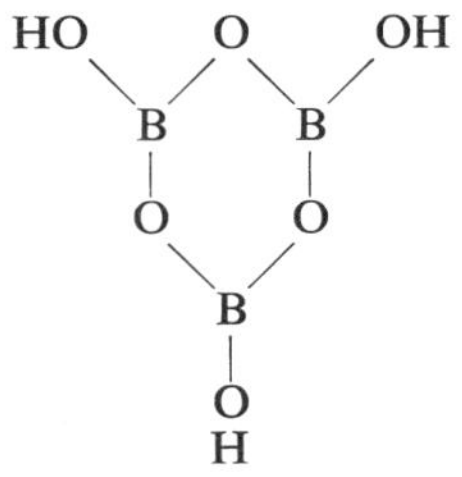

The repeating unit in this structure has the formula HBO_2. Since H_3BO_3 is the formal hydrate of HBO_2, H_3BO_3 has been called orthoboric acid and cyclotriboric acid has been called *trimetaboric acid.* Salts are known for this acid; they contain the trimetaborate ion, $B_3O_6^{3-}$.

Borax is the sodium salt of a highly polymerized hydroxy acid of boron. The acid consists of repeating units with the formula $H_2B_4O_7$. Borax is the hydrate of the sodium salt of this acid, $Na_2B_4O_7 \cdot 10H_2O$. The following is the structure of a portion of the molecule of this polymeric hydroxy acid:

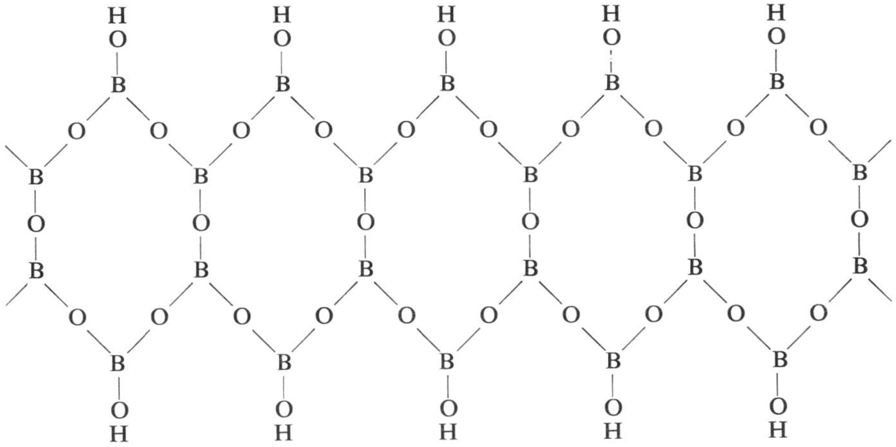

SUMMARY

The structures, formulas, and nomenclature of complex molecules can be understood in terms of a modular approach to molecular architecture.

The alkyl and phenyl halides can be related to the hydrogen halides; alcohols, ethers, and hydroxy acids and their derivatives can be related to water; amines and amides can be related to ammonia.

The hydroxy acids are parent compounds from which a great variety of compounds may be derived, including the polymeric hydroxy acids, salts, esters, acid halides, amides, nitro compounds, and ketones. Many of the hydroxy acids are so unstable that they have not been isolated and some probably do not exist. In general this instability can be attributed to a tendency toward dehydration.

Although the parent hydroxy acids may be unstable, their salts, esters, acid halides, and to some extent their amides, tend to be more stable. In addition, where polymerization is feasible, greater stability is found among the polymerized acids, which may be considered to be the partially dehydrated parent acids.

QUESTIONS AND PROBLEMS

9.1 Epsom salt is magnesium sulfate-7-hydrate. Suggest how the water molecules are incorporated into magnesium sulfate.

9.2 The following are some typical formulas for carbohydrates: $C_{12}H_{22}O_{11}$, $C_5H_{10}O_5$, $C_6H_{10}O_5$, $C_{18}H_{32}O_{16}$, and $C_6H_{12}O_6$. What is the significance of the name *carbohydrate*?

9.3 The usual commercial form of cobalt(II) sulfate is $CoSO_4 \cdot 7H_2O$. Suggest the structure for this salt.

9.4 When $MgCl_2 \cdot 6H_2O$ is heated, MgO and HCl form.
a) Write an equation for this reaction.
b) What does this suggest about the bonding in the hydrate?

9.5 The following compounds all form 6-hydrates: $Mg(BrO_3)_2$, $MgBr_2$, $Mg(ClO_3)_2$, $MgCl_2$, $Mg(NO_3)_2$, $Mg(ClO_4)_2$, $Mg(MnO_4)_2$, $MgSeO_4$, $MgSO_3$, and MgS_2O_3. What does this suggest?

9.6 What is the formal anhydride for each of the following?
a) NaOH
b) $Al(OH)_3$
c) H_6TeO_6
d) $H_2S_2O_7$
e) $Na_2S_2O_3 \cdot 5H_2O$

9.7 When a quantity of $Na_2SO_4 \cdot 10H_2O$ is sold, what percent of the purchase is water?

9.8 How is it possible to prepare esters of acids that do not exist, such as sulfurous acid?

9.9 Why are salts usually solids at room conditions?

9.10 How does bonding differ in esters and salts?

9.11 Why are the hydrogens in the methyl groups in dimethyl sulfate not hydrogen bonded?

9.12 Illustrate the hydrogen bonding in sulfuric diamide.

9.13 In chlorous acid, how is it possible for oxygen to make available a vacant orbital with which to accept an electron pair from chlorine?

9.14 Suggest possible bonding and geometries for the xenate and perxenate ions.

9.15 Suggest the general structural formulas for the hypohalous acids and the hypohalite ions.

9.16 Predict which has the higher dielectric constant at 25°C: $CHCl_3$ or CCl_4. Explain your choice.

9.17 When orthoperiodic acid is heated it dehydrates completely.
a) Suggest the formula for the anhydride and write an equation for the dehydration.
b) What is the oxidation state of iodine in orthoperiodic acid? In the anhydride of orthoperiodic acid?

9.18 What is the anticipated dehydration product for perchloric acid?

9.19 Write the structural formula for the orthoperiodate ion and the empirical formula for barium orthoperiodate.

9.20 Why is orthoperiodic acid more stable than metaperiodic acid?

9.21 Give two meanings for each of these prefixes: *meta-* and *ortho-*.

9.22 Perchloric acid forms a hydrate, $HClO_4 \cdot H_2O$, which is believed to contain water of coordination. Explain how this can be.

9.23 Write the structural formulas for eight isomeric alcohols with the formula $C_5H_{11}OH$, and name each.

9.24 An alternate nomenclature system has been used for alcohols. In this system, ethanol is hydroxyethane and 2-propanol is 2-hydroxypropane. Write the formula for 2,5-dihydroxy-3,4-dimethyl-5-ethyloctane, and name it also by the system used in this chapter.

9.25 Why should alcohols and ethers have lower electrostatic dipole moments than water?

9.26 Diethyl ether has a normal boiling point of 35°C, whereas 2-butanol has a normal boiling point of 108°C. Why is there such a great difference between these two compounds?

9.27 Criticize the following statement: Ethanol and dimethyl ether are isomers.

9.28 The following dielectric constants have been measured at 25°C. What does the trend suggest about these molecules?

$CH_3—OH$	32.6
$CH_3—CH_2—OH$	24.3
$CH_3—CH_2—CH_2—OH$	20.1
$CH_3—CH_2—CH_2—CH_2—OH$	17.1
$CH_3—CH_2—CH_2—CH_2—CH_2—OH$	13.9
$CH_3—CH_2—CH_2—CH_2—CH_2—CH_2—OH$	13.3

9.29 Write structural formulas and describe the geometries of dimethyl sulfite and sulfurous diamide.

9.30 The diacid chloride of sulfurous acid often is called thionyl chloride. Write a molecular and a structural formula for thionyl chloride. To what does the word *thionyl* probably refer?

9.31 What is the anhydride of sulfuric acid? When heated, sulfuric acid dehydrates. Write the equation for the dehydration reaction.

9.32 Why is the sulfate ion shaped like a regular tetrahedron, whereas the sulfuric acid molecule is distorted?

9.33 Write the structural formulas for the disulfate ion, trisulfate ion, and the diacid chloride of disulfuric acid.

9.34 For each of the following write a correct name and indicate the oxidation number of sulfur.

a) $K_2S_2O_7$ b) $Na_2S_3O_{10}$ c) $K_2S_2O_8$ d) Na_2SO_5

9.35 Suggest a reason why there appears to be no persulfuric acid.

9.36 The hydrogen sulfite ion sometimes is called the *bisulfite ion.* A compound with the formula $Na_2S_2O_5$ has been prepared, its common name is sodium metabisulfite.

a) What structure does this name imply?

b) Recently, interpretation of diffraction data led to the suggestion that there is a sulfur-to-sulfur bond in the metabisulfite ion. What structure does this imply?

c) What is the oxidation state of sulfur in metabisulfite ion?

9.37 Criticize the following statement: Peroxodisulfuric acid may be regarded as the product of the partial dehydration between two molecules of peroxomonosulfuric acid.

9.38 Suggest a structural formula and geometry for orthotelluric acid.

9.39 What is the formula for the acid chloride of selenic acid?

9.40 Explain the order in the following values for electrostatic dipole (Values are in debyes.)

H_2O	1.84
CH_3OH	1.70
NH_3	1.44
CH_3NH_2	1.25

9.41 Amides often have higher electrostatic dipole moments than amines. Suggest a reason for this.

9.42 Write structural formulas for phenyl amine, aminoethane, and 2-aminopropane.

9.43 Describe the geometry of the ammonium ion.

9.44 Suggest a structural formula for the methyl ammonium ion.

9.45 What are the advantages of coordinate covalent bond formation between ammonia and the hydrogen ion in ammonium ion formation?

9.46 Why is there no πd bonding indicated in the structures for nitrite and nitrate ions?

9.47 Assign formal charges to the atoms in the nitrite and nitrate ions and the nitric acid molecule.

9.48 Draw the probable structural formula for nitrous acid.

9.49 Compare the electronic structure of nitrite ion to the electronic structure of ozone.

9.50 The acid halides of nitrous acid and nitric acid are called nitrosyl and nitryl halides respectively. Write structural formulas for nitrosyl fluoride and nitryl chloride.

9.51 Why is it possible to isolate acid halides and esters of nitrous acid, but not the acid itself?

9.52 What are the anhydrides of nitrous and nitric acids?

9.53 Suggest why the bond angle in the nitrite ion is less than 120° but more than 109°28′.

9.54 What is the difference between the nitro group and the nitrite ion?

9.55 Account for the high polarities of the nitro compounds.

9.56 Nitromethane has a dielectric constant of 36 at 30°C. Does this appear to be in accord with its structure? Explain.

9.57 Consider orthophosphoric acid, phosphorous acid, and hypophosphorous acid.
a) What is the oxidation state of phosphorus in each of these acids?
b) What are the formulas of the anhydrides of these three acids?

9.58 Suggest why phosphorus hydroxy acids are more stable than the nitrogen hydroxy acids.

9.59 Write the structural formula for dimethyl phosphite.

9.60 What is the formula for the triacid halide of arsenious acid? Give several names for this compound.

9.61 Phosgene is the diacid chloride of carbonic acid and urea is the diamide of carbonic acid. Write the structural formulas for phosgene and urea.

9.62 Acetic acid undergoes partial dehydration by splitting out a molecule of water between two acid molecules. The product is called acetic anhydride. Write the structural formula for acetic anhydride.

9.63 Acetone has a dielectric constant of about 21 at 25°C. Is this to be expected from its structure? Explain.

9.64 Why are the polymerized silicic acids more stable than the simple hydroxy acids of silicon?

9.65 The structural formulas for two explosives, nitroglycerine and TNT, are given below. Classify each of these compounds by type.

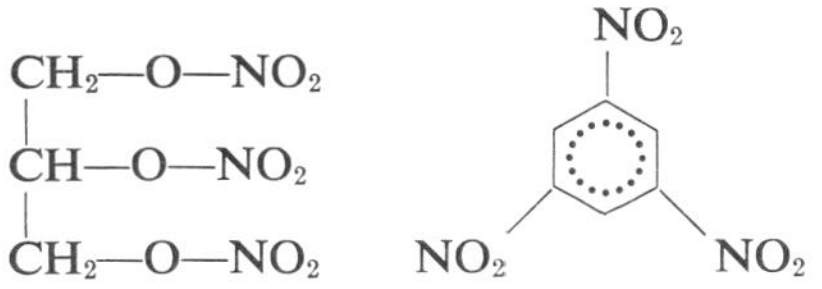

nitroglycerine TNT

9.66 The following are names used for some common chemical substances. Indicate the probable structural formula for each.

a) formamide
b) acetyl chloride
c) ethyl acetate
d) methyl ethyl ketone
e) sodium benzoate

9.67 Distinguish among the following with appropriate structural formulas:

a) beryl
b) pyroxene
c) amphibole
d) mica
e) asbestos

9.68 On paper, the amphibole structure resembles the polymeric $H_2B_4O_7$ structure. How do they differ?

9.69 What is the difference between orthoboric acid and metaboric acid?

9.70 What is the structure of $H_3B_3O_6$?

9.71 Write the structural formula and describe the geometry for each of the following:

a) 2,3-dibromo-2-butene
b) bromate ion
c) ethyl hypochlorite
d) sulfurous diamide
e) tetrathionate ion
f) amide of nitric acid
g) diphosphate ion
h) oxalate ion

9.72 Write a correct formula for each of the following:

a) sodium monohydrogen phosphate
b) ammonium peroxodisulfate
c) potassium cyanide
d) barium hydroxide
e) copper(II) bromate
f) magnesium hydrogen carbonate
g) arsenious acid
h) barium sulfate
i) magnesium perchlorate-6-hydrate
j) calcium nitrate
k) cyclohexasilicate ion
l) iron(III) phosphate
m) calcium hypophosphite
n) barium perxenate
o) 2-hexanol
p) dimethyl sulfite
q) sodium nitrite
r) barium phosphate
s) sodium thiosulfate-5-hydrate
t) lanthanum(III) carbonate-8-hydrate
u) ammonium iodide

9.73 Name each of the following compounds:

a) KOH
b) $MnSO_4 \cdot 4H_2O$
c) Na_6XeO_6
d) $CH_3—O—CH_2—CH_3$
e) $K_2Se_2O_7$
f) $Ca(H_2PO_4)_2$
g) $Na_2SO_4 \cdot 10H_2O$
h) NH_4NCS
i) $HIO_4 \cdot 2H_2O$
j) $(NH_4)_2S_3O_6$
k) $Ca(ClO)_2$
l) $KHSO_4$
m) $NaAsO_2$
n) $Sr(CN)_2$
o) NH_4NO_3
p) $ON—O—CH_2—CH_3$
q) $Na_3P_3O_9$
r) CaC_2O_4

10

The Metals and Some of Their Compounds

INTRODUCTION This chapter is an extension to the metals of the topic of Chapters 8 and 9. After a summary of some of the properties of the elements, the discussion covers some typical types of compounds formed by the metals. The chapter concludes with a consideration of coordination compounds, which are characteristic of many metallic ions.

THE ELEMENTS

In the elemental form, the metals consist of identical atoms. In the solid state, these atoms are packed into crystals, which usually can be described as hexagonal close-packed, face-centered cubic, or body-centered cubic. Approximately 70 percent of the metals crystallize with coordination number 12, that is, in one of the two close-packed arrangements. The remaining 30 percent crystallize with coordination number 8, that is, in the slightly less closely packed body-centered cubic arrangement. Like other substances, many metals crystallize in more than one crystal form; that is, they display crystal allotropy.

The crystal structures exhibited by the metals are shown in Table 10.1. Usually the stability of crystal allotropes is temperature dependent, and often it is possible to detect specific temperatures at which transitions occur from one form to another. For example, the hexagonal structure of titanium occurs below 880°C, whereas the body-centered form exists above this temperature. Similarly, the low-temperature hexagonal form of zirconium is converted to the body-centered cubic form above 865°C.

Because of the increased thermal agitation, it might be expected that at high temperatures the less closely packed body-centered cubic structure is favored. In some metals, such as titanium, zirçonium, scandium, lanthanum, and hafnium this appears to be the case. Molybdenum defies this rule, the more closely packed hexagonal form being favored at high temperatures. In other metals, a change in temperature leads to a transition from one closely packed arrangement to the other. The reason for this is not clear.

TABLE 10.1

Crystal Forms† of the Metals

Li bcc hx	*Be* hx													
Na bcc hx	*Mg* hx											*Al* fcc‡		
K bcc	*Ca* fcc hx bcc	*Sc* hx fcc	*Ti* hx bcc	*V* bcc	*Cr* bcc	*Mn*	*Fe* bcc fcc	*Co* hx fcc	*Ni* fcc hx	*Cu* fcc	*Zn* hx‡	*Ga*		
Rb bcc	*Sr* fcc hx bcc	*Y* hx	*Zr* hx bcc	*Nb* bcc	*Mo* bcc hx	*Tc* hx	*Ru* hx	*Rh* fcc	*Pd* fcc	*Ag* fcc	*Cd* hx	*In* fcc	*Sn*	
Cs bcc	*Ba* bcc	*La* hx fcc	*Hf* hx bcc	*Ta* bcc	*W* bcc	*Re* hx	*Os* hx	*Ir* fcc	*Pt* fcc	*Au* fcc	*Hg* fcc‡	*Tl* fcc‡ hx	*Pb* fcc	*Bi*
Fr	*Ra*	*Ac* fcc												

Ce fcc hx	*Pr* hx	*Nd* hx	*Pm*	*Sm*	*Eu* bcc	*Gd* hx	*Tb* hx	*Dy* hx	*Ho* hx	*Er* hx	*Tm* hx	*Yb* fcc	*Lu* hx
Th fcc	*Pa*	*U* bcc	*Np* bcc	*Pu* fcc bcc	*Am* hx								

† Key: fcc, face-centered cubic; bcc, body-centered cubic; hx, hexagonal close packed.
‡ Distorted

Iron undergoes one of the more interesting transitions. The body-centered cubic form is stable below 906°C and above 1401°C, and the closely packed face centered cubic form is stable in the intermediate temperature range.

In Table 10.1, the first form listed for each metal is the stable form at room temperature. From the table, some trends may be discerned among the room-temperature forms. The alkali metals adopt the more open, body-centered cubic structure. Among the transition metals there are similarities within families, with some exceptions such as iron, cobalt, and manganese.

Vacancies in Table 10.1 suggest that either structural information is not available or the structures are something other than the three close-packed arrangements. The distortions of the metals which border on the nonmetals probably are due to some covalent bonding that is superimposed on the metallic bonding. This probably imparts some directional characteristics to the bonds other than that imposed by the closest possible packing.

Density

In general, the solid nonmetals have a short range of densities. However, the metals tend to have a very wide range of values for the density, from the very low value of 0.54 g/ml for lithium to 22.70 g/ml for osmium at 25°C. The densities for some solid elements are shown in Table 10.2.

The general trend toward increasing density across a period of the metals corresponds to a decreasing atomic radius coupled with an increasing mass. Within the transition elements, the trend toward increasing density down the family corresponds to a rather large increase in mass coupled with a small increase or no increase in atomic size.

Certainly the density is a consequence of the atom packing within the crystals. Close packing leads to high densities. Thus the more open, body-centered cubic crystals tend to be less dense than comparable face-centered cubic or hexagonal arrangements. The alkali metals have low densities, not only because they have relatively large atoms for their masses, but also because they crystallize in the body-centered cubic arrangement.

Melting and Boiling Points and Heats of Fusion and Vaporization

Properties such as melting and boiling points and heats of fusion and vaporization depend upon the strength of the metallic bonds. On the whole, the metallic bond is moderately strong, and it appears to be strongest when there are maximum numbers of delocalized electrons in the sea surrounding the metallic ions. As a first, rough approximation, the number of delocalized electrons an atom might contribute to the delocalized sea is the total of the *ns*, *np*, and *unpaired* $(n-1)d$ electrons. On this basis, it is to be expected that maximum melting and boiling points should occur near the middle of the transition metal series in the vicinity of family 7B. Some data are given in Table 10.3 and a part of these are plotted in Fig. 10.1. Considering the lack of detail in our picture of the metallic bond, the tendency toward a maximum near the middle of the transition metal series is gratifying.

TABLE 10-2

Elemental Densitites in g/ml at 20°C (Solids Only)

	1A	2A	3B	4B	5B	6B	7B	8			1B	2B	3A	4A	5A	6A	7A	0
1	1 **H**																1 **H**	2 **He**
2	3 **Li** 0.53	4 **Be** 1.85											5 **B** 2.5	6 **C** 2.3	7 **N**	8 **O**	9 **F**	10 **Ne**
3	11 **Na** 0.97	12 **Mg** 1.74											13 **Al** 2.70	14 **Si** 2.4	15 **P** 1.8	16 **S** 2.0	17 **Cl**	18 **Ar**
4	19 **K** 0.86	20 **Ca** 1.55	21 **Sc** 3.02	22 **Ti** 4.50	23 **V** 5.9	24 **Cr** 7.14	25 **Mn** 7.3	26 **Fe** 6.03	27 **Co** 8.71	28 **Ni** 8.90	29 **Cu** 8.93	30 **Zn** 7.14	31 **Ga** 5.91	32 **Ge** 5.36	33 **As** 5.7	34 **Se** 4.5	35 **Br**	36 **Kr**
5	37 **Rb** 1.53	38 **Sr** 2.6	39 **Y** 4.47	40 **Zr** 6.4	41 **Nb** 8.57	42 **Mo** 9.01	43 **Tc** 11.5	44 **Ru** 12.2	45 **Rh** 12.5	46 **Pd** 12.0	47 **Ag** 10.5	48 **Cd** 8.65	49 **In** 7.3	50 **Sn** 5.8	51 **Sb** 6.68	52 **Te** 6	53 **I** 4.93	54 **Xe**
6	55 **Cs** 1.87	56 **Ba** 3.5	57 **La** 6.15	72 **Hf** 11.4	73 **Ta** 16.6	74 **W** 19.3	75 **Re** 20.5	76 **Os** 22.5	77 **Ir** 22.4	78 **Pt** 21.5	79 **Au** 19.3	80 **Hg**	81 **Tl** 11.8	82 **Pb** 11.3	83 **Bi** 9.78	84 **Po** 9.5	85 **At**	86 **Rn**
7	87 **Fr**	88 **Ra** 5	89 **Ac**															
					58 **Ce** 6.9	59 **Pr** 6.77	60 **Nd** 7.01	61 **Pm**	62 **Sm** 7.54	63 **Eu** 5.17	64 **Gd** 7.87	65 **Tb** 8.25	66 **Dy** 8.56	67 **Ho** 8.80	68 **Er** 9.06	69 **Tm** 9.32	70 **Yb** 6.96	71 **Lu** 9.85
					90 **Th** 11.0	91 **Pa** 15.4	92 **U** 18.9	93 **Np** 20.4	94 **Pu** 19	95 **Am**	96 **Cm** 7	97 **Bk**	98 **Cf**	99 **Es**	100 **Fm**	101 **Md**	102 **No**	103 **Lw**

TABLE 10.3

Melting and Boiling Points† of Elements in °K

†Below each element Symbol, the melting point is shown above the boiling point.

	1A	2A	3B	4B	5B	6B	7B	8	8	8	1B	2B	3A	4A	5A	6A	7A	0
1	1 **H** 14 20																1 **H** 14 20	2 **He** 1 4
2	3 **Li** 453 1540	4 **Be** 1551 3243											5 **B** 2570	6 **C** 3823 5100	7 **N** 63 77	8 **O** 54 90	9 **F** 53 86	10 **Ne** 24 27
3	11 **Na** 371 1165	12 **Mg** 922 1373											13 **Al** 933 2740	14 **Si** 1680 2628	15 **P** 317 553	16 **S** 392 718	17 **Cl** 172 238	18 **Ar** 84 87
4	19 **K** 337 1035	20 **Ca** 1083 1760	21 **Sc** 1812 3000	22 **Ti** 1953 3530	23 **V** 2160 3300	24 **Cr** 2173 2755	25 **Mn** 1523 2370	26 **Fe** 1812 3173	27 **Co** 1765 3173	28 **Ni** 1726 3005	29 **Cu** 1356 2853	30 **Zn** 692 1183	31 **Ga** 303 2673	32 **Ge** 1231 3100	33 **As**	34 **Se** 490 958	35 **Br** 266 332	36 **Kr** 116 121
5	37 **Rb** 312 973	38 **Sr** 1043 1657	39 **Y** 1773 3200	40 **Zr** 2123 3851	41 **Nb** 2740 5200	42 **Mo** 2893 5830	43 **Tc** 2470	44 **Ru** 2520 4173	45 **Rh** 2233 4000	46 **Pd** 1825 3200	47 **Ag** 1234 2453	48 **Cd** 594 1051	49 **In** 428 2270	50 **Sn** 505 2533	51 **Sb** 903 1653	52 **Te** 723 1260	53 **I** 387 457	54 **Xe** 161 165
6	55 **Cs** 303 963	56 **Ba** 998 1413	57 **La** 1193 3742	72 **Hf** 2423 5673	73 **Ta** 3273 6273	74 **W** 3653 5973	75 **Re** 3443 5898	76 **Os** 3270 5270	77 **Ir** 2683 4800	78 **Pt** 2042 4073	79 **Au** 1336 3239	80 **Hg** 234 630	81 **Tl** 577 1740	82 **Pb** 600 2017	83 **Bi** 544 1833	84 **Po** 527 1235	85 **At**	86 **Rn** 202 211
7	87 **Fr**	88 **Ra** 973 2010	89 **Ac** 1323 3500															
					58 **Ce** 1077 3741	59 **Pr** 1205 3400	60 **Nd** 1297 3358	61 **Pm** 1308 3000	62 **Sm** 1325 2170	63 **Eu** 1098 1712	64 **Gd** 1598 3300	65 **Tb** 1633 3100	66 **Dy** 1680 2870	67 **Ho** 1734 2870	68 **Er** 1773 3170	69 **Tm** 1818 2000	70 **Yb** 1097 1700	71 **Lu** 1925 3600
					90 **Th** 2023 4300	91 **Pa** 1500	92 **U** 1405 4091	93 **Np** 913	94 **Pu** 913 3508	95 **Am** 1125	96 **Cm**	97 **Bk**	98 **Cf**	99 **Es**	100 **Fm**	101 **Md**	102 **No**	103 **Lw**

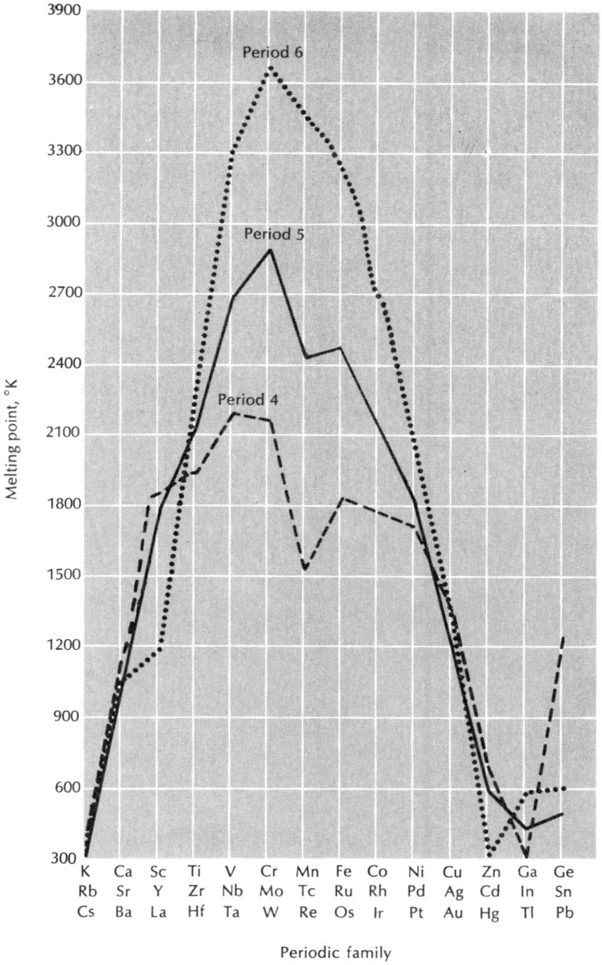

Fig. 10.1. Plot of melting point against periodic family for the metals of periods 4, 5, and 6

Table 10.4 and Fig. 10.2 show some similar data for heats of fusion and vaporization. Approximately, the trends follow those of the melting and boiling points. The extremely high heats of vaporization are good indicators of the strength of the metallic bond. The fact that the heats of vaporization are much higher than the heats of fusion indicate that the liquid state is more like the solid state than it is like the vapor state. In a sense, even the solid metal has some properties similar to the liquid state, in that the planes of ions in the crystals may slip past each other

TABLE 10-4

Heats of Fusion and Vaporization† of the Elements in kcal/mole

†Below each element symbol, the heat of fusion is shown above the heat of vaporization.

	1A	2A	3B	4B	5B	6B	7B	8	8	8	1B	2B	3A	4A	5A	6A	7A	0
1	1 **H** 0.014 0.108																1 **H** 0.014 0.108	2 **He** 0.005 0.020
2	3 **Li** 0.723 35.4	4 **Be** 2.99 70.2											5 **B** 5.30 129	6 **C** 143	7 **N** 0.086 0.66	8 **O** 0.053 0.82	9 **F** 0.186 0.78	10 **Ne** 0.080 0.421
3	11 **Na** 0.622 21.3	12 **Mg** 2.14 31.5											13 **Al** 2.56 70.2	14 **Si** 11.1 94.2	15 **P** 4.85 2.97	16 **S** 0.085 2.7	17 **Cl** 0.76 2.44	18 **Ar** 0.281 1.56
4	19 **K** 0.554 18.5	20 **Ca** 2.07 35.8	21 **Sc** 3.85 72.9	22 **Ti** 3.70 103	23 **V** 4.18 109	24 **Cr** 3.49 83.4	25 **Mn** 3.49 5.37	26 **Fe** 3.70 84.6	27 **Co** 3.66 91.5	28 **Ni** 4.25 90.9	29 **Cu** 3.11 72.6	30 **Zn** 1.74 27.4	31 **Ga** 1.33 60.7	32 **Ge** 7.12 79.8	33 **As** 7.75	34 **Se** 1.29 22.8	35 **Br** 1.26 3.59	36 **Kr** 0.391 2.16
5	37 **Rb** 0.526 16.5	38 **Sr** 2.20 33.2	39 **Y** 4.11 93.9	40 **Zr** 4.78 139	41 **Nb** 6.40 166	42 **Mo** 6.59 142	43 **Tc**	44 **Ru** 5.2 148	45 **Rh** 5.2 127	46 **Pd** 4.11 89	47 **Ag** 2.69 60.7	48 **Cd** 1.53 23.9	49 **In** 0.78 54.0	50 **Sn** 1.69 69.4	51 **Sb** 4.8 46.6	52 **Te** 4.18 27.2	53 **I** 1.87 5.2	54 **Xe** 0.548 3.02
6	55 **Cs** 0.521 15.7	56 **Ba** 1.83 36.0	57 **La** 2.70 75.6	72 **Hf** 5.21 158	73 **Ta** 7.50 180	74 **W** 8.41 191	75 **Re** 9.08 169	76 **Os** 6.4 162	77 **Ir** 6.6 152	78 **Pt** 5.18 107	79. **Au** 3.05 77.5	80 **Hg** 0.548 14.1	81 **Tl** 1.00 38.8	82 **Pb** 1.14 42.9	83 **Bi** 2.61 36.2	84 **Po** 24.6	85 **At**	86 **Rn** 0.638 3.92
7	87 **Fr**	88 **Ra** 2.0 32.7	89 **Ac** 3.39 94.8															

58 **Ce** 3.08 75.0	59 **Pr** 2.70 79	60 **Nd** 3.49	61 **Pm**	62 **Sm** 2.6 46	63 **Eu** 2.5 40	64 **Gd** 3.7 72	65 **Tb** 3.9 70	66 **Dy** 4.1 67	67 **Ho** 4.1 67	68 **Er** 4.1 70	69 **Tm** 4.4 56	70 **Yb** 2.2 40	71 **Lu** 4.6 59
90 **Th** 3.74	91 **Pa** 130	92 **U** 4.71 98.4	93 **Np**	94 **Pu**	95 **Am** 56.9	96 **Cm**	97 **Bk**	98 **Cf**	99 **Es**	100 **Fm**	101 **Md**	102 **No**	103 **Lw**

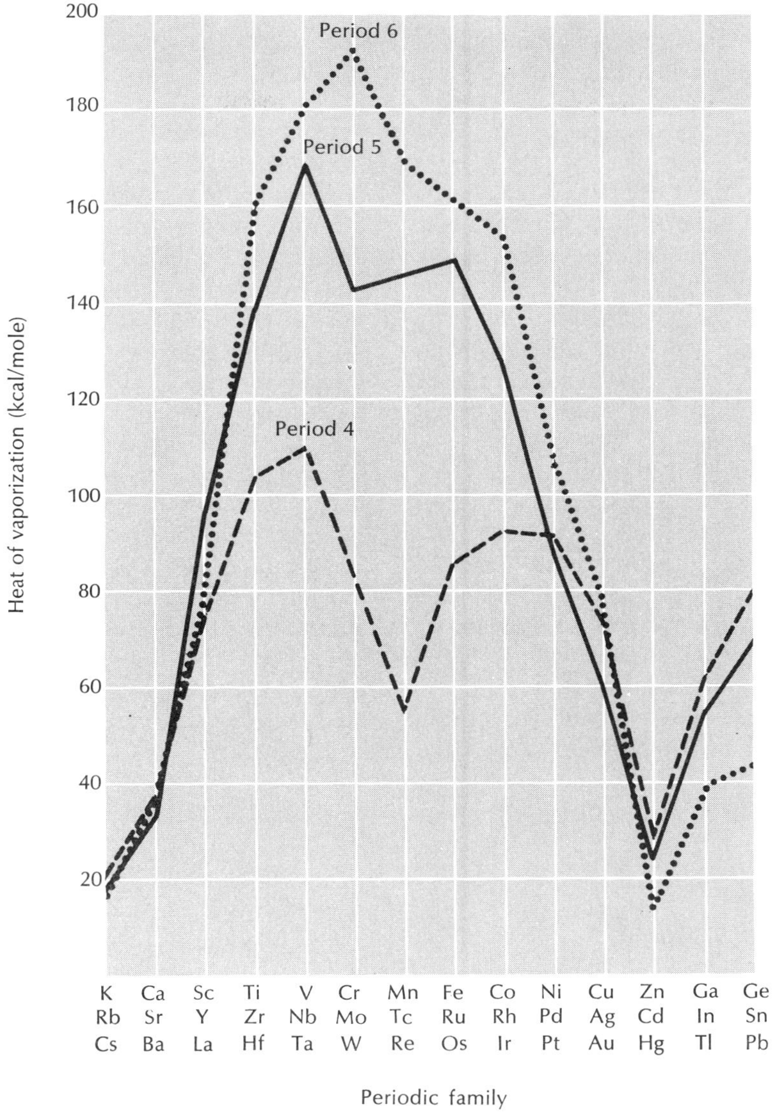

Fig. 10.2. Plot of heat of vaporization against periodic family for the metals of periods 4, 5, and 6

and the solids may be deformed. The liquid state is a semistructured state, and an expansion of the solid crystal structure and some slight disordering may be all that is necessary to transform the metal to a liquid.

Thus the metallic bond probably is still operative in the liquid state; relatively little energy is required to make the transformation from solid to liquid, since relatively few bonds must be broken. On the other hand, vaporization requires the breaking of most of the bonds to put the metal in the atomic state or at least a state of small aggregates in the vapor.

Electrical Conductivity

One of the characteristics of metals is their ability to conduct electricity, which means that metals permit the flow of electrons. This is reasonable in view of the metallic bond, in which electrons are delocalized and thus are somewhat free to move.

If we imagine an electron pump attached across the ends of a metal rod, then electrical conduction amounts to pumping electrons from one end and onto the other end of the rod. This is comparable to pumping water into the Atlantic Ocean in New York City while someone pumps water from the ocean at the same rate on the west coast of Africa.

In general, it is found that the electrical conductivity of metals decreases as temperature increases, whereas the electrical conductivity of nonmetals increases as temperature increases. In the metals the increase in temperature increases the vibrations of the ions in the lattice and these increased vibrations impede the flow of electrons. However, in the nonmetals, the initial problem is that electrons are not free to move because they are strongly localized between atoms. With increasing temperature, the covalent bonds tend to be broken, and as these bonds break, electrons become free to move for the first time. Thus, regardless of the fact that the lattice may be vibrating more than before, the conductance improves because of the increased mobility of the electrons.

Thermal Conductivity

Not only are metals good conductors of electricity, but they are good conductors of heat as well. Apparently, the close packing of atoms in the metallic lattice is responsible for the ability of the metals to conduct heat.

Thus when two ends of a metal bar are at different temperatures, the cooler end becomes warmer and the warmer end becomes cooler. The more quickly this happens, the greater is the thermal conductivity. The atoms in the warmer end have greater kinetic energy. Since the atoms are closely packed in the metal structure, these atoms collide with other atoms quickly and impart some of their kinetic energy to them. In turn these atoms collide with others and kinetic energy is transferred quickly.

BINARY COMPOUNDS

The metals tend to form the usual kinds of binary compounds with very little tendency to catenate in any usual covalent sense. They often exhibit a variety of oxidation states, and this is especially true of the transition metals. A summary of the common oxidation numbers was given in Table 6.5.

The alkali metals always exhibit an oxidation state of 1+, and the alkaline earths always exhibit an oxidation state of 2+, corresponding to the involvement of the *ns* electrons. In family 3B, scandium, yttrium, lanthanum, and actinium all show an oxidation state of 3+ corresponding to the involvement of the two *ns* electrons and the one $(n - 1)d$ electron. In addition the inner transition elements of period 6 all show the 3+ oxidation state, again corresponding to the involvement of the two *ns* electrons and the one $(n - 1)d$ electron.

Among the transition elements not in family 3A, the 2+ state of oxidation is rather common, corresponding to the involvement of the two *ns* electrons. In addition the 3+ state is found quite often and this involves one $(n - 1)d$ electron in addition to the two *ns* electrons. The involvement of the additional $(n - 1)d$ electrons for higher oxidation states appears possible early in the transition metals but seems to diminish later within each series in each period. The maximum positive oxidation state still normally appears to correspond to the family number in the periodic table, although there are a number of elements which do not exhibit this maximum. In general, it is found that oxidation states normally do not exceed the sum of the number of *ns* electrons and *unpaired* $(n - 1)d$ electrons.

Generally speaking, we tend to consider most of the common binary compounds of these metals as ionic or mostly so when the oxidation states are reasonably low, say 2+ or 3+. As the oxidation state increases, the tendency for a given element to form ionic bonds decreases and covalency predominates.

The metals form halides, including fluorides, chlorides, bromides, and iodides. However, the abundance of the compounds appears to fall off in the order fluorides, chlorides, bromides, and iodides. Iodides are found rarely among the elements in high oxidation states. For the most part the fluorides are considered to be ionic as are most chlorides. Bromides and iodides, which are somewhat more scarce, are considered to be much more covalent as a general rule.

Oxides and sulfides are common, the sulfides tending to be much more covalent and somewhat less prevalent among the high oxidation states. It is probably safe to say that oxides exist in nearly all of the oxidation states listed in Table 6.5.

TERNARY AND HIGHER COMPOUNDS

Salts

The metals are found frequently as salts of ternary hydroxy acids of the nonmetals. Thus the perchlorates, sulfates, nitrates, phosphates, carbonates, and others, constitute a large portion of the compounds of the metals. For example, manganese often is found as manganese(II) sulfate and copper often is found as copper(II) sulfate and copper(II) nitrate.

Hydroxy Compounds

The hydroxy compounds of the metals are common. Among the metals in low oxidation states these compounds are considered to be ionic hydroxides. For metals in higher oxidation states, the hydroxy compounds tend to be somewhat unstable with respect to dehydration. Some of them form hydroxy acids or at least derivatives of hypothetical hydroxy acids.

For example, vanadium in the 5+ oxidation state tends to resemble phosphorus, in the 5+ oxidation state. Although vanadic acid, H_3VO_4, has not been isolated salts of vanadic acid are known. In addition, vanadic acid tends to polymerize in the same fashion that phosphoric acid polymerizes. Thus salts are known which contain divanadate ions, $V_2O_7^{4-}$, tetravanadate ions, $V_4O_{13}^{6-}$, pentavanadate ions, $V_5O_{16}^{7-}$, and even in the extreme, polymetavanadate ions, $(VO_3^-)_n$.

Similarly, chromium in the 6+ state of oxidation resembles sulfur in the 6+ state of oxidation. The simple chromic acid, H_2CrO_4, is not known, but salts and acid halides are well known for this parent compound. Even to a greater extent than sulfuric acid, this hydroxy acid polymerizes. Chromate ions, CrO_4^{--}, dichromate ions, $Cr_2O_7^{--}$, and trichromate ions, $Cr_3O_{10}^{--}$, are well known in salts.

In the 7+ state of oxidation, manganese tends to resemble chlorine in the 7+ state. The permanganic acid, $HMnO_4$, has not been isolated, but salts containing the permanganate ion, MnO_4^-, are well known.

Coordination Complexes

Metal cations are highly electronegative and also have vacant orbitals suitable for bonding. Thus these metal cations may form coordinate covalent bonds by sharing pairs of otherwise nonbonding electrons in other molecules and ions. From this point of view we refer to the metallic cations as *electron-pair acceptors*. Those ions and molecules which serve as *electron-pair donors* often are called *ligands*. The covalently bonded molecule or ion formed by an electron pair acceptor and one or more ligands is called a *coordination complex*.

We have mentioned examples of such complexes at least twice before. Water of coordination is an example in which water molecules serve as ligands and a typical coordination complex may be represented as $[Co(H_2O)_6]^{++}$. The ammonium ion, NH_4^+, is a coordination complex in which the ammonia molecule is the ligand and the hydrogen ion is the electron-pair acceptor. Different cations may accept different numbers of electron pairs. In fact, under different circumstances, the same cation may accept different numbers of electron pairs. The number of electron pairs which a cation accepts in the formation of a coordination complex is called the *coordination number* of the cation.

CHARACTERISTICS OF ELECTRON PAIR ACCEPTORS. The more electronegative the metal cations are, the greater is their attraction for electrons and the greater is the potential-energy advantage of sharing electron pairs with the metal cation. In general, the smaller, more highly charged cations are the most electronegative, and therefore the more apt to form strong coordinate covalent bonds.

For example, the alkali metal and alkaline earth cations tend not to form strong coordinate covalent bonds, since they are relatively large ions with rather low ionic charges. However, within each of the families, the smallest ions, Li^+, Be^{++}, and Mg^{++}, have the greatest tendencies to form coordinate covalent bonds.

Similarly, within a given period, among similarly charged cations of the transition elements, the tendency toward coordinate covalent bond formation increases with atomic number as the ionic size decreases. Thus Fe^{3+} tends to coordinate more strongly than Sc^{3+}.

Also, as a general rule, more highly charged cations of the same element tend to form stronger covalent bonds than cations with a smaller ionic charge. Thus Fe^{3+} tends to coordinate more strongly than Fe^{++}.

Most metal cations have the one *ns* and three *np* orbitals available for electron pair acceptance. However, in many cases it is found that the coordination number of the cation exceeds four. In these cases, it is assumed that either the $(n - 1)d$ or *nd* orbitals are used for bonding.

CHARACTERISTICS OF LIGANDS. Ligands must have electron pairs available for coordinate covalent bond formation. This is true of most anions as well as many molecules. However, for ions and molecules to serve as good ligands, the electrons to be shared should not be attracted too strongly by the ligand. This is to say that if the electronegativity of the ligand is too high, covalent bond formation will not be possible, since the ligand will not permit the electron pairs to be close enough to the cation nucleus, on the average, to provide adequate potential-energy decrease for covalent bond formation. This does not prevent an electrostatic interaction between a cation and an anionic or polar ligand, but it does not lead to good covalent bond formation.

For example, Sanderson has estimated the electronegativities of some anions. Some of these are shown in Table 10.5. Note the perchlorate ion. Despite the fact that it is an anion, it has an electronegativity comparable to elemental fluorine. This is not unexpected, since the ion is composed of five very highly electronegative atoms. However, it means that perchlorate tends to form highly ionic compounds and that it does not form covalent bonds readily.

Thus the perchlorate ion is not a good ligand in coordinate covalent bond formation. In practice, it is found that those ions listed in Table 10.5, with electronegativities in excess of about 3.5, do not serve as good ligands in coordinate covalent bond formation. Of course, the simple halide ions do serve as ligands since they have low electronegativities.

There is an additional factor to be considered, however. Often the more highly electronegative oxygen and nitrogen atoms serve as electron-pair donors more readily than the less electronegative atoms such as sulfur and phosphorus. The oxygen and nitrogen atoms are among those tiny atoms on which electron-pair repulsions are significant. By localizing these electron pairs, repulsions by nonbonding electron pairs are decreased.

Taking these factors into consideration, we may summarize some of the characteristic species which serve as ligands. Among the halogens, the halide ions are the most common electron pair donors.

Oxygen serves as an electron-pair donor in the hydroxide ion, the oxide ion, and in many oxy anions such as carbonate, sulfite, and the carboxylates. In addition, many molecular species containing oxygen also serve as ligands, such as water, alcohols, ethers, ketones, and hydroxy acids. The sulfide ion does serve as an

TABLE 10.5
Electronegativities of Some Anions

ClO_4^-	perchlorate	4.2
HSO_4^-	hydrogen sulfate ion	4.0
NO_3^-	nitrate	3.9
$H_2PO_4^-$	dihydrogen phosphate ion	3.8
HSO_3^-	hydrogen sulfite ion	3.7
HCO_3^-	hydrogen carbonate ion	3.6
NO_2^-	nitrite	3.4
CH_3COO^-	acetate	3.4
SO_4^{--}	sulfate	3.1
HPO_4^{--}	monohydrogen phosphate ion	3.1
NCS^-	thiocyanate	2.7
ClO^-	hypochlorite	2.7
SO_3^{--}	sulfite	2.6
PO_4^{3-}	orthophosphate	2.1
CN^-	cyanide	2.0
BO_3^{3-}	orthoborate	1.2
F^-	fluoride	0.8
Cl^-	chloride	0.3
Br^-	bromide	0.1
O^{--}	oxide	<0
I^-	iodide	<0
S^{--}	sulfide	<0
Se^{--}	selenide	<0

electron-pair donor, but other sulfur-containing species tend to form relatively few coordination complexes.

In the nitrogen family, nitrogen is an electron-pair donor, as it appears in ammonia, amines, hydrazine, and the thiocyanate ion. Phosphine and arsine also appear as ligands.

In the carbon family, there are relatively few good examples in which the elements serve as electron-pair donors. The best example appears to be the cyanide ion, CN^-, in which the coordination normally proceeds through the carbon rather than the nitrogen.

Some ligands can donate more than one pair of electrons through different atoms within the molecule or anion. Such ligands are called *chelating agents* and the complexes are called *chelates*. The chelating agents are classified by the number of electron pairs which they donate when they serve as chelating agents. This is designated by the appropriate numerical prefix added to the root, *dentate*. Thus a chelating agent that donates two pairs of electrons is called a bidentate chelating agent. Chelating agents are known to be bidentate, terdentate, quadridentate, quiquidentate, and sexidentate. Some examples of a few common chelating agents are shown in Table 10.6.

ION-LIGAND SIZES AND COMPLEX GEOMETRY. In addition to the number of available orbitals on the metal cation, another factor affects the value of the coordination number which a cation displays. This is the relative size of the ligand with respect to the size of the central cation.

TABLE 10.6

Some Common Chelating Agents

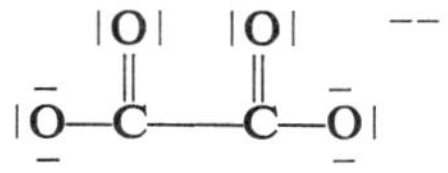	oxalate	bidentate
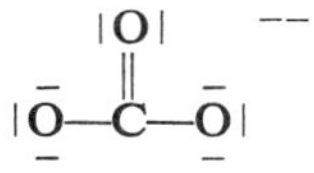	carbonate	bidentate
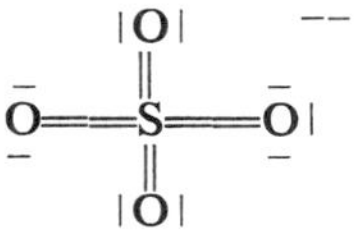	sulfate	bidentate
$H_2\bar{N}—CH_2—CH_2—\bar{N}H_2$	ethylenediamine	bidentate
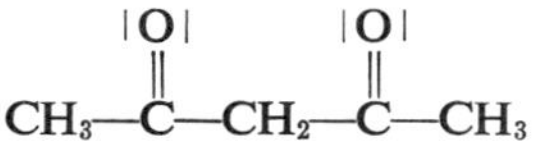	acetyl acetone	bidentate
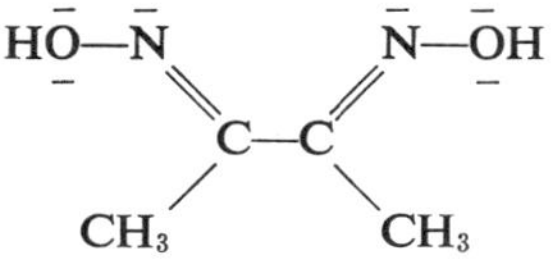	dimethyl glyoxime	bidentate
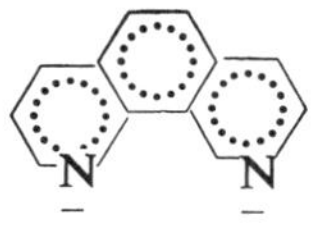	1,10-phenanthroline	bidentate
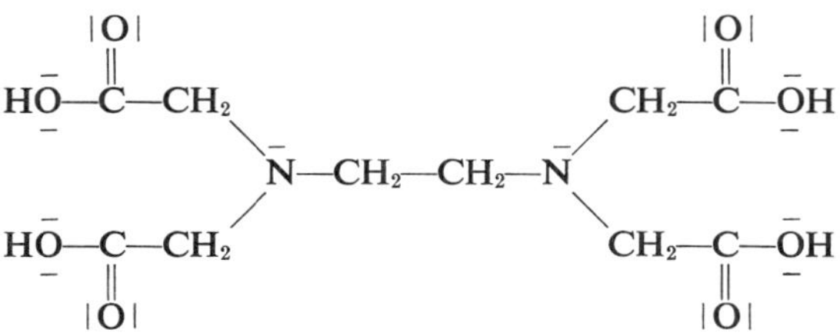	ethylenediamine-tetraacetic acid	up to sexidentate

If the central cation is quite small and the ligands are quite large, very few ligands can cluster around the central cation before the ligands interfere with each other. On the other hand, a larger number of small ligands can cluster around a large cation before there is major interference among the ligands.

Essentially, this is the same problem we faced in talking about fitting ions into holes in crystals. Again, the radius ratio indicates the feasible coordination numbers of the metal cations based on size relationships only. For our use here, the radius ratio is defined in slightly different words, as follows:

$$\text{radius ratio} = \frac{\text{radius of cation}}{\text{radius of ligand}}$$

Table 10.7 shows the ranges of radius ratio values that are possible for various coordination numbers. It should be noted that the geometries of the coordination complexes are exactly what is expected since the bonding electrons have maximum separation on the surface of the central metal cation.

Of course, for coordination numbers 1 and 2, there are no radius ratio limits, since the complexes normally can be linear. However, when the coordination number is three, the three ligands lie in a plane with the metal cation. They lie at the corners of an equilateral triangle circumscribed around the central metal cation. If the radius ratio is less than 0.155, the ligands are large enough to touch each other and thus interfere with each other without touching the central metal cation. Or, we might say that the hole in which the central cation fits is too large, and it "rattles around." Under these conditions, the central cation cannot form covalent bonds with all of the ligands.

TABLE 10.7
Radius Ratios Related to Coordination Numbers

COORDINATION NUMBER	GEOMETRY	RADIUS RATIO
3	trigonal	>0.155
4	tetrahedral	>0.225
5	trigonal bipyramidal	>0.414
6	octahedral	>0.414

Similarly, if there are four ligands, they tend to be arranged tetrahedrally around the central cation, but to provide adequate bonding to the central metal cation, the radius ratio must exceed 0.225. If there are five ligands, they are arranged at the vertices of a trigonal bipyramid. Similarly, six ligands are arranged at the vertices of an octahedron. In either case, the radius ratio must exceed 0.414 to allow adequate bonding between the central cation and the ligands.

FORMULAS AND NOMENCLATURE. In writing formulas for coordination complexes, it is common practice to put the central cation and its surrounding ligands within square brackets. The square brackets then enclose the coordination complex, which sometimes is called the *coordination sphere*. The complex, or coordination sphere, may be a cation such as $[Cu(NH_3)_4]^{++}$ where four neutral ammonia molecules are coordinated to a copper(II) ion. The coordination sphere may be an electrically neutral molecule such as $[PtCl_2(NH_3)_2]$, where two neutral ammonia molecules and two chloride ions have been coordinated to a platinum(II) ion. The coordination sphere may be an anion, as in the case of $[Fe(CN)_6]^{3-}$ where six cyanide ions are coordinated to an iron(III) ion.

In formulas for coordination spheres, the central cation is written first, followed by the anionic ligands and neutral ligands in that order. Thus the complex consisting of one cobalt(III) ion bonded to four ammonia molecules and two chloride ions is written as $[CoCl_2(NH_3)_4]^+$ rather than $[Co(NH_3)_4Cl_2]^+$ or $[(NH_3)_4CoCl_2]^+$.

The nomenclature of these coordination spheres is based upon the following rules:

1. Species in the coordination sphere are named in the following order: anionic ligands, neutral ligands, central cation.
2. *Anionic ligands* have the suffix *-o;* for example, Cl^- is *chloro*, OH^- is *hydroxo*, CN^- is *cyano*, CO_3^{--} is *carbonato*, SO_4^{--} is *sulfato*, etc.
3. Neutral, molecular ligands retain their names unchanged with the two important exceptions: water, H_2O, is *aquo* and ammonia, NH_3, is *ammine.*
4. The usual numerical prefixes are used to designate the number of each kind of ligand except that *mono-* usually is omitted. The numerical prefixes *bis-*, *tris-*, and *tetrakis-* are used for two, three, and four respectively when the name of the ligand is complex. When these prefixes are used, the name of the ligand is set off in parentheses immediately after the prefix.
5. The names of *anionic* coordination spheres end with the suffix *-ate,* whereas molecular and cationic coordination spheres have no special suffixes.
6. The final part of the name of any coordination sphere is the parenthetical Roman numeral indicating the oxidation state of the *central cation.*
7. The total name of the coordination sphere is written as one word.

For example, the complex cation, $[Co(NH_3)_6]^+$ is named as hexaamminecobalt(III) ion. The chloride of this ion has the formula $[Co(NH_3)_6]Cl_3$ and is named hexaamminecobalt(III) chloride. Similarly, the cation, $[CoCl_2(NH_3)_4]^+$, is named dichlorotetraamminecobalt(III) ion and its sulfate, $[CoCl_2(NH_3)_4]_2SO_4$ is named dichlorotetraamminecobalt(III) sulfate.

An example of an anionic coordination sphere is $[Co(C_2O_4)_3]^{3-}$, whch is named trioxalatocobaltate(III) ion, or sometimes tris(oxalato)cobaltate(III) ion. The prefix *tris-* is used in case there might be some single species which might be called the trioxalate ion.

An example of a molecular coordination sphere is $[PtBrCl(H_2O)(NH_3)]$, which is named bromochloroaquoammineplatinum(II). In some cases a compound may consist of a complex cation and a complex anion. $[Cr(NH_3)_6][Co(CN)_6]$ is an example. The main difficulty in naming compounds of this type is in knowing what the oxidation states of the two central cations are. Once it has been established that they are both in the 3+ state, the compound may be named as hexaamminechromium(III) hexacyanocobaltate(III).

ISOMERISM. It is characteristic of the compounds of the coordination complexes that they exhibit many kinds of isomerism. We shall limit our discussion here to the octahedral complexes for which many examples of isomerism have been established. However, this does not imply that isomerism is limited to the octahedral complexes.

Several types of isomers do not depend directly upon the geometry of the complex. For example, *hydrate isomerism* depends upon the way in which water of hydration is incorporated into the compound. Thus hexaaquochromium(III) chloride, $[Cr(H_2O)_6]Cl_3$, has six waters coordinated to the central chromium(III) ion with three chlorides outside the coordination sphere. Chloropentaaquochromium(III) chloride-1-hydrate, $[CrCl(H_2O)_5]Cl_2 \cdot H_2O$, in which one of the waters of hydration is not coordinated to the central chromium(III) ion, is an isomer of this compound. However, the coordination number of six is maintained for the chromium(III) ion, since one of the chlorides is coordinated and only two remain outside the coordination sphere.

Somewhat similar to hydrate isomerism is *ionization isomerism*, in which the difference between isomers depends upon which anion is in the coordination sphere and which is not. Dichlorotetraammineplatinum(IV) bromide, $[PtCl_2(NH_3)_4]Br_2$ is an isomer of dibromotetraammineplatinum(IV) chloride, $[PtBr_2(NH_3)_4]Cl_2$, for example. In the first case, the bromide ions are outside the coordination sphere as ions, and in the second case, they are inside the coordination sphere, where they are covalently bonded to the platinum(IV).

Cis-trans isomerism is common among the octahedral coordination complexes. In the octahedral structure there are both opposite and adjacent positions. If we pick any one position, there is one *trans* position and four *cis* positions with respect to it. For example, if we view the octahedron as follows, we may look at the various positions by number:

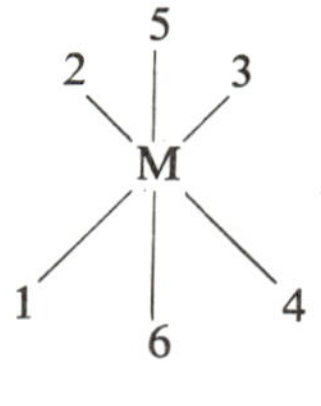

The plane formed by the metal cation, *M*, and positions 1, 2, 3, and 4 is viewed as perpendicular to the sheet of the paper, with positions 1 and 4 in front of the page and positions 2 and 3 behind the page. Positions 5 and 6 lie in the plane of the paper along with the central metal cation, with position 5 above and position 6 below the plane formed by positions 1, 2, 3, and 4.

Looking at position 1, position 3 is opposite, *trans*, and positions 2, 4, 5, and 6 are all adjacent, *cis*. Looking at position 5, position 6 is opposite, *trans*, and positions 1, 2, 3, and 4 are all adjacent, *cis*.

The dichlorotetraammine cobalt(III) ion exhibits *cis-trans* isomerism, in which the two chloride ions may be bonded to the cobalt(III) in either the *cis* or *trans* positions as follows:

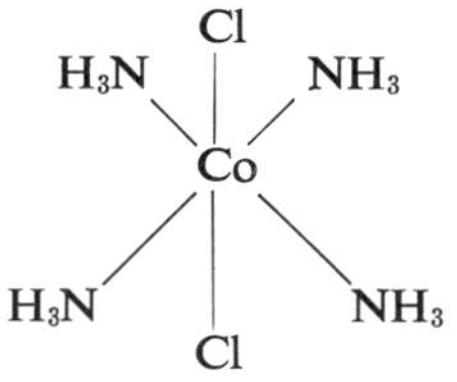

trans

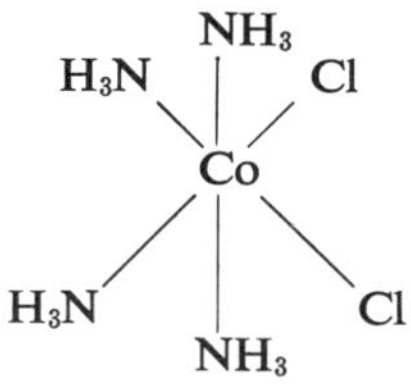

cis

Similarly, trichlorotriamminecobalt(III) exhibits *cis-trans* isomerism:

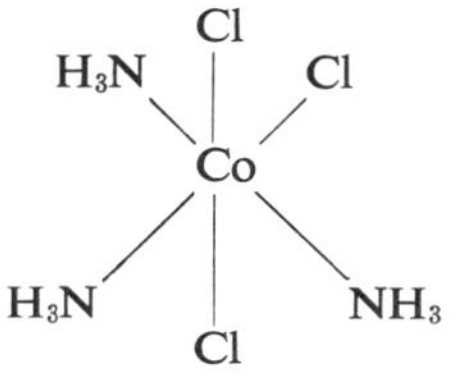

trans

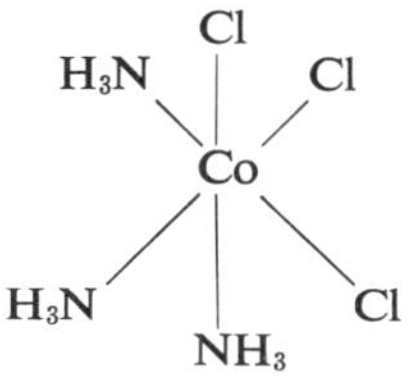

cis

In the *trans* isomer, two of the chlorides and two of the ammonias are *trans* to each other. In the *cis* isomer all three chlorides are *cis* to each other and all three ammonias are *cis* to each other.

By no means do these few examples exhaust the types of isomerism displayed by the coordination complexes, but they illustrate a few of the common types.

LIGAND FIELD THEORY. Many coordination complexes are paramagnetic and many are highly colored. Of course, the paramagnetism is attributed to unpaired electrons. In general, color in chemical species is attributed to the ability of the species to absorb light, allowing the promotion of an electron from a state of low potential energy to a state of higher potential energy. To understand both the magnetic properties and the colors of the complexes, as well as the bonding that occurs, a new approach to bonding in the complexes has proved useful. This is called the *ligand field theory.*

The point of view of the ligand field theory is that the approach of ligands affects the potential energy of the electrons in the $(n - 1)d$ orbitals of the central cation. The easiest situation to visualize is the formation of an octahedral complex of the following type, where M is the metal cation and L_1 through L_6 represent the six ligands:

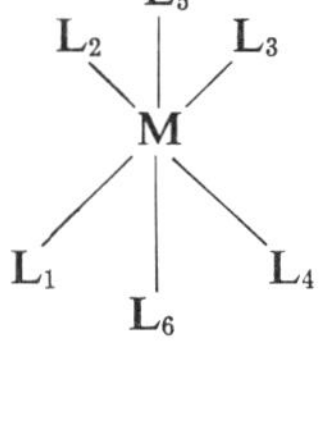

These six ligands lie on three mutually perpendicular axes with L_1 and L_3 on one axis, L_2 and L_4 on another axis, and L_5 and L_6 on the third axis. These three axes cross at the central metal cation.

Recalling the geometrical arrangement for the charge distributions of the electrons in the five *d* orbitals shown in Fig. 4.8, we note that the d_{z^2} and the $d_{x^2-y^2}$ charge distributions lie along three mutually perpendicular axes with the d_{xy}, d_{xz}, and d_{yz} charge distributions directed between the axes. With the six ligands approaching along the three axes, electrons in the d_{z^2} or $d_{x^2-y^2}$ orbitals experience stronger repulsions than any electrons in the d_{xy}, d_{xz}, or d_{yz} orbitals, due to the approach of the ligands and their electron pairs. As a result, the electrons in the d_{z^2} and $d_{x^2-y^2}$ orbitals are at higher potential energy than electrons in the d_{xy}, d_{xz}, and d_{yz} orbitals.

In the isolated metal cation, all five $(n-1)d$ orbitals are at the same potential energy. However, upon the approach of the six ligands at the octahedral positions, the five $(n-1)d$ orbitals are split into two potential-energy groups, with two of the orbitals at a higher potential-energy level than the other three, as suggested by Fig. 10.3.

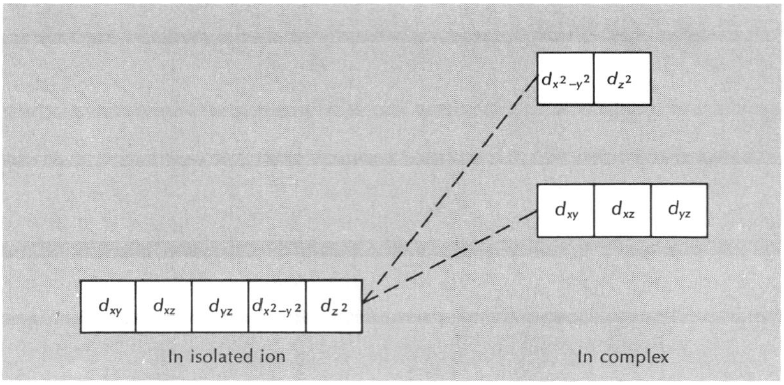

Fig. 10.3. Representation of the energy level split for the (*n*-1)*d* orbitals in an octahedral complex

The energy difference between the two groups of $(n-1)d$ orbitals is called 10Dq. The magnitude of 10Dq depends upon many factors, but it is customary to attribute the size of the energy split in the $(n-1)d$ orbitals to the electron-repelling ability of the ligands. Thus ligands which cause large values for 10Dq are called *strong-field ligands* and those which lead to relatively small values for 10Dq are called *weak-field ligands*. The effective difference between strong- and weak-field ligands is shown in Fig. 10.4.

Thus the approach of the six ligands affects the distribution of the $(n-1)d$ electrons in the central metal cation prior to bond formation. This, in turn, affects the orbitals that are available for electron pair acceptance, that is, the orbitals that will be involved in molecular bonding orbital formation.

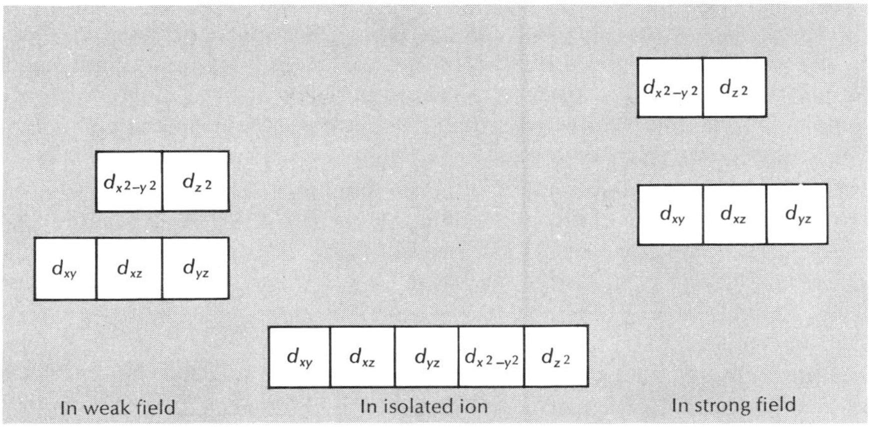

Fig. 10.4. Comparison of the effects of weak and strong field ligands on the energy level split for the $(n\text{-}1)d$ orbitals in an octahedral complex

To understand the distribution of the $(n - 1)d$ electrons, it is necessary to recall two competing effects. First, according to the rule of maximum multiplicity, electrons in the $(n - 1)d$ orbitals tend toward maximum unpairing of electron spins. On the other hand, the electrons seek positions of lowest potential energy.

The problem, then, may be stated as follows: Given a number of electrons in the $(n - 1)d$ level of a central metal cation of an octahedral complex, will these electrons be found in a state of maximum unpairing so that they are distributed over all five of the orbitals, or will they pair within the three orbitals of lower potential energy before any are located in the two orbitals of higher potential energy?

The answer to this question is not a simple one, but it appears that both of these situations may occur, depending upon the strength of the ligands and consequently the value of 10Dq.

For example, if the difference in energy between the high-energy group of orbitals and the low-energy group of orbitals is small, as it is for a weak-field ligand, then maximum multiplicity prevails, and the electrons distribute themselves over all five of the $(n - 1)d$ orbitals just as they normally would in the isolated ion. This is equivalent to saying that the increase in potential energy due to the pairing of electrons is greater than 10Dq, so that electrons go to the lower energy state of maximum unpairing.

On the other hand, if the difference in energy between the high-energy group and the low-energy group of orbitals is large, as it is for strong-field ligands, then electrons fill the three orbitals of lower energy prior to occupying any of the orbitals of higher energy. This is equivalent to saying that the increase in potential energy required for pairing is less than the increase in potential energy required to move from the lower energy group of orbitals to the higher energy group of orbitals.

The differences in effect of strong-field and weak-field ligands in octahedral complexes is shown in Table 10.8. In this table, the split in the $(n-1)d$ level is shown; the two upper boxes stand for the two higher $(n-1)d$ orbitals and the bottom three boxes for the lower $(n-1)d$ orbitals. The obvious difference between a complex containing weak- and strong-field ligands is the number of unpaired electrons that can be measured by magnetic dipole moments.

In Table 10.8, note that metal cations containing 1, 2, 3, 8, 9, or 10 $(n-1)d$ electrons have the same electronic configurations regardless of the nature of the ligands. However, the number of unpaired electrons differs when the central metal cation has 4, 5, 6, or 7 $(n-1)d$ electrons.

TABLE 10.8 • *Octahedral Electron Distributions for Strong- and Weak-Field Ligands*

d ELECTRONS	EXAMPLES	STRONG FIELD	WEAK FIELD
1	Ti^{3+}, V^{4+}	[ ][ ] [↓][ ][ ]	[ ][ ] [↓][ ][ ]
2	V^{3+}	[ ][ ] [↓][↓][ ]	[ ][ ] [↓][↓][ ]
3	Cr^{3+}, V^{++}, Mn^{4+}	[ ][ ] [↓][↓][↓]	[ ][ ] [↓][↓][↓]
4	Cr^{++}, Mn^{3+}	[ ][ ] [↑↓][↓][↓]	[↓][ ] [↓][↓][↓]
5	Mn^{++}, Fe^{3+}	[ ][ ] [↑↓][↑↓][↓]	[↓][↓] [↓][↓][↓]
6	Fe^{++}, Co^{3+}	[ ][ ] [↑↓][↑↓][↑↓]	[↓][↓] [↑↓][↓][↓]
7	Co^{++}	[↓][ ] [↑↓][↑↓][↑↓]	[↓][↓] [↑↓][↑↓][↑↓]
8	Ni^{++}	[↓][↓] [↑↓][↑↓][↑↓]	[↓][↓] [↑↓][↑↓][↑↓]
9	Cu^{++}	[↑↓][↓] [↑↓][↑↓][↑↓]	[↑↓][↓] [↑↓][↑↓][↑↓]
10	Zn^{++}, Cu^{+}	[↑↓][↑↓] [↑↓][↑↓][↑↓]	[↑↓][↑↓] [↑↓][↑↓][↑↓]

To take a specific case, consider a complex in which the central metal cation has six $(n-1)d$ electrons, such as a complex of iron(II) or cobalt(III). If the ligands are weak-field ligands, then the difference in energy between the upper two orbitals and the lower three orbitals is small, and the preference of the electrons is maximum unpairing, with one pair of electrons in one of the lower three orbitals and unpaired electrons in the other four orbitals. This leads to a magnetic dipole moment corresponding to four unpaired electrons.

On the other hand, if the ligands are strong-field ligands, then the energy difference between the upper and lower orbitals is too great to allow maximum unpairing, and the result is that the six electrons preferentially fill the lower three orbitals with all electrons paired. Such a complex does not display any paramagnetism.

Colors of the complexes have been attributed to the ability of the complexes to absorb electromagnetic radiation of the correct wavelength so as to promote an electron from one of the lower energy orbitals to one of the higher energy orbitals. Of course, the frequency of the radiation absorbed is related directly to 10Dq, since the energy of the electron transition is directly proportional to the frequency.

Thus by measuring *absorption spectra*, that is the wavelengths and frequencies of light absorbed, and using the magnetic criteria, it has been possible to arrange some common ligands into an approximate order of increasing field strength as follows:

low field strength high field strength

$$I^- < Br^- < Cl^- < F^- < H_2O < C_2O_4^{--} < NH_3 < NO_2^- < CN^-$$

A given ligand may be a strong-field ligand with one central metal cation and a weak-field ligand with another central metal cation. For example, water is a strong-field ligand with cobalt(III) and a weak-field ligand with iron(III). For this reason, rather than identify the ligand as a weak- or strong-field ligand, we often identify the total complex as a *low-spin* complex when the ligands act as strong-field ligands, thus leading to maximum electron pairing, and as a *high-spin* complex when the ligands act as weak-field ligands, thus leading to maximum electron unpairing.

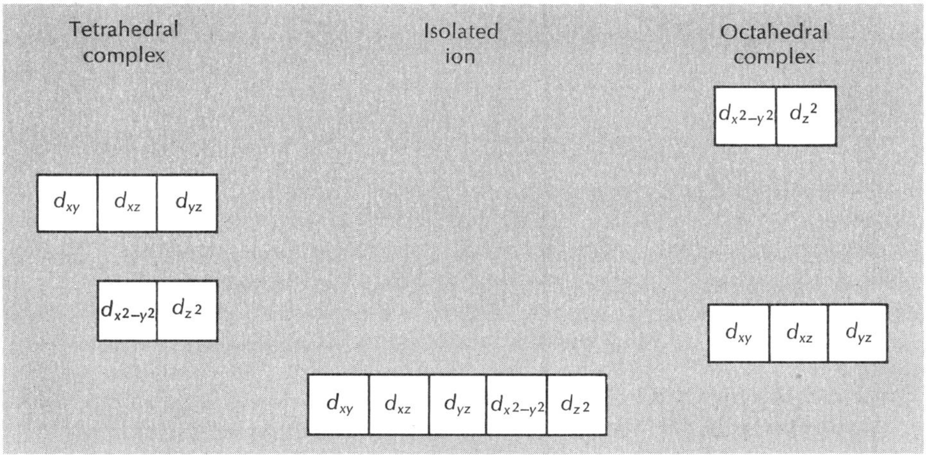

Fig. 10.5. Comparison of tetrahedral and octahedral geometry effects on the energy level split for the (*n*-1)*d* orbitals

As a general rule, the value of 10Dq is higher for cations of the fifth-period transition metals than it is for cations of the fourth-period transition metals. Similarly, the value usually is higher for cations of the sixth period than it is for cations of the fifth period. In fact this trend is sufficiently pronounced that nearly all metal cations in the fifth and sixth periods form low-spin complexes.

It is possible to make an analysis for the tetrahedral complexes similar to the foregoing discussion of the octahedral complexes. The derivation for the tetrahedral complexes is more difficult, however. The results of the derivation are shown in Fig. 10.5. In this case three orbitals are increased in energy more than the other two as the four ligands approach to form the tetrahedral complex. In addition the value for 10Dq is smaller in tetrahedral complexes than in octahedral complexes as a general rule.

The effect of the ligand field splitting of the $(n-1)d$ orbitals upon the bonding is to make available or inaccessible the $(n-1)d$ orbitals for bonding by electron pair acceptance. Those orbitals partially occupied by single unpaired electrons are not available for coordinate covalent bond formation, whereas completely vacant $(n-1)d$ orbitals are available for bonding.

A few examples should illustrate the application of ligand field theory to bonding problems in coordination complexes. For example, consider the hexaammine-cobalt(III) ion, which is a low-spin complex. This means that ammonia serves as a strong-field ligand. From Table 10.8 it is clear that in the presence of a strong-field ligand, the six $(n-1)d$ electrons in cobalt(III) are all paired in three orbitals, leaving vacant two of the $(n-1)d$ electrons for bonding purposes. This permits the formation of the octahedral complex employing the 4*s*, three 4*p*, and two 3*d* orbitals of cobalt(III) for bonding. In addition the complex does not display any paramagnetism.

On the other hand, hexachlorocobaltate(III) ion, $[CoCl_6]^{3-}$, is a high-spin complex. This means that chloride is a weak-field ligand. From Table 10.8 it is clear that in the presence of a weak-field ligand, the six $(n-1)d$ electrons in the cobalt(III) ion occupy all five of the $(n-1)d$ orbitals so that these are not available for bond formation. Thus this complex must employ the 4*s*, three 4*p*, and two 4*d* orbitals for covalent bond formation, and the complex is paramagnetic.

Or consider manganese(II) as another example. For manganese(II), water is a weak-field ligand. From Table 10.8 it is clear that in the presence of a weak-field ligand, all five $(n-1)d$ electrons in the manganese(II) occupy different 3*d* orbitals and none of these orbitals is available for bond formation. Thus manganese tends to form the tetrahedral tetraaquomanganese(II) ion, $[Mn(H_2O)_4]^{++}$, in which the 4*s*, and three 4*p* orbitals are used for bonding instead of the hexaaquomanganese(II) ion, $[Mn(H_2O)_6]^{++}$, which would require the use of less favorable 4*d* orbitals for bonding.

On the other hand the cyanide ion, CN^-, is a strong-field ligand with respect to manganese(II). Again, it is clear from Table 10.8, that in the presence of a strong-field ligand there are two vacant 3*d* orbitals in manganese(II), and correspondingly cyanide forms the hexacyanomanganate(II) ion, $[Mn(CN)_6]^{4-}$, in which the 4*s*, three 4*p*, and two 3*d* orbitals of manganese are involved in the bonding. It should be noted that this complex is paramagnetic, having one unpaired electron.

SUMMARY

The metals differ from the nonmetals in several respects. First they have low electronegativities and thus tend to form simple cations. In addition, the bonds between metal atoms do not appear to be covalent. As a result, metals do not form discrete molecules in the elemental state, but their atoms are close packed in crystals. Unlike the nonmetals, the metals are good electrical conductors, their conductivity decreasing as temperature increases. They also are good thermal conductors.

The metals tend to have relatively few valence electrons of the *ns* and *np* types, and their cations often are devoid of these electrons altogether. Thus the ions have available *ns* and *np* orbitals in addition to $(n - 1)d$ orbitals for coordinate covalent bond formation. Thus the coordination complexes tend to be characteristic species formed by metallic cations, the cations serving as electron-pair acceptors.

QUESTIONS AND PROBLEMS

10.1 Why is the body-centered cubic structure generally preferred at higher temperatures over the hexagonal or face-centered cubic structures?

10.2 From Table 10.1 it is clear that lithium has two crystal allotropes. In addition, the stable form is the body-centered cubic structure. Would you expect the transition temperature between these two allotropes to be above or below room temperature? Why?

10.3 Calcium has two transition temperatures for crystal allotropes at 250°C and 450°C. Suggest what the transitions are at each of these temperatures.

10.4 What is peculiar about the iron crystal structures?

10.5 What factors contribute to the very high density of osmium? To the very low density of lithium?

10.6 The metals in family 6B are very hard. Suggest a reason for this.

10.7 In each of the following pairs of metals, which probably has the higher melting point and why? a) Cr, Co; b) Pd, Pt.

10.8 Why are the heats of vaporization of the metals so high?

10.9 Silicon, normally considered a nonmetal, has the same boiling point as chromium, a metal. Explain why this is true.

10.10 Both xenon and rhodium form face-centered cubic crystals. However, the heat of fusion of rhodium is more than nine times that of xenon. Why?

10.11 Elemental fluorine has a heat of vaporization of 0.78 kcal/mole of molecules and elemental cesium has a heat of vaporization of 15.7 kcal/mole of atoms. At their respective boiling points, which requires more heat to vaporize, 1.00 g of fluorine or 1.00 g of cesium?

10.12 The molar heat of vaporization of sodium fluoride is more than twice the molar heat of vaporization of elemental sodium. Why?

10.13 Why does the electrical conductivity of metals decrease with increasing temperature, while the reverse is true for the nonmetals?

10.14 The electrical conductivity of arsenic is approximately one million times the electrical conductivity of phosphorus. What does this suggest about phosphorus and arsenic?

10.15 Name each of the following compounds:

a) TiO_2
b) V_2O_3
c) Mn_2O_7
d) FeS
e) Zn_3N_2
f) HgS
g) $CoBr_2$
h) $CrCl_2$
i) CuCl
j) Hg_2Cl_2
k) CdS
l) $Sc(NO_3)_3$
m) $Cr_2(SO_4)_3$
n) $MnCO_3$
o) $FeSO_4 \cdot 7H_2O$
p) $CoCrO_4$
q) $Ni(CH_3COO)_2$
r) NiC_2O_4
s) $CuSeO_4 \cdot 5H_2O$
t) $ZnS_2O_6 \cdot 6H_2O$
u) Ag_3AsO_4
v) $AgClO_3$
w) $Cd(CN)_2$
x) $Mn_2P_2O_7$
y) $H_4V_2O_7$

10.16 Write the correct formula for each of the following:

a) iron(III) oxide
b) cobalt(II) sulfide
c) nickel(II) sulfide
d) copper(I) oxide
e) zinc iodide
f) silver iodide
g) mercury(II) chloride
h) scandium bromide
i) titanium(III) sulfate
j) chromium(III) hydroxide
k) manganese(II) sulfate-4-hydrate
l) iron(III) nitrate
m) cobalt(II) iodate
n) nickel(II) sulfite
o) copper(II) phosphate
p) zinc carbonate
q) cadmium acetate
r) cadmium sulfite
s) mercury(I) bromate
t) sodium vanadate
u) potassium dichromate
v) silver permanganate

10.17 Suggest a structural formula for each of the following:

a) vanadate ion
b) divanadate ion
c) tetravanadate ion
d) chromate ion
e) dichromate ion
f) trichromate ion
g) permanganate ion

10.18 The diacid chloride of chromic acid is known as chromyl chloride. Write the formula for chromyl chloride.

10.19 The term "coordination number" has now been used in two different situations. How do the two definitions differ? How are they alike?

10.20 The sulfate ion may be viewed as a coordination complex. How?

10.21 Barium chloride-2-hydrate has crystal water, whereas magnesium chloride-6-hydrate has water of coordination. Suggest reasons for the difference.

10.22 Compounds of copper(I) tend not to be hydrated whereas the compounds of copper(II) tend to be hydrated. Why?

10.23 Why does the perchlorate ion have a higher electronegativity than the hypochlorite ion?

10.24 Why does the dihydrogen phosphate ion have a higher electronegativity than the phosphate ion?

10.25 Why does methane *not* serve as a ligand, whereas both water and ammonia do?

10.26 Sulfide ion tends to be a better ligand than hydrogen sulfide. Suggest reasons for this.

10.27 When the thiocyanate ion serves as a ligand, the electron pair usually is donated from the nitrogen. Suggest possible reasons why this occurs instead of electron-pair donation from either carbon or sulfur.

10.28 Use a dictionary to determine the derivation of the word "chelate." What is the significance of this derivation in coordination chemistry? In this connection, what is the significance of the root *dentate* in representing the chelating agents?

10.29 Fluoride ion tends to serve as a ligand best with cations which are small and highly charged. Suggest a reason for this.

10.30 The tetraiodomercurate(II) ion is a more stable complex than the tetraiodocadmate(II) ion which is more stable than the tetraiodozincate(II) ion. Why?

10.31 Iron(III) forms a very stable hexafluoro complex, a less stable tetrachloro complex, and a still less stable dibromo complex. Suggest reasons for these observations.

10.32 Is there any difference between the definition of radius ratio in this chapter and in Chapter 7? If so, what is it?

10.33 Using a radius-ratio argument, suggest why complexes with coordination number five tend to be rare.

10.34 The radii of several ions in angstrom units are: antimony(V) = 0.62; O^{--} or OH^- = 1.40; Arsenic(V) = 0.47; F^- = 1.33; Cr^{3+} = 0.62; Zn^{++} = 0.74. For each of the following species, indicate the most probable formula:

a) antimonate: SbO_3^-, SbO_4^{3-}, $[Sb(OH)_6]^-$
b) arsenate: AsO_3^-, AsO_4^{3-}, $[As(OH)_6]^-$
c) fluorochromate(III): $[CrF_4]^-$, $[CrF_6]^{3-}$
d) chromite: CrO_2^-, $[Cr(OH)_4]^-$, $[Cr(OH)_6]^{3-}$
e) zincate: ZnO_2^{--}, $[Zn(OH)_4]^{--}$, $[Zn(OH)_6]^{4-}$

10.35 Write the correct name for each of the following:

a) $K_2[TiCl_6]$
b) $[V(CN)_6]^{4-}$
c) $[VF_6]^-$
d) $[CrCl(H_2O)_5]Cl_2 \cdot H_2O$
e) $(NH_4)_4[Mn(CN)_6]$
f) $K_3[Fe(CN)_6]$
g) $Na_3[Co(CN)_6]$
h) $[CuCl_4]^{--}$
i) $[Ag(NH_3)_2]^+$
j) $(NH_4)_2[Hg(C_2O_4)_2(H_2O)_2]$
k) $[Pt(Br)(Cl)(NO_2)(NH_3)_3]^+$
l) $[Co(CN)_2(NO_2)_2(NH_3)_2]^-$
m) $[Ag(NH_3)_2]_2[PtBr_4]$
n) $[CoCl(BrO_3)(NH_3)_4]NO_3$
o) $NH_4[Cr(NCS)_4(NH_3)_2]$
p) $[Co(OH)(Cl)(NH_3)_3H_2O]Br$

10.36 Write the correct formula for each of the following:

a) tetraamminecopper(II) ion
b) copper(II) tetrahydroxoplatinate(II)
c) dichlorodiaquocopper(II)
d) tetraamminezinc(II) orthophosphate
e) hexafluoroferrate(II) ion
f) potassium tetracyanonickelate(II)
g) dichlorotetraaquochromium(III) sulfate
h) tris(oxalato)iron(III) ion
i) calcium hexafluorovanadate(V)
j) trihydroxotriaquoaluminum(III)
k) ammonium hexafluorotitanate(III)
l) trifluorotriaquovanadium(III)
m) potassium hexathiocyanatochromate(III)
n) dichlorotetraamminechromium(III) ion
o) tris(oxalato)manganate(III) ion
p) potassium hexacyanoferrate(II)
q) potassium bis(thiosulfato)argentate(I)
r) tetrahydroxozincate(II) ion
s) potassium tetraiodomercurate(II)

10.37 What other name might be given to dichlorodiaquocopper(II)?

10.38 Solid phosphorus(V) halides form ions which resemble coordination complexes. Thus phosphorus(V) chloride is $[PCl_4][PCl_6]$ and phosphorus(V) bromide is $[PBr_4]Br$. Name each of these as compounds containing coordination complexes.

10.39 Internationally approved nomenclature rules call for the naming of oxy anions derived from the simple hydroxy acids as coordination complexes in which the ligands are oxide ions which are given the name *oxo*. Thus carbonate ion is named trioxocarbonate(IV) ion.
a) What is the formula for hexaoxotellurate(VI) ion?
b) What is the formula for dioxochlorate(III) ion?
c) What is the more common name for each of the above species?
d) In this system, what is the name for the dihydrogen phosphate ion?

10.40 There are no known *cis-trans* isomers among simple tetrahedral complexes. Suggest a reason for this.

10.41 Criticize the following statement: The compounds $[Co(NH_3)_3(NO_2)_3]$ and $[Co(NH_3)_6][Co(NO_2)_6]$ are isomers.

10.42 Draw structures for all isomers of each of the following:
a) $[Co(NO_2)_3(NH_3)_3]$
b) $[CoSO_4(NH_3)_5]Br$
c) $[Co(NO_2)_4(NH_3)_2]^-$
d) $NH_4[PtBr_2Cl_3(NH_3)]$
e) $[Co(NO_2)_2(H_2O)_2(NH_3)_2]Cl$

10.43 Write the structures for at least eleven isomers of $[Cr(NH_3)_6][Co(CN)_6]$ in which both chromium and cobalt are in the 3+ oxidation state.

10.44 Describe the probable bonding in $[AlF_6]^{3-}$.

10.45 Why are the compounds of the transition metals often colored and paramagnetic?

10.46 Explain the fact that cobalt(II) forms $[CoCl_4]^{--}$ with a coordination number of four but also $[Co(NH_3)_6]^{++}$ with a coordination number of six. In addition, the hexaamminecobalt(II) ion is easily converted to the hexaamminecobalt(III) ion.

10.47 Coordination complexes of zinc(II) seldom are colored. Suggest a reason for this.

10.48 Discuss the bonding and magnetic properties for the tetrahedral and octahedral coordination complexes of iron(II) and iron(III) with strong-field ligands such as cyanide and weak-field ligands such as fluoride.

10.49 Hexacyanoferrate(III) ion is colorless, whereas both hexachloroferrate(III) ion and tetrachloroferrate(III) ion are colored. Why?

10.50 Prepare a table analogous to Table 10.8 for the electron distribution in metal cations with one to ten $(n - 1)d$ electrons in tetrahedral coordination complexes.

11

Chemical Reactions: Stoichiometry

INTRODUCTION At this point we shall begin to change our point of view somewhat. Except for our discussions of the three states of matter, we have dealt so far with individual molecules and atoms. Now we shall consider systems containing very large numbers of particles and the way in which the particles in these systems interact with each other.

In this chapter, we shall deal with the relative amounts of interacting substances in chemical reactions, an area that is often called chemical *stoichiometry*. Since we will be dealing with large numbers of particles (atoms and molecules), we will use mole quantities almost exclusively.

In the process, the mole notation will be abbreviated slightly. Since the mole is the number 6.023×10^{23}, it is necessary to designate the kinds of things represented by this number. Thus for water, H_2O, we might want to designate a mole of water molecules or a mole of hydrogen-oxygen bonds.

However, it is common practice to use the term *mole* by itself to mean *one mole of molecules*. When something else is intended, the objects involved are specified. Thus a *mole of water* is a *mole of water molecules* and a *mole of oxygen* is a *mole of oxygen molecules*. For substances in which specific molecules cannot be depicted, the mole represents the moles of particles represented in the simplest formula of the compound. Thus a mole of ionic sodium fluoride, NaF, is that amount of sodium fluoride that contains a mole of sodium ions and a mole of fluoride ions. Similarly, a mole of aluminum sulfate, $Al_2(SO_4)_3$, is the amount of aluminum sulfate that contains 2 moles of aluminum ions and 3 moles of sulfate ions.

Silicon(IV) oxide is a highly polymeric substance that might best be represented as $(SiO_2)_x$ where SiO_2 is the repeating polymer unit rather than an individual molecule. In this case, however, a mole of silicon(IV) oxide is the amount of silicon(IV) oxide that contains 1 mole of silicon atoms and 2 moles of oxygen atoms.

The usual metric prefixes and abbreviations may be applied to the mole as well as other units. Thus the *millimole*, which is abbreviated as *mmole*, is 10^{-3} mole and the *kilomole*, which is abbreviated as *kmole*, is 10^3 moles.

Chemical reactions are described by chemical equations, and the writing and interpretation of these chemical equations is the subject of this chapter. We shall look at numerous examples of useful quantitative interpretations of chemical equations after taking a brief look at the fundamental meaning of chemical equations.

CHEMICAL EQUATIONS

Fundamental Equation Structure

It is a fact that chemical reactions occur. For example, liquid benzene burns in air, and if a cold object is held over the burning benzene, water condenses on the object and a deposit of soot (carbon) also forms. By controlling the environment, we can show that the reaction is between benzene, C_6H_6, and atmospheric oxygen, O_2, to form water, H_2O, and carbon, C. This may be described symbolically as follows:

$$C_6H_6(l) + O_2(g) = C(s) + H_2O(l) \qquad (11.1)$$

This statement correctly indicates that benzene and oxygen react to form carbon and water. The parenthetic letters designate the state in which each of the reactants and products exists, that is, solid, liquid, or gas. Depending upon the use to which the statement is to be put, these symbols often are omitted.

Note that statement 11.1 is written as an equation, but it is not an equation, since it does not represent an equality. Many years of chemical experience have shown that the most important aspect of nonnuclear chemical reactions is the conservation of atoms. Thus atoms are neither created nor destroyed in these reactions, they are just rearranged. Thus for statement 11.1 to be an equation, there must be as many atoms of carbon represented among the products as among the reactants, and the same may be said for the atoms of hydrogen and oxygen.

The equality in carbon atoms among the reactants and products may be shown by indicating 6 moles of carbon atoms among the products:

$$C_6H_6 + O_2 = 6C + H_2O \qquad (11.2)$$

Statement 11.2 is not yet an equation, since there are 6 moles of hydrogen atoms among the reactants and only 2 moles of hydrogen atoms among the products. This may be corrected by representing 3 moles of water in the products:

$$C_6H_6 + O_2 = 6C + 3H_2O \tag{11.3}$$

However, statement 11.3 still is not an equation, since there are 2 moles of oxygen atoms among the reactants and 3 moles of oxygen atoms among the products. This may be corrected by representing $1\frac{1}{2}$ moles of oxygen among the reactants:

$$C_6H_6 + 1\tfrac{1}{2}\ O_2 = 6C + 3H_2O \tag{11.4}$$

Statement 11.4 is a *chemical equation* for a reaction of benzene with oxygen. It indicates the kinds of substances that undergo reaction and the kinds of substances that result from the reaction. In addition, it indicates the relative quantities of reactants and products involved in the reaction. Thus, from Equation (11.4), it is clear that for every mole of benzene that undergoes this reaction, $1\frac{1}{2}$ moles of oxygen are consumed and 6 moles of carbon atoms and 3 moles of water are produced.

As a matter of practice, we often avoid the use of fractional coefficients in chemical equations, although fractional coefficients are not incorrect. Equation (11.4) may be multiplied by 2 to eliminate the fractional coefficient:

$$2C_6H_6 + 3O_2 = 12C + 6H_2O \tag{11.5}$$

According to Equation (11.5), for every 2 moles of benzene reacting in this way, 3 moles of oxygen are consumed and 12 moles of carbon atoms and 6 moles of water are produced.

Equation Interpretation

It is important to realize that the chemical equation represents a *net* chemical reaction, since it does not tell us anything about why or how the reaction occurs or whether or not it might be the sum of several intermediate reactions that simply add up to this result.

In addition, the equation represents only the amounts of substances that actually participate in the stated reaction. Thus not all of the benzene in a given container burns instantaneously. Rather it burns over a period of time. At a given moment during the burning process, only a fraction of the benzene present will have burned, and it is this amount of benzene that reacted according to Equation (11.5).

In addition, there always is the possibility that not all of the benzene undergoes the same reaction. For example, some of the benzene may burn to form water and carbon dioxide. This reaction is represented by Equation (11.6).

$$2C_6H_6 + 15O_2 = 12CO_2 + 6H_2O \tag{11.6}$$

In this case, *each* mole of benzene that undergoes this particular reaction consumes 7.5 moles of oxygen to produce 6 moles of carbon dioxide and 3 moles of water. This equation describes only that portion of the over-all burning process in which benzene is converted to carbon dioxide and water.

We shall now consider several interpretative problems involving Equations (11.5) and (11.6).

MASS INTERPRETATIONS. For the moment consider only the reaction represented by Equation (11.5). Suppose that we know that 35.0 g of benzene have been burned, can we compute how much oxygen has been consumed in the process? From the statement of the problem, the chemical equation, and formulas, we know the following:

1. 35.0 g of benzene have been burned.
2. From the formula for benzene, C_6H_6, the molecular weight is determined to be 78.1, which means that 1 mole of benzene has a mass of 78.1 g.
3. From Equation (11.5), it is clear that 3 moles of oxygen are consumed for every 2 moles of benzene that burn.
4. From the formula for oxygen, the molecular weight of oxygen is determined to be 32.0, which means that 1 mole of oxygen has a mass of 32.0 g.

These four relationships may be put together as follows to solve the problem:

$$35.0 \text{ g } C_6H_6 \times \frac{1 \text{ mole } C_6H_6}{78.1 \text{ g } C_6H_6} \times \frac{3 \text{ moles } O_2}{2 \text{ moles } C_6H_6} \times \frac{32.0 \text{ g } O_2}{1 \text{ mole } O_2} = 21.5 \text{ g } O_2$$

This shows a step-by-step breakdown of the solution to this problem. Often several steps may be combined mentally. For example, we might reason as follows:

1. The experimental fact is that 35.0 g of benzene burned.
2. From Equation (11.5) it is clear that 2 moles of benzene consume 3 moles of oxygen. Since each mole of benzene has a mass of 78.1 g and each mole of oxygen has a mass of 32.0 g, this is equivalent to saying that 2×78.1 g of benzene consume 3×32.0 g of oxygen or 156 g of benzene consume 96 g of oxygen.

Thus the solution of the problem may be shown in more abbreviated form:

$$35.0 \text{ g } C_6H_6 \times \frac{96.0 \text{ g } O_2}{156 \text{ g } C_6H_6} = 21.5 \text{ g } O_2$$

As another similar example, compute the number of grams of water produced when 46.0 g of benzene burns in the reaction represented by Equation (11.5). According to this equation, 6 moles of water are produced for every 2 moles of benzene burned, or 6×18 g of water are produced for every 2×78.1 g of benzene burned. Thus the solution is

$$46.0 \text{ g } C_6H_6 \times \frac{108 \text{ g } H_2O}{156 \text{ g } C_6H_6} = 31.8 \text{ g } H_2O$$

Now suppose that the burning of benzene is limited to the two reactions represented by Equations (11.5) and (11.6), but that both reactions may occur simultaneously. Experimentally it is determined that the burning of 125 g of benzene produces 75.9 g of carbon. What percent of the benzene burned by each reaction?

The only reaction in which carbon is deposited is that represented by Equation (11.5). According to this equation, 12 moles of carbon atoms are produced from the burning of 2 moles of benzene, or 144 g of carbon is produced from 156 g of benzene. Thus the amount of benzene that burned to produce carbon may be computed:

$$75.9 \text{ g C} \times \frac{156 \text{ g } C_6H_6}{144 \text{ g C}} = 82.2 \text{ g } C_6H_6 \text{ burned by reaction (11.5)}$$

Since a total of 125 g of benzene burned and only 82.2 g of it burned by the reaction represented by Equation (11.5), the percent of the total that burned by this reaction may be found:

$$100 \text{ g } C_6H_6 \text{ total} \times \frac{82.2 \text{ g } C_6H_6 \text{ by reaction (11.5)}}{125 \text{ g } C_6H_6 \text{ total}}$$

$$= 65.8 \text{ percent of the benzene burned by reaction (11.5)}$$

Since there was only one other reaction by which the benzene could burn, the remainder or 34.2 percent must have burned according to reaction (11.6).

A similar problem may be stated in slightly different terms. If 125 g of benzene burned with the consumption of 317 g of oxygen, what percent of the benzene burned according to reaction (11.5)?

In this case oxygen is consumed by both reactions, and we know only the total amount of oxygen consumed. The strategy for solving this problem is as follows:

1. If we let x equal the number of grams of benzene burned by reaction (11.5), then the remainder of the benzene, or $125 - x$ g of benzene, must have burned by reaction (11.6).
2. Using Equations (11.5) and (11.6), the number of grams of oxygen consumed by each reaction may be computed in terms of x.
3. The sum of these two quantities of oxygen must equal the experimentally determined quantity of oxygen consumed. This provides an algebraic equation in x that may be solved to indicate the number of grams of benzene burned by reaction (11.5).
4. This quantity of benzene may be compared to the total quantity of benzene burned to determine the percent of the total that burned by reaction (11.5).

This strategy is carried out as follows:

Let x = grams of benzene burned by reaction (11.5).

Then $125 - x$ = grams of benzene burned by reaction (11.6).

Using Equation (11.5) the quantity of oxygen consumed in burning x grams of benzene may be computed:

$$x \text{ g } C_6H_6 \times \frac{96 \text{ g } O_2}{156 \text{ g } C_6H_6} = 0.615x \text{ g } O_2 \text{ consumed by reaction (11.5)}$$

Using Equation (11.6), the quantity of oxygen consumed in burning $125 - x$ g of benzene may be computed:

$$(125 - x) \text{ g } C_6H_6 \times \frac{480 \text{ g } O_2}{156 \text{ g } C_6H_6} = 3.08(125 - x) \text{ g } O_2 \text{ consumed by reaction (11.6)}$$

Since the total oxygen consumption was 317 g, the sum of these two separate oxygen consumptions by the individual reactions may be set equal to 317 g, and the resulting equation may be solved for x.

$$0.615x + 3.08(125 - x) = 317$$
$$0.615x + 385 - 3.08x = 317$$
$$2.46x = 68$$
$$x = 68/2.46 = 28 \text{ g } C_6H_6 \text{ burned by reaction (11.5)}$$

Since 125 g of benzene was burned and only 28 g was burned by reaction (11.5), the percent of the total benzene burned by reaction (11.5) may be computed:

$$100 \text{ g } C_6H_6 \text{ total} \times \frac{28 \text{ g } C_6H_6 \text{ by reaction (11.5)}}{125 \text{ g } C_6H_6 \text{ total}}$$
$$= 22 \text{ percent burned by reaction (11.5)}$$

VOLUME INTERPRETATIONS. If we assume that gaseous substances are ideal gases, we may use the ideal gas equation in conjunction with chemical equations to compute volumes of gaseous substances involved in chemical reactions.

For example, using Equation (11.6), compute the number of liters of carbon dioxide produced at 27°C at a pressure of 650 mm Hg by burning 3.10 g of benzene.

With the pressure and temperature given and the volume to be computed, the chemical equation may be used to compute the number of moles of CO_2 produced. The volume may then be computed from the ideal gas equation, $PV = nRT$.

From Equation (11.6), it is clear that the burning of 2 moles of benzene leads to the production of 12 moles of CO_2. Since the molecular weight of benzene is 78.1, the 2 moles of benzene have a mass of 156 g. Thus the number of moles of CO_2 formed during the burning of 3.10 g of benzene may be computed:

$$3.10 \text{ g } C_6H_6 \times \frac{12 \text{ moles } CO_2}{156 \text{ g } C_6H_6} = 0.239 \text{ mole } CO_2$$

From the ideal gas equation, the volume of CO_2, V_{CO_2}, may be found:

$$V_{CO_2} = \frac{nRT}{P} = 0.239 \text{ mole} \times \frac{0.0821 \text{ l} \times \text{atm}}{\text{mole} \times {}^\circ\text{K}} \times 300^\circ\text{K} \times \frac{760 \text{ mm Hg}}{650 \text{ mm Hg} \times \text{atm}}$$
$$= 6.86 \text{ liters of } CO_2$$

If the density of liquids or solids is known, the volumes of liquid and solid substances participating in chemical reactions may be computed. For example, the density of benzene is 0.879 g/ml at 25°C. How many ml of benzene at 25°C have been burned by reaction (11.6) when 21.4 ml of water at 25°C have been formed?

Recalling that the density of water at 25°C is 1.00 g/ml, we find that the 21.4 ml of water formed has a mass of 21.4 grams. Knowing that 21.4 g of water is formed, we may use the chemical equation to find the corresponding number of grams of benzene that must have been burned to form 21.4 g of water. Then using the density of benzene, we may compute the volume of that mass of benzene:

$$21.4 \text{ g } H_2O \times \frac{156 \text{ g } C_6H_6}{108 \text{ g } H_2O} \times \frac{1 \text{ ml } C_6H_6}{0.879 \text{ g } C_6H_6} = 35.2 \text{ ml } C_6H_6$$

SOME APPLICATIONS

Preparative Reactions

Often chemical reactions are used to prepare otherwise unavailable compounds. For example, barium nitrite may be prepared by treating barium nitrate with lead metal. The reaction is described by Equation (11.7).

$$Ba(NO_3)_2 + 2Pb = Ba(NO_2)_2 + 2PbO \tag{11.7}$$

CHOICE OF QUANTITIES OF REACTANTS. Suppose that 10.0 g of barium nitrite is needed for some purpose. Theoretically, how much barium nitrate and how much lead are needed to carry out the reaction to produce it?

According to Equation (11.7), 1 mole of barium nitrate is required to produce 1 mole of barium nitrite. Since the molecular weights of barium nitrate and barium nitrite are 261 and 229 respectively, theoretically, 229 g of barium nitrite can be prepared from 261 grams of barium nitrate. Thus the mass of barium nitrate required to produce 10.0 g of barium nitrite may be computed as follows:

$$10.0 \text{ g } Ba(NO_2)_2 \times \frac{261 \text{ g } Ba(NO_3)_2}{229 \text{ g } Ba(NO_2)_2} = 11.4 \text{ g } Ba(NO_3)_2$$

Similarly, using the atomic weight of lead as 207, the mass of lead required to react with 11.4 g of barium nitrate to produce 10.0 g of barium nitrite may be computed as follows:

$$10.0 \text{ g } Ba(NO_2)_2 \times \frac{414 \text{ g Pb}}{229 \text{ g } Ba(NO_2)_2} = 18.1 \text{ g Pb}$$

Thus, theoretically, 11.4 g of barium nitrate should react exactly with 18.1 g of lead to produce 10.0 g of barium nitrite.

EXCESS REACTANTS. Sometimes it is advantageous to carry out reactions with one of the reactants in excess of that required for the reaction. For example, in the preparation of barium nitrite, one set of directions calls for the use of 10.0 g of barium nitrate with 32 g of lead.

From Equation (11.7), it appears that 1 mole of barium nitrate reacts with 2 moles of lead atoms, or using the molecular and atomic weights, 261 g of barium nitrate reacts with 414 g of lead. Thus the theoretical mass ratio is $Pb:Ba(NO_3)_2 = 414\ g:261\ g = 1.59$. From the recorded directions for carrying out this preparation, the recommended mass ratio is $Pb:Ba(NO_3)_2 = 32\ g:10\ g = 3.2$. Thus in carrying out the reaction, a large excess of lead is recommended.

THEORETICAL YIELDS. If it is known how much of each reactant has been used for a particular reaction, it is possible to calculate how much of a particular product theoretically can be expected. In the case of the preparation of barium nitrite from 10.0 g of barium nitrate and 32 g of lead, it has been pointed out that the lead is used in excess of that needed to react with all of the barium nitrate used. It should be clear that any calculation of the theoretical yield of barium nitrite must be based on the amount of barium nitrate used, and not on the amount of lead used, since it is not possible to have a reaction in which all of the lead reacts. This is to say, in general, that theoretical yields cannot be based upon quantities of reactants that have been used in excess.

However, using Equation (11.7), it is possible to compute the expected amount of barium nitrite that can be prepared from 10.0 g of barium nitrate with lead in excess:

$$10.0\ g\ Ba(NO_3)_2 \times \frac{229\ g\ Ba(NO_2)_2}{261\ g\ Ba(NO_3)_2} = 8.77\ g\ Ba(NO_2)_2$$

PERCENT YIELD. Often in carrying out preparative reactions, the experimental yields are lower than the theoretical yields. This may be due to any of a variety of causes, such as mechanical losses of reactants or products, losses due to competing reactions, failure of the given reaction to go to completion, or failure to follow directions.

Whatever the cause, it is common practice to express the results of a preparation as a percent yield, which is the percent of the theoretical yield actually obtained. For example, 5.72 g of barium nitrite was produced in one instance by reacting 10.0 g of barium nitrate with 32 g of lead. Since the theoretical yield is 8.77 g of barium nitrite, the percent yield may be found as follows:

$$100\ g\ Ba(NO_2)_2\ \text{expected} \times \frac{5.72\ g\ Ba(NO_2)_2\ \text{obtained}}{8.77\ g\ Ba(NO_2)_2\ \text{expected}} = 65.2\ \text{percent yield}$$

CONSECUTIVE REACTIONS. Often preparations are carried out in steps, so that a reactant needed is prepared in a separate step. For example, nitric acid is prepared commercially by the following reactions.

$$N_2 + 3H_2 = 2NH_3 \tag{11.8}$$

$$4NH_3 + 7O_2 = 4NO_2 + 6H_2O \tag{11.9}$$

$$3NO_2 + H_2O = 2HNO_3 + NO \tag{11.10}$$

Assuming that hydrogen, oxygen, and water are made available in excess, let us consider what the theoretical yield of nitric acid in moles would be from 1 mole of nitrogen.

The three equations indicate the following three relationships:

1. 1 mole of nitrogen leads to 2 moles of ammonia.
2. 4 moles of ammonia lead to 4 moles of nitrogen(IV) oxide.
3. 3 moles of nitrogen(IV) oxide lead to 2 moles of nitric acid.

These relationships may be used to calculate the theoretical molar yield of nitric acid from 1 mole of nitrogen as follows:

$$1 \text{ mole } N_2 \times \frac{2 \text{ moles } NH_3}{1 \text{ mole } N_2} \times \frac{4 \text{ moles } NO_2}{4 \text{ moles } NH_3} \times \frac{2 \text{ moles } HNO_3}{3 \text{ moles } NO_2} = 1.33 \text{ moles } HNO_3$$

In a similar way, we might ask what the theoretical yield of nitric acid might be, starting with 5.00 kg of nitrogen and 1.00 kg of hydrogen, assuming that oxygen and water are made available in excess.

From Equation (11.8), the mass ratio $N_2:H_2$ is 28.0:6.06 = 4.6. In this situation in this problem, the mass ratio $N_2:H_2$ is 5.00:1.00 = 5.00. Since the actual ratio is higher than that required theoretically, nitrogen is in excess and all calculations must be based upon hydrogen.

Thus beginning with 1.00 kg of hydrogen, the theoretical yield of nitric acid may be found as follows:

$$1.00 \text{ kg } H_2 \times \frac{1 \text{ kmole } H_2}{2.02 \text{ kg } H_2} \times \frac{4 \text{ kmole } NO_2}{6 \text{ kmole } H_2} \times \frac{2 \text{ kmole } HNO_3}{3 \text{ kmole } NO_2} \times \frac{63.0 \text{ kg } HNO_3}{1 \text{ kmole } HNO_3}$$
$$= 13.9 \text{ kg } HNO_3$$

Reaction Progress

As indicated earlier, chemical reactions do not occur instantaneously. They occur over a period of time. In following chemical reactions, when the chemical equation for a reaction is known, it is helpful to observe how far the reaction has progressed.

For example, commercially available sodium hydroxide comes in pellets about the size of a pea, which look very much like shiny hailstones. If these pellets are placed in a dry atmosphere containing carbon dioxide, they begin to lose their luster and a fine white powder appears on the surface. When enough of this powder has formed, it can be scraped off the surface of the pellet and studied to show that it is sodium hydrogen carbonate. The equation for the reaction may be written as follows:

$$NaOH(s) + CO_2(g) = NaHCO_3(s) \qquad (11.11)$$

Suppose that in a controlled experiment, 3.50 g of sodium hydroxide is placed in a 5.00-l closed container with 7.50 g of carbon dioxide at 25°C. After a certain time has elapsed, it is found that 6.25 g of sodium hydrogen carbonate, $NaHCO_3$, has been produced. This is sufficient information to indicate how much of each substance was present at the end of this time, assuming that no other reactions occur.

Thus from Equation (11.11), the amounts of NaOH and CO_2 consumed may be calculated:

$$6.25 \text{ g } NaHCO_3 \times \frac{40.0 \text{ g NaOH}}{84.0 \text{ g } NaHCO_3} = 2.98 \text{ g NaOH consumed}$$

$$6.25 \text{ g } NaHCO_3 \times \frac{44.0 \text{ g } CO_2}{84.0 \text{ g } NaHCO_3} = 3.27 \text{ g } CO_2 \text{ consumed}$$

Therefore, at the end of this interval of time there must be (3.50 − 2.98) g of NaOH = 0.52 g NaOH remaining and (7.50 − 3.27) g of CO_2 = 4.23 g CO_2 remaining along with the 6.25 g of $NaHCO_3$.

According to Equation (11.11), the required mass ratio CO_2:NaOH is 44.0:40.0 = 1.10. The ratio at the outset of this experiment was CO_2:NaOH = 7.50:3.50 = 2.14. Since this is higher than theoretical, CO_2 is in excess. Thus the progress of the reaction may be expressed in terms of the percent of NaOH that has disappeared. Thus when 0 percent of the NaOH has disappeared, the reaction is 0 percent complete. Similarly, when all of the NaOH has disappeared, the reaction is 100 percent complete.

In this case, when the 6.25 g of $NaHCO_3$ was formed, 2.98 g of the original 3.50 g of NaOH was consumed. From the following calculation, it is clear that 85.1 percent of the NaOH disappeared and that the reaction was 85.1 percent complete at this point.

$$100 \text{ g NaOH total} \times \frac{2.98 \text{ g NaOH consumed}}{3.50 \text{ g NaOH total}} = 85.1 \text{ percent NaOH consumed}$$

As a result of the experimental determination of the quantity of a substance present after a reaction has proceeded for a period of time, it is possible to follow the progress of a chemical reaction. The experimental problem is one of measuring the quantity of a substance present.

In reactions involving gases, often it is convenient to trace the course of the reaction by monitoring the pressure of the system. In the absorption of gaseous CO_2 by solid sodium hydroxide to form solid sodium hydrogen carbonate with the temperature and volume constant, the pressure should decrease as the reaction proceeds. Assuming that CO_2 behaves ideally, the ideal gas equation, $PV = nRT$, may be used to show that the pressure of the system is directly proportional to the number of moles of CO_2 present, with RT/V as the proportionality constant.

$$P = \frac{RT}{V}(n_{CO_2}) \qquad (11.12)$$

The number of moles of CO_2 present is related to the completeness of the reaction. For example, in the previous problem, at the outset, the 5.00-l container enclosed 7.50 g of CO_2 at 25°C. At this point the number of moles of CO_2 is 7.50 g/(44.0 g/mole) = 0.170 mole of CO_2. Thus the initial pressure may be found as follows:

$$P = \frac{RTn}{V} = \frac{0.0821 \text{ l} \times \text{atm}}{\text{mole} \times {}^\circ\text{K}} \times \frac{298^\circ\text{K}}{5.00 \text{ l}} \times 0.170 \text{ mole} = 0.832 \text{ atm}$$

When the reaction is complete, all 3.50 g of NaOH will have reacted with 0.0875 mole of CO_2 as shown by the following computation:

$$3.50 \text{ g NaOH} \times \frac{1 \text{ mole } CO_2 \text{ consumed}}{40.0 \text{ g NaOH}} = 0.0875 \text{ mole } CO_2 \text{ consumed}$$

Since there was 0.170 mole of CO_2 present initially, when the reaction is complete, there is (0.170 − 0.088) mole CO_2 = 0.082 mole CO_2 remaining. Using the ideal gas equation, the pressure at completion may be computed as follows:

$$P = \frac{RTn}{V} = \frac{0.0821 \text{ l} \times \text{atm}}{\text{mole} \times {}^\circ\text{K}} \times \frac{298^\circ\text{K}}{5.00 \text{ l}} \times 0.082 \text{ mole} = 0.401 \text{ atm}$$

Using this approach it is possible to plot the system pressure against the percent completion of the reaction, as shown in Fig. 11.1. Thus, if the pressure of a system is measured, the percent completion may be read directly from the graph.

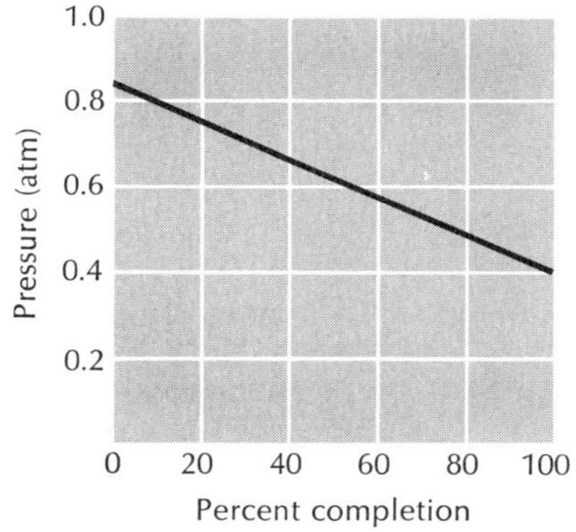

Fig. 11.1. Plot of pressure against percent completion for the reaction of sodium hydroxide with carbon dioxide

More complex reactions may be followed similarly. For example, consider the following reaction between solid iron(II) sulfide and gaseous oxygen:

$$4FeS(s) + 7O_2(g) = 2Fe_2O_3(s) + 4SO_2(g) \quad (11.13)$$

Suppose that 10.0 g of FeS is dropped into a 20.0-l container of oxygen gas at 27°C and a pressure of 700 mm Hg. After an interval of time it is found that 4.00 g of iron(III) oxide has been produced. What is the pressure in the container after this time interval, assuming that the temperature remains constant at 27°C?

We may formulate our strategy as follows. The initial pressure is due to the oxygen present initially. Using the ideal gas equation, the original number of moles of gas may be found. The final pressure is due to the remaining unreacted oxygen and the sulfur(IV) oxide produced. Knowing that 4.00 g of iron(III) oxide is produced, the number of moles of oxygen consumed and the number of moles of sulfur(IV) oxide produced can be computed using chemical Equation (11.13). Then the total number of moles of gas may be computed, and this may be used in the ideal gas equation to find the final pressure.

Using the subscripts i and f respectively for the initial state before the reaction started and the final state after the lapse of the time interval, the ideal gas equation may be written for each of these states:

$$P_iV = n_iRT \tag{11.14}$$

$$P_fV = n_fRT \tag{11.15}$$

Subscripts are omitted from V, R, and T since they remain constant.

Equation (11.14) may be divided by Equation (11.15) as shown in the following steps:

$$\frac{P_iV}{P_fV} = \frac{n_iRT}{n_fRT} \tag{11.16}$$

$$\frac{P_i}{P_f} = \frac{n_i}{n_f} \tag{11.17}$$

Equation (11.17) may be solved for P_f.

$$P_f = \frac{P_in_f}{n_i} \tag{11.18}$$

Values for P_i, n_i, and n_f may be determined as follows:

$P_i = 700$ mm Hg from the problem statement

$$n_i = \frac{P_iV}{RT} = \frac{700 \text{ mm Hg} \times \text{atm}}{760 \text{ mm Hg}} \times \frac{20.0 \text{ l}}{300°\text{K}} \times \frac{\text{mole} \times °\text{K}}{0.0821 \text{ l} \times \text{atm}} = 0.748 \text{ mole}$$

n_f = number of moles O_2 remaining + number of moles SO_2 produced

$$\text{moles } O_2 \text{ consumed} = \frac{4.00 \text{ g } Fe_2O_3}{} \times \frac{7 \text{ moles } O_2}{2 \times 160 \text{ g } Fe_2O_3} = 0.088 \text{ mole}$$

$$\begin{aligned}\text{moles } O_2 \text{ remaining} &= \text{moles } O_2 \text{ originally} - \text{moles } O_2 \text{ consumed} \\ &= (0.748 - 0.088) \text{ mole} \\ &= 0.660 \text{ mole } O_2 \text{ remaining}\end{aligned}$$

$$\text{moles } SO_2 \text{ produced} = \frac{4.00 \text{ g } Fe_2O_3}{} \times \frac{4 \text{ moles } SO_2}{2 \times 160 \text{ g } Fe_2O_3} = 0.050 \text{ mole}$$

$$n_f = 0.660 \text{ mole } O_2 + 0.050 \text{ mole } SO_2 = 0.710 \text{ mole gas}$$

These values may be used in Equation (11.18) to find P_f as follows:

$$P_f = \frac{P_i n_f}{n_i} = \frac{700 \text{ mm Hg}}{} \times \frac{0.710 \text{ mole}}{0.748 \text{ mole}} = 664 \text{ mm Hg}$$

Analysis

Often it is desirable to know the purity of a compound or the percent composition of a compound or mixture of compounds. By observing chemical reactions under conditions in which they readily go to completion, it is possible to make an analysis.

For example, compounds containing chloride ions can be converted readily to silver chloride when treated with silver nitrate, and the silver chloride can be isolated and weighed. Thus, if calcium chloride is treated with silver nitrate, the following reaction occurs:

$$2AgNO_3 + CaCl_2 = 2AgCl + Ca(NO_3)_2 \tag{11.19}$$

The important aspect of this reaction for our purposes is the reaction of the chloride ion of the $CaCl_2$ with the silver ion from the $AgNO_3$ to form the AgCl. This may be represented by an ionic equation which indicates only the important reacting species:

$$Ag^+ + Cl^- = AgCl \tag{11.20}$$

This is a general equation indicating that any source of silver ions will react with any source of chloride ions to form silver chloride.

Suppose that a 0.679-g sample of a pure compound containing chloride ions is treated with excess silver nitrate to form 1.32 g of silver chloride. What is the percent of chloride in the compound?

From Equation (11.20) it is clear that one mole of AgCl is formed for every mole of chloride ions, and thus since the molecular weight of AgCl is 143 and the atomic weight of chlorine is 35.5, there must be 143 g of AgCl formed for every 35.5 g of chloride ion in the sample. Thus the mass of chloride in the sample may be found as follows.

$$1.32 \text{ g AgCl} \times \frac{35.5 \text{ g Cl}^-}{143 \text{ g AgCl}} = 0.328 \text{ g Cl}^-$$

Thus 0.328 g of the 0.679 g of sample is chloride, and the percent chloride in the sample may be found as follows:

$$100 \text{ g sample} \times \frac{0.328 \text{ g Cl}^-}{0.679 \text{ g sample}} = 48.3 \text{ percent chloride}$$

Consider another related example in which a mixture is known to contain only sodium chloride, NaCl, and calcium chloride, $CaCl_2$. A 1.11-g sample of this mixture is treated with an excess of $AgNO_3$ so that all of the chloride in both compounds is converted to 2.75 g of AgCl. What is the percent composition of this mixture?

We may formulate our strategy as follows: Since only NaCl and $CaCl_2$ are present, we may let x represent the number of grams of NaCl present. Then $(1.11 - x)$ must be the number of grams of $CaCl_2$ present. The mass of AgCl formed from each of these constituents may be formulated in terms of x and the sum must be 2.75 g. This provides an equation in x which may be solved, and from the solution, the percent of NaCl may be determined.

This plan may be carried out as follows:

Let x = grams of NaCl in the mixture
Then $(1.11 - x)$ = grams of $CaCl_2$ in the mixture

Since each mole of NaCl has a mass of 58.4 g and contains 1 mole of chloride ions, the number of grams of AgCl formed from x grams of NaCl may be found:

$$x \text{ g NaCl} \times \frac{143 \text{ g AgCl}}{58.4 \text{ g NaCl}} = 2.45x \text{ g AgCl}$$

Since each mole of $CaCl_2$ has a mass of 111 g and contains 2 moles of chloride ions, the number of grams of AgCl formed from $1.11 - x$ g of $CaCl_2$ may be found:

$$(1.11 - x) \text{ g CaCl}_2 \times \frac{2 \times 143 \text{ g AgCl}}{111 \text{ g CaCl}_2} = 2.58(1.11 - x) \text{ g AgCl}$$

Since the total amount of AgCl formed is 2.75 g, the equation in x may be formulated and solved:

$$\begin{aligned} 2.45x + 2.58(1.11 - x) &= 2.75 \\ 2.45x + 2.86 - 2.58x &= 2.75 \\ 0.13x &= 0.11 \\ x &= 0.11/0.13 = 0.85 \text{ g NaCl} \end{aligned}$$

This means that only 0.85 g of the 1.11 g of mixture is NaCl. Thus the percent of NaCl in the sample is found:

$$100 \text{ g mixture} \times \frac{0.85 \text{ g NaCl}}{1.11 \text{ g mixture}} = 77 \text{ percent NaCl}$$

By difference, the mixture must be 23 percent $CaCl_2$.

Sometimes gas volumes may be measured and used to carry out analyses. For example, compounds containing the nitrite ion react with the monoamide of sulfuric acid to produce gaseous nitrogen. Thus, barium nitrite reacts as shown by Equation (11.21).

$$Ba(NO_2)_2 + 2NH_2SO_3H = 2N_2 + BaSO_4 + H_2SO_4 + 2H_2O \qquad (11.21)$$

The only gaseous substance in this system is the nitrogen. If the volume, pressure, and temperature of the liberated nitrogen are determined experimentally, it is possible to discover how much barium nitrite has reacted.

For example, suppose that barium nitrite is prepared by the reaction represented earlier by Equation (11.7). To check the purity of the barium nitrite produced, a 176-mg sample of the barium nitrite was treated with an excess of the monoamide of sulfuric acid so that all of the barium nitrite in the sample would react. In the process, 36.7 ml of nitrogen gas was formed at 20°C and 710 mm Hg. What was the percent of barium nitrite in the sample?

Using the ideal gas equation, it is possible to calculate the number of mmoles of nitrogen formed:

$$n = \frac{PV}{RT} = \frac{710 \text{ mm Hg} \times \text{atm}}{760 \text{ mm Hg}} \times \frac{36.7 \text{ ml}}{293°\text{K}} \times \frac{\text{mmole} \times °\text{K}}{0.0821 \text{ ml} \times \text{atm}}$$
$$= 1.43 \text{ mmole } N_2$$

From the number of mmoles of nitrogen formed, Equation (11.21) may be used to compute the number of milligrams of barium nitrite which reacted.

$$1.43 \text{ mmoles } N_2 \times \frac{229 \text{ mg } Ba(NO_2)_2}{2 \text{ mmoles } N_2} = 164 \text{ mg } Ba(NO_2)_2$$

Assuming that for all intents and purposes all of the barium nitrite reacted, then only 164 mg of the 176-mg sample was really barium nitrite, or only 93.2 percent of the sample was barium nitrite, as shown by the following computation:

$$100 \text{ mg sample} \times \frac{164 \text{ mg } Ba(NO_2)_2}{176 \text{ mg sample}} = 93.2 \text{ percent } Ba(NO_2)_2$$

SUMMARY

The chemical equation is a very useful and powerful tool in describing the interactions among atoms, ions, and molecules in systems containing many such particles. The symbols and formulas in equations may be interpreted as moles of particles that may be related to masses and volumes.

Illustrative of the many uses to which chemical equations may be put are the computations of relative amounts of reactants necessary to carry out reactions, the determination of theoretical and actual yields in chemical preparations, the tracking of the progress of chemical reactions, and analyses of substances and mixtures.

As powerful and useful as chemical equations may be, it must be remembered that they describe only the gross, over-all picture of the chemical reaction, and do not indicate why or how particular reactions occur.

QUESTIONS AND PROBLEMS

11.1 Write a complete equation for each of the following chemical reactions:

a) Sodium hypochlorite produces sodium chloride and sodium chlorate when heated.

b) Carbon disulfide burns in air to give carbon dioxide and sulfur(IV) oxide.
c) Metallic sodium reacts with water to produce sodium hydroxide and hydrogen.
d) Metallic aluminum reacts with iron(III) oxide to produce metallic iron and aluminum oxide.
e) Potassium chlorate may be heated to produce potassium chloride and potassium perchlorate.
f) Calcium carbonate yields calcium oxide and carbon dioxide when heated.
g) Elemental mercury and oxygen are produced when mercury(II) oxide is heated.
h) Ammonia burns in oxygen to produce water and nitrogen dioxide.
i) Manganese(IV) oxide reacts with hydrogen chloride to produce water, manganese(II) chloride, and elemental chlorine.
j) Silica reacts with hydrogen fluoride to produce silicon(IV) fluoride and water.
k) Ammonia is prepared by the reaction of elemental nitrogen and hydrogen.
l) Sodium chromate, carbon dioxide, and iron(III) oxide result when iron(II) chromite, $Fe(CrO_2)_2$, is heated with sodium carbonate and elemental oxygen.
m) Upon heating, ammonium dichromate decomposes to chromium(III) oxide, nitrogen, and water.
n) Iodine reacts with sodium thiosulfate to produce sodium iodide and sodium tetrathionate.
o) Uranium metal reacts with elemental iodine to form uranium(III) iodide.
p) Ethanol is produced by the hydration of ethene.
q) Nitric acid reacts with benzene to produce nitrobenzene and water.
r) Tetrachloromethane reacts with steam to produce the diacid chloride of carbonic acid and hydrogen chloride.
s) Copper metal reacts with nitric acid to produce nitrogen(II) oxide, copper(II), nitrate, and water.
t) Magnesium chloride-6-hydrate yields magnesium oxide, hydrogen chloride, and water upon heating.
u) Hydrogen peroxide reacts with potassium hydroxide and potassium hexacyanoferrate(III) to produce water, oxygen, and potassium hexacyanoferrate(II).
v) Peroxydisulfuric acid reacts with water to produce sulfuric acid and hydrogen peroxide.
w) Tin(II) chloride and hydrogen chloride react with ozone to produce tin(IV) chloride and water.
x) Hydrazine is produced, along with sodium chloride and water, when ammonia reacts with sodium hypochlorite.
y) Hypophosphorous acid and hydrogen iodide are produced by the reaction of phosphine and iodine with water.
z) Iron(III) oxide reacts with carbon monoxide to produce metallic iron and carbon dioxide.

11.2 When ammonium chloride and calcium oxide are heated together, ammonia, calcium chloride, and water are produced.
a) Write the equation for this reaction.
b) How many moles of calcium oxide must react in order to produce 5.0 moles of water?

c) How many grams of calcium chloride are produced if 100 g of ammonium chloride undergoes this reaction?
d) How many moles of ammonia are produced if 3.00 g of water are produced simultaneously by this reaction?
e) A sample of 200 g of ammonium chloride was mixed with 200 g of calcium oxide and heated until 25 g of ammonia had been produced. How much calcium oxide was left unreacted at this point?

11.3 Hexane is a liquid with a density of 0.659 g/ml at room temperature. It burns in air to produce carbon dioxide and water.
a) Write the chemical equation for this reaction.
b) How many moles of oxygen are required to burn 5.0 moles of hexane?
c) How many moles of carbon dioxide are produced when 10.0 g of hexane is burned?
d) How many liters of oxygen at 25°C and 740 mm Hg must be consumed in this reaction in order to produce 95.0 g of water?
e) How many milliliters of carbon dioxide gas at 400°C and 0.38 atm can be produced by this reaction when 25.0 ml of hexane is burned? (The volume of the hexane is measured at room temperature.)

11.4 Silver chloride reacts with potassium cyanide to produce potassium chloride and potassium dicyanoargentate(I).
a) Write the chemical equation for this reaction.
b) What is the maximum amount of potassium chloride that can be produced by this reaction if 0.10 mole of silver chloride is added to 0.10 mole of potassium cyanide?
c) A 1.0 mole sample of silver chloride was mixed with 0.5 mole of potassium cyanide. After a certain time had elapsed, it was found that the mixture contained 0.15 mole KCl. Indicate what species were present in the mixture and how much of each.

11.5 Copper metal reacts in two ways with nitric acid. In reaction I, the products are copper(II) nitrate, nitrogen(II) oxide, and water. In reaction II, the products are copper(II) nitrate, nitrogen dioxide, and water. At room temperature, the only gases in the system are nitrogen(II) oxide and nitrogen dioxide.
a) Write the equations for reactions I and II.
b) In which reaction is there more water produced per gram of copper consumed?
c) In which reaction is there more copper(II) nitrate produced per gram of copper consumed?
d) In which reaction does one gram of copper require the more nitric acid with which to react?
e) If 5.13 g of copper metal react with excess nitric acid to produce 0.471 g of nitrogen(II) oxide, what percent of the copper reacts by each of the two reactions?
f) If 3.79 g of copper metal reacts with excess nitric acid to produce 2.49 l of gas measured at 150°C and 684 mm Hg, what percent of the copper reacts by each of the two reactions? (Assume no association of the gases and assume that they behave ideally.)
g) If the purpose of reacting copper with nitric acid is to produce copper(II) nitrate, which of the two reactions has the higher theoretical yield of copper(II) nitrate per g of Cu used? Which reaction is more economical in that it requires relatively less nitric acid per gram of copper(II) nitrate produced?

11.6 Hydrazine is prepared by the reaction of ammonia with sodium hypochlorite. However, there are two competing reactions. In reaction I, the products are hydrazine, sodium chloride, and water. In reaction II, the products are ammonium chloride, sodium chloride, sodium hydroxide, nitrogen, sodium hydroxide, and water.

a) Write the equations for these two reactions.
b) How does the relative degree to which these two reactions occur affect the yield of hydrazine?
c) In one preparation of hydrazine using this reaction, ammonia was kept in excess and the reaction of 46.8 g of sodium hypochlorite led to the production of 6.70 g of ammonium chloride.
   1) What percent of the sodium hypochlorite reacted by each of the two reactions?
   2) Theoretically, how much hydrazine was produced?
   3) How much ammonia was consumed?
d) In another preparation, the chemist was unaware of the possibility of reaction II. He also kept ammonia in excess and on the basis of the equation for reaction I and the amount of sodium hypochlorite used, he calculated a 38.4 percent yield of hydrazine. Assuming that there were no other possible reactions or losses of product, what percent of the sodium hypochlorite underwent each of these reactions?
e) In still another preparation with ammonia in excess, the reaction of 42.8 g of sodium hypochlorite led to the formation of 16.1 g of sodium chloride.
   1) What percent of the sodium hypochlorite reacted by each of these two reactions?
   2) How much hydrazine was produced?

11.7 Potassium bis(oxalato)diaquochromate(III)-2-hydrate is prepared by the reaction of potassium dichromate with oxalic acid-2-hydrate. Carbon dioxide and water are formed also.

a) Write the chemical equation for this reaction.
b) According to this equation, how many grams of potassium dichromate are required to react with 12.0 g of oxalic acid-2-hydrate?
c) Assuming that the actual yield is near the theoretical yield, how many grams of potassium dichromate and oxalic acid-2-hydrate are required to produce 15.0 g of potassium bis(oxalato)diaquochromate(III)-2-hydrate?
d) What is the theoretical yield of potassium bis(oxalato)diaquochromate(III)-hydrate when 6.28 g of potassium dichromate reacts with 15.9 g of oxalic acid-2-hydrate?
e) In one preparation, a yield of 5.08 g of potassium bis(oxalato)diaquochromate(III)-2-hydrate was obtained from 5.08 g of potassium dichromate and 20.3 g of oxalic acid-2-hydrate. What was the percent yield?

11.8 Potassium hexafluorophosphate, KPF_6, is prepared along with potassium chloride from phosphorus(V) chloride and potassium fluoride.

a) Write the equation for the reaction.
b) Theoretically, how much potassium fluoride is required to react with 9.74 g of phosphorus(V) chloride?
c) What is the theoretical yield of potassium hexafluorophosphate when 6.00 g of phosphorus(V) chloride is mixed with 15.6 grams of potassium fluoride?
d) If 1.93 g of potassium hexafluorophosphate is isolated from the reaction between 3.14 g of phosphorus(V) chloride and 4.92 g of potassium fluoride, what is the percent yield?

11.9 Acetyl chloride, which is the acid chloride of acetic acid, may be prepared from acetic acid and phosphorus(III) chloride. Phosphorous acid also is formed.

a) Write the equation for this reaction.

b) A 67 percent yield of acetyl chloride has been reported using this reaction. Considering this, how much acetic acid and phosphorus(III) chloride are required to prepare 25.0 ml of acetyl chloride, which has a density of 1.10 g/ml?

c) Acetyl chloride reacts quantitatively with ammonia to produce ammonium chloride and acetamide, which is the simple amide of acetic acid.

  1) Write the equation for this reaction.

  2) Considering the reported yield of acetyl chloride from phosphorus(III) chloride and acetic acid, how much acetamide could be prepared from 10.0 g of acetic acid?

11.10 The Solvay process is an industrial process for the production of sodium hydrogen carbonate and sodium carbonate.

a) This process consists of the following five steps. Write the equation for each step.

  1) Calcium carbonate is heated to produce calcium oxide and carbon dioxide.

  2) The carbon dioxide from step 1) is treated with ammonia and water to produce ammonium hydrogen carbonate.

  3) The ammonium hydrogen carbonate from step 2) is treated with sodium chloride to produce sodium hydrogen carbonate and ammonium chloride.

  4) The sodium hydrogen carbonate from step 3) is heated to produce sodium carbonate, water, and carbon dioxide with the carbon dioxide reused in step 2).

  5) The ammonium chloride from step 3) is heated with the calcium oxide from step 1) to produce calcium chloride, water, and ammonia with the ammonia reused in step 2).

b) What are the over-all products of this process?

c) If the process is carried through to produce sodium carbonate, theoretically, what percent of the carbon dioxide consumed in step 2) is recovered in step 4)?

d) Theoretically, does the company have either excess calcium oxide or ammonium chloride for sale?

e) Theoretically, what percent of the ammonia consumed in step 2) is recovered in step 5)?

11.11 Although silver metal often is found as the element in nature, it is necessary to separate it from many other substances with which it is mixed. One technique often employed is the cyaniding process.

a) The cyaniding process is carried out in two steps as follows. Write an equation for each step.

  1) Elemental silver is converted to sodium dicyanoargentate(I) by treatment with sodium cyanide, oxygen, and water. Sodium hydroxide is the other product.

  2) The sodium dicyanoargentate(I) is treated with powdered zinc metal to form elemental silver metal and sodium tetracyanozincate(II).

b) In order to produce 50.0 g of silver metal,

  1) how much sodium cyanide is required?

  2) how much zinc metal is required?

c) If the reaction between elemental silver and sodium cyanide is carried out in a closed container at constant volume and temperature, what happens to the pressure as the reaction proceeds?

11.12 Solid sodium cyanide often is prepared by treating solid sodium carbonate and solid carbon with gaseous nitrogen. Gaseous carbon monoxide is a by-product.
a) Write the equation for this reaction.
b) Theoretically, how much carbon is required for the preparation of 11.8 g of sodium cyanide?
c) If this reaction is carried out in a closed container at constant volume and constant temperature, what happens to the pressure?
d) If this reaction is carried out at constant pressure and constant temperature, what happens to the volume of the system?

11.13 The reaction between sodium hydroxide and carbon dioxide to form sodium hydrogen carbonate was discussed in this chapter. Could the progress of this reaction be followed by monitoring the volume of the system if the reaction were carried out at constant temperature and pressure? What would a plot of volume against percent completion of the reaction look like?

11.14 Suppose that the reaction between iron(II) sulfide and oxygen as described by Equation (11.13) was carried out at constant pressure and constant temperature. If 10.0 g of iron(II) sulfide is dropped into 20.0 l of oxygen at 27°C and at a pressure of 700 mm Hg, what is the volume when 4.00 g of iron(III) oxide has been formed?

11.15 For each of the following reactions carried out at constant temperature and constant volume, indicate whether the pressure increases, decreases, or stays the same as the reaction proceeds.
a) Gaseous ozone forms gaseous oxygen.
b) Gaseous nitrogen(II) oxide reacts with gaseous oxygen to form gaseous nitrogen dioxide.
c) Gaseous hydrogen sulfide burns in gaseous oxygen to form gaseous water and gaseous sulfur(IV) oxide.
d) Gaseous hydrogen and gaseous oxygen from liquid water.
e) Gaseous hydrogen and gaseous chlorine form gaseous hydrogen chloride.
f) Gaseous methane burns in gaseous oxygen to form gaseous carbon dioxide and gaseous water.
g) Liquid bromine reacts with gaseous hydrogen sulfide to form solid sulfur and gaseous hydrogen bromide.

11.16 Compounds containing sulfate ion react with barium chloride to form barium sulfate as represented by the following ionic equation:

$$Ba^{++} + SO_4^{--} = BaSO_4$$

a) A 1.69-g sample of a mixture containing sodium sulfate and other nonsulfate-type substances was treated with excess barium chloride to produce 1.88 g of barium sulfate. What was the percent of sodium sulfate in the sample?
b) A mixture is composed of only lithium sulfate and potassium sulfate. How much barium chloride-2-hydrate must be used to treat a 1.50-g sample of this mixture to be sure that an excess of barium chloride has been added?
c) A mixture is composed of only ammonium sulfate and epsom salt, which is magnesium sulfate-7-hydrate. If 1.68 g of barium sulfate is produced when an excess of barium chloride is added to 1.24 g of the mixture, what is the percent composition of the mixture?

11.17 When a sample containing magnesium ions is treated with reagents containing ammonium and phosphate ions, magnesium ammonium phosphate is formed according to the following ionic equation:

$$Mg^{++} + NH_4^+ + PO_4^{3-} = MgNH_4PO_4$$

When magnesium ammonium phosphate is heated above 600°C, it quantitatively decomposes to magnesium diphosphate, ammonia, and water.

a) Write the equation for the ignition of magnesium ammonium phosphate.

b) How much ammonium dihydrogen phosphate must be used to be sure that an excess of ammonium and phosphate ions are provided for the treatment of an impure 2.00-g sample of magnesium chloride-6-hydrate?

c) A 0.109-g sample of a metal alloy containing magnesium was treated with hydrochloric acid so that the magnesium was converted to magnesium chloride with the evolution of hydrogen. The magnesium chloride was converted to magnesium ammonium phosphate which was ignited to 0.425-g of magnesium diphosphate. What was the percent magnesium in the alloy?

11.18 When materials containing calcium ions are treated with sodium oxalate, the calcium is converted to calcium oxalate as represented by the following ionic equation:

$$Ca^{++} + C_2O_4^{--} = CaC_2O_4$$

When the calcium oxalate is ignited at temperatures in excess of 850°C, it is converted to calcium oxide, carbon monoxide, and carbon dioxide.

a) Write the equation for the ignition of calcium oxalate.

b) A compound is known to be a hydrate of calcium chloride. When a 1.92-g sample of this compound is treated with excess sodium oxalate, the resulting calcium oxalate is ignited to form 0.733 g of calcium oxide.

 1) What are the percents of calcium, chlorine, and water in the compound?

 2) What is the formula of the compound?

c) A 2.61-g sample of calcium oxalate was placed in a furnace at 1000°C. After ten minutes, the sample was removed, cooled, and found to weigh 1.58 g. What percent of the calcium oxalate had been converted to calcium oxide in this time interval?

11.19 Carbonates react with hydrochloric acid to form chlorides, water, and carbon dioxide.

a) Write the equations for the reactions of calcium carbonate and sodium carbonate with hydrochloric acid.

b) An impure 234-mg sample of calcium carbonate was treated with excess hydrochloric acid to produce 40.8 ml of carbon dioxide at 28°C and 690 mm Hg. What was the percent of calcium carbonate in the sample?

c) A mixture contains only calcium carbonate and sodium carbonate-10-hydrate. A 312-mg sample of this mixture is treated with excess hydrochloric acid to produce 57.8 ml of carbon dioxide at 20°C and 731 mm Hg. What is the percent composition of the mixture?

11.20 Both potassium chlorate and potassium perchlorate form oxygen and potassium chloride when heated.

a) Write an equation for each of these ignition reactions.

b) A 1.84-g sample of a mixture of potassium chlorate and potassium perchlorate was heated in an open container until it reached the constant mass of 1.09 g. What was the percent composition of the mixture?

c) A 99.4-mg sample of another mixture of potassium chlorate and potassium perchlorate was heated and the oxygen, which was collected at 29°C and 753 mm Hg, had a volume of 31.4 ml. What was the percent composition of the mixture?

d) Suppose that 1.00-g samples of various mixtures of potassium chlorate and potassium perchlorate are to be decomposed by heating. Prepare a graph in which the mass of the potassium chloride residue is plotted against the percent of potassium chlorate in the sample.

11.21 When 1.09 g of dichlorotetraamminecobalt(III) chloride reacts with 3.00 g of silver chloride, 0.672 g of silver chloride is produced. What does this suggest about dichlorotetraamminecobalt(III) chloride and its reaction with silver nitrate?

12

Chemical Reactions: Why They Occur

INTRODUCTION In the last chapter we used the chemical equation to describe chemical changes that can be shown to occur. No attention was paid to the question of why such changes occur. In this chapter we shall attempt to discover some of the driving forces behind these changes.

In previous chapters, we put great stress upon the tendency to achieve minimum potential energy as the driving force in the aggregation of subatomic particles into atoms and atoms into molecules. Certainly this same driving force exists in chemical reactions, and we shall find it useful to explore it in a little more detail.

To make this study, it is convenient to think of the atoms, molecules, and ions of the reactants and products of a chemical reaction as a system of particles. The complete description of this system is called the *state* of the system, which may be specified by indicating the values of a number of properties of the system.

As a macroscopic system contains large numbers of particles, the properties specified are those gross properties of the system that can be observed and measured easily. The choice of these properties is somewhat arbitrary. For example, it is possible to specify the state of a system of particles by specifying the pressure, volume, and temperature. If the system is a gaseous one, which behaves ideally, the ideal gas equation relates these three properties.

The properties used to specify the state of a system are called *state functions.* It is characteristic that the changes in state functions that accompany system changes depend only upon the initial and final state of the system and not on the method by which that change is brought about. Thus the pressure change accompanying the system change does not depend upon the way in which the pressure change or the system change is brought about, but only on the initial and final state of the system. However, the rate at which the change occurs depends upon the manner in which the change is brought about.

In this chapter we shall establish some useful state functions that suggest the driving forces behind system changes. Although the development is general and independent of the nature of the particles involved, we shall attempt to show how changes in state functions may be interpreted through the atomic and molecular models developed earlier in this book.

SYSTEM ENERGY

The total energy E of a system composed of many particles may be described as the sum of the following energies:

1. The external kinetic energy E_k, which the system as a whole possesses due to the motion of the entire system;
2. The external potential energy E_p, which the system as a whole possesses due to the position of the entire system in a force field;
3. The internal energy U, which is the aggregate energy associated with the internal makeup of the system, which, in the case of a system of atoms, molecules, and ions, includes the kinetic energies of the individual particles, the energies associated with the bonding and intermolecular forces, the energies associated with vibrations of atoms in molecules and with the rotation and spinning of molecules, as well as the energies associated with the electrons within the molecules.

This may be expressed by Equation (12.1).

$$E = E_k + E_p + U \tag{12.1}$$

The only way in which the total energy of the system can change is by interaction with the environment with which it is in contact. If the energy of the system increases, the energy of the environment must decrease, and if the energy of the system decreases, the energy of the environment must increase. This is in accord with the principle of energy conservation.

Energy is transferred between the system and its environment in several ways. One way is through mechanical work performed on or by the system. For example, if the system expands in volume against a constraining environment, the system expends energy in performing work on the environment, and the energy of the environment increases by the amount of work performed on it. Similarly, if the environment performs work on the system, as in a contraction of the system, then the energy of the environment decreases by the amount of work performed and the energy of the system increases by a corresponding amount.

A second method of energy transfer between the system and its environment is by heat transfer, which results from a temperature difference between the system and its environment. Thus, if the system is cooler than the surroundings with which it is in contact, heat flows from the environment to the system, and the energy of the system increases. If the system is warmer than its environment, then heat flows from the system to the surroundings, and the energy of the system decreases.

A third method of energy transfer is by transfer of electrical energy, which we shall not consider at this time.

The study of this energy transfer between a system and its environment is called *thermodynamics*. From the thermodynamic point of view, the energy change in a system ΔE, is the difference between the heat q absorbed by the system from its environment and the work w done by the system on its environment, as shown by Equation (12.2).

$$\Delta E = q - w \tag{12.2}$$

By convention, q is positive when heat is absorbed by the system from its environment and negative when heat is lost by the system to its environment. Similarly, w is positive when the system performs work on its environment, and negative when the environment performs work on the system.

From these conventions and Equation (12.2), the following relationships should be clear:

1. ΔE is positive when heat is absorbed by the system *and* work is done on the system, that is, when q is positive and w is negative.
2. ΔE is negative when heat is lost by the system *and* the system performs work on the environment, that is, when q is negative and w is positive.
3. Depending on the relative magnitudes of q and w, ΔE may be either positive or negative when heat is absorbed by the system *and* the system performs work on the environment, *or* when heat is lost by the system and the environment performs work on the system, that is, when q and w are both positive *or* both negative.

Work

The work done by an expanding system against a constraining environment may be understood with the aid of Fig. 12.1. A gas is confined to a container by a movable, weightless, and frictionless piston. If the gas expands against the surrounding atmosphere, then the gas must do work to push away the atmosphere. The work is the product of the force F exerted by the surrounding atmosphere, and the distance d through which the piston moves, as shown by Equation (12.3).

$$w = Fd \tag{12.3}$$

Fig. 12.1. Representation of a gas in a movable, weightless, frictionless piston

Since the external atmospheric pressure is the force exerted per unit area of the piston, the force F against which the system does work is the product of the constant external pressure P_{ex} and the area A of the piston as shown by Equation (12.4).

$$F = P_{ex}A \tag{12.4}$$

The expression for F in Equation (12.4) may be substituted into Equation (12.3) as follows.

$$w = P_{ex}Ad \tag{12.5}$$

However, the product of the area of the piston and the distance d through which it moves, Ad, is the increase in volume ΔV of the system, as shown by Equation (12.6), which may be substituted into Equation (12.5) to give Equation (12.7).

$$Ad = \Delta V \tag{12.6}$$

$$w = P_{ex}\Delta V \tag{12.7}$$

Equation (12.7) shows that the work done by the system on the environment is the product of a constant pressure against which the work is done, and the change in volume of the system. This is called *pressure-volume work*.

As defined here, the pressure-volume work is not a state function. This may be seen in the following example. Suppose that 1 l of an ideal gas at 1.0 atm pressure undergoes the following changes at constant temperature:

	ΔV	$P_{ex}\Delta V$
Step 1: Expansion to 2.0 l against 0.5 atm	1.0 l	+0.5 (l × atm)
Step 2: Contraction to 0.5 l against 2.0 atm	−1.5 l	−3.0 (l × atm)
Step 3: Expansion to 1.0 l against 1.0 atm	0.5 l	+0.5 (l × atm)
Total Pressure-Volume Work		−2.0 (l × atm)

Although the system has changed in a cyclic process back to its original state, the net work, $P_{ex}\Delta V$, is not zero as it should be for a state function. In this case 2.0 (l × atm) of work has been performed on the system without a net change of state. What is more, the net work depends on the path by which the change occurs, as illustrated by the following three-step process also carried out at constant temperature, and on the same system.

	ΔV	$P_{ex}\Delta V$
Step 1: Contraction to 0.5 l against 2 atm	−0.5 l	−1.0 (l × atm)
Step 2: Expansion to 2.0 l against 0.5 atm	+1.5 l	+0.75 (l × atm)
Step 3: Contraction to 1.0 l against 1.0 atm	−1.0 l	−1.0 (l × atm)
Total Pressure-Volume Work		−1.25 (l × atm)

In this case, the net work performed on the system was 1.25 (l × atm) despite the fact that the initial and final states were the same within this series of changes and were also the same as the initial and final states for the previous series of changes, where the net work performed on the system was 2.0 (l × atm).

However, the internal pressure of a gas is a state property of the gas, and if this pressure could be used in the $P\Delta V$ expression, then $P\Delta V$ would be a state function of the system. The internal pressure equals the external pressure only if the changes in the system are carried out in infinitesimally small steps so that the internal pressure of the gas always equals the external pressure. Under these conditions, the process is said to be thermodynamically *reversible*. In reality, it is not possible to carry out changes in a perfectly reversible manner, although close approximations can be made. However, it is convenient to use the concept of thermodynamic reversibility in order to use the state property of internal gas pressure.

For reversible processes, Equation (12.7) may be rewritten as Equation (12.8), where w is the reversible work and P is the internal gas pressure.

$$w = P\Delta V \tag{12.8}$$

Of course it is not necessary that changes occur at constant internal pressure. The product of the pressure and volume is a state function, since both pressure and volume are state properties. Thus the reversible work may be expressed by Equation (12.9).

$$w = \Delta(PV) = P_fV_f - P_iV_i \tag{12.9}$$

Some Elementary System Changes

Imagine a sample of hydrogen gas confined by a movable, weightless, and frictionless piston at a pressure of 1 atm and a temperature of 25°C, as shown in Fig. 12.2. Suppose that the gas expands to double its volume so that the final pressure is 0.5 atm. There are two ways in which this expansion may be carried out.

The expansion is said to be *adiabatic* if heat transfer between the system and

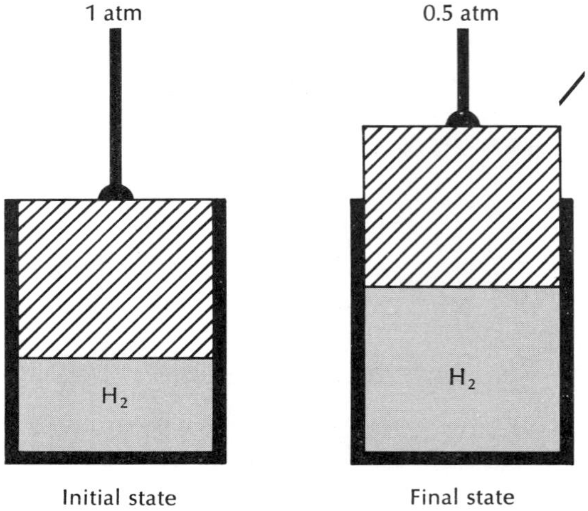

Fig. 12.2. Initial and final states for the expansion of hydrogen

its environment is made impossible with a suitable insulator. Under these circumstances, q is zero and Equation (12.2) may be rewritten as Equation (12.10) with the aid of Equation (12.9).

$$\Delta E = -w = -\Delta(PV) \tag{12.10}$$

This equation shows that for an *adiabatic* process, the pressure-volume work performed by the system on its environment is done at the expense of the total energy of the system.

If the gas behaves ideally, the individual molecules do not interact with each other beyond the transfer of kinetic energy through elastic collisions. The expansion of the gas must occur because the gas molecules impart some of their kinetic energy to the environment, and as a result, the average kinetic energy of the molecules in the system decreases with an adiabatic expansion. Consequently, the temperature of the system decreases.

The expansion is said to be *isothermal* if the temperature remains constant. To maintain this condition, the gas absorbs heat from the environment in an amount equivalent to the work the system performs on the environment. Thus, under *isothermal* conditions, the total energy of the system remains constant, that is, $\Delta E = 0$, and Equation (12.2) reduces to Equation (12.11) with the aid of Equation (12.9).

$$q = w = \Delta(PV) \tag{12.11}$$

This equation shows that an *isothermal* expansion requires the system to absorb just enough heat from the environment to carry out the pressure-volume work.

If the piston in Fig. 12.2 is replaced by a rigid wall so that system changes occur at constant volume, there can be no expansion and contraction, and there can be no pressure-volume work, and $w = 0$ and Equation (12.2) reduces to Equation (12.12).

$$q = \Delta E \tag{12.12}$$

This equation shows that the addition of heat to a system at constant volume leads to an equivalent increase in the total energy of the system.

ENTHALPY

For reversible changes, Equation (12.2) may be combined with Equation (12.9) as follows:

$$q = \Delta E + \Delta(PV) \tag{12.13}$$

Since the delta notation indicates the difference between initial and final states Equation (12.13) may be rewritten and rearranged:

$$q = E_f - E_i + (P_f V_f - P_i V_i)$$

$$q = E_f + P_f V_f - (E_i + P_i V_i) \tag{12.14}$$

The right side of Equation (12.14) appears to be the difference between an initial and final state, and it is convenient to define a new property, H, called the *enthalpy*, as shown in Equation (12.15).

$$H = E + PV \tag{12.15}$$

With Equation (12.15), Equation (12.14) may be rewritten:

$$q = H_f - H_i = \Delta H \tag{12.16}$$

Equation (12.16) shows that the addition of heat to a system from the environment leads to an increase in the enthalpy of the system, and the loss of heat by the system to the environment leads to a decrease in the enthalpy of the system. In addition, the magnitude of the enthalpy change is the same as the magnitude of the heat transfer.

As defined by Equation (12.15), the enthalpy is the sum of the energy of the system and the product of the pressure and volume of the system. Since the enthalpy is defined in terms of three state properties, it is a state property as well. Thus the changes in enthalpy which accompany system changes depend only on the initital and final states of the system.

The Components of Enthalpy Changes

Since the change in enthalpy really is absorption or loss of heat in a reversible process, Equation (12.2) may be rewritten as Equation (12.17) for a reversible process.

$$\Delta H = \Delta E + w \tag{12.17}$$

Then for changes which involve only pressure-volume work, Equation (12.9) may be used to rewrite Equation (12.17) as follows.

$$\Delta H = \Delta E + \Delta(PV) \tag{12.18}$$

From Equation (12.18), we might ask about the nature of the contributions to the enthalpy change that accompanies a system change in terms of the energy change and the change in the pressure-volume work.

From Equation (12.1) it is clear that any change in system energy must be the sum of the changes in external kinetic energy, external potential energy, and internal energy. In most common situations, there are no significant changes in external or potential energies, and thus the changes in system energy usually are changes in internal energy. In addition, if a system change is carried out isothermally, then the average kinetic energy of the particles in the system remains constant. Thus at constant temperature, the predominant component in the system energy change is the change in potential energy among the particles of the system. Although there are several contributions to the potential energy change, the most significant are changes due to changes in bonding and changes in states of matter. Thus one important contribution to the change in enthalpy accompanying a system change is the potential energy change for the particles of which the system is composed.

In addition to the system energy change, the change in the pressure-volume product also contributes to the enthalpy change. Often this contribution is small. Usually volume changes are small among systems containing only liquids and solids, and if pressure changes are small also, the change in the pressure-volume product may be negligible for many considerations.

However, volume changes can be significant in gaseous systems. Some notion of the magnitude of this contribution to the enthalpy change can be obtained from the ideal gas equation, which is shown for the final and initial states by Equations (12.19) and (12.20) respectively.

$$PV_f = n_fRT \tag{12.19}$$

$$PV_i = n_iRT \tag{12.20}$$

The subscripts have been omitted for the pressure and temperature, which we shall assume are constant. Under these conditions the volume changes only if the number of moles of particles changes.

Equations (12.19) and (12.20) may be subtracted to form the following equation, which may be rearranged to Equations (12.22) and (12.23).

$$PV_f - PV_i = n_fRT - n_iRT \tag{12.21}$$

$$P(V_f - V_i) = (n_f - n_i)RT \tag{12.22}$$

$$P\Delta V = RT\Delta n \tag{12.23}$$

Thus, assuming ideal gas conditions, the value for $P\Delta V$ may be computed from $RT\Delta n$. A computed value for R is 1.987 calories/(mole × °K). As an approximation, this is about 2 calories/(mole × °K), and if the temperature is room temperature of about 300°K, this amounts to a $P\Delta V$ product of about 600 calories/mole of gaseous particles changed during the reaction. Although this is a significant energy, it often is minor in comparison to system energy changes, which often are more than 20,000 calories. However, it should be remembered that there are situations in which the pressure-volume product is more significant than the change in system energy.

Enthalpy Change as a Driving Force

It has just been shown that in many cases, the enthalpy change that accompanies a system change reflects the change in potential energy of the particles in the system. Equation (12.16) indicated that the change in enthalpy corresponds to the heat absorbed or lost by the system during a reversible change. Those changes in which heat is lost by the system are called *exothermic* changes, and those changes in which heat is absorbed by the system are called *endothermic* changes. Thus for exothermic changes, heat is lost by the system, q is negative, and ΔH is negative. The negative value for ΔH indicates that the final enthalpy is less than the initial enthalpy, so that the enthalpy decreases during an exothermic change.

It has been observed that there is a tendency for exothermic reactions to occur spontaneously. This is only a general trend and there are exceptions, but this suggests that there is a driving force leading to the loss of heat by the system. This has been translated into enthalpy terms by the depiction of a decrease in enthalpy as a driving force behind system changes. In those cases in which the enthalpy change reflects the potential energy change among the particles in the system, the decrease in enthalpy corresponds to the decrease in potential energy that we have already mentioned in a number of situations as an indication of spontaneity of change.

Thus, system changes that occur with a decrease in enthalpy are favored over those changes that occur with an increase in enthalpy.

Standard States

For the chemist, chemical reactions constitute thermodynamic systems of great interest. In these systems, the reactants are present in the initial state and the products are present in the final state. Since the enthalpy difference between the initial and final states is a measure of part of the driving force behind the reaction, it is of interest to know what this change will be so that some predictions may be made about the feasibility of the reaction. However, the task of listing enthalpy changes for all conceivable reactions under all possible sets of conditions is impossible. For this reason, several conventions have been adopted for the recording and listing of enthalpy changes

First, the state of each reactant and each product needs to be indicated if the initial and final states of the system are to be specified. For this purpose, a set of *standard conditions* is agreed upon. *Each* substance is understood to be in its most stable form at one atmosphere of pressure. In addition the temperature must be specified in descriptions of state. Here we shall use 25°C unless otherwise noted.

In addition, *standard* enthalpy changes are designated for specific reactions; these reactions may be represented by chemical equations, and there must be some understanding that the quantities of reactants and products in the initial and final states of the system are those represented by the equation.

For example, consider the reaction between hydrogen and chlorine to produce hydrogen chloride as represented by Equation (12.24).

$$H_2(g) + Cl_2(g) = 2HCl(g) \tag{12.24}$$

The reaction is represented at standard conditions in the very hypothetical and schematic drawing in Fig. 12.3.

The initial state consists of the system in which H_2, Cl_2, and HCl are present, each at a partial pressure of 1 atm, which is the standard pressure. As the reaction occurs in the central chamber, H_2 and Cl_2 are consumed and more H_2 and Cl_2 must be forced in to maintain the pressure of H_2 and Cl_2 at 1 atm throughout the reaction. At the same time, HCl is formed and must be bled out to keep the HCl partial pressure down to 1 atm. At the final state, the system still consists of H_2, Cl_2, and HCl, each at a partial pressure of 1 atm. However, in making the change from the

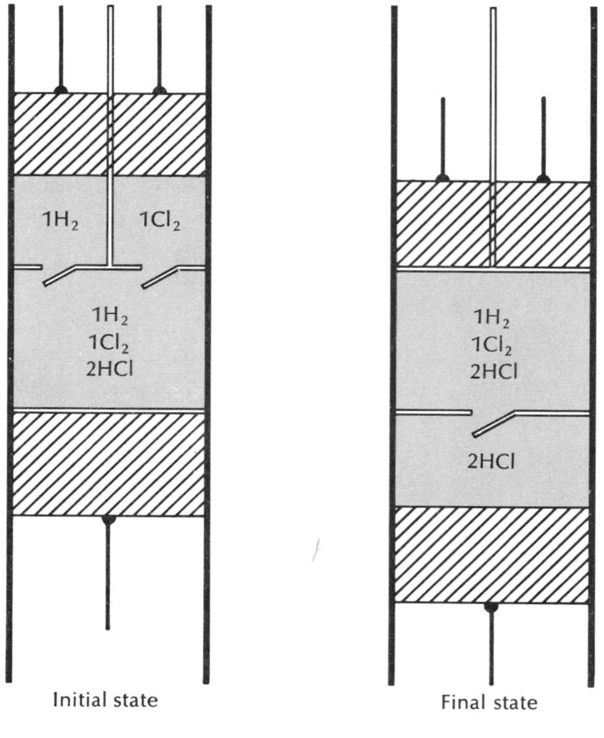

Fig. 12.3. Representation of the initial and final states at standard conditions for the reaction of hydrogen with chlorine to form hydrogen chloride

initial to the final state, 1 mole of H_2 and 1 mole of Cl_2 are consumed and 2 moles of HCl are produced.

The meaning of the enthalpy change for this reaction is specified by Equation (12.25), which shows the enthalpy change as the sum of the enthalpies of the reactants subtracted from the enthalpy of the products.

$$\Delta H = H_{HCl} - (H_{H_2} + H_{Cl_2}) \tag{12.25}$$

Using Equation (12.15), Equation (12.25) may be rewritten as follows:

$$\Delta H = E_{HCl} + P_{HCl}V_{HCl} - (E_{H_2} + P_{H_2}V_{H_2}) - (E_{Cl_2} + P_{Cl_2}V_{Cl_2})$$

This equation may be rewritten as follows by noting that P_{HCl}, P_{H_2}, and P_{Cl_2} have the same constant value of 1 atm:

$$\Delta H = E_{HCl} - (E_{H_2} + E_{Cl_2}) + P[V_{HCl} - (V_{H_2} + V_{Cl_2})]$$

This equation may be put into the delta notation:

$$\Delta H = \Delta E + P\Delta V \tag{12.26}$$

Equation (12.26) shows that this is a constant pressure reaction when the partial pressures of each of the components of the reaction are held constant.

Although it is not true in all chemical reactions, in this case the reaction also occurs at constant volume, since the initial volume of the reactants is the same as the final volume of the products. Since $\Delta V = 0$, from Equation (12.26) it can be seen that $\Delta H = \Delta E$. Thus the enthalpy change corresponds to the system energy change, and the most important components of the energy change are the changes in bonds. One mole of H—H bonds and 1 mole of Cl—Cl bonds have been converted to 2 moles of H—Cl bonds. Using the bond energies in Table 6.9, the potential energies associated with the mole of H—H bonds, the mole of Cl—Cl bonds and the 2 moles of H—Cl bonds are -104 kcal, -58 kcal, and -206 kcal respectively. Thus the change from initial to final state at standard conditions leads to a decrease in potential energy of about 44 kcal. The accepted value for the enthalpy change is -44.12 kcal.

The change in enthalpy that occurs when the substances in both the initial and final states are at standard conditions is called the *standard enthalpy change;* it is designated with a superscript circle, $\Delta H°$. Thus for the conversion of hydrogen and chlorine to hydrogen chloride at standard conditions in the gas phase, $\Delta H° = -44.12$ kcal.

Enthalpies of Formation

Since the enthalpy is a state property, the enthalpy change associated with a system change is the sum of the enthalpy changes associated with all of the intermediate changes in the system. For example, suppose that a system changes from state 1 to state 4 through intermediate states 2 and 3 under standard conditions. The three steps involved in this process may be outlined as follows:

$$\begin{array}{rl}
\text{Step 1: From state 1 to state 2, } \Delta H^\circ_{12} &= H^\circ_2 - H^\circ_1 \\
\text{Step 2: From state 2 to state 3, } \Delta H^\circ_{23} &= H^\circ_3 - H^\circ_2 \\
\text{Step 3: From state 3 to state 4, } \Delta H^\circ_{34} &= H^\circ_4 - H^\circ_3 \\
\hline
\text{Over-all: From state 1 to state 4, } \Delta H^\circ_{14} &= H^\circ_4 - H^\circ_1 \\
&= \Delta H^\circ_{12} + \Delta H^\circ_{23} + \Delta H^\circ_{34}
\end{array}$$

Thus the over-all enthalpy change is the sum of the enthalpy changes for each of the steps in the process. What is more, the nature of the steps is of no consequence. For example, the same system may have proceeded from state 1 to state 4 through intermediate states *a*, *b*, and *c* as follows:

$$\begin{array}{rl}
\text{Step 1: From state 1 to state } a, \ \Delta H^\circ_{1a} &= H^\circ_a - H^\circ_1 \\
\text{Step 2: From state } a \text{ to state } b, \ \Delta H^\circ_{ab} &= H^\circ_b - H^\circ_a \\
\text{Step 3: From state } b \text{ to state } c, \ \Delta H^\circ_{bc} &= H^\circ_c - H^\circ_b \\
\text{Step 4: From state } c \text{ to state 4, } \Delta H^\circ_{4c} &= H^\circ_4 - H^\circ_c \\
\hline
\text{Over-all: From state 1 to state 4, } \Delta H^\circ_{14} &= H^\circ_4 - H^\circ_1 \\
&= \Delta H^\circ_{1a} + \Delta H^\circ_{ab} + \Delta H^\circ_{bc} + \Delta H^\circ_{4c}
\end{array}$$

Thus regardless of the path employed, ΔH°_{14} has the same value.

In dealing with chemical reactions, we may choose a convenient set of steps without regard for their actual validity. The procedure which may be employed for any chemical reaction is as follows:

1. The first steps in the change are the conversions of all reactants to the elements of which they are composed.
2. The final steps in the change are the conversions of the elements to the products.

For example, suppose that we wish to study the reaction of carbon tetrachloride vapor with water vapor to produce gaseous carbon dioxide and hydrogen chloride as represented by the following equation:

$$CCl_4(g) + 2H_2O(g) = CO_2(g) + 4HCl(g) \tag{12.27}$$

We may imagine the conversion of carbon tetrachloride to carbon and chlorine, and of water to hydrogen and oxygen, followed by the conversion of carbon and oxygen to carbon dioxide, and of hydrogen and chlorine to hydrogen chloride, as follows:

$$CCl_4(g) = C(s) + 2Cl_2(g) \tag{12.28}$$

$$2H_2O(g) = 2H_2(g) + O_2(g) \tag{12.29}$$

$$C(s) + O_2(g) = CO_2(g) \tag{12.30}$$

$$2H_2(g) + 2Cl_2(g) = 4HCl(g) \tag{12.31}$$

$$\overline{CCl_4(g) + 2H_2O(g) = CO_2(g) + 4HCl(g)}$$

If we wish to find the over-all enthalpy change, we need to know the enthalpy changes for each of the steps. Each of these steps involves a reaction of a compound to form its elements or a reaction of the elements to form a compound. Since any chemical reaction may be written as the sum of a series of changes of this sort, it has become common practice to list *standard enthalpies of formation*. These are the standard enthalpy changes when 1 mole of a compound is formed from its elements. Thus Equation (12.30) represents the formation of 1 mole of CO_2 molecules from 1 mole of C atoms and 1 mole of oxygen molecules. The enthalpy change associated with this change at standard conditions is called the standard enthalpy of formation of CO_2 and is represented as ΔH_f° or $\Delta H_{f\text{CO}_2}^\circ$.

Similarly, Equation (12.31) represents the formation of hydrogen chloride from its elements. However, the equation represents the formation of 4 moles of HCl molecules. The standard enthalpy of formation of HCl is the enthalpy change associated with the following reaction:

$$\tfrac{1}{2}\text{H}_2(\text{g}) + \tfrac{1}{2}\text{Cl}_2(\text{g}) = \text{HCl(g)} \qquad (12.32)$$

The standard enthalpy of formation of HCl is represented as $\Delta H_{f\text{HCl}}^\circ$. However, since Equation (12.31) represents the formation of 4 moles of HCl molecules, the enthalpy change represented by this equation is four times the standard enthalpy of formation, that is, $4\Delta H_{f\text{HCl}}^\circ$.

Equation (12.28) is the reverse of the equation for the formation of carbon tetrachloride from its elements. The formation of CCl_4 is represented by Equation (12.33).

$$\text{C(s)} + 2\text{Cl}_2(\text{g}) = \text{CCl}_4(\text{g}) \qquad (12.33)$$

The difference between the systems represented by Equations (12.28) and (12.33) is that the initial and final states have been reversed. This simply leads to a change in sign for the standard enthalpy change since the magnitude of the difference between the two states remains unchanged. Thus the standard enthalpy change for the reaction represented by Equation (12.28) is the negative of the standard enthalpy of formation for carbon tetrachloride, that is, $-\Delta H_{f\text{CCl}_4}^\circ$.

By similar arguments, the standard enthalpy change for the reaction represented by Equation (12.28) is $-2\Delta H_{f\text{H}_2\text{O}}^\circ$.

In this way, the standard enthalpy changes associated with the reactions represented by Equations (12.27) through (12.31) may be given in terms of the standard enthalpies of formation as follows:

$$\begin{array}{rcll}
\text{CCl}_4(\text{g}) &=& \text{C(s)} + 2\text{Cl}_2(\text{g}) & -\Delta H_{f\text{CCl}_4}^\circ \\
2\text{H}_2\text{O(g)} &=& 2\text{H}_2(\text{g}) + \text{O}_2(\text{g}) & -2\Delta H_{f\text{H}_2\text{O}}^\circ \\
\text{C(s)} + \text{O}_2(\text{g}) &=& \text{CO}_2(\text{g}) & +\Delta H_{f\text{CO}_2}^\circ \\
2\text{H}_2(\text{g}) + 2\text{Cl}_2(\text{g}) &=& 4\text{HCl(g)} & +4\Delta H_{f\text{HCl}}^\circ \\
\hline
\text{CCl}_4(\text{g}) + 2\text{H}_2\text{O(g)} &=& \text{CO}_2(\text{g}) + 4\text{HCl(g)} &
\end{array}$$

$$\Delta H^\circ = \Delta H_{f\text{CO}_2}^\circ + 4\Delta H_{f\text{HCl}}^\circ - (\Delta H_{f\text{CCl}_4}^\circ + 2\Delta H_{f\text{H}_2\text{O}}^\circ)$$

The important point to see here is that for this reaction, as any other, the standard enthalpy change may be found by subtracting the sum of the standard enthalpies of formation of the reactants from the sum of the standard enthalpies of formation of the products. Thus by using a table of standard enthalpies of formation, the standard enthalpy changes for any reactions involving the substances in the table can be ascertained.

A few standard enthalpies of formation are given in Table 12.1, and additional values are listed in Table 12.2 later in this chapter. From Table 12.1, the standard

TABLE 12.1

Some Standard Enthalpies of Formation

COMPOUND	FORMATION REACTION	ΔH_f° kcal/mole
$CCl_4(g)$	$C(s) + 2Cl_2(g) = CCl_4(g)$	-25.50
$H_2O(g)$	$H_2(g) + \frac{1}{2}O_2(g) = H_2O(g)$	-57.80
$CO_2(g)$	$C(s) + O_2(g) = CO_2(g)$	-94.05
$HCl(g)$	$\frac{1}{2}H_2(g) + \frac{1}{2}Cl_2(g) = HCl(g)$	-22.06
$NH_3(g)$	$\frac{1}{2}N_2(g) + \frac{3}{2}H_2(g) = NH_3(g)$	-11.04
$NO(g)$	$N_2(g) + O_2(g) = NO(g)$	$+21.60$

enthalpy change for the reaction represented by Equation (12.27) may be computed as follows:

$$\begin{aligned}\Delta H^\circ &= \Delta H^\circ_{f\,CO_2} + 4\Delta H^\circ_{f\,HCl} - (\Delta H^\circ_{f\,CCl_4} + 2\Delta H^\circ_{f\,H_2O}) \\ &= [-94.05 - 4(22.06) + 25.50 + 2(57.80)] \text{ kcal} \\ &= [-94.05 - 88.24 + 25.50 + 115.60] \text{ kcal} \\ &= -41.19 \text{ kcal}\end{aligned}$$

Thus the enthalpy driving force favors this reaction of 1 mole of carbon tetrachloride vapor and 2 moles of water vapor to form 1 mole of carbon dioxide gas and 4 moles of hydrogen chloride gas at standard conditions.

As a final example, we might inquire about the standard enthalpy change for the reaction of gaseous nitrogen(II) oxide and water vapor to form ammonia and oxygen as shown by the following equation:

$$4NO(g) + 6H_2O(g) = 4NH_3(g) + 5O_2(g) \qquad (12.34)$$

The standard enthalpies of formation are listed in Table 12.1 for all components of this system except oxygen. Of course, oxygen already is in its elemental form, which is stable at standard conditions, and so it has no standard enthalpy of formation. This is true of all elements that appear in their standard states.

Using the values in Table 12.1, the standard enthalpy change for the reaction shown in Equation (12.34) may be computed as follows:

$$\begin{aligned}\Delta H^\circ &= 4\Delta H^\circ_{f\,NH_3} - 4\Delta H^\circ_{f\,NO} - 6\Delta H^\circ_{f\,H_2O} \\ &= [4(-11.04) - 4(21.60) - 6(-57.80)] \text{ kcal} \\ &= [-44.16 - 86.40 + 346.80] \text{ kcal} \\ &= +216.24 \text{ kcal}\end{aligned}$$

Clearly the enthalpy change does not favor the conversion of 4 moles of gaseous nitrogen(II) oxide and 6 moles of water vapor into 4 moles of gaseous ammonia and 5 moles of gaseous oxygen at standard conditions, since the value for the standard enthalpy change is positive.

It should be clear that a limited amount of standard enthalpy difference information can go a long way in allowing the computation of standard enthalpy changes for many other reactions. This is a part of the power of thermodynamics in allowing predictions about the feasibility of chemical reactions.

ENTROPY

A somewhat more subtle driving force often competes with the enthalpy change as the controlling force behind chemical reactions. Consider Fig. 12.4, which shows a large container divided into two compartments of equal volume, with a valve in the neck connecting the two compartments. With the valve closed, 1 mmole of each of two ideal gases, *A* and *B*, are introduced into the respective sides. Throughout the change to be discussed here the temperature remains constant.

If the valve is opened, the two gases begin to mix, and after a period of time it is found that both of the gases are equally divided between the two compartments. From your experience, this is probably what you would have expected, but what

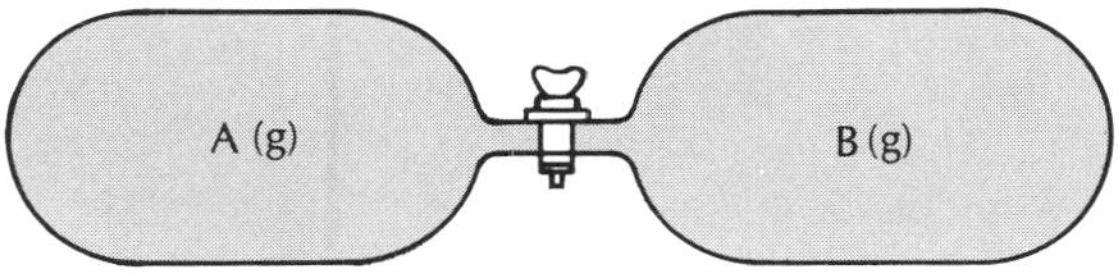

Fig. 12.4. Apparatus for the mixing of two gases

is the driving force behind this mixing? Also, from your experience you might predict that it is unlikely that over a long period of time the two gases will be ordered so that all molecules of gas *A* are in one compartment and all of the molecules of gas *B* are in the other compartment.

In this example, the mixing process occurs at constant volume and constant pressure as well as constant temperature. There is no change in internal energy and there is no change in the pressure-volume product. There is no enthalpy change to serve as a driving force, and yet this is a spontaneous change.

This driving force is not unfamiliar in changing systems. Suppose you start with a well-ordered deck of 52 playing cards, and that you know the order of these cards. If you then shuffle the cards and determine the order once more, in general you do not find the original arrangement.

The reason for this is not that the system "dislikes" the particular order in which the cards were placed originally, but rather that there are so many other possible arrangements that are equally satisfactory, that the probability that the cards will fall into the original order is extremely small. With 52 cards there are more than 10^{30} possible arrangements.

For the two ideal gases, the number of possible arrangements is much larger, and what we would call the ordered arrangement in which all molecules of gas *A* are in one compartment and all molecules of gas B are in the other compartment allows only two alternatives among the very large number of possibilities.

In general, the meaning of order implies at best a few limited arrangements, and in physical systems the number of such arrangements usually is so small in comparison to the total number of possible arrangements for particles that it appears that there is a tendency toward the formation of unordered arrangements.

The measure of the disorder of a system of particles is called the *entropy* of the system and is represented by the symbol S. It really is the measure of the improbability of finding the particles in an ordered arrangement. The entropy driving force behind a system change appears to be toward increasing entropy, that is toward increasing disorder. Thus if the final entropy of a system is given by S_f and the initial entropy of the system is given by S_i, then the change in entropy is ΔS, which is defined by Equation (12.35).

$$\Delta S = S_f - S_i \tag{12.35}$$

It can be shown that the entropy of a system is a state property of the system, and therefore ΔS depends only upon the initial and final states of the system. In terms of the entropy, spontaneous system changes occur with an increase in entropy, that is a positive value for ΔS. Similarly, nonspontaneous changes occur with a decrease in entropy, that is, a negative value for ΔS.

FREE ENERGY

The conclusion is that two driving forces must be considered in chemical systems. One is the tendency to change toward lower enthalpy and the other is the tendency to change toward higher entropy. The total driving force behind chemical changes is the sum of these two driving forces.

However, if enthalpy and entropy values are to be compared, they must be expressed in the same units. Entropy usually is expressed as energy per degree Kelvin, and thus the entropy must be multiplied by the absolute temperature. Then in order to sum up the two driving forces, a new state property is defined acording to Equation (12.36), where the new property, G, is called the *free energy* of the system.

$$G = H - TS \tag{12.36}$$

To use the usual delta notation, Equations (12.37) and (12.38) may be written for the final and initial states respectively and may be subtracted to provide Equation (12.39), which may be rearranged to Equations (12.40) and (12.41). Throughout, the temperature is held constant.

$$G_f = H_f - TS_f \tag{12.37}$$

$$G_i = H_i - TS_i \tag{12.38}$$

$$G_f - G_i = H_f - H_i - TS_f + TS_i \tag{12.39}$$

$$G_f - G_i = H_f - H_i - T(S_f - S_i) \tag{12.40}$$

$$\Delta G = \Delta H - T\Delta S \tag{12.41}$$

Equation (12.41) is the important one in this series, which sums up the total driving force for isothermal system changes. The difference in free energy between the final and initial states, ΔG, is the measure of the actual driving force; it equals the difference between the enthalpy driving force, ΔH, and the entropy driving force, $T\Delta S$.

It should be clear from Equation (12.41) that if the value for ΔH is negative corresponding to a decrease in enthalpy, and the value for ΔS is positive corresponding to an increase in entropy, then the value for ΔG will be negative and will be more negative than the value for ΔH. Thus the direction of change in free energy for a spontaneous process is the same as the direction change in enthalpy for a spontaneous process, that is, it decreases. Thus if the free-energy change for a proposed reaction is shown to be negative, the reaction is thermodynamically spontaneous, whereas if the change is positive, the reaction is thermodynamically nonspontaneous.

When ΔH is negative and ΔS is positive, then ΔG must be negative, and the reaction is spontaneous. When ΔH is positive and ΔS is negative, then ΔG must be positive, and the reaction is nonspontaneous. When ΔH is negative and ΔS is negative, or when ΔH is positive and ΔS is positive, then whether or not ΔG is positive or negative depends upon the relative sizes of the ΔH and ΔS values.

The Relative Sizes of Enthalpy and Entropy Changes

Computations with free-energy changes may be handled in exactly the same way that computations with enthalpy changes were handled earlier. The same standard conditions are used to define standard free-energy changes and it is profitable to tabulate standard free energies of formation which may be used in the same way that standard enthalpies of formation were used. The standard free energies and standard enthalpies of formation for a large number of compounds are listed in Table 12.2.

Standard entropy changes may be computed from the data in Table 12.2 by using Equation (12.41). For example, consider the data for liquid water. The reaction represented by the standard free energy and standard enthalpy of formation is the following:

$$H_2(g) + \tfrac{1}{2}O_2(g) = H_2O(l) \tag{12.42}$$

Equation (12.41) may be solved for $T\Delta S$:

$$T\Delta S_f^\circ = \Delta H_f^\circ - \Delta G_f^\circ \tag{12.43}$$

Using the values for ΔH_f° and ΔG_f° given in Table 12.2, the value for $T\Delta S_f^\circ$ may be computed as follows:

$$T\Delta S_f^\circ = (-68.23 - [-56.69])\ \text{kcal} = -11.63\ \text{kcal}$$

TABLE 12.2 Some Standard Enthalpies and Free Energies of Formation in kcal/Mole

	ΔH_f°	ΔG_f°		ΔH_f°	ΔG_f°
$AlBr_3(s)$	−125.8	−120.7	$CH_3{-}O{-}CH_3(l)$	−65.2	−27.75
$Al_2Cl_6(g)$	−303.6		$CH_3OH(g)$	−48.10	−38.70
$AlCl_3(s)$	−166.2	−152.2	$CH_3OH(l)$	−57.04	−39.75
$AlCl_3 \cdot 6H_2O(s)$	−641.1	−542.4	$CH_4(g)$	−17.89	−12.14
$AlF_3(s)$	−311	−294	$C_6H_6(g)$	+19.82	+30.99
$AlI_3(s)$	−75.2	−75.0	$C_6H_6(l)$	+11.63	+29.40
$Al(NO_3)_3 \cdot 6H_2O(s)$	−680.7	−525.8	$CO(g)$	−26.42	−32.81
$Al_2O_3(s)$	−399.1	−376.8	$CO_2(g)$	−94.05	−94.26
$Al_2S_3(s)$	−121.6	−117.7	$CS_2(g)$	+27.55	+15.55
$Al_2(SO_4)_3(s)$	−821.0	−739.0	$CS_2(l)$	+21.0	+15.2
$BBr_3(g)$	−44.6	−51.0	$CaBr_2(s)$	−161.3	−156.8
$BBr_3(l)$	−52.8	−52.4	$CaBr_2 \cdot 6H_2O(s)$	−597.2	
$BCl_3(g)$	−94.5	−90.9	$CaCO_3(s)$	−288.4	−269.8
$BCl_3(l)$	−100.0	−90.6	$CaCl_2(s)$	−190.0	−179.3
$BF_3(g)$	−265.4	−261.3	$CaCl_2 \cdot 6H_2O(s)$	−623.2	
$B_2H_6(g)$	+7.5	+19.8	$CaF_2(s)$	−290	−277.7
$B_2O_3(s)$	−302.0	−283.0	$CaI_2(s)$	−127.8	−126.6
$BaBr_2(s)$	−180.4		$Ca(NO_3)_2(s)$	−224.0	−177.3
$BaBr_2 \cdot 2H_2O(s)$	−326.3		$Ca(NO_3)_2 \cdot 4H_2O(s)$	−509.4	−406.5
$BaCO_3(s)$	−291.3	−272.2	$CaO(s)$	−151.9	−144.4
$BaCl_2(s)$	−205.6	−193.8	$Ca(OH)_2(s)$	−235.8	−214.3
$BaCl_2 \cdot 2H_2O(s)$	−349.4	−309.7	$CaS(s)$	−115.3	−114.1
$BaF_2(s)$	−286.9		$CaSO_4(s)$	−342.4	−315.6
$BaI_2(s)$	−144.0		$CaSO_4 \cdot \frac{1}{2}H_2O(s)$	−376.5	−343.0
$BaI_2 \cdot 2H_2O(s)$	−290.9		$CaSO_4 \cdot 2H_2O(s)$	−483.1	−429.2
$Ba(NO_3)_2(s)$	−237.1	−190.0	$ClO_2(g)$	+24.1	+27.5
$BaO(s)$	−133.4	−126.3	$Cl_2O(g)$	+18.20	+22.40
$BaS(s)$	−106.0		$FeBr_2(s)$	−60.02	
$BaSO_4(s)$	−350.2	−323.4	$FeCO_3(s)$	−178.7	−161.1
$CCl_4(g)$	−25.50	−15.35	$FeCl_2(s)$	−81.5	−72.2
$CCl_4(l)$	−33.34	−16.43	$FeCl_2 \cdot 4H_2O(s)$	−370.7	
$CF_4(g)$	−162.5	−151.8	$FeCl_3(s)$	−96.8	
$CH{\equiv}CH(g)$	+54.19	+50.00	$FeCl_3 \cdot 6H_2O(s)$	−532.0	
$CHCl_3(g)$	−24	−16	$Fe(NO_3)_3 \cdot 9H_2O(s)$	−784.4	
$CHCl_3(l)$	−31.5	−17.1	$FeO(s)$	−63.7	−58.4
$CH_2{=}CH_2(g)$	+12.50	+16.28	$Fe_2O_3(s)$	−196.5	−177.4
$CH_3Br(g)$	−8.2	−5.9	$Fe_3O_4(s)$	−267.0	−242.4
$CH_3Cl(g)$	−19.58	−13.96	$Fe(OH)_2(s)$	−135.8	−115.6
$CH_3{-}CH_3(g)$	−20.24	−7.86	$Fe(OH)_3(s)$	−197.0	−166.3
$CH_3{-}CO{-}CH_3(g)$	−51.79	−36.50	$FeSO_4(s)$	−220.5	−195.5
$CH_3{-}CO{-}CH_3(l)$	−59.32	−37.22	$FeSO_4 \cdot 7H_2O(s)$	−718.7	
$CH_3COOH(g)$	−104.7	−91.2	$H_3BO_3(s)$	−260.2	−230.2
$CH_3{-}NH_2(g)$	−6.7	+6.6	$HBr(g)$	−8.66	−12.72
$CH_3{-}NO_2(l)$	−26.7	−3.1	$HCl(g)$	−22.06	−22.77
$CH_3{-}O{-}CH_3(g)$		−28.00	$HClO_4(l)$	−11.1	

	ΔH_f°	ΔG_f°
HF(g)	−64.2	−64.7
HI(g)	+6.20	+0.31
HIO_3(s)	−57.03	
H_2O(g)	−57.80	−54.64
H_2O(l)	−68.32	−56.69
H_2O_2(g)	−31.83	
H_2S(g)	−4.82	−7.89
H_2Se(g)	+20.5	+17
ICl(g)	+9.75	+0.91
ICl_3(g)	+4.20	+1.32
$KAl(SO_4)_2$(s)	−589.24	−534.3
$KAl(SO_4)_2 \cdot 12H_2O$(s)	−1447.7	−1227.8
KBr(s)	−93.73	−90.63
K_2CO_3(s)	−273.9	
KCl(g)	−51.6	−56.2
KCl(s)	−104.2	−97.59
$KClO_3$(s)	−93.50	−69.29
$KClO_4$(s)	−103.6	−72.7
KF(s)	−134.5	−127.4
KHF_2(s)	−220.0	−203.7
KI(s)	−78.31	−77.03
KNH_2(s)	−28.3	
KNO_3(s)	−117.8	−93.96
K_2O(s)	−86.4	
KOH(s)	−101.8	
K_2SO_4(s)	−342.7	−314.6
$MgBr_2$(s)	−123.7	
$MgBr_2 \cdot 6H_2O$(s)	−575.4	−491.0
$MgCl_2$(s)	−153.4	−141.6
$MgCl_2 \cdot 6H_2O$(s)	−597.4	−305.6
MgF_2(s)	−263.5	−250.8
MgI_2(s)	−86.0	
$Mg(NO_3)_2$(s)	−188.7	−140.6
$Mg(NO_3)_2 \cdot 6H_2O$(s)	−624.4	
MgO(s)	−143.8	−136.1
$Mg(OH)_2$(s)	−221.0	−199.3
$MgSO_4$(s)	−305.5	−280.5
$MgSO_4 \cdot 7H_2O$(s)	−808.7	
NF_3(g)	−27.2	
NH_3(g)	−11.04	−3.98
$NH_4Al(SO_4)_2 \cdot 12H_2O$(s)	−1419	−1179
NH_4Br(s)	−64.61	
NH_4Cl(s)	−75.38	−48.73
NH_4F(s)	−111.6	
NH_4I(s)	−48.30	
NH_4NO_3(s)	−87.27	
$(NH_4)_2SO_4$(s)	−281.86	−215.19
NO(g)	+21.60	+20.72
NO_2(g)	+8.09	+12.39
N_2O(g)	+19.49	+24.76
N_2O_4(g)	+2.31	+23.49
NOCl(g)	+12.57	+15.86
NaBr(s)	−86.03	
$NaBr \cdot 2H_2O$(s)	−227.2	
Na_2CO_3(s)	−270.3	−250.4
$Na_2CO_3 \cdot 10H_2O$(s)	−975.6	
NaCl(g)	−43.50	
NaCl(s)	−98.23	−91.79
$NaClO_3$(s)	−92.18	
NaF(s)	−136.0	−129.3
$NaHF_2$(s)	−216.6	
NaI(s)	−68.84	
$NaNO_3$(s)	−101.5	−87.45
Na_2O(s)	−99.4	−90.0
Na_2O_2(s)	−120.6	−105.0
NaOH(s)	−102.0	−90.60
Na_2SO_4(s)	−330.9	−302.8
$Na_2SO_4 \cdot 10H_2O$(s)	−1033	−870.9
O_3(g)	−34.4	+32.4
PBr_3(g)	−35.9	−41.2
PBr_3(l)	−47.5	
PBr_5(s)	−66	
PCl_3(g)	−73.72	−68.42
PCl_3(l)	−81.0	−63.3
PCl_5(g)	−95.35	−77.59
PCl_5(s)	−110.7	
PH_3(g)	+2.21	+4.36
SF_6(g)	−262	−237
SO_2(g)	−70.76	−71.79
SO_3(g)	−94.45	−88.52
SO_3(l)	−103.0	−88.28
$SOCl_2$(l)	−49.2	
SO_2Cl_2(l)	−93.0	−75
SeO_2(s)	−55.0	
$SiCl_4$(g)	−145.7	−136.2
$SiCl_4$(l)	−153.0	−136.9
SiF_4(g)	−370	−360
SiH_4(g)	−14.8	−9.4
SiI_4(s)	−31.6	
SiO_2(quartz)(s)	−205.4	−192.4

Since the temperature cannot be negative, this means that the standard entropy change must be negative and the entropy change is unfavorable. The actual value for ΔS_f° may be computed as follows:

$$\Delta S_f^\circ = \frac{\Delta H_f^\circ - \Delta G_f^\circ}{T} = \frac{-11.63 \text{ kcal/mole}}{298°\text{K}} = -39 \text{ cal/(mole} \times °\text{K)}$$

However, since the enthalpy change is so highly favorable, the rather small unfavorable entropy effect decreases the free-energy change in comparison to the enthalpy change, but still leaves the free-energy change negative, so that the reaction of 2 moles of gaseous hydrogen with 1 mole of gaseous oxygen to form 2 moles of liquid water at standard conditions is a thermodynamically spontaneous change.

This example, along with a study of Table 12.2, should show that entropy changes at standard conditions tend to be relatively small. Thus, in reactions in which the standard enthalpy change is quite large, the enthalpy change tends to be the dominant factor in the free-energy change, and the direction of change in enthalpy and free energy often are the same. This is the basis for the frequently stated rule that exothermic reactions tend to be spontaneous. However, there are cases in which the enthalpy change for a reaction is relatively small, and under these circumstances the entropy change can be decisive in determining the direction of the free-energy change.

In this connection it is worth while to look at several factors that are quite important in determining the nature of the free-energy change for chemical reactions.

Bonding. Attractive forces among particles in any system are important in determining the potential energy of the particles in the system. As indicated earlier, these forces tend to dictate the magnitude of the enthalpy changes. However, the attractive forces lead to more orderly arrays of particles and allow them less random distribution. In this way they oppose the spontaneous tendency toward greater entropy.

Earlier in this book we considered a number of different kinds of bonds, including covalent bonds, hydrogen bonds, ionic bonds, metallic bonds and van der Waals bonds. Those earlier discussions also pointed out that there is a great geometrical orderliness, not only among the atoms in molecules, but also among the particles in solid crystals. In general, however, the decrease in entropy brought about by these attractive forces is not sufficient to counterbalance the decrease in enthalpy due to these same forces.

As a general rule, there is a driving force in chemical reactions that favors the replacement of weak bonds with strong bonds; the factors that affect bond strength influence the enthalpy changes and consequently the free-energy changes. A quick review of some of these factors is given below.

1. Polar covalent bonds tend to be stronger than nonpolar covalent bonds.
2. Ionic bonds allow the formation of ionic aggregates in which electrically charged ions may surround another ion of opposite charge, leading to great potential-energy stabilization.

3. Very small, highly electronegative atoms have a high capacity for multiple-bond formation, and multiple bonds are stronger than single bonds between the same two atoms.
4. Covalent bonds tend to be strongest when formed between atoms whose valence electrons are of the same potential energy.
5. Coordinate covalent bonds in complexes are favored by cations of high electronegativity and species that can serve as electron-pair donors.
6. Bonds in any complex molecule or ion are weakened by the presence of large, cumbersome ligands surrounding a small central atom or ion so that the ligands interfere with each other; that is, bonds are weakened by very small radius ratios.
7. Hydrogen bonds tend to be formed among molecules that have hydrogen bonded to fluorine, oxygen, or nitrogen.
8. Metallic bonds are strongest when formed among metals with maximum *ns*, *np*, and unpaired $(n-1)d$ electrons.

Changes in state are included in these changes involving changes in bonds. In general, the enthalpy and entropy driving forces oppose each other in changes from solid to liquid, liquid to gas, and so on. The more condensed states are favored by the enthalpy driving force since the condensed state represents the state of lowest energy. However, since the condensed state is more ordered, the less condensed states are favored by the entropy driving force.

Some entropy effects are recorded in Table 12.3. In this table, $\Delta S^\circ_{\text{fusion}}$ represents the difference between the entropy of the liquid and the solid and $\Delta S^\circ_{\text{vapor}}$ represents the difference between the entropy of the gas and the liquid, all at standard conditions. These values show that the difference in entropy is greater between the liquid and gaseous states than it is between the solid and liquid states. This suggests that the liquid state is still a very highly ordered state, and in terms of the order and disorder in the arrangement of particles, the liquid state resembles the solid state more than it resembles the gaseous state.

A careful examination of the data in Table 12.2 reveals many examples in which the state of matter plays an important role in the enthalpy and free-energy

TABLE 12.3

Some Standard Entropy Changes for Fusion and Vaporization Cal/(Mole x °K) at 298°K

SUBSTANCE	$\Delta S^\circ_{\text{fusion}}$	$\Delta S^\circ_{\text{vapor}}$
B_2O_3	+5.64	+35.87
BeO	+7.12	+36.71
$HgBr_2$	+6.09	+29.71
Li_2O	+0.88	+46.17
Na	+1.53	+22.88
NaCl	+2.89	+34.66
$PbCl_2$	+5.84	+38.29
PbO	+4.96	+36.80
Si	+6.78	+28.91
SiO_2	+1.35	+43.27
TiO_2	+3.42	+41.01

changes. For example, the standard enthalpy of formation for potassium sulfate is more negative than the standard free energy of formation. This suggests that in terms of enthalpy, the all-solid potassium sulfate system is preferred to a system containing solid potassium, solid sulfur, and gaseous oxygen. However, in terms of entropy, the system of solid potassium, solid sulfur, and gaseous oxygen is preferred over the all-solid system of potassium sulfate. In the initial state, the gaseous oxygen provides for greater randomness in the arrangement of particles, whereas the final state is entirely solid.

Or, consider carbon tetrachloride in the gaseous and liquid states as recorded in Table 12.2. Clearly the enthalpy difference between the carbon tetrachloride and solid carbon and gaseous chlorine is more favorable for the formation of liquid carbon tetrachloride than it is for the formation of gaseous carbon tetrachloride. This is in agreement with the lower potential energy for the liquid state.

However, the entropy change is unfavorable for the formation of carbon tetrachloride either as a liquid or as a gas. In formation of liquid carbon tetrachloride, the all-liquid final state is less randomly oriented than the partially gaseous initial state. The reason for the unfavorable entropy change for the formation of gaseous carbon tetrachloride may be found in the equation for this formation:

$$C(s) + 2Cl_2(g) = CCl_4(g)$$

In the initial state, the system contains 2 moles of gas, and in the final state it contains only 1 mole of gas. Clearly the particles have been more compactly bonded in the final state and the arrangement of particles is less random than in the initial state.

TEMPERATURE. Although changes in temperature normally do not affect the enthalpy differences greatly, the driving force as expressed by the entropy function, $T\Delta S$, is directly dependent on temperature. Clearly, as the temperature increases, the entropy driving force increases, and theoretically, for any system, the entropy driving force will predominate at some sufficiently high temperature. To state it somewhat differently, all chemical reactions that occur with an increase in entropy will occur spontaneously at sufficiently high temperatures.

This helps to account for the fact that if temperatures are raised sufficiently, all chemical bonds will break eventually. This means that at very high temperatures, the species are quite different from those at normal room temperature or even temperatures of several hundred degrees Celsius. Much interest is being shown currently in the study of high temperature systems, and the areas of plasma physics and plasma chemistry are revealing many new properties of matter at very high temperatures.

Chemical Equilibrium

REACTIONS AT NONSTANDARD PRESSURES. Experimentally, the partial pressures of components of gaseous systems often are not 1 atm. Thus it is desirable to have a relationship with which conversions can be made for different partial pressures of the components. Making the assumption that each of the gases behaves ideally, it is possible to show that the free energy of formation, ΔG_f, of a compound at some partial pressure P other than the standard pressure $P°$ is related to the standard free energy of formation, ΔG_f°, by the following equation:

$$\Delta G_f = \Delta G_f^\circ + 2.3RT \log \frac{P}{P^\circ} \tag{12.44}$$

Since $P°$ has been chosen as 1 atm, it is possible to express this relationship as Equation (12.45), in which the units for pressure must be atmospheres.

$$\Delta G_f = \Delta G_f^\circ + 2.3RT \log P \tag{12.45}$$

Thus the free energy of formation of a compound is the standard free energy of formation which has been corrected for nonstandard conditions. The corrective term is $2.3RT \log P$. Specifically, if the pressure of the compound being formed in the final state is greater than the standard pressure, that is, $P > P°$, then the fraction $P/P°$ exceeds unity and $\log P/P°$ is positive. Under these conditions, $\Delta G_f > \Delta G_f^\circ$, and the driving force to form the compound from its elements is less at this pressure than at standard pressure.

Similarly, if the pressure of the compound being formed in the final state is less than the standard pressure, that is, $P < P°$, then the fraction $P/P°$ is less than 1 and $\log P/P°$ is negative. Under these circumstances, $\Delta G_f < \Delta G_f^\circ$ and the driving force to form the compound from its elements is greater at this pressure than at standard pressure.

Now suppose that we apply this to a chemical system we studied earlier, which was represented by Equation (12.27), reproduced here:

$$CCl_4(g) + 2H_2O(g) = CO_2(g) + 4HCl(g)$$

In the initial state, the carbon tetrachloride has a partial pressure of P_{CCl_4} and water has a partial pressure of P_{H_2O}. In the final state, carbon dioxide and hydrogen chloride have the respective pressures P_{CO_2} and P_{HCl}.

Using the free energies of formation, the free-energy change ΔG for this system can be expressed in terms of the free energies of formation for the component compounds as follows:

$$\Delta G = \Delta G_{fCO_2} + 4\Delta G_{fHCl} - (\Delta G_{fCCl_4} + 2\Delta G_{fH_2O}) \tag{12.46}$$

However, each of the free energies of formation can be expressed in terms of the standard free energy of formation and the partial pressure of the compound, using Equation (12.45) as follows:

$$\Delta G_{fCO_2} = \Delta G^\circ_{fCO_2} + 2.3RT \log P_{CO_2} \tag{12.47}$$

$$4\Delta G_{fHCl} = 4\Delta G^\circ_{fHCl} + 4 \times 2.3RT \log P_{HCl}$$

$$= 4\Delta G^\circ_{fHCl} + 2.3RT \log P^4_{HCl} \tag{12.48}$$

$$\Delta G_{f\mathrm{CCl_4}} = \Delta G^\circ_{f\mathrm{CCl_4}} + 2.3RT \log P_{\mathrm{CCl_4}} \tag{12.49}$$

$$2\Delta G_{f\mathrm{H_2O}} = 2\Delta G^\circ_{f\mathrm{H_2O}} + 2 \times 2.3RT \log P_{\mathrm{H_2O}}$$

$$= 2\Delta G^\circ_{f\mathrm{H_2O}} + 2.3RT \log P^2_{\mathrm{H_2O}} \tag{12.50}$$

If the relationships shown by Equations (12.47) through (12.50) are substituted into Equation (12.46), Equation (12.51) results.

$$\Delta G = \Delta G^\circ_{f\mathrm{CO_2}} + 2.3RT \log P_{\mathrm{CO_2}} + 4\Delta G^\circ_{f\mathrm{HCl}} + 2.3RT \log P^4_{\mathrm{HCl}}$$

$$- (\Delta G^\circ_{f\mathrm{CCl_4}} + 2.3RT \log P_{\mathrm{CCl_4}} + 2\Delta G^\circ_{f\mathrm{H_2O}} + 2.3RT \log P^2_{\mathrm{H_2O}}) \tag{12.51}$$

Equation (12.52) results when terms are collected in Equation (12.51):

$$\Delta G = \Delta G^\circ_{f\mathrm{CO_2}} + 4\Delta G^\circ_{f\mathrm{HCl}} - (\Delta G^\circ_{f\mathrm{CCl_4}} + 2\Delta G^\circ_{f\mathrm{H_2O}})$$

$$+ 2.3RT[\log P_{\mathrm{CO_2}} + \log P^4_{\mathrm{HCl}} - (\log P_{\mathrm{CCl_4}} + \log P^2_{\mathrm{H_2O}})] \tag{12.52}$$

The first four collected terms in Equaton (12.52) correspond to ΔG° for the reaction represented by Equation (12.27) and the last group of terms can be collected, remembering that the sum of two logarithms is the same as the logarithm of the product of the two numbers, and that the difference of two logarithms is the same as the logarithm of the quotient of the two numbers. In this way, Equation (12.53) may be written:

$$\Delta G = \Delta G^\circ + 2.3RT \log \frac{P_{\mathrm{CO_2}}\, P^4_{\mathrm{HCl}}}{P_{\mathrm{CCl_4}}\, P^2_{\mathrm{H_2O}}} \tag{12.53}$$

In the trivial case in which the partial pressure of each of the four substances involved is 1 atm, the logarithm of the quotient in Equation (12.53) reduces to log 1, which is zero, and $\Delta G = \Delta G^\circ$, which is to be expected since the stated conditions are standard conditions.

THE PRESSURE EQUILIBRIUM CONSTANT. Note that in Equation (12.53), in the logarithm term, the pressures of the products appear in the numerator and the pressures of the reactants appear in the denominator. Now consider what happens as the chemical reaction occurs. The reactants are consumed and thus their pressures are reduced. This means that the logarithm term becomes larger and the value for ΔG becomes larger, that is, more positive. Thus as the chemical reaction proceeds, the driving force behind it decreases.

In a similar fashion, as the reaction proceeds, the pressures of the products increase. Such an increase in the product pressures leads also to an increase in the logarithm term in Equation (12.53) and this means that ΔG becomes more positive; the over-all driving force of the reaction decreases. An alternative interpretation is that there are two competing reactions in the chemical system described by chemical Equation (12.27). One of these is the so-called forward reaction in which carbon tetrachloride vapor reacts with water vapor to produce carbon dioxide and hydrogen chloride. The driving force behind this forward reaction decreases as the reaction proceeds and the reactants are consumed.

The other reaction involved in the system is the *reverse* reaction, in which CO_2 and HCl react to produce carbon tetrachloride and water vapors. This reverse reaction is represented by chemical Equation (12.54).

$$CO_2(g) + 4HCl(g) = CCl_4(g) + 2H_2O(g) \tag{12.54}$$

In a fashion analogous to Equation (12.53), the free-energy change for this reaction may be represented by Equation (12.55), where the subscript r represents the "reverse" reaction.

$$\Delta G_r = \Delta G_r^\circ + 2.3RT \log \frac{P_{CCl_4} P_{H_2O}^2}{P_{CO_2} P_{HCl}^4} \tag{12.55}$$

As the forward reaction proceeds, carbon tetrachloride and water are consumed, decreasing the driving force behind this reaction, and carbon dioxide and hydrogen chloride are produced. From Equation (12.55), it is clear that an increase in the pressures of CO_2 and HCl leads to a decrease in the logarithm term with a corresponding decrease in ΔG_r. Thus, as the forward reaction proceeds, the driving force behind the reverse reaction increases.

It should be clear, then, that the system tends to reach the condition in which these two driving forces are equal. When the two driving forces are equal, there is no net change in the system, since there is as great a tendency for CCl_4 and water to form CO_2 and HCl as there is for CO_2 and HCl to form CCl_4 and water. When this condition has been reached, the system is said to be at *equilibrium*.

This does not mean that at equilibrium the driving force for each of the reactions is zero, but only that the driving force leading to a net change in the system is zero. In fact, the equilibrium condition in chemical reactions is dynamic, with both the forward and reverse reactions occurring simultaneously.

If ΔG for the over-all reaction is interpreted as the driving force leading to a net change in the system, then equilibrium is reached when $\Delta G = 0$. Under this condition, the two opposing driving forces are the same, and no net change is occurring. This equilibrium condition may be described mathematically by setting ΔG equal to zero in Equation (12.53):

$$0 = \Delta G^\circ + 2.3RT \log \frac{(P_{CO_2})(P_{HCl})^4}{(P_{CCl_4})(P_{H_2O})^2}$$

$$\Delta G^\circ = -2.3RT \log \frac{(P_{CO_2})(P_{HCl})^4}{(P_{CCl_4})(P_{H_2O})^2} \tag{12.56}$$

The parentheses enclosing the pressure terms indicate that the pressures are *equilibrium pressures*.

Of the terms in Equation (12.56), ΔG°, R, and T are all constant for an isothermal process. This means that the logarithm of the quotient shown there must be constant, and therefore the quotient itself must be constant at equilibrium. This constant is called the *equilibrium constant*, or, since it involves the partial pressures of the reactants and products, it sometimes is called the *pressure equilibrium constant*. The pressure equilibrium constant is given the symbol K_p. Equation (12.57) defines the pressure equilibrium constant for the system in which carbon tetrachloride reacts with water to produce carbon dioxide and hydrogen chloride.

$$K_p = \frac{(P_{\mathrm{CO_2}})(P_{\mathrm{HCl}})^4}{(P_{\mathrm{CCl_4}})(P_{\mathrm{H_2O}})^2} \tag{12.57}$$

The pressure equilibrium constant is related to the standard free-energy change through Equation (12.56):

$$\Delta G^\circ = -2.3RT \log K_p \tag{12.58}$$

Like the free-energy change, the pressure equilibrium constant is a measure of the driving force behind a chemical reaction, since it is a thermodynamic function directly related to the standard free-energy change.

For example, a very negative value for ΔG°, which corresponds to a strong standard driving force, also indicates a very positive value for the logarithm of K_p and consequently a high value for K_p. Similarly, a very positive value for ΔG°, which corresponds to a weak driving force, indicates a very negative value for the logarithm of K_p and consequently a very small value for K_p.

It should be clear from the derivation of the equilibrium-constant expression, that the exponents associated with the pressure terms in the expression are the coefficients for the species in chemical Equation (12.27). Thus in general, the pressure equilibrium constant for a reaction in the gaseous phase may be defined as the product of the equilibrium pressures of the gaseous products, each raised to the power of the corresponding coefficient in the chemical equation, divided by the product of the equilibrium pressures of the gaseous reactants, each raised to the power of the appropriate coefficient in the chemical equation.

Thus for the general chemical reaction described by Equation (12.59), where the capital letters represent chemical species and the small letters represent numerical coefficients, the pressure equilibrium constant may be expressed as Equation (12.60):

$$a\mathrm{A} + b\mathrm{B} = y\mathrm{Y} + z\mathrm{Z} \tag{12.59}$$

$$K_p = \frac{(P_{\mathrm{Y}})^y(P_{\mathrm{Z}})^z}{(P_{\mathrm{A}})^a(P_{\mathrm{B}})^b} \tag{12.60}$$

Depending upon the values for *a*, *b*, *y*, and *z* in Equation (12.60), the pressure equilibrium constant may be unitless or may have units associated with it. In the reaction described by Equation (12.27), the pressure equilibrium constant has the units *atm*^{2} when the pressures are given in atmospheres as shown below:

$$K_p = \frac{(P_{\mathrm{CO_2}})(P_{\mathrm{HCl}})^4}{(P_{\mathrm{CCl_4}})(P_{\mathrm{H_2O}})^2} = \frac{\mathrm{atm} \times \mathrm{atm}^4}{\mathrm{atm} \times \mathrm{atm}^2} = \mathrm{atm}^2$$

However, in the case of the reaction shown in Equation (12.61), the pressure equilibrium constant is formulated by Equation (12.62).

$$H_2(g) + CO_2(g) = H_2O(g) + CO(g) \tag{12.61}$$

$$K_p = \frac{(P_{H_2O})(P_{CO})}{(P_{H_2})(P_{CO_2})} \tag{12.62}$$

That the pressure equilibrium constant is unitless can be demonstrated:

$$K_p = \frac{(P_{H_2O})(P_{CO})}{(P_{H_2})(P_{CO_2})} = \frac{\text{atm} \times \text{atm}}{\text{atm} \times \text{atm}} = 1$$

The pressure equilibrium constant is unitless whenever the sum of the coefficients of the reactants in the chemical equation equals the sum of the coefficients of the products. Another example is given by chemical Equation (12.63), and the mathematical formulation of the pressure equilibrium constant along with its units follows:

$$C_2H_4(g) + 3O_2(g) = 2CO_2(g) + 2H_2O(g) \tag{12.63}$$

$$K_p = \frac{(P_{CO_2})^2(P_{H_2O})^2}{(P_{C_2H_4})(P_{O_2})^3} = \frac{\text{atm}^2 \times \text{atm}^2}{\text{atm} \times \text{atm}^3} = 1$$

It has become customary to omit units when writing pressure equilibrium constants and it is assumed that all pressures are expressed in atmospheres when this custom is observed.

THE CONCENTRATION EQUILIBRIUM CONSTANT. When the ideal gas equation is rearranged into the form of Equation (12.64), it is evident that at isothermal conditions, the partial pressure P_X of a gaseous constituent "X" in a system is directly proportional to the number of moles of this constituent, n_X, per unit volume V of the system.

$$P_X = RT\frac{n_X}{V} \tag{12.64}$$

The proportionality constant is RT. The number of moles of constituent per liter of system is called the *molar concentration* of the constituent and is represented by M or M_X. Thus Equation (12.65) serves to define the molar concentration of constituent X.

$$M_X = \frac{n_X}{V} \tag{12.65}$$

The *equilibrium* molar concentration is represented by the symbol or formula for the constituent in brackets, as shown in Equation (12.66).

$$\text{at equilibrium } M_X = [X] \tag{12.66}$$

This permits rewriting Equation (12.64) as Equation (12.67) and *at equilibrium* as Equation (12.68).

$$P_X = RTM_X \tag{12.67}$$

$$(P_X) = RT[X] \tag{12.68}$$

Referring now to chemical Equation (12.59) and its associated pressure equilibrium constant expressed by Equation (12.60), we note that it is possible to use the relationship in Equation (12.68) to express the constant in terms of the molar concentrations of the components of the system. Thus

$$(P_Y) = RT[Y]$$

$$(P_Z) = RT[Z]$$

$$(P_A) = RT[A]$$

$$(P_B) = RT[B]$$

These relationships may be substituted into Equation (12.60) to produce Equation (12.69).

$$K_p = \frac{(RT[Y])^y(RT[Z])^z}{(RT[A])^a(RT[B])^b} \tag{12.69}$$

Equation (12.69) may be rearranged as follows:

$$K_p = \frac{(RT)^y(RT)^z}{(RT)^a(RT)^b} \times \frac{[Y]^y[Z]^z}{[A]^a[B]^b} \tag{12.70}$$

$$K_p = (RT)^{(y+z)-(a+b)}\frac{[Y]^y[Z]^z}{[A]^a[B]^b} \tag{12.71}$$

Thus the pressure equilibrium constant is directly proportional to the product of the *equilibrium* molar concentrations of the products each raised to the power of its coefficient in the chemical equation divided by the product of the *equilibrium* molar concentrations of the reactants each raised to the power of its coefficient in the chemical equation. This new quotient is called the *concentration equilibrium constant;* it is denoted by K_c, and defined by Equaton (12.72).

$$K_c = \frac{[Y]^y[Z]^z}{[A]^a[B]^b} \tag{12.72}$$

Like the pressure equilibrium constant, the concentration equilibrium constant may or may not have units associated with it. When the sum of the coefficients of the reactants in the chemical equation equals the sum of the coefficients of the products, then the concentration equilibrium constant, like the pressure equilibrium constant, is unitless. Under other conditions it has units. For example, the units for the concentration equilibrium constant associated with the reaction formulated by chemical Equation (12.27) are derived as follows:

$$K_c = \frac{[CO_2][HCl]^4}{[CCl_4][H_2O]^2} = \frac{(\text{mole}/1)(\text{mole}/1)^4}{(\text{mole}/1)(\text{mole}/1)^2} = \frac{\text{mole}^2}{\text{liter}^2}$$

Also, like the pressure equilibrium constant, the concentration equilibrium constant normally is stated without units, on the understanding that all concentrations are in moles per liter.

The concentration equilibrium constant K_c permits the extension of the equilibrium constant concept to nongaseous systems, and this constant will be used generally in the later chapters of this book.

SOME PROPERTIES OF EQUILIBRIUM CONSTANTS. It is important to realize that the equilibrium constant is the description of the equilibrium state of a chemical system at the stated temperature. Thus, regardless of the initial conditions of the system, the changes in the system tend to bring the system back to equilibrium.

For example, consider the system described by Equation (12.73).

$$H_2(g) + CO_2(g) = H_2O(g) + CO(g) \tag{12.73}$$

Suppose that the pressure equilibrium constant for this system is determined to be 2.95×10^{-4} at 298°K. From this information, it is possible to calculate the standard free energy change for this reaction at this temperature, from Equation (12.58) as follows:

$$\begin{aligned} \Delta G^0 &= -2.3RT \log K_p \\ &= -2.3 \times 1.99 \times 10^{-3} \times 298 \log(2.95 \times 10^{-4}) \\ &= -2.3 \times 1.99 \times 10^{-3} \times 298 \times (-3.53) \\ &= +4.82 \text{ kcal} \end{aligned}$$

This indicates that at standard conditions and this temperature, the reaction is not spontaneous. In addition, this calculation shows how free-energy change data may be computed if equilibrium constants can be evaluated independently.

Since the sum of the coefficients for the reactants is the same as the sum of the coefficients for the products in Equation (12.73), the pressure and concentration equilibrium constants have the same values, as follows:

$$K_p = \frac{(P_{H_2O})(P_{CO})}{(P_{H_2})(P_{CO_2})} = K_c = \frac{[H_2O][CO]}{[H_2][CO_2]} = 2.95 \times 10^{-4} \tag{12.74}$$

To illustrate some of the relationships involving equilibrium constants, we shall consider a number of different problems involving chemical Equation (12.73) and the equilibrium constant relationships in Equation (12.74).

Suppose we start with a system containing only hydrogen and carbon dioxide, both of which are at a pressure of 1.00 atm. From the pressure equilibrium constant, we may determine the condition of the system at equilibrium. From chemical Equation (12.73), we expect that the change in the system will occur with H_2 and CO_2 reacting to produce water vapor and carbon monoxide.

Since equal numbers of moles of water and CO_2 molecules are produced by this reaction, at equilibrium, their pressures must be equal, as shown in the following relationship:

$$(P_{H_2O}) = (P_{CO}) \tag{12.75}$$

Since the pressures of H_2 and CO_2 were equal initially, and since 1 mole of H_2 disappears every time 1 mole of CO_2 disappears, the equilibrium pressures of H_2 and CO_2 must be the same at equilibrium.

$$(P_{H_2}) = (P_{CO_2}) \tag{12.76}$$

In addition, from chemical Equation (12.73), it is clear that 1 mole of water molecules and 1 mole of CO_2 molecules are produced for each mole of H_2 molecules and CO_2 molecules that react. This means that the initial pressure of the H_2 and CO_2 must decrease by the same amount that the water and CO_2 pressures increase, that is,

$$(P_{H_2}) = (P_{CO_2}) = 1.00 - (P_{H_2O}) = 1.00 - (P_{CO}) \tag{12.77}$$

These relationships may be inserted into the pressure equilibrium constant expression:

$$K_p = \frac{(P_{H_2O})(P_{CO})}{(P_{H_2})(P_{CO_2})} = \frac{(P_{H_2O})(P_{H_2O})}{[1.00 - (P_{H_2O})][1.00 - (P_{H_2O})]} = 2.95 \times 10^{-4} \tag{12.78}$$

This expression may be solved for (P_{H_2O}), beginning with the following rearrangements:

$$\frac{(P_{H_2O})^2}{1.00 - 2(P_{H_2O}) + (P_{H_2O})^2} = 2.95 \times 10^{-4}$$

$$(P_{H_2O})^2 = 2.95 \times 10^{-4} - 5.90 \times 10^{-4}(P_{H_2O}) + 2.95 \times 10^{-4}(P_{H_2O})^2$$

$$(P_{H_2O})^2 + 5.90 \times 10^{-4}(P_{H_2O}) - 2.95 \times 10^{-4} = 0 \tag{12.79}$$

Equation (12.79) is a quadratic equation of the form shown in Equation (12.80).

$$ax^2 + bx + c = 0 \tag{12.80}$$

The solution to quadratic equations of this type is given by Equation (12.81).

$$x = \frac{-b \pm \sqrt{b^2 - 4ac}}{2a} \tag{12.81}$$

Equation (12.81) may be used to solve Equation (12.79) where x is (P_{H_2O}), a is 1, b is 5.90×10^{-4}, and c is -2.95×10^{-4}

$$(P_{H_2O}) = \frac{-5.90 \times 10^{-4} \pm \sqrt{(5.90 \times 10^{-4})^2 - 4(-2.95 \times 10^{-4})}}{2}$$

$$= \frac{-5.90 \times 10^{-4} \pm \sqrt{11.8 \times 10^{-4}}}{2} = \frac{-5.90 \times 10^{-4} \pm 3.44 \times 10^{-2}}{2}$$

The choice between the plus and minus sign can be made easily, since the negative sign leads to a negative answer, and thus a negative pressure, which has no meaning. Thus the final solution is as follows:

$$(P_{H_2O}) = \frac{-5.90 \times 10^{-4} + 3.44 \times 10^{-2}}{2} = 1.69 \times 10^{-2} \text{ atm}$$

Returning now to Equations (12.75) and (12.77), we note that the equilibrium state may be described by the following pressures for the components of the system:

$$(P_{H_2O}) = 1.69 \times 10^{-2} \text{ atm}$$

$$(P_{CO}) = 1.69 \times 10^{-2} \text{ atm}$$

$$(P_{H_2}) = 9.8 \times 10^{-1} \text{ atm}$$

$$(P_{CO_2}) = 9.8 \times 10^{-1} \text{ atm}$$

The results of this calculation show that very little H_2 and CO_2 react before equilibrium is attained. At the equilibrium condition, most of the system is composed of H_2 and CO_2, the starting reactants. This is consistent with the small value for the equilibrium constant and the related positive value for the standard free-energy change.

It is possible to reach an equilibrium condition also by starting with the products, water and carbon monoxide, in which case the reverse reaction is the one of importance initially. For example, suppose we start with the molarities of water and CO equal at 5.00000 *M*. Thus at the outset, the following relationship holds:

$$M_{H_2O} = M_{CO} = 5.00000 \text{ moles/liter} \tag{12.82}$$

At equilibrium, equal numbers of moles of H_2 and CO_2 will have been produced, according to chemical Equation (12.73), and, therefore, the equilibrium molarity of H_2 and CO_2 will be the same, as indicated by the following relationship:

$$[H_2] = [CO_2] \tag{12.83}$$

Similarly, since the molarities of H_2 and CO_2 are the same at the beginning, and since they decrease by equal amounts in attaining equilibrium, their equilibrium concentrations must be the same as shown by the following relationship:

$$[H_2O] = [CO] = 5.0000 - [H_2] = 5.0000 - [CO_2] \tag{12.84}$$

The concentration equilibrium constant may then be used to determine the equilibrium concentrations of all components of the system as follows:

$$K_c = \frac{[H_2O][CO]}{[H_2][CO_2]} = \frac{(5.0000 - [H_2])^2}{[H_2]^2} = 2.95 \times 10^{-4}$$

$$25.00 - 10.000[H_2] + [H_2]^2 = 2.95 \times 10^{-4}[H_2]^2$$

$$0.99970[H_2]^2 - 10.000[H_2] + 25.00 = 0$$

$$[H_2] = \frac{10.00 \pm \sqrt{(10.000)^2 - 4(0.99970)(25.00)}}{2(0.99970)}$$

$$= \frac{10.00 \pm \sqrt{100.000 - 99.970}}{2.00}$$

$$= \frac{10.00 \pm \sqrt{0.030}}{2.00} = \frac{10.00 \pm 0.17}{2.00}$$

The choice between the plus and minus in this expression may be made by considering the nature of the system. If the plus sign is chosen, the $[H_2]$ will exceed 5.00 *M*. This is impossible since the original system contained no hydrogen, and the maximum concentration of hydrogen possible if all of the water and carbon monoxide react is 5.00 *M*. Thus the minus sign is the only one that makes sense for this system, and the calculations may be completed as follows:

$$[H_2] = \frac{10.00 - 0.17}{2.00} = \frac{9.83}{2.00} = 4.92\ M$$

By referring to Equations (12.83) and (12.84) the equilibrium state may be described as follows:

$$[H_2] = 4.92\ M$$

$$[CO_2] = 4.92\ M$$

$$[H_2O] = 0.08\ M$$

$$[CO] = 0.08\ M$$

In this case the reverse reaction is very nearly complete and relatively little of the initial components of the system are left at equilibrium. This is consistent with the small equilibrium constant for the forward reaction, which is equivalent to a large equilibrium constant for the reverse reaction, and with the positive standard free energy change for the forward reaction, which is equivalent to a negative standard free energy change for the reverse reaction.

Of course, it is not necessary to start with some of the equilibrium components absent. It is possible to start with a nonequilibrium mixture of all of the eventual components and ask what the equilibrium condition is. Suppose we start with a gaseous mixture of hydrogen, carbon dioxide, water, and carbon monoxide such that the molar concentrations are as follows:

$$M_{H_2} = 1.250$$

$$M_{CO_2} = 1.100$$

$$M_{H_2O} = 0.300$$

$$M_{CO} = 0.200$$

By formulating a concentration quotient, Q_c, which resembles the concentration equilibrium constant but employs the nonequilibrium initial concentrations, it is possible to predict how the system will change as it approaches equilibrium.

$$Q_c = \frac{M_{H_2O}\, M_{CO}}{M_{H_2}\, M_{CO_2}} = \frac{(0.300)(0.200)}{(1.250)(1.100)} = 4.38 \times 10^{-2} \tag{12.85}$$

This value must be compared to the value of the equilibrium constant, K_c, which is formulated as follows:

$$K_c = \frac{[H_2O][CO]}{[H_2][CO_2]} = 2.95 \times 10^{-4} \tag{12.86}$$

Since Q_c represents the initial state and K_c represents the final state of this system, it is clear that the value of the quotient must decrease from the initial state to the final state. For this to occur, water and carbon monoxide must react to form hydrogen and carbon dioxide, thus decreasing the values in the numerator of the expression in Equation (12.85) and increasing the values in the denominator.

The concentrations of each of the components at equilibrium may be computed by letting x be the increase in molar concentration for hydrogen. In this way, the concentrations of the four components at equilibrium may be expressed in terms of x as follows:

$$[H_2] = 1.250 + x$$

$$[CO_2] = 1.100 + x$$

$$[H_2O] = 0.300 - x$$

$$[CO] = 0.200 - x$$

These expressions may be substituted into Equation (12.86) as follows:

$$K_c = \frac{[H_2O][CO]}{[H_2][CO_2]} = \frac{(0.300 - x)(0.200 - x)}{(1.250 + x)(1.100 + x)} = 2.95 \times 10^{-4}$$

$$\frac{0.0600 - 0.500x + x^2}{1.375 + 2.350x + x^2} = 2.95 \times 10^{-4}$$

$$0.0600 - 0.500x + x^2 = 4.056 \times 10^{-4} + 6.932 \times 10^{-4}x + 2.95 \times 10^{-4}x^2$$

$$0.9997x^2 - 0.5007x + 0.0596 = 0$$

$$x = \frac{0.501 \pm \sqrt{(-0.5007)^2 - 4(0.9997)(0.0596)}}{2(0.9997)}$$

$$= \frac{0.501 \pm 0.111}{2.00}$$

The choice between the plus and minus sign can be made by considering the system. The maximum possible value for x is 0.200 since the increase in hydrogen concentration is made at the expense of carbon monoxide and the initial concentration of carbon monoxide is 0.200 *M*. If the plus sign is chosen, x will exceed 0.200, and therefore the plus sign has no valid meaning for this system.

$$x = \frac{0.501 - 0.111}{2.00} = \frac{0.390}{2.00} = 0.195\ M$$

With this value of x, the equilibrium concentrations of the components of this system are determined as follows:

$$[H_2] = 1.250 + 0.195 = 1.445\ M$$

$$[CO_2] = 1.100 + 0.195 = 1.295\ M$$

$$[H_2O] = 0.300 - 0.195 = 0.105\ M$$

$$[CO] = 0.200 - 0.195 = 0.005\ M$$

Again it is apparent that the equilibrium condition exists with relatively little of the products present, consistent with the small value of the equilibrium constant.

CHANGES IN PRESSURE OR CONCENTRATION OF INDIVIDUAL COMPONENTS. Often systems at equilibrium are put into a nonequilibrium condition by some change, and it is desirable to know what happens to the system as it returns to the equilibrium condition.

Suppose that a system at equilibrium contains hydrogen, carbon dioxide, carbon monoxide, and water vapor at 25°C and the partial pressures are as follows:

$$(P_{H_2O}) = 4.40 \times 10^{-2}\ \text{atm}$$

$$(P_{CO}) = 1.00 \times 10^{-2}\ \text{atm}$$

$$(P_{CO_2}) = 6.00 \times 10^{-1}\ \text{atm}$$

From the equilibrium constant value of 2.95×10^{-4}, the partial pressure of hydrogen may be computed as follows:

$$K_p = \frac{(P_{H_2O})(P_{CO})}{(P_{H_2})(P_{CO_2})} = 2.95 \times 10^{-4}$$

$$(P_{H_2}) = \frac{(P_{H_2O})(P_{CO})}{(P_{CO_2})\,K_p} = \frac{(4.40 \times 10^{-2})(1.00 \times 10^{-2})}{(6.00 \times 10^{-1})(2.95 \times 10^{-4})} = 2.49 \text{ atm}$$

Thus the equilibrium state may be described by the four pressures: $(P_{H_2O}) = 4.40 \times 10^{-2}$ atm; $(P_{CO}) = 1.00 \times 10^{-2}$ atm; $(P_{H_2}) = 2.49$ atm; $(P_{CO_2}) = 6.00 \times 10^{-1}$ atm. Now suppose that while this system is at this equilibrium condition, more CO_2 is forced into the container so that the partial pressure of CO_2 is 5.00 atm. At this moment, the CO_2 pressure exceeds the original equilibrium pressure and the system no longer is at equilibrium. That the pressure quotient now exceeds the equilibrium constant can be shown as follows:

$$Q_p = \frac{P_{H_2O}\,P_{CO}}{P_{CO_2}\,P_{H_2}} = \frac{(4.40 \times 10^{-2})(1.00 \times 10^{-2})}{5.00(2.49)} = 3.55 \times 10^{-5}$$

Under these conditions, Q_p is less than K_p, so to reach a new equilibrium condition, the value for Q_p must increase; this may be accomplished if carbon dioxide and hydrogen react to produce water and carbon monoxide. This net reaction continues until equilibrium is reached once more. It is possible to compute the pressures of the components at the new equilibrium condition by letting x equal the increase in the water vapor pressure. Then the new equilibrium pressures may be expressed in terms of x as follows; where the prime sign on the pressures indicates the new equilibrium pressure, not to be confused with the original equilibrium pressure:

$$(P'_{H_2O}) = 4.40 \times 10^{-2} + x$$

$$(P'_{CO}) = 1.00 \times 10^{-2} + x$$

$$(P'_{CO_2}) = 5.00 - x$$

$$(P'_{H_2}) = 2.49 - x$$

These expressions may be substituted into the expression for the pressure equilibrium constant, which then may be solved for x as follows:

$$K_p = \frac{(P'_{H_2O})(P'_{CO})}{(P'_{H_2})(P'_{CO_2})} = \frac{(4.40 \times 10^{-2} + x)(1.00 \times 10^{-2} + x)}{(2.49 - x)(5.00 - x)} = 2.95 \times 10^{-4}$$

$$\frac{4.40 \times 10^{-4} + 5.40 \times 10^{-2}x + x^2}{13.0 - 7.49x + x^2} = 2.95 \times 10^{-4}$$

$$x^2 + 5.56 \times 10^{-2}x - 3.39 \times 10^{-3} = 0$$

$$x = \frac{-5.56 \times 10^{-2} \pm \sqrt{(5.56 \times 10^{-2})^2 - 4(-3.39 \times 10^{-3})}}{2.00}$$

$$= \frac{-5.56 \times 10^{-2} \pm 1.29 \times 10^{-1}}{2.00}$$

The choice between the plus and minus sign in this equation is easy since the choice of the minus sign would lead to a negative pressure.

$$x = \frac{-5.56 \times 10^{-2} + 1.29 \times 10^{-1}}{2.00} = \frac{7.3 \times 10^{-2}}{2.00} = 3.6 \times 10^{-2}$$

The new equilibrium state may be described by the following four pressures:

$$(P'_{H_2O}) = 4.42 \times 10^{-2} + 3.6 \times 10^{-2} = 8.0 \times 10^{-2} \text{ atm}$$

$$(P'_{CO}) = 1.00 \times 10^{-2} + 3.6 \times 10^{-2} = 4.6 \times 10^{-2} \text{ atm}$$

$$(P'_{CO_2}) = 5.00 - 3.6 \times 10^{-2} = 4.96 \text{ atm}$$

$$(P'_{H_2}) = 2.49 - 3.6 \times 10^{-2} = 2.45 \text{ atm}$$

In comparing these values with the initial equilibrium pressures, note that the partial pressures of water and CO_2 are greater than in the initial equilibrium. This is to be expected, since both water and CO were produced in attaining the second equilibrium. The partial pressure of hydrogen is less in the new equilibrium state than in the original. This is expected also, since no H_2 was added and some was consumed in the process of attaining the new equilibrium state. The new equilibrium partial pressure of CO_2 is higher than the original equilibrium partial pressure, but less than the total partial pressure of CO_2 immediately after it was added to the system.

Qualitatively, this over-all effect is summarized by LeChatelier's principle, which may be stated as follows: *When a stress is applied to a system at equilibrium, the system reacts to relieve the stress.* In the example we have been considering, the stress was the excess CO_2 pressure (or concentration), and this stress was relieved by consuming some of the excess CO_2 through reaction with H_2 to produce more water and CO.

For convenience, Equation (12.73) is repeated here:

$$H_2(g) + CO_2(g) = H_2O(g) + CO(g)$$

In describing this equilibrium system, we say that the addition of CO_2 to the system causes the equilibrium to shift toward the formation of the products. This means that the addition of excess reactant is a stress applied to the system, and to relieve the stress the system reacts to remove some of this excess reactant; in so doing it forms more product. The same effect could be produced by adding additional hydrogen to the system so that the H_2 partial pressure is increased.

It is to be noted however, from the last calculation, that not all of the additional CO_2 added is consumed in returning to equilibrium. It is generally true that the increase in concentration or pressure is not all deleted in the return to the new equilibrium state. This is a required condition if the equilibrium constant is to be a constant.

In a similar way, if either water or CO is added to the equilibrium system described by Equation (12.73), the stress on the system is an excess concentration of one of the products. To relieve this stress the reverse reaction is favored, with water and CO consumed to produce more reactants, namely H_2 and CO_2. In this case, the equilibrium shifts toward the formation of the reactants.

If the concentration of one of the reactants can be removed in some way, then the stress on the equilibrium system is a deficiency of the reactant, and the system reacts to produce more reactants, that is, it shifts toward the production of more reactants. Similarly, if the concentration of one of the products can be reduced, the stress on the equilibrium system is a low concentration of products, and the system reacts to increase the concentration of the products, that is, the equilibrium shifts toward the formation of the products.

SYSTEMIC CHANGES IN PRESSURE AND CONCENTRATION. Rather than change the pressure or concentration of one component of a system, it is possible in gaseous systems to change all of the pressures or concentrations of the components equally. This may be accomplished by changing the total volume of the system. Thus, if the volume of the system suddenly is doubled, the total pressure of the system is halved, as is the total concentration. At the same time, each partial pressure and each concentration is cut in half. Similarly, if the volume of the system suddenly is cut in half, the total system pressure is doubled as is the total system concentration, and each partial pressure and concentration is doubled likewise.

For example, suppose that an equilibrium system described by Equation (12.73) suddenly has its volume cut in half so that each pressure and concentration is doubled. The concentration quotient then takes on the following form:

$$Q_c = \frac{2[H_2O] \times 2[CO]}{2[H_2] \times 2[CO_2]} = \frac{[H_2O][CO]}{[H_2][CO_2]} = K_c \tag{12.87}$$

In this case, a change in system volume does not affect the equilibrium of the system.

However, consider the equilibrium system represented by Equation (12.88).

$$C_2H_4(g) + H_2(g) = C_2H_6(g) \tag{12.88}$$

The equilibrium condition is described by the concentration equilibrium constant expression shown by Equation (12.89).

$$K_c = \frac{[C_2H_6]}{[H_2][C_2H_4]} \tag{12.89}$$

Now, if the volume of the entire system is cut in half suddenly so that each concentration is doubled, the concentration quotient may be formulated as follows:

$$Q_c = \frac{2[C_2H_6]}{2[H_2] \times 2[C_2H_4]} = \frac{1}{2}\frac{[C_2H_6]}{[H_2][C_2H_4]} = \frac{K_c}{2} \tag{12.90}$$

In this case the nonequilibrium quotient is less than the equilibrium constant. This means that in returning to equilibrium, the system will consume hydrogen and ethene to produce ethane; that is, the equilibrium will shift toward the production of the product.

From the point of view of LeChatelier's principle, the stress applied to this equilibrium system is excess pressure. If possible, the system will react to reduce this pressure. The only way in which this may be done in a constant-volume container is by the reduction of the number of particles (molecules) in the system. This is possible when the chemical equation shows a difference between the number of moles of molecules of reactants and the number of moles of molecules of products. This is the case with reaction (12.88), where there are two moles of reactant molecules and one mole of product molecules represented by the equation. Thus excess pressure can be relieved if reactants are converted to products. For every 2 moles of reactant molecules consumed, only 1 mole of product molecules is formed. Thus with increased pressure the equilibrium shifts toward the formation of product in this example.

In the case of reaction (12.73), each 2 moles of gaseous reactants consumed leads to 2 moles of gaseous products, and there is no way this system can react toward an over-all pressure change; the equilibrium is thus unaffected by over-all pressure changes.

Temperature effects. For systems of ideal gases, changes in pressure and concentration do not affect the values of the equilibrium constants. However, changes in temperature do affect the values of the equilibrium constants. We shall not derive the relationship between the equilibrium constant and temperature, but qualitatively the effect of temperature may be understood from LeChatelier's principle.

Consider a hypothetical gaseous system described by Equation (12.91).

$$\mathrm{A + B = C + D + Heat} \qquad (12.91)$$

As written this is an exothermic reaction. The concentration equilibrium constant is represented as follows:

$$K_c = \frac{[\mathrm{C}][\mathrm{D}]}{[\mathrm{A}][\mathrm{B}]} \qquad (12.92)$$

If the equilibrium system is heated, the stress applied is excess heat and the system reacts to consume heat. This means that the reverse reaction is favored over the forward reaction, and some of the products are consumed and more reactants are produced. This corresponds to a numerical value decrease for the numerator in the quotient of Equation (12.92) and a numerical value increase for the denominator. This means that the value of the equilibrium constant must decrease for an exothermic reaction as the temperature increases.

Similarly, it can be shown that the value of the equilibrium constant for an endothermic reaction increases as temperature increases.

Additional examples. The reaction between hydrogen and carbon dioxide to produce water vapor and carbon monoxide, as described in the foregoing discussion is fairly simple to handle mathematically because all of the coefficients in the chemical equation are unity and there are equal numbers of moles of reactant molecules and product molecules. Although the following examples are more complex, the reasoning required is the same throughout. This section provides additional practice in the use of equilibrium constants.

At elevated temperatures nitrogen dioxide decomposes according to the reaction represented by the following equation:

$$2NO_2(g) = 2NO(g) + O_2(g) \tag{12.93}$$

Specifically at 279°C and a total system pressure of 1.00 atm, NO_2 is 13 percent decomposed at equilibrium. This is sufficient information to compute the values for the pressure and concentration equilibrium constants.

For every 100 moles of NO_2 molecules present originally, there are 87 moles of molecules left at equilibrium since 13 percent of the NO_2 decomposed. The other 13 moles of NO_2 molecules have been converted to 13 moles of nitrogen(II) oxide molecules and 6.5 moles of oxygen molecules. This means that for every 100 moles of gas present orginally, there are 106.5 moles of gas present at equilibrium. These 106.5 moles of gas must be in a volume such that they exert 1.00 atm of pressure. Thus the equilibrium pressure of each component may be computed:

$$(P_{NO}) = 13.0 \text{ moles} \times \frac{1.00 \text{ atm}}{106.5 \text{ moles}} = 0.12 \text{ atm}$$

$$(P_{O_2}) = 6.5 \text{ moles} \times \frac{1.00 \text{ atm}}{106.5 \text{ moles}} = 0.061 \text{ atm}$$

$$(P_{NO_2}) = 87.0 \text{ moles} \times \frac{1.00 \text{ atm}}{106.5 \text{ moles}} = 0.82 \text{ atm}$$

These equilibrium pressures may be used to compute the value for the pressure equilibrium constant:

$$K_p = \frac{(P_{NO})^2(P_{O_2})}{(P_{NO_2})^2} = \frac{(0.12)^2(0.061)}{(0.82)^2} = \frac{(0.014)(0.061)}{0.67} = 1.3 \times 10^{-3}$$

The expression for the concentration equilibrium constant may then be formulated:

$$K_c = \frac{[NO]^2[O_2]}{[NO_2]^2} \tag{12.94}$$

If the gases are assumed to behave ideally, then the ideal gas equation may be used to relate the moles per liter (n/V) to the pressure:

$$\frac{n}{V} = M = \frac{P}{RT} \tag{12.95}$$

In this way the equilibrium concentrations of the components may be expressed in terms of the equilibrium pressures:

$$[NO] = \frac{(P_{NO})}{RT}$$

$$[O_2] = \frac{(P_{O_2})}{RT}$$

$$[NO_2] = \frac{(P_{NO_2})}{RT}$$

These expressions may be substituted into Equation (12.94) to determine the value for the concentration equilibrium constant:

$$K_c = \frac{(P_{NO})^2(P_{O_2})}{(P_{NO_2})^2} \times \frac{(RT)^2}{(RT)^2(RT)} = \frac{K_p}{RT} = \frac{1.3 \times 10^{-3}}{0.082 \times 552} = 2.9 \times 10^{-5}$$

In this case the numerical value for the concentration equilibrium constant is smaller than the value for the pressure equilibrium constant, although they describe identical sets of systems.

Finally, consider the decomposition of carbonyl fluoride into carbon dioxide and carbon tetrafluoride, described by the following equation:

$$2COF_2(g) = CO_2(g) + CF_4(g) \tag{12.96}$$

At 1000°C, it is found that the pressure equilibrium constant is 4.00 and thus the concentration equilibrium constant also is 4.00 at this temperature.

If 5.00 moles of carbonyl fluoride is placed in a 1.500-l container at this temperature, what percent of the carbonyl fluoride has decomposed when equilibrium is reached if the volume of the system has not changed?

The initial concentration of carbonyl fluoride, M_{COF_2}, is 3.33 moles/l as shown below:

$$M_{COF_2} = \frac{5.00 \text{ moles}}{1.50 \text{ l}} = 3.33 \text{ moles/l}$$

As the system undergoes the change to the equilibrium state, the molar concentration of COF_2 decreases. If we allow x to represent this decrease in molar concentration of COF_2, then the equilibrium concentration of carbonyl fluoride may be expressed as follows:

$$[COF_2] = 3.33 - x \tag{12.97}$$

Since 1 mole of CO_2 is produced for every 2 moles of COF_2 consumed, the equilibrium concentration of CO_2 must be one-half of the concentration decrease for the COF_2.

$$[CO_2] = \frac{x}{2} \tag{12.98}$$

In addition, since CO_2 and CF_4 are produced in equal quantities, the equilibrium concentration of CF_4 must be the same as the equilibrium concentration of CO_2.

$$[CF_4] = [CO_2] = \frac{x}{2} \tag{12.99}$$

These expressions in x may be substituted into the expression for the concentration equilibrium constant, which may be solved for x as follows:

$$K_c = \frac{[CO_2][CF_4]}{[COF_2]^2} = \frac{(x/2)(x/2)}{(3.33 - x)^2} = 4.00$$

$$\frac{x^2}{4(11.1 - 6.66x + x^2)} = 4.00$$

$$15x - 107x + 178 = 0$$

$$x = \frac{107 \pm \sqrt{(-107)^2 - 4(15)(178)}}{2(15)} = \frac{107 \pm \sqrt{1.14 \times 10^4 - 1.07 \times 10^4}}{30}$$

$$= \frac{107 \pm \sqrt{7.0 \times 10^2}}{30} = \frac{107 \pm 27}{30}$$

The choice between the plus and minus sign in this expression can be made after a little reflection on the problem. If the plus sign is chosen, then x will exceed 4, which is impossible since x is the decrease in concentration of COF_2 and original concentration was only 3.33 M. Thus the value for x may be determined as follows:

$$x = \frac{107 - 27}{30} = \frac{80}{30} = 2.7\ M$$

With the value of x, the equilibrium concentrations of each of the components may be determined as follows:

$$[COF_2] = 3.3 - 2.7 = 0.6\ M$$

$$[CO_2] = 2.7/2 = 1.4\ M$$

$$[CF_4] = 2.7/2 = 1.4\ M$$

From the equilibrium concentration of COF_2, the number of moles of COF_2 in the 1.5-l container may be computed as follows:

$$1.5\ \text{l} \times \frac{0.6\ \text{mole}}{1.0\ \text{l}} = 1.0\ \text{mole}$$

Thus the percent of COF_2 remaining may be computed as follows:

$$100\ \text{moles } COF_2 \text{ originally} \times \frac{1.0\ \text{mole } COF_2 \text{ left}}{5.0\ \text{moles } COF_2 \text{ originally}} = 20\ \text{percent } COF_2 \text{ left}$$

SUMMARY

A study of thermodynamics provides some insight into the reasons why chemical reactions occur. The driving force behind chemical changes is a free-energy decrease from the initial to the final state.

The free-energy change really is the sum of the changes in the enthalpy and the product of the temperature and the change in the entropy. The enthalpy, which often is closely related to the potential energy of the atoms and molecules in the system, spontaneously decreases. The entropy, which is the measure of the disorder among the particles in the system, spontaneously increases.

From a related point of view, the driving force behind chemical reactions is the tendency to attain equilibrium. The equilibrium state is described by the equilibrium constant, which is related to the standard free-energy change for the reaction. To some extent, the degree to which a chemical reaction occurs depends upon how far the initial condition differs from the equilibrium condition. Equilibrium constants may be used to compute this difference and to describe the composition of the equilibrium state, which then may be compared to the initial state. Qualitatively, LeChatelier's principle is useful in predicting how chemical systems will change.

QUESTIONS AND PROBLEMS

12.1 Describe some conditions under which the external potential energy of a system may change.

12.2 A circular piston with a radius of 7.50 cm is forced out 12.7 cm by an expanding gas against a constant external pressure of 710 mm Hg. At the same time, the gas absorbs 120 calories of heat.
   a) Compute the number of kilocalories of work done by the gas in this expansion.
   b) Compute the change in system energy, ΔE.
   c) Did the energy of the system increase or decrease? Why?

12.3 A sample of an ideal gas was compressed by a constant pressure of 5.16 atm from 45.8 l to 1.60 l with the evolution of 141 calories of heat.
   a) How many kilocalories of work were performed on the gas?
   b) What was the change in energy of the system?
   c) Did the energy of the system increase or decrease? Why?

12.4 Calculate the values for q, ΔE, and w when 4.00 l of helium is compressed adiabatically to 0.680 l at constant pressure of 3.0 atm.

12.5 Calculate the values for q, ΔE, w, and ΔH when 5.00 l of nitrogen at 7.00 atm expands isothermally at 20°C to a pressure of 1.75 atm against a constant pressure of 1.75 atm.

12.6 In Equation (12.12) it is suggested that for constant volume changes in a gaseous system, $q = \Delta E$. Describe some types of changes that might possibly occur at constant volume.

12.7 Under what conditions does $q = \Delta H$?

12.8 Using the bond energies from Tables 6.9 and 6.10, calculate $\Delta H°$ values for each of the following reactions and compare them to the results computed using values from Table 12.2.

a) $2H_2(g) + O_2(g) = 2H_2O(g)$
b) $N_2(g) + 3H_2(g) = 2NH_3(g)$
c) $H_2(g) + O_2(g) = H_2O_2(g)$
d) $CH_4(g) + 2O_2(g) = CO_2(g) + 2H_2O(g)$
e) $4NH_3(g) + 5O_2(g) = 4NO(g) + 6H_2O(g)$

12.9 Characterize the *standard* initial and final states for the following chemical systems.

a) $4NH_3(g) + 5O_2(g) = 4NO(g) + 6H_2O(g)$
b) $2NOCl(g) + CO(g) = COCl_2(g) + 2NO(g)$
c) $CS_2(g) + 4NOCl(g) = CCl_4(g) + 2SO_2(g) + 2N_2(g)$
d) $SO_3(g) + 6HF(g) = SF_6(g) + 3H_2O(g)$

12.10 The enthalpy of combustion for a substance is the enthalpy change associated with the complete burning of the substance in oxygen. Thus to say that the enthalpy of combustion of methane is −212.8 kcal/mole is to say that $\Delta H° =$ −212.8 kcal for the following reaction:

$$CH_4(g) + 2O_2(g) = CO_2(g) + 2H_2O(l)$$

a) Show that the standard enthalpy of combustion for methane may be computed from values in Table 12.2.
b) The standard enthalpy of combustion of sucrose, $C_{12}H_{22}O_{11}(s)$, is −1349 kcal/mole. What is the standard enthalpy of formation for sucrose?
c) The standard enthalpy of combustion for glycerol, $C_3H_5(OH)_3(l)$ is −3970 kcal/mole. What is the standard enthalpy of formation for glycerol?
d) The standard enthalpy of combustion of ethanol, $CH_3CH_2OH(l)$, is −327.6 kcal/mole. What is the standard enthalpy change for the following reaction?

$$CH_2{=}CH_2(g) + H_2O(l) = CH_3CH_2OH(l)$$

12.11 Compute the standard enthalpy change for each of the following reactions.

a) $CS_2(l) + 4NOCl(g) = CCl_4(l) + 2SO_2(g) + 2N_2(g)$
b) $2H_2S(g) + 3O_2(g) = 2H_2O(l) + 2SO_2(g)$
c) $B_2H_6(g) + 2NF_3(g) = 2BF_3(g) + 2NH_3(g)$
d) $BaBr_2 \cdot 2H_2O(s) = BaBr_2(s) + 2H_2O(l)$
e) $FeO(s) + 2H_2O(l) = Fe(OH)_2(s)$
f) $NH_4NO_3(s) = N_2O(g) + 2H_2O(g)$

12.12 What is the difference between enthalpy and entropy?

12.13 Where necessary, use values from Table 12.2 to answer the following:

a) Why is $\Delta G_f°$ more positive than $\Delta H_f°$ for $BF_3(g)$? For $NaCl(s)$?
b) Why is $\Delta G_f°$ more negative than $\Delta H_f°$ for $BBr_3(g)$? For $H_2S(g)$?
c) Why is the value of $\Delta G_f°$ about the same as the value of $\Delta H_f°$ for each of the following?
   1) $BaSO_4(s) + 2C(s) = BaS(s) + 2CO_2(g)$
   2) $MgF_2(s) + CaCl_2(s) = CaF_2(s) + MgCl_2(s)$
   3) $3CaSO_4(s) + CaCl_2 \cdot 6H_2O(s) = 3CaSO_4 \cdot 2H_2O(s) + CaCl_2(s)$
   4) $2Fe(OH)_3(s) = Fe_2O_3(s) + 3H_2O(l)$

12.14 Why do $\Delta H°$ and $\Delta G°$ have very nearly the same value for the following reaction?

$$CH_4(g) + 2O_2(g) = CO_2(g) + 2H_2O(g)$$

12.15 Why is $\Delta H°$ positive while $\Delta G°$ is negative for the following reaction?

$$CH_3OH(l) + NH_3(g) = CH_3NH_2(g) + H_2O(g)$$

12.16 Compute the standard free-energy change and the standard entropy change for each of the following reactions.

a) $CaSO_4 \cdot 2H_2O(s) = CaSO_4(s) + 2H_2O(g)$
b) $CaSO_4 \cdot 2H_2O(s) = CaSO_4(s) + 2H_2O(l)$
c) $CH_4(g) + 2O_2(g) = CO_2(g) + 2H_2O(g)$
d) $CH_4(g) + Cl_2(g) = CH_3Cl(g) + HCl(g)$
e) $MgCl_2 \cdot 6H_2O(s) = Mg(OH)_2(s) + 2HCl(g) + 4H_2O(l)$
f) $NH_4Cl(s) + NaOH(s) = NH_3(g) + NaCl(s) + H_2O(l)$
g) $MgCl_2 \cdot 6H_2O(s) = MgCl_2(s) + 6H_2O(l)$
h) $2N_2O(g) + N_2O_4(g) = 6NO(g)$

12.17 Write the equilibrium-constant expression in both the pressure and concentration forms for each of the following reactions:

a) $N_2(g) + 3H_2(g) = 2NH_3(g)$
b) $4HCl(g) + O_2(g) = 2H_2O(g) + 2Cl_2(g)$
c) $2O_3(g) = 3O_2(g)$
d) $SO_3(g) + 6HF(g) = SF_6(g) + 3H_2O(g)$
e) $2NO(g) + Cl_2(g) = 2NOCl(g)$

12.18 From the data in Table 12.2, compute the values for K_p and K_c for each of the reactions in Problem 12.17. Also indicate the units in which each should be expressed.

12.19 Amounts of HCl, O_2, and Cl_2 are mixed in a constant-volume container at a fixed temperature and allowed to come to equilibrium as described by the following equation:

$$4HCl(g) + O_2(g) = 2H_2O(g) + 2Cl_2(g)$$

Complete the following table, using the information already given and then compute the equilibrium constant for this reaction.

	INITIAL MOLARITY	CHANGE IN MOLARITY	MOLARITY AT EQUILIBRIUM
HCl	0.08	________	________
O_2	0.04	________	________
H_2O	0.00	________	________
Cl_2	0.01	________	0.03

12.20 For each of the following equilibrium systems, use LeChatelier's principle to predict the effect of

a) a decrease in temperature; and
b) an increase in over-all system pressure brought about by a decrease in over-all volume.

1) $2H_2(g) + O_2(g) = 2H_2O(g) + 115$ kcal
2) $2O_3(g) = 3O_2(g) + 61.4$ kcal
3) $PCl_5(g) + 30$ kcal $= PCl_3(g) + Cl_2(g)$
4) $N_2(g) + O_2(g) + 43.2$ kcal $= 2NO(g)$
5) $3CF_4(g) + 4NH_3(g) = 4NF_3(g) + 3CH_4(g)$
6) $CS_2(g) + 4NOCl(g) = CCl_4(g) + 2SO_2(g) + 2N_2(g)$

12.21 Should the value of the equilibrium constant increase, decrease, or remain the same for each change shown in Question 12.20?

12.22 An inert gas, helium, is forced into a strong cylinder of fixed volume that is at constant temperature and contains the equilibrium system represented by the following equation:

$$2SO_2(g) + O_2(g) = 2SO_3(g)$$

What effect will this change have on the equilibrium yield of SO_3?

12.23 It is found that 3×10^{-3} mole of carbon monoxide, 2×10^{-3} mole of chlorine, and 8×10^{-3} mole of carbonyl chloride, $COCl_2$, are in equilibrium as described by the following equation:

$$COCl_2(g) = CO(g) + Cl_2(g)$$

Is it possible to compute the concentration equilibrium constant for this reaction from this information? If not, why not? If so, calculate the value for the concentration equilibrium constant.

12.24 It is found that 0.8 mole of carbon dioxide, 0.5 mole of carbon tetrafluoride, and 0.3 mole of carbonyl fluoride, COF_2, are in equilibrium as described by the following equation:

$$2COF_2(g) = CO_2(g) + CF_4(g)$$

Is it possible to compute the concentration equilibrium constant for this reaction from this information? If not, why not? If so, calculate the value for the concentration equilibrium constant.

12.25 Ammonia is produced commercially by the Haber process, employing the following equilibrium reaction:

$$N_2(g) + 3H_2(g) = 2NH_3(g) + \text{Heat}$$

a) Discuss the effect of a temperature increase on the value of the equilibrium constant and upon the equilibrium yield of ammonia.
b) Discuss the effect of a system pressure increase brought about by compressing the system on the value of the equilibrium constant and on the equilibrium yield of ammonia.

12.26 At a given temperature the equilibrium system represented by the following equation was analyzed:

$$SO_2(g) + NO_2(g) = SO_3(g) + NO(g)$$

The following equilibrium concentrations were found: $[SO_3] = 6M$; $[NO] = 4M$; $[SO_2] = 8M$; $[NO_2] = 1M$. Without changing the temperature or the volume, more nitrogen(II) oxide was added to the container. When the system regained equilibrium, it was found that $[NO_2] = 3M$. Calculate the amount of nitrogen(II) oxide per liter that must have been *added* to produce this equilibrium shift.

12.27 The reaction represented by the following equation has a concentration equilibrium constant of 1.60 at 986°C:

$$H_2(g) + CO_2(g) = H_2O(g) + CO(g)$$

In several experiments the gases were mixed in the following initial concentrations. In each case calculate the final equilibrium concentration of each component at 986°C.

a) $M_{H_2} = 0.50\ M$; $M_{CO_2} = 0.50\ M$.
b) $M_{H_2O} = 0.50\ M$; $M_{CO} = 0.50\ M$.
c) $M_{H_2} = M_{CO_2} = M_{H_2O} = M_{CO} = 0.50\ M$.

12.28 Using the same equilibrium system described in Problem 12.27, calculate the concentration of each component in the final equilibrium state when the system in the initial state contains 1.00 mole of hydrogen, 2.00 moles of carbon dioxide, 3.00 moles of water vapor, and 4.00 moles of carbon monoxide in a 10.00 liter container at 986°C.

12.29 The concentration equilibrium constant for the following reaction is 3 at 1000°C.

$$2SO_2(g) + O_2(g) = 2SO_3(g)$$

An equilibrium system at 1000°C contains 18 moles of sulfur(VI) oxide and 6 moles of sulfur(IV) oxide in a 3 liter container. How many moles of oxygen are present in this equilibrium system?

12.30 Hydrogen iodide is 20 percent dissociated at a given temperature as described by the following equation:

$$2HI(g) = H_2(g) + I_2(g)$$

Determine the number of moles of each component present in the mixture if 0.5 mole of hydrogen iodide is placed in a vessel at this temperature.

12.31 At 2000°K, the pressure equilibrium constant is 10^{-3} for the following reaction.

$$CO_2(g) = CO(g) + \tfrac{1}{2}O_2(g)$$

What is the value of the pressure equilibrium constant for the following reaction at 2000°K?

$$2CO_2(g) = 2CO(g) + O_2(g)$$

12.32 The value for the pressure equilibrium constant decreases as temperature increases for the equilibrium system described by the following equation.

$$N_2(g) + 3H_2(g) = 2NH_3(g)$$

What does this suggest about this reaction?

12.33 At elevated temperatures, water may be decomposed to hydrogen and oxygen. At 2530°K, water is 1.77 percent decomposed at equilibrium and 1.00 atm total pressure. What are the values for the pressure and concentration equilibrium constants for this reaction at 2530°K?

12.34 At 500°C, the following values are found for the pressure equilibrium constant at the indicated total-system pressure for the Haber-process reaction for the preparation of ammonia.

$$N_2(g) + 3H_2(g) = 2NH_3(g)$$

P(atm)	K_p
10	0.145
50	0.150
100	0.162
300	0.248

How might you account for this observation?

13

Solutions and Colloids

INTRODUCTION Not all chemical reactions occur in the gas phase. In fact, many important reactions occur in liquid solutions. It is the purpose of this chapter to study some of the properties of solutions in general with special emphasis on liquid solutions. This will set the stage for the remaining chapters, which will deal largely with reactions in liquid solutions.

Solutions are homogeneous mixtures of two or more substances. Homogeneity is an important property of solutions, and the meaning of this term must be clearly understood. Of course, it implies a uniformity of composition throughout. On the other hand, no system is perfectly uniform in the minutest detail. If it were possible to reach into a system composed of two substances and extract individual molecules, we would conclude that the system was not homogeneous, for molecules of different substances might be extracted. On the other hand, if we dip into a solution and extract a cupful of material, for example, we expect that the composition of all cupsful will be the same. Then it is logical to ask how small an amount of a solution we might take and still expect each sample to have the same composition. Ultimately this depends upon the sizes of the particles within the mixture.

TYPES OF SOLUTIONS AND COLLOIDS

Solutions

True solutions are limited to homogeneous mixtures of particles of relatively small molecular size. For spherical particles, the diameters may range up to approximately 10 Å.

Since a solution is homogeneous, it must consist of one continuous phase, that is, it cannot have any boundaries within it, such as that which exists between a gas and a liquid, or between a solid and a liquid, for example. On the other hand, it is possible to have substances in a solution that normally exist in different states (solid, liquid, or gas) at the temperature and pressure of the solution. However, in the solution, only one phase is exhibited.

One of the substances in the solution normally can be considered as the medium in which the other substances are dispersed. This substance is called the *solvent*. All other substances are called *solutes*. The solutes are said to be *dissolved* in the solvent. Usually, the state in which the solution exists is the state of the solvent, and it is possible to classify solutions according to the states of the solvents and solutes separately as shown in Table 13.1.

All gases mix homogeneously and any such mixture is a solution. The gaseous systems discussed in Chapter 12 were solutions. Gases may dissolve in liquids to form liquid solutions. Gases from the atmosphere dissolve in water, and when water is heated, bubbles of these gases rise prior to the boiling of the liquid. Gases dissolve in solids rather rarely, although there are some examples of gases that dissolve in solid metals.

It is not useful to define a solution of a liquid dissolved in a gas, since such solutions are mixtures of the vapor of the liquid with the gas. As such they are solutions of two gases. However, solutions of a liquid in a liquid are quite common. Examples include alcoholic beverages, antifreeze in automobile radiators, and gasoline. Examples of liquids dissolved in solids are not numerous, but some *amalgams* are in this category. Amalgams are solutions involving liquid mercury and other metals. These range from solids in which liquid mercury is dissolved in solid metal, to liquids in which a solid metal is dissolved in liquid mercury.

TABLE 13.1
Types of Solutions

SOLUTE	SOLVENT	EXAMPLES
Gas	gas	air
Gas	liquid	oxygen in water
Gas	solid	H_2 in Pd
Liquid	liquid	alcohol in water
Liquid	solid	amalgams
Solid	liquid	sugar in water
Solid	solid	brass

It is not useful to consider solutions in which solids are dissolved in a gas. The only examples would be the vapors from highly volatile solids which tend to sublime, and these solutions really are solutions of two gases. However, there are many common examples of solids dissolved in liquids. Examples include sugar or salt dissolved in water or greases dissolved in gasoline, benzene, or carbon tetrachloride. Solutions in which a solid is dissolved in another solid are known among the alloys; examples are brass, in which zinc is dissolved in copper, and coinage metal, in which nickel is dissolved in copper.

It should be noted that many solutions studied in this chapter are liquids, with water as the solvent. This is due to the great importance of water as a solvent. These are called *aqueous solutions*. It should not be concluded, however, that only aqueous solutions are important.

Colloids

Larger particles, in the range of 10 to 10,000 Å in diameter, may be dispersed within a continuous medium and still be stable. Such a system is called a *colloidal dispersion* or a *colloidal suspension*. At the lower size limit, these systems appear to be uniformly mixed like solutions and sometimes they are incorrectly called colloidal solutions. At the upper size limit, these systems appear to be less uniformly mixed and in this respect they resemble suspensions, in which very large particles are dispersed in a medium.

The continuous phase in which the particles are dispersed is called the *dispersion medium*. The particles dispersed constitute the *dispersed phase*. There is a detectable surface boundary between the dispersed phase particles and the dispersion medium. Thus the system is not homogeneous in the same sense that solutions are homogeneous.

Some examples of colloidal dispersions are given in Table 13.2. The dispersed phase may consist of either of two kinds of particles. Commonly, these are aggre-

TABLE 13.2
Some Types of Colloidal Dispersions

DISPERSED PHASE	DISPERSION MEDIUM	COMMON NAMES	EXAMPLES
Gas	liquid	foam	whipped cream, soap suds, beaten egg whites
Gas	solid	solid foam	marshmallow, pumice
Liquid	gas	liquid aerosol	fog, clouds, mists
Liquid	liquid	emulsion	mayonnaise
Liquid	solid	solid emulsion	cheese, butter, jellies, opal (H_2O in SiO_2), pearl (H_2O in $CaCO_3$)
Solid	gas	solid aerosol	smoke, dust
Solid	liquid	sol	paints, some inks, milk of magnesia
Solid	solid	solid sol	some alloys, ruby glass, some gems

gates of smaller atoms, molecules, or ions. For example, the dispersed phase in foams may be bubbles of gas containing many molecules and the dispersed phase in solid aerosols consists of large solid particles that are aggregates of many smaller particles. However, other colloids contain huge polymeric single molecules as the particles in the dispersed phase. Particularly near the lower limit of particle size, there are polymeric species, such as the silicates and proteins, which have single molecules of the appropriate sizes to form colloidal dispersions.

CONCENTRATION

It is characteristic of solutions and colloidal dispersions that their compositions may vary widely. This is not true of pure compounds, of course. For example, water always contains 2 g of hydrogen and 16 g of oxygen in each 18 g of water. However, solutions may be prepared, of water and sugar, for example, which vary greatly in the relative amounts of water and sugar they contain. A number of ways of expressing the compositions of solutions and colloidal dispersions are discussed in the following sections. Henceforth, solution terminology will be defined, but in general the same concentration notations may be used for colloidal dispersions as well.

Concentrated and Dilute

The terms *concentrated* and *dilute* are qualitative terms used to compare solutions. A concentrated solution contains a greater proportion of solute than a dilute solution. In general these terms do not represent specific concentrations.

Mass of Solute per Unit Mass of Solvent

Handbooks often record concentration as the number of grams of solute per hundred grams of solvent. Often this is used with reference to aqueous solutions. Since the density of water is approximately 1 g/ml, it is possible from such information to make a quick *estimate* of how much solute must be dissolved in 100 ml of water to make a specified solution.

Percent

Unless specified to the contrary, percent concentration is expressed in mass or weight units. Thus the percent concentration of a component of a solution is the number of mass units of this component in 100 mass units of *solution*. In more commonly used terms, the percent concentration is the number of grams of component per 100 g of *solution*. Thus a 25-percent aqueous solution of sugar contains 25 g of sugar and 75 g of water in each 100 g of solution. It is customary, but not mandatory, to have the percentage expressed for the solute. However, the percent solvent may also be given. Consequently the 25 percent aqueous sugar solution also is 75 percent water.

Sometimes percent by volume is specified. The volume percent is the number of volume units of constituent per 100 volume units of *solution*. In more commonly used terms, the volume percent is the number of milliliters of solution component per 100 ml of *solution*. This concentration unit can be difficult to use. For example, it is not necessarily true that 100 ml of solution will result from the mixing of 50 ml of ethanol with 50 ml of water. For this reason, it is better to use mass-weight percentages since it is true that 100 g of solution results when 50 g of ethanol is mixed with 50 g of water.

The following two illustrative problems show how percent concentration may be used in computations:

The density of a 25.0 percent aqueous solution of ammonia is 0.911 g/ml. What volume of this solution must be taken to provide 15.0 g of NH_3*?*

The number that determines the amount of solution required is the amount of ammonia desired. Beginning with this quantity, 15.0 g, the number of grams of solution required may be computed, using the percentage composition information. In turn this may be used with the density to determine the required solution volume.

$$15.0 \text{ g NH}_3 \times \frac{100 \text{ g soln}}{25 \text{ g NH}_3} \times \frac{1.00 \text{ ml soln}}{0.911 \text{ g soln}} = 65.9 \text{ ml soln}$$

If 58.7 g of acetic acid is dissolved in 100.0 g of water, what is the percent of acetic acid in the solution?

By definition, the percent of acetic acid is the number of grams of acetic acid in 100 g of solution. Therefore there must be 58.7 g of acetic acid in each (100.0 + 58.7) g or 159 g of solution. Thus,

$$100 \text{ g soln} \times \frac{58.7 \text{ g CH}_3\text{COOH}}{159 \text{ g soln}} = 36.9 \text{ g CH}_3\text{COOH}$$

$$= 36.9 \text{ percent CH}_3\text{COOH}$$

Molarity

The *molarity* is the number of moles of a substance per liter of *solution*. This concentration expression was introduced with gaseous solutions in Chapter 12 and is one of the most widely used concentration units in chemistry. Usually it is designated by the symbol M. Thus a 4.0 M sugar solution contains 4.0 moles of sugar per liter of solution.

Since both terms in the molarity definition can be divided by 1000 without changing the concentration, molarity may also be defined as the number of millimoles of a substance per milliliter of solution. Thus a 4.0 M sugar solution also contains 4.0 mmoles of sugar per milliliter of solution.

Since the molarity is based on the *volume* of solution, molarity is temperature sensitive. As a result, the molarity of a single solution often changes as the temperature changes, because the volume of solution changes while the number of moles of solution component does not. For fairly dilute aqueous solutions, this is not very significant, particularly over short ranges of temperature change.

The following six problems illustrate some ways in which the molarity may be used in computations.

If 75 g of $CaCl_2$ is dissolved in enough solvent to make 350 ml of solution, what is the molarity of the solution?

Since molarity is the number of moles per liter of solution, the problem is to find the number of moles of calcium chloride in one liter of solution.

$$1.0 \text{ l soln} \times \frac{75 \text{ g } CaCl_2}{0.35 \text{ l soln}} \times \frac{1 \text{ mole } CaCl_2}{111 \text{ g } CaCl_2} = 1.9 \text{ moles } CaCl_2$$

Therefore this solution is 1.9 *M* in calcium chloride.

How many grams of ammonia are there in 760 ml of a 0.70 M aqueous ammonia solution?

If the volume of a solution is multiplied by the molarity, the number of moles (or mmoles) of solute is computed. The molecular weight of the solute may be used to compute the number of grams.

$$760 \text{ ml soln} \times \frac{0.70 \text{ mole } NH_3}{1000 \text{ ml soln}} \times \frac{17 \text{ g } NH_3}{1 \text{ mole } NH_3} = 9.0 \text{ g } NH_3$$

If 40.0 ml of 0.60 M H_2SO_4 is added to 75.0 ml of 1.20 M H_2SO_4, what is the resulting molarity of sulfuric acid?

Since the molarity is the number of moles per liter or the number of mmoles per ml, the number of mmoles in each of the original solutions may be found by multiplying the number of ml by the molarity. The sum of these two must be the number of mmoles of sulfuric acid in the final solution. Finally, some assumption must be made about the final volume. It is assumed that the final volume is the sum of the two separate volumes of the original solutions. Usually this is not a bad assumption if the solutions are somewhat dilute and have the same solvent.

$$\text{Total mmoles } H_2SO_4 = 40.0 \text{ ml} \times \frac{0.60 \text{ mmole}}{1.0 \text{ ml}} + 75.0 \text{ ml} \times \frac{1.20 \text{ mmole}}{1.0 \text{ ml}}$$

$$= 24 \text{ mmoles} + 90 \text{ mmoles} = 114 \text{ mmoles}$$

$$\text{Total Volume} = 75.0 \text{ ml} + 40.0 \text{ ml} = 115 \text{ ml}$$

$$\text{Molarity} = \frac{114 \text{ mmoles } H_2SO_4}{115 \text{ ml soln}} = 0.99\ M$$

How can 50 ml of 2.3 M aqueous hydrochloric acid be prepared from a 5.5 M aqueous hydrochloric acid solution?

This is a problem of preparing a dilute solution from a more concentrated one. The problem is to determine the number of milliliters of the concentrated solution that must be used to provide the requisite number of millimoles of hydrochloric acid. This then may be diluted with the necessary amount of solvent to make the dilute solution. In this case, the product of the volume and molarity of the dilute solution indicates the requisite number of millimoles of hydrochloric acid in that solution. This may be divided by the molarity of the more concentrated solution to compute the required number.

$$50 \text{ ml dil soln} \times \frac{2.3 \text{ mmole HCl}}{1 \text{ ml dil soln}} \times \frac{1 \text{ ml conc soln}}{5.5 \text{ mmole HCl}} = 20 \text{ ml conc soln}$$

Thus 20 ml of the concentrated solution must be used and diluted to 50 ml with the appropriate amount of water.

Commercially, nitric acid is available as a 69.5 percent aqueous solution with a density of 1.42 g/ml. What is the molarity of this solution?

Since the molarity of a solution is the number of moles per liter, the question may be asked also in the following way: How many moles of nitric acid are there in one liter (1000 ml) of solution? Beginning with this 1000 ml of solution, the density may be used to determine the number of grams of solution in 1000 ml. Then the percent of nitric acid may be used to determine the number of grams of nitric acid in the liter of solution. Finally the molecular weight of nitric acid may be used to find the number of moles of nitric acid present.

$$1000 \text{ ml soln} \times \frac{1.42 \text{ g soln}}{1.0 \text{ ml soln}} \times \frac{69.5 \text{ g } HNO_3}{100 \text{ g soln}} \times \frac{1 \text{ mole } HNO_3}{63 \text{ g } HNO_3} = 15.7 \text{ mole } HNO_3$$

Thus the solution is 15.7 *M*.

How many grams of barium chloride are needed to react with 40 ml of 0.72 M H_2SO_4 according to the following reaction?

$$BaCl_2 + H_2SO_4 = 2HCl + BaSO_4$$

The number of moles of sulfuric acid to be treated is determined from the volume and molarity of the sulfuric acid solution to be used. From the equation, the number of grams of barium chloride needed to react with the prescribed number of moles of sulfuric acid can be determined and used to find the final number of grams of barium chloride.

$$40 \text{ ml soln} \times \frac{0.72 \text{ moles } H_2SO_4}{1000 \text{ ml soln}} \times \frac{208 \text{ g } BaCl_2}{1 \text{ mole } H_2SO_4} = 6.0 \text{ g } BaCl_2$$

Molality

The molality is the number of moles of solution component per kilogram of *solvent.* In other words, molality is the number of millimoles of solution component per gram of solvent. Carefully note the difference between molality and molarity. The molarity is based on the *volume* of *solution*. The molality is based on the *mass* of *solvent*. Since the molality is based entirely on masses, it is not temperature sensitive, provided the solute in question is not a gas that is evolved upon heating. Thus the molality is used commonly when measurements must be made over some temperature range.

Symbolically the molality is designated by m. Thus a 4.0 m aqueous solution of sugar contains 4.0 moles of sugar per kilogram of water.

The following problem illustrates a typical computation involving the molality: *What is the molality of a 25 percent solution of potassium nitrate?*

Since the molality is the number of moles of potassium nitrate in 1000 g of solvent, it is possible to find the grams of potassium nitrate by using the percentage composition. Finally the molecular weight of potassium nitrate may be used to determine the number of moles of potassium nitrate in the 1000 grams of solvent, which is the molality.

$$1000 \text{ g solv} \times \frac{25 \text{ g } KNO_3}{75 \text{ g solv}} \times \frac{1 \text{ mole } KNO_3}{101 \text{ g } KNO_3} = 3.3 \text{ mole } KNO_3$$

Thus the solution is 3.3 m in KNO_3.

Mole Fraction and Mole Percent

The mole fraction is the ratio of the number of moles of a solution constituent to the total number of moles of solution, where the number of moles of solution is the sum of the number of moles of the solvent and all solutes. The mole percent is 100 times the mole fraction, or is the number of moles of constituent per 100 moles of solution.

The following two problems illustrate computations involving mole fraction and mole percent:

If 27 g of water, 40 g of hydrogen fluoride, and 40 g of hydrogen bromide are mixed, what is the mole fraction of each constituent?

The number of moles of each constituent of the solution may be determined as follows:

$$27 \text{ g } H_2O \times \frac{1 \text{ mole } H_2O}{18 \text{ g } H_2O} = 1.5 \text{ moles } H_2O$$

$$40 \text{ g HF} \times \frac{1 \text{ mole HF}}{20 \text{ g HF}} = 2.0 \text{ moles HF}$$

$$40 \text{ g HBr} \times \frac{1 \text{ mole HBr}}{80 \text{ g HBr}} = 0.5 \text{ moles HBr}$$

The total moles in the solution must be the sum of these three quantities:

$$\text{Total moles of solution} = (1.5 + 2.0 + 0.5) \text{ moles} = 4.0 \text{ moles}$$

The individual mole fractions may be computed as follows:

$$\text{mole fraction } H_2O = \frac{1.5 \text{ moles } H_2O}{4.0 \text{ moles soln}} = 0.38$$

$$\text{mole fraction HF} = \frac{2.0 \text{ moles HF}}{4.0 \text{ moles soln}} = 0.50$$

$$\text{mole fraction HBr} = \frac{0.50 \text{ moles HBr}}{4.0 \text{ moles soln}} = 0.12$$

It should be obvious that the sum of the mole fractions of the components of a solution must be unity. In addition the mole percent of each of the components is the mole fraction multiplied by 100. In this case the solution is 38 mole percent water, 50 mole percent hydrogen fluoride, and 12 mole percent hydrogen bromide.

What is the mole fraction composition of a 0.50 m solution of iodine in carbon tetrachloride?

The fact that the solution is 0.50 *m* in iodine indicates that there is 0.50 mole of iodine in 1000 g of carbon tetrachloride. The number of moles of carbon tetrachloride in the 1000 g may be computed as follows:

$$1000 \text{ g } CCl_4 \times \frac{1 \text{ mole } CCl_4}{154 \text{ g } CCl_4} = 6.49 \text{ moles } CCl_4$$

Therefore, the total number of moles of solution is (0.50 + 6.49) moles or 6.99 moles. The mole fractions may be computed as follows:

$$\text{mole fraction } I_2 = \frac{0.50 \text{ mole } I_2}{6.99 \text{ mole soln}} = 0.072$$

$$\text{mole fraction } CCl_4 = \frac{6.49 \text{ moles } CCl_4}{6.99 \text{ moles soln}} = 0.928$$

LIQUID SOLUTIONS

Among the most frequently encountered solutions are those which contain a liquid solvent. Of these, the most common are those in which water is the solvent.

Although the solution, which is the end product of the dissolving process, is homogeneous, the system during the dissolving process is heterogeneous. By this we mean that the solvent and solute are in two different phases, and the dissolving process amounts to the incorporation of the two phases into a single phase.

For example, when a solid is placed in a liquid, there is a boundary, or *interface*, between them. If the solid is soluble in the liquid, this interface may disappear as a single phase is formed, that is, as the solid dissolves in the liquid. However, during the dissolving process, the boundary exists and the process is said to be heterogeneous.

If the interface between the two phases disappears, we say that the solute is soluble in the solvent. If the interface does not disappear, then we say that the solute is insoluble in the solvent. For example, we say that in general oxygen, carbon tetrachloride, and sand are insoluble in water, or at least they are not very soluble. However, ammonia, ethanol, and sugar are soluble in water.

The question of why solutes dissolve in a liquid solvent is answered best by thermodynamics. Like other changes, the spontaneous dissolving process is accompanied by a decrease in free energy. As seen before, this free-energy change is the sum of the enthalpy change and the product of the temperature and the entropy change.

The major contributions to enthalpy changes are changes in bonding, or more generally, the changes in potential energy among the system particles. A decrease in enthalpy occurs in the dissolving process when there are stronger attractions between the solvent and solute particles than there are between the solvent particles and the solute particles separately. Thus the increase in interparticle attractions helps to account for an enthalpy decrease. In this way solute particles become intermixed with solvent particles in the liquid phase. The forces between the solvent and the solute are called *solvation forces,* and the solute that is in the solvent and attracted to the solvent particles is said to be *solvated.* When the solvent is water, the attractive forces are called *hydration forces* and the solute is said to be *hydrated.* From the enthalpy change point of view, soluble solutes are those which have strong attraction for the solvent, and insoluble solutes are those which do not have strong attractions for the solvent.

For the most part, the forces of attraction in solutions are electrostatic. They may be due to van der Waals forces, intermolecular attractions between polar molecules, interionic attractions between oppositely charged ions, or attractions between ions and polar molecules. These are the same kinds of attractions that exist between individual solvent particles or between individual solute particles. In this respect the dielectric constant of the solvent plays an important role in reducing the forces of attraction between individual solute particles. Recall that the dielectric constant is a measure of the insulating ability of a medium. Thus a liquid solvent with a high dielectric constant reduces the electrostatic forces of attraction between the solute particles in the solution.

Some typical dielectric constants are recorded in Table 13.3. Note that water has a high dielectric constant and thus is able to reduce the electrostatic attractions between solute molecules.

Several factors affect the entropy change accompanying the solution process. One is the change of state for the solute. When a solid solute dissolves in a liquid, the solute changes from the solid state to the liquid state. This constitutes a favorable entropy change. However, when a gas dissolves in a liquid, the solute changes from the gaseous state to the liquid state. This is accompanied by an unfavorable entropy change. Inasmuch as the dissolving process involves the mixing of the solute and solvent, this one factor suggests a favorable entropy change. Thus when one liquid dissolves in another liquid, there is no change of state, but there is an increase in randomness. Finally, if there is a very strong interaction between the solute and the solvent, solvent molecules may be attracted in clusters to the solute

TABLE 13.3
Dielectric Constants of Some Solvents at Room Temperature

Hydrogen cyanide	HCN	115
Formamide	$HCONH_2$	110
Water	H_2O	78.5
Hydrazine	$NH_2—NH_2$	58.5
Formic acid	$HCOOH$	57.9
Cyanomethane	$CH_3—CN$	37.5
Nitrobenzene	$C_6H_5—NO_2$	35
Methanol	CH_3OH	33.6
Ethanol	CH_3CH_2OH	25.0
Ammonia	NH_3	22.4(−33°C)
Phosphoryl chloride	$POCl_3$	22
Acetone	$CH_3—CO—CH_3$	21.5
Dichloromethane	CH_2Cl_2	9.1
Acetic acid	CH_3COOH	6.2
Ethyl acetate	$CH_3COOCH_2CH_3$	6.0
Chloroform	$CHCl_3$	4.8
Phosphorus(III) chloride	PCl_3	4.7
Diethyl ether	$CH_3CH_2—O—CH_2CH_3$	4.3
Carbon disulfide	CS_2	2.6
Silicon(IV) chloride	$SiCl_4$	2.4
Benzene	C_6H_6	2.3
Carbon tetrachloride	CCl_4	2.2
Hexane	C_6H_{14}	1.9

molecules. This clustering is accompanied by a decrease in the freedom to mix, and the entropy decreases.

Clearly, many factors affect the dissolving process. To explain these more fully, several kinds of solutions are examined in the following sections.

Gas Dissolved in a Liquid

When both solvent and solute are polar, the two polar species interact electrostatically through their separated charge centers. For example, when gaseous nitrogen(II) oxide comes in contact with water, the polar water molecules interact with the polar nitrogen(II) oxide molecules as represented in Fig. 13.1. This is a planar representation of a three-dimensional situation in which the water mole-

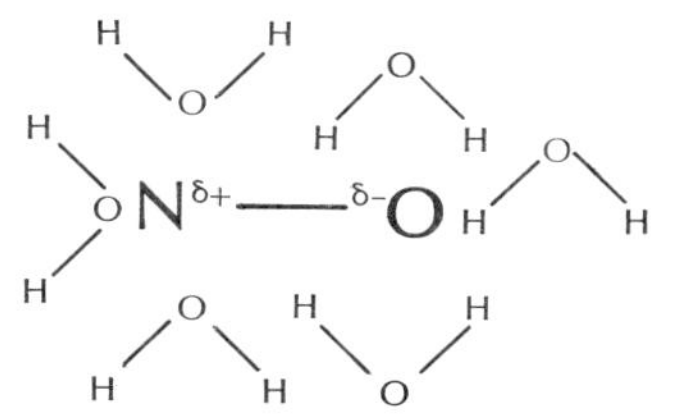

Fig. 13.1. Representation of the hydration of nitrogen(II) oxide in solution

cules are clustered around the nitrogen(II) oxide molecules, with the negative centers of the water molecules oriented to the positive ends of the nitrogen(II) oxide molecules, and the positive ends of the water molecules oriented to the negative ends of the nitrogen(II) oxide molecules.

Figure 13.1 is a representation of hydrated nitrogen(II) oxide. However, the actual degree of solvation usually cannot be specified exactly. Certainly it depends upon how polar the respective molecules are. In addition, there are relative size and geometry factors that determine how many solvent molecules can cluster around a solute molecule.

The hydration of nitrogen(II) oxide may be represented by either Equation (13.1) or (13.2).

$$NO(g) + xH_2O(l) = NO(H_2O)_x \tag{13.1}$$

$$NO(g) = NO(aq) \tag{13.2}$$

Equation (13.2) is an abbreviated form of Equation (13.1) in which the solvent is omitted from the equation. In Equation (13.2), the NO(aq) term is read as aqueous nitrogen(II) oxide, nitrogen(II) oxide aqueous, or hydrated nitrogen(II) oxide.

A qualitative picture of the dissolving process is depicted by Fig. 13.2. Initially, the system is composed of pure liquid water with an atmosphere of gaseous nitrogen(II) oxide over it. As the nitrogen(II) oxide molecules move about randomly

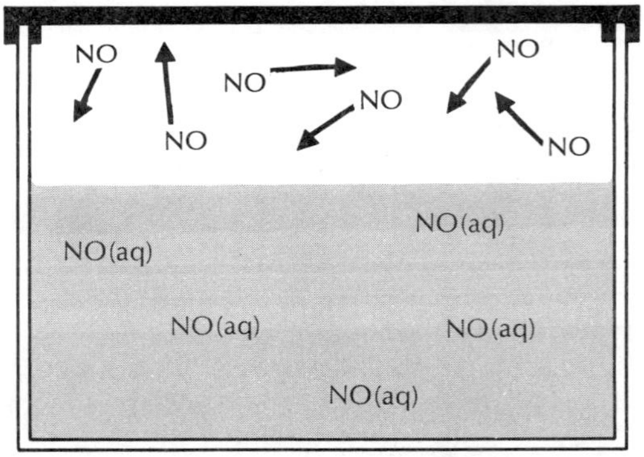

Fig. 13.2. Representation of the attainment of solubility equilibrium for gaseous nitrogen(II) oxide in water

in the gaseous phase, some of them strike the surface of the water, are attracted electrostatically by the polar water molecules, and are incorporated into the liquid phase.

Within the solution, nitrogen(II) oxide molecules are completely surrounded by water molecules, and thus are strongly attracted to the solvent, that is, they are at a state of low potential energy. However, these hydrated nitrogen(II) oxide molecules move about in the solution dragging hydrating water molecules with them. Occasionally these molecules strike the surface of the liquid and break away from the hydrating water molecules. As a result, some of the hydrated nitrogen(II) oxide molecules escape from the solvent and reenter the gas phase.

Initially, all of the change is from the gaseous phase to the hydrated state. However, as molecules of nitrogen(II) oxide enter the solution, the concentration of nitrogen(II) oxide above the liquid decreases. This means that fewer molecules of nitrogen(II) oxide strike the surface of the liquid per unit time, and, as the process proceeds, the rate of dissolving decreases.

However, as the concentration of hydrated nitrogen(II) oxide increases, there is an increasing probability that hydrated nitrogen(II) oxide molecules will strike the surface and enter the gaseous phase. Thus as the dissolving process proceeds, the rate of escape of nitrogen(II) oxide molecules from the solution increases.

Eventually, the rate of dissolving and the rate of escape are equal, and the system then is at dynamic equilibrium, in which there is no apparent change in the over-all concentration of the nitrogen(II) oxide in solution. Such a solution is said to be *saturated.* A saturated solution is one in which the dissolved solute is capable of being in equilibrium with undissolved solute. The concentration of solute in the saturated solution is called the *solubility* of the solute.

Note that the solubility equilibrium is heterogeneous, with the dissolved solute in one phase and the undissolved solute in another. Such heterogeneous equilibria may be described by concentration equilibrium constants in which the concentrations are expressed for the respective phases in which a particular component exists. For example, the concentration equilibrium constant K_c for the nitrogen(II) oxide solubility equilibrium in water may be written as follows from Equation (13.1).

$$K_c = \frac{[NO(H_2O)_x]}{[NO(g)][H_2O(l)]^x} \tag{13.3}$$

In this expression, $[NO(H_2O)_x]$ is the molarity of nitrogen(II) oxide in the liquid phase; $[NO(g)]$ is the molarity of nitrogen(II) oxide in the gaseous phase; and $[H_2O(l)]$ is the molarity of the solvent water in the liquid phase. This latter molarity is for the water that is not clustered around the solute molecules as water of solvation.

The great experimental difficulty in determining the value of x in Equations (13.2) and (13.3) makes the concentration equilibrium constant difficult to determine and use. However, in some special conditions it is possible to make some assumptions that permit the simplification of the equilibrium constant expression so that it may be used to make approximate calculations, which often are quite adequate.

The special requirement is that the saturated solution must be very dilute. In other words, the solute must not be very soluble. Under these conditions, the addition of the solute does not affect the volume of the liquid phase appreciably nor does it tie up an appreciable fraction of water molecules in the hydration of the nitrogen(II) oxide.

For example, at 0°C and a pressure of 1 atm, the solubility of nitrogen(II) oxide in water is about 3.3×10^{-4} moles in 5.5 moles of water. Since the number of molecules of water associated with each nitrogen(II) oxide molecule is unknown, we might make a guess at a reasonable number. Suppose that on the average, 10 molecules of water are used to hydrate each nitrogen(II) oxide molecule. This would tie up 3.3×10^{-3} moles of water in the hydration compared to the original total of 5.5 moles of water. This is insignificant to two significant figures and, in fact, considerably more water could be used in the hydration before it affected the over-all amount of solvent water. Thus the number of moles of water is virtually unaffected by the hydration of nitrogen(II) oxide in the saturated solution.

Since the volume does not change appreciably in the dissolving process, the value of $[H_2O(l)]$ in Equation (13.3) is a constant, and Equation (13.3) may be rewritten as Equation (13.4).

$$K_c[H_2O(l)]^x = \frac{[NO(H_2O)_x]}{[NO(g)]} = K_s \tag{13.4}$$

Since $[H_2O(l)]^x$ is essentially constant under these conditions, the product of K_c and $[H_2O(l)]^x$ is essentially constant also. This constant is called the solubility constant, K_s.

Note that whereas K_c is the equilibrium constant for the dissolving process depicted by Equation (13.1), K_s is a modified equilibrium constant that is valid only under limited conditions and appears to be related to Equation (13.2), which is the modified form of Equation (13.1). Thus the expression for the solubility constant may be written as follows:

$$K_s = \frac{[NO(aq)]}{[NO(g)]} \tag{13.5}$$

It should be clear that the solubility of gases in general will increase as the pressure of the gas above the solvent increases. From the point of view of LeChatelier's principle, if a system is in equilibrium and the pressure of nitrogen(II) oxide is increased, then to relieve this pressure increase, more nitrogen(II) oxide dissolves.

From the point of view of the rates of dissolving and escape, an increase in pressure of the nitrogen(II) oxide is an increase in concentration also. This increases the rate at which nitrogen(II) oxide dissolves over the existing equilibrium rate at which nitrogen(II) oxide escapes from solution. As a result, the system is not at equilibrium; more nitrogen(II) oxide dissolves to decrease the concentration of gaseous nitrogen(II) oxide and consequently the rate of dissolving decreases also. As more nitrogen(II) oxide dissolves, the concentration of nitrogen(II) oxide in solution increases and the rate at which nitrogen(II) oxide escapes from solution increases, until the rates of solution and escape are equal and a new equilibrium is established.

From the point of view of the solubility constant K_s, as expressed in Equation (13.5), an increase in concentration of gaseous nitrogen(II) oxide makes the ratio $M_{NO(aq)}/M_{NO(g)}$ less than the solubility constant. To regain equilibrium, the concentration of NO(g) decreases and the concentration of NO(aq) increases until the concentration ratio equals the value of the solubility constant.

Although the entropy change accompanying the dissolving of a gas in a liquid tends to be unfavorable, the trends in solubility often may be understood in terms of the enthalpy changes. In turn, these enthalpy changes are related to the bond breaking and bond making that occurs in the dissolving process. In general, in order to fit the solute molecules into the liquid solvent structure, some of the bonds between solvent molecules must be broken. In water, these include the hydrogen bonds. However, bonds may be formed between the solvent and solute molecules. Trends in solubility often may be understood in these terms.

For example, the saturated solution of nitrogen(II) oxide in water at 1 atm of pressure and 0°C is about 0.006 mole percent nitrogen(II) oxide. This low solubility probably is due to the low polarity of the nitrogen(II) oxide and the failure of nitrogen(II) oxide to form strong bonds with water.

Even nonpolar gases dissolve to some extent. At 0°C and 1 atm of pressure, the solubility of nitrogen in water is about 0.002 mole percent, and the solubility of oxygen is about 0.004 mole percent. The interactions between these gases and water can be attributed to van der Waals forces.

On the other hand, the solubility of sulfur(IV) oxide in water at 0°C and 1 atm of pressure is about 7.0 mole percent. This higher solubility probably can be attributed to the higher polarity of sulfur(IV) oxide, which accounts for a stronger interaction between the sulfur(IV) oxide and water.

The solubility of ammonia in water at 0°C and one atmosphere of pressure is almost 50 mole percent. Not only is ammonia highly polar, but it is also capable of forming hydrogen bonds with the water.

Liquid Dissolved in a Liquid

For liquids dissolved in liquids, the rough rule of thumb is that like dissolves like. Polar solutes dissolve in polar solvents and nonpolar solutes dissolve in nonpolar solvents. Polar solvent molecules have strong attractions for each other, and a solute must have as strong or stronger attractions for the solvent. This is possible for polar solutes that have strong attractions for solvent molecules. Nonpolar solutes are "squeezed out" of the semistructured state of the aggregated solvent molecules.

In many cases it is possible for polar solute molecules to replace solvent molecules in the aggregated liquid state of the solvent. In this case the enthalpy change is not very great but there is an appreciable increase in entropy.

Since nonpolar molecules have relatively small attractions for each other, nonpolar liquids tend to mix well. The dissolving of one nonpolar liquid in another nonpolar liquid occurs primarily because of entropy effects.

Methanol, CH_3OH, and water, HOH, are examples of two polar, highly hydrogen-bonded substances. When mixed they may form hydrogen bonds with each other rather than with themselves. These two molecules are sufficiently similar that methanol molecules may replace water molecules in the semistructured liquid water without much enthalpy change. In fact, these two substances are infinitely soluble in each other. Such liquids, which can be mixed homogeneously in all proportions, are said to be *miscible*. With miscible liquids it is impossible to establish a solubility equilibrium and it is impossible to have a saturated solution.

Alcohol solubility in water varies with the alkyl group attached to the hydroxy group. Ethanol, CH_3CH_2OH, is miscible with water, and both 1-propanol, $CH_3CH_2CH_2OH$, and 2-propanol, $CH_3CH(OH)CH_3$, are miscible with water. However, 1-butanol, $CH_3CH_2CH_2CH_2OH$, is not miscible with water. The long alkyl group serves as a nonpolar tail that decreases solubility by requiring the breaking of solvent bonds to make room for it in the solvent structure, without providing any bonding possibilities with the solvent. Certainly 1-butanol is polar, but the polarity is at the hydroxy end of the molecule, and at this end, hydrogen bonding occurs with water.

If several smaller alkyl groups are clustered around the carbon to which the hydroxy group is bonded, then the alcohol is miscible with water. Thus 2-methyl-2-propanol, an isomer of 1-butanol,

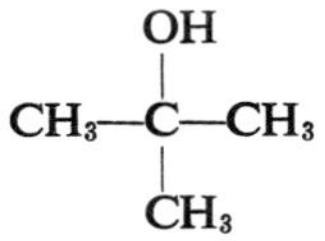

is miscible with water. If a second hydroxy group is attached to the end of the 1-butanol chain, then 1,4-dihydroxybutane, HO—CH_2—CH_2—CH_2—CH_2—OH, is miscible with water. The mole percent solubilities of some hydroxy butanes are shown in Table 13.4.

Long chain alcohols are insoluble in water, although many of them form thin films on the water surface. This they do by immersing their polar hydroxy heads in the water while keeping their nonpolar and insoluble tails out of the water. This film, called a *monomolecular layer*, is represented schematically in Fig. 13.3. Such films adhere to the water surface and can be used to decrease the evaporation of water. They have been used to coat the surfaces of lakes and reservoirs to decrease losses by evaporation.

Acetone, with the following structural formula,

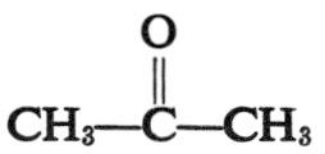

is another polar liquid that is miscible with water. Although it is quite polar, acetone apparently does not form hydrogen bonds. However, its polarity is sufficient to allow it to interact strongly with water, although higher ketones are only slightly soluble in water.

Nonpolar liquids such as carbon tetrachloride, CCl_4, and benzene, C_6H_6, have very low solubilities in water since there is very little tendency for them to be attracted to the polar solvent. The hydrogen bonds between the water molecules themselves are too strong to allow the nonpolar molecules to break the bonding

TABLE 13.4

Solubilities of Some Hydroxy Butanes in Water at Room Temperature

		MOLE %
$CH_3—CH_2—CH_2—CH_2—OH$	1-butanol	2
$CH_3—CH_2—CH(OH)—CH_3$	2-butanol	3
$CH_3—CH(CH_3)—CH_2—OH$	2-methyl-1-propanol	2.5
$CH_3—C(OH)(CH_3)—CH_3$	2-methyl-2-propanol	miscible
$CH_3—C(OH)(CH_3)—CH_2—OH$	2-methyl-1,2-dihydroxypropane	30
$CH_3—CH_2—CH(OH)—CH_2—OH$	1,2-dihydroxybutane	very soluble
$CH_3—CH(OH)—CH_2—CH_2—OH$	1,3-dihydroxybutane	miscible
$HO—CH_2—CH_2—CH_2—CH_2—OH$	1,4-dihydroxybutane	miscible
$CH_3—CH(OH)—CH(OH)—CH_3$	2,3-dihydroxybutane	miscible

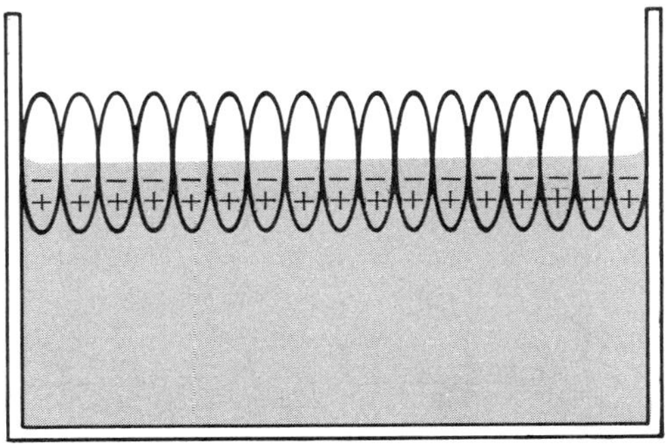

Fig. 13.3. Representation of a monomolecular layer on the surface of water

in order to fit into the semistructured liquid state. However, these two nonpolar liquids are miscible in each other. In this case, there are relatively weak interactions among all of the molecules, and the major effect probably is an entropy effect, bringing about maximum randomness.

Solid Dissolved in a Liquid

FEATURES OF THE DISSOLVING PROCESS. The rule that like dissolves like applies also to the dissolving of solids in liquids. A nonpolar solid such as iodine, which exists as discrete I_2 molecules in the solid state, is soluble in such nonpolar solvents as carbon tetrachloride and benzene. The molecules on the surface of the crystal are not entirely surrounded by other iodine molecules and in a sense are in a state of higher potential energy than those molecules buried in the interior of the crystal. Under the influence of van der Waals forces and a favorable entropy change, these exterior molecules drift into the solvent.

Iodine is much less soluble in water than it is in carbon tetrachloride. However, it does have some solubility in water, probably due to van der Waals forces between the iodine molecules and the water molecules.

Many polar solids have high solubilities in water. A saturated solution of sucrose (table sugar) in water at 25°C is over 67 percent sucrose by weight.

Many ionic solids are soluble in polar solvents such as water. The highly polar nature of water, along with its high dielectric constant, makes it a generally good solvent for ionic substances. Some of the features of dissolving ionic substances in water can be understood by referring to Fig. 13.4. The polar water molecules approach the charged ions in the crystal lattice; the attraction of the positively charged ions for the negative portion of the water dipole, and of the negatively charged ions for the positive portion of the water dipole, allows the ions to separate from the crystal lattice and move into the liquid phase, where they are each surrounded by layers of water molecules.

The high degree of orientation of the water molecules around the solute ions

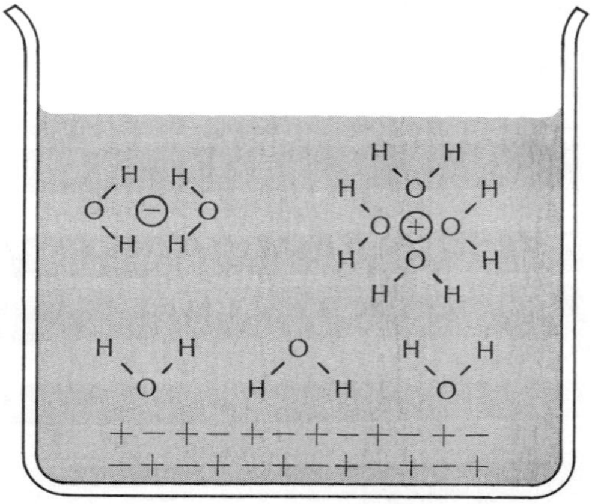

Fig. 13.4. Representation of the dissolving of an ionic substance in water

in the solution explains the unfavorable entropy change that accompanies the dissolving of many ionic solids in water.

The enthalpy change for this transformation includes the enthalpy change accompanying the removal of ions from the crystal lattice and the enthalpy change accompanying the solvation of the ions. This may be represented by lattice enthalpy, solvation enthalpy, and solution enthalpy. The lattice enthalpy, ΔH_{lat}, is the decrease in enthalpy that accompanies the formation of 1 mole of solid crystalline compound from the gaseous ions. Thus ΔH_{lat} normally is negative, corresponding to a decrease in enthalpy. The solvation enthalpy, ΔH_{solv}, is the decrease in enthalpy accompanying the dissolving of the gaseous ions in the solvent. The enthalpy of solution, ΔH_{soln}, is the change in enthalpy when 1 mole of the solid compound dissolves in the solvent.

The energetics for the dissolving of solid sodium hydroxide in water may be explained in terms of the following hypothetical enthalpy cycle:

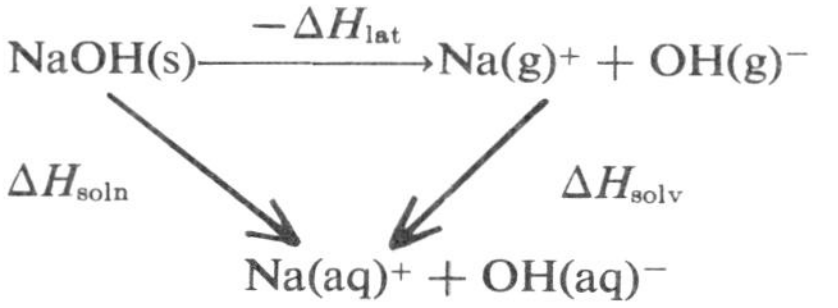

The enthalpy of solution represents the over-all enthalpy change that accompanies the dissolving of solid sodium hydroxide in water. In this cycle, the change is represented as the sum of the changes for converting solid sodium hydroxide to its gaseous ions and then dissolving these gaseous ions in the water.

$$\Delta H_{\text{soln}} = \Delta H_{\text{solv}} - \Delta H_{\text{lat}} \tag{13.6}$$

Since the solvation enthalpy and the lattice enthalpy normally are both negative, the sign of the solution enthalpy depends upon the relative sizes of these two enthalpies. The most favorable enthalpy of solution occurs when the lattice enthalpy is small and the solvation enthalpy is large.

Among the factors affecting the lattice enthalpy are the ionic charges. The higher the ionic charges, the more negative the lattice enthalpy. This reflects the greater attractions between the ions when they have greater charges. In addition, small ions in a lattice have greater attractions and lead to a more negative lattice enthalpy. A much more difficult factor to evaluate is the geometry of the crystal, which indicates the coordination numbers of the ions. Generally, the greater the crystal coordination number, the more negative the lattice enthalpy is.

The solvation enthalpy is affected somewhat similarly. Cations of high charge and small size are those which tend to form the best coordination complexes, and they are also the ones most heavily hydrated. This is reasonable simply on the basis of electrostatic forces. Anions tend to be more highly hydrated the larger they are and the more negative their charges are. In the case of the anions, the large size favors polarizability by the polar solvent molecules.

The concentration of the solution into which the dissolving solute must disperse also affects the solvation enthalpy. If the solution already is moderately concentrated, a large portion of the solvent molecules are involved in the solvation of other solute ions or molecules. This may mean that there are insufficient water molecules remaining to hydrate the new solute adequately, so that the solvation enthalpy is less than it would be if the solute were dissolving into pure water. Thus at high concentrations, solvation enthalpy may decrease somewhat from what it is at lower concentrations. Therefore, it is possible that when a substance is dissolved initially in pure water, the solution process is exothermic, whereas near the saturation point the solution process is endothermic.

Unfortunately for comparison purposes, some of the factors affecting solution enthalpy affect both solvation and lattice enthalpies. Thus cations of high charge and small size tend to form stable crystals but also tend to hydrate readily. It is therefore necessary for us to make a cursory survey now of some common solid, ionic substances and their solubilities in water.

SOLUBILITY TRENDS AMONG IONIC SOLIDS IN WATER. *Halogen Species.* Fluorides are not easily hydrated and are generally insoluble except for those of the alkali metals, ammonium, silver, aluminum, mercury, thallium, and tin. Other halides normally are soluble except those of silver, mercury(I), tin(IV), and lead(II), as well as mercury(II) bromide and iodide, and bismuth(III) iodide.

The more highly electronegative oxyanions tend to hydrate readily. As a result, the chlorates, perchlorates, and bromates are generally soluble while the iodates and periodates are generally insoluble except for the alkali metals and ammonium.

Chalcogen Species. Oxides, sulfides, selenides, and tellurides all tend to be insoluble, consistent with high lattice energies associated with these highly charged anions. Exceptions are the compounds of the alkali metals and ammonium. Alkaline earth, chromium(III) and aluminum sulfides all decompose in water.

Sulfates generally are soluble except those of lead(II), barium, and strontium, and to some extent, calcium. Sulfites, which are less electronegative than sulfates, tend to be insoluble, except those of the alkali metals and ammonium. Thiosulfates are soluble except for those of lead(II) and silver.

Nitrogen Family Species. Binary compounds of this family, nitrides, phosphides, etc., all decompose in water. The nitrate ion has low charge and tends to form crystals of low lattice enthalpy. With relatively high electronegativity it hydrates readily. Thus all nitrates are soluble. The more highly charged phosphate, arsenate, and arsenite ions tend to form crystals of higher lattice enthalpy, and these are all insoluble except those of the alkali metals and ammonium. The compounds containing the hydrogen phosphates, hydrogen arsenates, and the hydrogen arsenites are more soluble. The ammonium ion is a rather large cation of low charge and like the alkali metal ions tends to form crystals of rather low lattice enthalpy. All compounds of ammonium ion are soluble.

Carbon Family Species. Binary compounds such as carbides and silicides generally decompose in water, and if they do not, they are insoluble. Carbonate and silicate anions are of relatively low electronegativity. In addition, the silicates tend to form highly polymeric species. All of their compounds are insoluble except the compounds with the alkali metals and ammonium. Acetates are soluble and oxalates are insoluble except those of the alkali metals and ammonium.

Thiocyanates, NCS^-, generally are soluble except for those of silver, mercury(I), tin(IV), and lead(II). Cyanides generally are insoluble except those of the alkali metals and ammonium.

Other Anionic Species. Like the oxyanions of low electronegativity, the borates tend to be insoluble except those of the alkali metals and ammonium. Chromates tend to be insoluble except for those of the alkali metals, ammonium, magnesium, calcium, and copper(II). The dichromates, which have the same charge spread over a larger ion, tend not to have such high lattice enthalpies and are more soluble. Permanganates, which have a low charge, have low lattice enthalpies and are soluble.

Some Cationic Species. The alkali metals tend to form compounds with low lattice enthalpies, and thus all of their compounds are soluble. Ammonium compounds resemble the alkali metal compounds and are soluble. The more highly charged cations of the alkaline earths form crystals with higher lattice enthalpies, and as a result, the compounds of the alkaline earths are much more insoluble than the compounds of the alkali metals. Compounds of silver(I) tend to be insoluble, with the exception of the fluoride, perchlorate, chlorate, sulfate, nitrate, and acetate.

SOLUBILITY EQUILIBRIA. When an ionic solid is dissolved in water, the surface ions of the crystal are attracted to the water molecules and break away, as shown earlier in Fig. 13.4. However, as the concentration of dissolved ions increases in the solution, there is increasing probability that some of the ions in the solution will strike the surface of the remaining undissolved solid and adhere to it. In this way, some of the ions in solution escape from the liquid solution to the solid phase.

When the rates of dissolving and escape from solution are equal, the system is in dynamic equilibrium. This equilibrium condition may be described by a concentration equilibrium constant.

For a general case, consider the compound $A_xB_y(s)$, which is composed of the ions A^{+y} and B^{-x}. The dissolving process may be described as follows:

$$A_xB_y(s) + (r + s)H_2O(l) = xA(H_2O)_r{}^{+y} + yB(H_2O)_s{}^{-x} \tag{13.7}$$

In words, Equation (13.7) indicates that solid A_xB_y is in equilibrium with its hydrated ions in solution. These ions are written separately since they apparently have some degree of independent motion, the cations being insulated from the anions somewhat by the surrounding sheaths of hydrating water molecules.

If this equation describes the equilibrium, then the concentration equilibrium constant may be formulated as follows:

$$K_c = \frac{[A(H_2O)_r{}^{+y}]^x[B(H_2O)_s{}^{-x}]^y}{[A_xB_y(s)][H_2O(l)]^{(r+s)}} \tag{13.8}$$

As in the case of nitrogen(II) oxide discussed earlier, it is difficult experimentally to determine values for the average number of hydrating water molecules for each of the ions. In this case, this means that r and s cannot be determined easily.

To simplify this equilibrium constant expression, the same limitations are imposed here that were used before in relation to the solubility equilibrium between a gaseous solute and a liquid solution. Thus when the equilibrium solution is very dilute, the concentration of water, which is the solvent, remains essentially constant, since the relatively few solute ions tie up an insignificant number of the water molecules in solvating action. Under these circumstances, the term $[H_2O(l)]^{(r+s)}$ is a constant, and may be transposed to the other side of Equation (13.8):

$$K_c[H_2O(l)]^{(r+s)} = \frac{[A(H_2O)_r{}^{+y}]^x[B(H_2O)_s{}^{-x}]^y}{[A_xB_y(s)]} \qquad (13.9)$$

In addition, since the solid compound A_xB_y, constitutes the entire solid phase in this heterogeneous equilibrium, its *molarity* is constant. In fact, the molarity is directly related to the density of the solid. Thus Equation (13.9) may be rewritten as Equation (13.10):

$$K_c[H_2O(l)]^{(r+s)}[A_xB_y(s)] = [A(H_2O)_r{}^{+y})^x[B(H_2O)_s{}^{-x}]^y = K_{sp} \qquad (13.10)$$

The product $K_c[H_2O(l)]^{(r+s)}[A_xB_y(s)]$ is a constant under the stated conditions; it is called the *solubility product constant*, or sometimes just the *solubility product*, and is given the symbol K_{sp}. Note that the solubility product constant is expressed as the product of the molarities of the solvated ions, each raised to the power of its coefficient in the solubility equilibrium equation. When using the conditions under which the solubility product is constant, it is customary to write the chemical equilibrium equation as follows:

$$A_xB_y(s) = xA(aq)^{+y} + yB(aq)^{-x} \qquad (13.11)$$

From this the solubility product constant expression may be written:

$$K_{sp} = [A(aq)^{+y}]^x[B(aq)^{-x}]^y \qquad (13.12)$$

Note carefully that the solubility product constant is a modified equilibrium constant and that it may be used only under the rigid condition that at equilibrum the concentration of dissolved ions in the solution is very low. This really means that the solubility product constant may be applied only to those substances which we have been calling insoluble, or at best to those substances which have very low solubilities. This also means that values for the solubility product constant must be small. Some solubility product constants at room temperature are listed in Table 13.5.

TABLE 13.5

Some Solubility Product Constants at Room Temperature

ARSENATE	
Ag_3AsO_4	1.1×10^{-20}
$Ba_3(AsO_4)_2$	7.8×10^{-51}
$Ca_3(AsO_4)_2$	6.8×10^{-19}
$Cu_3(AsO_4)_2$	7.6×10^{-36}
$Sr_3(AsO_4)_2$	8.1×10^{-19}
$Zn_3(AsO_4)_2$	1.1×10^{-27}
AZIDE	
AgN_3	2.6×10^{-9}
CuN_3	4.9×10^{-9}
$Hg_2(N_3)_2$	7.1×10^{-10}
$Pb(N_3)_2$	2.6×10^{-9}
TlN_3	2.2×10^{-4}
BROMATE	
$AgBrO_3$	5.2×10^{-5}
BROMIDE	
AgBr	4.6×10^{-13}
CuBr	5.3×10^{-9}
Hg_2Br_2	5.8×10^{-23}
$PbBr_2$	2.8×10^{-5}
TlBr	1.5×10^{-5}
CARBONATE	
Ag_2CO_3	6.4×10^{-12}
$CaCO_3$	7.2×10^{-9}
$CoCO_3$	1.4×10^{-13}
$CuCO_3$	2.3×10^{-10}
$FeCO_3$	2.1×10^{-11}
$MgCO_3$	3.5×10^{-8}
$MnCO_3$	1.8×10^{-11}
$NiCO_3$	1.4×10^{-7}
$PbCO_3$	7.2×10^{-14}
$SrCO_3$	1.6×10^{-9}
$ZnCO_3$	1.7×10^{-11}
CHLORIDE	
AgCl	1.8×10^{-10}
CuCl	1.9×10^{-7}
Hg_2Cl_2	1.3×10^{-17}
$PbCl_2$	1.7×10^{-5}
TlCl	1.9×10^{-4}
CHROMATE	
Ag_2CrO_4	2.4×10^{-12}
$BaCrO_4$	1.2×10^{-10}
$CuCrO_4$	3.6×10^{-6}
Hg_2CrO_4	2.0×10^{-9}
$PbCrO_4$	2.8×10^{-13}
$SrCrO_4$	2.2×10^{-5}
Tl_2CrO_4	9.8×10^{-13}
CYANIDE	
AgCN	1.2×10^{-16}
CuCN	3.2×10^{-20}
$Hg_2(CN)_2$	5.0×10^{-40}
FLUORIDE	
BaF_2	1.0×10^{-6}
CaF_2	4.0×10^{-11}
MgF_2	6.5×10^{-9}
SrF_2	2.5×10^{-9}
HEXACYANOFERRATE(II)	
$Ag_4[Fe(CN)_6]$	1.6×10^{-41}
$Cu_2[Fe(CN)_6]$	1.3×10^{-16}
$Zn_2[Fe(CN)_6]$	4.1×10^{-16}
HYDROXIDE	
$Cd(OH)_2$	1.2×10^{-14}
$Co(OH)_2$	1×10^{-15}
$Cu(OH)_2$	1×10^{-19}
$Fe(OH)_2$	9.8×10^{-15}
$Fe(OH)_3$	4.5×10^{-37}
$Mg(OH)_2$	1.2×10^{-11}
$Mn(OH)_2$	2×10^{-13}
$Zn(OH)_2$	3.3×10^{-17}
IODATE	
$AgIO_3$	3.0×10^{-8}
$Ba(IO_3)_2$	1.5×10^{-9}
$Ca(IO_3)_2$	7.1×10^{-7}
$Ce(IO_3)_3$	3.2×10^{-10}
$Cu(IO_3)_2$	7.4×10^{-8}
$Hg_2(IO_3)_2$	1.9×10^{-14}
$La(IO_3)_3$	6.2×10^{-12}
$Pb(IO_3)_2$	3.2×10^{-13}
$TlIO_3$	3.1×10^{-6}
IODIDE	
AgI	8.3×10^{-17}
Hg_2I_2	4.9×10^{-29}
PbI_2	7.1×10^{-9}
TlI	6.5×10^{-8}

TABLE 13.5 (cont.)

SELENATE		SULFIDE	
$BaSeO_4$	3.5×10^{-8}	Ag_2S	6.3×10^{-50}
$CaSeO_4$	4.0×10^{-5}	CoS	3.1×10^{-23}
$PbSeO_4$	1.5×10^{-7}	Cu_2S	7.2×10^{-49}
SELENITE		CuS	4.0×10^{-36}
$CaSeO_3$	3.0×10^{-6}	FeS	5.1×10^{-18}
Hg_2SeO_3	6×10^{-15}	HgS	1.9×10^{-53}
$MgSeO_3$	1.3×10^{-5}	MnS	1.1×10^{-15}
$PbSeO_3$	3×10^{-12}	NiS	1.0×10^{-24}
$SrSeO_3$	8.5×10^{-7}	PbS	6.8×10^{-29}
SULFATE		ZnS	1.2×10^{-28}
Ag_2SO_4	1.6×10^{-5}	THIOCYANATE	
$BaSO_4$	9.8×10^{-11}	$AgNCS$	1.1×10^{-12}
$CaSO_4$	9.1×10^{-6}	$CuNCS$	4.8×10^{-15}
Hg_2SO_4	4.8×10^{-7}	$Hg_2(NCS)_2$	1.8×10^{-20}
$PbSO_4$	1.6×10^{-8}	$Pb(NCS)_2$	2.0×10^{-5}
$RaSO_4$	4.2×10^{-15}	$TlNCS$	1.7×10^{-4}
$SrSO_4$	6.2×10^{-7}		

From Table 13.5, it is possible to set up the solubility product expressions for the substances listed there. For example, for calcium arsenate, the solubility equilibrium equation can be written as follows:

$$Ca_3(AsO_4)_2(s) = 3Ca(aq)^{++} + 2AsO_4(aq)^{3-} \quad (13.13)$$

The solubility product constant expression then may be written:

$$K_{sp} = [Ca(aq)^{++}]^3[AsO_4(aq)^{3-}]^2 = 6.8 \times 10^{-19} \quad (13.14)$$

SAMPLE SOLUBILITY-PRODUCT CALCULATIONS. For the following calculations, we shall assume that the solubility equilibrium is not complicated by any additional reactions. Systems involving multiple equilibria will be considered later.

Solubility data may be used to compute values for the solubility product constant. For example, it has been found that the molar solubility of calcium fluoride is 2.15×10^{-4} *M*. The solubility equilibrium equation may be written as follows:

$$CaF_2(s) = Ca(aq)^{++} + 2F(aq)^{-} \quad (13.15)$$

From Equation (13.15) it should be clear that for each mole of calcium fluoride that dissolves, 1 mole of hydrated calcium ions and 2 moles of hydrated fluoride ions are produced. Since 2.15×10^{-4} moles of calcium fluoride dissolved per liter, at equilibrium the concentrations of hydrated calcium ions and hydrated fluoride ions must be 2.15×10^{-4} *M* and $2 \times 2.15 \times 10^{-4}$ *M* respectively. Thus the value for the solubility product constant may be computed as follows:

$$K_{sp} = [Ca(aq)^{++}][F(aq)^{-}]^2 = (2.15 \times 10^{-4})(4.30 \times 10^{-4})^2 = 4.0 \times 10^{-11}$$

Similarly, it is possible to calculate the molar solubility of a substance from its solubility product constant. For example, compute the molar solubility of lead(II) iodate at room temperature.

The solubility equilibrium may be represented as follows:

$$Pb(IO_3)_2(s) = Pb(aq)^{++} + 2IO_3(aq)^- \qquad (13.16)$$

The solubility product expression may be written as follows using the value recorded in Table 13.5.

$$K_{sp} = [Pb(aq)^{++}][IO_3(aq)^-]^2 = 3.2 \times 10^{-13}$$

From Equation (13.16) it can be seen that for each mole of lead(II) iodate that dissolves, 1 mole of hydrated lead(II) ions and 2 moles of hydrated iodate ions are formed. Thus if x is the molar solubility of lead(II) iodate, the equilibrium concentrations of hydrated lead(II) ions and hydrated iodate ions may be expressed in terms of x:

$$[Pb(aq)^{++}] = x$$

$$[IO_3(aq)^-] = 2x$$

These expressions for the ion concentrations may be substituted into the solubility product equation, which then may be solved for x:

$$(x)(2x)^2 = 3.2 \times 10^{-13}$$

$$4x^3 = 3.2 \times 10^{-13}$$

$$x^3 = 0.8 \times 10^{-13} = 80 \times 10^{-15}$$

$$x = \sqrt[3]{80 \times 10^{-15}} = 4.3 \times 10^{-5}\ M$$

Therefore a saturated solution of lead(II) iodate at room temperature is 0.000043 *M*.

The solubility equilibrium is subject to the same effects as other equilibria. Considering the equilibrium described by Equation (13.16), suppose that the iodate ion concentration suddenly is increased in the equilibrium system. This produces a stress on the equilibrium that may be attributed to the excess iodate. To relieve this stress, the equilibrium shifts toward the production of more solid lead(II) iodate. Thus lead(II) iodate is less soluble in a solution of sodium iodate, say, than it is in water alone. Commonly this is called the *common ion effect*, which may be described as follows: the solubility of a relatively insoluble substance is decreased by the presence of one or more of the ions common to that substance in the solution.

Suppose that we wish to calculate the molar solubility of lead(II) iodate in 0.010 *M* sodium iodate solution. If we let x be the molar solubility of lead(II) iodate in 0.010 *M* sodium iodate solution, then at equilibrium the concentration of lead(II) ions in solution must be x, since 1 mole of hydrated lead(II) ions is produced for each mole of lead(II) iodate that dissolves.

However, there are two sources of iodate ion in the solution. One is the sodium iodate that is present at the outset, and the other is the dissolved lead(II) iodate. The concentration of iodate ion from lead(II) iodate must be $2x$, since two moles of iodate ion are produced in solution for each mole of lead(II) iodate that dissolves. Thus the total molar concentration of iodate ion in solution at equilibrium must be $(0.010 + 2x)$. However, it was shown in the previous problem that the molar solubility of lead(II) iodate in pure water is only 0.000043 *M*, and it has already been shown that the solubility will be even less than that in 0.010 *M* sodium iodate. Therefore, the value of $2x$ must be somewhat less than 0.000086 *M*, and this is insignificant with respect to 0.010 *M*. Consequently, the concentration of hydrated iodate ion may be expressed as 0.010 *M*. Then the equilibrium concentrations of hydrated lead(II) ions and iodate ions may be expressed as follows:

$$[Pb(aq)^{++}] = x$$

$$[IO_3(aq)^-] = 0.010 = 10^{-2}\ M$$

These may be substituted into the expression for the solubility product constant, which then may be solved for x.

$$[Pb(aq)^{++}][IO_3(aq)^-]^2 = 3.2 \times 10^{-13}$$

$$(x)(10^{-2})^2 = 3.2 \times 10^{-13}$$

$$x = 3.2 \times 10^{-9}\ M$$

Obviously, the solubility of lead(II) iodate is greatly reduced by the presence of excess iodate in the solution. Whereas it is soluble to the extent of 4.3×10^{-5} *M* in water, it is soluble only to the extent of 3.2×10^{-9} *M* in 0.010 *M* sodium iodate solution.

Like other equilibrium constants, the solubility product constant is not constant with a temperature change. The solubilities of most substances change with temperature, and the way they change can be predicted from the energetics of the dissolving reaction at equilibrium. It was pointed out earlier that at equilibrium the dissolving reaction often is endothermic. The reverse reaction, in which the solute is separated from the solution, then, must be exothermic. If this is the case, an increase in temperature favors the endothermic reaction. Since the endothermic reaction is the dissolving reaction for many substances, their solubilities increase with increasing temperature. However, there are substances for which the reverse is true, so that the dissolving process is exothermic at equilibrium. For these substances, the solubility decreases as temperature increases.

PRECIPITATION. When the ions of a relatively insoluble substance are brought together in solution so that the product of their concentrations raised to the powers of the coefficients in the solubility equilibrium equation exceeds the solubility product constant, the solid, insoluble substance should leave the solution and form the solid phase. The process by which insoluble substances are removed from the solution is called *precipitation.*

Suppose that 50 ml of 0.01 M $AgNO_3$ is mixed with 50 ml of 0.001 M NaCl. At the moment of mixing, the solution will consist of a mixture of the following ions in water: $Na(aq)^+$, $Cl(aq)^-$, $Ag(aq)^+$, and $NO_3(aq)^-$. Considering the possible combinations of ions, the following combinations represent possible compounds that might be formed: NaCl, $NaNO_3$, AgCl, $AgNO_3$. Of these, only AgCl, silver chloride, is insoluble. Thus the AgCl should precipitate according to the following equation:

$$Ag(aq)^+ + Cl(aq)^- = AgCl(s) \tag{13.17}$$

Note that this is a simple ionic equation and that not all of the ions in the solution are represented, but only those which can be expected to react.

Considering the solubility product constant value in Table 13.5, it is expected that if the product $M_{Ag(aq)^+}M_{Cl(aq)^-}$ exceeds the value of 1.8×10^{-10}, then AgCl should precipitate until the concentrations of the ions have decreased sufficiently so that the concentration product will equal the solubility product constant. If, at the moment of mixing, the concentration product is less than the solubility product constant, then something less than a saturated solution is formed. Such a solution is said to be *unsaturated.* Under such conditions precipitation is not expected. If the concentration product equals the solubility product constant, the solution is saturated with respect to AgCl, and precipitation is not expected.

By computing the concentrations of the silver(I) and chloride ions at the moment of mixing, it is possible to predict whether or not silver chloride will precipitate. To compute the hydrated ion concentrations, note that two solutions are mixed, and that the volume after mixing is therefore greater than either of the volumes before mixing. Whereas each hydrated ion is in 50 ml of solution before mixing, it is in 100 ml of solution after mixing, assuming that the volumes are additive. This has the effect of halving each concentration, so that at the moment of mixing the ion concentrations are as follows:

$$M_{Ag(aq)^+} = 0.005\ M$$

$$M_{Cl(aq)^-} = 0.0005\ M$$

The concentration product then is

$$M_{Ag(aq)^+}M_{Cl(aq)^-} = (5 \times 10^{-3})(5 \times 10^{-4}) = 25 \times 10^{-7} = 2.5 \times 10^{-6}$$

Clearly, at the moment of mixing, the hydrated ion concentration product exceeds the solubility product constant of 1.8×10^{-10} as recorded in Table 13.5. Thus it is anticipated that precipitation will occur as depicted in Equation (13.17).

However, the precipitation process is fraught with hazards. For example, the simple collision of one silver ion, $Ag(aq)^+$, with one chloride ion, $Cl(aq)^-$, hardly constitutes precipitation. First, the two ions must break through the surrounding sheaths of water molecules. Once together, however, they still do not represent a solid precipitate. A two-ion aggregate hardly represents a crystal lattice. After all, the potential-energy advantage of the crystal lattice is that a very large number of ions have come together so that most of them are surrounded by oppositely charged ions. When single silver and chloride ions come together they are still under strong attack by water molecules.

Thus it is important that these two ions stay close together until additional ions collide with them to form a larger, more stable aggregate. During this process of growth of the central aggregate or nucleus upon which crystallization may occur, there is no evident precipitation. This period of time is called the period of *nucleation*. Thus precipitation does not occur until the nucleus has reached a certain critical size and stability at which its continued existence is assured. It is believed that the critical size of the nucleus is about 100 atoms or ions.

Even when solute concentrations are sufficiently high to cause precipitation in theory, it is possible to have conditions under which the nucleus does not grow to adequate size to provide a route through which precipitation may begin. A solution containing a higher concentration of solute than the saturated solution is called a *supersaturated solution*. Supersaturated solutions are metastable systems, and often precipitation may be induced in them by adding a small crystal of the solute to serve as a suitable nucleus. Sometimes agitation or other "shock treatments" are used.

Once the solid precipitates, the surface ions of the solid still attract ions in the surrounding solution. For example, silver chloride tends to attract chloride ions from the solution, and these adhere to the surface of the crystal. This produces a negative surface charge on the solid. Somewhat more loosely associated with the layer of negative chloride ions is a layer of counter ions of opposite charge, as shown in Fig. 13.5.

During their growth, precipitating particles pass through the size range characteristic of particles held in colloidal dispersions. If the growth stops at this point, the result is a colloidal dispersion instead of a precipitate. In silver chloride, these small colloid-size particles may have a layer of negative chloride ions adhering to

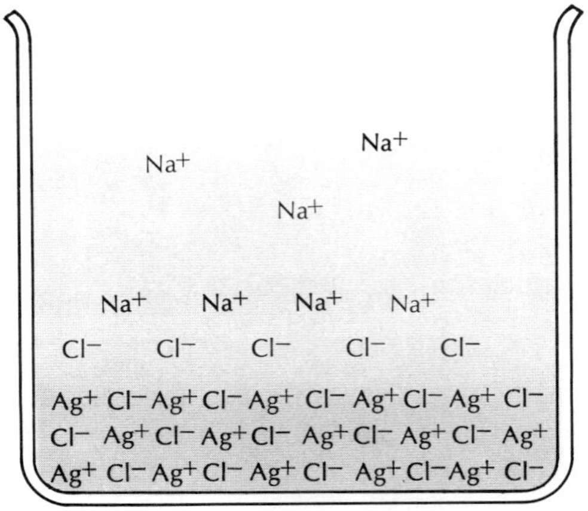

Fig. 13.5. Representation of surface adsorption for silver chloride

them, so that the particles appear to be negative to each other and thus repel each other. In this way further coagulation to form larger particles that will precipitate is hindered.

Some insoluble substances have a great affinity for water. This is true particularly of hydroxides and sulfides. Often water forms the adsorbed layer on the particles. When these substances finally do precipitate, they again form colloids, with the liquid solvent dispersed in the solid precipitate. Such precipitates are not crystalline but are gelatinous colloids, and are usually difficult to handle.

These factors help to account for the production of impure precipitates. Almost all precipitates have impurities adhering to their surfaces. In addition, impurities often are imbedded in the interior of the solid matrix. Adhering impurities usually can be washed off, but those which are incorporated into the precipitate are more difficult to remove. One standard purification technique for such precipitates is *recrystallization*. A solvent is chosen in which the precipitate has a markedly different solubility at different temperatures. The precipitate and its impurities are dissolved in the solvent at high temperature. Upon cooling, the solubility decreases, and solid precipitates. Since the impurities now are rather widely dispersed in the solvent there is less chance for them to be incorporated into the precipitate and the precipitate is more pure than it was before.

Quantitative precipitations may be used to analyze unknown substances. This area of quantitative analysis is called *gravimetric analysis*. To carry out such an analysis, the substance is dissolved and then treated with a reagent that forms a very insoluble precipitate with the constituent of the sample to be measured. The precipitate must be sufficiently insoluble so that only insignificant amounts of the constituent being measured remain behind in the solution. Sometimes the common-ion effect can be used to good advantage to decrease the solubility of the precipitate, although large excesses of reagents normally must be avoided lest the precipitate be highly contaminated. Then the precipitate is filtered, dried, and weighed. From the mass of the original sample and the mass of the precipitate, the percent of the desired constituent in the sample may be computed.

For example, a sample contains chlorine in the form of chloride. A 0.4321-g sample is dissolved in water and enough silver nitrate is added to precipitate all of the chloride. The mass of the clean, dry silver chloride is 0.6144 g. What is the percent of chlorine in the sample?

The equation for the precipitation is as follows:

$$\mathrm{Ag(aq)^+ + Cl(aq)^- = AgCl(s)}$$

We wish to find how many grams of chlorine there are in 100 g of sample. From the mass of AgCl precipitated and the mass of the sample used, the number of grams of AgCl that would be precipitated from 100 g of sample may be computed. From this, the number of grams of chlorine may be computed, using the atomic weight of chlorine and the molecular weight of AgCl.

$$100\text{ g sample} \times \frac{0.6144\text{ g AgCl}}{0.4321\text{ g sample}} \times \frac{35.46\text{ g Cl}}{143.3\text{ g AgCl}} = 35.18\text{ percent Cl}$$

COLLIGATIVE PROPERTIES OF SOLUTIONS

Vapor Pressure, Boiling Point, and Freezing Point

Experimentally it is found that *nonvolatile, nonion producing* solutes cause a decrease in the vapor pressure of the solvent in which they are dissolved. Specifically, in *dilute solutions*, the vapor pressure of the solvent over the solution is directly proportional to the mole fraction of the solvent in the solution. The proportionality constant is the vapor pressure of the *pure* solvent at the temperature of the measurement. This is expressed by Equation (13.18), in which P is the vapor pressure of the solvent over the solution, $P°$ is the vapor pressure of the pure solvent, and X_{solv} is the mole fraction of the solvent in the solution.

$$P = P° X_{\text{solv}} \tag{13.18}$$

To explain this observation, think of a closed container half full of liquid solvent in equilibrium with its vapor. Now suppose that a solute is added to this solvent. When it dissolves, some of the solvent is used to solvate the solute, and the *effective* concentration of the solvent decreases. Since some of the solvent molecules are thus tied up, they are less likely to strike the liquid surface and escape from the solution, and the rate of vaporization decreases. At this instant, the rate of condensation exceeds the rate of vaporization, and some of the vapor condenses to the liquid to reestablish the equilibrium. When the new equilibrium is established, the pressure of the vapor is less than it was before.

Using the relationship in Equation (13.18), computations may be made for specific systems. For example, the vapor pressure of ethanol at 30°C is 78.2 mm Hg. If 50 g of hydroxybenzene, C_6H_5OH, is added to 250 g of ethanol, CH_3CH_2OH, at 30°C, what is the vapor pressure of ethanol over the solution?

To make the calculation, the mole fraction of ethanol in the solution must be determined. From the masses and molecular weights of the two constituents, the number of moles of each one and the total number of moles of solution may be found:

$$250 \text{ g } CH_3CH_2OH \times \frac{1 \text{ mole } CH_3CH_2OH}{46.1 \text{ g } CH_3CH_2OH} = 5.42 \text{ moles } CH_3CH_2OH$$

$$50 \text{ g } C_6H_5OH \times \frac{1 \text{ mole } C_6H_5OH}{94 \text{ g } C_6H_5OH} = 0.53 \text{ mole } C_6H_5OH$$

$$\text{Total moles soln} = 5.42 + 0.53 = 5.95 \text{ moles soln}$$

The mole fraction of solvent ethanol is found as follows:

$$X_{CH_3CH_2OH} = \frac{5.42}{5.95} = 0.91$$

From Equation (13.18), the vapor pressure of ethanol over the solution may be computed:

$$P = P° X_{CH_3CH_2OH} = (78.2 \text{ mm Hg})(0.91) = 71.2 \text{ mm Hg}$$

Sometimes the vapor pressure depression relationship is expressed in a slightly different way. This new relationship may be derived by writing the identity (13.19) and subtracting from that Equation (13.18) to obtain Equation (13.20).

$$P^\circ = P^\circ \tag{13.19}$$

$$P = P^\circ X_{\text{solv}} \tag{13.18}$$

$$P^\circ - P = P^\circ - P^\circ X_{\text{solv}} \tag{13.20}$$

The usual delta notation may be introduced through Equation (13.21) which may be substituted into Equation (13.20).

$$\Delta P = P - P^\circ \tag{13.21}$$

$$-\Delta P = P^\circ(1 - X_{\text{solv}}) \tag{13.22}$$

The expression $(1 - X_{\text{solv}})$ is the mole fraction of the *solute*, X_{solu}, so that Equation (13.22) may be rewritten thus:

$$-\Delta P = P^\circ X_{\text{solu}} \tag{13.23}$$

The mole fraction of solute is proportional to the molality of the solution. This is expressed by Equation (13.24), which may be substituted into Equation (13.23).

$$X_{\text{solu}} = km \tag{13.24}$$

$$-\Delta P = kP^\circ m \tag{13.25}$$

Since the vapor pressure of a liquid is constant at a fixed temperature, the product of the proportionality constant k and the vapor pressure of the solvent, P°, is a constant called the *molal vapor pressure constant*, which is represented by the symbol k_{vp}. Thus Equation (13.25) may be rewritten thus:

$$-\Delta P = k_{vp}m \tag{13.26}$$

This equation indicates that the vapor-pressure depression caused by adding solute to a solvent is directly proportional to the molality of the solution.

It follows that a lowering of the vapor pressure of a solvent by addition of a solute corresponds to an elevation of the boiling point for the solvent in solution over that of the pure solvent. In fact, the quantitative relationship is just like the vapor-pressure expression in that the boiling-point elevation ΔT_{bp} is directly proportional to the molality of the solution, as shown by Equation (13.27).

$$\Delta T_{bp} = k_{bp}m \tag{13.27}$$

The proportionality constant k_{bp} is called the *molal boiling-point constant*, or the *ebullioscopic constant*.

Corresponding to the boiling-point elevation is the freezing-point depression.

When solute is added to a solvent, the solvent in the solution freezes at a lower temperature than the pure solvent. This probably is associated with the interference of the foreign molecules with the orderly organization required of the solvent molecules to form the solid lattice.

The form of the quantitative relationship is the same as that for the vapor-pressure depression and the boiling-point elevation, namely, the freezing-point depression $-\Delta T_{fp}$ is directly proportional to the molality of the solution.

$$-\Delta T_{fp} = k_{fp}m \tag{13.28}$$

The proportionality constant k_{fp} is called the *molal freezing-point constant*, or the *cryoscopic constant.*

The properties of vapor-pressure depression, boiling-point elevation, and freezing-point depression are called *colligative properties* of solutions. These properties depend only upon the relative number of solute particles per quantity of solvent. The proportionality constants are properties of the solvent, and ideally the nature of the solute is of little consequence. Values for some ebullioscopic and cryoscopic constants are given in Table 13.6.

TABLE 13.6 Some Ebullioscopic and Cryoscopic Constants for Solvents (deg x kg/Mole)

NAME	FORMULA	k_{fp}	k_{bp}
Acetic acid	CH_3COOH	3.9	3.07
Acetone	CH_3COCH_3		1.71
Benzene	C_6H_6	5.12	2.53
Bromobenzene	C_6H_5Br		6.26
Bromoform	$CHBr_3$	14.4	
Camphor	$C_{10}H_{16}O$	40.0	5.95
Carbon disulfide	CS_2		2.34
Carbon tetrachloride	CCl_4		5.03
Chlorobenzene	C_6H_5Cl		4.15
Chloroform	$CHCl_3$		3.63
Cyclohexane	C_6H_{12}	20.0	2.79
1,2-Dibromoethane	$BrCH_2CH_2Br$	12.5	6.44
Diethyl ether	$CH_3CH_2OCH_2CH_3$		2.02
1,4-Dimethylbenzene	$C_6H_4(CH_3)_2$	4.3	
Ethanol	CH_3CH_2OH		1.22
Ethyl Acetate	$CH_3COOCH_2CH_3$		2.77
Formamide	$HCONH_2$	3.8	
Formic Acid	$HCOOH$	2.77	
Hexane	C_6H_{14}		2.75
Methanol	CH_3OH		0.83
Methylbenzene	$C_6H_5CH_3$		3.33
Naphthalene	$C_{10}H_8$	6.9	5.65
Nitrobenzene	$C_6H_5NO_2$	8.1	5.24
Sulfuric acid	H_2SO_4	6.81	
Water	H_2O	1.86	0.512

The following problems illustrate some types of calculations that may be made with the colligative properties of solutions.

What concentration of ethylene glycol, $HOCH_2CH_2OH$, *in water is needed for water to freeze at* $-4.00°C$? The molal freezing-point constant for water from Table 13.6 is $1.86°C \times kg/mole$. From the statement of the problem, $\Delta T_{fp} = -4.00°C$. Equation (13.28) may be solved for the molality as follows:

$$m = \frac{-\Delta T_{fp}}{k_{fp}} = \frac{4.00°C \times \text{mole}}{1.86°C \times \text{kg}} = 2.15\ m$$

When 20.0 g of nitrobenzene, $C_6H_5NO_2$, *is dissolved in 150 g of naphthalene,* $C_{10}H_8$, *the freezing point is lowered by 7.6°C. What is the cryoscopic constant for naphthalene?* Equation (13.28) may be solved for the cryoscopic constant, k_{fp}.

$$k_{fp} = \frac{-\Delta T_{fp}}{m}$$

The value for ΔT_{fp} is given in the problem statement. The molality must be computed. The molality is the number of moles of solute in 1000 g of solvent. Starting with the 1000 g of solvent naphthalene, the data allow the determination of the number of grams of solute nitrobenzene in the 1000 g of solvent naphthalene. Then the molecular weight can be used to determine the number of moles of nitrobenzene.

$$1000\ \text{g}\ C_{10}H_8 \times \frac{20\ \text{g}\ C_6H_5NO_2}{150\ \text{g}\ C_{10}H_8} \times \frac{1\ \text{mole}\ C_6H_5NO_2}{123\ \text{g}\ C_6H_5NO_2} = 1.1\ m$$

$$k_{fp} = \frac{-\Delta T_{fp}}{m} = \frac{7.6°C \times \text{kg}}{1.1\ \text{mole}} = 6.9°C \times \text{kg/mole}$$

The most useful application of freezing-point depression and boiling-point elevation is the determination of molecular weights. For example, *if 15.0 g of compound X is added to 200 g of benzene, and the boiling point of the benzene is elevated by 1.25°C, what is the molecular weight of X?*

From Table 13.6, the ebullioscopic constant for benzene is found to be 2.53 $°C \times kg/mole$. The value of ΔT_{bp} is given as 1.25°C. With these two values, the molality of the solution may be computed from Equation (13.27). However, the number of moles of X in the solution may be determined by multiplying the molality by the number of kilograms of solvent.

$$m = \frac{\Delta T_{bp}}{k_{bp}} = \frac{\text{moles } X}{\text{kg } C_6H_6}$$

$$\text{moles } X = \frac{\Delta T_{bp} \times \text{kg } C_6H_6}{k_{bp}} = \frac{1.25°C \times 0.200\ \text{kg } C_6H_6 \times \text{moles } X}{2.53°C \times \text{kg } C_6T_6}$$

$$= 0.0988\ \text{mole } X$$

The sample used had a mass of 15.0 g.

$$\frac{15 \text{ g } X}{0.0988 \text{ mole } X} = 152 \text{ g/mole } X$$

Solutions of Electrolytes

The colligative properties of solutions appear to depend only upon the concentration of the solute particles. Electrolytes are solutes that produce ions in solution. Since the ions are produced by the splitting of aggregates into many ions, it is to be expected that electrolytes have a greater effect upon the colligative properties than nonelectrolytes do.

Thus, the freezing point of a 1.0 m aqueous solution of a nonelectrolyte is $-1.86°C$. Experimentally, it has been shown that the freezing point of a 1.0 m solution of sodium chloride, NaCl, in water is $-3.37°C$. That the freezing point for 1.0 m NaCl is lower than the freezing point for 1.0 m nonelectrolyte is attributed to the fact that each mole of NaCl that dissolves produces two moles of ions in solution. However, it might be expected that the freezing point of 1.0 m NaCl should be $2(-1.86°C) = -3.72°C$. That the freezing-point depression caused by electrolytes often is less than expected is a significant indication of the nature of electrolyte solutions.

To pursue this problem further, it is convenient to define the *mole number i* as follows.

$$i = \frac{\text{Colligative property value for electrolyte}}{\text{Colligative property value for nonelectrolyte}} \tag{13.29}$$

Thus in the example, the freezing-point depression for 1.0 m NaCl is 3.37°C and the freezing-point depression for 1.0 m nonelectrolyte is 1.86°C. In this case, i has the following value:

$$i = \frac{3.37}{1.86} = 1.81$$

This indicates that the effect of NaCl upon the freezing point of water is 1.81 times the effect of a nonelectrolyte upon the freezing point of water. Since a mole of NaCl forms 2 moles of ions in solution, it is expected that the value for i in this case should be 2.00. In other words, the mole number i roughly indicates the *effective* number of moles of ions produced per mole of solute. In this case, it appears that each mole of NaCl dissolved effectively produces 1.81 moles of ions. Why the effective number of moles of ions differs from the actual number of moles of ions remains to be seen.

Some additonal data for a few other 1.0 m aqueous solutions are provided in Table 13.7. From this data it is clear that the values of i are less than predicted on the basis of the number of moles of ions available per mole of electrolyte. It appears that values for i between 1 and 2 correspond to electrolytes providing 2

TABLE 13.7 Some Freezing-point Depression Data for 1.0 m Aqueous Solutions of Electrolytes†

ELECTROLYTE	$-\Delta T_{fp}$ (Theor.)	$-\Delta T_{fp}$ (Exptl.)	i
KNO_3	3.72°C	2.56°C	1.38
$AgNO_3$	3.72	2.63	1.41
NH_4NO_3	3.72	2.92	1.57
$NaNO_3$	3.72	3.02	1.62
KCl	3.72	3.25	1.75
KBr	3.72	3.29	1.77
NH_4Cl	3.72	3.33	1.79
KI	3.72	3.37	1.81
NaCl	3.72	3.37	1.81
NaBr	3.72	3.51	1.89
$Sr(NO_3)_2$	5.58	3.90	2.10
$Ca(NO_3)_2$	5.58	4.59	2.47
$BaCl_2$	5.58	5.20	2.80
$Cd(NO_3)_2$	5.58	5.42	2.91

† $-\Delta T_{fp}$ for nonelectrolytes = 1.86°C

moles of ions per mole of solute. Similarly, values for i between 2 and 3 correspond to electrolytes providing 3 moles of ions per mole of solute.

It is interesting to see how these values for i deviate from the predicted as the molal concentration of the solution varies. It is possible to find the freezing points of solutions of varying concentrations and then to see how these various values correspond to what we might predict. The graph in Fig. 13.6 shows these variations for KCl, $MgSO_4$, K_2SO_4, and $BaCl_2$. The graph includes the theoretical lines that may be compared to the experimental results.

In studying the graph, it is clear that there is minimal deviation from the expected at very high dilutions, since all of these curves converge on the origin. Also it is obvious that of these four electrolytes, KCl deviates least and $MgSO_4$ deviates most from the expected values. The $BaCl_2$ and K_2SO_4 are intermediate in this respect.

Figure 13.7 shows another way of looking at the data for these same four electrolytes. There the values of the mole number i are plotted against solute molality. Again it is clear that in very dilute solutions the mole numbers approach the theoretical values. Again the KCl has the smallest deviation from the theoretical and $MgSO_4$ has the greatest deviation from the theoretical, with $BaCl_2$ and K_2SO_4 intermediate.

Solutions of electrolytes are electrical conductors. The mode of conductance is shown in Fig. 13.8. The battery serves as an electron pump pulling electrons from the electrode labeled as positive and pushing them on to the electrode labeled negative. The ions in the solution are able to move to the oppositely charged electrodes to give up or pick up the electrons. It is this motion of the ions in the solution that accounts for these solutions serving as conductors.

Thus the ability of an electrolyte solution to serve as an electrical conductor depends upon the charges on the ions and mobility of the ions, that is, the ability of the ions to migrate in the solution. In order to distinguish between these effects

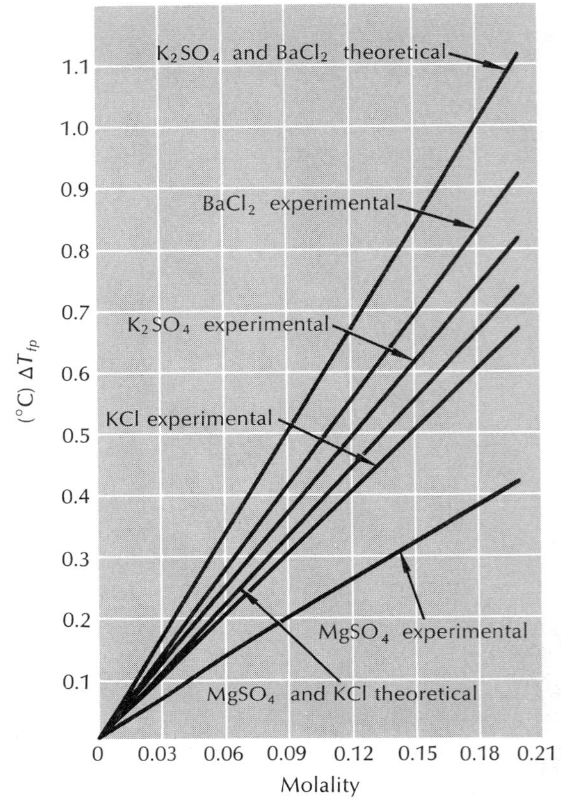

Fig. 13.6. Plot of freezing-point depression against molality for several electrolytes

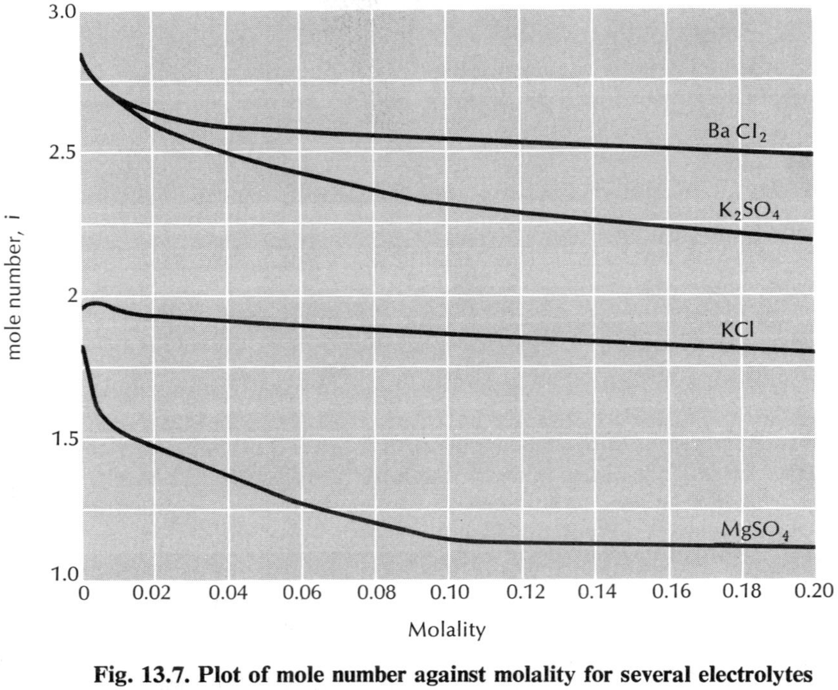

Fig. 13.7. Plot of mole number against molality for several electrolytes

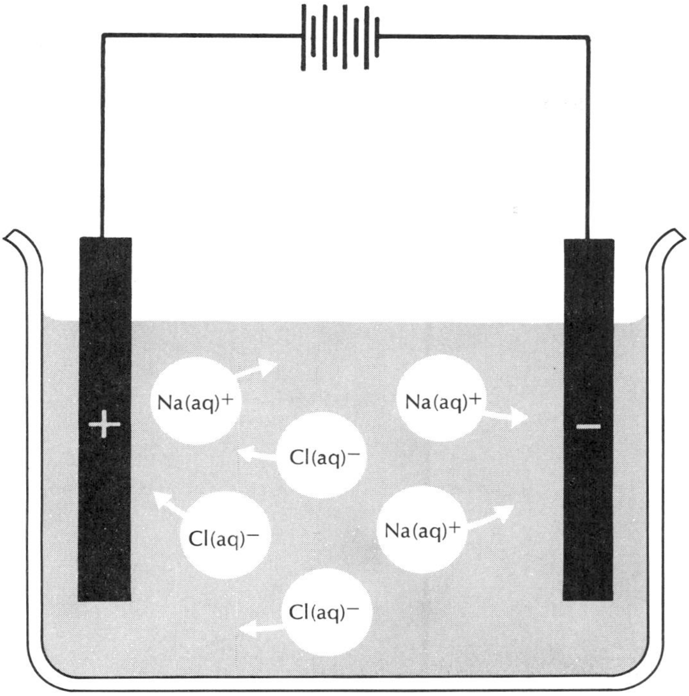

Fig. 13.8. Schematic diagram of a cell in which an aqueous solution of sodium chloride acts as a conductor

it is convenient to measure the conductance in such a way that the number of charges between the electrodes remain constant. Under these conditions the measurement of the conductance serves as a measure of the ion mobility.

The *equivalent conductance* is the conductance in $ohms^{-1}$ (or mhos) for a solution containing 1 mole each of positive and negative ionic quantum charges between two electrode plates that are 1.00 cm apart. It should be clear that any concentration of solution may be used, and the only requirement is that the plates be sufficiently large for very dilute solutions, so that the total volume of solution between the plates, large as it may be, will contain 1 mole each of positive and negative ionic quantum charges The symbol for the equivalent conductance is Λ.

Consequently, in measuring the equivalent conductance, the total amount of ionic charge between the electrodes is always the same, and the values for the equivalent conductance indicate the ability of the ions to migrate between the electrodes. In Fig. 13.9, some equivalent conductance data are plotted against the concentration of ionic charges in the solution for KCl, K_2SO_4, and $MgSO_4$. From the graphs it is clear that KCl is the best conductor, $MgSO_4$ is the worst, and K_2SO_4 is intermediate. In addition it is evident that the equivalent conductances of all three of these solutes increase with dilution.

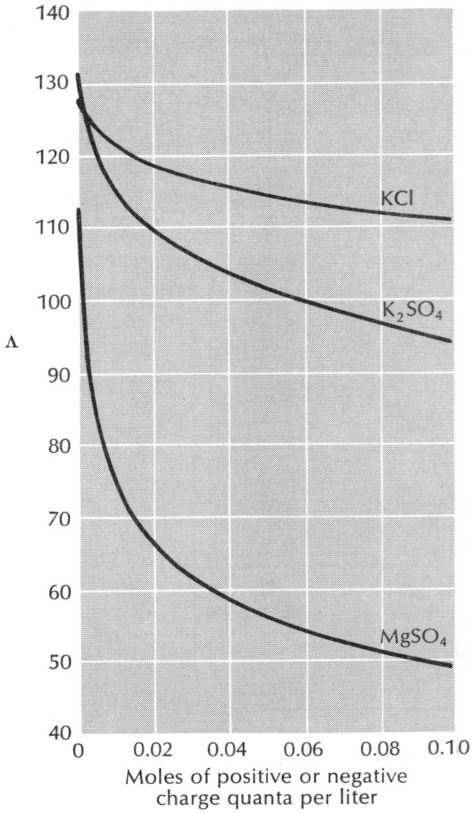

Fig. 13.9. The effect of concentration of ionic charge on the equivalent conductance of several electrolytes

The deviations in colligative-property values for electrolytes with concentration changes, and the effects of dilution upon conductance, along with the observations of the variations in all of these properties with respect to the nature of the electrolyte, has led to a picture of electrolytes in solution that is extremely important. This is the Debye-Hückel interionic attraction theory.

According to this theory ions in solution retain strong attractions for each other despite the fact that they are surrounded by sheaths of water molecules. Thus the ions are not totally independent of each other, and there is a tendency toward clustering of hydrated ions. As a result of this clustering, the mobility of the ions is decreased. Thus those electrolytes with ions of small charge are the best conductors and those with the largest charges are the worst conductors. In addition the clustering is more pronounced as the concentration increases. As a result electrolytes tend to conduct best in dilute solutions.

Insofar as the colligative properties are concerned, a cluster of several hydrated ions may act as a single particle within the solution. The result is that the *effective* concentration of the solute is somewhat less than the predicted concentration. This effect is less severe with ions of low charge and in dilute solutions. This circumstance helps to account for the fact that the mole numbers for electrolytes tend to be less than predicted, the deviation decreasing as concentration decreases.

If this picture is correct, it seems reasonable to expect that the solubility of a substance may be greater if it is dissolving into an electrolyte solution rather than into pure water. Our picture of the dissolving process involved the attraction between the polar water molecules and the ions in the crystal. If there are other ions in the water, they may also attract ions from the crystal into the solution. If this is the case, then the solubility product constant should not be constant over a range of different concentrations of other ions in solution.

To account for the attracting ability of a solution due to the presence of solvated ions, the *ionic strength* I is defined as a new concentration unit. The ionic strength is one-half the sum of the products of the molar concentrations of each of the ions times the square of its charge. This is expressed by Equation (13.30) where Σ represents the sum and the subscript i represents the ith ion over which the summation is taken; z is the ionic charge.

$$I = \tfrac{1}{2}\Sigma M_i z_i^2 \qquad (13.30)$$

The computation of the ionic strength of a solution may be illustrated by the following example: A solution has the following ions at the indicated molarity: $M_{\mathrm{Na(aq)^+}} = 0.20\ M$; $M_{\mathrm{Ba(aq)^{++}}} = 0.15\ M$; $M_{\mathrm{Ca(aq)^{++}}} = 0.30\ M$; $M_{\mathrm{NO_3(aq)^-}} = 0.20\ M$; $M_{\mathrm{Cl(aq)^-}} = 0.90\ M$. The ionic strength is computed as follows:

$$I = \tfrac{1}{2}[(0.20 \times 1^2) + (0.15 \times 2^2) + (0.30 \times 2^2) + (0.20 \times 1^2) + (0.90 \times 1^2)]$$

$$= \tfrac{1}{2}[0.20 + 0.60 + 1.20 + 0.20 + 0.90] = 1.55$$

Figure 13.10 shows a plot of the solubility product of thallium(I) iodate against the ionic strength of the solution into which the thallium(I) iodate is dissolving. The ions used to create the solution of specified ionic strength are unimportant as long as they do not react in some other way with the thallium iodate. Thus we

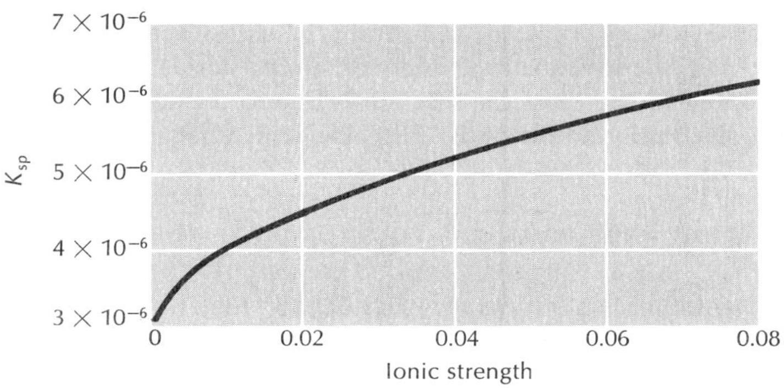

Fig. 13.10. The effect of ionic strength on the solubility product constant for thallium(I) iodate

must modify our notion of the solubility product constant to recognize that it is constant only at constant ionic strength.

In addition, the clustering effect of hydrated ions affects equilibrium constant computations in general. Whereas we have used the apparent molar concentrations for ions in solution for solubility product constant calculations, for example, in fact the *effective* concentration of these ions should be used.

The effective concentration is called the *activity* of the ion and is given the symbol a. In order to obtain the activity, we imagine multiplying the apparent concentration of the species by a corrective factor called the *activity coefficient,* which is given the symbol γ. Thus the activity may be defined by Equation (13.31)

$$a = M\gamma \tag{13.31}$$

In dilute solutions the activity coefficient is nearly unity, and the molar concentration may be used interchangeably with the molar activity without any appreciable error. However, strictly speaking, equilibrium constants for reactions in solution are correct only when activities are used instead of concentrations. It is customary to use brackets for equilibrium molar concentrations and parentheses for activities. Thus for the solubility equilibrium depicted by Equation (13.32), the correct solubility product expression is given by Equation (13.33).

$$MgF_2(s) = Mg(aq)^{++} + 2F(aq)^- \tag{13.32}$$

$$K_{sp} = (Mg(aq)^{++})(F(aq)^-)^2 = [Mg(aq)^{++}][F(aq)^-]^2\gamma_{Mg(aq)^{++}}\gamma^2_{F(aq)^-} \tag{13.33}$$

One of the difficulties in using activities is finding the appropriate value for the activity coefficients. For moderately dilute solutions, the empirical relationship shown in Equation (13.34) may be used.

$$\log\gamma_{ion} = -0.509z^2\sqrt{I} \tag{13.34}$$

In this equation z is the charge on the ion in question.

Consider the following problem to illustrate the use of activity coefficients in computations: Calculate the molar solubility of silver chloride in 0.03 M $NaNO_3$.

Most of the ionic strength of the solution is due to the sodium nitrate, since relatively little silver chloride dissolves. Thus the ionic strength I may be calculated as follows:

$$I = \tfrac{1}{2}[(0.03 \times 1^2) + (0.03 \times 1^2) = 0.03$$

Then the activity coefficients for $Ag(aq)^+$ and $Cl(aq)^-$ may be computed through Equation (13.34):

$$\log \gamma_{Ag(aq)^+} = -0.509 \times 1^2\sqrt{0.03} = -0.0875$$

$$\gamma_{Ag(aq)^+} = \text{antilog}\,(-0.0875) = 0.82$$

$$\log \gamma_{Cl(aq)^-} = -0.509 \times (-1)^2\sqrt{0.03} = -0.0875$$

$$\gamma_{Cl(aq)^-} = 0.82$$

The solubility product constant expression then may be developed:

$$K_{sp} = (Ag(aq)^+)(Cl(aq)^-) = [Ag(aq)^+][Cl(aq)^-]\gamma_{Ag(aq)^+}\gamma_{Cl(aq)^-} = 1.8 \times 10^{-10}$$

$$[Ag(aq)^+][Cl(aq)^-] = \frac{1.8 \times 10^{-10}}{\gamma_{Ag(aq)^+}\gamma_{Cl(aq)^-}} = \frac{1.8 \times 10^{-10}}{(0.82)^2}$$

Let the molar solubility of AgCl(s) $= x$

Then $[Ag(aq)^+] = [Cl(aq)^-] = x$

$$x^2 = \frac{1.8 \times 10^{-10}}{(0.82)^2}$$

$$x = 1.6 \times 10^{-5}\ M$$

SUMMARY

Solubility and the dissolving process show some applications of thermodynamic principles to heterogeneous systems. In addition, some manipulations of equilibrium constants illustrate how some computations with equilibrium constants may be simplified by the use of legitimate approximations based upon the nature of the specific system involved. Equilibrium calculations are useful in computing solubilities, predicting precipitations, and discovering the magnitude of the common-ion effect.

Solutions are very common systems in which chemical reactions occur, and in some respects there are similarities between solutions and colloidal dispersions. Although the colloidal dispersions often are useful in their own right, they also cause problems, especially in the precipitation process.

Some of the forces responsible for colloidal dispersion formation also contribute to the apparent anomalies among the colligative properties of solutions of electrolytes. In addition, the tendency for ionic solutes to aggregate somewhat accounts for increased solubility of electrolytes in the presence of other electrolytes, a circumstance that requires a modification also of the solubility product constant expression.

QUESTIONS AND PROBLEMS

13.1 Distinguish between gaseous, liquid, and solid solutions, giving examples.

13.2 How many grams of each of the following substances should be dissolved in water to produce one liter of 0.10 *M* solution?

a) HCl
b) sodium hydroxide
c) H_2SO_4
d) calcium nitrate-4-hydrate
e) $Al_2(SO_4)_3$
f) H_3PO_4

13.3 How many grams of each of the following substances should be dissolved in water to produce 310 ml of a solution that is 0.10 *M* in sodium ion?
a) NaCl
b) Na_2CO_3
c) Na_3PO_4
d) $Na_2SO_4 \cdot 10H_2O$

13.4 How many grams of each of the following substances is contained in 475 ml of a 0.23 *M* solution of each substance?
a) $Ba(OH)_2$
b) $Cr_2(SO_4)_3$
c) $CaCl_2$

13.5 If 423 ml of an aqueous solution of potassium chloride contains 5.69 g of potassium chloride, what is the molarity of the solution?

13.6 If 1600 ml of aqueous solution contains 413 g of H_2SO_4, what is the molarity of the solution?

13.7 How many liters of 1.27 *M* sodium chloride solution can be made from 956 g of sodium chloride?

13.8 What is the molality of a 5.0 percent aqueous solution of ammonium chloride? What additional information, if any, is needed to compute the molarity?

13.9 A solution is prepared by mixing 100 ml of each of the following liquids: ethanol (density 0.79 g/ml), diethyl ether (density 0.71 g/ml), and carbon tetrachloride (density 1.60 g/ml). Assuming that the volumes are additive, compute each of the following:
a) volume percent of each substance
b) mass percent of each substance
c) mole fraction and mole percent of each substance
d) molarity of carbon tetrachloride

13.10 Often the concentrations of commercially available solutions are indicated by mass percent. For each of the following substances, the mass percent and the density of the aqueous solution available commercially is indicated. Compute the molarity of these solutions.
a) ammonia, 58.6 percent, 0.90 g/ml;
b) hydrochloric acid, 36.0 percent, 1.19 g/ml;
c) sulfuric acid, 96.0 percent, 1.84 g/ml;
d) acetic acid, 99.5 percent, 1.06 g/ml.

13.11 Using the data in Problem 13.10, compute the number of milliliters of commercially available concentrated hydrochloric acid that must be used to prepare 500 ml of 0.55 *M* aqueous hydrochloric acid solution.

13.12 What is the molarity of a sulfuric acid solution made by diluting 100 ml of the commercially available concentrated acid solution to 300 ml? (See Problem 13.10 for data on the commercially available solution.)

13.13 Make each of the following conversions:
a) 2.5 *M* barium chloride to milligrams of $BaCl_2$ per milliliter of solution
b) 5.0 g sodium sulfate per liter of aqueous solution to molarity
c) 0.58 mmole Na_2HPO_4 per milliliter of aqueous solution to grams of Na_2HPO_4 per liter of solution
d) 15% $CaCl_2$ in water to molality
e) 0.60 *m* NH_4NO_3 in water to percent ammonium nitrate

13.14 How many moles of H_3PO_4 are there in 150 ml of 6.0 *M* solution?

13.15 How many mmoles of potassium permanganate are there in 350 ml of 0.65 *M* aqueous solution? How many grams?

13.16 To carry out a reaction, 35 g of Na_3PO_4 is required. How many milliliters of a 2.5 *M* solution of sodium phosphate must be used?

13.17 To carry out a reaction, 8.4 mmoles of hydrogen bromide are needed. How many milliliters of 0.70 *M* solution of hydrobromic acid must be used?

13.18 How can you prepare 35.0 ml of a 0.65 *M* solution from a 3.4 *M* solution?

13.19 If 35.0 ml of 2.00 *M* $MgCl_2$ is mixed with 85 ml of 3.10 *M* $(NH_4)_2SO_4$, what is the concentration of each compound in the final solution? What is the concentration of each ion in the final solution?

13.20 How many grams of manganese are there in 150 ml of 0.25 *M* manganese(II) sulfate?

13.21 How can you make 500 ml of a solution containing 10 mg of iron per milliliter from a solution that is 3.0 *M* in iron(III) nitrate?

13.22 For each of the following substances, comment on the probable solubility in water and in carbon tetrachloride:

a) $H_2(g)$
b) $HBr(g)$
c) $I_2(s)$
d) $Na_2SO_4(s)$
e) $SiO_2(s)$
f) $CO_2(g)$
g) $N_2O(g)$
h) $CH_3NH_2(g)$
i) $H_2S(g)$
j) $CHCl_3(l)$

13.23 Why do perchlorate ions hydrate more readily than periodate ions?

13.24 Why does the enthalpy of solution depend upon the concentration of the solution? Sometimes we talk about the enthalpy of dilution, that is, the enthalpy change occurring when a solution is made more dilute by the addition of more solvent. Why should there be any enthalpy change in the dilution process?

13.25 For each of the following gases, the aqueous solubility is indicated in grams of gas per 100 g of water at 25°C and a partial pressure of 726 mm Hg for the gas. Compute the value of the solubility constant and the molar solubility for each gas, assuming that the final volume of each solution is 100.0 ml and that each gas is ideal.

a) acetylene, 0.105
b) carbon monoxide, 0.00260
c) methane, 0.00209
d) ethene, 0.0131

13.26 What is the solubility-constant expression for one liquid dissolving in another liquid?

13.27 How can you account for the fact that ethanol is miscible with both water and carbon tetrachloride?

13.28 Sometimes in the laboratory, wet glassware is rinsed first with ethanol to remove the water and then with diethyl ether to remove the ethanol. What would happen if the order of the rinses was reversed?

13.29 A saturated aqueous solution of nitromethane at approximately room conditions is 1.5 *M*, whereas a saturated aqueous solution of nitrobenzene is about 0.015 *M* under similar conditions. Suggest a reason for this difference.

13.30 Would you expect *m*-dinitrobenzene or *p*-dinitrobenzene to be the more soluble in water under comparable conditions? Why?

13.31 Chlorobenzene is miscible with ethanol, but iodobenzene is not. Suggest a reason for this.

13.32 There appear to be two different solubilities possible for zinc sulfide in water at the same temperature. Suggest a reason for this.

13.33 How is the solubility of silver(I) acetate affected by the addition of a) silver(I) nitrate; b) sodium acetate?

13.34 Suppose that AZ, BZ, and CX are three slightly soluble salts each composed of univalent cations and anions. The corresponding molecular weights are 120,

130, and 140. If the respective solubilities are 0.1080, 0.1040, and 0.0980 g/l of solution, what are the solubility product constants for these three salts?

13.35 Criticize the following statement. In comparing solubility product constants for two substances, the substance with the higher solubility product constant has the greater solubility.

13.36 The solubility equilibrium for the hypothetical compound RQ_2 in water is given by the following equation.

$$RQ_2(s) = R(aq)^{++} + 2Q(aq)^-$$

The solubility product constant at room temperature is 3.2×10^{-11}. Compute the number of moles of $RQ_2(s)$ which can dissolve in

a) 50.0 ml pure water;

b) 50.0 ml of an aqueous solution in which the concentration of $R(aq)^{++}$ already is 0.020 *M*;

c) 50.0 ml of a solution in which the concentration of $Q(aq)^-$ already is 0.020 *M*.

13.37 Criticize the following statement: Lead(II) iodate is more soluble in 0.010 *M* sodium iodate than it is in 0.010 *M* lead(II) nitrate.

13.38 Using the data in Table 13.5, compute the molar solubilities of the following in water:

a) MgF_2 b) CuCl c) $AgIO_3$ d) $Hg_2(IO_3)_2$ e) $La(IO_3)_3$

13.39 It is found that 1.6×10^{-6} mole of lead(II) sulfate is all that can be dissolved in 100 ml of 0.0010 *M* sodium sulfate.

a) Compute the solubility product constant for lead(II) sulfate and compare it to the value recorded in Table 13.5.

b) Should a precipitate be formed if 20.0 ml of 2.0×10^{-4} *M* lead(II) nitrate is mixed with 80.0 ml of 1.0×10^{-4} *M* sodium sulfate? Justify your answer.

13.40 The solubility of each of the following compounds was reported to be as indicated in grams per 100 ml of solution. In each case compute the value of the solubility product constant and compare to the value recorded in Table 13.5.

a) $BaSO_4$, 2.4×10^{-4} b) Hg_2SO_4, 0.25 c) Hg_2Cl_2, 8.5×10^{-5}

13.41 In each of the following cases show whether or not a precipitate will form.

a) 1 ml of 0.1 *M* silver(I) nitrate is added to 1 l of 0.01 *M* potassium sulfate.

b) 1 g of thallium(I) nitrate is put into 100 ml of 0.01 *M* hydrochloric acid.

c) 1 ml of 1.0 *M* sodium hydroxide is added to 1 l of 10^{-4} *M* magnesium chloride.

13.42 What iodate ion concentration is needed to commence precipitation of each of the following metals as the iodate if each is present to the extent of 1 mg of metal ion per 10 ml of solution?

a) $Ag(aq)^+$ b) $Pb(aq)^{++}$ c) $La(aq)^{3+}$

13.43 What concentration of hydrated bromide ion is needed to start precipitation of silver bromide from a saturated solution of each of the following? a) silver sulfate; b) silver chloride.

13.44 Without doing the complete formal calculation, but by comparing solubility product constant values from Table 13.5, indicate which compound in each of the following pairs is the *least soluble:*

a) AgI, AgBr

b) $BaSO_4$, $PbSO_4$

c) MgF_2, BaF_2

d) $Ba(IO_3)_2$, $BaSO_4$

e) $BaSO_4$, $La(IO_3)_3$

13.45 Compute the molar solubility of barium sulfate in a) 0.0005 *M* aluminum sulfate; b) 0.0005 *M* barium chloride.

13.46 Compute the molar solubility of silver sulfate in a) 0.0005 *M* aluminum sulfate; b) 0.0005 *M* silver nitrate. Be sure that any assumptions you make are valid. At least show how to solve these problems.

13.47 Consider 75 ml of 0.21 *M* silver nitrate.
a) What mass of silver chloride can be obtained by precipitating all of the silver as silver chloride?
b) What mass of calcium chloride is required to precipitate all of the silver as silver chloride?
c) What volume of 0.32 *M* sodium chloride is required to precipitate all of the silver as silver chloride?

13.48 When 50.0 ml of 1.00 *M* calcium chloride is added to a solution in which a 1.00 g sample of an unknown has been dissolved, 0.444 g of calcium fluoride precipitates. What is the percent of fluoride in the unknown sample?

13.49 If 20 g of a compound with the formula $C_6H_{10}O_5$ is dissolved in 250 g of water, what are the boiling and freezing points of the solution? What is the vapor pressure of the water at the boiling point?

13.50 Calculate the molecular weights for the following nonelectrolytes:
a) Water boils at 100.3°C when 2.4 g of solute is dissolved in 80 g of water.
b) Water freezes at −0.50°C when 12 g of solute is dissolved in 20 g of water.
c) The freezing point of benzene is lowered 0.53°C when 0.372 g of solute is dissolved in 20 g of benzene.
d) The freezing point of camphor is lowered 22°C when 1.5 g of solute is dissolved in 30 g of camphor.

13.51 For each of the following pairs of conditions, which solution will have the higher freezing point?
a) aqueous 0.5 *m* sucrose, $C_{12}H_{22}O_{11}$; or aqueous 0.5 *m* ethylene glycol, $HOCH_2CH_2OH$?
b) 10 g of naphthalene, $C_{10}H_8$, in 100 ml of benzene; or 10 g of camphor, $C_{10}H_{16}O$, in 100 ml of benzene?
c) 10 g of naphthalene, $C_{10}H_8$, in 100 g of benzene; or 10 g of naphthalene in 100 g of 1,2-dibromoethane?
d) 2 g of iodobenzene, C_6H_5I, in 35 g of nitrobenzene; or 14 g of dibromobenzene, $C_6H_4Br_2$, in 89 g of naphthalene, $C_{10}H_8$?

13.52 A 4.25-g sample of sulfur dissolved in 15.0 g of benzene lowered the freezing point of benzene by 0.76°C. What was the formula of the sulfur?

13.53 A compound is 38.7 percent carbon, 9.7 percent hydrogen, and 51.6 percent oxygen. A solution of 3.32 g of this compound in 100 g of water was nonelectrolytic and froze at −1.00°C. Calculate the approximate molecular weight, the formula for the compound, and the exact molecular weight.

13.54 Explain why it is difficult to use colligative properties to determine the molecular weights of electrolytes.

13.55 Explain why a 1.0 *m* aqueous solution of sodium chloride freezes at about −3.0°C.

13.56 Calculate the mole number *i* for a 0.2 *m* aqueous solution of an electrolyte that has a freezing point of −0.62°C.

13.57 If a 0.1 *m* aqueous solution of an electrolyte has a boiling point of 100.135°C, how many ions are produced per mole of compound?

13.58 Discuss the Debye-Hückel interionic attraction theory of electrolytes in solution and explain how this theory explains "abnormalities" in data on freezing-point lowering.

13.59 Explain why silver chloride is more soluble in 0.1 *M* KNO_3 than in water but less soluble in 0.1 *M* $AgNO_3$ than in water.

13.60 Compute the ionic strength of a solution made by mixing 50 ml 0.2 *M* barium chloride, 50 ml 0.4 *M* magnesium nitrate, and 50 ml 0.5 *M* sodium chloride.

14

Chemical Reactions: How They Occur

INTRODUCTION The study of thermodynamics leads to an understanding of why chemical reactions occur. They occur in order to decrease the free energy of the system. On this basis, it is clear that reactions cannot be carried out spontaneously if they are accompanied by a free-energy increase.

On the other hand, some reactions that would be accompanied by a free-energy decrease do not appear to occur, or occur so slowly that no change is really observed. There appears to be no simple relationship between the over-all free-energy decrease for a reaction and the speed with which it occurs. Also recall that the free-energy change for a reaction is independent of the path by which the change occurs.

In this chapter, we shall take a brief look at *chemical kinetics*, the study of the rates of chemical reactions. Kinetics provides information about the paths by which chemical reactions occur—something that thermodynamics cannot provide. These paths are called the *mechanisms* of the reactions.

We begin this chapter with a brief summary of some of the experimental results of kinetic studies. This is followed by theoretical interpretations and specific examples.

THE MEANING OF REACTION RATE

The rate of a chemical reaction is expressed as the number of moles per liter of reactant particles disappearing per unit time, or as the number of moles per liter of product particles produced per unit time. Thus the rate of reaction is the change in concentration per unit time for a reactant or a product.

As a matter of convention, the rate of disappearance of a reactant is labeled as a negative rate and the rate of appearance of a product is labeled as a positive rate.

For example, suppose that a generalized reaction may be represented by Equation (14.1).

$$A + 2B = 3C + D \tag{14.1}$$

The rate of this reaction may be expressed as the rate of disappearance of either of the reactants, A or B, or as the rate of appearance of either of the products, C or D. In this case, we may express the rate of disappearance of A as $-r_A$ and of B as $-r_B$, and the rates of appearance of C and D as r_C and r_D respectively.

From Equation (14.1), it should be clear that reactant B disappears twice as fast as reactant A, since 2 moles of B disappear every time 1 mole of A disappears. Similarly, the rate of appearance of product D is numerically the same as the rate of disappearance of reactant A, since 1 mole of D is produced for each mole of A that disappears. Also, the rate at which product C appears is three times the rate at which product D appears. This is summarized in Equations (14.2) through (14.5).

$$-r_B = 2(-r_A) \tag{14.2}$$

$$r_D = -r_A \tag{14.3}$$

$$r_C = 3r_D \tag{14.4}$$

$$r_C = 3r_D = -\frac{3}{2}r_B = -3r_A \tag{14.5}$$

This means that if we wish to record absolute reaction rate values, we must specify the chemical species upon which the rate is based. However, it is clear from Equation (14.5) that each of the rates for a particular reaction is proportional to each of the other rates. Therefore, if we are interested in proportional relationships, we can think of a generalized reaction rate. For example, if something happens to double the reaction rate based upon reactant A, it will also double the reaction rates based upon reactant B and products C and D.

These definitions establish the base from which reaction rates may be determined experimentally. To determine a reaction rate, it is necessary to measure the concentration of one or more components at various intervals of time. It is the job of the analytical chemist to provide techniques for measuring the concentration of a component of a system. Gravimetric analyses may be used, for example, or the changing intensity of the color of a colored chemical species might be used. Some other techniques will be discussed in later chapters.

In the following sections, we shall summarize some of the general observations that have emerged from the study of reaction rates.

FACTORS AFFECTING REACTION RATES

Temperature

An increase in temperature increases the rate of reaction. Do not confuse this effect with the effect of temperature upon an equilibrium. In an equilibrium, there are two reactions occurring simultaneously, namely, the forward and reverse reactions. An increase in temperature increases the rate of both reactions, but usually not equally. The increase in temperature increases the rate of the endothermic reaction more than it increases the rate of the exothermic reaction This leads to a shift in the equilibrium in favor of the endothermic reaction.

Catalysts

Catalysts are substances that increase the rate of chemical reaction without themselves appearing among the over-all products of the reaction. For example, aqueous hydrogen peroxide decomposes slowly according to the following equation:

$$2H_2O_2(aq) = 2H_2O(l) + O_2(g) \tag{14.6}$$

However, if solid manganese(IV) oxide is added to the solution, the oxygen is released rapidly enough to cause vigorous effervescence. However, at the end of the reaction, there is as much manganese(IV) oxide present as there was initially. The manganese(IV) oxide served as a catalyst for this reaction. The only net effect of a catalyst is to increase the rate of reaction.

Reactant Concentration

In general, it is found that the rate of a reaction depends upon the concentration of the reactants. This is not always true, and in some instances, where there are several reactants, the rate of reaction may depend on the concentrations of only one or two of the reactants while remaining independent of the concentrations of the other reactants.

The mathematical expression relating the rate of reaction to the concentrations of the chemical components is called the *rate equation* for the reaction.

Often it is found that the reaction rate is directly proportional to the concentrations of some reactants, each raised to some integral power.

For example, nitrogen(V) oxide dissolved in carbon tetrachloride decomposes according to Equation (14.7), where (sv) means solvated.

$$2N_2O_5(sv) = 4NO_2(g) + O_2(g) \tag{14.7}$$

Experimentally it has been found that the generalized reaction rate r_f for the forward reaction is directly proportional to the molarity of the nitrogen(V) oxide. This is expressed by Equation (14.8), where k_f is the proportionality constant (or rate constant) and $M_{N_2O_5}$ is the molarity of nitrogen(V) oxide.

$$r_f = k_f M_{N_2O_5} \tag{14.8}$$

For another example, consider the gas-phase reaction of nitrogen(II) oxide with oxygen:

$$2NO(g) + O_2(g) = 2NO_2(g) \tag{14.9}$$

The generalized rate equation for the forward reaction is

$$r_f = k_f M_{NO}^2 M_{O_2} \tag{14.10}$$

The experimentally determined rate equation for the reverse reaction is

$$r_r = k_r M_{NO_2}^2 \tag{14.11}$$

There are cases, however, in which fractional exponents are found for the reactant concentrations. For example, chlorine and carbon monoxide combine to form carbonyl chloride:

$$Cl_2(g) + CO(g) = COCl_2(g) \tag{14.12}$$

The experimentally determined rate equation is

$$r_f = k_f M_{CO} M_{Cl_2}^{3/2} \tag{14.13}$$

There are still other cases in which the rate equations are much more complex. For example, hydrogen and bromine combine to form hydrogen bromide:

$$H_2(g) + Br_2(g) = 2HBr(g) \tag{14.14}$$

The corresponding rate equation is

$$r_f = \frac{k_f M_{H_2} M_{Br_2}^{3/2}}{M_{Br_2} + 0.1\ M_{HBr}} \tag{14.15}$$

Equations (14.8), (14.10), (14.11), (14.13), and (14.15) are empirical equations, that is, they have been derived experimentally. They represent the experimentally determined relationship between the rates of reactions and the concentrations of components of the reacting systems.

It is customary to define *reaction order* as the sum of the exponents associated with the concentrations in the rate equation. The total reaction order is the sum of all exponents on all concentrations in the rate equation. On the other hand, the reaction order *with respect to* a particular component in the reacting system is the exponent associated with the concentration of that component in the rate equation.

For example, the reaction represented by Equation (14.7) is first order over-all and first order with respect to nitrogen(V) oxide. The reaction represented by Equation (14.9) is third order over-all, second order with respect to nitrogen(II) oxide and first order with respect to oxygen. The reverse reaction of Equation (14.9) is second order over-all and second order with respect to nitrogen dioxide. The reaction represented by Equation (14.12) is 5/2 order over-all, 3/2 order with respect to chlorine, and first order with respect to carbon monoxide.

INTERPRETATION OF REACTION RATES

Collision Theory

In order to account for the observations made on reaction rates, a theoretical model for the manner in which chemical reactions occur has developed.

According to this theory, in order for a reaction to occur, there must be collisions among the ions, atoms, and molecules in the system, or among all three. These collisions must induce new bond formation or the breaking of existing bonds, or both. Not all collisions are effective in this way.

For example, as two atoms approach, there is an initial repulsion among electron clouds that must be overcome (Chapter 6). This means that there is an increase in potential energy, and in a large-scale system this is reflected as an increase in enthalpy. Thus as atomic, molecular, or ionic particles approach each other, there is an enthalpy increase. In order for the particles to come together sufficiently so that their electron clouds merge, which probably is necessary for reaction to occur, they must have sufficient kinetic energy to overcome this potential-energy barrier. Thus collisions between two particles having low kinetic energy will not tend to be effective collisions insofar as bond-making and bond-breaking are concerned.

In some cases, the collisions may be effective only if the particles collide in the right way. Thus, if a molecule such as 1-propanol, $CH_3CH_2CH_2OH$, participates in a reaction that involves the breaking of the O—H bond, the collision of another particle with the number three carbon probably will not be very effective. This means that in some cases collisions must occur in some orderly geometric pattern in order to be effective.

The course of a simple chemical reaction may be described graphically as in Fig. 14.1. In this graph, the reaction coordinate simply is a convenient axis on which to record the progress of the reaction. Thus movements from left to right on this axis describe the progress of the reaction. The free energy of the isolated reactants is labeled G_r. As the particles of reactants approach, the free energy increases to a maximum, G_a. This increase in free energy is due in part to the increase in enthalpy associated with electron cloud repulsions, and in part to the decrease in entropy associated with the geometrical requirements for effective collisions.

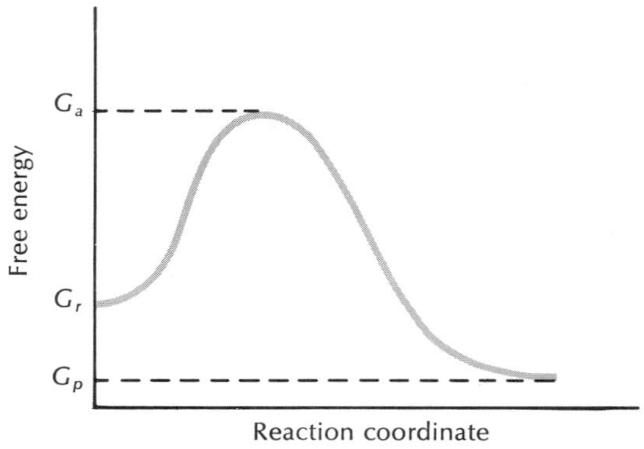

Fig. 14.1. Representation of a typical free-energy change as a chemical reaction occurs

At the maximum free energy G_a the electron clouds of the colliding particles merge and a new composite particle called the *activated complex* forms. This activated complex is like a molecule or ion, except that it is very unstable; it is in the process of coming apart. The activated complex exists for such a short time that it cannot be isolated.

Within the activated complex, old bonds break and new ones form. As this occurs, the free energy decreases as represented by the right side of the "hill" in Fig. 14.1, until the free energy of the products is reached at G_p. This drawing represents the so-called free-energy hill or barrier over which the chemical reaction must pass. The increase in free energy from the initial reactants to the activated complex is called the *free energy of activation*, ΔG_a, and it is defined by Equation (14.16).

$$\Delta G_a = G_a - G_r \tag{14.16}$$

If the free energy of activation is small, that is, if the hill in Fig. 14.1 is low, then collision efficiency is high and the reaction rate is high. However, if the free energy of activation is large, then collision efficiency is low and the reaction rate is low.

The change in free energy for the over-all reaction ΔG is given simply by the difference in free energies for the reactants and products:

$$\Delta G = G_p - G_r \tag{14.17}$$

Thus the reaction represented in Fig. 14.1 has a favorable free-energy change, and the products are at lower free energy than the reactants. However, if the free energy of activation is very large, the free-energy hill is very high, and the reaction may proceed so slowly that for all intents and purposes it does not occur at all.

Collision Frequency and Collision Efficiency

Reaction rate depends on both collision frequency and collision efficiency. The more collisions there are per unit time, the greater is the chance that an effective collision will occur, and the greater the reaction rate will be. Also if a high percentage of the collisions that do occur actually lead to a change, the reaction rate will tend to be high. The free energy of activation already has been mentioned as an important factor affecting collision efficiency. However, temperature and concentration of reactants also are important factors affecting collision frequency and collision efficiency.

TEMPERATURE. Temperature changes are related to both collision frequency and collision efficiency. An increase in temperature corresponds to an increase in average kinetic energy. This means that there is an increase in the average speed of the particles and therefore an increase in collision frequency.

In addition, the increase in average kinetic energy accompanying the temperature increase leads to an increase in collision efficiency, for more particles have sufficient kinetic energy to overcome the potential-energy barrier that is due to electron cloud repulsions. Thus an increase in temperature leads to an increase in reaction rate as observed.

CONCENTRATION. Concentration changes affect the collision frequency. As the concentration of one of the particles increases, the probability of collision increases also per unit of time. In fact, this appears to lend itself to quantitative interpretations.

For example, suppose that some volume of solution contains one A particle and four B particles. The A particle then has the chance to collide with each of the four B particles. However, if the number of B particles is doubled to eight, each A particle has the chance of colliding with either of the eight B particles. Thus, the doubling of the concentration of B particles doubles the possibilities for collisions. Similarly, if this is followed by a doubling of the concentration of A particles from one to two in this volume, then the possibilities for collisions between A and B particles are doubled from eight to sixteen.

This suggests that the rate of a reaction is directly proportional to the concentration of each of the colliding particles. In turn, this suggests a way to derive rate equations for chemical reactions. However, the fact that not all reactants need be represented in the rate equation indicates that some modifications are necessary.

It is unlikely that large numbers of particles will come together in one grand collision. Rather it is more likely that two particles will collide and later the products of this collision will undergo further collisions with other particles. In this way, complex reactions may be viewed as a series of consecutive, simple, single-step reactions. The series of steps through which an over-all reaction proceeds is called the *mechanism* of the reaction.

Some Types of Single-Step Reactions

A single-step reaction is viewed as the reaction resulting from the simultaneous collision of several particles. The number of particles involved in this single collision, and consequently the number of particles combined in the activated complex, is called the *molecularity* of the reaction. Note that the molecularity is defined for single-step reactions. When two particles collide, the reaction is said to be bimolecular, and when three particles collide, the reaction is said to be termolecular. The unimolecular reaction is one in which it appears that a single particle undergoes a change without requiring a collision with another particle. This definition is a bit deceptive, and the unimolecular reaction will be discussed a little later.

The following sections describe briefly three commonly postulated single-step reactions.

THE IRREVERSIBLE REACTION. Suppose that a single-step reaction may be described as follows:

$$A + B \rightarrow C + D \tag{14.18}$$

The arrow indicates that the reaction is irreversible; that is, C does not react with D to produce A and B. This is not an equilibrium reaction. The equation indicates that there is a collision between one A molecule and one B molecule.

Since the rate of this single-step reaction is proportional to the concentration of each of the colliding particles, the rate equation for this reaction may be written as follows:

$$r_f = k_f M_A M_B \tag{14.19}$$

Clearly, as this reaction proceeds, the concentrations of A and B decrease and the rate of reaction decreases.

Such irreversible reactions may occur rapidly or slowly. Normally, if the mechanism of a reaction consists of several consecutive irreversible steps, each step will occur at its own distinctive rate. Then the over-all rate of reaction will be the rate of the slowest step, for the reaction cannot occur any more rapidly than the slowest step in the series of consecutive steps. In principle, this resembles a stream of heavy highway traffic, in which the cars cannot proceed at a greater rate than the rate of the slowest moving vehicles.

For example, suppose that the over-all reaction represented by Equation (14.23) consists of three steps represented by Equations (14.20) through (14.22).

$$A + B \rightarrow C + D \quad \text{(fast)} \tag{14.20}$$

$$C + E \rightarrow F + G \quad \text{(slow)} \tag{14.21}$$

$$D + F \rightarrow H + I \quad \text{(fast)} \tag{14.22}$$

$$A + B + E \rightarrow G + H + I \tag{14.23}$$

In this case, reactants A and B are consumed quickly to produce C and D. However, C is consumed slowly in reacting with E to produce F and G. Then F is consumed quickly as it is produced by reacting with D, which was produced in the first step. The rate of the over-all reaction represented by Equation (14.23) is the rate of the second step, represented by Equation (14.21). The rate equation for this second step of the mechanism is given by Equation (14.24).

$$r_f = k_f M_C M_E \tag{14.24}$$

However C has only intermediate existence. It is not one of the original reactants, nor is it one of the final products. It is produced in the first step of the reaction and consumed in the seond step. Thus if we determine the rate equation for the over-all reaction experimentally, the rate will appear to depend only upon the concentration of E, since the existence of C probably will not be known. Thus the experimentally determined rate equation for the over-all reaction as represented by Equation (14.23) is given by Equation (14.25).

$$r_f = k_f M_E \tag{14.25}$$

The reaction represented by Equation (14.23) should thus be found experimentally to be a first-order reaction. Each of the steps in the mechanism of this reaction is second order and bimolecular. Often the over-all reaction is assigned a molecularity equal to the molecularity of the rate-determining step, that is, the slowest step. Thus the over-all reaction is said to be bimolecular.

As another hypothetical example, consider the over-all reaction represented by Equation (14.28), which occurs in two steps indicated by Equations (14.26) and (14.27).

$$A + B \rightarrow C + D \qquad \text{(slow)} \tag{14.26}$$

$$A + C \rightarrow E + F \qquad \text{(fast)} \tag{14.27}$$

$$2A + B \rightarrow D + E + F \tag{14.28}$$

In this case A and B react slowly to produce C and D and as quickly as C is produced it reacts with A to produce E and F. Since the initial step is the slow step, it is the rate-determining step, and the rate of this initial step is the rate of the over-all reaction. Thus the rate of the over-all reaction may be expressed as follows:

$$r_f = k_f M_A M_B \tag{14.29}$$

In this example the over-all reaction is second order and bimolecular.

As a final hypothetical reaction, consider the reaction represented by Equation (14.33), the mechanism for which consists of three steps represented by Equations (14.30), (14.31), and (14.32).

$$A + B \rightarrow C + D \qquad \text{(fast)} \tag{14.30}$$

$$B + D \rightarrow 2E \qquad \text{(fast)} \tag{14.31}$$

$$C + E \rightarrow 2F \qquad \text{(slow)} \tag{14.32}$$

$$A + 2B \rightarrow E + 2F \tag{14.33}$$

Here reactants A and B react rapidly to produce C and D. Then reactant B consumes D rapidly to produce E. Finally C and E react slowly to produce F. This final step is the rate-determining step, and the rate equation for it may be expressed as

$$r_f = k_f M_C M_E \tag{14.34}$$

However, the rate-determining step does not involve any of the reactants for the over-all reaction. Thus in a determination of the relationship between the rate of the over-all reaction and the concentrations of the reactants A and B, the experimental results will show that the rate of the reaction is independent of the concentrations of A and B. The rate equation will then be

$$r_f = k_f \tag{14.35}$$

In this example the reaction is said to be zero order, although it is bimolecular in the rate-determining step.

THE RAPID EQUILIBRIUM REACTION. Suppose that molecules A and B collide to produce molecules C and D. Then molecules C and D collide in the reverse reaction to produce original reactants A and B. This equilibrium is established rapidly and the equilibrium shifts in response to stresses very rapidly also. This equilibrium is represented by Equation (14.36), where the double arrows indicate the equilibrium condition.

$$\mathrm{A} + \mathrm{B} \rightleftharpoons \mathrm{C} + \mathrm{D} \tag{14.36}$$

The rate of the forward reaction in this equilibrium, r_{fe}, is proportional to the concentrations of A and B:

$$r_{fe} = k_{fe} M_\mathrm{A} M_\mathrm{B} \tag{14.37}$$

The rate of the reverse reaction in this equilibrium, r_{re}, is proportional to the concentrations of C and D:

$$r_{re} = k_{re} M_\mathrm{C} M_\mathrm{D} \tag{14.38}$$

At equilibrium, the rates of the forward and reverse reactions are equal, as indicated in Equation (14.39), where the concentrations are in brackets to indicate that they are equilibrium concentrations.

$$r_{fe} = r_{re} = k_{fe}[\mathrm{A}][\mathrm{B}] = k_{re}[\mathrm{C}][\mathrm{D}] \tag{14.39}$$

Equation (14.39) may be rearranged as shown in Equation (14.40):

$$\frac{k_{fe}}{k_{re}} = \frac{[\mathrm{C}][\mathrm{D}]}{[\mathrm{A}][\mathrm{B}]} = K_c \tag{14.40}$$

Thus the equilibrium constant for this and other single-step equilibria is the ratio of the rate constant for the forward reaction to the rate constant for the reverse reaction.

Now suppose that such a rapid equilibrium step occurs before a slow rate-determining irreversible reaction. An example is given in the two steps represented by Equations (14.41) and (14.42) for the over-all reaction depicted by Equation (14.43).

$$A + B \rightleftharpoons C \quad \text{(rapid)} \tag{14.41}$$

$$A + C \rightarrow D \quad \text{(slow)} \tag{14.42}$$

$$2A + B \rightarrow D \tag{14.43}$$

In this case, reactants A and B establish a rapid equilibrium with C, and then C is slowly consumed by reaction with A to produce D. With the initial equilibrium adjusted rapidly, the equilibrium is maintained as C is consumed.

The second step, the irreversible reaction, is the rate-determining step, so that the rate of this step as well as the rate of the over-all reaction may be expressed as follows:

$$r_f = k_f M_A M_C \tag{14.44}$$

However, the concentrations of A and C are the equilibrium concentrations since both of these are involved in the initial equilibrium reaction. Thus Equation (14.44) may be rewritten as Equation (14.45).

$$r_f = k_f[A][C] \tag{14.45}$$

Although C is not one of the original reactants, its concentration is related to the concentrations of the original reactants through the equilibrium constant expression. Thus Equation (14.40) may be solved for the concentration of C:

$$[C] = \frac{k_{fe}[A][B]}{k_{re}} \tag{14.46}$$

The expression for the concentration of C in Equation (14.46) may be substituted into Equation (14.45) to produce Equation (14.47), which may be rearranged to form Equation (14.48).

$$r_f = \frac{k_f k_{fe}}{k_{re}}[A]^2[B] \tag{14.47}$$

$$r_f = k_f'[A]^2[B] \tag{14.48}$$

Equation (14.48) is the form of the anticipated experimental rate equation in which k_f' is the generalized rate constant. This reaction is third order and bimolecular.

THE UNIMOLECULAR REACTION. It is possible that collisions among molecules may occur without immediate reaction. In the collision process, certain molecules of a substance may be activated; that is, the vibrations and oscillations of the atoms within the molecules may be disturbed and at a later time these activated molecules may break apart. It has been shown that the rate of such a reaction, as represented by Equation (14.49) should be directly proportional to the concentration of the disintegrating species, as shown in Equation (14.50).

$$A \rightarrow B + C \tag{14.49}$$

$$r_f = k_f M_A \tag{14.50}$$

This type of reaction is first order and is called a unimolecular reaction.

SUMMARY. Complex reactions may be understood in terms of a series of simple single-step reactions. For the most part, these single-step reactions have low molecularities, most of them being unimolecular or bimolecular. However, reactions with a variety of over-all orders may be understood in terms of these simple reactions.

One of the areas of current research interest is reaction mechanisms. Although many techniques are employed in this endeavor, kinetic studies play a very important role. Once the experimental rate equations have been established, attempts are made to conceive of possible mechanisms that are consistent with the rate equations and with the other known properties of the systems and their components.

In the following section we look briefly at some specific reactions and interpretations that have been given to kinetic data.

Some Illustrative Reactions and Interpretations

HYDROLYSIS OF ETHYL ACETATE. Ethyl acetate reacts slowly with large amounts of water according to Equation (14.51).

$$CH_3COOCH_2CH_3 + H_2O = CH_3COOH + CH_3CH_2OH \qquad (14.51)$$

It is found that the rate of reaction may be expressed as follows:

$$r_f = k_f M_{CH_3COOCH_2CH_3} \qquad (14.52)$$

This is a first-order reaction. It might appear to be a stepwise reaction in which there is a rate-determining step that is unimolecular or involves only the ethyl acetate. However, it develops that this reaction only appears to be first order because of the large excess of water present. Thus the concentration of water in the solution is affected only very, very slightly by the small amount consumed by the reaction. Thus, for all intents and purposes, the water concentration is not changed as the reaction occurs and the reaction appears to be independent of the water concentration. However, if smaller amounts of water are used in other solvents, the reaction is second order, depending upon both the water and the ethyl acetate concentrations.

DECOMPOSITION OF NITROGEN(V) OXIDE. It was pointed out earlier that the decomposition of nitrogen(V) oxide as represented by Equation (14.53) is first order, as indicated by rate Equation (14.54).

$$2N_2O_5 = 4NO_2 + O_2 \qquad (14.53)$$

$$r_f = k_f' M_{N_2O_5} \qquad (14.54)$$

The following three-step mechanism has been proposed.

$$2(N_2O_5 \rightleftharpoons NO_2 + NO_3) \qquad \text{(rapid)} \qquad (14.55)$$

$$NO_2 + NO_3 \rightarrow NO + O_2 + NO_2 \qquad \text{(slow)} \qquad (14.56)$$

$$NO + NO_3 \rightarrow 2NO_2 \qquad \text{(fast)} \qquad (14.57)$$

The first step is the rapid equilibrium in which nitrogen(V) oxide is in equilibrium with nitrogen dioxide and nitrogen(VI) oxide, a hypothetical species that has not been identified. Then nitrogen dioxide and nitrogen(VI) oxide react slowly to produce nitrogen(II) oxide, oxygen, and nitrogen dioxide. Finally the nitrogen(II) oxide and nitrogen(VI) oxide react to produce more nitrogen dioxide. The second step, represented by Equation (14.56), is the rate-determining step, and the rate equation for this step is

$$r_f = k_f M_{NO_2} M_{NO_3} = k_f[NO_2][NO_3] \qquad (14.58)$$

However, for the equilibrium step represented by Equation (14.55), Equations (14.59) and (14.60) may be written:

$$\frac{k_{fe}}{k_{re}} = \frac{[NO_2][NO_3]}{[N_2O_5]} \qquad (14.59)$$

$$[NO_2][NO_3] = \frac{k_{fe}[N_2O_5]}{k_{re}} \qquad (14.60)$$

Equation (14.60) may be used to rewrite Equation (14.58) as Equation (14.61), which agrees with the experimentally determined rate equation.

$$r_f = \frac{k_f k_{fe}}{k_{re}}[N_2O_5] = k_f'[N_2O_5] \qquad (14.61)$$

Decomposition of hydrogen iodide. At elevated temperatures, hydrogen iodide decomposes into hydrogen and iodine, as indicated by Equation (14.62).

$$2HI(g) = H_2(g) + I_2(g) \qquad (14.62)$$

The rates of the forward and reverse reactions are indicated by Equations (14.63) and (14.64) respectively.

$$r_f = k_f M_{HI}^2 \qquad (14.63)$$

$$r_r = k_r M_{H_2} M_{I_2} \qquad (14.64)$$

Each of these equations agrees with a single-step, second-order, bimolecular reaction, and there is no need to postulate a more complicated mechanism.

Nitration of benzene. When benzene is treated with a large excess of nitric acid in acetic acid solvent, nitrobenzene is formed:

$$C_6H_6 + HNO_3 = C_6H_5NO_2 + H_2O \tag{14.65}$$

Experimentally it is found that this reaction is a zero-order reaction. The following two-step mechanism has been proposed to be consistent with this observation.

$$2HNO_3 \rightarrow NO_2^+ + NO_3^- + H_2O \quad \text{(slow)} \tag{14.66}$$

$$C_6H_6 + NO_2^+ \rightarrow C_6H_5NO_2 + H^+ \quad \text{(fast)} \tag{14.67}$$

The first step is the rate-determining step, and the rate equation for this step should be the rate equation for the over-all reaction:

$$r_f = k_f M_{HNO_3}^2 \tag{14.68}$$

However, with a large excess of nitric acid present, the concentration of nitric acid will not vary appreciably during the course of the reaction. For all intents and purposes the molarity of nitric acid is constant, and thus the rate is constant. This is consistent with a zero-order reaction.

HYDROGEN PEROXIDE AND HYDROGEN IODIDE. Hydrogen peroxide reacts with hydrogen iodide:

$$H_2O_2 + 2HI = 2H_2O + I_2 \tag{14.69}$$

The experimentally determined rate equation is

$$r_f = k_f M_{HI} M_{H_2O_2} \tag{14.70}$$

Thus this reaction is second order. The following two-step mechanism has been proposed to account for this observation.

$$H_2O_2 + HI \rightarrow HIO + H_2O \quad \text{(slow)} \tag{14.71}$$

$$HIO + HI \rightarrow H_2O + I_2 \quad \text{(fast)} \tag{14.72}$$

In this case the over-all rate equation is the same as the rate equation for the initial step, which is the same as the experimentally determined rate equation.

NITROGEN(II) OXIDE AND OXYGEN. Nitrogen(II) oxide undergoes a third-order reaction with oxygen to produce nitrogen(IV) oxide, as indicated by Equations (14.73) and (14.74).

$$2NO + O_2 = 2NO_2 \tag{14.73}$$

$$r_f = k_f' M_{NO}^2 M_{O_2} \tag{14.74}$$

Two different mechanisms have been proposed for this reaction.

The first of these is shown in the following two steps:

$$2NO \rightleftharpoons (NO)_2 \quad \text{(rapid)} \tag{14.75}$$

$$(NO)_2 + O_2 \rightarrow 2NO_2 \quad \text{(slow)} \tag{14.76}$$

The first step involves the association of two nitrogen(II) oxide molecules, which seems reasonable when it is recalled that nitrogen(II) oxide molecules each have an unpaired electron. However, the second step is the rate-determining step for the over-all reaction. Thus the rate equation for the over-all reaction may be expressed as follows:

$$r_f = k_f M_{(NO)_2} M_{O_2} = k_f[(NO)_2][O_2] \tag{14.77}$$

However, from the initial equilibrium, Equations (14.78) and (14.79) may be written.

$$\frac{k_{fe}}{k_{re}} = \frac{[(NO)_2]}{[NO]^2} \tag{14.78}$$

$$[(NO)_2] = \frac{k_{fe}}{k_{re}}[NO]^2 \tag{14.79}$$

Using Equation (14.79), Equation (14.77) may be rewritten:

$$r_f = \frac{k_f k_{fe}}{k_{re}}[NO]^2[O_2] = k_f'[NO]^2[O_2] \tag{14.80}$$

Thus, according to this mechanism, the reaction should be third order and bimolecular.

The second mechanism proposed for this reaction is outlined in the following two steps:

$$NO + O_2 \rightleftharpoons NO \cdot O_2 \qquad \text{(rapid)} \tag{14.81}$$

$$NO \cdot O_2 + NO \rightarrow 2NO_2 \qquad \text{(slow)} \tag{14.82}$$

The species indicated as $NO \cdot O_2$ is a molecule formed by the association of a molecule of nitrogen(II) oxide and oxygen, with no particular bonding implied. However, since both nitrogen(II) oxide and oxygen are paramagnetic, it is possible that the unpaired electrons in each of these molecules are involved in the bonding.

2-Bromo-2-methylpropane with water. When shaken with a large excess of water, 2-bromo-2-methylpropane forms 2-methyl-2-propanol:

$$CH_3—\underset{\displaystyle CH_3}{\overset{\displaystyle CH_3}{\underset{|}{\overset{|}{C}}}}—Br + 2H_2O = CH_3—\underset{\displaystyle CH_3}{\overset{\displaystyle CH_3}{\underset{|}{\overset{|}{C}}}}—OH + H_3O^+ + Br^- \tag{14.83}$$

As experimentally determined, the reaction is first order.

$$r_f = k_f M_{C_4H_9Br} \tag{14.84}$$

The postulated mechanism consists of the following three steps:

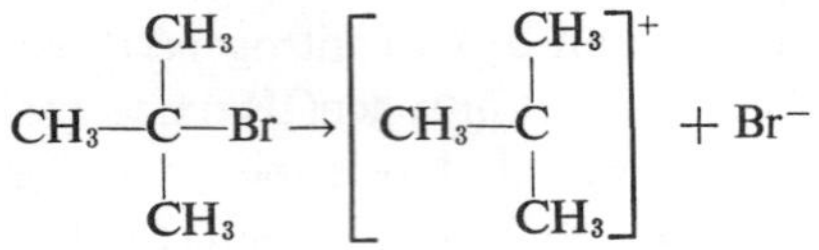
(slow) (14.85)

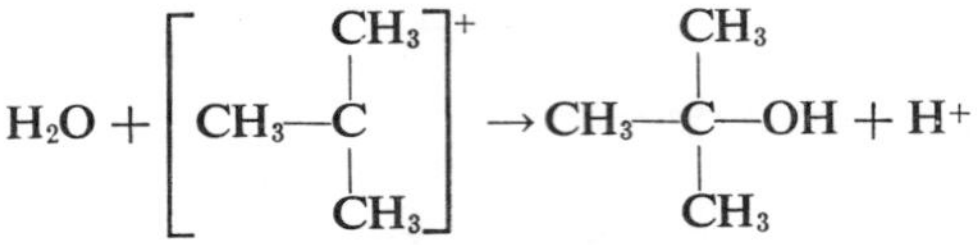
(fast) (14.86)

$$H^+ + H_2O \rightarrow H_3O^+ \qquad \text{(fast)} \tag{14.87}$$

The initial step in this sequence is a unimolecular reaction. Undoubtedly the collisions with solvent molecules have something to do with this reaction, but in terms of the kinetics it is unimolecular. Since this first step is the rate-determining step, the over-all reaction is first order and the over-all rate equation is the same as the rate equation for the reaction represented by Equation (14.85)

IODOMETHANE AND AQUEOUS HYDROXIDE. Aqueous hydroxide reacts with iodomethane to replace the iodine with a hydroxy group to form methanol.

$$CH_3I + OH^- = CH_3OH + I^- \tag{14.88}$$

The over-all reaction is second order.

$$r_f = k_f M_{CH_3I} M_{OH^-} \tag{14.89}$$

This is the rate equation that would be anticipated if Equation (14.88) represented a simple, single step, bimolecular reaction. No further mechanism needs to be postulated.

OZONE DECOMPOSITION. Ozone is unstable and decomposes to oxygen.

$$2O_3(g) = 3O_2(g) \tag{14.90}$$

It has been shown experimentally that the over-all reaction is first order and the rate equation is as follows:

$$r_f = k_f' \frac{M_{O_3}^2}{M_{O_2}} \tag{14.91}$$

To account for this rate equation, the following two-step mechanism has been proposed:

$$O_3 \rightleftharpoons O_2 + O \qquad \text{(rapid)} \tag{14.92}$$

$$O + O_3 \rightarrow 2O_2 \qquad \text{(slow)} \tag{14.93}$$

The first step is a rapid equilibrium between ozone and oxygen and atomic oxygen. The second step is the rate-determining step between atomic oxygen and ozone. The rate of the reaction thus may be expressed by Equation (14.94).

$$r_f = k_f' M_O M_{O_3} = k_f'[O][O_3] \tag{14.94}$$

From the initial equilibrium, Equations (14.95) and (14.96) may be written:

$$\frac{k_{fe}}{k_{re}} = \frac{[O_2][O]}{[O_3]} \tag{14.95}$$

$$[O] = \frac{k_{fe}[O_3]}{k_{re}[O_2]}$$

$$r_f = \frac{k_f k_{fe}[O_3]^2}{k_{re}[O_2]} = k_f' \frac{[O_3]^2}{[O_2]} \tag{14.96}$$

HYDROGEN WITH BROMINE. As a final example, consider the reaction of hydrogen with bromine in the gas phase. As indicated earlier in this chapter, the rate equation for this reaction is quite complex.

$$r_f = \frac{k_f M_{H_2} M_{Br_2}^{1/2}}{1 + \dfrac{0.1\ M_{HBr}}{M_{Br_2}}} \tag{14.97}$$

To explain this complex rate equation, it has been suggested that the reaction occurs in a series of five steps, not all of which need to be in a particular sequence. These five steps are as follows:

Step 1	$Br_2 \rightarrow 2Br$	(14.98)
Step 2	$Br + H_2 \rightarrow HBr + H$	(14.99)
Step 3	$H + Br_2 \rightarrow HBr + Br$	(14.100)
Step 4	$H + HBr \rightarrow H_2 + Br$	(14.101)
Step 5	$2Br \rightarrow Br_2$	(14.102)

This is conceived as a chain reaction in which the initial step consists of the breaking of bromine molecules into their atoms. This may be accomplished by the absorption of electromagnetic radiation of the same energy as the Br—Br bond energy.

Once the bromine atoms are formed, they are very reactive, and either of two things may happen to them. Two bromine atoms may recombine to form molecular bromine as indicated in Step 5. In this case, there is no net reaction. Thus Step 5 is a termination step that brings the chain reaction to a halt.

On the other hand, bromine atoms may collide with hydrogen molecules to produce hydrogen bromide and atomic hydrogen as shown in Step 2. The atomic hydrogen is very reactive; once it is formed, there are two important reactions it may undergo. First, it may react with bromine molecules to form hydrogen bromide and bromine atoms, as shown in Step 3. The bromine atoms formed in this step may react in either of the two ways mentioned before. Second, atomic hydrogen may react with hydrogen bromide formed in Steps 2 and 3 to form elemental hydrogen and atomic bromine, as shown in Step 4.

Step 1 is the initiation step, for without it, the bromine atoms are not produced to carry out the other steps. Steps 2, 3, and 4 are the propagation steps, for in each of these steps very highly reactive species are formed, and these proceed to react with other species. These reactive species are atomic hydrogen in Step 2 and atomic bromine in Steps 3 and 4. As indicated earlier, Step 5 is the termination step.

It is assumed that during the course of this reaction, the concentrations of atomic hydrogen and atomic bromine remain constant, since they are consumed in some reactions and produced by others. No distinction is made for speed among these reactions since they all are relatively fast. Once this assumption is made, it is possible to write two useful rate equations, one for the rate of formation of atomic bromine and one for the rate of formation of atomic hydrogen. Since there is no net change in concentration of each of these species, each of these rates is zero.

The net rate of formation of bromine atoms, r_{Br}, is the sum of the rate of formation of bromine atoms in Steps 1, 3, and 4 minus the rate of consumption of bromine atoms in Steps 2 and 5. This is expressed as Equation (14.103), where numerical subscripts on the rate constants refer to step numbers.

$$r_{Br} = k_1 M_{Br_2} - k_2 M_{Br} M_{H_2} + k_3 M_H M_{Br_2} + k_4 M_H M_{HBr} - k_5 M^2_{Br} = 0 \quad (14.103)$$

Similarly, the rate of formation of atomic hydrogen, r_H, may be expressed as Equation (14.104).

$$r_H = k_2 M_{Br} M_{H2} - k_3 M_H M_{Br2} - k_4 M_H M_{HBr} = 0 \quad (14.104)$$

If Equation (14.104) is added to Equation (14.103), Equation (14.105) is the result, and this equation may be solved for the concentration of atomic bromine as shown by Equation (14.106).

$$k_1 M_{Br_2} = k_5 M^2_{Br} \quad (14.105)$$

$$M_{Br} = \left(\frac{k_1}{k_5}\right)^{1/2} M^{1/2}_{Br2} \quad (14.106)$$

The expression for the concentration of atomic bromine may be substituted into Equation (14.104) to produce Equation (14.108), which is solved for the concentration of atomic hydrogen.

$$k_2 M_{Br} M_{H_2} = M_H(k_3 M_{Br_2} + k_4 M_{HBr}) \quad (14.107)$$

$$M_H = \frac{k_2 M_{Br} M_{H_2}}{k_3 M_{Br_2} + k_4 M_{HBr}} = \frac{k_2 (k_1/k_5)^{1/2} M_{H_2} M^{1/2}_{Br2}}{k_3 M_{Br_2} + k_4 M_{HBr}} \quad (14.108)$$

The rate of formation of hydrogen bromide is a suitable measure of the rate of the over-all reaction. This rate of formation of hydrogen bromide must be the sum of the rates of formation in Steps 2 and 3 minus the rate of consumption in Step 4. This is shown in Equation (14.109), which may be rearranged to Equation (14.110).

$$r_{HBr} = k_2 M_{Br} M_{H_2} + k_3 M_H M_{Br_2} - k_4 M_H M_{HBr} \tag{14.109}$$

$$r_{HBr} = k_2 M_{Br} M_{H_2} + M_H (k_3 M_{Br_2} - k_4 M_{HBr}) \tag{14.110}$$

The expression for the concentration of atomic bromine from Equation (14.106) and the expression for the concentration of atomic hydrogen from Equation (14.108) may be substituted into Equation (14.110).

$$r_{HBr} = k_2 \left(\frac{k_1}{k_5}\right)^{1/2} M_{Br_2}{}^{1/2}\, M_{H_2} + (k_3 M_{Br_2} - k_4 M_{HBr}) \frac{k_2 (k_1/k_5)^{1/2}\, M_{H_2} M_{Br_2}^{1/2}}{k_3 M_{Br_2} + k_4 M_{HBr}}$$

This equation may be rearranged to the following form:

$$r_{HBr} = k_2 \left(\frac{k_1}{k_5}\right)^{1/2} M_{Br_2}^{1/2} M_{H_2} \left(1 + \frac{k_3 M_{Br_2} - k_4 M_{HBr}}{k_3 M_{Br_2} + k_4 M_{HBr}}\right)$$

Finally, this equation may be rearranged.

$$r_{HBr} = \frac{2 k_2 k_3 (k_1/k_5)^{1/2}\, M_{H_2} M_{Br_2}^{1/2}}{1 + \dfrac{k_4 M_{HBr}}{k_3 M_{Br_2}}} \tag{14.111}$$

Equation (14.111) corresponds to the experimental rate Equation (14.97), where $2k_2k_3(k_1/k_5)^{1/2}$ is k_f and k_4/k_3 is 0.1.

This is a good illustration of the way in which simple stepwise reactions may be combined to account for rather complex reaction kinetics.

CATALYSIS

As indicated earlier, catalysts are substances that increase the speed of reactions without themselves appearing among the over-all products of the reaction. A hypothetical catalyzed reaction is represented by Equation (14.112), where X is the catalyst.

$$\mathrm{A + B + X = C + D + X} \tag{14.112}$$

Since the catalyst appears among the products as well as the reactants, the net equation for the reaction may be written as follows:

$$\mathrm{A + B = C + D}$$

Since the catalyst is the same before and after the reaction, the free energy of the catalyst does not change and thus the catalyst does not contribute to the over-all reaction free-energy change. As a result the catalyst does not alter the value of the equilibrium constant for the reaction, which is closely related to the standard free-energy change. Thus the addition of a catalyst does not alter the equilibrium position or the value of the equilibrium constant. Rather, the catalyst allows the equilibrium to be attained more quickly. It also follows that a catalyst increases the rate of both the forward and reverse reactions equally.

In general, the catalytic activity is explained by a change in the mechanism for the reaction. The mechanism for the uncatalyzed reaction involves steps with high free energies of activation, whereas the mechanism for the catalyzed reaction involves steps with lower free energies of activation. With the lower free energy of activation, collision efficiency is increased and the rate of reaction increases. In addition, this lower free energy of activation accounts for the increase in rates for both the forward and reverse reactions.

A simplified picture of this free energy of activation decrease is seen in Fig. 14.2, a graphical representation of a single-step uncatalyzed reaction and the same single-step catalyzed reaction. The solid curve represents the uncatalyzed reaction and the dashed curve represents the catalyzed reaction, The two curves represent two different mechanisms.

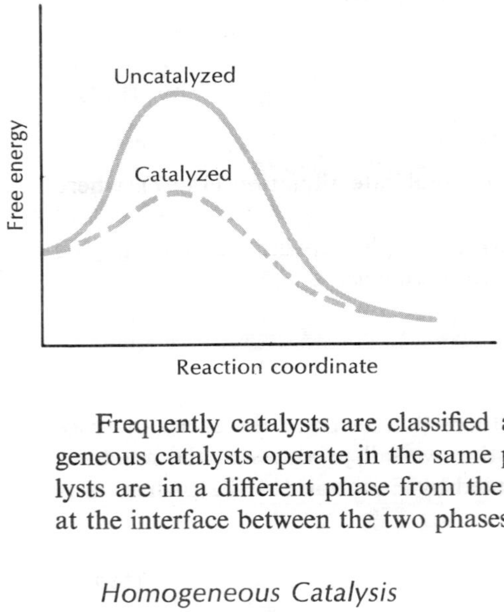

Fig. 14.2. Comparison of the free-energy changes for a catalyzed and an uncatalyzed reaction

Frequently catalysts are classified as homogeneous or heterogeneous. Homogeneous catalysts operate in the same phase as the reactants. Heterogeneous catalysts are in a different phase from the reactants, and the catalytic activity occurs at the interface between the two phases.

Homogeneous Catalysis

The reaction between sulfur(IV) oxide and elemental oxygen to form sulfur(VI) oxide is very slow.

$$2SO_2(g) + O_2(g) = 2SO_3(g) \tag{14.113}$$

The mechanism for this reaction is unknown, but the fact that it is slow suggests that it may be a single-step termolecular reaction; or it may be unimolecular or bimolecular with a very high free energy of activation.

However, gaseous nitrogen(II) oxide serves as a catalyst. The following is a simplified view of a complex process:

$$2NO \rightleftharpoons (NO)_2 \qquad (14.114)$$

$$(NO)_2 + O_2 \rightarrow 2NO_2 \qquad (14.115)$$

$$2(NO_2 + SO_2 \rightarrow NO + SO_3) \qquad (14.116)$$

The first two steps were discussed previously as a possible mechanism for the reaction of nitrogen(II) oxide with oxygen. The third step is the reaction of nitrogen dioxide, formed from nitrogen(II) oxide and oxygen, with sulfur(IV) oxide. In this last step, the nitrogen(II) oxide catalyst is regenerated and the over-all reaction may be represented by Equation (14.113).

Thus the catalyst is not a magical potion that increases the speed of reaction by its mere proximity to the reacting system. Actually, the catalyst participates in the chemical reaction and in this way provides another mechanism with lower free energy of activation, so that the over-all reaction may occur more rapidly.

Another example is the thermal decomposition of acetaldehyde, which, as an uncatalyzed reaction, appears to be a single-step unimolecular reaction as suggested by the following equation:

$$CH_3{-}\overset{\overset{\large O}{\|}}{C}{-}H(g) = CH_4(g) + CO(g) \qquad (14.117)$$

This reaction is catalyzed by iodine and the following two-step catalytic mechanism has been proposed.

$$CH_3{-}\overset{\overset{\large O}{\|}}{C}{-}H + I_2 \rightarrow CH_3I + HI + CO \qquad (14.118)$$

$$CH_3I + HI \rightarrow CH_4 + I_2 \qquad (14.119)$$

Again the catalyst participates in the reaction and is regenerated so that the over-all reaction may be represented by Equation (14.117).

This reaction is of interest as an example of a multiple-step mechanism that has a higher rate of reaction than a single-step reaction.

Heterogeneous Catalysis

A very common type of heterogeneous catalysis involves a solid catalyst and reactant in the gas phase. In general, those substances which serve as catalysts for this type of reaction are substances upon which the gaseous reactants are readily adsorbed. This suggests that bonds are formed between the catalyst and the reactants. The formation of these bonds may weaken one or more bonds within the reactant molecules, thus decreasing the bond energy within these molecules. In effect, this makes the molecule more reactive. Molecules of reactants that are adsorbed in adjacent locations on the surface of the catalyst may react to form the products, or the activated adsorbed molecules may be bombarded by other molecules from the gas phase to form the products. After this reaction has occurred, the products are desorbed and the catalyst surface is regenerated and may be used to repeat this process.

For example, hydrogen and oxygen react slowly to form water.

$$2H_2(g) + O_2(g) = 2H_2O(g) \qquad (14.120)$$

Metallic platinum is a catalyst for this reaction. The catalytic process may be described in the following steps:

Step 1. Hydrogen and oxygen molecules are adsorbed on the surface of the platinum, somewhat as depicted in Fig. 14.3. The formation of bonds with the catalytic surface weakens the H—H and O═O bonds.

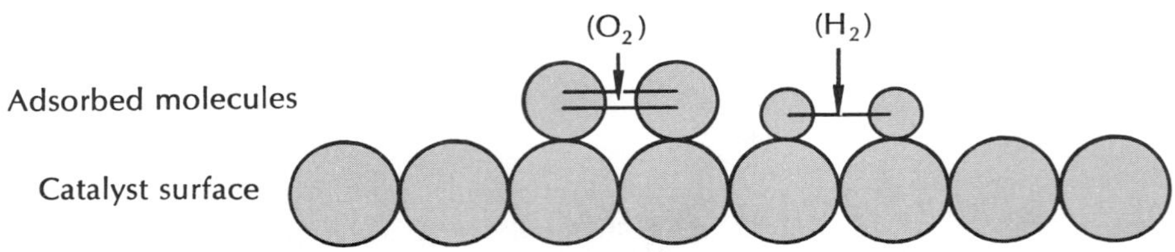

Fig. 14.3. Schematic representation of the adsorption of two reactants on the surface of a heterogeneous catalyst

Step 2. Adjacent hydrogen and oxygen molecules on the catalytic surface react to form H—O bonds and water molecules, which remain adsorbed on the platinum surface.

Step 3. The adsorbed water molecules are desorbed, leaving the platinum surface ready to repeat this reaction.

SUMMARY

The mechanisms of chemical reactions are interpreted in terms of series of single-step reactions. Some of the evidence for the nature of these steps comes from the experimental determination of the rate equations for the over-all reactions.

In general, the rate of any single-step reaction may be conceived as the result of the collision frequency and the collision efficiency among the reacting particles.

The collision frequency is related to the temperature and the concentration of the reacting particles. At high temperatures, the average kinetic energy of the particles is high and thus the average speed is high. This means that particles have relatively great opportunity to collide with other particles. Similarly, at high concentration, particles are relatively close together and collisions are frequent.

Collision efficiency is related to the free energy of activation and the temperature. The free energy of activation serves as a barrier to chemical reactions. In large measure, the enthalpy of activation is a result of the increase in potential energy occasioned by the repulsions of electron clouds as molecular and atomic particles approach each other. To some degree, the entropy of activation is due to the geometrical requirements in the formation of the activated complex. Thus the entropy of the activated complex often is lower than the entropy of the isolated particles.

The free energy of activation is the sum of the enthalpy of activation and the entropy of activation effects. A high value for the free energy of activation leads to a slow reaction. In this case, relatively few particles collide with sufficient kinetic energy to overcome the electrostatic repulsions among electron clouds, or in the correct geometry to form the activated complex. An increase in temperature corresponds to an increase in average kinetic energy, so that more of the collisions occur between particles of sufficient kinetic energy to surmount the potential energy portion of the free energy of activation barrier.

Collision efficiency also is related to reaction molecularity. A termolecular reaction may be slow, because three particles must collide simultaneously and this will occur rather rarely in comparison to a bimolecular collision, for example.

In some sense, catalysts improve collision efficiency, although, in the presence of a catalyst, the effective collisions are between particles different from those involved in the uncatalyzed reaction. The catalyst provides an alternate mechanism with a lower over-all free energy of activation.

QUESTIONS AND PROBLEMS

14.1 Criticize the following statement: The mechanism by which a reaction occurs is more important than the free-energy change in determining whether or not a reaction will occur.

14.2 Could Equations (14.2) through (14.5) be any different if Equation (14.1) does not represent a single-step reaction?

14.3 For each of the following chemical equations and their corresponding rate equations indicate the over-all order and the order with respect to each of the reactants:

a) $5Br(aq)^- + BrO_3(aq)^- + 6H_3O(aq)^+ = 3Br_2(l) + 9H_2O(l)$

$$r_f = k_f M_{Br^-} M_{BrO_3^-} M^2_{H_3O^+}$$

b) $3I(aq)^- + H_3AsO_4(aq) + 2H_3O(aq)^+ = I_3(aq)^- + H_3AsO_3(aq) + 3H_2O(l)$

$$r_f = k_f M_{I^-} M_{H_3AsO_4} M_{H_3O^+}$$

c) $CH_3{-}\overset{\overset{\displaystyle O}{\|}}{C}{-}CH_3 + I_2 = CH_3{-}\overset{\overset{\displaystyle O}{\|}}{C}{-}CH_2{-}I + HI$

$$r_f = k_f$$

14.4 The following reaction is found to be third order over-all. Write all of the possible rate equations that are consistent with this fact.

$$2A + 3B + C = D + 2E$$

14.5 What is the order of a radioactive decay process?

14.6 What units are associated with the rate constants for first, second, and third-order reactions?

14.7 Gaseous hydrogen iodide decomposes according to the following equation:

$$2HI(g) = H_2(g) + I_2(g)$$

The activated complex often is represented as follows, where the solid lines are the original bonds in the hydrogen iodide molecules and the dotted lines are the new bonds that are forming:

```
H····H
|    |
I····I
```

a) Draw a free-energy graph like Fig. 14.1, assuming the free energy of the products is lower than the free energy of the reactants.
b) Draw a free-energy graph like Fig. 14.1, assuming the free energy of the products is greater than the free energy of the reactants.
c) Discuss the factors that contribute to the free energy of activation.
d) Compare the entropy of the activated complex to the entropy of the reactants and then to the entropy of the products.
e) Assuming the reaction is a single step, write the rate equation. What is the order of the reaction? The molecularity?
f) Suggest the nature of the activated complex for the reverse reaction.
g) Discuss how the rate of this reaction might be increased.

14.8 How can the collision frequency for a reaction be decreased? How can the collision efficiency for a reaction be increased?

14.9 Is it possible to increase the collision frequency while at the same time decreasing the collision efficiency? If so, how? If not, why not?

14.10 The rate-determining step for a homogeneous reaction involves one molecule of A and one molecule of B as indicated in the following equation:

$$A + B \rightarrow 2C$$

Suggest four ways to increase the rate of formation of C.

14.11 A hypothetical reaction occurs in the following steps.

$$A + B \rightarrow C + D \qquad \text{(fast)}$$
$$C + D \rightarrow E + F \qquad \text{(slow)}$$
$$E + B \rightarrow D \qquad \text{(fast)}$$

a) What is the equation for the over-all reaction?
b) Write the rate equation for the over-all reaction.
c) What is the order of this reaction? The molecularity?

14.12 Dilute aqueous nitrous acid decomposes according to the following equation.

$$3HNO_2(aq) = 2NO(g) + H_3O(aq)^+ + NO_3(aq)^-$$

The following mechanism has been proposed:

$$2HNO_2 \rightleftharpoons NO + NO_2 + H_2O \qquad \text{(rapid)}$$
$$2NO_2 \rightleftharpoons N_2O_4 \qquad \text{(rapid)}$$
$$N_2O_4 + H_2O \rightarrow HNO_2 + H^+ + NO_3^- \qquad \text{(slow)}$$
$$H^+ + H_2O \rightarrow H_3O^+ \qquad \text{(fast)}$$

a) Suggest the form of the rate equation for the over-all reaction.
b) What is the order of this reaction?

14.13 The rate-determining step for a homogeneous gaseous reaction is as follows:

$$A(g) + 2B(g) \rightarrow C(g)$$

a) Write the rate equation for the reaction.
b) How does the rate change as each of the reactant concentrations is doubled separately?
c) How does the rate change if the enclosing volume suddenly is halved?
d) What is the molecularity for the reaction?

14.14 The following equation represents an over-all net reaction.

$$2A + 2B = C + D$$

The rate law is found to be as follows:

$$r_f = k_f M_A M_B$$

Propose a mechanism for the reaction that is consistent with this rate equation.

14.15 An over-all reaction has a two-step mechanism with a slowly attained equilibrium followed by a fast irreversible reaction as follows:

$$A + B \rightleftharpoons C + D \qquad \text{(slow)}$$
$$C + E \rightarrow F \qquad \text{(fast)}$$

a) What is the equation for the over-all reaction?
b) Derive the rate equation for the over-all reaction.
c) What is the order of this reaction?

14.16 What is meant by a zero-order reaction? Can there be a reaction that has zero molecularity? If so, how? If not, why not?

14.17 Criticize the following statement: The molecularity of a reaction depends upon the mechanism of the reaction, whereas the order of the reaction does not.

14.18 In aqueous solution, iodide ions react with hypochlorite ions as follows:

$$I(aq)^- + ClO(aq)^- = IO(aq)^- + Cl(aq)^-$$

Experimentally the rate equation is found to be

$$r_f = k_f \frac{M_{I^-} M_{ClO^-}}{M_{OH^-}}$$

a) How can you account for the fact that the concentration of hydroxide ion appears in the rate equation but not in the over-all chemical equation?
b) Is this reaction a single-step reaction? Explain.

14.19 Some of the evidence supporting the nitrogen(V) oxide decomposition shown in Equations (14.55) through (14.57) has been the way in which the presence of large excesses of nitrogen(II) oxide affects the reaction. On the basis of this mechanism, predict the effect of a large excess of nitrogen(II) oxide on this reaction.

14.20 For the second mechanism for the reaction of nitrogen(II) oxide with oxygen as shown by Equations (14.81) and (14.82), derive the over-all rate equation and show that it corresponds to Equation (14.74).

14.21 Nitrogen(II) oxide reacts with hydrogen as indicated by the following over-all reaction equation.

$$2NO(g) + H_2(g) = N_2O(g) + H_2O(g)$$

The over-all reaction is third order; it is second order with respect to nitrogen(II) oxide and first order with respect to hydrogen. Suggest mechanisms for the reaction consistent with this data.

14.22 Hydrogen and bromine may be mixed in a brown glass bottle and not react. However, if mixed in a colorless transparent bottle they may react explosively. Explain.

14.23 Equation (14.91) is the rate equation for the decomposition of ozone and is reproduced as follows:

$$r_f = k_f \frac{M_{O_3}^2}{M_{O_2}}$$

On the basis of this equation, the decomposition of ozone has been called a first-order reaction. What is the justification for calling this a first-order reaction? What is the order with respect to oxygen?

14.24 Criticize the following statement: Since catalysts do not appear in the over-all net chemical equation, it is not possible for the catalyst concentration to appear in the rate equation.

14.25 Ethyl acetate reacts slowly with water to form acetic acid and ethanol as indicated by the following equation:

$$CH_3\text{—}\overset{\overset{\displaystyle O}{\|}}{C}\text{—}O\text{—}CH_2\text{—}CH_3 + H_2O = CH_3\text{—}\overset{\overset{\displaystyle O}{\|}}{C}\text{—}OH + CH_3\text{—}CH_2\text{—}OH$$

The decomposition of the ester is more rapid in the presence of hydroxide ion in the water. The reaction may be represented by the following equation:

$$CH_3\text{—}\overset{\overset{\displaystyle O}{\|}}{C}\text{—}O\text{—}CH_2\text{—}CH_3 + OH^- = CH_3\text{—}\overset{\overset{\displaystyle O}{\|}}{C}\text{—}O^- + CH_3\text{—}CH_2\text{—}OH$$

Is the hydroxide ion a catalyst in this reaction? Why?

14.26 Chloroethane and benzene by themselves do not appear to undergo reaction. However, in the presence of aluminum chloride, ethylbenzene is produced. The mechanism which has been proposed for this reaction is as follows:

$$CH_3CH_2Cl + AlCl_3 \rightarrow CH_3CH_2^+ + [AlCl_4]^-$$
$$CH_3CH_2^+ + C_6H_6 \rightarrow C_6H_5CH_2CH_3 + H^+$$
$$H^+ + [AlCl_4]^- \rightarrow HCl + AlCl_3$$

Is the aluminum chloride a catalyst in this reaction? Why?

14.27 Carbon monoxide reacts with oxygen to form carbon dioxide.

a) The reaction is catalyzed by nitrogen(II) oxide. Suggest a mechanism for the catalysis reaction.

b) The reaction also is catalyzed by solid quartz. How does the mechanism of this catalysis differ from the mechanism of the nitrogen(II) oxide catalysis?

14.28 The decomposition of acetaldehyde as represented by Equation (14.117) is slow at low temperatures. How can you account for this fact when the reaction appears to be a simple, unimolecular, single-step reaction?

14.29 In practice, solid, heterogeneous catalysts often are very finely divided. Suggest the reason for this.

15

Acids and Bases: An Introduction

INTRODUCTION Among the many classifications of chemical reactions, one of the most useful is the group called acid-base reactions. Not only are acid-base reactions common, but they often serve as single-step reactions in the mechanisms for complex reactions.

In the overview of some concepts of acids and bases which follows this introduction, note that the original basis for classifying substances as acids or bases was experimental. Thus substances were classified according to the type of reaction they produced with certain other kinds of substances. However, once the classification was made, there were attempts to find the common "cause" of acidity and basicity among the often diverse substances in each category.

Following the overview, an introduction is given to the Bronsted-Lowry concept of acids and bases, with some attempt to relate acid-base behavior to molecular and ionic structure. The Bronsted-Lowry concept has been singled out for emphasis because it has proved useful in a wide variety of situations, particularly in aqueous-solution reactions. However, in many instances, the Lewis concept, which is described briefly in the overview, has proved even more fruitful.

In considering this chapter, it is helpful to keep in mind the Lewis approach and compare it to the Bronsted-Lowry approach. The resulting practice in trying to develop an idea from a few fundamental postulates will lead to a clearer view of both the Bronsted-Lowry and the Lewis concepts.

OVERVIEW OF SOME ACID-BASE CONCEPTS

Early chemical workers were familiar with certain materials, such as spirit of vinegar or oil of vitriol, which changed the color of vegetable dyes, tasted sour, and acted on various metals, earths, and alkalies, often with effervescence. Experimenters realized that in such processes the acid principle was neutralized to produce a "neutral salt," that is, a soluble, crystalline substance with a characteristic taste, such as sea salt.

Lavoisier was among the first to propose a theory of acids. He postulated that all acids are compounds of an "acidifiable base" (an element such as sulfur or phosphorus) with the "acidifying principle" (oxygen). Although it was a popular theory for many years, opponents of this model successfully showed that many acids contain no oxygen—HCl, H_2S, Cl_2 for example—while many oxygen-containing compounds are not acidic—for example, K_2O, and CaO.

In 1887 Arrhenius postulated the existence of ions in aqueous solutions. According to the Arrhenius theory of acids and bases, an acid is a hydrogen-containing compound that produces hydrogen ion, H^+, in water, and a base is a hydroxide that dissociates to form hydroxide ion, OH^-, in water. For example, hydrogen chloride, HCl, sulfuric acid, H_2SO_4, and acetic acid, CH_3COOH, are acids, whereas sodium hydroxide, NaOH, calcium hydroxide, $Ca(OH)_2$, and nickel(II) hydroxide, $Ni(OH)_2$, are bases.

Although this was the most important theory of acids and bases for many years, it was too restrictive. There are many nonhydrogen-containing substances whose aqueous solutions taste sour, will change the colors of vegetable dyes, and will react with so-called active metals to produce hydrogen, all of which experimental properties are those by which acids are identified. Thus aqueous solutions of iron(III) nitrate, aluminum chloride, and tin(IV) chloride are acidic. Similarly, there are many substances whose aqueous solutions taste bitter, feel slippery, and neutralize acids, which are experimental properties by which bases are identified. Examples are potassium sulfide, sodium carbonate, ammonia, and sodium oxide. Furthermore, the Arrhenius definitions restricted acids and bases to aqueous solutions.

In 1923, Bronsted and Lowry independently suggested a set of definitions of acids and bases which together are now known as the *Bronsted-Lowry* theory of acids and bases. An acid-base reaction is viewed as a competition between two chemical species for a hydrogen ion, H^+. The winner of this competition is the base and the loser is the acid. Sometimes the base is called a hydrogen-ion acceptor, and the acid is called a hydrogen-ion donor. Since the most common isotope of hydrogen is hydrogen-1, and the hydrogen ion produced from this isotope is a proton, an acid may be defined as a proton loser (or donor) and a base may be defined as a proton winner (or acceptor). Thus from the Bronsted-Lowry point of view, an acid-base reaction involves the transfer of a proton from an acid to a base.

The reaction between gaseous hydrogen chloride and gaseous ammonia is an example of a Bronsted-Lowry acid-base reaction.

$$NH_3(g) + HCl(g) = NH_4^+Cl^-(s) \quad (15.1)$$

This reaction is viewed as the loss of a proton, H^+, by the HCl, and a gain of a proton by the NH_3, which uses its pair of nonbonding electrons to bond the hydrogen ion covalently into the ammonium ion, NH_4^+. Thus the NH_3 acts as a base or proton winner and HCl acts as an acid or proton loser.

In comparison to the Arrhenius theory, this theory has a broader concept of a base, so that a base is not restricted to the hydroxide-ion-containing substances. However, acidic properties still are identified with the hydrogen ion.

A broader view of acids and bases was presented by Franklin and Germann in the *solvent system* concept of acids and bases. According to this view, solvents tend to undergo self-dissociation into ions. For example, water dissociates as follows:

$$H_2O + H_2O = H_3O^+ + OH^- \quad (15.2)$$

Equation (15.2) represents a Bronsted-Lowry acid-base reaction in which one molecule of water serves as a base to remove a proton from another molecule of water, the acid. Then an acid is any solute which increases the concentration of the cation produced by the self-dissociation of the solvent: H_3O^+ in this case. Similarly, a base is any solute which increases the concentration of the anion produced by the self-dissociation of the solvent: OH^- in this case.

For protonic solvents, the self-dissociation involves the transfer of a proton from one molecule to another. Thus ammonia undergoes the following dissociation:

$$NH_3 + NH_3 = NH_4^+ + NH_2^- \quad (15.3)$$

In liquid ammonia, an acid increases the concentration of ammonium ion, NH_4^+, and a base increases the concentration of amide ion, NH_2^-.

Proponents of the solvent-system theory have extended it to nonprotonic solvents such as liquid sulfur(IV) oxide and carbonyl chloride. The self-dissociations of these two substances are as follows:

$$SO_2 + SO_2 = SO^{++} + SO_3^{--} \quad (15.4)$$

$$COCl_2 = CO^{++} + 2Cl^- \quad (15.5)$$

Thus in liquid SO_2, an acid increases the concentration of thionyl ion, SO^{++}, and a base increases the concentration of sulfite ion, SO_3^{--}. In $COCl_2$, acids increase the concentration of carbonyl ion, CO^{++}, and bases increase the concentration of chloride ion, Cl^-.

Opponents of the solvent-system theory point out that relatively few solvents are ionizing. Only those solvents of high dielectric constant have a tendency to dissolve ionic substances or to permit the existence of ions in solution. There is evidence for the self-dissociation of water and ammonia, but there is no evidence for the self-dissociation of sulfur(IV) oxide and carbonyl chloride. Thus this theory tends to degenerate into a formalism which may be useful but has little factual support.

A still broader concept of acids and bases was proposed by G. N. Lewis. From the Bronsted-Lowry point of view, a base is a proton winner because it has a pair of nonbonding electrons with which to form a coordinate covalent bond with the proton. Lewis suggested that an acid-base reaction might be defined as any reaction in which a coordinate covalent bond is formed. From this point of view an acid-base reaction does not depend upon the chemical nature of the species involved, but only on the type of bond formed. Thus any species that serves as an electron-pair donor is a base, and any species that can accept a pair of electrons is a potential acid.

The following are some examples of Lewis acid-base reactions.

Acid *Base*

$$H^+ + OH^- = H{-}OH \qquad (15.6)$$

$$H^+ + NH_3 = H{-}NH_3^+ \qquad (15.7)$$

$$\begin{array}{c} F \\ | \\ F{-}B \\ | \\ F \end{array} + NH_3 = \begin{array}{c} F \\ | \\ F{-}B{-}NH_3 \\ | \\ F \end{array} \qquad (15.8)$$

$$Pb^{++} + 4HOH = [Pb(H_2O)_4]^{++} \qquad (15.9)$$

SOME ILLUSTRATIVE ACID-BASE REACTIONS IN AQUEOUS MEDIA

Consider the reaction of hydrogen chloride with water, as shown in the following equation:

$$HCl(g) + H_2O = H_3O(aq)^+ + Cl(aq)^- \qquad (15.10)$$

Although HCl is not ionic, it reacts with water to produce ions in solution.

From the Bronsted-Lowry view, HCl is an acid, since it loses a hydrogen ion to water. From the Lewis view, the hydrogen ion is an acid, since it accepts a pair of electrons from water to form a coordinate covalent bond. Technically, the Lewis approach indicates that hydrogen ion, H^+, is the acid, whereas the Bronsted-Lowry approach indicates that HCl is the acid. However, since HCl is the source of H^+, the HCl often is called the acid in Lewis approach terminology. From the solvent system view, HCl is an acid since it reacts to increase the concentration of the hydronium ion, which is the cation from the self-dissociation of water.

In the reaction represented by Equation (15.10), water is a base from the Bronsted-Lowry view, since it wins a proton from HCl. Water is a base also from the Lewis view, since it donates a pair of electrons to the H^+ to form a coordinate covalent bond.

Aqueous solutions consist of molecules, cations, and anions. Each of these kinds of species may serve as acids and bases. In the following section these are briefly discussed.

Molecules and Precipitates

Molecular species may serve as either acids or bases. For example, aqueous acetic acid acts as an acid as shown by the following equation:

$$CH_3COOH(aq) + H_2O = H_3O(aq)^+ + CH_3COO(aq)^- \qquad (15.11)$$

From the Bronsted-Lowry view, acetic acid is an acid since it loses a proton to water. Also it is an acid from the Lewis view, since the proton it provides accepts a pair of electrons from water in the formation of a coordinate covalent bond. Finally it is an acid from the solvent system view, since the reaction increases the concentration of hydronium ion, which is the cation produced in the self-dissociation of water, the solvent. From the point of view of the Bronsted-Lowry concept and the Lewis concept, water is a base since it wins a proton from water and provides the pair of electrons by which the proton is bound covalently to the water molecules in the hydronium ions.

Molecular ammonia also reacts with water, as follows.

$$NH_3(aq) + H_2O = NH_4(aq)^+ + OH(aq)^- \qquad (15.12)$$

Ammonia is a base in this reaction since it wins a proton from water, provides a pair of electrons for coordinate covalent bond formation with the proton from water, and increases the concentration of hydroxide ion, which is the anion from the self-dissociation of water. At the same time, water is an acid, since it loses a proton to ammonia and provides a proton which accepts a pair of electrons from ammonia in coordinate covalent bond formation.

From comparison of the reactions represented by Equations (15.11) and (15.12), it is clear that in one, water acts as a base and in the other, it acts as an acid. Any substance or species that acts as an acid under some conditions and as a base under other conditions is said to be *amphoteric*. Specifically, species which are amphoteric in the Bronsted-Lowry sense are said to be *amphiprotic*.

A number of amphoteric substances are insoluble in water but may be dissolved by either acids or bases. For example, zinc hydroxide is insoluble in water but in the presence of hydronium ion, an acid, it dissolves as follows:

$$Zn(OH)_2(s) + 2H_3O(aq)^+ = Zn(aq)^{++} + 4H_2O \qquad (15.13)$$

In the Bronsted-Lowry sense, $Zn(OH)_2$ is a base in this reaction, since the hydroxide of the $Zn(OH)_2$ wins a proton from the hydronium ion. From the Lewis view, it is a base, because the hydroxide ion donates a pair of electrons to the formation of a coordinate covalent bond with the proton from the hydronium ion.

However, $Zn(OH)_2$ also dissolves in an aqueous solution containing excess hydroxide ions. From the Bronsted-Lowry view, the reaction occurs as follows.

$$Zn(OH)_2(s) + 2OH(aq)^- = ZnO_2(aq)^{--} + 2H_2O \qquad (15.14)$$

Thus $Zn(OH)_2$ is an acid since it loses two protons to hydroxide ions, which then act as bases.

From the Lewis view, $Zn(OH)_2$ may react as an acid in an alternative manner:

$$Zn(OH)_2(s) + 2OH(aq)^- = [Zn(OH)_4]^{--} \qquad (15.15)$$

In this case, $Zn(OH)_2$ is an acid since the zinc(II) ions accept pairs of electrons from hydroxide ions to form coordinate covalent bonds.

Formally, tetrahydroxozincate(II) ion is the dihydrate of the zincate ion, ZnO_2^{--}. As a result, in aqueous solution, it has not been possible to distinguish between the two species. Thus the reaction of zinc hydroxide with hydroxide ions is represented equally well by Equations (15.14) and (15.15).

The reactions represented by Equation (15.13) and either (15.14) or (15.15) indicate the amphoteric character of zinc hydroxide. More specifically, these equations demonstrate the amphiprotic character of zinc hydroxide.

The following set of reaction pairs shows examples of amphoterism that are not examples of amphiprotism.

$$AlF_3 + 3H_3O^+ = Al^{3+} + 3HF + 3H_2O \qquad (15.16)$$

$$AlF_3 + 3F^- = [AlF_6]^{3-} \qquad (15.17)$$

$$Fe(CN)_2 + 2H_3O^+ = Fe^{++} + 2HCN + 2H_2O \qquad (15.18)$$

$$Fe(CN)_2 + 4CN^- = [Fe(CN)_6]^{4-} \qquad (15.19)$$

$$Sb_2S_3 + 6H_3O^+ = 2Sb^{3+} + 3H_2S + 6H_2O \qquad (15.20)$$

$$Sb_2S_3 + S^{--} = 2[SbS_2]^- \qquad (15.21)$$

Cations

Since cations are good electron-pair acceptors in coordination complexes, substances that produce cations are acids. For example, salts of iron(III) contain the iron(III) ion. When placed in water, this ion hydrates readily, with an indeterminate coordination number, as follows:

$$Fe^{3+} + xH_2O = [Fe(H_2O)_x](aq)^{3+} \qquad (15.22)$$

In the Lewis sense, this is an acid-base reaction with iron(III) ions as the acid and water as the base. However, the simple iron(III) ion contains no hydrogen and thus cannot serve as an acid in the Bronsted-Lowry sense.

The hydrated iron(III) ion, however, contains hydrogens in the water molecules and may serve as a Bronsted-Lowry acid as follows:

$$[Fe(H_2O)_x](aq)^{3+} + H_2O = [Fe(OH)(H_2O)_{x-1}](aq)^{++} + H_3O(aq)^+ \qquad (15.23)$$

From the Bronsted-Lowry view, the hydrated iron(III) ion is an acid with solvent water serving as a base. Of course this reaction is an acid-base reaction also from the solvent system and Lewis views as well.

Thus cations frequently serve as acids.

Anions

Nonhydrogen-containing anions often serve as bases. For example, the phosphate ion acts as a base in the following reaction:

$$PO_4(aq)^{3-} + H_2O = HPO_4(aq)^{--} + H_3O(aq)^+ \qquad (15.24)$$

In the Bronsted-Lowry sense, phosphate ion is a base, winning a proton from water. Of course phosphate is a base from the solvent-system and Lewis views as well.

Hydrogen-containing anions may be either acids or bases and are often amphoteric. Thus the dihydrogen phosphate ion undergoes the following two reactions, which demonstrate its amphoteric nature:

$$H_2PO_4(aq)^- + H_3O(aq)^+ = H_3PO_4(aq) + H_2O \qquad (15.25)$$

$$H_2PO_4(aq)^- + OH(aq)^- = HPO_4(aq)^{--} + H_2O \qquad (15.26)$$

SOME FEATURES OF THE BRONSTED-LOWRY CONCEPT

Conjugate Acids and Bases

Although the reactants in a Bronsted-Lowry acid-base reaction are an acid and a base, the products of the reaction also are a potential acid and a potential base. For example, consider the addition of hydrazoic acid, HN_3, to water with the reaction represented here:

$$HN_2(aq) + H_2O = H_3O(aq)^+ + N_3(aq)^- \qquad (15.27)$$

For the forward reaction, hydrazoic acid is the acid and water is the base. However, potentially, the reaction might be reversed so that hydronium ion is the acid and azide ion is the base. Thus the components of the system in which an acid-base reaction occurs consists of two acid-base pairs. In this case one pair is the hydrazoic acid and water and the other pair is hydronium ion and azide ion. If one of these pairs is chosen as the primary pair, say hydrazoic acid and water, then the other pair is called the *conjugate* acid-base pair, in this case, hydronium ion and azide ion.

Since the azide ion is derived from the hydrazoic acid, it is called the *conjugate base* of hydrazoic acid. Since the hydronium ion is derived from the water, it is called the *conjugate acid* of water.

Any Bronsted-Lowry acid has a conjugate base that is derived by removing one hydrogen ion from the acid. In the case of HN_3 the conjugate base is N_3^- and in the case of $H_2PO_4^-$ the conjugate base is HPO_4^{--}. Similarly, any Bronsted-Lowry base has a conjugate acid that is derived by adding one hydrogen ion to the base. In the case of water, the conjugate acid is H_3O^+, and in the case of $H_2PO_4^-$, the conjugate acid is H_3PO_4.

Formally, it is possible to write conjugate acids and bases for many different species. For example, it is possible to think of methane, CH_4, as an acid, although it rarely acts that way. However, if CH_4 acts as an acid, its conjugate base must be CH_3^-. It is improbable that CH_4 can act as a base, since there are no nonbonding electrons available to form a coordinate covalent bond to a proton.

As another example, HCl readily acts as an acid and its conjugate base is Cl^-. On the other hand, it is conceivable that HCl might act as a base, since there are electron pairs available on the chlorine to form a coordinate covalent bond with a proton. If HCl acts as a base, its conjugate acid is H_2Cl^+.

Acid-Base Equilibria

The conjugate acid-base pair picture suggests that acid-base reactions are reversible and thus establish chemical equilibria. This appears to be correct. These equilibria, like others, may be described by equilibrium constants. In the following sections, we shall discuss some typical acid-base equilibria.

THE WATER CONSTANT AND pH. The solvent system concept is based on the self-dissociation of the solvent, water, as described earlier by Equation (15.2), which is reproduced here.

$$H_2O + H_2O = H_3O(aq)^+ + OH(aq)^-$$

The equilibrium constant for this reaction, K_c, may be written as follows:

$$K_c = \frac{(H_3O(aq)^+)(OH(aq)^-)}{(H_2O)^2} \tag{15.28}$$

In dilute solutions, the activities may be replaced by molarities. In addition, the concentration of water remains essentially constant, since water is the solvent that is present in great excess. Thus Equation (15.28) may be rewritten as follows:

$$K_c[H_2O]^2 = [H_3O(aq)^+][OH(aq)^-] = K_w \qquad (15.29)$$

Since the water concentration is constant, $K_c[H_2O]^2$ is constant, and is called the water constant, K_w. It is the product of the equilibrium concentrations of hydronium and hydroxide ions. The experimentally determined value for K_w at 25°C is 1.00×10^{-14}. This means that in pure water or dilute aqueous solutions, the product of the hydronium and hydroxide ion molarities is 1.00×10^{-14}.

From Equation (15.2), it is clear that the equilibrium concentrations of hydronium and hydroxide ions must be equal in pure water, that is, $[H_3O(aq)^+] = [OH(aq)^-]$, so that Equation (15.29) may be rewritten as follows:

$$[H_3O(aq)^+]^2 = 1.00 \times 10^{-14}$$

$$[H_3O(aq)^+] = \sqrt{1.00 \times 10^{-14}} = 1.00 \times 10^{-7}\ M$$

$$[OH(aq)^-] = [H_3O(aq)^+] = 1.00 \times 10^{-7}\ M$$

According to the solvent-system concept, an acid is a substance that increases the hydronium ion concentration. However, from Equation (15.29) it is clear that the product of the hydronium and hydroxide ion molarities must remain constant. When an acid is placed in water, the increase in hydronium ion concentration is accompanied by a decrease in hydroxide ion concentration, so that the product of the two concentrations remains constant. This may be viewed also as a shift in equilibrium for the reaction represented by Equation (15.2). An increase in hydronium ion concentration is a stress that is relieved by the consumption of both hydronium and hydroxide ions to produce water. In the process, the hydroxide ion concentration falls below 10^{-7} *M*.

Similarly, a base increases the hydroxide ion concentration at the expense of the hydronium ion concentration. In a basic solution the hydroxide ion concentration exceeds 10^{-7} *M* and the hydronium ion concentration is less than 10^{-7} *M*.

This provides the means for the experimental determination of acidic and basic solutions. An acidic solution is one in which the hydronium ion concentration exceeds 10^{-7} *M*, and a basic solution is one in which the hydronium ion concentration is less than 10^{-7} *M*. A solution in which the hydronium ion concentration equals 10^{-7} *M* is called a neutral solution.

Solution acidities often are expressed on the pH scale. The pH of a solution is defined as the negative logarithm of the molar concentration of hydronium ion, as follows:

$$pH = -\log[H_3O(aq)^+] \qquad (15.30)$$

The equilibrium concentration is used in the definition since the self-dissociation of water is a rapidly attained and adjusted equilibrium.

To get our bearings on the pH scale, first consider pure water for which $[H_3O(aq)^+] = 1.00 \times 10^{-7}$. By application of the pH definition, it can be seen that the pH of pure water is 7.

$$\text{pH} = -\log[\text{H}_3\text{O(aq)}^+] = -\log(1.00 \times 10^{-7}) = -(\log 1.00 + \log 10^{-7})$$
$$= -(0 + [-7]) = -[-7] = 7$$

A value of 7 for the pH of a solution is called the neutral pH.

For an acidic solution, the hydronium ion concentration exceeds 10^{-7} *M*. For example, if the hydronium ion concentration is 3.0×10^{-5} *M*, the pH of the solution is 4.5, as shown by the following computation:

$$\text{pH} = -\log[\text{H}_3\text{O(aq)}^+] = -\log(3.0 \times 10^{-5}) = -(\log 3.0 + \log 10^{-5})$$
$$= -(0.5 + [-5]) = -0.5 - [-5] = -0.5 + 5 = 4.5$$

Similarly, for a basic solution, the hydronium ion concentration is less than 10^{-7} *M*. For example, if the concentration is 3.0×10^{-9} *M*, the pH is 8.5.

$$\text{pH} = -\log[\text{H}_3\text{O(aq)}^+] = -\log(3.0 \times 10^{-9}) = -(\log 3.0 + \log 10^{-9})$$
$$= -(0.5 + [-9]) = -0.5 - [-9] = 0.5 + 9 = 8.5$$

It should be clear that neutral solutions have a pH of 7, whereas *acidic solutions* have pH values *less than 7* and *basic solutions* have pH values *greater than 7*.

Often the p symbolism is extended to other values. When used in front of symbols or formulas for chemical species, it indicates the negative logarithm of the molar concentration of that species:

$$\text{pOH} = -\log[\text{OH(aq)}^-] \qquad (15.31)$$

$$\text{pNa} = -\log M_{\text{Na}^+} \qquad (15.32)$$

Often this symbolism is used with equilibrium constants to indicate the negative logarithm of the constant, as follows:

$$\text{p}K_c = -\log K_c \qquad (15.33)$$

$$\text{p}K_w = -\log K_w \qquad (15.34)$$

A useful relationship may be derived from the water constant expression.

$$K_w = [\text{H}_3\text{O(aq)}^+][\text{OH(aq)}^-]$$
$$-\log K_w = -\log\{[\text{H}_3\text{O(aq)}^+][\text{OH(aq)}^-]\}$$
$$= -\log[\text{H}_3\text{O(aq)}^+] - \log[\text{OH(aq)}^-]$$
$$\text{p}K_w = \text{pH} + \text{pOH} \qquad (15.35)$$

Since K_w is 1.00×10^{-14} at 25°C, pK_w is 14 and Equation (15.35) may be rewritten for aqueous solutions at 25°C.

$$pH + pOH = 14 \tag{15.36}$$

The following problems illustrate computations with pH and pOH.

What is the pH and pOH for a solution that is 4 × 10^{-8} M in hydroxide ion? This may be solved in two slightly different but related ways. First, the pOH may be found directly as follows:

$$pOH = -\log[OH(aq)^-] = -\log(4 \times 10^{-8}) = 7.4$$

Then the pH may be found by using Equation (15.36) as follows:

$$pH = 14 - pOH = 14 - 7.4 = 6.6$$

Alternately, the pH may be found from the definition, using the water constant to find the hydronium ion concentration from the given value for the hydroxide ion concentration.

$$[H_3O(aq)^+] = \frac{K_w}{[OH(aq)^-]} = \frac{1.0 \times 10^{-14}}{4.0 \times 10^{-8}} = 2.5 \times 10^{-7}\ M$$

$$pH = -\log[H_3O(aq)^+] = -\log(2.5 \times 10^{-7}) = 6.6$$

Often, when the pH of a solution is determined experimentally, we desire to know the hydronium ion concentration. For example, *What is the hydronium ion concentration when the pH is 3.84?*

To solve this, recall the following relationship:

If $\log Y = X$, then $Y = 10^X$.

The problem may be solved as follows.

$$\log[H_3O(aq)^+] = -pH$$

$$[H_3O(aq)^+] = 10^{-pH} = 10^{-3.84} = 10^{(-4.00+0.16)} = 10^{(0.16-4)}$$

$$= 10^{0.16} \times 10^{-4} = 1.4 \times 10^{-4}\ M$$

ACID-BASE CONSTANTS. Consider the reaction of aqueous hydrogen fluoride with water as represented by the following equation:

$$HF(aq) + H_2O = H_3O(aq)^+ + F(aq)^- \tag{15.37}$$

This is a typical, Bronsted-Lowry equilibrium acid-base reaction. The equilibrium constant expression for this reaction may be written as follows:

$$K_c = \frac{(H_3O(aq)^+)(F(aq)^-)}{(HF(aq))(H_2O)} \tag{15.38}$$

Again, if we limit our considerations to dilute aqueous solutions, activities may be replaced by molarities and the water concentration remains essentially constant, since water is the solvent present in large excess. Thus Equation (15.38) may be rewritten as Equation (15.39):

$$K_c[H_2O] = \frac{[H_3O(aq)^+][F(aq)^-]}{[HF(aq)]} = K_a \tag{15.39}$$

This modified equilibrium constant expression is called the *acid constant*, K_a.

In a similar way, a *base constant* expression may be written for aqueous solutions of bases. For example, consider ammonia as a typical base that undergoes the following reaction with water:

$$NH_3(aq) + H_2O = NH_4(aq)^+ + OH(aq)^- \tag{15.40}$$

The expression for the base constant K_b may be written as follows:

$$K_b = \frac{[NH_4(aq)^+][OH(aq)^-]}{[NH_3(aq)]} \tag{15.41}$$

Together, acid and base constants often are called *ionization constants*. The numerical values for these ionization constants serve as a measure of the strengths of acids and bases.

For example, a high value for a base constant indicates that a relatively high percentage of the dissolved base molecules or ions have reacted with water at equilibrium to form the conjugate acid and base. This means that a base with a high base-constant value is able to remove a proton from water rather easily. Such a base is called a *strong base*.

However, if the base-constant value is low, only a small percentage of the basic molecules or ions has removed protons from water at equilibrium, indicating that this base does not remove protons from water very easily. Such a base is called a *weak base*.

In a similar fashion, acids with high acid-constant values are called *strong acids*, and acids with small acid-constant values are called *weak acids*.

In Table 15.1, some acid and base-constant values are listed. The acids are listed in the order of decreasing strength, whereas the bases are listed in the order of increasing strength. The reason for this can be seen in the relationship between an acid constant and the base constant for the conjugate base of the acid.

For example, nitrous acid reacts with water as follows:

$$HNO_2(aq) + H_2O = H_3O(aq)^+ + NO_2(aq)^- \tag{15.42}$$

The acid-constant expression is given by the following equation:

$$K_a = \frac{[H_3O(aq)^+][NO_2(aq)^-]}{[HNO_2(aq)]} \tag{15.43}$$

Nitrite ion is the conjugate base of nitrous acid. The base-constant expression shown in Equation (15.45) is based on the following reaction:

$$NO_2(aq)^- + H_2O = HNO_2(aq) + OH(aq)^- \tag{15.44}$$

$$K_b = \frac{[HNO_2(aq)][OH(aq)^-]}{[NO_2(aq)^-]} \tag{15.45}$$

Equation (15.45) may be solved for $[NO_2(aq)^-]/[HNO_2(aq)]$ as follows:

$$\frac{[NO_2(aq)^-]}{[HNO_2(aq)]} = \frac{[OH(aq)^-]}{K_b} \tag{15.46}$$

This expression may be substituted into Equation (15.45).

$$K_a = \frac{[H_3O(aq)^+][OH(aq)^-]}{K_b} \tag{15.47}$$

However, the numerator of the expression in Equation (15.47) is the water constant. Therefore the equation may be rewritten.

$$K_aK_b = K_w \tag{15.48}$$

Thus the product of the acid constant and the base constant for the conjugate base of the acid is constant. This means that the conjugate base of a weak acid is a strong base, and the conjugate base of a strong acid is a weak base. The bases listed in Table 15.1 are the conjugate bases of the acids listed there.

TABLE 15.1

Some Acid and Base Constants at Room Temperature

ACID CONSTANTS	ACID	CONJUGATE BASE	BASE CONSTANTS
very large	$HClO_4(aq)$	$ClO_4(aq)^-$	very small
very large	$H_2SO_4(aq)$	$HSO_4(aq)^-$	very small
very large	$HCl(aq)$	$Cl(aq)^-$	very small
very large	$HNO_3(aq)$	$NO_3(aq)^-$	very small
very large	$H_3O_2(aq)^+$	$H_2O_2(aq)$	very small
very large	$NH_2SO_3H(aq)$	$NH_2SO_3(aq)^-$	very small
9.1	$H_3O(aq)^+$	H_2O	1.1×10^{-15}
5.6×10^{-2}	$H_2C_2O_4(aq)$	$HC_2O_4(aq)^-$	1.8×10^{-13}
1.7×10^{-2}	$SO_2(H_2O)_x$	$HSO_3(aq)^-$	5.9×10^{-13}
1.2×10^{-2}	$HSO_4(aq)^-$	$SO_4(aq)^{--}$	8.3×10^{-13}
1×10^{-2}	$[Co(H_2O)_x]^{3+}$	$[Co(OH)(H_2O)_{x-1}]^{++}$	1×10^{-12}
8.5×10^{-3}	$[Sn(H_2O)_x]^{++}$	$[Sn(OH)(H_2O)_{x-1}]^+$	1.2×10^{-12}
7.1×10^{-3}	$H_3PO_4(aq)$	$H_2PO_4(aq)^-$	1.4×10^{-12}
6.8×10^{-3}	$[Fe(H_2O)_x]^{3+}$	$[Fe(OH)(H_2O)_{x-1}]^{++}$	1.5×10^{-12}
6.0×10^{-3}	$H_3AsO_4(aq)$	$H_2AsO_4(aq)^-$	1.7×10^{-12}
1×10^{-3}	$H_3VO_3(aq)$	$H_2VO_3(aq)^-$	1×10^{-11}

TABLE 15.1 (cont.)

ACID CONSTANTS	ACID	CONJUGATE BASE	BASE CONSTANTS
6.8×10^{-4}	$HF(aq)$	$F(aq)^-$	1.5×10^{-11}
5.1×10^{-4}	$HNO_2(aq)$	$NO_2(aq)^-$	2.0×10^{-11}
3×10^{-4}	$H_2VO_3(aq)^-$	$HVO_3(aq)^{--}$	3×10^{-11}
1.6×10^{-4}	$[Cr(H_2O)_x]^{3+}$	$[Cr(OH)(H_2O)_{x-1}]^{++}$	6.3×10^{-11}
6.3×10^{-5}	$C_6H_5COOH(aq)$	$C_6H_5COO(aq)^-$	1.6×10^{-10}
5.2×10^{-5}	$HC_2O_4(aq)^-$	$C_2O_4(aq)^{--}$	1.9×10^{-10}
1.9×10^{-5}	$HN_3(aq)$	$N_3(aq)^-$	5.3×10^{-10}
1.7×10^{-5}	$CH_3COOH(aq)$	$CH_3COO(aq)^-$	5.9×10^{-10}
4.4×10^{-7}	$CO_2(H_2O)_x$	$HCO_3(aq)^-$	2.3×10^{-8}
1.1×10^{-7}	$H_2AsO_4(aq)^-$	$HAsO_4(aq)^{--}$	9.1×10^{-8}
8.9×10^{-8}	$H_2S(aq)$	$HS(aq)^-$	1.1×10^{-7}
6.3×10^{-8}	$H_2PO_4(aq)^-$	$HPO_4(aq)^{--}$	1.6×10^{-7}
6.2×10^{-8}	$HSO_3(aq)^-$	$SO_3(aq)^{--}$	1.6×10^{-7}
1.6×10^{-8}	$[Pb(H_2O)_x]^{++}$	$[Pb(OH)(H_2O)_{x-1}]^+$	6.3×10^{-7}
1.0×10^{-8}	$NH_2—NH_3(aq)^+$	$NH_2—NH_2(aq)$	9.8×10^{-7}
1.0×10^{-8}	$[Cu(H_2O)_x]^{++}$	$[Cu(OH)(H_2O)_{x-1}]^+$	1.0×10^{-6}
7.9×10^{-9}	$[Ni(H_2O)_x]^{++}$	$[Ni(OH)(H_2O)_{x-1}]^+$	1.3×10^{-6}
8.3×10^{-10}	$H_3AsO_3(aq)$	$H_2AsO_3(aq)^-$	1.2×10^{-5}
7.8×10^{-10}	$H_3BO_3(aq)$	$H_2BO_3(aq)^-$	1.3×10^{-5}
5.9×10^{-10}	$NH_4(aq)^+$	$NH_3(aq)$	1.7×10^{-5}
4.8×10^{-10}	$HCN(aq)$	$CN(aq)^-$	2.1×10^{-5}
1.5×10^{-10}	$[Zn(H_2O)_x]^{++}$	$[Zn(OH)(H_2O)_{x-1}]^+$	6.7×10^{-5}
1.1×10^{-10}	$C_6H_5OH(aq)$	$C_6H_5O(aq)^-$	9.1×10^{-5}
4.8×10^{-11}	$HCO_3(aq)^-$	$CO_3(aq)^{--}$	2.1×10^{-4}
3.0×10^{-12}	$HAsO_4(aq)^{--}$	$AsO_4(aq)^{3-}$	3.3×10^{-3}
2.6×10^{-12}	$H_2O_2(aq)$	$HO_2(aq)^-$	3.8×10^{-3}
1×10^{-12}	$[Cd(H_2O)_x]^{++}$	$[Cd(OH)(H_2O)_{x-1}]^+$	1×10^{-2}
7.4×10^{-13}	$H_2AsO_3(aq)^-$	$HAsO_3(aq)^{--}$	1.4×10^{-2}
6.3×10^{-13}	$[Co(H_2O)_x]^{++}$	$[Co(OH)(H_2O)_{x-1}]^+$	1.6×10^{-2}
4.4×10^{-13}	$HPO_4(aq)^{--}$	$PO_4(aq)^{3-}$	2.3×10^{-2}
1.3×10^{-13}	$HS(aq)^-$	$S(aq)^{--}$	7.7×10^{-2}
4.0×10^{-14}	$HAsO_3(aq)^{--}$	$AsO_3(aq)^{3-}$	2.5×10^{-1}
1.1×10^{-15}	H_2O	$OH(aq)^-$	9.1
very small	$CH_3CH_2OH(aq)$	$CH_3CH_2O(aq)^-$	very large
very small	$NH_3(aq)$	$NH_2(aq)^-$	very large
very small	$OH(aq)^-$	$O(aq)^{--}$	very large
very small	$CH_4(aq)$	$CH_3(aq)^-$	very large

The effect of the solvent upon the acid-base character of particular species must be considered. Acid-base strength is related directly to the acidic and basic characteristics of the solvent used.

At the top of Table 15.1 there is a group of six acids, all of which are listed with very large acid constants. In effect, these ionization constants are so high that it is impossible to measure them, and for all intents and purposes, when one of these acids dissolves in water, ionization is complete. Thus all six acids appear to be equally strong in water, and it is impossible to differentiate among them since they are completely ionized. This is because water is a sufficiently good base to make all of these acids appear to be strong. This phenomenon is called the *leveling effect* of the solvent. It is evident also among the bases at the end of Table 15.1. For the four strong bases listed there, water is sufficiently acidic to make these bases react completely in removing protons from water. The bases are leveled to a common strength by the solvent in the same way strong acids are leveled to a common strength by the solvent.

However, if the strong acids listed in Table 15.1 are dissolved in a more acidic solvent, such as liquid acetic acid, they do not appear to be so strong. Here the solvent is less basic and does not remove the protons from the acid as readily. Some acid constant values are recorded in Table 15.2 for three of the strong acids of Table 15.1 when these acids are dissolved in acetic acid.

TABLE 15.2 Some Acid Constants in Acetic Acid at 25°C

$HClO_4$	2.0×10^{-3}
H_2SO_4	5.6×10^{-5}
HCl	5.0×10^{-6}

POLYPROTIC ACIDS AND BASES. There are acids capable of losing more than one proton per molecule or ion. For example, oxalic acid, $H_2C_2O_4$, may lose two protons per molecule. If one proton is lost, the conjugate base is the hydrogen oxalate ion, $HC_2O_4^-$. If this species acts as an acid, its conjugate base is the oxalate ion, $C_2O_4^{--}$. Acids that may lose more than one proton per molecule or ion are called *polyprotic acids*. Oxalic acid is a diprotic acid, since it may lose two protons per molecule; arsenic acid, H_3AsO_4, is a triprotic acid which may lose three protons per molecule.

Proton loss appears to occur in a stepwise fashion. Generally, the second proton is lost less readily than the first, and the third is lost less readily than the second. The over-all equilibrium for the loss of several protons may be viewed as the sum of several single-step equilibria, so that the over-all ionization constant is the product of the ionization constants for the single steps.

For example, the two steps in the ionization of oxalic acid may be represented as follows:

$$H_2C_2O_4(aq) + H_2O = HC_2O_4(aq)^- + H_3O(aq)^+ \quad (15.49)$$

$$HC_2O_4(aq)^- + H_2O = C_2O_4(aq)^{--} + H_3O(aq)^+ \quad (15.50)$$

$$H_2C_2O_4(aq) + 2H_2O = C_2O_4(aq)^{--} + 2H_3O(aq)^+ \quad (15.51)$$

The over-all reaction is represented by Equation (15.51), which is the sum of Equations (15.49) and (15.50). The ionization represented by Equation (15.49) is called the *first ionization* of oxalic acid, and the equilibrium is described by the *first ionization constant* of oxalic acid, K_{a1}. The ionization represented by Equation (15.50) is called the *second ionization* of oxalic acid, and the equilibrium is described by the second ionization constant of oxalic acid, K_{a2}. The expressions for the first and second ionization constants follow:

$$K_{a1} = \frac{[HC_2O_4(aq)^-][H_3O(aq)^+]}{[H_2C_2O_4(aq)]} = 5.6 \times 10^{-2} \tag{15.52}$$

$$K_{a2} = \frac{[C_2O_4(aq)^{--}][H_3O(aq)^+]}{[HC_2O_4(aq)^-]} = 5.2 \times 10^{-5} \tag{15.53}$$

The product of these two constants is the over-all ionization constant, K_{aT}, which describes the over-all reaction represented by Equation (15.51).

$$K_{a1}K_{a2} = \frac{[HC_2O_4(aq)^-][H_3O(aq)^+][C_2O_4(aq)^{--}][H_3O(aq)^+]}{[H_2C_2O_4(aq)][HC_2O_4(aq)^-]}$$

$$= \frac{[C_2O_4(aq)^{--}][H_3O(aq)^+]^2}{[H_2C_2O_4(aq)]} = K_{aT} = 2.9 \times 10^{-6}$$

One important observation to be made is that the equilibrium constant for a reaction that may be conceived as the sum of several equilibria is the product of the separate equilibrium constants for the individual steps. Thus it is possible also to find over-all base constants for polyprotic bases, that is, bases that win more than one proton per molecule or ion. For example, when phosphate ions are dissolved in water, the following reactions occur:

$$PO_4(aq)^{3-} + H_2O = HPO_4(aq)^{--} + OH(aq)^- \tag{15.54}$$

$$HPO_4(aq)^{--} + H_2O = H_2PO_4(aq)^- + OH(aq)^- \tag{15.55}$$

$$H_2PO_4(aq)^- + H_2O = H_3PO_4(aq) + OH(aq)^- \tag{15.56}$$

$$PO_4(aq)^{3-} + 3H_2O = H_3PO_4(aq) + 3OH(aq)^- \tag{15.57}$$

The total base constant for phosphate ion may be computed as follows:

$$K_{bT} = \frac{[H_3PO_4(aq)][OH(aq)^-]^3}{[PO_4(aq)^{3-}]} = K_{b1}K_{b2}K_{b3}$$

$$= (2.3 \times 10^{-2})(1.6 \times 10^{-7})(1.4 \times 10^{-12})$$

$$= 5.1 \times 10^{-21}$$

OTHER ACID-BASE EQUILIBRIA. From the ionization constant values in Table 15.1, it is possible to determine over-all equilibrium constants for many different acid-base reactions by the technique of adding equations and multiplying equilibrium constants.

For example, hydrofluoric acid is an acid, and cyanide ion is a base, as both are listed in Table 15.1. They might be expected to react according to the following equation:

$$HF(aq) + CN(aq)^- = HCN(aq) + F(aq)^- \qquad (15.58)$$

To learn whether or not this reaction takes place to a large extent it is useful to compute the equilibrium constant. A large value of the constant suggests a complete reaction and a small value suggests an incomplete reaction. Since the equilibrium constant is a thermodynamic function, its value depends only upon the initial and final states. Thus it is possible to construct a set of hypothetical steps through which the reaction might proceed, without worrying about whether or not these are actual steps in a mechanism. Since values are recorded for acid and base constants and the water constant, it is convenient to use the reactions corresponding to these values for the hypothetical steps, as follows:

$$HF(aq) + H_2O = H_3O(aq)^+ + F(aq)^- \qquad K_a = 6.8 \times 10^{-4} \qquad (15.59)$$

$$CN(aq)^- + H_2O = HCN(aq) + OH(aq)^- \qquad K_b = 2.1 \times 10^{-5} \qquad (15.60)$$

$$H_3O(aq)^+ + OH(aq)^- = 2H_2O \qquad 1/K_w = 1.0 \times 10^{14} \qquad (15.61)$$

$$HF(aq) + CN(aq)^- = HCN(aq) + F(aq)^- \qquad K_c = 1.4 \times 10^{6} \qquad (15.62)$$

Clearly the high value for the equilibrium constant indicates that this reaction is a complete one; that is, there is a strong driving force behind it.

In this reaction, the acid HF reacts with the base CN^-, which is a stronger base than the conjugate base of the HF. Looking at it from another point of view, the base reacts with an acid that is stronger than the conjugate acid of the base. In all such cases, the equilibrium constant exceeds unity. When an acid reacts with a base that is weaker than its conjugate base, or when a base reacts with an acid that is weaker than its conjugate acid, the equilibrium constant is less than unity.

FACTORS AFFECTING ACID-BASE STRENGTH

In order for species to serve as Bronsted-Lowry acids, they must contain hydrogen. Most acids may be classified into two groups: the hydroxy acids and the binary acids. Typical hydroxy acids are sulfuric acid, H_2SO_4; nitrous acid, HNO_2; perchloric acid, $HClO_4$; and acetic acid, CH_3COOH. In addition, the hydrated cations belong in this category. Typical binary acids are hydrofluoric acid, HF, and hydrogen sulfide, H_2S. Often hydrocyanic acid, HCN, and ammonium ion, NH_4^+, are included among the binary acids.

Since acid and base constants are related to the concentration equilibrium constants, they are related also to the standard free energy changes for the reactions of the acids or bases with water. Since this free-energy change may be explained in terms of the enthalpy and entropy changes, the trends in acid-base strength expressed in Table 15.1 may be understood in these same terms.

Enthalpy Changes

For our purposes, we shall consider a hypothetical molecular acid, HA, which undergoes the following reaction with water:

$$HA(aq) + H_2O = H_3O(aq)^+ + A(aq)^- \qquad (15.63)$$

The enthalpy change for this reaction may be understood from the following hypothetical, stepwise cycle, known as a Born-Haber cycle.

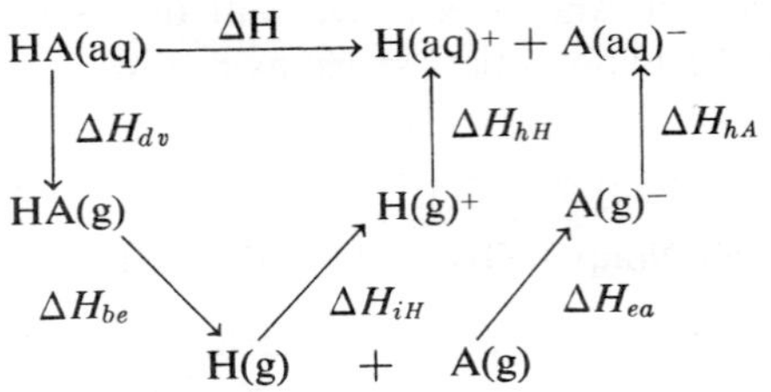

The over-all enthalpy change for the reaction, ΔH, is the sum of the enthalpy changes for each of the steps, as follows:

$$\Delta H = \Delta H_{dv} + \Delta H_{be} + \Delta H_{iH} + \Delta H_{ea} + \Delta H_{hH} + \Delta H_{hA} \qquad (15.64)$$

The individual steps are discussed briefly in the following sections.

ENTHALPY OF DEHYDRATION AND VAPORIZATION, ΔH_{dv}. Generally, this enthalpy change is positive, indicating an energy input to separate the acid and the waters of hydration and to put the solute into the gaseous phase. The most significant differences among various acids lies in the degree of association the acid particles have with themselves. When association is high, the enthalpy change is quite positive and thus unfavorable for the reaction.

ENTHALPY OF INTRAMOLECULAR BONDING, ΔH_{be}. In this step, the bond between the acidic hydrogen and the rest of the molecule is broken. For convenience, we shall call the rest of the acid molecule the acid substrate. Thus in the hypothetical acid, HA, the acid substrate is A; in HCl it is Cl, and in $HClO_4$ it is ClO_4. The acid substrate differs from the conjugate base in having one less negative-charge quantum per particle.

The enthalpy of intramolecular bonding is positive and represents the input of energy equivalent to the bond dissociation energy for the H—A bond. Among the hydroxy acids this is always the H—O bond, and thus does not differ greatly from one hydroxy acid to another. However, there can be great variances among the binary acids, depending upon the element to which hydrogen is bonded.

ENTHALPIES OF IONIZATION, ΔH_{iH} AND ΔH_{eA}. The enthalpy of ionization of hydrogen, ΔH_{iH}, is positive and corresponds to the ionization potential for hydrogen. The value for this enthalpy change is the same for all Bronsted-Lowry acids.

The enthalpy of ionization of the acid substrate, A, ΔH_{eA}, usually is negative (although it may be positive if A is an anion) and corresponds to the electron affinity of the acid substrate. Since the trends in electron affinity are essentially the same as those for electronegativity, we shall frequently use that term in discussing these trends.

Thus when the acid substrate is highly electronegative, the value of ΔH_{eA} is highly negative, representing a highly favorable enthalpy change. For example, the fact that perchloric acid, $HClO_4$, is a stronger acid than hypochlorous acid, HClO, can be attributed in part to the high electronegativity of the ClO_4 substrate compared to the electronegativity of the ClO substrate.

ENTHALPIES OF HYDRATION, ΔH_{hH} AND ΔH_{hA}. The final steps in this Born-Haber cycle consist of the hydrations of the proton and the conjugate base of the acid. Both of these enthalpies are negative, representing the energy released due to the bond formation between the proton and water molecules and the conjugate base and water molecules. In general, very negative enthalpies of hydration are associated with small, highly charged particles. Certainly, the high charge leads to strong electrostatic interactions with polar water molecules and thus large decreases in enthalpy. The small size of the particle keeps the electrostatic charge intense and does not permit it to be diffusely spread throughout a large volume. Thus solvent water molecules may closely approach the electrostatic charge and then be attracted strongly.

This means that the hydration of the proton, which is a very small particle, is accompanied by a sizable decrease in enthalpy, which constitutes a favorable driving force. This decrease is identical for all Bronsted-Lowry acids. However, the hydration of the conjugate base varies among acids, the most favorable enthalpy changes occurring for very negative anions of small size. In many instances, the enthalpy of hydration of the conjugate base is very negative when the enthalpy of ionization of the acid substrate is very negative, since the same factors tend to be operative.

Entropy Changes

From Equation (15.63), it may appear at first that a typical reaction of an acid with water may not lead to any appreciable entropy change. However, it must be remembered that ions in aqueous solution lead to much clustering of water molecules around the ions. This clustering represents a more highly ordered state for the solvent, and an unfavorable situation from the entropy point of view. As written, Equation (15.63) represents an unfavorable entropy change, since neutral molecules are converted to ions.

Although acids need not be neutral molecules, it is true generally that the entropy changes accompanying the reactions described by the acid constants are unfavorable. The production of the hydronium ion is constant in all of these reactions, and thus in any comparison of acids, the entropy effects brought about by the conjugate bases are important. For example, the fact that hydrofluoric acid is a weaker acid than hydrochloric acid can be attributed in part to the small fluoride ion, which is more highly hydrated than the larger chloride ion and thus decreases the entropy more than the chloride ion.

In general, those conjugate bases which are more highly hydrated and have the more negative enthalpies of hydration also have the more unfavorable entropy effects associated with them.

Bases

Although the same free-energy considerations apply to bases, it is more common to give the arguments for the acids and then to relate the base strength to that of its conjugate acid. Thus a base is strong if it has a weak conjugate acid, and is weak if it has a strong conjugate acid.

A SURVEY OF SOME ACIDS AND BASES

In the remainder of this chapter some typical acids and bases are reviewed in the light of the foregoing discussion of factors affecting acid and base strengths. The arrangement is by periodic family of the characteristic elements in the acids and bases.

Family 7A: The Halogens

Some of the halogen-containing acids and bases are listed in Table 15.3 along with their acid and base constants.

Among the binary hydrohalic acids, it is clear that hydrofluoric acid is the weakest, and the remaining acids are leveled by the water solvent to the category of strong acids. Some studies in other solvents have suggested that hydroiodic is the strongest, followed by hydrobromic and hydrochloric acids in the order shown in Table 15.3. It is clear, however, that the big difference occurs between hydrofluoric acid and the other three hydrohalic acids.

The unfavorable free-energy change for hydrofluoric acid probably is due to both an unfavorable enthalpy change and an unfavorable entropy change. The reaction in question is represented by Equation (15.65).

$$HF(aq) + H_2O = H_3O(aq)^+ + F(aq)^- \qquad (15.65)$$

In large measure, the unfavorable enthalpy change is due to the strong hydrogen-bonding properties of hydrogen fluoride, which make the dehydration and vaporization enthalpy change highly positive. Not only does hydrogen fluoride form hydrogen bonds to itself, but undoubtedly it also forms hydrogen bonds with the hydrating water molecules. The unfavorable entropy change is due to the small size of the fluoride ion, which is highly hydrated and thus produces a high degree of organization of the solvent water molecules as they cluster around these ions.

TABLE 15.3 Acid and Base Constants at Room Temperature for Some Halogen Species

ACID CONSTANTS	ACID	CONJUGATE BASE	BASE CONSTANTS
very large	$HClO_4(aq)$	$ClO_4(aq)^-$	very small
very large	$HI(aq)$	$I(aq)^-$	very small
very large	$HBr(aq)$	$Br(aq)^-$	very small
very large	$HCl(aq)$	$Cl(aq)^-$	very small
very large	$HClO_3(aq)$	$ClO_3(aq)^-$	very small
very large	$HBrO_3(aq)$	$BrO_3(aq)^-$	very small
1.6×10^{-1}	$HIO_3(aq)$	$IO_3(aq)^-$	6.2×10^{-14}
1.1×10^{-2}	$HClO_2(aq)$	$ClO_2(aq)^-$	9.1×10^{-13}
6.8×10^{-4}	$HF(aq)$	$F(aq)^-$	1.5×10^{-11}
6.0×10^{-8}	$HClO(aq)$	$ClO(aq)^-$	1.6×10^{-7}
5.4×10^{-9}	$H_4IO_6(aq)^-$	$H_3IO_6(aq)^{--}$	1.8×10^{-6}
2.5×10^{-9}	$HBrO(aq)$	$BrO(aq)^-$	4.0×10^{-6}
4.5×10^{-13}	$HIO(aq)$	$IO(aq)^-$	2.2×10^{-2}
very small	$H_3IO_6(aq)^{--}$	$H_2IO_6(aq)^{3-}$	very large

Following are some comparable standard free energy changes in kcal/mole:

	X = *F*	*Cl*	*Br*	*I*
$HX(aq) + H_2O = H_3O(aq)^+ + X(aq)^-$	+4.33	−8.6	−11.8	−12.7

The most important factor governing the acidity of the hydroxy acids appears to be the electronegativity of the acid substrate. For the very highly electronegative substrates, the enthalpies of ionization of the substrates and hydration of the conjugate bases are quite negative and so tend to be important factors in making the enthalpy change favorable. Thus there is a difference in electronegativity of the substrate due to the nature of the central halogen. Thus hypochlorous acid is a stronger acid than hypobromous acid, since the chlorine is more electronegative than the bromine.

In addition, the number of oxygens attached to the central halogen helps to determine the electronegativity of the acid substrate. The more oxygen atoms there are, the more electronegative the substrate and the stronger the hydroxy acid. For example, perchloric acid, $HOClO_3$, is much stronger than hypochlorous acid, HOCl. The obvious difference is that perchloric acid molecules have three additional oxygens bonded to the central chlorine.

Family 6A: The Chalcogens

Some acid and base constants for chalcogen-containing species are listed in Table 15.4.

Although all of the binary acids are fairly weak, it is clear that water is much weaker as an acid than the other three binary acids. The situation is comparable to that of hydrogen fluoride, in which the water molecules are hydrogen-bonded and the conjugate base, OH^-, is relatively small and highly hydrated and thus highly organizing for solvent water molecules.

Among the hydroxy acids, the electronegativity of the acid substrate again appears to be a dominant factor. Thus sulfuric acid is a stronger acid than sulfurous acid and also a stronger acid than selenic acid. Even when one of the hydroxy groups of sulfuric acid is replaced by an amido group, NH_2, this group is sufficiently electronegative to allow amidosulfuric acid, NH_2SO_2OH, to be a strong acid.

Sulfurous acid is listed in Table 15.4 as $SO_2(H_2O)_x$. It is probable that the actual form of sulfur dioxide in water is mostly hydrated sulfurous acid, H_2SO_3, although

TABLE 15.4 Acid and Base Constants at Room Temperature for Some Chalcogen Species

ACID CONSTANTS	ACID	CONJUGATE BASE	BASE CONSTANTS
very large	$H_2SO_4(aq)$	$HSO_4(aq)^-$	very small
very large	$H_3O_2(aq)^+$	$H_2O_2(aq)$	very small
very large	$NH_2SO_2OH(aq)$	$NH_2SO_3(aq)^-$	very small
9.1	$H_3O(aq)^+$	H_2O	1.1×10^{-15}
2.4×10^{-1}	$H_2S_2O_3(aq)$	$HS_2O_3(aq)^-$	4.2×10^{-14}
1.9×10^{-2}	$HS_2O_3(aq)^-$	$S_2O_3(aq)^{--}$	5.3×10^{-13}
1.7×10^{-2}	$SO_2(H_2O)_x$	$HSO_3(aq)^-$	5.9×10^{-13}
1.2×10^{-2}	$HSO_4(aq)^-$	$SO_4(aq)^{--}$	8.3×10^{-13}
1×10^{-2}	$HSeO_4(aq)^-$	$SeO_4(aq)^{--}$	1×10^{-12}
2.3×10^{-3}	$H_2Te(aq)$	$HTe(aq)^-$	4.3×10^{-12}
2.4×10^{-3}	$H_2SeO_3(aq)$	$HSeO_3(aq)^-$	4.2×10^{-12}
1.3×10^{-4}	$H_2Se(aq)$	$HSe(aq)^-$	7.7×10^{-11}
8.9×10^{-8}	$H_2S(aq)$	$HS(aq)^-$	1.1×10^{-7}
6.2×10^{-8}	$HSO_3(aq)^-$	$SO_3(aq)^{--}$	1.5×10^{-7}
2.5×10^{-8}	$H_6TeO_6(aq)$	$H_5TeO_6(aq)^-$	4.0×10^{-7}
4.8×10^{-9}	$HSeO_3(aq)^-$	$SeO_3(aq)^{--}$	2.1×10^{-6}
1.0×10^{-11}	$HSe(aq)^-$	$Se(aq)^{--}$	1.0×10^{-3}
1.0×10^{-11}	$HTe(aq)^-$	$Te(aq)^{--}$	1.0×10^{-3}
6.5×10^{-12}	$H_5TeO_6(aq)^-$	$H_4TeO_6(aq)^{--}$	1.5×10^{-3}
2.6×10^{-12}	$H_2O_2(aq)$	$HO_2(aq)^-$	3.8×10^{-3}
1.3×10^{-13}	$HS(aq)^-$	$S(aq)^{--}$	7.7×10^{-2}
1.1×10^{-15}	H_2O	$OH(aq)^-$	9.1

there may be some SO_2 dissolved in a more loosely hydrated form. The value of the acid constant for sulfurous acid is for the total dissolved sulfur(IV) oxide, regardless of its form in solution.

Family 5A: The Nitrogen Family

Acid and base constants for species of the nitrogen family are shown in Table 15.5.

The diversity of the species in this family makes comparisons somewhat difficult. For example, nitric acid is a metaacid, whereas phosphoric acid is an orthoacid. Phosphorous acid has the peculiar structure in which one hydrogen is bonded directly to phosphorus. In addition, phosphoric acid tends to form highly poly-

TABLE 15.5 *Acid and Base Constants at Room Temperature for Some Nitrogen Family Species*

ACID CONSTANTS	ACID	CONJUGATE BASE	BASE CONSTANTS
very large	$HNO_3(aq)$	$NO_3(aq)^-$	very small
1×10^{-1}	$H_3PO_2(aq)$	$H_2PO_2(aq)^-$	1×10^{-13}
3.0×10^{-2}	$H_4P_2O_7(aq)$	$H_3P_4O_7(aq)^-$	3.3×10^{-13}
1.0×10^{-2}	$H_3PO_3(aq)$	$H_2PO_3(aq)^-$	1.0×10^{-12}
1×10^{-2}	$[Bi(H_2O)_x]^{3+}$	$[Bi(OH)(H_2O)_{x-1}]^{++}$	1×10^{-12}
7.1×10^{-3}	$H_3PO_4(aq)$	$H_2PO_4(aq)^-$	1.4×10^{-12}
6.0×10^{-3}	$H_3AsO_4(aq)$	$H_2AsO_4(aq)^-$	1.7×10^{-12}
5.0×10^{-3}	$H_3P_3O_{10}(aq)^{--}$	$H_2P_3O_{10}(aq)^{3-}$	2.0×10^{-12}
4.4×10^{-3}	$H_3P_2O_7(aq)^-$	$H_2P_2O_7(aq)^{--}$	2.3×10^{-12}
5.1×10^{-4}	$HNO_2(aq)$	$NO_2(aq)^-$	2.0×10^{-11}
1.9×10^{-5}	$HN_3(aq)$	$N_3(aq)^-$	5.3×10^{-10}
1.0×10^{-6}	$NH_3OH(aq)^+$	$NH_2OH(aq)$	1.0×10^{-8}
5.5×10^{-7}	$H_2P_3O_{10}(aq)^{3-}$	$HP_3O_{10}(aq)^{4-}$	1.8×10^{-8}
2.5×10^{-7}	$H_2P_2O_7(aq)^{--}$	$HP_2O_7(aq)^{3-}$	4.0×10^{-8}
2.0×10^{-7}	$H_2PO_3(aq)^-$	$HPO_3(aq)^{--}$	5.0×10^{-8}
1.1×10^{-7}	$H_2AsO_4(aq)^-$	$HAsO_4(aq)^{--}$	9.1×10^{-8}
6.3×10^{-8}	$H_2PO_4(aq)^-$	$HPO_4(aq)^{--}$	1.6×10^{-7}
1.0×10^{-8}	$NH_2—NH_3(aq)^+$	$NH_2—NH_2(aq)$	9.8×10^{-7}
1.3×10^{-9}	$HP_3O_{10}(aq)^{4-}$	$P_3O_{10}(aq)^{5-}$	7.7×10^{-6}
8.3×10^{-10}	$H_3AsO_3(aq)$	$H_2AsO_3(aq)^-$	1.2×10^{-5}
5.9×10^{-10}	$NH_4(aq)^+$	$NH_3(aq)$	1.7×10^{-5}
5.6×10^{-10}	$HP_2O_7(aq)^{3-}$	$P_2O_7(aq)^{4-}$	1.8×10^{-5}
3.0×10^{-12}	$HAsO_4(aq)^-$	$AsO_4(aq)^{3-}$	3.3×10^{-3}
7.4×10^{-13}	$H_2AsO_3(aq)^-$	$HAsO_3(aq)^{--}$	1.4×10^{-2}
4.4×10^{-13}	$HPO_4(aq)^{--}$	$PO_4(aq)^{3-}$	2.3×10^{-2}
4.0×10^{-14}	$HAsO_3(aq)^{--}$	$AsO_3(aq)^{3-}$	2.5×10^{-1}
very small	$NH_3(aq)$	$NH_2(aq)^-$	very large

merized species. Later members in the family tend to be more metallic, and the characteristic species of bismuth is the hydrated bismuth(III) ion.

Hypophosphorous acid, H_3PO_2, is a stronger acid than phosphorous acid, H_3PO_3, which is a stronger acid than phosphoric acid, H_3PO_4. Although this trend may be associated with the direct bonding of the hydrogen to phosphorus in H_3PO_2 and H_3PO_3, the reason for the trend is not clear.

The highly polymeric hydroxy acids of phosphorus tend to be strong acids and this is not surprising. For example, diphosphoric acid, $H_4P_2O_7$, is a tetraprotic acid that may be considered as orthophosphoric acid, H_3PO_4, in which one of the hydroxy groups has been replaced by the PO_4H_2 group, which is more highly electronegative than the hydroxy group.

It is not especially surprising that the hydrated bismuth(III) ion is rather highly acidic. The acid substrate for this species is $[Bi(OH)(H_2O)_{x-1}]^{3+}$, and as a highly charged cation it has a high electronegativity that helps to account for its acidity.

It is possible to compare the ammonium ion, NH_4^+, the hydrazinium ion, $NH_2—NH_3^+$, and the hydroxylammonium ion, $NH_3—OH^+$, since the latter two may be considered as substituted ammonium ions. The hydroxylammonium ion is an ammonium ion in which the highly electronegative hydroxy group has replaced a hydrogen and thus the hydroxylammonium ion is a stronger acid than the ammonium ion. The hydrazinium ion is an ammonium ion in which a hydrogen has been replaced by an amino group which is less electronegative than the hydroxy group but more electronegative than a hydrogen. Thus the hydrazinium ion is more acidic than the ammonium ion but less acidic than the hydroxylammonium ion.

Family 4A: The Carbon Family

Acid and base constants for some species of the carbon family are shown in Table 15.6.

The binary hydrogen compounds of the carbon-family elements are such weak acids that they do not play an important role in aqueous acid-base chemistry. For the most part, this can be attributed to the low electronegativity of the acid substrates.

The simple hydroxy acids in this family are somewhat difficult to compare since the hydroxy acid of carbon, carbonic acid, is unstable and the hydroxy acids of both silicon and germanium tend to form high polymers.

The somewhat unstable carbonic acid is formed by dissolving CO_2 in water as shown by the following equation:

$$CO_2(g) + H_2O = H_2CO_3(aq) \tag{15.66}$$

However, not all CO_2 dissolved in water is converted to carbonic acid. Much of it is more loosely hydrated and might be represented more accurately by the formula $CO_2(H_2O)_x$, in which the discrete H_2CO_3 molecule may not be discernible. In Table 15.1 and Table 15.6, the species listed as $CO_2(H_2O)_x$ with an acid constant of 4.4×10^{-7} represents all of the dissolved CO_2 in some kind of hydrated form, regardless of the specific species present. On the other hand, the species listed as H_2CO_3 with an acid constant of 1.7×10^{-4} is the true carbonic acid in aqueous solution. It is this constant that must be used to compare the strength of carbonic acid to the strengths of other acids. On the other hand, to make computations involving the acidity of solutions in which CO_2 has been dissolved, the former value must be used since some of the hydrated CO_2 that cannot be identified as true carbonic acid may also have protons removed from the hydrating water molecules.

Aside from formic acid, HCOOH, the simplest carboxylic acid is acetic acid, CH_3COOH. This acid is somewhat weaker than carbonic acid. Acetic acid may be viewed as carbonic acid in which one hydroxy group has been replaced by a less electronegative methyl group. The nature of the alkyl group makes rather little difference, and propionic acid, CH_3CH_2COOH, butyric acid, $CH_3CH_2CH_2COOH$, etc., all have acid constants with the approximate value of 10^{-5}. However, if other electronegative groups are substituted on the methyl group of acetic acid, the

TABLE 15.6
Acid and Base Constants at Room Temperature for Some Carbon Family Species

ACID CONSTANTS	ACID	CONJUGATE BASE	BASE CONSTANTS
very large	$CH_3SO_3H(aq)$	$CH_3SO_3(aq)^-$	very small
very large	$HNCS(aq)$	$NCS(aq)^-$	very small
2×10^{-1}	$Cl_3C—COOH(aq)$	$Cl_3C—COO(aq)^-$	5×10^{-14}
5.6×10^{-2}	$HOOC—COOH(aq)$	$HOOC—COO(aq)^-$	1.8×10^{-13}
5×10^{-2}	$NO_2—CH_2—COOH(aq)$	$NO_2—CH_2—COO(aq)^-$	2×10^{-13}
3.3×10^{-2}	$Cl_2CH—COOH(aq)$	$Cl_2CH—COO(aq)^-$	3.0×10^{-13}
8.5×10^{-3}	$[Sn(H_2O)_x]^{++}$	$[Sn(OH)(H_2O)_{x-1}]^+$	1.2×10^{-12}
5.1×10^{-3}	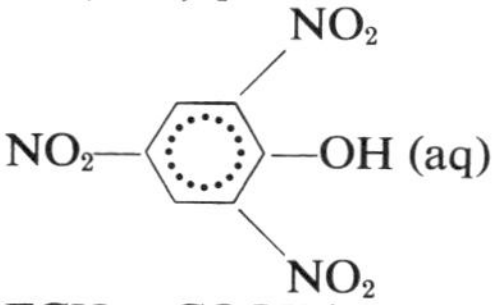	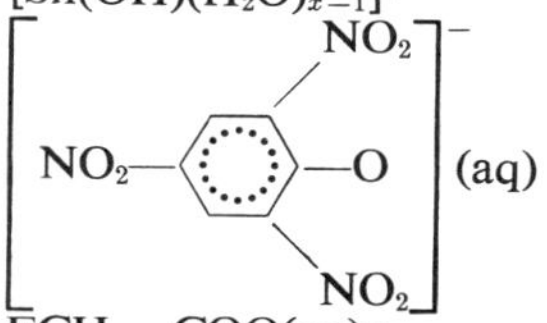	2.0×10^{-12}
2.6×10^{-3}	$FCH_2—COOH(aq)$	$FCH_2—COO(aq)^-$	3.8×10^{-12}
1.4×10^{-3}	$HOOC—CH_2—COOH(aq)$	$HOOC—CH_2—COO(aq)^-$	7.1×10^{-12}
1.4×10^{-3}	$ClCH_2—COOH(aq)$	$ClCH_2—COO(aq)^-$	7.1×10^{-12}
1.3×10^{-3}	$BrCH_2—COOH(aq)$	$BrCH_2—COO(aq)^-$	7.7×10^{-12}
6.7×10^{-4}	$ICH_2—COOH(aq)$	$ICH_2—COO(aq)^-$	1.5×10^{-11}
2.2×10^{-4}	$CH_3CH_2CH—COOH(aq)$ | OH	$CH_3CH_2CH—COO(aq)^-$ | OH	4.5×10^{-11}
1.8×10^{-4}	$HCOOH(aq)$	$HCOO(aq)^-$	5.6×10^{-11}
1.7×10^{-4}	$H_2CO_3(aq)$	$HCO_3(aq)^-$	5.9×10^{-11}
1.5×10^{-4}	$HOCH_2—COOH(aq)$	$HOCH_2—COO(aq)^-$	6.7×10^{-11}
6.3×10^{-5}	$C_6H_5COOH(aq)$	$C_6H_5COO(aq)^-$	1.6×10^{-10}
5.2×10^{-5}	$HOOC—COO(aq)^-$	$OOC—COO(aq)^{--}$	1.9×10^{-10}
4.1×10^{-5}	$CH_3CHCH_2—COOH(aq)$ | OH	$CH_3CHCH_2—COO(aq)^-$ | OH	2.4×10^{-10}
2.4×10^{-5}	$C_6H_5NH_3(aq)^+$	$C_6H_5NH_2(aq)$	4.2×10^{-10}
1.7×10^{-5}	$CH_3COOH(aq)$	$CH_3COO(aq)^-$	5.9×10^{-10}
1.3×10^{-5}	$CH_3CH_2—COOH(aq)$	$CH_3CH_2—COO(aq)^-$	7.7×10^{-10}
1.0×10^{-5}	$[Sn(OH)(H_2O)_{x-1}]^+$	$[Sn(OH)_2(H_2O)_{x-2}]$	1.0×10^{-9}
1.4×10^{-6}	$HOOC—CH_2—COO(aq)^-$	$OOC—CH_2—COO(aq)^{--}$	7.1×10^{-9}
4.4×10^{-7}	$CO_2(H_2O)_x$	$HCO_3(aq)^-$	2.3×10^{-8}
3.0×10^{-7}	$NH_3—CH_2CH_2—NH_3(aq)^{++}$	$NH_3—CH_2CH_2—NH_2(aq)^+$	3.3×10^{-8}
1.6×10^{-8}	$[Pb(H_2O)_x]^{++}$	$[Pb(OH)(H_2O)_{x-1}]^+$	6.3×10^{-7}
1.1×10^{-9}	O O ‖ ‖ $CH_3CCH_2CCH_3(aq)$	O O ‖ ‖ $CH_3CCHCCH_3(aq)^-$	9.1×10^{-6}
4.8×10^{-10}	$HCN(aq)$	$CN(aq)^-$	2.1×10^{-5}
2.8×10^{-10}	$[Sn(OH)_2(H_2O)_{x-2}]$	$[Sn(OH)_3(H_2O)_{x-3}]^-$	3.6×10^{-5}
1.6×10^{-10}	$(CH_3)_3—NH(aq)^+$	$(CH_3)_3N(aq)$	6.3×10^{-5}
1.1×10^{-10}	$C_6H_5OH(aq)$	$C_6H_5O(aq)^-$	9.1×10^{-5}
4.8×10^{-11}	$HCO_3(aq)^-$	$CO_3(aq)^{--}$	2.1×10^{-4}
2.4×10^{-11}	$CH_3—NH_3(aq)^+$	$CH_3—NH_2(aq)$	4.2×10^{-4}
1.3×10^{-11}	$NH_3—CH_2CH_2—NH_2(aq)^+$	$NH_2—CH_2CH_2—NH_2(aq)$	7.7×10^{-4}
1.7×10^{-11}	$(CH_3)_2—NH_2(aq)^+$	$(CH_3)_2—NH(aq)$	6.0×10^{-4}
1×10^{-11}	$NO_2CH_3(aq)$	$NO_2CH_2(aq)^-$	1×10^{-3}
very small	$CH_3OH(aq)$	$CH_3O(aq)^-$	very large
very small	$CH_3CH_2OH(aq)$	$CH_3CH_2O(aq)^-$	very large
very small	$CH_4(aq)$	$CH_3(aq)^-$	very large

acidity increases markedly. The acid constant values in Table 15.6 show that the following acids are in the order of decreasing acid strength: Cl_3C—COOH > NO_2CH_2—COOH > Cl_2CH—COOH > FCH_2—COOH > $ClCH_2$—COOH > $BrCH_2$—COOH > $HOCH_2$—COOH > ICH_2—COOH. Thus three highly electronegative chlorines on the carbon adjacent to the carboxyl group makes the acid rather strong, and certainly stronger than dichloroacetic acid, which is stronger than monochloroacetic acid. Similarly, the acid strengths of the monohaloacetic acids decrease in the order of decreasing electronegativity of the halogen. It is not surprising that nitroacetic acid is fairly strong since the nitro group is rather highly electronegative.

Alcohols such as methanol, CH_3OH, and ethanol, CH_3CH_2OH, may be considered to consist of water molecules in which one hydrogen has been replaced by alkyl groups. If anything, the alkyl groups may be slightly less electronegative than the hydrogens, and alcohols are in general very weak acids, unless some other electronegative atom or group is involved also. However, the phenyl group, C_6H_5, appears to have a strong electron-withdrawing effect; when hydrogen of water is replaced by a phenyl group to form hydroxybenzene, otherwise known as phenol or carbolic acid, the acid constant jumps from 1.1×10^{-15} to 1.1×10^{-10}. If three highly electronegative nitro groups are substituted on the phenyl group of phenol to form 2,4,6-trinitrophenol, otherwise called picric acid, the acid constant jumps to 5.1×10^{-3}.

Generally hydrogens bonded to carbon are not very acidic since the acid substrate consists of carbon and hydrogen and thus is not extremely electronegative. However, if other highly electronegative atoms or groups are involved to make the acid substrate more electronegative, the hydrogens bonded to carbon may become acidic. One example is hydrogen cyanide, HCN, where the rather highly electronegative nitrogen is included in the acid substrate. Another example is 2,4-pentandione,

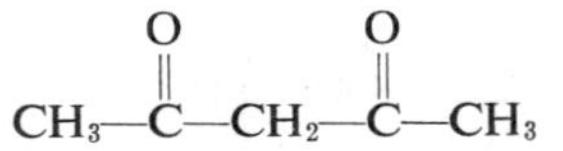

where the electronegative carbonyl groups help to account for an acid constant of about 10^{-9}. Although methane has negligible acidity in water, when a nitro group replaces one hydrogen the resulting nitromethane displays small but perceptible acidic properties toward water with an acid constant of about 10^{-11}.

When a hydroxy acid may be visualized as the result of replacing one hydroxy group in a polyhydroxy acid with an alkyl group, the resulting acid usually is only slightly weaker than the parent acid. For example, sulfuric acid, HO—SO_2—OH, is a strong dihydroxy acid. If one of the hydroxy groups is replaced by an alkyl group, the resulting acid is one of the sulfonic acids. Methyl sulfonic acid, CH_3—SO_2—OH, is an example. Like sulfuric acid, the sulfonic acids are strong acids.

Amines, such as aminomethane, CH_3NH_2, and aminobenzene, $C_6H_5NH_2$, are bases. They may be considered to be substituted ammonias, and their conjugate acids, $CH_3NH_3^+$ and $C_6H_5NH_3^+$, may be thought of as substituted ammonium ions. Like ammonium ion, these substituted ammonium ions act as acids. That the ammonium ion itself is a better acid than water can be attributed in part to the lower N—H bond energy and to the greater electronegativity of the positively charged NH_3^+ acid substrate for the ammonium ion. The substitution for hydrogen of groups which have electronegativities similar to that of hydrogen alters the acidity of the ammonium ion only slightly.

Species Containing the Metals

Acid and base constants for some species of the metals are shown in Table 15.7.

For the most part these species are coordination complexes in which at least one of the ligands is water. In these cases it is a hydrogen from one of the water molecules which is lost in the acid-base reaction and these species are hydroxy acids. All the complexes listed in Table 15.7 as acids are cationic, and this corresponds to a high electronegativity for the acid substrate, helping to account for their acidity in terms of the Born-Haber cycle. In this light it can be seen from Table 15.7 that the strongest acids tend to be those complexes with the highest ionic charges.

When similar complexes that differ only in the metallic cation present are compared, the relative acidity of the two species often can be related to the relative electronegativity of the two central metallic cations. For example, $[Co(H_2O)_x]^{3+}$ is more acidic than $[La(H_2O)_x]^{3+}$, where cobalt is more electronegative than lanthanum.

Color Indicators

Many dyes are pH sensitive, and their colors change as the pH varies. These substances are acids and bases, which are called *acid-base color indicators*. In the acid form they are frequently represented by the simplified formula, HInd, the conjugate base being represented by Ind^-. The equilibrium reaction in water is shown by Equation (15.67).

$$HInd(aq) + H_2O = H_3O(aq)^+ + Ind(aq)^- \qquad (15.67)$$

The acid form, HInd, and the conjugate base form, Ind^-, have different colors and the shifting of the equilibrium represented by Equation (15.67) can be observed as a color change as acidity changes. With increasing acidity, the equilibrium shifts toward the acid form, HInd, and the color of the acid form predominates in highly acidic solutions. If the solution is made basic, the equilibrium shifts toward the basic form, Ind^-, and the color becomes that of the basic form of the indicator.

TABLE 15.7
Acid and Base Constants at Room Temperature for Some Metal-containing Species

ACID CONSTANTS	ACID	CONJUGATE BASE	BASE CONSTANTS
very large	$HMnO_4(aq)$	$MnO_4(aq)$	very small
6×10^{-1}	$[In(H_2O)_x]^{3+}$	$[InOH(H_2O)_{x-1}]^{++}$	2×10^{-14}
1×10^{-2}	$[Co(H_2O)_x]^{3+}$	$[CoOH(H_2O)_{x-1}]^{++}$	1×10^{-12}
1×10^{-2}	$[Tl(H_2O)_x]^{3+}$	$[TlOH(H_2O)_{x-1}]^{++}$	1×10^{-12}
1×10^{-2}	$[Bi(H_2O)_x]^{3+}$	$[BiOH(H_2O)_{x-1}]^{++}$	1×10^{-12}
8.5×10^{-3}	$[Sn(H_2O)_x]^{++}$	$[SnOH(H_2O)_{x-1}]^{+}$	1.2×10^{-12}
6.8×10^{-3}	$[Fe(H_2O)_x]^{3+}$	$[FeOH(H_2O)_{x-1}]^{++}$	1.5×10^{-12}
1.3×10^{-3}	$[V(H_2O)_x]^{3+}$	$[VOH(H_2O)_{x-1}]^{++}$	7.7×10^{-12}
2.5×10^{-4}	$[Co(NH_3)_2(H_2O)_4](aq)^{3+}$	$[CoOH(NH_3)_2(H_2O)_3](aq)^{++}$	4.0×10^{-11}
2.2×10^{-4}	$[Hg(H_2O)_x]^{++}$	$[HgOH(H_2O)_{x-1}]^{+}$	4.5×10^{-11}
1.6×10^{-4}	$[Cr(H_2O)_x]^{3+}$	$[CrOH(H_2O)_{x-1}]^{++}$	6.3×10^{-11}
1×10^{-4}	$[Be(H_2O)_x]^{++}$	$[BeOH(H_2O)_{x-1}]^{+}$	1×10^{-10}
7.1×10^{-5}	$H_2CrO_4(aq)$	$HCrO_4(aq)^{-}$	1.4×10^{-10}
5.1×10^{-5}	$[Co(NH_3)_3(H_2O)_3](aq)^{3+}$	$[CoOH(NH_3)_3(H_2O)_2](aq)^{++}$	2.0×10^{-10}
2.5×10^{-5}	$[Sc(H_2O)_x]^{3+}$	$[ScOH(H_2O)_{x-1}]^{++}$	4.0×10^{-10}
1.1×10^{-5}	$[Al(H_2O)_x]^{3+}$	$[AlOH(H_2O)_{x-1}]^{++}$	9.1×10^{-10}
1×10^{-5}	$[Hg_2(H_2O)_x]^{++}$	$[Hg_2OH(H_2O)_{x-1}]^{+}$	1×10^{-9}
8.3×10^{-6}	$[Cr(NH_3)_4(H_2O)_2](aq)^{3+}$	$[CrOH(NH_3)_4(H_2O)](aq)^{++}$	1.2×10^{-9}
5.0×10^{-6}	$[Cr(NH_3)_5(H_2O)](aq)^{3+}$	$[CrOH(NH_3)_5](aq)^{++}$	2.0×10^{-9}
2.5×10^{-6}	$[La(H_2O)_x]^{3+}$	$[LaOH(H_2O)_{x-1}]^{++}$	4.0×10^{-9}
2.0×10^{-6}	$[Co(NH_3)_5(H_2O)](aq)^{3+}$	$[CoOH(NH_3)_5](aq)^{++}$	5.0×10^{-9}
1.6×10^{-6}	$[Co(NH_3)_4(H_2O)_2](aq)^{3+}$	$[CoOH(NH_3)_4H_2O](aq)^{++}$	6.3×10^{-9}
1.4×10^{-6}	$[Rh(NH_3)_5(H_2O)](aq)^{3+}$	$[RhOH(NH_3)_5](aq)^{++}$	7.1×10^{-9}
1.2×10^{-6}	$[Fe(H_2O)_x]^{++}$	$[FeOH(H_2O)_{x-1}]^{+}$	8.3×10^{-9}
1.0×10^{-6}	$[CrCl_2(H_2O)_4](aq)^{+}$	$[CrCl_2(OH)(H_2O)_3](aq)$	1.0×10^{-8}
4.4×10^{-8}	$[CrOH(NH_3)_4H_2O](aq)^{++}$	$[Cr(OH)_2(NH_3)_4](aq)^{+}$	2.3×10^{-7}
1.6×10^{-8}	$[Pb(H_2O)_x]^{++}$	$[PbOH(H_2O)_{x-1}]^{+}$	6.3×10^{-7}
1.2×10^{-8}	$[Pt(NH_3)_6](aq)^{4+}$	$[Pt(NH_2)(NH_3)_5](aq)^{3+}$	8.3×10^{-7}
1.0×10^{-8}	$[Cu(H_2O)_x]^{++}$	$[CuOH(H_2O)_{x-1}]^{+}$	1.0×10^{-6}
7.9×10^{-9}	$[Ni(H_2O)_x]^{++}$	$[NiOH(H_2O)_{x-1}]^{+}$	1.3×10^{-6}
1.5×10^{-10}	$[Zn(H_2O)_x]^{++}$	$[ZnOH(H_2O)_{x-1}]^{+}$	6.7×10^{-5}
8.3×10^{-11}	$[Pt(NH_2)(NH_3)_5](aq)^{3+}$	$[Pt(NH_2)_2(NH_3)_4](aq)^{++}$	1.2×10^{-4}
2.5×10^{-11}	$[Mn(H_2O)_x]^{++}$	$[MnOH(H_2O)_{x-1}]^{+}$	4.0×10^{-4}
7.2×10^{-12}	$[Rh(NH_3)_6](aq)^{3+}$	$[Rh(NH_2)(NH_3)_5](aq)^{++}$	1.4×10^{-3}
2.0×10^{-12}	$[Ag(H_2O)_x]^{+}$	$[AgOH(H_2O)_{x-1}]$	5.0×10^{-3}
1×10^{-12}	$[Cd(H_2O)_x]^{++}$	$[CdOH(H_2O)_{x-1}]^{+}$	1×10^{-2}
6.3×10^{-13}	$[Co(H_2O)_x]^{++}$	$[CoOH(H_2O)_{x-1}]^{+}$	1.6×10^{-2}
1.6×10^{-13}	$[Mg(H_2O)_x]^{++}$	$[MgOH(H_2O)_{x-1}]^{+}$	6.3×10^{-2}
very small	$[Ca(H_2O)_x]^{++}$	$[CaOH(H_2O)_{x-1}]^{+}$	very large
very small	$[Sr(H_2O)_x]^{++}$	$[SrOH(H_2O)_{x-1}]^{+}$	very large
very small	$[Ba(H_2O)_x]^{++}$	$[BaOH(H_2O)_{x-1}]^{+}$	very large
very small	$[Li(H_2O)_x]^{+}$	$[LiOH(H_2O)_{x-1}]$	very large
very small	$[Na(H_2O)_x]^{+}$	$[NaOH(H_2O)_{x-1}]$	very large
very small	$[K(H_2O)_x]^{+}$	$[KOH(H_2O)_{x-1}]$	very large
very small	$[Rb(H_2O)_x]^{+}$	$[RbOH(H_2O)_{x-1}]$	very large

Although small changes in color may be detected with instruments, certain conditions must be met before small color changes may be detected by the eye. For example, if the indicator is in a highly acidic solution and the acid form has an intense color, the small amount of basic form present is not detectable and small changes in acidity will go undetected by the human eye. However, in solutions containing nearly equal concentrations of acid and base forms which are of similar intensity, the indicator is visibly sensitive to pH changes.

The indicator equilibrium described by Equation (15.67) may also be described by an ionization constant, called the indicator constant, K_{ind}, which is defined by the following equation.

$$K_{ind} = \frac{[H_3O(aq)^+][Ind(aq)^-]}{[HInd(aq)]} \tag{15.68}$$

Equation (15.68) may be solved for $[H_3O(aq)^+]$ as follows:

$$[H_3O(aq)^+] = \frac{K_{ind}[HInd(aq)]}{[Ind(aq)^-]} \tag{15.69}$$

In general, if the intensities of the acid and base forms of the indicator are about the same, the most sensitive pH for color-change observation is the pH at which $[HInd(aq)] = [Ind(aq)^-]$, so that any change in pH leads to a predominance of one of the two forms. Using Equation (15.69), it is clear that under these circumstances the most sensitive conditions for color change are described by Equations (15.68) and (15.71).

$$[H_3O(aq)^+] = K_{ind} \tag{15.70}$$

$$pH = pK_{ind} \tag{15.71}$$

In practice, it is found that visible color changes occur in indicators over a short pH range, often over about 2 pH units; the pH value lies in the middle of this range at the approximate value of pK_{ind}.

The value of the indicator constant, and consequently, the pH range over which visible color changes occur, depend upon the molecular structure of the indicator. Generally, the useful indicators are complex carbon-containing species that have highly delocalized electron systems, such as those found in benzene type compounds, which allow electron transitions that account for the color. These systems are affected by addition of or loss of a proton so that the energy of electron transitions changes and consequently the wavelength of light needed to bring about the transitions changes.

A few common indicators are listed in Table 15.8, along with their common names, structures, colors, and pH ranges for visible detection.

TABLE 15.8
Some Common Acid-Base Color Indicators

COMMON NAME	CHEMICAL STRUCTURE	ACID COLOR	BASE COLOR	COLOR CHANGE pH RANGE
picric acid	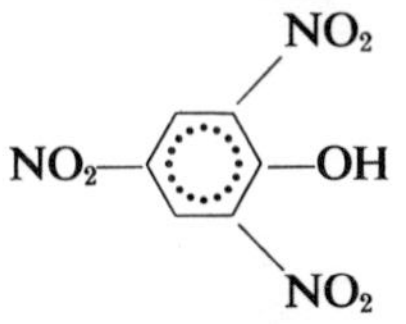	colorless	yellow	0.0-1.3
p-nitrophenol	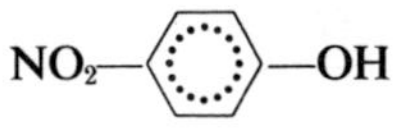	colorless	yellow	5.6-7.6
nitramine	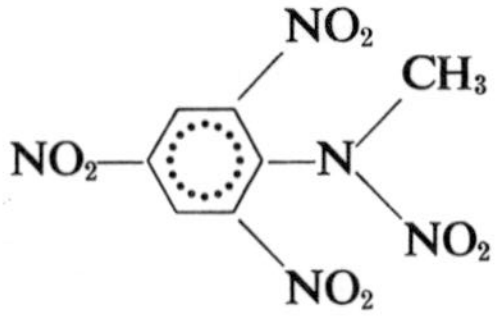	colorless	brown	10.8-13.0
methyl orange	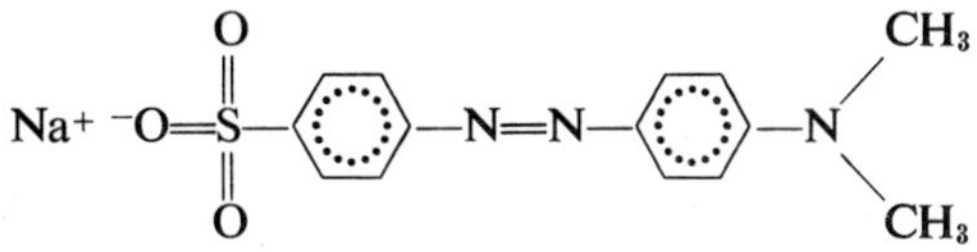	red	yellow-orange	3.1-4.4
methyl red	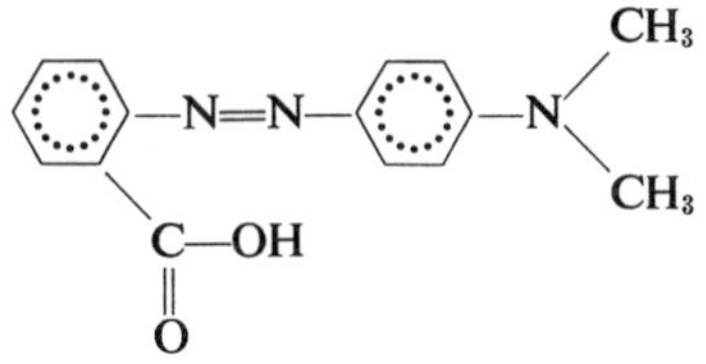	red	yellow	4.2-6.2
alizarin yellow R	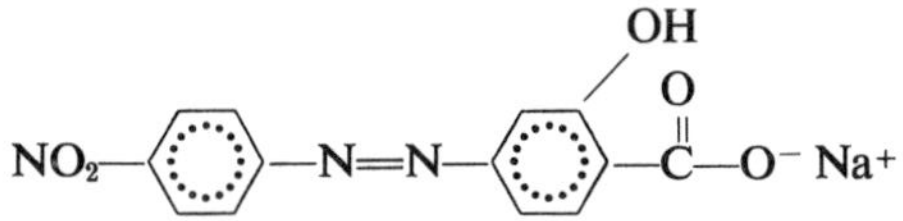	yellow red	violet yellow	11.0-12.0 1.9-3.3

congo red	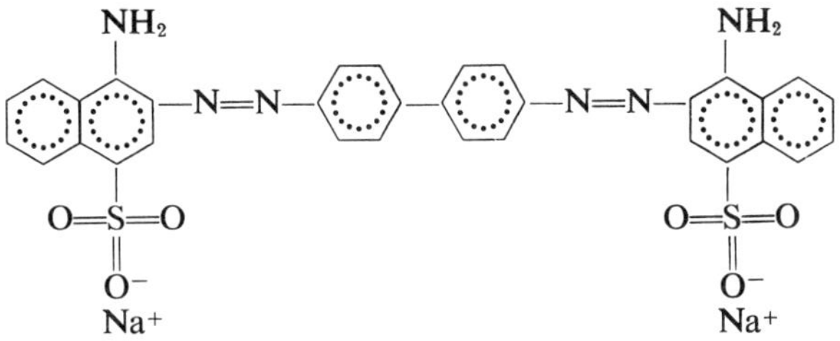	blue	red	3.0-5.0
methyl violet	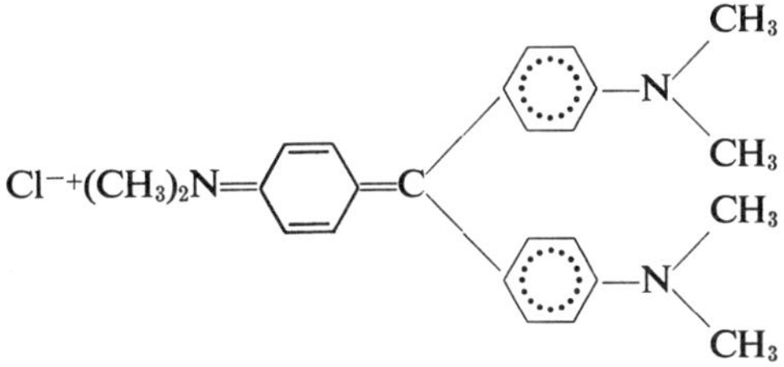	yellow	violet	2.0-3.1
phenolphthalein	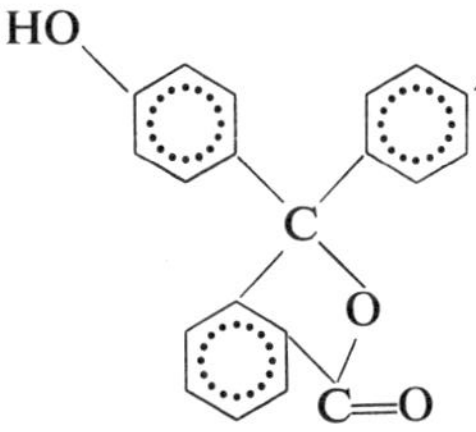	colorless	red	8.0-9.8
bromphenol blue	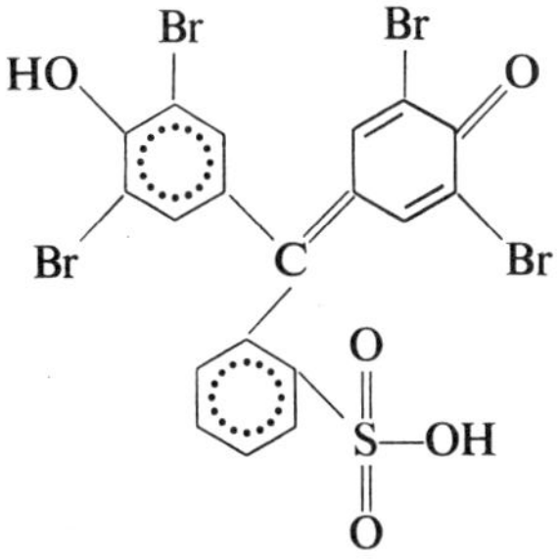	yellow	purple	3.0-4.6

TABLE 15.8 (cont.)

COMMON NAME	CHEMICAL STRUCTURE	ACID COLOR	BASE COLOR	COLOR CHANGE pH RANGE
phenol red	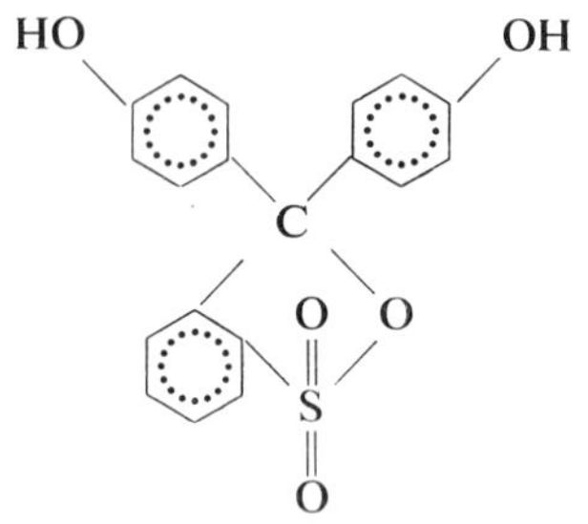	yellow	red	6.8-8.4
alizarin sodium sulfonate	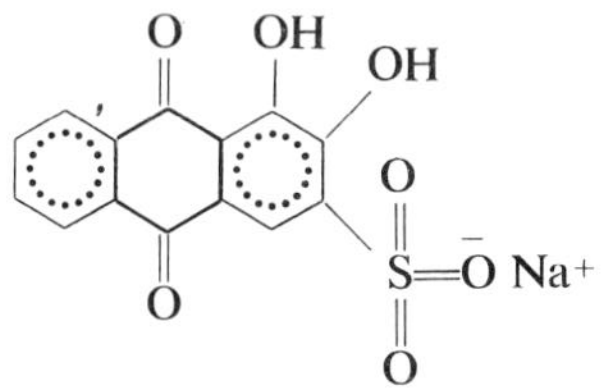	yellow	purple	3.7-5.2
indigo carmine	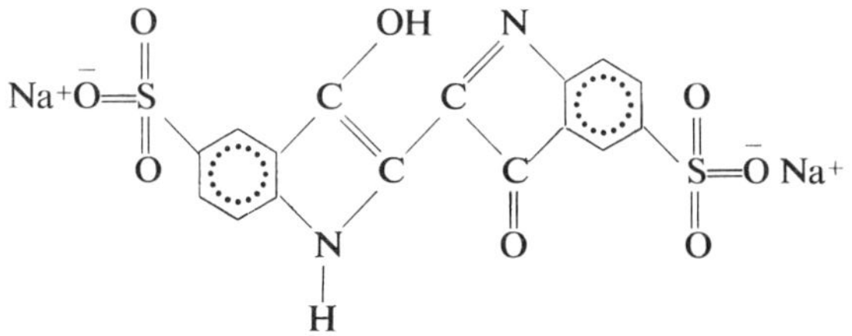	blue	yellow	11.6-14.0

SUMMARY

Experimental observations have been used to class substances as acids, bases, or neither acids nor bases. A study of these substances reveals characteristics which acids have in common along with those which bases have in common.

Of central importance to us is the Bronsted-Lowry concept, by which an acid-base reaction is viewed as a competition for a hydrogen ion, the winner being called a base and the loser an acid. The relative competitive strengths of acids and bases can be discerned by comparison of the acid and base constants for various species.

In water, the acidity or basicity of solutions is based on the relative concentrations of hydronium and hydroxide ions; a 10^{-7} M concentration of each is taken as the neutral point. This corresponds to a pH of 7. Basic substances increase the hydroxide ion concentration at the expense of the hydronium ion concentration, and thus basic solutions have hydronium ion concentrations less than 10^{-7} M, corresponding to pH values greater than 7. Similarly, acidic substances increase the hydronium ion concentration at the expense of hydroxide ion concentration, so that acidic solutions have hydronium ion concentrations greater than 10^{-7} M corresponding to pH values less than 7.

The effect of the addition of a substance to water can be predicted by separate evaluation of the acidic and basic species it produces in solution. For example, if sodium hydroxide, NaOH, is dissolved in water, sodium and hydroxide ions are produced. Sodium ion, Na^+, is neutral, but the addition of hydroxide ion, OH^-, increases the hydroxide ion concentration and the solution is basic. Similarly, if sodium carbonate, Na_2CO_3, is added to water, sodium and carbonate ions are produced. Although the sodium ion is neutral, the carbonate ion, CO_3^{--}, is basic as indicated by its base constant value. Thus the solution is basic.

On the other hand, if iron(III) chloride is added to water, the chloride ion and hydrated iron(III) ion, $[Fe(H_2O)_x]^{3+}$, are produced. The chloride ion is neutral, but the iron(III) ion is acidic, as indicated by its acid constant value. Consequently the solution is acidic.

If sodium hydrogen carbonate, $NaHCO_3$, is added to water, sodium ions, Na^+, and hydrogen carbonate ions, HCO_3^-, are produced. Again the sodium ions are neutral, but the hydrogen carbonate ions are amphoteric. However, since the acid constant for the hydrogen carbonate ion is 4.8×10^{-11} and the base constant is slightly larger at 2.3×10^{-8}, it is clear that the hydrogen carbonate ion is a stronger base than it is an acid and an aqueous solution of sodium hydrogen carbonate is slightly basic.

In a similar fashion, when ammonium fluoride, NH_4F, is added to water, ammonium ions, NH_4^+, and fluoride ions, F^- are produced. The ammonium ion is an acid, with an acid constant of 5.9×10^{-10}, and the fluoride ion is a base, with a base constant of 1.5×10^{-11}. The slightly higher value for the acid constant shows that an aqueous solution of ammonium fluoride is slightly acidic.

For the most part, the acid-base strengths of chemical species can be understood in terms of free-energy changes, which in turn can be understood in terms of entropy and enthalpy changes. Among the factors affecting the enthalpy changes are the bond energy for the bond between hydrogen and the acid substrate, and the electronegativity of the acid substrate.

QUESTIONS AND PROBLEMS

15.1 Define the terms "acid," "base," and "acid-base reaction" according to each of the following concepts: a) Arrhenius; b) Bronsted-Lowry; c) solvent system; d) Lewis.

15.2 How does each of the following species satisfy each of the definitions of an acid given in Question 15.1?
a) HSO_4^- in water
b) copper(II) ions in water
c) ammonium ions in water
d) ammonium ions in liquid ammonia
e) HCN in water
f) water in liquid ammonia

15.3 Write an equation for the self-dissociation of each of the following: a) CH_3OH; b) HCN; c) SO_3; d) $SOCl_2$.

15.4 Criticize each of the following statements:
a) all Bronsted-Lowry acids and bases are also included as acids and bases in the solvent-system approach.
b) The solvent-system theory presents a broader, more inclusive, picture of acids and bases than the Bronsted-Lowry theory.
c) Both the Bronsted-Lowry and solvent-system theories define acids and bases only for systems in liquid solution.
d) The Bronsted-Lowry theory is the most correct theory of acids and bases.

15.5 Write equations for the reactions of the following Bronsted-Lowry acids and bases with water. Label the conjugate acid-base pairs.
a) $HS(aq)^-$
b) $H_2PO_4(aq)^-$
c) $NH_4(aq)^+$
d) $[Fe(H_2O)_x]^{3+}$
e) $CO_3(aq)^{--}$
f) $CN(aq)^-$
g) $O(aq)^{--}$
h) $H_3PO_3(aq)$
i) $[Mn(H_2O)_x]^{3+}$
j) $H_2O_2(aq)$
k) $NH_2SO_3H(aq)$
l) $NH_2—NH_2(aq)$
m) $NH_2—NH_3(aq)^+$
n) HF(aq)

15.6 How might each of the following species react as acids or bases when placed in liquid acetic acid as solvent? Write equations for any reactions. a) CN^-; b) NH_4^+; c) CO_3^{--}; d) HF; e) HS^-.

15.7 Consider the reactions represented by Equations (15.10), (15.11), (15.22) and (15.23). Are these acid-base reactions from the solvent system point of view? Explain.

15.8 Show how the terms "acid," "base," "conjugate acid," and "conjugate base" in the Bronsted-Lowry sense apply to the following reactions in liquid sulfuric acid as solvent; (sa) = solvation by sulfuric acid.
a) $HF(sa) + H_2SO_4 = H_2F(sa)^+ + HSO_4(sa)^-$
b) $HClO_4(sa) + H_2SO_4 = H_3SO_4(sa)^+ + ClO_4(sa)^-$

15.9 From the following list, select those species which may act as Lewis acids and those which may act as Lewis bases.

a) S^{--} d) NH_3 g) BF_3
b) H^+ e) CH_4 h) Al^{3+}
c) H_2O f) NH_2^- i) O^{--}

15.10 Magnesium hydroxide is insoluble in water but is soluble in aqueous ammonium chloride. Why?

15.11 Use equations to suggest why silver iodide is soluble in aqueous sodium thiosulfate, but not in water.

15.12 Most water-insoluble carbonates dissolve in acidic solutions. Why?

15.13 In Equations (15.16) through (15.21), identify all acids and bases.

15.14 Each of the following species is amphiprotic. Use equations to suggest reactions that illustrate this amphiprotism.

a) $HCO_3(aq)^-$ b) $Pb(OH)_2(s)$ c) $[FeOH(H_2O)_{x-1}]^{++}$

15.15 For each of the following write the formula for the conjugate base and/or the conjugate acid:

a) C_6H_6 c) NH_3 e) O^{--} g) H_2Se
b) H_3PO_4 d) H_2O f) $[CrCl_2(H_2O)_4]^+$

15.16 Assume that the typical Bronsted-Lowry acid-base reaction is a single-step equilibrium reaction represented by the following equation:

$$HA(aq) + H_2O = H_3O(aq)^+ + A(aq)^-$$

Write the rate equations for the forward and reverse reactions and derive the acid constant expression for HA.

15.17 Consider the following equilibrium:

$$NH_3(aq) + H_2O = NH_4(aq)^+ + OH(aq)^-$$

Indicate the changes in equilibrium and hydroxide ion concentration which occur upon the addition of each of the following substances to the solution.

a) $HCl(g)$ b) $NH_4Cl(s)$ c) $NaOH(s)$ d) $NaCl(s)$ e) $NH_3(g)$

15.18 Each of the following compounds is dissolved in water. Indicate whether the resulting solution will be acidic, basic, or neutral and write equations to justify your answers.

a) Na_2CO_3 d) $NaCl$ g) $(NH_4)_2SO_3$
b) $AlCl_3$ e) K_2HPO_4 h) $Fe(CH_3COO)_3$
c) NH_4Cl f) $(NH_4)_2SO_4$ i) NH_4CH_3COO

15.19 Which of the following species are amphiprotic in water? For each one that is, write two chemical equations as illustrations.

a) HCl c) $H_2PO_4^-$ e) S^{--}
b) NH_4CN d) NH_3 f) $[CrOH(H_2O)_{x-1}]^{++}$

15.20 Compute the pH and pOH values for the following aqueous solutions at room temperature.

a) 3.45×10^{-3} *M* $HClO_4$ d) 4.04×10^{-4} *M* Na_2O
b) 7.75×10^{-4} *M* HI e) 1.0×10^{-9} *M* HCl
c) 8.86×10^{-3} *M* NaOH

15.21 Where possible, convert the following pH values to hydronium ion molarity:

a) 4 c) 2.50 e) −0.44
b) 7 d) 8.20 f) 16.11

15.22 Criticize the following statements:

a) It is not possible to have solutions with negative pH values.
b) It is not possible to have solutions with pH values in excess of 15.

15.23 The water constant, K_w, has the value of 10^{-12} at 100°C.

a) What does this suggest about the effect of temperature on the self-dissociation of water?
b) What is the pH of water at its boiling point?
c) What happens to the pH of an aqueous solution of sodium hydroxide as it is heated?

15.24 Discuss the leveling effect.

15.25 Using the following acid-constant values, compute the value for the base constant for the conjugate base of the listed acid.
a) $HClO(aq)$, 6.0×10^{-8}
b) $HTe(aq)^-$, 1.0×10^{-11}
c) $H_3P_3O_{10}(aq)^{--}$, 5.0×10^{-3}
d) $CH_3NH_3(aq)^+$, 1.9×10^{-11}
e) $[In(H_2O)_x]^{3+}$, 6×10^{-4}

15.26 Show how formic acid, HCOOH, may act as either a strong, weak, or very weak acid depending upon whether it is placed in water, liquid ammonia, or pure acetic acid as a solvent.

15.27 Choose several acid-base pairs in Table 15.1 and compute their pK_a and pK_b values. Then criticize the following statement: If acid HA has a higher pK_a value than acid HB, then acid HA is stronger than acid HB.

15.28 Criticize the following statement: An aqueous solution with pH of 6 is more acidic than an aqueous solution with pH of 5.

15.29 In Table 15.1, why is the base constant for oxide ion listed as "very large," without a numerical value?

15.30 How might one go about distinguishing between the base strengths of oxide and amide ions?

15.31 In Table 15.1, the acid constant listed for water is not the typical water constant, K_w. For comparative purposes, the self-dissociation of water is treated like other acid-base reactions. The reaction equation may be written as follows:

$$H_2O(aq) + H_2O = H_3O(aq)^+ + OH(aq)^-$$

The $H_2O(aq)$ is viewed as the acidic species and H_2O is viewed as the solvent. Thus the acid constant for water may be written as follows:

$$K_a = \frac{[H_3O(aq)^+][OH(aq)^-]}{[H_2O]}$$

a) Recalling that the density of water is 1.00 g/ml, find the molarity of water in pure water. (That is, how many moles of water are there per liter of water?)
b) What is the relationship between the values for K_a and K_w for water?
c) Compute the value for K_a and compare it to the value in Table 15.1.
d) What is the relationship between K_a for water and K_b for hydroxide ion?
e) Write the chemical equation from which the acid constant for hydronium ion may be written.
f) What is the relationship between K_a for water and K_a for hydronium ion?
g) What is the relationship between K_a and K_b for water?

15.32 What would happen to the values in Table 15.1 if liquid acetic acid were used as the solvent instead of water? If liquid ammonia were used as the solvent?

15.33 Criticize the following statement: The base constant for ammonia is the modified equilibrium constant for the following reaction:

$$NH_3(g) + H_2O = NH_4(aq)^+ + OH(aq)^-$$

15.34 Criticize the following statement: The acid constant for ammonia is the modified equilibrium constant for the following reaction:

$$NH_4(aq)^+ + H_2O = NH_3(aq) + H_3O(aq)^+$$

15.35 Consider the diprotic acid H_2B.

a) Write the chemical equation for the following:
   1) First ionization of H_2B in water with the expression for K_{a1}
   2) Second ionization of H_2B in water with the expression for K_{a2}
   3) First ionization of B^{--} in water with the expression for K_{b1}
   4) Second ionization of B^{--} in water with the expression for K_{b2}

b) Note that equations 2) and 4) in a) describe the amphiprotic behavior of HB^-. Show that by combining the ionization constant expressions for these two reactions the following expression gives the hydronium ion concentration at equilibrium.

$$[H_3O(aq)^+] = \frac{K_w K_{a2}[H_2B(aq)]}{K_{b2}[B(aq)^{--}]}$$

c) Suppose that NaHB is added to water; under what conditions will the pH be as follows?
   1) exactly 7.0
   2) greater than 7.0
   3) less than 7.0

15.36 Compute the total ionization constant value for each of the following polyprotic acids and bases.
a) diprotic $H_3O_2(aq)^+$
b) triprotic $H_3AsO_4(aq)$
c) triprotic $PO_4(aq)^{3-}$

15.37 Compute the equilibrium constant value for each of the following:
a) $HSO_4(aq)^- + NO_2(aq)^- = HNO_2(aq) + SO_4(aq)^{--}$
b) $[Fe(H_2O)_x]^{3+} + CO_3(aq)^{--} = [FeOH(H_2O)_{x-1}]^{++} + HCO_3(aq)^-$
c) $H_3AsO_3(aq) + H_2PO_4(aq)^- = H_2AsO_3(aq)^- + H_3PO_4(aq)$
d) $H_2C_2O_4(aq) + S(aq)^{--} = H_2S(aq) + C_2O_4(aq)^{--}$

15.38 Use the Born-Haber cycle to explain why the second ionization constant for an acid such as H_2S is smaller than the first ionization constant.

15.39 Suggest a possible Born-Haber cycle for bases.

15.40 In each of the following pairs of acids, choose the one you expect to be the strongest and indicate the reason for your choice.
a) H_2Se, H_2Te
b) H_2O, HF
c) HClO, HBrO
d) HClO, $HClO_2$
e) H_3AsO_3,H_3AsO_4
f) H_2SO_4, H_2SeO_4
g) $HClO_4$, $HClO_3$

15.41 On theoretical grounds, arrange the following acids in the order of decreasing acid strength:
a) CH_3COOH
b) $BrCH_2COOH$
c) $Cl_2CHCOOH$
d) $ClCH_2COOH$
e) ICH_2COOH

15.42 Why is water more acidic than either ammonia or hydrogen sulfide?

15.43 Why is an amine such as CH_3NH_2 more basic than an alcohol such as CH_3OH?

15.44 Why is an amine such as $CH_3CH_2NH_2$ more basic than a similar amide such as

$$CH_3\text{—}\overset{\overset{\large O}{\|}}{C}\text{—}NH_2?$$

15.45 What does the fact that aminobenzene, $C_6H_5NH_2$, is a much weaker base than aminomethane, CH_3NH_2, imply about the electron-attracting ability of the phenyl group as compared to the methyl group?

15.46 Pauling has proposed a very approximate rule of thumb for estimating the values of acid constants for hydroxy acids of the form $XO_m(OH)_n$, where m is the number of oxygen atoms bonded to the central X atom but not bonded to hydrogen. According to the Pauling rule, K_a is about 10^{-7} when $m = 0$; K_a is about 10^{-2} when $m = 1$; K_a is large when $m = 2$; and K_a is very large when $m = 3$. How can you account for this trend with the Born-Haber cycle?

15.47 Often it is stated that among aquo-coordination complexes of a given metal ion, the complex having the metal in the higher oxidation state is the more acidic. Why should this be true?

15.48 According to Table 15.7, $[Co(H_2O)_x]^{3+}$ is more acidic than $[Co(NH_3)_2(H_2O)_4]$-$(aq)^{3+}$. Suggest a reason for this.

15.49 For each of the following pairs of species indicate your choice for the strongest acid, and the reasons for your choice.

a) $[Fe(H_2O)_x]^{3+}$, $[Fe(H_2O)_x]^{++}$
b) $[Na(H_2O)_x]^{+}$, $[Li(H_2O)_x]^{+}$
c) $[Na(H_2O)_x]^{+}$, $[Mg(H_2O)_x]^{++}$

15.50 The pK_{ind} for bromthymol blue is 7.0. Compute the pH of the solution when the following percentages of the indicator are in the basic form:

a) 1.0 b) 10.0 c) 25.0 d) 50.0 e) 80.0

15.51 Two typical Lewis acid-base reactions are described by the following two equations:

$$Cu(aq)^{++} + 4NH_3(aq) = [Cu(NH_3)_4](aq)^{++}$$
$$Zn(aq)^{++} + 4NH_3(aq) = [Zn(NH_3)_4](aq)^{++}$$

a) Suggest the appropriate forms of the acid-base equilibrium constant expressions for these reactions.
b) How would the two constants be related if zinc(II) ions were stronger acids than copper(II) ions?
c) What would you suggest about the molecularity of such reactions?

16

Quantitative Treatment of Some Acid-Base Equilibria

INTRODUCTION In the last chapter acids and bases were defined and the concept of acid and base strength was clarified. This chapter treats some acid-base equilibria quantitatively, beginning with a single acid or base in water and concluding with systems in which an acid and a base react with each other in aqueous solution.

Throughout this chapter we shall intermix specific and general examples of acid-base equilibria. It is important to keep both points of view in mind. The general viewpoint is essential so that each specific example does not appear to be an entirely new problem, but the ability to apply these considerations to specific examples is also necessary.

AQUEOUS SOLUTIONS OF ACIDS AND BASES

Strong Acids and Bases

When a strong acid, such as hydrochloric acid, dissolves in water, it ionizes completely as indicated by Equation (16.1).

$$HCl(aq) + H_2O = H_3O(aq)^+ + Cl(aq)^- \qquad \textbf{(16.1)}$$

In addition, the self-dissociation of water occurs, as indicated by the following equilibrium:

$$H_2O(aq) + H_2O = H_3O(aq)^+ + OH(aq)^- \quad (16.2)$$

From these two reactions, it is clear that the solution is composed of the following species:

1. $Cl(aq)^-$ from the ionization of hydrochloric acid
2. $H_3O(aq)^+$ from the ionizations of both water and hydrochloric acid
3. $OH(aq)^-$ from the ionization of water
4. H_2O left over from the *partial* ionization of water

Thus the hydronium ion concentration, $[H_3O(aq)^+]$, is the sum of the hydronium ion concentration derived from water, $[H_3O(aq)^+]_W$, and the hydronium ion concentration derived from the hydrochloric acid, $[H_3O(aq)^+]_{HCl}$ as indicated in the following equation:

$$[H_3O(aq)^+] = [H_3O(aq)^+]_W + [H_3O(aq)^+]_{HCl} \quad (16.3)$$

However, the maximum concentration of hydronium ion that can be secured from the ionization of water is 10^{-7} *M*, which occurs in pure water. When an acid is added to the water, the hydronium ions produced from the acid serve as a stress on the self-ionization equilibrium of water, so that the equilibrium shifts from the hydronium and hydroxide ions to the molecular water. The result is that in the presence of another acid, the concentration of hydronium ion from the ionization of water is less than 10^{-7} *M*.

This means that the contribution of the water ionization to the total hydronium ion concentration is insignificant for most concentrations of a strong acid in water. Only when the concentraton of the acid approaches 10^{-7} *M* is it necessary to consider the contribution of water to the hydronium ion concentration. Of course at acid concentrations less than 10^{-7} *M* the water is the major contributor to the hydronium ion concentration.

Thus in 0.10 *M* HCl, the hydronium ion concentration is 0.10 *M* and the pH is 1.0.

Weak Monoprotic Acids

When the generalized weak acid, HA, is placed in water, it undergoes an equilibrium reaction as follows:

$$HA(aq) + H_2O = H_3O(aq)^+ + A(aq)^- \quad (16.4)$$

In addition, the water ionization is present also as described earlier in Equation (16.2).

Consequently, the following species are present in this aqueous solution:

1. H_2O left from the *partial* ionization of water;

2. $H_3O(aq)^+$ from the partial ionizations of both HA and H_2O;
3. $OH(aq)^-$ from the partial ionization of water;
4. $A(aq)^-$ from the partial ionization of HA;
5. HA(aq) left from the *partial* ionization of the original HA used to make the solution.

Keep this picture of the composition of the solution in mind as we now look at several different types of problems related to the weak acid equilibrium.

CALCULATIONS OF ACID CONSTANTS. Recalling Equation (16.4), we may write the acid constant expression for HA as follows:

$$K_a = \frac{[H_3O(aq)^+][A(aq)^-]}{[HA(aq)]} \tag{16.5}$$

To find a value for K_a requires values for $[H_3O(aq)^+]$, $[A(aq)^-]$, and $[HA(aq)]$, or at least expressions for these in terms of K_a.

One way to secure data for determining the acid constant is to prepare a solution of HA which has a known concentration. The measurement of the pH of that solution provides enough information to evaluate the acid constant. For example, suppose that a 3.0×10^{-3} M solution of HA has a pH of 4.2. What is the value of K_a for HA?

Our strategy is to find numerical values for the molarities of the three species shown in Equation (16.5). First, $[H_3O(aq)^+]$ may be determined directly from the pH as follows:

$$[H_3O(aq)^+] = 10^{-pH} = 10^{-4.2} = 10^{-5.0+0.8} = 6.3 \times 10^{-5}\ M$$

The only source of $A(aq)^-$ is the ionization of HA(aq), as described by Equation (16.4). From the stoichiometry of that reaction, the molarity of $A(aq)^-$ must be the same as the molarity of $H_3O(aq)^+$ derived from the ionization of HA(aq). In turn the molarity of $H_3O(aq)^+$ derived from the ionization of HA(aq) must be the difference between the total equilibrium molarity of $H_3O(aq)^+$ and the molarity of $H_3O(aq)^+$ derived from the ionization of water. This may be expressed as follows:

$$[A(aq)^-] = [H_3O(aq)^+]_{HA} = [H_3O(aq)^+] - [H_3O(aq)^+]_W$$

The over-all equilibrium molarity of $H_3O(aq)^+$ has been computed from the pH. The molarity of $H_3O(aq)^+$ derived from water is the same as the molarity of $OH(aq)^-$ derived from water. However, water is the only source of $OH(aq)^-$, so the molarity of $H_3O(aq)^+$ from water is the same as the molarity of $OH(aq)^-$ in the solution. This total molarity of $OH(aq)^-$ is related to the total molarity of $H_3O(aq)^+$ through K_w as follows:

$$[OH(aq)^-] = \frac{K_w}{[H_3O(aq)^+]}$$

Then the over-all relationship may be expressed as follows:

$$[H_3O(aq)^+]_W = [OH(aq)^-]_W = [OH(aq)^-] = \frac{K_w}{[H_3O(aq)^+]}$$

$$= \frac{1.0 \times 10^{-14}}{6.3 \times 10^{-5}} = 1.6 \times 10^{-10}\ M$$

Thus the molarity of $A(aq)^-$ may be determined as follows:

$$[A(aq)^-] = [H_3O(aq)^+] - [H_3O(aq)^+]_W = 6.3 \times 10^{-5}\ M - 1.6 \times 10^{-10}\ M$$

$$= 6.3 \times 10^{-5}\ M$$

Finally, the equilibrium concentration of HA(aq) must be the difference between the original concentration of HA(aq), M_{HA}, as it was added, and the concentration of HA which ionized, $M_{HA\,ion}$. In addition, the concentration of HA which ionized must be the same as the equilibrium concentration of $A(aq)^-$. This is shown as follows:

$$[HA(aq)] = M_{HA} - M_{HA\,ion} = M_{HA} - [A(aq)^-] = 3.0 \times 10^{-3}\ M - 6.3 \times 10^{-5}\ M$$

$$= 2.9 \times 10^{-3}\ M$$

These numerical values may be substituted into the acid-constant expression to find the numerical value for K_a:

$$K_a = \frac{[H_3O(aq)^+][A(aq)^-]}{[HA(aq)]} = \frac{(6.3 \times 10^{-5})^2}{2.9 \times 10^{-3}} = 1.4 \times 10^{-6}$$

Two things should be noted in this calculation. First, acid HA is strong enough and concentrated enough that the contribution of the water ionization to the hydronium ion concentration is negligible. In this case, it is safe to assume that effectively all of the hydronium ion comes from the HA. This would not be the case, however, if the acid were extremely weak or extremely dilute.

Second, acid HA is weak enough to allow only a small percentage of the acid to ionize. This means that the equilibrium concentration of HA(aq) is nearly the same as the original concentration before any ionization occurred. In fact, from the data available in this problem, it is possible to compute the percent of the original acid ionized at equilibrium. The original concentration of HA(aq) was 3.0×10^{-3} moles/l, and the concentration of HA(aq) which ionized was 6.3×10^{-5} moles/l. The percent ionization may be computed as follows:

$$100 \text{ moles orig HA} \times \frac{1 \text{ l soln}}{3.0 \times 10^{-3} \text{ mole orig HA}} \times \frac{6.3 \times 10^{-5} \text{ mole HA ionized}}{1 \text{ l soln}}$$

$$= 2.1 \text{ percent ionized}$$

In fact experimental data sometimes is expressed as percent ionization, and from this it is desired to find the acid constant. For example, suppose that it is found that acid HA is 15 percent ionized in a 0.030 *M* solution. What is the value for K_a?

In this case the expression for K_a is the same as that shown in Equation (16.5). Since the only source of $A(aq)^-$ is the ionization of HA(aq), the equilibrium molarity of $A(aq)^-$ must be 15 percent of the over-all concentration.

$$[A(aq)^-] = \frac{0.030 \text{ total moles HA}}{1 \text{ soln}} \times \frac{15 \text{ moles HA ionized}}{100 \text{ total moles HA}} = 4.5 \times 10^{-3}\ M$$

As before, the hydronium ion comes from both the HA and water, so the equilibrium hydronium ion concentration may be expressed as follows:

$$[H_3O(aq)^+] = [H_3O(aq)^+]_W + [H_3O(aq)^+]_{HA}$$

However, the concentration of $H_3O(aq)^+$ from HA is the same as the concentration of $A(aq)^-$ from HA, which is 4.5×10^{-3} *M*. Since the hydronium ion concentration from the water ionization cannot exceed 10^{-7} *M*, there is no point in even computing a value for this since it must be insignificant in comparison to 4.5×10^{-3} *M*. Therefore, the hydronium ion concentration in the solution is 4.5×10^{-3} *M*.

Since 15 percent of the original acid ionized, only 85 percent of it remains unionized at equilibrium.

$$[HA(aq)] = \frac{3.0 \times 10^{-2} \text{ total moles HA}}{1 \text{ soln}} \times \frac{85 \text{ moles HA unionized}}{100 \text{ total moles HA}}$$

$$= 2.6 \times 10^{-2}\ M$$

The value for K_a may be computed from these three values:

$$K_a = \frac{[H_3O(aq)^+][A(aq)^-]}{[HA(aq)]} = \frac{(4.5 \times 10^{-3})^2}{2.6 \times 10^{-2}} = 7.7 \times 10^{-4}$$

Calculation of Concentrations from Acid-constant Data

Suppose that a solution of known concentration of acid HA is prepared and the value of K_a is known for this acid. With this information it is possible to compute the concentrations of all species in the solution, and in particular it is possible to compute the hydronium ion molarity from which the pH may be found.

We shall proceed by developing a rigorous algebraic expression in which the hydronium ion molarity is related to the acid constant of HA, the water constant, and the over-all molarity of the acid, all of which are known. We shall see that this equation is cumbersome and tedious to solve and we shall show how the solution of the equation may be estimated by making some reasonable approximations. Then we shall apply these approximations to some specific examples to see how good they are. Finally, we shall return to the rigorous equation to see how a graphical interpretation can aid us in evaluating the approximations.

DERIVATION OF EQUATION RELATING $[H_3O(aq)^+]$, K_a AND K_w. The hydronium ions in the solution come from two sources, namely the ionizations of HA and water. The total hydronium ion molarity is the sum of the molarities of hydronium ion from these two sources:

$$[H_3O(aq)^+] = [H_3O(aq)^+]_{HA} + [H_3O(aq)^+]_W \tag{16.6}$$

Since hydronium and hydroxide ions are produced in equal numbers by the ionization of water [see Equation (16.2)], and since the ionization of water is the only source of hydroxide ions, Equation (16.7) may be written and that expression for $[H_3O(aq)^+]_W$ may be substituted into Equation (16.6):

$$[H_3O(aq)^+]_W = [OH(aq)^-]_W = [OH(aq)^-] \tag{16.7}$$

$$[H_3O(aq)^+] = [H_3O(aq)^+]_{HA} + [OH(aq)^-] \tag{16.8}$$

By use of the water constant, the hydroxide ion molarity may be expressed in terms of the over-all hydronium ion molarity by Equation (16.9) which may be used to rewrite Equation (16.8) as Equation (16.10).

$$[OH(aq)^-] = \frac{K_w}{[H_3O(aq)^+]} \tag{16.9}$$

$$[H_3O(aq)^+] = [H_3O(aq)^+]_{HA} + \frac{K_w}{[H_3O(aq)^+]} \tag{16.10}$$

From Equation (16.4) it is clear that in the ionization of HA(aq), a mole of $A(aq)^-$ ions is produced for each mole of $H_3O(aq)^+$ ions. This permits us to write Equation (16.11), and this expression may be substituted into Equation (16.10) to form Equation (16.12).

$$[H_3O(aq)^+]_{HA} = [A(aq)^-] \tag{16.11}$$

$$[H_3O(aq)^+] = [A(aq)^-] + \frac{K_w}{[H_3O(aq)^+]} \tag{16.12}$$

The acid-constant expression was given earlier as Equation (16.5), and this may be solved for $[A(aq)^-]$ as shown in Equation (16.13).

$$[A(aq)^-] = \frac{K_a[HA(aq)]}{[H_3O(aq)^+]} \tag{16.13}$$

However, the equilibrium concentration of HA(aq) is the difference between the over-all original concentration of acid, M_{HA}, and the concentration of acid that ionizes, $M_{HA\,ion}$, as shown by Equation (16.14).

$$[HA(aq)] = M_{HA} - M_{HA\,ion} \tag{16.14}$$

Since each mole of HA(aq) that ionizes produces a mole of $A(aq)^-$, then $M_{HAion} = [A(aq)^-]$ and Equation (16.14) may be rewritten as Equation (16.15).

$$[HA(aq)] = M_{HA} - [A(aq)^-] \tag{16.15}$$

The expression for [HA(aq)] in Equation (16.15) may be substituted into Equation (16.13).

$$[A(aq)^-] = \frac{K_a(M_{HA} - [A(aq)^-])}{[H_3O(aq)^+]} \tag{16.16}$$

If Equation (16.16) is solved for $[A(aq)^-]$, the following expression results.

$$[A(aq)^-] = \frac{K_a M_{HA}}{[H_3O(aq)^+] + K_a} \tag{16.17}$$

This may be substituted into Equation (16.12) as follows:

$$[H_3O(aq)^+] = \frac{K_a M_{HA}}{[H_3O(aq)^+] + K_a} + \frac{K_w}{[H_3O(aq)^+]} \tag{16.18}$$

Fractions may be cleared and Equation (16.18) may be put into the more standard cubic equation form.

$$[H_3O(aq)^+]^3 + K_a[H_3O(aq)^+]^2 - (K_a M_{HA} + K_w)[H_3O(aq)^+] - K_a K_w = 0 \tag{16.19}$$

If values for K_a, M_{HA}, and K_w are known, Equation (16.19) may be solved for $[H_3O(aq)^+]$. However, the solution of such a cubic equation is tedious and after performing the necessary operations a few times, one begins to look for short-cuts in self defense. The following section indicates two approximations that may be made in many commonly encountered situations.

APPROXIMATIONS. Earlier, approximations in the computations of acid constant values from solution pH data and from percent ionization data were mentioned.

The first of these approximations is valid if the acid is sufficiently strong and concentrated to make the contribution of the water ionization to the hydronium ion concentration negligible in comparison to the contribution made by the ionization of the acid. In this case, Equation (16.6) may be rewritten as Equation (16.20).

$$[H_3O(aq)^+] = [H_3O(aq)^+]_{HA} \tag{16.20}$$

Since hydronium ions and $A(aq)^-$ ions are produced in equal numbers by the ionization of HA(aq), the relationship shown in Equation (16.11) may be used to rewrite Equation (16.20).

$$[H_3O(aq)^+] = [A(aq)^-] \tag{16.21}$$

When this relationship holds, the acid-constant expression may be written as follows:

$$K_a = \frac{[H_3O(aq)^+][A(aq)^-]}{[HA(aq)]} = \frac{[H_3O(aq)^+]^2}{[HA(aq)]} \qquad (16.22)$$

Since the maximum concentration of hydronium ion which may be produced by the water ionization is 10^{-7} *M*, a rather dilute solution of a very weak acid is required to provide a smaller concentration than this. As a result, this approximation tends to be a good one for many acids over a wide range of concentrations.

The second approximation is valid if the acid is sufficiently weak and concentrated so that only a very small percentage of the original acid is ionized at equilibrium. If the concentration of ionized acid is small compared to the original over-all acid concentration, then Equation (16.14) may be rewritten:

$$[HA(aq)] = M_{HA} \qquad (16.23)$$

When this relationship is valid, the acid-constant expression in Equation (16.22) may be rewritten:

$$K_a = \frac{[H_3O(aq)^+]^2}{M_{HA}} \qquad (16.24)$$

This is a much simplified version of Equation (16.19)

Whether or not these two approximations are suitable in any particular case depends upon the requirements of the investigator making the calculations. If only very approximate answers are desired, the chances are good that these approximations may be used most of the time. However, where moderately accurate results are demanded, some care must be used. Of the two approximations, the second usually leads to the largest errors. Even a moderately weak acid in dilute solution often will ionize by several percent, and this will cause a corresponding error in the calculated value for the hydronium ion concentration.

In the next section we shall perform some sample calculations involving these approximations and use these to secure some feeling for the approximate size of the errors involved.

Sample calculations using approximations. As an example of a situation in which the approximations may be used successfully, consider an 0.080 *M* solution of NH_4Cl. When dissolved, the solute produces $NH_4(aq)^+$ and $Cl(aq)^-$ ions. The $NH_4(aq)^+$ ion is an acid, whereas the $Cl(aq)^-$ does not undergo a perceptible acid-base reaction. The known data and relationships are summarized by Equations (16.25) through (16.29).

$$NH_4(aq)^+ + H_2O = NH_3(aq) + H_3O(aq)^+ \qquad (16.25)$$

$$K_a = \frac{[NH_3(aq)][H_3O(aq)^+]}{[NH_4(aq)^+]} = 5.9 \times 10^{-10} \qquad (16.26)$$

$$H_2O(aq) + H_2O = H_3O(aq)^+ + OH(aq)^- \qquad (16.27)$$

$$K_w = [H_3O(aq)^+][OH(aq)^-] = 1.0 \times 10^{-14} \tag{16.28}$$

$$M_{NH_4^+} = 8.0 \times 10^{-2}\ M \tag{16.29}$$

Using the approximation that the only significant concentration of hydronium ion comes from $NH_4(aq)^+$ and not from water, Equation (16.26) may be written as Equation (16.30).

$$\frac{[H_3O(aq)^+]^2}{[NH_4(aq)^+]} = 5.9 \times 10^{-10} \tag{16.30}$$

Using the approximation that very little $NH_4(aq)^+$ actually ionizes, Equation (16.30) may be rewritten as Equation (16.31).

$$\frac{[H_3O(aq)^+]^2}{[NH_4(aq)^+]} = \frac{[H_3O(aq)^+]^2}{M_{NH_4^+}} = \frac{[H_3O(aq)^+]^2}{8.0 \times 10^{-2}} = 5.9 \times 10^{-10} \tag{16.31}$$

Equation (16.31) may be solved for the hydronium ion concentration as follows:

$$[H_3O(aq)^+]^2 = 47 \times 10^{-12}$$

$$[H_3O(aq)^+] = \sqrt{47 \times 10^{-12}} = 6.9 \times 10^{-6}\ M$$

Using this approximate value for the hydronium ion concentration, it is possible to compute the approximate values for the other species in the solution as well. For example, the hydroxide ion concentration may be computed by using the water constant.

$$[OH(aq)^-] = \frac{K_w}{[H_3O(aq)^+]} = \frac{1.0 \times 10^{-14}}{6.9 \times 10^{-6}} = 0.14 \times 10^{-8} = 1.4 \times 10^{-9}\ M$$

The molarity of $NH_3(aq)$ is the same as the molarity of $H_3O(aq)^+$ produced from the ionization of $NH_4(aq)^+$. However we have assumed that all of the $H_3O(aq)^+$ has come from this ionization and thus

$$[NH_3(aq)] = [H_3O(aq)^+] = 6.9 \times 10^{-6}\ M$$

Finally, since very little of the ammonium ion actually ionizes, the equilibrium concentration of the ammonium ion is the same as the original concentration, which in turn is the same as the chloride ion concentration, or $8.0 \times 10^{-2}\ M$. All of these concentrations determined by use of the approximations are summarized as follows:

$$[H_3O(aq)^+] = 6.9 \times 10^{-6}\ M$$

$$[OH(aq)^-] = 1.4 \times 10^{-9}\ M$$

$$[NH_3(aq)] = 6.9 \times 10^{-6}\ M$$

$$[NH_4(aq)^+] = 8.0 \times 10^{-2}\ M$$

$$[Cl(aq)^-] = 8.0 \times 10^{-2}\ M$$

With these concentrations available, it is possible to check the validity of the approximations. According to the first, the concentration of hydronium ion derived from the water ionization is insignificant with respect to the concentration of hydronium ion derived from ammonium ion. Since the concentration of hydronium ion derived from water is the same as the concentration of hydroxide ion in the solution, $1.4 \times 10^{-9}\ M$ may be compared to $6.9 \times 10^{-6}\ M$, whereupon it is clear that to two significant figures the concentration of hydronium ion derived from water is insignificant.

According to the second approximation, the concentration of ammonium ion ionized at equilibrium is insignificant with respect to the total ammonium ion concentration originally present. Since the concentration of ammonium ion ionized is the same as the equilibrium concentration of $NH_3(aq)$, $6.9 \times 10^{-6}\ M$ may be compared to $8.0 \times 10^{-2}\ M$, whereupon it is clear that to two significant figures, the concentration of ammonium ion ionized is insignificant.

From this example, it is clear that the calculation of the concentrations of all species in solution is relatively simple when the approximations are valid. However, when the approximations are used, the answers obtained should be used to check the validity of the approximations.

As another example, consider a case in which the approximations are not valid. What are the concentrations of all species in 0.0010 M HF? The following equations summarize the appropriate relationships and information with which to make the calculations.

$$HF(aq) + H_2O = H_3O(aq)^+ + F(aq)^- \tag{16.32}$$

$$K_a = \frac{[H_3O(aq)^+][F(aq)^-]}{[HF(aq)]} = 6.8 \times 10^{-4} \tag{16.33}$$

$$M_{HF} = 1.0 \times 10^{-3}\ M \tag{16.34}$$

Using the approximations, we find that the concentrations of $H_3O(aq)^+$ and $F(aq)^-$ are the same, since they are produced in equal concentrations by the ionization of HF(aq). In addition the equilibrium concentration of HF(aq) is 1.0×10^{-3} M, the concentration of the original solution prior to any ionization of HF(aq). Equation (16.33) may be rewritten and solved as follows:

$$\frac{[H_3O(aq)^+]^2}{1.0 \times 10^{-3}} = 6.8 \times 10^{-4}$$

$$[H_3O(aq)^+]^2 = 6.8 \times 10^{-7} = 68 \times 10^{-8}$$

$$[H_3O(aq)^+] = \sqrt{68 \times 10^{-8}} = 8.2 \times 10^{-4}\ M$$

The concentrations of other species in the solution are summarized in the following section:

$$[\text{OH(aq)}^-] = \frac{K_w}{[\text{H}_3\text{O(aq)}^+]} = \frac{1.0 \times 10^{-14}}{8.2 \times 10^{-4}} = 1.2 \times 10^{-11}\ M$$

$$[\text{F(aq)}^-] = [\text{H}_3\text{O(aq)}^+] = 8.2 \times 10^{-4}\ M$$

$$[\text{HF(aq)}] = M_{\text{HF}} = 1.0 \times 10^{-3}\ M$$

In checking the approximations, it appears that the ionization of water still is insignificant with respect to the acid ionization, since the concentration of hydronium ion (same as concentration of hydroxide ion) from the water ionization is about 10^{-11} M, compared to the total hydronium ion concentration estimated to be about 10^{-4} M. However, the second approximation, that the percent ionization of HF(aq) is small, is a poor approximation. Since the only source of F(aq)^- in the solution is the ionization of HF(aq), the concentration of HF(aq) ionized at equilibrium should be 8.2×10^{-4} M. When this is compared to the over-all acid concentration of 1.0×10^{-3}, it appears that 82 percent of the acid has ionized.

$$100 \text{ moles HF orig} \times \frac{1 \text{ l soln}}{1.0 \times 10^{-3} \text{ moles HF orig}} \times \frac{8.2 \times 10^{-4} \text{ moles HF ionized}}{1 \text{ l soln}} = 82 \text{ percent ionized}$$

This apparent percent of ionization is not the actual percent of ionization, since we are using approximate concentrations that obviously are faulty. However, the apparent 82 percent ionization for HF(aq) is sufficiently high to indicate that this second approximation is not very good and some other technique should be used to find a better answer.

PROCEDURE USED WHEN APPROXIMATIONS FAIL. When the approximations are inadequate, it is possible to return to the more exact relationship of Equation (16.19), which can be used not only to calculate the hydronium ion concentration of the solution, but also as a guide to indicate when the approximations are adequate.

One way to find a satisfactory value for $[\text{H}_3\text{O(aq)}^+]$ for a particular concentration of a given acid is to make a graph in which $[\text{H}_3\text{O(aq)}^+]$ is plotted against M_{HA}. Then, if the value for M_{HA} is known, the value for $[\text{H}_3\text{O(aq)}^+]$ may be read from the graph. The easiest way in which to secure the data for preparing such a graph is to assign various values to $[\text{H}_3\text{O(aq)}^+]$ and then calculate the corresponding values for M_{HA}. To do this, Equation (16.19) usually is solved for M_{HA} as follows:

$$M_{\text{HA}} = \frac{[\text{H}_3\text{O(aq)}^+]^2}{K_a} + [\text{H}_3\text{O(aq)}^+] - \frac{K_w}{[\text{H}_3\text{O(aq)}^+]} - \frac{K_w}{K_a} \quad (16.35)$$

Some typical values for M_{HA} are given in Table 16.1. The values in each vertical column are values for M_{HA} corresponding to an acid with the value of K_a indicated at the head of the column. The assigned hydronium ion concentrations are listed along the left side of the table.

When we consider graphing these values, it is clear that there is a very wide range of values for both $[H_3O(aq)^+]$ and M_{HA}. In order to facilitate the graphing, we elect to use the negative logarithm of the two variables so the graph will be a plot of $-\log[H_3O(aq)^+]$ (or pH) against $-\log M_{HA}$. This means that the numbers plotted are negative values of the exponents shown in Table 16.1. The graph is shown as Fig. 16.1.

It is clear from the graph that the pH of these acidic solutions levels off at 7.0 at different concentrations depending upon the value for the acid constant. Thus an acid with $K_a = 10^{-1}$, does not level off at pH 7 until its concentration is about 10^{-7} *M* or lower. This is true also for acids with K_a values down to about 10^{-7}. For these acids, when the concentration is 10^{-7} *M* or less, the contribution of the acid to the hydronium ion concentration becomes insignificant with respect to the contribution made by the ionization of water.

However, the contribution of water to the hydronium ion concentration becomes significant before pH 7 is reached, and as an approximation, we can say that it becomes significant at pH 6 or greater. Thus for acids with K_a values of 10^{-5} or greater, the water ionization is significant only when the acid concentration is 10^{-6} *M* or less. However, for weaker acids, the water ionization is significant at higher concentrations. Thus for an acid with $K_a = 10^{-7}$, the water ionization is significant at acid concentrations of 10^{-5} *M* or less. For a very weak acid with $K_a = 10^{-13}$, the water ionization is significant at virtually all concentrations, at least all greater than 1.0 *M*.

If we say, somewhat arbitrarily, that the commonly used second approximation is valid for our purposes when 10 percent or less of the acid is ionized at equilibrium, then Fig. 16.1 may be used to indicate the range for the validity of this approximation. Thus for a 0.1 *M* solution of HA, the approximation is valid for $[H_3O(aq)^+]$ less than 10^{-2} *M*, that is, for pH greater than 2. From the graph it can

TABLE 16.1

Values for M_{HA} at Various Values for $[H_3O(aq)^+]$ and K_a

	K_a						
$[H_3O(aq)^+]$	10^{-1}	10^{-3}	10^{-5}	10^{-7}	10^{-9}	10^{-11}	10^{-13}
10^{-1}	$10^{-0.7}$						
10^{-2}	$10^{-2.0}$	$10^{-1.0}$					
10^{-3}	$10^{-3.0}$	$10^{-2.7}$	$10^{-1.0}$				
10^{-4}	$10^{-4.0}$	$10^{-4.0}$	$10^{-3.0}$	$10^{-1.0}$			
10^{-5}	$10^{-5.0}$	$10^{-5.0}$	$10^{-4.7}$	$10^{-3.0}$	$10^{-1.0}$		
10^{-6}	$10^{-6.0}$	$10^{-6.0}$	$10^{-6.0}$	$10^{-5.0}$	$10^{-3.0}$	$10^{-1.0}$	
$10^{-6.2}$	$10^{-6.2}$	$10^{-6.2}$	$10^{-6.2}$	$10^{-5.4}$	$10^{-3.4}$	$10^{-1.4}$	$10^{+0.6}$
$10^{-6.4}$	$10^{-6.4}$	$10^{-6.4}$	$10^{-6.4}$	$10^{-5.7}$	$10^{-3.8}$	$10^{-1.8}$	$10^{+0.2}$
$10^{-6.5}$	$10^{-6.5}$	$10^{-6.5}$	$10^{-6.5}$	$10^{-5.9}$	$10^{-4.0}$	$10^{-2.0}$	$10^{-0.0}$
$10^{-6.6}$	$10^{-6.7}$	$10^{-6.7}$	$10^{-6.7}$	$10^{-6.1}$	$10^{-4.3}$	$10^{-2.3}$	$10^{-0.3}$
$10^{-6.7}$	$10^{-6.8}$	$10^{-6.8}$	$10^{-6.8}$	$10^{-6.4}$	$10^{-4.5}$	$10^{-2.5}$	$10^{-0.5}$
$10^{-6.8}$	$10^{-7.0}$	$10^{-7.0}$	$10^{-7.0}$	$10^{-6.6}$	$10^{-4.8}$	$10^{-2.8}$	$10^{-0.8}$
$10^{-6.9}$	$10^{-7.3}$	$10^{-7.3}$	$10^{-7.3}$	$10^{-6.9}$	$10^{-5.2}$	$10^{-3.2}$	$10^{-1.2}$

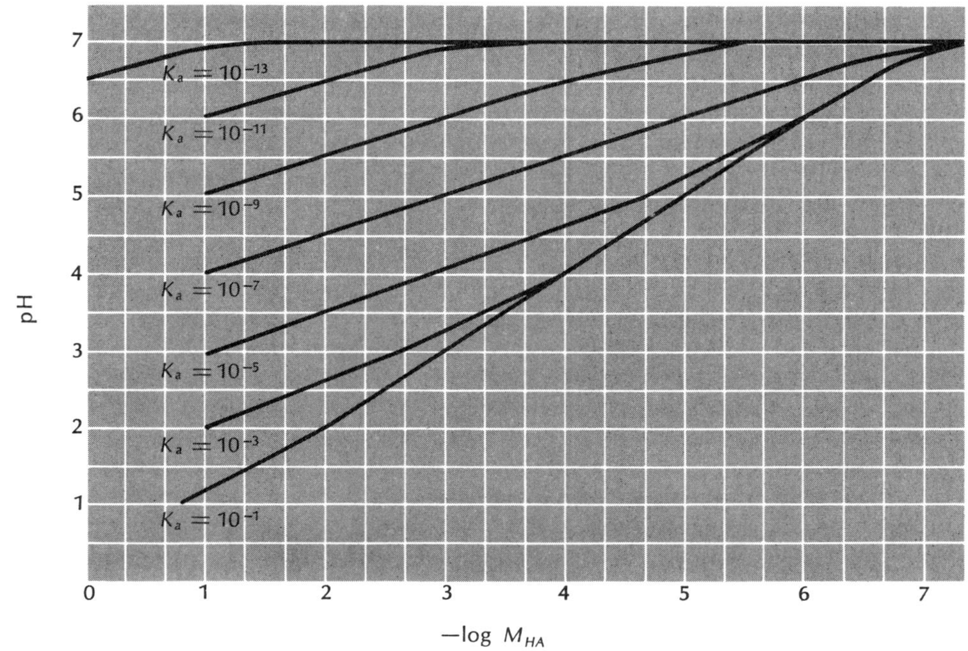

Fig. 16.1. pH plotted against $-\log M_{HA}$ for various values of K_a

be seen that this is true for acids with K_a less than 10^{-3}. Similarly, for a 10^{-3} M solution of HA, the approximation is valid if $[H_3O(aq)^+]$ is less than 10^{-4} M or if the pH is greater than 4. This is true for acids with K_a values less than 10^{-5}.

Weak Bases

The approach to monoprotic weak bases is the same as that used for monoprotic weak acids. An example should suffice to illustrate the similarities. Compute the concentrations of all species present in a 0.050 M solution of sodium azide, NaN_3.

The appropriate equilibria and relationships are summarized in the following list of equations:

$$N_3(aq)^- + H_2O = HN_3(aq) + OH(aq)^- \quad (16.36)$$

$$K_b = \frac{[HN_3(aq)][OH(aq)^-]}{[N_3(aq)^-]} = 5.3 \times 10^{-10} \quad (16.37)$$

$$H_2O(aq) + H_2O = H_3O(aq)^+ + OH(aq)^- \quad (16.38)$$

$$K_w = [H_3O(aq)^+][OH(aq)^-] = 1.0 \times 10^{-14} \quad (16.39)$$

$$M_{N_3^-} = 5.0 \times 10^{-2}\ M \quad (16.40)$$

Clearly, the species present in the solution are the following:

H_2O from the solvent left unionized
$OH(aq)^-$ from reaction of $N_3(aq)^-$ with water and from the ionization of water
$HN_3(aq)$ from reaction of $N_3(aq)^-$ with water
$H_3O(aq)^+$ from the ionization of water
$N_3(aq)^-$ left unreacted with water, from NaN_3
$Na(aq)^+$ from the NaN_3

Although $OH(aq)^-$ is produced from the reaction of $N_3(aq)^-$ with water as well as from the ionization of water, the principal source must be the former, since the concentration from the ionization of water cannot exceed 10^{-7} *M*. Since hydroxide ions and hydrazoic acid molecules are produced in equal numbers by the reaction of azide ion with water, as shown in Equation (16.36), the concentrations of these two species must be nearly equal, and Equation (16.37) may be rewritten thus:

$$\frac{[OH(aq)^-]^2}{[N_3(aq)^-]} = 5.3 \times 10^{-10} \tag{16.41}$$

If we assume that azide is a sufficiently weak base that only a small percentage of it reacts with water at equilibrium, then $[N_3(aq)^-]$ essentially is the same as $M_{N_3^-}$ or 5.0×10^{-2} *M*. This permits the rewriting of Equation (16.41) as Equation (16.42), which may be solved for $[OH(aq)^-]$ as indicated in the following section:

$$\frac{[OH(aq)^-]^2}{5.0 \times 10^{-2}} = 5.3 \times 10^{-10} \tag{16.42}$$

$$[OH(aq)^-]^2 = 26 \times 10^{-12}$$

$$[OH(aq)^-] = \sqrt{26 \times 10^{-12}} = 5.1 \times 10^{-6}\ M$$

With this value the other concentrations may be computed:

$$[HN_3(aq)] = [OH(aq)^-] = 5.1 \times 10^{-6}\ M$$

$$[H_3O(aq)^+] = \frac{K_w}{[OH(aq)^-]} = \frac{1.0 \times 10^{-14}}{5.1 \times 10^{-6}} = 2.0 \times 10^{-9}\ M$$

$$[N_3(aq)^-] = M_{N_3^-} = 5.0 \times 10^{-2}\ M$$

$$M_{Na^+} = 5.0 \times 10^{-2}\ M$$

Finally, the approximations *must* be checked. Since the hydroxide ion concentration from the ionization of water is the same as the hydronium ion concentration, $[OH(aq)^-]_W$ is 2.0×10^{-9} *M* which is insignificant to two significant figures when compared to the computed hydroxide ion concentration at 5.1×10^{-6} *M*. Thus the first approximation is valid.

The equilibrium concentration of $HN_3(aq)$ represents the concentration of azide ion which reacted with water, and this value of 5.1×10^{-6} *M* is insignificant when compared to the original azide ion concentration of 5.0×10^{-2} *M*. Thus the second approximation is valid also.

Polyprotic Species

Phosphoric acid, H_3PO_4, is an example of a polyprotic acid. When phosphoric acid is dissolved in water, it undergoes the following three successive reactions:

$$H_3PO_4(aq) + H_2O = H_2PO_4(aq)^- + H_3O(aq)^+ \qquad K_{a1} = 7.1 \times 10^{-3} \quad (16.43)$$

$$H_2PO_4(aq)^- + H_2O = HPO_4(aq)^{--} + H_3O(aq)^+ \qquad K_{a2} = 6.3 \times 10^{-8} \quad (16.44)$$

$$HPO_4(aq)^{--} + H_2O = PO_4(aq)^{3-} + H_3O(aq)^+ \qquad K_{a3} = 4.4 \times 10^{-13} \quad (16.45)$$

Although it is possible to add Equations (16.43), (16.44), and (16.45) to obtain over-all Equation (16.46), some care must be exercised in its interpretation.

$$H_3PO_4(aq) + 3H_2O = 3H_3O\,(aq)^+ + PO_4(aq)^{3-} \qquad K_{aT} = K_{a1}K_{a2}K_{a3} \qquad (16.46)$$

There is a temptation to think that the hydronium ion concentration is three times the phosphate ion concentration and also that the species, $H_2PO_4(aq)^-$ and $HPO_4(aq)^-$, are absent from the solution. This is not true.

Since the first ionization constant is much larger than the second, very little of the $H_2PO_4(aq)^-$ produced in the first ionization goes on to react according to Equation (16.44). This means that most of the dihydrogen phosphate ion produced in the first ionization is still in solution when the final equilibrium is attained. Finally, as indicated by the very small third ionization constant, only a very small percentage of the monohydrogen phosphate produced in the second ionization goes on to form phosphate ion. Thus we may construct a list of species present in an aqueous solution of phosphoric acid:

- $H_3PO_4(aq)$: What is left of the original acid after partial ionization.
- $H_2PO_4(aq)^-$: What is left after that produced by the first ionization undergoes partial ionization.
- $HPO_4(aq)^{--}$: What is left after that produced by the second ionization undergoes partial ionization.
- $PO_4(aq)^{3-}$: What is produced by the third ionization.
- $H_3O(aq)^+$: What is produced by the ionization of water and by the three ionizations of phosphoric acid.
- $OH(aq)^-$: What is produced by the ionization of water.
- H_2O: What is left of the solvent after it undergoes partial ionization.

Since the calculation of the concentrations of all species in solutions of polyprotic acids (or bases) is somewhat more laborious than the corresponding calculations for solutions of monoprotic acids (or bases), it is even more important that we be able to make some reasonable approximations.

There are four sources of hydronium ion in the solution: the ionization of water, and the three ionizations of phosphoric acid. Since the first ionization constant of phosphoric acid is rather high, for most common solutions we may assume that the ionization of water leads to a negligible concentration of hydronium ion. In addition, since the second and third ionization constants of phosphoric acid are much smaller than the first ionization constant, these two ionizations also are negligible in their contributions to the hydronium ion concentration in the solution. Thus the first approximation is that the only major source of hydronium ion is the first ionization of the phosphoric acid. It should be noted that this approximation is valid only when the first ionization constant is much larger than the later ionization constants. This is not always true, and care must be exercised in making this approximation for polyprotic acids in general.

The only source of dihydrogen phosphate ion, $H_2PO_4^-$, is the first ionization of phosphoric acid. A small percentage of this is consumed in the second ionization, but since the second constant is much smaller than the first, the concentration thus consumed is negligible compared to the concentration produced in the first ionization. Thus the second approximation is that the final equilibrium concentration of $H_2PO_4^-$ is the same as the concentration of that ion from the first ionization of phosphoric acid. This leads to the relationship expressed in Equation (16.47), in which the equilibrium concentrations of $H_3O(aq)^+$ and $H_2PO_4(aq)^-$ are the same since these species are produced in equal numbers in the first ionization.

$$[H_3O(aq)^+] = [H_2PO_4(aq)^-] \qquad (16.47)$$

The only source of monohydrogen phosphate ion, HPO_4^{--}, is the second ionization, and since the third ionization is small, only a negligible concentration of this ion is consumed in the final step.

Finally, the equilibrium concentration of H_3PO_4 must be the original over-all concentration minus the concentration of ionizing H_3PO_4. If it is assumed, as it was for weak monoprotic acids, that the acid is sufficiently weak, then the concentration of ionizing acid will be negligible compared to the over-all original concentration, and the relationship shown in Equation (16.48) will be valid.

$$[H_3PO_4(aq)] = M_{H_3PO_4} \qquad (16.48)$$

Now let us apply these approximations to the calculation of the concentrations of all species in 1.00 M H_3PO_4.

The hydronium ion concentration may be found by using the approximations with the first ionization constant.

$$K_{a1} = \frac{[H_3O(aq)^+][H_2PO_4(aq)^-]}{[H_3PO_4(aq)]} = 7.1 \times 10^{-3}$$

$$\frac{[H_3O(aq)^+]^2}{1.00} = 7.1 \times 10^{-3}$$

$$[H_3O(aq)^+]^2 = 7.1 \times 10^{-3} = 71 \times 10^{-4}$$

$$[H_3O(aq)^+] = \sqrt{71 \times 10^{-4}} = 8.4 \times 10^{-2}\ M$$

From this hydronium ion concentration, the concentration of hydroxide ion may be found by utilizing the water constant, and the concentration of the dihydrogen phosphate ion is the same as the hydronium ion concentration if the approximation is used.

$$[OH(aq)^-] = \frac{K_w}{[H_3O(aq)^+]} = \frac{1.0 \times 10^{-14}}{8.4 \times 10^{-2}} = 1.2 \times 10^{-13}\ M$$

$$[H_2PO_4(aq)^-] = [H_3O(aq)^+] = 8.4 \times 10^{-2}\ M$$

The values just computed for $[H_2PO_4(aq)^-]$ and $[H_3O(aq)^+]$ may be used in the expression for the second ionization constant to compute the concentration of monohydrogen phosphate ion. However, the approximations are sufficient to carry out this calculation without the use of actual numerical values.

$$K_{a2} = \frac{[H_3O(aq)^+][HPO_4(aq)^{--}]}{[H_2PO_4(aq)^-]}$$

$$[HPO_4(aq)^{--}] = \frac{K_{a2}[H_2PO_4(aq)^-]}{[H_3O(aq)^+]}$$

However, since $[H_2PO_4(aq)^-] = [H_3O(aq)^+]$ from the approximations, this relationship simplifies to the following one:

$$[HPO_4(aq)^{--}] = K_{a2} = 6.3 \times 10^{-8}\ M$$

The values for $[HPO_4(aq)^{--}]$ and $[H_3O(aq)^+]$ may be used in the third ionization constant expression to find the value for $[PO_4(aq)^{3-}]$.

$$K_{a3} = \frac{[PO_4(aq)^{3-}][H_3O(aq)^+]}{[HPO_4(aq)^{--}]}$$

$$[PO_4(aq)^{3-}] = \frac{K_{a3}[HPO_4(aq)^{--}]}{[H_3O(aq)^+]} = \frac{(4.4 \times 10^{-13})(6.3 \times 10^{-8})}{8.4 \times 10^{-2}} = 3.3 \times 10^{-19}\ M$$

The computed concentrations may be summarized in the following way:

$$[H_3O(aq)^+] = 8.4 \times 10^{-2}\ M$$

$$[OH(aq)^-] = 1.2 \times 10^{-13}\ M$$

$$[H_2PO_4(aq)^-] = 8.4 \times 10^{-2}\ M$$

$$[HPO_4(aq)^{--}] = 6.3 \times 10^{-8}\ M$$

$$[PO_4(aq)^{3-}] = 3.3 \times 10^{-19}\ M$$

$$[H_3PO_4(aq)] = 1.00\ M$$

Finally it is necessary to check the validity of the approximations. Certainly the very low value for $[PO_4(aq)^{3-}]$ compared to $[HPO_4(aq)^{--}]$ indicates that a negligible portion of the monohydrogen phosphate ionizes. Similarly, the low value for $[HPO_4(aq)^{--}]$ compared to $[H_2PO_4(aq)^{-}]$ indicates that a negligible portion of the dihydrogen phosphate ionizes. The concentrations of $H_2PO_4(aq)^{-}$, $HPO_4(aq)^{--}$, and $PO_4(aq)^{3-}$ also indicate the relative concentrations of hydronium ion contributed by the three ionizations of phosphoric acid and it is clear that the first ionization is the only major contributor. In addition, the concentration of hydronium ion from the first ionization is so much higher than the possible contribution of the ionization of water, that we are quite safe in making this approximation.

However, the comparison of $[H_3O(aq)^+] = 8.4 \times 10^{-2}\ M$ to $[H_3PO_4(aq)] = 1.00\ M$ indicates an apparent 8.4 percent ionization of the acid, which is a bit high. Whether or not this approximation—that the ionization of the acid is negligible—is suitable depends upon our demands. If we say somewhat arbitrarily that we are willing to accept apparent ionizations up to about 10 percent, then this is within reason and the approximation is a valid one.

Mixtures of Strong and Weak Acids

The addition of a strong acid to a solution of a weak acid decreases the pH, that is, increases the hydronium ion molarity. However, the hydronium ion concentration increase is less than the increase caused by the strong acid, since the addition of the strong acid increases the hydronium ion concentration, which then produces a shift in the equilibrium for the weak acid.

For example, in a solution of hydrazoic acid, HN_3, the following equilibrium exists:

$$HN_3(aq) + H_2O = H_3O(aq)^+ + N_3(aq)^- \tag{16.49}$$

If a strong acid, such as hydrochloric acid, is added, the hydronium ion concentration increases due to the ionization of the HCl, and equilibrium (16.49) shifts to relieve the stress by reacting some of the $H_3O(aq)^+$ with $N_3(aq)^-$ to form $HN_3(aq)$ and H_2O. Thus the addition of the strong acid suppresses the ionization of the weak acid.

As an example of a typical calculation, we shall compute the concentrations of all species in a solution which is 0.020 *M* in HN_3 and 0.00050 *M* in HCl.

Clearly the hydronium ion comes from three sources, namely the ionizations of HCl, HN_3, and H_2O.

$$[H_3O(aq)^+] = [H_3O(aq)^+]_W + [H_3O(aq)^+]_{HN_3} + [H_3O(aq)^+]_{HCl} \tag{16.50}$$

Since hydrochloric acid is ionized completely, the concentration of hydronium ion from this source is $5.0 \times 10^{-4}\ M$. This in itself guarantees that the contribution from the ionization of water is negligible, so that Equation (16.50) may be rewritten as Equation (16.51).

$$[H_3O(aq)^+] = [H_3O(aq)^+]_{HN_3} + 5.0 \times 10^{-4} \qquad (16.51)$$

However, since the ionization of HN_3 produces equal numbers of hydronium and azide ions, the hydronium ion concentration from this ionization must be the same as the equilibrium concentration of azide ions, and Equation (16.51) may be rewritten as Equation (16.52).

$$[H_3O(aq)^+] = [N_3(aq)^-] + 5.0 \times 10^{-4}$$

$$[N_3(aq)^-] = [H_3O(aq)^+] - 5.0 \times 10^{-4} \qquad (16.52)$$

The final approximation is the same one used several times before, namely that the percent ionization of $HN_3(aq)$ is negligible compared to the total original concentration. With the hydrochloric acid now suppressing this ionization, this would appear to be a better approximation than it has been in past calculations.

These relationships may be used with the acid-constant expression for hydrazoic acid.

$$K_a = \frac{[H_3O(aq)^+][N_3(aq)^-]}{[HN_3(aq)]} = 1.9 \times 10^{-5} \qquad (16.53)$$

The expression for $[N_3(aq)^-]$ shown as Equation (16.52), along with the original concentration of hydrazoic acid, may be substituted into Equation (16.53).

$$\frac{[H_3O(aq)^+]([H_3O(aq)^+] - 5.0 \times 10^{-4})}{2.0 \times 10^{-2}} = 1.9 \times 10^{-5}$$

This may be put into standard quadratic form:

$$[H_3O(aq)^+]^2 - 5.0 \times 10^{-4}[H_3O(aq)^+] - 3.8 \times 10^{-7} = 0$$

Recall that the roots for the quadratic $ax^2 + bx + c = 0$ may be found as follows:

$$x = \frac{-b \pm \sqrt{b^2 - 4ac}}{2a}$$

Using this formula, the hydronium ion concentration may be found.

$$[H_3O(aq)^+] = \frac{(5.0 \times 10^{-4}) \pm \sqrt{(5.0 \times 10^{-4})^2 + 4(3.8 \times 10^{-7})}}{2}$$

$$= \frac{(5.0 \times 10^{-4}) \pm \sqrt{(2.5 \times 10^{-7}) + 15 \times 10^{-7}}}{2}$$

$$= \frac{(5.0 \times 10^{-4}) \pm \sqrt{1.8 \times 10^{-6}}}{2} = \frac{5.0 \times 10^{-4} \pm 1.3 \times 10^{-3}}{2}$$

To make the correct $\pm$ option, note that 1.3×10^{-3} is larger than 5.0×10^{-4} and if subtraction is performed the hydronium ion concentration is negative and that is meaningless. Therefore addition is performed.

$$[H_3O(aq)^+] = \frac{5.0 \times 10^{-4} + 1.3 \times 10^{-3}}{2} = \frac{1.8 \times 10^{-3}}{2} = 9.0 \times 10^{-4}\ M$$

From this value of $[H_3O(aq)^+]$ the remaining concentrations may be computed. The hydroxide ion concentration is found using the water constant.

$$[OH(aq)^-] = \frac{K_w}{[H_3O(aq)^+]} = \frac{1.0 \times 10^{-14}}{9.0 \times 10^{-4}} = 1.1 \times 10^{-10}\ M$$

The azide ion concentration is found through the relationship in Equation (16.52).

$$[N_3(aq)^-] = [H_3O(aq)^+] - 5.0 \times 10^{-4}$$

$$= 9.0 \times 10^{-4} - 5.0 \times 10^{-4} = 4.0 \times 10^{-4}\ M$$

The hydrazoic acid concentration is the same as the original and the chloride ion concentration is the same as the hydrochloric acid concentration. The following list summarizes the concentrations of all species in the solution.

$$[H_3O(aq)^+] = 9.0 \times 10^{-4}\ M$$

$$[OH(aq)^-] = 1.1 \times 10^{-11}\ M$$

$$[N_3(aq)^-] = 4.0 \times 10^{-4}\ M$$

$$[HN_3(aq)] = 2.0 \times 10^{-2}\ M$$

$$[Cl(aq)^-] = 5.0 \times 10^{-4}\ M$$

The approximation that the water ionization is an insignificant source of hydronium ion is verified by the very small hydroxide ion concentration compared to the hydronium ion concentration. When the azide ion concentration is compared with the hydrazoic acid concentration, there is an apparent 2 percent ionization, which is not very large and is well within an arbitrarily agreed-upon maximum of 10 percent.

Buffers

The action of a weak acid equilibrium to resist changes in hydronium ion concentration by consuming at least some of the added hydronium ion is related to much more efficient systems which resist pH change. Such efficient systems for the resistance of pH changes are called *buffer systems, buffer solutions, or* commonly just *buffers.* Usually a buffer is a mixture of a weak acid and its conjugate base, or conversely a weak base and its conjugate acid. Thus, if an acid is added to the system, it reacts with the base so that, although the pH decreases, it decreases much less than it would if the buffer were absent. Similarly, if a base is added, it reacts with the acid present so that the pH rise is less than it would be otherwise.

An example of a buffer is hydrazoic acid and sodium azide in an aqueous solution. This may be viewed either from the point of view of the acidic reaction of hydrazoic acid as shown by Equation (16.54), or from the point of view of the basic reaction of azide ion, as shown by Equation (16.55).

$$HN_3(aq) + H_2O = H_3O(aq)^+ + N_3(aq)^- \qquad (16.54)$$

$$N_3(aq)^- + H_2O = HN_3(aq) + OH(aq)^- \qquad (16.55)$$

From the point of view of Equation (16.54), the addition of acid shifts the equilibrium toward the production of $HN_3(aq)$ and the addition of base shifts the equilibrium toward the ionization of $HN_3(aq)$. From the point of view of Equation (16.55), the addition of acid shifts the equilibrium toward the production of $HN_3(aq)$ and the addition of base shifts the equilibrium toward the ionization of $HN_3(aq)$.

As a sample calculation, we shall find the pH of an aqueous solution that is 0.050 M in HN_3 and 0.020 M in NaN_3. As a first approach, consider this from the point of view of the ionization of the acid, as shown in Equation (16.54). The ionization constant expression may be written in the following way:

$$K_a = \frac{[H_3O(aq)^+][N_3(aq)^-]}{[HN_3(aq)]} = 1.9 \times 10^{-5} \qquad (16.56)$$

The azide ion molarity is the sum of the molarities of azide ion derived from the ionizations of $HN_3(aq)$ and NaN_3. The sodium azide is completely ionized, and the azide ion concentration from the sodium azide suppresses the ionization of hydrazoic acid. Since hydrazoic acid is a weak acid anyway, we make the approximation that the equilibrium concentration of azide ion is the same as the concentration of sodium azide, that is, 2.0×10^{-2} M.

In addition, since the presence of the sodium azide suppresses the ionization of the hydrazoic acid, the equilibrium concentration of hydrazoic acid is nearly the same as the original concentration of hydrazoic acid, namely, 5.0×10^{-2} M.

With these approximations Equation (16.56) takes on the following form and the concentration of hydronium ion may be computed as indicated.

$$\frac{[H_3O(aq)^+](2.0 \times 10^{-2})}{5.0 \times 10^{-2}} = 1.9 \times 10^{-5}$$

$$[H_3O(aq)^+] = \frac{(1.9 \times 10^{-5})(5.0 \times 10^{-2})}{(2.0 \times 10^{-2})} = 4.8 \times 10^{-5}\ M$$

$$pH = -\log[H_3O(aq)^+] = -\log(4.8 \times 10^{-5}) = 4.3$$

Since hydrazoic acid is the main source of hydronium ion, the azide ion concentration derived from hydrazoic acid should be the same as the hydronium ion concentration, or 4.8×10^{-5} *M*. This is insignificant compared to 5.0×10^{-2} *M*, which is the azide ion concentration derived from sodium azide. Thus the first approximation is satisfactory.

The hydronium ion concentration of 4.8×10^{-5} *M* is the same as the concentration of hydrazoic acid ionized, and this is about 0.05 percent of the original concentration of hydrazoic. For most purposes, then, the second approximation is satisfactory also.

This problem may be solved just as well from the point of view of Equation (16.55) and the base constant for azide ion. The same approximations may be used.

$$K_b = \frac{[HN_3(aq)][OH(aq)^-]}{[N_3(aq)^-]} = 5.3 \times 10^{-10}$$

$$[OH(aq)^-] = \frac{K_b[N_3(aq)^-]}{[HN_3(aq)]} = \frac{(5.3 \times 10^{-10})(2.0 \times 10^{-2})}{5.0 \times 10^{-2}} = 2.1 \times 10^{-10}\ M$$

$$pOH = -\log[OH(aq)^-] = -\log(2.1 \times 10^{-10}) = 9.7$$

$$pH = 14.0 - pOH = 14.0 - 9.7 = 4.3$$

Often, in preparing buffers, it is most important to know the ratio of concentrations of acid and conjugate base necessary to make a solution with a specified pH. Within the limits of the approximations that the original concentrations of the acid and its conjugate base are the same as the equilibrium concentrations, this information is obtained easily from the acid constant expression. As a general case, consider the weak acid HA and its conjugate base, A^-.

$$HA(aq) + H_2O = H_3O(aq)^+ + A(aq)^-$$

$$K_a = \frac{[H_3O(aq)^+][A(aq)^-]}{[HA(aq)]} = \frac{[H_3O(aq)^+]M_{A^-}}{M_{HA}}$$

$$\frac{M_{HA}}{M_{A^-}} = \frac{[H_3O(aq)^+]}{K_a} \tag{16.57}$$

Thus, if it is desired to make a buffer with pH = 6, from hydrazoic acid and sodium azide, the molar ratio of hydrazoic acid to sodium azide may be found using Equation (16.57).

$$\frac{M_{HN_3}}{M_{N_3^-}} = \frac{[H_3O(aq)^+]}{K_a} = \frac{1.0 \times 10^{-6}}{1.9 \times 10^{-5}} = 0.053$$

Thus 0.053 moles of HN_3 must be used for each mole of NaN_3 used.

Usually buffer action is most efficient when the acid and base are present in equal concentrations. Under these conditions the buffer resists increases and decreases in pH almost equally well. From the acid constant expression for hydrazoic acid in Equation (16.56), it is clear that $[H_3O(aq)^+] = K_a$ when $[HN_3(aq)] = [N_3(aq)^-]$. Thus the most efficient buffer pH can be computed from the acid constant for the acid, since pH = pK_a. In the case of hydrazoic acid and sodium azide, the most effective pH occurs at $[H_3O(aq)^+] = 1.9 \times 10^{-5}$ *M* or pH = 4.7.

Amphiprotic Species

Many species are amphiprotic, that is, they may either lose or win a proton. An example is the monohydrogen phosphate ion. The species present in an aqueous solution of Na_2HPO_4 include $Na(aq)^+$, H_2O, $H_3O(aq)^+$, $OH(aq)^-$, $HPO_4(aq)^{--}$, $PO_4(aq)^{3-}$, $H_2PO_4(aq)^-$, and $H_3PO_4(aq)$. The following equations represent the equilibrium reactions responsible for the presence of the several phosphorus-containing species.

$$HPO_4(aq)^{--} + H_2O = H_2PO_4(aq)^- + OH(aq)^- \tag{16.58}$$

$$H_2PO_4(aq)^- + H_2O = H_3PO_4(aq) + OH(aq)^- \tag{16.59}$$

$$HPO_4(aq)^{--} + H_2O = PO_4(aq)^{3-} + H_3O(aq)^+ \tag{16.60}$$

In reaction (16.58) the monohydrogen phosphate acts as a base, and in reaction (16.60) it acts as an acid. From Table 15.1 the acid constant of monohydrogen phosphate ion is found to be 4.4×10^{-13}, whereas the base constant is found to be 1.6×10^{-7}. Since the base constant is larger than the acid constant, the resulting solution should be slightly basic; that is, it should have a pH greater than 7.

Also, it is apparent that the monohydrogen phosphate ion may act as a buffer by itself, since it is both a weak acid and a weak base. If an acid is added to an aqueous solution of Na_2HPO_4, the $HPO_4(aq)^{--}$ acts as a base to consume hydronium ions, and if a base is added it acts as an acid.

Typical quantitative relationships may be illustrated by considering the amphiprotic species, HA^-, which has the conjugate acid, H_2A, and the conjugate base, A^{--}. When NaHA is dissolved in water, the system may be described by the following chemical equations and equilibrium relationships.

$$HA(aq)^- + H_2O = H_3O(aq)^+ + A(aq)^{--} \tag{16.61}$$

$$K_a = \frac{[H_3O(aq)^+][A(aq)^{--}]}{[HA(aq)^-]} \tag{16.62}$$

$$HA(aq)^- + H_2O = H_2A(aq) + OH(aq)^- \tag{16.63}$$

$$K_b = \frac{[H_2A(aq)][OH(aq)^-]}{[HA(aq)^-]} \tag{16.64}$$

$$H_2O(aq) + H_2O = H_3O(aq)^+ + OH(aq)^- \tag{16.65}$$

$$K_w = [H_3O(aq)^+][OH(aq)^-] \tag{16.66}$$

Clearly there are two sources of hydronium ion, namely the ionization of water and the acid reaction of HA^-.

$$[H_3O(aq)^+] = [H_3O(aq)^+]_W + [H_3O(aq)^+]_{HA^-} \tag{16.67}$$

The hydronium ion concentration derived from water is the same as the hydroxide ion concentration derived from water, and the hydronium ion concentration from HA^- is the same as the concentration of $A(aq)^{--}$, which is derived only from the acidic reaction of $HA(aq)^-$. This permits rewriting Equation (16.67) as Equation (16.68).

$$[H_3O(aq)^+] = [OH(aq)^-]_W + [A(aq)^{--}] \tag{16.68}$$

Since the basic reaction of $HA(aq)^-$ also produces hydroxide ion, the concentration of hydroxide ion derived from water must be the equilibrium concentration of hydroxide ion minus the concentration of hydroxide ion derived from $HA(aq)^-$. When this is incorporated into Equation (16.68), Equation (16.69) results.

$$[H_3O(aq)^+] = [OH(aq)^-] - [OH(aq)^-]_{HA^-} + [A(aq)^{--}] \tag{16.69}$$

From Equation (16.63) it is clear that the hydroxide ion concentration derived from HA^- is the same as the concentration of $H_2A(aq)$ derived from HA^-, which is the only source of $H_2A(aq)$. As a result, Equation (16.69) may be rewritten as follows:

$$[H_3O(aq)^+] = [OH(aq)^-] - [H_2A(aq)] + [A(aq)^{--}] \tag{16.70}$$

Expressions for the three terms on the right-hand side of Equation (16.70) may be obtained by solving Equation (16.66) for $[OH(aq)^-]$, Equation (16.64) for $[H_2A(aq)]$, and Equation (16.62) for $[A(aq)^{--}]$ as shown in the following equations.

$$[OH(aq)^-] = \frac{K_w}{[H_3O(aq)^+]} \tag{16.71}$$

$$[H_2A(aq)] = \frac{K_b[HA(aq)^-]}{[OH(aq)^-]} = \frac{K_b[HA(aq)^-][H_3O(aq)^+]}{K_w} \tag{16.72}$$

$$[A(aq)^{--}] = \frac{K_a[HA(aq)^-]}{[H_3O(aq)^+]} \tag{16.73}$$

The expressions in these three equations may be substituted into Equation (16.70).

$$[H_3O(aq)^+] = \frac{K_w}{[H_3O(aq)^+]} - \frac{K_b[HA(aq)^-][H_3O(aq)^+]}{K_w} + \frac{K_aK_w[HA(aq)^-]}{[H_3O(aq)^+]} \tag{16.74}$$

Equation (16.74) may be solved for $[H_3O(aq)^+]^2$ as follows:

$$[H_3O(aq)^+]^2 = \frac{K_w^2 + K_aK_w[HA(aq)^-]}{K_w + K_b[HA(aq)^-]} \tag{16.75}$$

Equation (16.75) represents an exact relationship, since no approximations were used in its derivation. However, under certain circumstances this relationship may be simplified by the use of some approximations. For example, if HA^- is sufficiently weak as both an acid and a base, the percentage of the original $HA(aq)^-$ concentration that is reacted at equilibrium is small and may be neglected, so that the equilibrium concentration of $HA(aq)^-$ is the same as the over-all concentration of the amphiprotic species, M_{HA^-}. Then Equation (16.75) may be written as Equation (16.76).

$$[H_3O(aq)^+]^2 = \frac{K_w^2 + K_aK_wM_{HA^-}}{K_w + K_bM_{HA^-}} \qquad (16.76)$$

Since K_w^2 is quite small, 10^{-28} specifically, this term may be ignored in the numerator when the product, $K_aM_{HA^-}$, is appreciably larger than 10^{-14}. Similarly, when the product, $K_bM_{HA^-}$, is appreciably larger than 10^{-14}, K_w may be neglected in the denominator. Under these circumstances, Equation (16.76) reduces to Equation (16.77) where it is apparent that the pH of the solution is independent of the original concentration of $HA(aq)^-$ *if* these conditions are met.

$$[H_3O(aq)^+]^2 = \frac{K_aK_wM_{HA^-}}{K_bM_{HA^-}} = \frac{K_aK_w}{K_b} \qquad (16.77)$$

For example, compute the pH of 0.10 M NaHS. For the amphiprotic $HS(aq)^-$, $K_a = 1.3 \times 10^{-13}$, and $K_b = 1.1 \times 10^{-7}$. It is evident that $K_aK_wM_{HS^-} = 1.3 \times 10^{-28}$ which cannot be ignored with respect to 1.0×10^{-28}. In this case the approximation fails and the hydronium ion concentration may be computed from Equation (16.76) in the following manner:

$$[H_3O(aq)^+] = \sqrt{\frac{(1.0 \times 10^{-14})^2 + (1.3 \times 10^{-13})(1.0 \times 10^{-14})(1.0 \times 10^{-1})}{(1.0 \times 10^{-14}) + (1.1 \times 10^{-7})(1.0 \times 10^{-1})}}$$

$$= 1.4 \times 10^{-10}\ M$$

$$pH = -\log[H_3O(aq)^+] = -\log(1.4 \times 10^{-10}) = 9.1$$

Thus the solution is basic as expected, since the base constant is higher than the acid constant for $HS(aq)^-$.

As a final example, compute the pH of 0.10 M $NaHCO_3$. For $HCO_3(aq)^-$, $K_b = 2.3 \times 10^{-8}$ and $K_a = 4.8 \times 10^{-11}$. In this case $K_aK_wM_{HCO_3^-} = 4.8 \times 10^{-6}$. To two significant figures, K_w^2 is insignificant by comparison, and the relationship shown in Equation (16.77) may be used.

$$[H_3O(aq)^+] = \frac{(4.8 \times 10^{-11})(1.0 \times 10^{-14})}{2.3 \times 10^{-8}} = 4.6 \times 10^{-9}\ M$$

$$pH = -\log[H_3O(aq)^+] = -\log(4.6 \times 10^{-9}) = 8.3$$

REACTIONS BETWEEN ACIDS AND BASES IN AQUEOUS SOLUTIONS

Reactions between acids and bases dissolved in water tend to be rapid, so that equilibria are established quickly. However, it is possible to follow changes in pH when one of the reacting species is added to the second in small increments.

We shall look at the way the pH changes as a solution of one species is added to a solution of another species. Then we shall see how an understanding of this relationship provides a powerful tool in analytical chemistry.

Addition of a Strong Base to a Strong Acid

Suppose that we begin with 50.0 ml of 0.100 *M* hydrochloric acid. Since hydrochloric acid is completely ionized, the concentration of hydronium ion from the HCl ionization is 0.100 *M*. This is much higher than any concentration that can be derived from the water ionization, so the over-all hydronium ion concentration is 0.100 *M*.

Next, suppose that 10.0 ml of 0.100 *M* sodium hydroxide is added to this solution. The $OH(aq)^-$ from the NaOH solution reacts with the $H_3O(aq)^+$ in the hydrochloric acid solution as follows:

$$H_3O(aq)^+ + OH(aq)^- = 2H_2O \qquad (16.78)$$

For our purposes, this reaction is complete. Thus each mmole of $OH(aq)^-$ ions added reacts with a mmole of $H_3O(aq)^+$ ions in the solution. The 10.0 ml of 0.100 *M* NaOH solution contains 1.00 mmole of $OH(aq)^-$ ions. The original 50.0 ml of 0.100 *M* HCl solution contained 5.00 mmoles of $H_3O(aq)^+$ ions. Thus in the resulting solution there are only 4.00 mmoles of $H_3O(aq)^+$ ions left. In addition the total volume of this solution now is 60.0 ml, so the hydronium ion concentration is 6.7×10^{-2} *M*.

$$[H_3O(aq)^+] = \frac{4.00 \text{ mmoles}}{60.0 \text{ ml}} = 6.7 \times 10^{-2}\ M$$

The pH of this solution is 1.2.

$$\text{pH} = -\log[H_3O(aq)^+] = -\log(6.7 \times 10^{-2}) = 1.2$$

Similarly, if another 10.0 ml of 0.100 *M* NaOH is added, this makes a total of 20.0 ml added, which contain 2.00 mmoles of NaOH, which in turn consume 2.00 mmoles of $H_3O(aq)^+$, leaving just 3.00 mmoles of $H_3O(aq)^+$ ions in a total volume of 70.0 ml. At this point the molarity of $H_3O(aq)^+$ is 4.3×10^{-2} *M* and the pH is 1.4.

$$[H_3O(aq)^+] = \frac{3.00 \text{ mmoles}}{70.0 \text{ ml}} = 4.3 \times 10^{-2}\ M$$

$$\text{pH} = -\log[H_3O(aq)^+] = -\log(4.3 \times 10^{-2}) = 1.4$$

Similar reasoning may be used to compute pH values after additional increments of NaOH have been added. It should be noted, however, that when 50.0 ml of 0.100 *M* NaOH has been added, 5.00 mmoles of NaOH have been added and this is exactly enough NaOH to react with the 5.00 mmoles of HCl originally present in the solution. At this point the solution is an aqueous solution of sodium chloride. This is evident in the following equation, where all ions are included to represent the total reaction:

$$Na(aq)^+ + OH(aq)^- + H_3O(aq)^+ + Cl(aq)^- = 2H_2O + Na(aq)^+ + Cl(aq)^-$$

Since neither $Na(aq)^+$ nor $Cl(aq)^-$ is acidic or basic, the solution is neutral, that is, it has pH = 7.0.

When just enough reagent has been added to react exactly with another substance, the system is said to have reached the *equivalence point*. In the example above, the equivalence point corresponds to the addition of 50.0 ml of 0.100 *M* NaOH to 50.0 ml of 0.100 *M* HCl. The pH at the equivalence point for this system is 7.0.

If sodium hydroxide is added beyond the equivalence point, the solution is just a more dilute solution of sodium hydroxide containing dissolved sodium chloride. Thus when 10.0 ml of 0.100 *M* NaOH is added beyond the equivalence point, the resulting solution contains 1.00 mmole of $OH(aq)^-$ in 110 ml of solution; the hydroxide ion concentration is 9.1×10^{-3} *M*, and the pH is 12.0.

$$[OH(aq)^-] = \frac{1.00 \text{ mmole}}{110 \text{ ml}} = 9.1 \times 10^{-3}\ M$$

$$pOH = -\log[OH(aq)^-] = -\log(9.1 \times 10^{-3}) = 2.0$$

$$pH = 14.0 - pOH = 14.0 - 2.0 = 12.0$$

Additional data for the addition of 0.100 *M* NaOH to 50.0 ml of 0.100 *M* HCl are given in Table 16.2. The values given in Table 16.2 are used to plot pH against milliliters of 0.100 *M* NaOH added in Fig. 16.2. Figure 16.2 clearly shows the very rapid change in pH very near the equivalence point, with very small changes in the amount of sodium hydroxide added. As a result it is evident that if one has some means of following pH changes, this is a very sensitive way in which to detect the equivalence point. One means of doing this is with a pH meter; another is with acid-base color indicators chosen for their color-change range near the pH at the equivalence point. With the rapid change in pH at the equivalence point, the indicator undergoes a sharp color change.

The laboratory process of measuring the amount of one reactant required to react completely with a given amount of a second reactant, as the first is added continuously to the second, is called *titration*. The added reactant is called the *titrant*. The reactant, to which the titrant is added, is called the *titree*. In any titration, the equivalence point is that point during the titration at which exactly enough titrant has been added to react with the titree. One of the technical problems in any titration is the detection of the equivalence point. The *end-point* of a titration is the point at which an *observable* change occurs to indicate the close approach or passage of the equivalence point. Ideally, the end-point and the equivalence point should be identical, but in practice it is rare that one can make

TABLE 16.2
Addition of 0.100 M NaOH to 50.0 ml of 0.100 M HC1

TOTAL ml 0.100 *M* NaOH ADDED	TOTAL SOLN VOLUME (ml)	mmoles $H_3O(aq)^+$ IN SOLN	mmoles $OH(aq)^-$ IN SOLN	$[H_3O(aq)^+]$	pH
0.0	50.0	5.00		1.0×10^{-1}	1.0
10.0	60.0	4.00		6.7×10^{-2}	1.2
20.0	70.0	3.00		4.3×10^{-2}	1.4
30.0	80.0	2.00		2.5×10^{-2}	1.6
40.0	90.0	1.00		1.1×10^{-2}	2.0
45.0	95.0	0.50		5.3×10^{-3}	2.3
46.0	96.0	0.40		4.2×10^{-3}	2.4
47.0	97.0	0.30		3.1×10^{-3}	2.5
48.0	98.0	0.20		2.0×10^{-3}	2.7
49.0	99.0	0.10		1.0×10^{-3}	3.0
50.0	100.0			1.0×10^{-7}	7.0
51.0	101.0		0.10	1.0×10^{-11}	11.0
52.0	102.0		0.20	5.0×10^{-12}	11.3
53.0	103.0		0.30	3.4×10^{-12}	11.5
54.0	104.0		0.40	2.6×10^{-12}	11.6
55.0	105.0		0.50	2.1×10^{-12}	11.7
60.0	110.0		1.00	1.1×10^{-12}	12.0

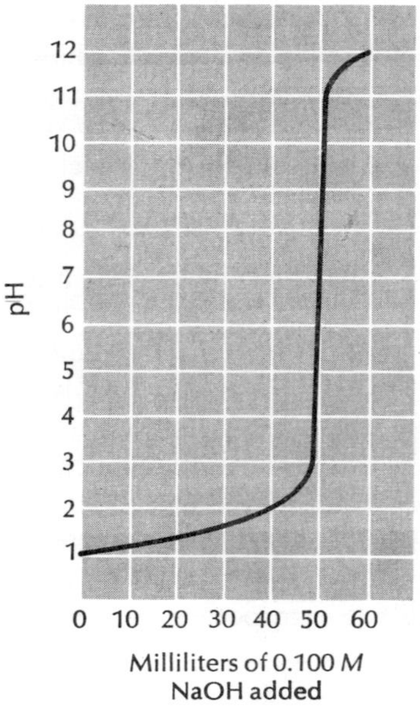

Fig. 16.2. Titration of 50.0 ml of 0.100 *M* HCl with 0.100 *M* NaOH

the two coincide exactly, although there are many instances in which the difference between the two is negligible.

For titrations in general a number of different techniques are employed for the observation of the end-point. However, for acid-base titrations the measurement of pH is one of the most common techniques. The graph in Fig. 16.2 often is called a *titration curve*, for it describes the pH change as titrant is added. Since the pH at the equivalence point is 7.0 in this case, if the volume of titrant is measured as it is added to the titree, when pH 7.0 is reached, the amount of titrant added is exactly the amount needed to react with the titree.

Some specific uses of titrations will be illustrated later in this chapter.

Addition of a Strong Acid to a Strong Base

From the previous discussion, it should be clear that if sodium hydroxide is used as the titree and hydrochloric acid as the titrant, the equivalence point pH should be 7.0. The solution at the equivalence point is just a solution of sodium chloride. The main difference between this example and the one described earlier is that the titrant has a high pH at the outset and a low pH beyond the equivalence point. Figure 16.3 shows the titration curve for the addition of 0.100 *M* HCl to 50.0 ml of 0.100 *M* NaOH.

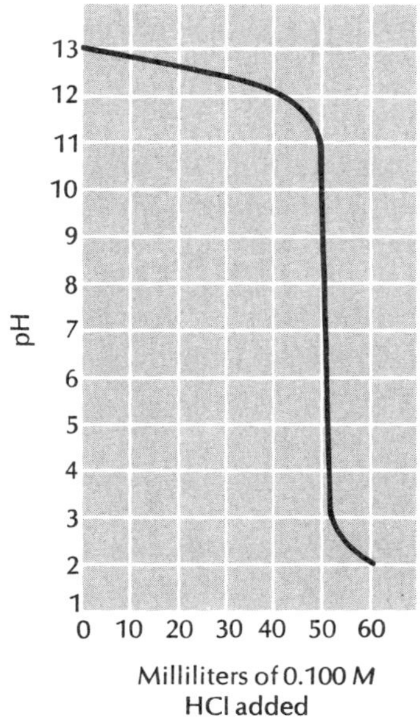

Fig. 16.3. Titration of 50.0 ml of 0.100 *M* NaOH with 0.100 *M* HCl

Addition of a Strong Titrant to a Weak Titree

As an example of a strong titrant and a weak titree, consider the titration of 50.0 ml of 0.100 *M* acetic acid by 0.100 *M* sodium hydroxide. Since the weak acid is only slightly ionized, we may write the over-all titration reaction in the following equation form:

$$CH_3COOH(aq) + OH(aq)^- = CH_3COO(aq)^- + H_2O \qquad (16.79)$$

Initially the titree is acidic, although not as acidic as 0.100 *M* HCl, since the acetic acid is only partially ionized. To determine the pH it is necessary to use the methods developed earlier for the determination of the pH of a weak acid. Employing the usual approximations, we may compute the initial pH as follows:

$$CH_3COOH(aq) + H_2O = H_3O(aq)^+ + CH_3COO(aq)^- \qquad (16.80)$$

$$K_a = \frac{[H_3O(aq)^+][CH_3COO(aq)^-]}{[CH_3COOH(aq)]} = 1.7 \times 10^{-5} = \frac{[H_3O(aq)^+]^2}{1.0 \times 10^{-1}}$$

$$[H_3O(aq)^+] = \sqrt{1.7 \times 10^{-6}} = 1.3 \times 10^{-3}\ M$$

$$pH = -\log(1.3 \times 10^{-3}) = 2.9$$

Upon the addition of titrant, acetic acid is consumed and converted to acetate ions, so that the concentration of acetic acid decreases and the concentration of acetate ions increases. However, the equilibrium represented by Equation (16.80) still is valid, and the hydronium ion concentration may be represented by the following expression:

$$[H_3O(aq)^+] = \frac{K_a[CH_3COOH(aq)]}{[CH_3COO(aq)^-]} \qquad (16.81)$$

According to Equation (16.79) each mmole of NaOH added consumes a mmole of acetic acid to form a mmole of acetate ions. During the titration most of the acetate ion concentration is due to the acetate ion produced by the reaction of acetic acid with NaOH. As the acetate ion concentration from this source increases, it suppresses the acetic acid ionization. Thus, in addition, it appears safe to assume that only a very small percentage of acetic acid present will be ionized.

The original titree, 50.0 ml of 0.100 *M* acetic acid, contains 5.00 mmoles CH_3COOH. If 10.0 ml of 0.100 *M* NaOH is added, this is equivalent to the addition of 1.00 mmole of NaOH, which consumes 1.00 mmole CH_3COOH to produce 1.00 mmole of CH_3COO^-. At the same time, the solution volume increases to 60.0 ml. Thus there are 4.00 mmoles CH_3COOH in 60.0 ml of solution and the molarity of CH_3COOH is $4.00/60.0 = 6.7 \times 10^{-2}\ M$. Similarly there is 1.00 mmole of CH_3COO^- in the 60.0 ml of solution, and the molarity of acetate ion is $1.00/60.0 = 1.7 \times 10^{-2}\ M$. These concentrations may be inserted into Equation (16.81) to compute the hydronium ion molarity, which then may be used to compute the pH.

$$[H_3O(aq)^+] = \frac{(1.7 \times 10^{-5})(6.7 \times 10^{-2})}{1.7 \times 10^{-2}} = 6.7 \times 10^{-5}\ M$$

$$pH = -\log(6.7 \times 10^{-5}) = 4.2$$

This approach may be used to compute pH values during the titration prior to the equivalence point. At the equivalence point, when 50.0 ml of 0.100 *M* NaOH has been added, 5.00 mmoles of NaOH have been added and these have reacted with the 5.00 mmoles of CH_3COOH to produce a solution containing 5.00 mmoles of sodium acetate, CH_3COONa. Since the total volume is 100.0 ml at this time, the concentration of CH_3COONa is 5.00/100.0 = 0.0500 *M*. Thus at the equivalence point, the pH is that of a 0.0500 *M* solution of CH_3COONa. This pH is greater than 7.0 since acetate ions are basic and sodium ions are neutral.

The problem of finding the pH at the equivalence point reduces to one of finding the pH of a solution of a single base. This type of problem has been solved before. Employing the usual approximations, we may compute the pH in the following manner:

$$CH_3COO(aq)^- + H_2O = CH_3COOH(aq) + OH(aq)^- \tag{16.82}$$

$$K_b = \frac{[CH_3COOH(aq)][OH(aq)^-]}{[CH_3COO(aq)^-]} = 5.9 \times 10^{-10} = \frac{[OH(aq)^-]^2}{5.0 \times 10^{-2}}$$

$$[OH(aq)^-] = \sqrt{30 \times 10^{-12}} = 5.5 \times 10^{-6}\ M$$

$$pOH = -\log(5.5 \times 10^{-6}) = 5.3$$

$$pH = 14.0 - pOH = 14.0 - 5.3 = 8.7$$

When more titrant is added beyond the equivalence point, the hydroxide ions from the NaOH suppress the basic reaction of the acetate ions, and the added NaOH is the main source of $OH(aq)^-$. Thus if 5.0 ml of NaOH titrant is added beyond the equivalence point, 0.50 mmole of NaOH is added and the total solution volume is 105.0 ml. Thus the hydroxide ion concentration is 0.50/105 = 4.8×10^{-3} *M*. The pOH is 2.3 and the pH is 11.7.

Table 16.3 shows some data for this titration; and these data are used to plot the titration curve shown in Fig. 16.4. The general shape of this curve is the same as that for the titration of a strong acid with a strong base. However, it differs in several respects. First, the rise in pH near the equivalence point is not as great as that for a strong acid with a strong base. Second, the equivalence point does not occur at pH 7.0, but at a more basic pH, since the product of the titration is a base. Third, the pH at the beginning of the titration is higher and the initial addition of titrant leads to a larger increase in pH, which then levels off as more of the conjugate base of the acid is produced and the titree solution becomes buffered by this mixture of a weak acid and its conjugate base.

If we consider the case in which the base is used as the titree and the acid as the titrant, the initial pH is high for the basic titree and the pH decreases as the acidic titrant is added. However, the equivalence-point pH still is greater than 7.0 if the base is strong and the acid is weak. However, if the titration is carried out between a weak base and a strong acid, the equivalence-point pH is less than 7.0, due to the reaction of the conjugate acid of the weak base. For example, consider the titration of aqueous ammonia with hydrochloric acid as represented in Equation (16.83).

$$NH_3(aq) + H_3O(aq)^+Cl(aq)^- = NH_4(aq)^+Cl(aq)^- + H_2O \tag{16.83}$$

TABLE 16.3

Titration of 50.0 ml of 0.100 M CH_3COOH with 0.100 M NaOH

TOTAL ml 0.100 *M* NaOH ADDED	TOTAL SOLN VOLUME (ml)	mmoles $CH_3COOH(aq)$	$[CH_3COOH(aq)]$	$[CH_3COO(aq)^-]$	$[H_3O(aq)^+]$	$[OH(aq)^-]$	pH
0.0	50.0	5.00	1.0×10^{-1}		1.3×10^{-3}		2.9
10.0	60.0	4.00	6.7×10^{-2}	1.7×10^{-2}	6.7×10^{-5}		4.2
20.0	70.0	3.00	4.3×10^{-2}	2.9×10^{-2}	2.5×10^{-5}		4.6
30.0	80.0	2.00	2.5×10^{-2}	3.7×10^{-2}	1.1×10^{-5}		5.0
40.0	90.0	1.00	1.1×10^{-2}	4.4×10^{-2}	4.2×10^{-6}		5.4
45.0	95.0	0.50	5.3×10^{-3}	4.7×10^{-2}	1.9×10^{-6}		5.7
50.0	100.0			5.0×10^{-2}		5.5×10^{-6}	8.7
55.0	105.0					4.8×10^{-3}	11.7
60.0	110.0					9.1×10^{-3}	12.0

In this case the product of the titration is ammonium chloride, which is acidic due to the acidic ammonium ions that are the conjugate acids of the weakly basic ammonia molecules. The pH at the equivalence point of this titration is less than 7.0.

Since the sharpness of the rise in pH near the equivalence point is less for a strong titrant and weak titree than it is for a strong titrant and strong titree, it might be expected that the use of a very weak titree might lead to an even poorer definition of the pH change at the equivalence point. In general, this is true, and very weak acids cannot be titrated very successfully with strong bases, especially using color indicators. The successful use of color indicators depends upon their passing through the color-change range upon the addition of a very small volume of titrant, say about one drop of solution. Thus the change in color is very sharp and often dramatic. However, if the pH changes rather slowly with addition of titrant, it is very difficult to decide which slight change in color corresponds to the equivalence point. As a rule of thumb, acids with K_a values less than 10^{-7} usually titrate rather poorly with color indicators.

As an example, the titration curve for ammonium ion, $K_a = 5.9 \times 10^{-10}$, with NaOH is given in Fig. 16.5. The lack of a sharp vertical rise in the pH at the equivalence point indicates the difficulty to be expected in carrying out this titration in the laboratory.

What has been said here for the titration of very weak acids with strong bases applies equally well to the titration of very weak bases with strong acids.

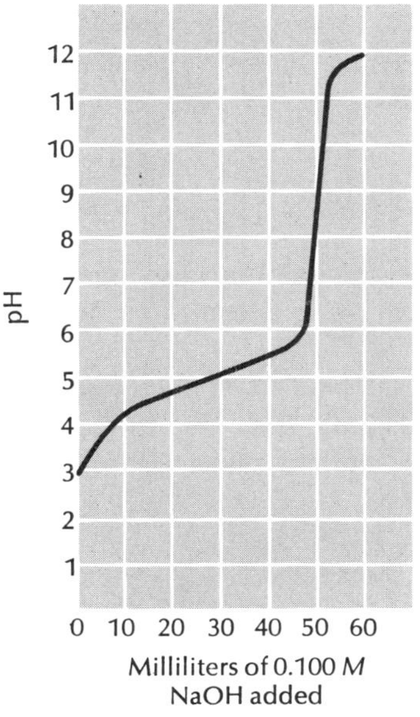

Fig. 16.4. Titration of 50.0 ml of 0.100 *M* CH_3COOH with 0.100 *M* NaOH

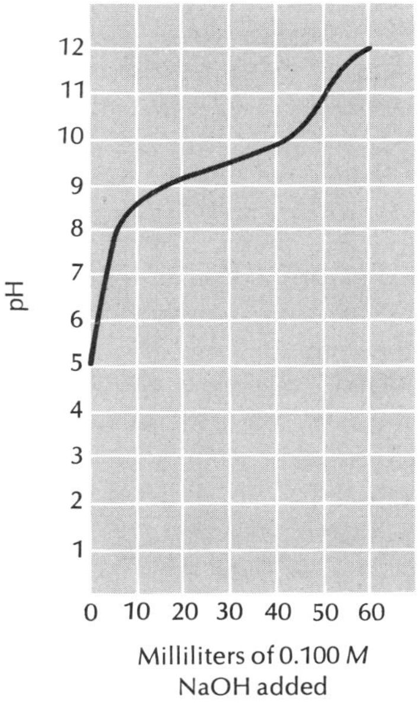

Fig. 16.5. Titration of 50.0 ml of 0.100 *M* NH_4Cl with 0.100 *M* NaOH

Addition of a Weak Titrant to a Weak Titree

From the previous discussion of the difficulties encountered in titrating a weak acid or base with a strong base or acid, it is not surprising that titrations involving both weak acids and weak bases have poorly defined end-points, since the pH rise near the equivalence point does not occur very sharply. A sample titration curve is given in Fig. 16.6 for the titration of acetic acid with ammonia.

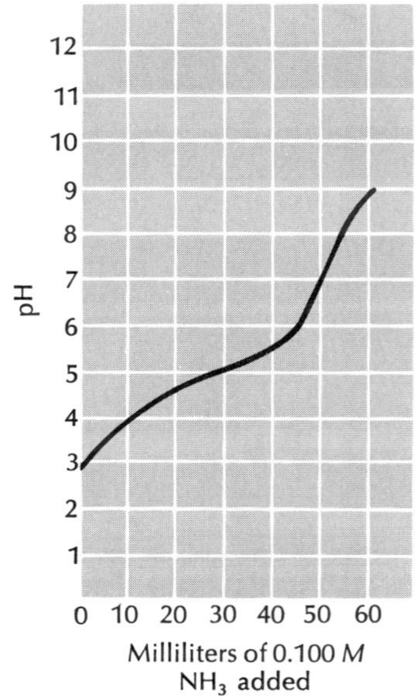

Fig. 16.6. Titration of 50.0 ml of 0.100 *M* CH_3COOH with 0.100 *M* NH_3

Titrations of Polyprotic Species

Polyprotic species with widely different successive ionization constants gain or lose protons stepwise. In such cases it is possible to detect more than one equivalence point.

For example, consider 50.0 ml of 0.100 *M* Na_2CO_3 as the titree to be titrated with 0.100 *M* HCl. At the outset the titree is quite basic due to the strongly basic carbonate. Carbonate is diprotic and in water undergoes the following two reactions:

$$CO_3(aq)^{--} + H_2O = HCO_3(aq)^- + OH(aq)^- \tag{16.84}$$

$$HCO_3(aq)^- + H_2O = CO_2(H_2O)(aq) + OH(aq)^- \tag{16.85}$$

Since the first ionization constant is much larger than the second, the main source of $OH(aq)^-$ ions is reaction (16.84) and the concentrations of hydroxide and hydrogen carbonate ions are approximately equal. In addition, assuming that very little carbonate actually reacts, it is possible to approximate the pH of the initial titree using K_{b1}.

$$K_{b1} = \frac{[HCO_3(aq)^-][OH(aq)^-]}{[CO_3(aq)^{--}]} = 2.1 \times 10^{-4} = \frac{[OH(aq)^-]^2}{1.0 \times 10^{-1}}$$

$$[OH(aq)^-] = \sqrt{21 \times 10^{-6}} = 4.6 \times 10^{-3}\ M$$

$$pOH = -\log(4.6 \times 10^{-3}) = 2.3$$

$$pH = 14.0 - pOH = 14.0 - 2.3 = 11.7$$

When 50.0 ml of 0.100 M HCl is added, this is just enough HCl to convert all of the carbonate into hydrogen carbonate, and this is the first equivalence point at which the titree is a solution of sodium hydrogen carbonate in water. At this point the 5.00 mmoles of added HCl have reacted with the 5.00 mmoles of carbonate ions to make 5.00 mmoles of hydrogen carbonate ions. The solution volume is 100.0 ml, so the solution is 0.050 M in $HCO_3(aq)^-$. Using Equation (16.75) developed for amphiprotic species such as $HCO_3(aq)^-$, it is possible to show that the pH is 8.9.

Beyond this, as more titrant is added, the hydrogen carbonate ions are converted to hydrated carbon dioxide, often called carbonic acid. When another 50.0 ml of titrant is added, all of the hydrogen carbonate ions are converted to carbonic acid. At this point the solution consists of 5.00 mmoles of carbonic acid in 150.0 ml of solution and the carbonic acid concentration is $5.00/150 = 3.3 \times 10^{-2}\ M$. This carbonic acid is a weak acid as shown by its K_a value of 4.4×10^{-7}. Using this value and the molarity, the pH can be shown to be 3.9 at this second equivalence point. The titration curve is shown in Fig. 16.7. The two breaks in the curve show the two end-points which can be detected by means of two different indicators, for example.

Polyprotic acids behave in a similar way. For example, phosphoric acid, H_3PO_4

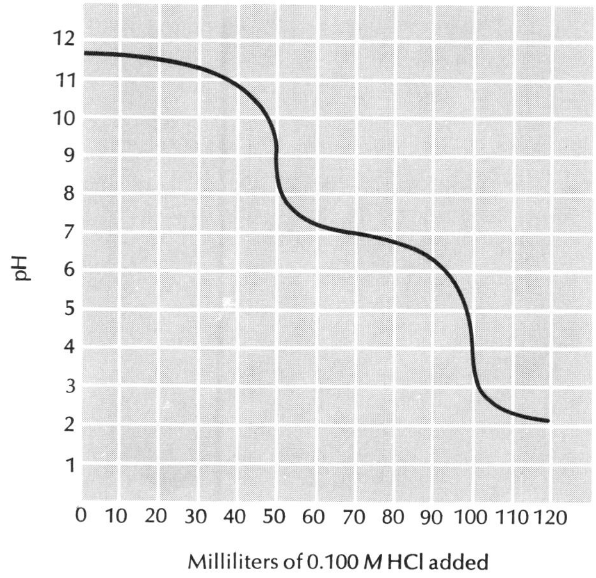

Fig. 16.7. Titration of 50.0 ml of 0.100 M Na_2CO_3 with 0.100 M HCl

theoretically is a triprotic acid. However, its third ionization constant is so small that the removal of the third hydrogen is very incomplete in aqueous solutions. Thus, for practical purposes, phosphoric acid is a diprotic acid in titrations with strong bases in water. Thus its titration curve shows two breaks.

Titration Calculations

The ways in which titration data may be used will be illustrated through several typical problems.

SAMPLE PROBLEM 1. If 32.0 ml of aqueous HF requires 24.3 ml of 0.149 *M* NaOH for titration, what is the molarity of the HF solution?

The chemical equation for the titration reaction is as follows:

$$HF(aq) + OH(aq)^- = F(aq)^- + H_2O \qquad (16.86)$$

Thus each mmole of NaOH added consumes 1.00 mmole of HF, converting it to 1.00 mmole of $F(aq)^-$. Since the question asks for the molarity of HF, it may be restated in the following way: How many mmoles of HF are there in 1.00 ml of the solution? Starting with 1.00 ml of solution, the number of mmoles present can be found in the following way:

$$1.00 \text{ ml HF} \times \frac{24.3 \text{ ml NaOH}}{32.0 \text{ ml HF}} \times \frac{0.149 \text{ mmole NaOH}}{1.00 \text{ ml NaOH}} \times \frac{1.00 \text{ mmole HF}}{1.00 \text{ mmole NaOH}} = 0.113\ M$$

In this solution, the first factor is derived from the data given in the problem, namely that 32.0 ml of the HF solution is chemically equivalent to 24.3 ml of the NaOH titrant. This provides a conversion from HF to NaOH, so that the molarity of the NaOH may be used in the second factor to put the units in terms of mmoles of NaOH actually added. Although the final factor may appear trivial, it is important since it establishes the molar relationship between HF and NaOH as described by Equation (16.86).

This problem illustrates a technique for producing *standard solutions*, that is, solutions of known concentration. The solution to be standardized is titrated against another standard solution.

SAMPLE PROBLEM 2. A 0.500-g sample of impure NaCN is dissolved in water and titrated with 35.2 ml of standard 0.050 *M* HCl. What is the percent NaCN in the sample?

Since the question is stated in percent terms, it may be restated as follows: How many grams of NaCN are there in 100 g of sample? Beginning with that 100 g of sample, the question may be answered as follows:

$$100 \text{ g samp} \times \frac{35.2 \text{ ml HCl}}{0.500 \text{ g samp}} \times \frac{0.050 \text{ mmole HCl}}{1.00 \text{ ml HCl}} \times \frac{1 \text{ mmole NaCN}}{1 \text{ mmole HCl}} \times \frac{0.049 \text{ g NaCN}}{1 \text{ mmole NaCN}} = 17.2 \text{ percent NaCN}$$

The first factor relates sample mass to volume of titrant. This relationship was established in the problem statement. The second factor is the molarity of the titrant as given in the problem statement. This puts the units on a molar basis. The third factor relates the numbers of moles of the two reactants through the following chemical equation:

$$CN(aq)^- + H_3O(aq)^+ = HCN(aq) + H_2O \quad (16.87)$$

The final factor relates mass to moles using the molecular weight of NaCN at 49 g/mole, or 0.049 g/mmole.

The analytical chemist often uses this technique to determine the composition of materials.

SAMPLE PROBLEM 3. A 0.620-g sample of impure Na_3AsO_3 is dissolved in water and titrated as a triprotic base with 47.8 ml of 0.0750 M H_2SO_4. What is the percent arsenic in the sample?

Potentially sulfuric acid is a diprotic acid. In fact it is a sufficiently strong acid, even in its second ionization, that in titration reactions in aqueous solutions it always serves as a diprotic acid. The chemical reaction for the titration is represented by Equation (16.88).

$$2Na(aq)_3{}^+AsO_3(aq)^{3-} + 3(H_3O(aq)^+)_2SO_4(aq)^{--} = 2H_3AsO_3(aq) + 3Na(aq)_2{}^+SO_4(aq)^{--} \quad (16.88)$$

To ask for the percent of arsenic in the sample is to ask for the number of grams of arsenic in 100 g of sample.

$$100 \text{ g samp} \times \frac{47.8 \text{ ml } H_2SO_4}{0.620 \text{ g samp}} \times \frac{0.0750 \text{ mmole } H_2SO_4}{1.00 \text{ ml } H_2SO_4} \times \frac{2 \text{ mmole As}}{3 \text{ mmole } H_2SO_4} \times \frac{0.0749 \text{ g As}}{1 \text{ mmole As}} = 32.4 \text{ percent arsenic}$$

The first factor relates sample mass to titrant volume as given in the problem statement. The second factor is the molarity of the titrant which puts the units on a mole basis. The third factor relates the number of mmoles of sulfuric acid used to the number of mmoles of arsenic atoms present. This is determined directly from the chemical equation. The final factor relates the moles of arsenic to grams of arsenic through the atomic weight.

It is common analytical practice to find the percent of some element in a sample although the element is not present in the elemental form. From formulas and equations it is possible to determine the relationship between the number of moles of atoms of the element and the number of moles of some reagent used in the reaction.

SAMPLE PROBLEM 4. In an organic synthesis, a mixture of monoprotic acids is produced. One of these is isolated and purified. Then 500 mg of the acid is dissolved in enough water to make 100 ml of solution. Titration of 50.0 ml of this solution requires 33.7 ml of 0.100 *M* NaOH. What is the molecular weight of this isolated acid?

The monoprotic acid may be represented schematically as HA. Since it is monoprotic, and the sodium hydroxide titrant is monoprotic as well, each mmole of NaOH added reacts with 1.00 mmole of HA. Since the question asks for the molecular weight, it may be restated as follows: How many milligrams of HA are there in 1.00 mmole of HA? The problem may then be solved.

$$1.00 \text{ mmole HA} \times \frac{1 \text{ mmole NaOH}}{1 \text{ mmole HA}} \times \frac{1 \text{ ml NaOH}}{0.100 \text{ mmole NaOH}} \times \frac{50.0 \text{ ml HA}}{33.7 \text{ ml NaOH}}$$

$$\times \frac{500 \text{ mg HA}}{100.0 \text{ ml HA}} = 74 \text{ mg/mmole}$$

The first factor is the molar relationship between the two reactants and puts the units on a molar basis. The second factor is the reciprocal of the molarity, which puts the units in terms of the volume of the titrant. The third factor relates the volume of titrant to the volume of titree actually used. The final factor relates the total volume of solution of HA actually prepared to the total number of mg of solid acid added.

This is another common application of titration data, to determine molecular weights of unknown substances.

SUMMARY

A knowledge of acid-base constant values is a powerful tool in predicting the concentrations of species in solutions of acids and bases. The stoichiometric relationships derived from the chemical equations describing the reactions may be used to interrelate the equilibrium concentrations of these species in solution.

The safest procedure in dealing with such solutions is to describe them as exactly as possible, including all species present in the solution and taking into account all of the sources for each of the species. Then, on the basis of the specific system under scrutiny, the relationships may be simplified by making appropriate approximations. However, when final solutions to the problems are obtained, the approximations should be checked to verify their validity.

QUESTIONS AND PROBLEMS

16.1 Consider an aqueous solution of HCl. What happens to the contribution of the water ionization to the equilibrium hydronium ion concentration as the solution is diluted by the addition of water?

16.2 What are the concentrations of all species in 0.100 *M* HNO_3? What is the pOH of the solution?

16.3 What is the approximate contribution of water ionization to the equilibrium hydronium ion concentration in 10^{-6} *M* $HClO_4$?

16.4 Criticize the following statement: The pH of 10^{-10} *M* HI is 10.

16.5 For dilute aqueous solutions of each of the following, indicate the species present in the solution and all of the sources for each of these species.

a) NaOH
b) HI
c) NH_2SO_3H
d) HCN
e) NH_3
f) CH_3COOH
g) CH_3COONa
h) NaF
i) H_3PO_4
j) Na_2HAsO_4

16.6 What is the contribution of the water ionization to the equilibrium hydroxide ion concentration in solutions with the following pH values?

a) 10.0 b) 12.78 c) 8.42

16.7 What is the contribution of the water ionization to the equilibrium hydronium ion concentration in solutions with the following pH values?

a) 2.00 b) 3.75 c) 5.60

16.8 The pH of a 0.100 *M* aqueous solution of monoprotic acid, HA, is 5.2.

a) Calculate the value of K_a for HA.
b) What is the pH of a 0.025 *M* solution of this acid?

16.9 When 0.100 *M* solutions are prepared for weak acids HA and HB, the pH of the solution of HA is greater than the pH of the solution of HB. Which of the acids has the higher value for K_a?

16.10 The pH of a 0.050 *M* solution of base B is 9.90. Calculate the value of K_b for base B.

16.11 Criticize the following statement: Since the ionization constant of weak acid HA is constant, the percent of ionization of this acid in water is the same regardless of the concentration of the solution.

16.12 Compute the percent ionization of CH_3COOH in each of the following solutions of the indicated concentration:

a) 0.200 *M* b) 0.140 *M* c) 0.0050 *M*

16.13 Propionic acid was found to be 1.14 percent ionized in 0.100 *M* aqueous solution.

a) Calculate the value for K_a for propionic acid and compare with the value given in Table 15.6.
b) What is the percent ionization of propionic acid in 0.0050 *M* solution?
c) What is the pH of 0.0050 *M* propionic acid solution?

16.14 The base constant for aqueous ammonia is 1.75×10^{-5}.

a) Calculate the percent completion of the reaction of $NH_3(aq)$ with water in an 0.080 *M* solution.
b) Calculate the hydroxide ion concentration in this 0.080 *M* solution.
c) What is the pH of this solution?

16.15 Consider the following generalized ionization for base B.

$$B(aq) + H_2O = HB(aq)^+ + OH(aq)^-$$

Derive an equation similar to Equation (16.19) to relate $[OH(aq)^-]$, M_B, K_b, and K_w.

16.16 Referring to Table 16.1, suggest why some values for [HA(aq)] are missing. For example, there is no value for [HA(aq)] when $K_a = 10^{-13}$ and $[H_3O(aq)^+] = 10^{-1}$ *M*.

16.17 Prepare data to add to Table 16.1 for $K_a = 10^2$, 10^6, and 10^{10} and plot these in the same manner as Fig. 16.1.

16.18 The values in the first two columns of Table 16.1 are quite similar. Why?

16.19 Calculate the pH and the concentrations for all species for the following aqueous solutions, using K_a and K_b values from Table 15.1.

a) 0.020 M NH_4Cl
b) 0.15 M $Ni(NO_3)_2$
c) 0.0060 M KCN
d) 0.0030 M CH_3COONa
e) 30.0 ml of 0.100 M CH_3COOH added to 30.0 ml of 0.100 M KOH

16.20 For which of the following solutions will it be impossible to use the standard approximations in computing $[H_3O(aq)^+]$?
a) 0.0020 M HF
b) 0.360 M HNO_2
c) 0.00750 M HCN
d) 0.0410 M H_2O_2
e) 0.160 M $CoCl_3$
f) 0.250 M $(NH_4)_2SO_4$

16.21 For each of the following solutions compute the pH and give the concentrations of all species in solution.
a) 0.0050 M H_3AsO_4
b) 0.040 M Na_2SO_3

16.22 Compute the concentrations of all species in each of the following:
a) A solution which is 0.200 M in HF and 0.00320 M in HBr.
b) A solution which is 0.065 M in CH_3COOH and 0.00051 M in HI.
c) A solution which is 0.100 M in $NaHO_2$ and 0.0400 M in NaOH.

16.23 Determine the pH of each of the following solutions:
a) A 0.0500 M solution of NH_3 to which sufficient NH_4Cl has been added to make the total equilibrium concentration of $NH_4(aq)^+$ equal to 0.100 M.
b) A liter of solution in which are dissolved 0.080 mole CH_3COOH and 0.100 mole CH_3COONa.
c) A solution prepared by mixing 10.0 ml of 0.100 M CH_3COOH with 20.0 ml of 0.100 M CH_3COONa.

16.24 Explain the operation of a buffer, showing how it is possible for a buffer to resist pH changes.

16.25 A buffer solution is to be prepared having a pH of 9.5. Calculate the concentration ratio required to give this pH using a) HCN and NaCN; b) NH_4Cl and NH_3.

16.26 In Equation (16.77), the product, K_aK_b, for the K_a and K_b used there does not equal 10^{-14}. Why?

16.27 Solution A, with a pH of 2.0, is made by adding 0.010 mole of monoprotic acid, HA, to pure water to make 1000 ml of solution. Solution B, with $[H_3O(aq)^+] = 1.0 \times 10^{-2}$ M, is made by adding 1.0 mole of the monoprotic acid, HB, to pure water to make 10.0 l of solution.
a) Which acid is the stronger?
b) Which acid has the stronger conjugate base?

16.28 For each of the following solutions, compute the pH and the concentrations of each of the species present in the solution.
a) 3.00 g NaH_2PO_4 dissolved per 100 ml of solution
b) 0.040 M $NaHSO_3$

16.29 In the titration of 50.0 ml of 0.100 M HCl with 0.100 M NaOH, the initial pH is 1.0 and the equivalence-point pH is 7.0. Also 50.0 ml of titrant is required to reach the equivalence point. For each of the following total volumes of titrant added, what percent of titrant required to reach the equivalence point has been added, and what percent of the pH range from the initial pH to the equivalence point pH has been covered?
a) 20.0 ml
b) 40.0 ml
c) 45.0 ml
d) 49.0 ml

16.30 For each of the following titrations, write the chemical equation for the titration reaction; compute the pH of the titree before the titration begins and at the equivalence point; indicate the most appropriate pH range for the indicator color change. In cases of polyprotic acids and bases assume that the reactions involve the maximum number of transferable protons.

TITREE	TITRANT
a) 50.0 ml of 0.100 M HN_3	0.100 M KOH
b) 50.0 ml of 0.100 M HN_3	0.0100 M KOH
c) 15.0 ml of 0.150 M NH_3	0.250 M HCl
d) 35.0 ml of 0.0200 M $Ca(OH)_2$	0.0500 M HNO_3
e) 75.0 ml of 0.20 M Na_3PO_4	0.150 M H_2SO_4

16.31 Verify the titration data in Tables 16.2 and 16.3.

16.32 Prepare titration curves for the following:
a) Titration of 30.0 ml of 0.250 M HN_3 with 0.200 M NaOH.
b) Titration of 40.0 ml of 0.100 M NaOH with 0.100 M H_2SO_4.
c) Titration of 50.0 ml of 0.150 M NaH_2BO_3 with 0.150 M NH_2SO_3H.

16.33 The following reaction may be used as a titration for the determination of phosphate in a sample of NaH_2PO_4:

$$NaH_2PO_4 + NaOH = Na_2HPO_4 + H_2O$$

If 27.0 ml of 0.100 M NaOH is required to titrate an 0.800-g sample, what is the percent phosphate, phosphorus, and NaH_2PO_4 in the sample?

16.34 A 1.000-g sample of pure oxalic acid-2-hydrate, $H_2C_2O_4 \cdot 2H_2O$, acts as a diprotic acid when titrated with 44.50 ml of aqueous sodium hydroxide. What is the molarity of the aqueous sodium hydroxide?

16.35 Vinegar is a dilute acetic acid solution. A 15.64-ml sample of vinegar with a density of 1.06 g/ml requires 24.80 ml of 0.469 M KOH for titration. What is the percent acetic acid in the vinegar?

16.36 If 32.0 ml of aqueous HCl reacts with 0.480 g of pure Na_2CO_3 and Na_2CO_3 acts as a diprotic base, compute
a) The molarity of the HCl solution.
b) The concentration of the HCl solution as grams of HCl per ml of solution.

16.37 A 22.0-ml sample of aqueous H_2SO_4 was titrated with 24.2 ml of aqueous NaOH. The sulfuric acid acted as a diprotic acid. Upon evaporation, 0.550 g of dry Na_2SO_4 was found. Calculate the molarity of each solution.

16.38 If 300 mg of a monoprotic acid is dissolved in pure water to make 100 ml of solution A, and 50.0 mmoles of NaOH is used to make 500 ml of solution B, and 33.3 ml of solution B is needed to titrate all of solution A, what is the molecular weight of the acid?

16.39 A 538-mg sample of a pure alkali metal hydroxide was dissolved in 50.0 ml of aqueous solution. If 35.0 ml of 0.150 M HCl was needed to titrate this solution, what alkali metal hydroxide was in the sample?

16.40 Suggest the general shapes of the titration curves for each of the following:
a) Titration of H_3PO_4 with NaOH.
b) Titration of Na_3PO_4 with HCl.
c) Titration of H_2S with KOH.

17

Oxidation and Reduction Reactions

INTRODUCTION Although the term "oxidation" originally implied a combination with oxygen, it has come to have a much broader meaning today. One obvious change in oxygen when it undergoes combination is the change in its oxidation number. Like any other element, elemental oxygen has an oxidation number of zero, but in combination it exhibits oxidation states of 1− and 2−.

Since the oxidation number concept was introduced as an electron bookkeeping device, and since electron absorption and emission is not found among nonradioactive processes, it is not surprising that the process of oxidation number reduction also is accompanied by the process of oxidation number increase, usually in some other species. It is these reactions which involve both an increase and a decrease in oxidation number which now are called *oxidation-reduction reactions* or *reduction-oxidation reactions* or in more abbreviated form, *redox reactions.*

In reactions involving combination with oxygen, it is the oxygen which has its oxidation number reduced. Thus the oxygen is said to be reduced. Concurrently, the oxidation number of some other element in another species is increased and it is this element which is said to be oxidized. The oxidation of this substance is caused by the oxygen which then is called the *oxidizing agent* and this substance is viewed as the cause of the reduction of the oxygen and thus is called the *reducing agent.* Thus, in general terms, the oxidizing agent is the substance reduced and the reducing agent is the substance oxidized.

In this chapter we shall discuss various aspects of redox reactions. We shall give most of our attention to a very important subset of redox reactions, namely the electron transfer reactions. As you study this chapter you may find it useful to compare and contrast redox reactions with acid-base reactions.

EQUATIONS OF REDOX REACTIONS

To write an equation for any reaction it is necessary to know the reactants and products. Normally this requires experimentally determined information or a background of experience upon which to draw. Thus in writing equations for some typical redox reactions in this section we shall provide typical experimentally verifiable information as a starting point.

For example, when elemental phosphorus is placed in aqueous nitric acid, phosphoric acid and nitrogen(II) oxide result. Clearly this is an oxidation-reduction reaction. The oxidation number of phosphorus in the elemental form is zero; in phosphoric acid, H_3PO_4, it is 5+. Clearly the phosphorus has been oxidized. Similarly the oxidation number for nitrogen in nitric acid, HNO_3, is 5+, whereas it is 2+ in NO. Thus the nitrogen in nitric acid has been reduced or, in common parlance, the nitric acid has been reduced. In this reaction nitric acid is the oxidizing agent and phosphorus is the reducing agent.

To write the equation for this reaction we begin with the experimental information placed in equation form, although we recognize that the statement is not yet an equation.

$$P + HNO_3 = H_3PO_4 + NO \tag{17.1}$$

Note that phosphorus has been represented by the symbol P instead of the formula P_4. Since phosphorus may exist in several modifications, not all of which consist of P_4 molecules, it has become common practice to represent it by its symbol rather than by a formula. In addition, nitric acid is represented by a molecular formula instead of by its ions, although as a strong acid the aqueous solution consists of $H_3O(aq)^+$ and $NO_3(aq)^-$ ions. For the time being we shall write this equation in the molecular form; later we shall discuss how it might be written in ionic form.

The second step in completing the equation for the reaction is to indicate the oxidation numbers for those elements which undergo a change and also to indicate by how much each of these oxidation numbers has changed.

$$\overbrace{\overset{0}{P} + H\underbrace{\overset{5+}{N}O_3 = H_3\overset{5+}{P}}_{\text{decreased } 3}O_4 + \overset{2+}{N}O}^{\text{increased } 5} \tag{17.2}$$

Since electrons are not transferred between the surroundings and the reaction, the increase in oxidation number should equal the decrease in oxidation number.

The increase and decrease in oxidation number may be equalized by the addition of appropriate coefficients to the species in the equation. The choice of coefficients is similar to the problem of finding a least common denominator. Thus if 3 P atoms are represented and 5 N atoms, both the increase and decrease in oxidation number will be 15.

$$\overset{\text{increased } 3\times 5}{\overbrace{\overset{0}{3P} + 5H\overset{5+}{N}O_3 = 3H_3\overset{5+}{P}O_4}} \quad \text{(with } 5\overset{2+}{N}O\text{)}$$

$$\overset{0}{3P} + 5H\overset{5+}{N}O_3 = 3H_3\overset{5+}{P}O_4 + 5\overset{2+}{N}O \qquad (17.3)$$

(N: decreased 5×3)

At this point the oxidation number change has been balanced, and it is clear that 3 moles of phosphorus atoms reacts with 5 moles of nitric acid molecules to produce 3 moles of phosphoric acid molecules and 5 moles of nitrogen(II) oxide molecules. However atomic balance has not been achieved, since 15 moles of O atoms are shown on the left while there are 17 moles of O atoms on the right. Similarly, there are only 5 moles of H atoms on the left while there are 9 moles of H atoms on the right. Clearly there is a deficiency of 4 moles of H atoms and 2 moles of O atoms on the left, equivalent to 2 moles of water molecules. Since the reaction is carried out in aqueous solution, it is quite possible that the solvent molecules participate in the reaction. Thus the equation is completed by the addition of "$2H_2O$" on the left.

$$3P + 5HNO_3 + 2H_2O = 3H_3PO_4 + 5NO \qquad (17.4)$$

As indicated earlier, Equation (17.4) is a molecular equation for this reaction. However, nitric acid is ionized in water and thus might be written more correctly as $H_3O(aq)^+ + NO_3(aq)^-$. Of course, phosphoric acid is slightly ionized in water, but its ionization is suppressed by the presence of the nitric acid, so that it is best represented in molecular form. The ionic equation for this reaction is shown as Equation (17.5).

$$3P(s) + 5H_3O(aq)^+ + 5NO_3(aq)^- = 3H_3PO_4(aq) + 5NO(g) + 3H_2O \qquad (17.5)$$

Note that water appears among the products in Equation (17.5) but among the reactants in Equation (17.4). The difference is that 5 moles of water molecules are represented in the 5 moles of hydronium ions in Equation (17.5), and this is 3 moles more than the 2 moles of water molecules represented in Equation (17.4).

As a final example, consider the reaction of potassium dichromate and hydrogen sulfide in dilute aqueous perchloric acid solution to produce chromium(III) perchlorate, sulfur, and potassium perchlorate. The skeleton of the molecular equation is represented as follows:

(S: increased 2)

$$K_2\overset{6+}{Cr_2}O_7 + H_2\overset{2-}{S} + HClO_4 = \overset{3+}{Cr}(ClO_4)_3 + \overset{0}{S} + KClO_4 \qquad (17.6)$$

(Cr: decreased 3)

As indicated, the chromium is reduced from the 6+ state to the 3+ state and sulfur is oxidized from the 2− state to the 0 state.

Note that 2 chromiums are shown on the left; together these two will decrease in oxidation number by 6 units. This requires a coefficient of 2 for $Cr(ClO_4)_3$ on the right. A corresponding increase of 6 in the oxidation number of sulfur may be obtained by using the coefficient 3 for both H_2S and S.

increased 3 × 2

$$\overset{6+}{K_2Cr_2O_7} + 3\overset{2-}{H_2S} + HClO_4 = 2\overset{3+}{Cr}(ClO_4)_3 + 3\overset{0}{S} + KClO_4 \qquad (17.7)$$

decreased 2 × 3

This establishes the stoichiometric relationship between the oxidizing and reducing agents, namely that 1 mole of $K_2Cr_2O_7$ reacts with 3 moles of H_2S to produce 2 moles of $Cr(ClO_4)_3$ and 3 moles of S atoms. In addition, the 2 potassiums on the left require a coefficient of 2 for $KClO_4$ on the right. This indicates a total of 8 perchlorates on the right, which requires a coefficient of 8 for $HClO_4$ on the left.

$$K_2Cr_2O_7 + 3H_2S + 8HClO_4 = 2Cr(ClO_4)_3 + 3S + 2KClO_4 \qquad (17.8)$$

Superficially this appears to be a complete equation. However there are 39 oxygens and 14 hydrogens on the left and only 32 oxygens on the right. Thus the right side of this statement is deficient by 14 hydrogens and 7 oxygens, which is equivalent to $7H_2O$. The final, complete equation may be written as Equation (17.9).

$$K_2Cr_2O_7 + 3H_2S + 8HClO_4 = 2Cr(ClO_4)_3 + 3S + 2KClO_4 + 7H_2O \qquad (17.9)$$

In aqueous solution $K_2Cr_2O_7$, $HClO_4$, $Cr(ClO_4)_3$, and $KClO_4$ are completely ionized. If these are represented in the ionic form, Equation (17.9) may be rewritten:

$$2K(aq)^+ + Cr_2O_7(aq)^{--} + 3H_2S(aq) + 8H_3O(aq)^+ + 8ClO_4(aq)^- = 2Cr(aq)^{3+}$$
$$+ 6ClO_4(aq)^- + 3S(s) + 2K(aq)^+ + 2ClO_4(aq)^- + 7H_2O \qquad (17.10)$$

In this form it is clear that the potassium and perchlorate ions do not participate in this reaction. They are present at the beginning and the end. Consequently they may be removed from the equation which then takes on the following simplified form:

$$Cr_2O_7(aq)^{--} + 3H_2S(aq) + 8H_3O(aq)^+ = 2Cr(aq)^{3+} + 3S(s) + 15H_2O \qquad (17.11)$$

This form of the reaction representation is called an ionic equation, since it contains only those ions which participate in the reaction. Those substances not predominantly ionic are not written ionically, that is, the very weak acid, H_2S, water, and elemental sulfur. Note that although the practice is not demonstrated in this equation, insoluble ionic substances are written in molecular form in ionic equations.

Equation (17.11) is much more general than Equation (17.10). According to Equation (17.11), the oxidizing agent is dichromate ion, and it is not important whether it comes from sodium dichromate, potassium dichromate, calcium dichromate, or some other source. Similarly, it is not important what acid is used as a source of hydronium ions.

Now let us look at this same reaction from a slightly different point of view. Suppose that the experimental facts had been stated in this way: Acidified dichromate solution reacts with hydrogen sulfide to produce sulfur and chromium(III) ions. To write the equation we begin with the following skeleton:

increased 2

$$\overset{6+}{Cr_2}O_7(aq)^{--} + H_2\overset{2-}{S}(aq) = \overset{3+}{Cr}(aq)^{3+} + \overset{0}{S}(s) \qquad (17.12)$$

decreased 3

According to the same arguments used before to balance increase and decrease in oxidation number, this skeleton may be transformed into the following:

increased 3 × 2

$$\overset{6+}{Cr_2}O_7(aq)^{--} + 3H_2\overset{2-}{S}(aq) = 2\overset{3+}{Cr}(aq)^{3+} + 3\overset{0}{S}(s) \qquad (17.13)$$

decreased 2 × 3

If no electrons may be added from the surroundings or lost to the surroundings, the total of the ionic charges on one side of the equation must be the same as the total of the ionic charges on the other side. In other words, there must be a charge balance as well as an atomic balance. In the case of this skeletal equation, there are 2 negative charges on the left and 6 positive charges on the right. Since this reaction is carried out in an acidic aqueous medium, water molecules and hydronium ions are available in abundance, and it is quite possible that they may participate in the over-all reaction. Of these two species only the hydronium ion is charged. If we think of inserting hydronium ions into skeletal Equation (17.13), to bring about a charge balance, 8 hydronium ions need to be added to the left side, producing a net ionic charge of 6+ on the left to balance the 6+ ionic charge on the right.

$$Cr_2O_7(aq)^{--} + 3H_2S(aq) + 8H_3O(aq)^{+} = 2Cr(aq)^{3+} + 3S(s) \qquad (17.14)$$

Finally, there are 30 hydrogens and 15 oxygens on the left and none on the right, and this corresponds to 15 waters which need to be added to the right to bring about atomic balance.

$$Cr_2O_7(aq)^{--} + 3H_2S(aq) + 8H_3O(aq)^{+} = 2Cr(aq)^{3+} + 3S(s) + 15H_2O \qquad (17.15)$$

Equation (17.15) is identical to Equation (17.11).

This same general reaction also illustrates the effect of pH upon the reaction and the corresponding equation. For example, suppose that the reaction were carried out in basic solution instead of acidic solution. First, dichromate is converted to chromate in basic solution.

$$Cr_2O_7(aq)^{--} + 2OH(aq)^- = 2CrO_4(aq)^{--} + H_2O \qquad (17.16)$$

Second, in basic solution the hydrogen sulfide is converted to sulfide ions.

$$H_2S(aq) + 2OH(aq)^- = S(aq)^{--} + 2H_2O \qquad (17.17)$$

Third, chromium(III) hydroxide is insoluble and thus precipitates in a basic environment.

$$Cr(aq)^{3+} + 3OH(aq)^- = Cr(OH)_3(s) \qquad (17.18)$$

Thus in basic solution the reaction may be stated as follows: Chromate reacts with sulfide in basic aqueous solution to produce sulfur and chromium(III) hydroxide. The skeleton of the equation may be written as follows:

increased 2

$$\overset{6+}{Cr}O_4(aq)^{--} + \overset{2-}{S}(aq)^{--} = \overset{3+}{Cr}(OH)_3(s) + \overset{0}{S}(s) \qquad (17.19)$$

decreased 3

The increase and decrease in oxidation numbers may be balanced as before.

increased 3×2

$$2\overset{6+}{Cr}O_4(aq)^{--} + 3\overset{2-}{S}(aq)^{--} = 2\overset{3+}{Cr}(OH)_3(s) + 3\overset{0}{S}(s) \qquad (17.20)$$

decreased 2×3

Since the solution is basic, hydroxide ions and water molecules are abundant. To bring about charge balance, $10OH(aq)^-$ ions may be added to the right to balance the net ionic charge of 10− on the left.

$$2CrO_4(aq)^{--} + 3S(aq)^{--} = 2Cr(OH)_3(s) + 3S(s) + 10OH(aq)^- \qquad (17.21)$$

Since there are 8 oxygens on the left, 16 oxygens on the right, no hydrogens on the left, and 16 hydrogens on the right, 8 waters may be added to the left to bring about atomic balance:

$$2CrO_4(aq)^{--} + 3S(aq)^{--} + 8H_2O = 2Cr(OH)_3(s) + 3S(s) + 10OH(aq)^- \qquad (17.22)$$

This ionic equation represents the over-all reaction in basic solution.

ELECTRON TRANSFER REACTIONS

Voltaic and Electrolytic Cells

Figure 17.1 shows an experimental setup using a strip of zinc metal immersed in a solution of zinc sulfate and a strip of copper metal in a solution of copper(II) sulfate. The two solutions are connected by a *salt bridge*, a tube filled with a concentrated solution of an electrolyte. The two metal strips are connected by a wire.

When such a setup is prepared and observed for a while, the copper strip becomes noticeably larger by the deposition of more copper on its surface and the zinc strip becomes smaller. In addition, if the wire connecting the two metal strips is broken and the two ends are touched to the base of a small flashlight bulb, the

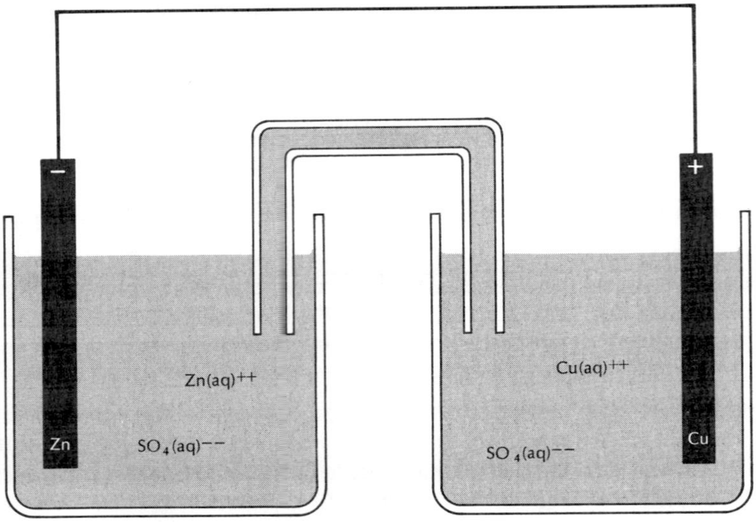

Fig. 17.1. Schematic diagram of a voltaic cell

bulb glows. Meters may be used also to show that there is a flow of electricity through the wire connecting the two metal strips. If the salt bridge is removed, the flow of electricity appears to stop almost immediately.

These observations may be interpreted with the aid of Fig. 17.2. According to this interpretation, the copper metal plus the copper(II) sulfate solution attracts electrons more strongly than the zinc metal plus the zinc sulfate solution. Thus

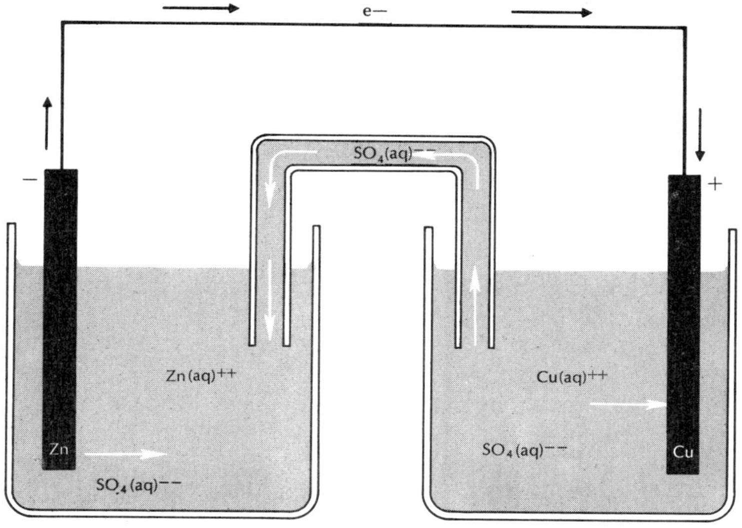

Fig. 17.2. Interpretation of the operation of a voltaic cell

electrons are attracted from the zinc metal through the wire to the copper metal. As the electrons leave the zinc metal, the zinc atoms are converted to zinc ions which enter the aqueous zinc sulfate solution. On the copper metal strip, the electrons are available to the copper(II) ions in the surrounding solution. These ions absorb the electrons and are converted to copper atoms on the strip of copper metal.

The two metal strips immersed in the solutions that serve as electrical conductors are called *electrodes.* Since the zinc atoms in the zinc electrode deposit electrons on that electrode while copper(II) ions remove electrons from the copper electrode, the zinc electrode is said to be negative with respect to the copper electrode. Thus there appears to be a spontaneous flow of negative electrons through the external circuit from the negative electrode to the positive electrode.

At the same time, positive zinc ions are added to the solution from the zinc electrode, so that momentarily there appears to be an excess of positive ionic charge in the zinc sulfate solution. However, as positive cations are converted to atoms at the copper electrode, the solution there appears to have a deficiency of cationic charge, or an excess of anionic charge, due to the excess sulfate ions. It is the purpose of the salt bridge to provide a means for maintaining electrical neutrality in these solutions by providing ions in a mobile medium. The net effect, electrostatically, is to allow sulfate ions from the copper(II) sulfate solution to migrate to the zinc sulfate solution. In practice, however, some other ions from the salt bridge solution instead of the sulfate ion from the copper(II) sulfate solution may be transferred into the zinc sulfate solution.

The chemical reaction occurring at the copper electrode is given by Equation (17.23).

$$Cu(aq)^{++} + 2e^- = Cu(s) \tag{17.23}$$

This reaction is a reduction, since the oxidation number of copper is reduced from 2+ to 0.

Similarly, the chemical reaction occurring at the zinc electrode is given by Equation (17.24).

$$Zn(s) = Zn(aq)^{++} + 2e^- \tag{17.24}$$

This reaction is an oxidation, since the oxidation number of zinc is increased from 0 to 2+.

Neither of these equations represents a complete chemical reaction, for one represents a reduction and the other an oxidation. The complete reaction consists of both processes. For this reason, these equations often are said to represent *half-reactions.* In addition, since they contain both ions and electrons, they are sometimes called *ion-electron equations.* The system described in Fig. 17.2, by which electrons are transferred through an external circuit between two chemical species, is called a *cell.* The two components of the cell are called *half-cells.* Thus Equations (17.23) and (17.24) represent the reactions that occur in the individual half-cells and often they are called *half-cell reaction equations.*

By convention, the electrode in the half-cell in which *reduction* occurs is called the *cathode* and the electrode in the half-cell in which *oxidation* occurs is called the *anode.* In the example discussed here, the strip of zinc metal is the anode and the copper electrode is the cathode.

Any cell that employs a spontaneous chemical reaction to generate an electrical current is called a *voltaic cell, galvanic cell,* or *primary cell;* the names are used interchangeably. In general, voltaic cells may be represented schematically as follows:

$$\text{anode}\left|\begin{matrix}\text{anodic}\\\text{electrolyte}\end{matrix}\right\|\begin{matrix}\text{cathodic}\\\text{electrolyte}\end{matrix}\left|\text{cathode}\right.$$

The half-cell containing the anode is represented first, with the composition of the anode separated from the composition of its surrounding electrolyte by a single vertical line. The anodic half-cell is separated from the cathodic half-cell by a double vertical line. Then the electrolyte composition is separated from the electrode composition by another single vertical line. As an example, the voltaic cell represented in Fig. 17.2 may be represented as follows.

$$Zn\left|\begin{matrix}\text{aqueous}\\ZnSO_4\end{matrix}\right\|\begin{matrix}\text{aqueous}\\CuSO_4\end{matrix}\left|Cu\right.$$

Sometimes only the reacting species within the electrolyte are listed so that the cell shown in Fig. 17.2 may be represented in the following manner also:

$$Zn|Zn(aq)^{++} \| Cu(aq)^{++}|Cu$$

The over-all chemical reaction associated with the voltaic cell may be represented by the sum of the equations for the half-cell reactions. Thus for the cell diagrammed in Fig. 17.2, the over-all chemical reaction is represented by the sum of Equations (17.23) and (17.24):

$$Cu(aq)^{++} + Zn(s) = Zn(aq)^{++} + Cu(s) \qquad (17.25)$$

From this equation, one is led to believe that copper metal and zinc ions may be produced if zinc metal is in contact with copper(II) ions in aqueous solution. In fact this is so. Clearly, Equation (17.25) represents a redox reaction. It also represents an electron transfer reaction as shown in the voltaic cell.

Electron transfer reactions constitute a large subset of the redox reactions. Characteristically, electron transfer reactions occur not only if the oxidizing and reducing agents are in direct contact, but also if they are separated as components of an electrical circuit, so that the electron transfer may occur through the circuit.

In many cases, it is possible to reverse spontaneous electron transfer reactions. Figure 17.3 shows schematically how this may be done for the zinc-copper(II)

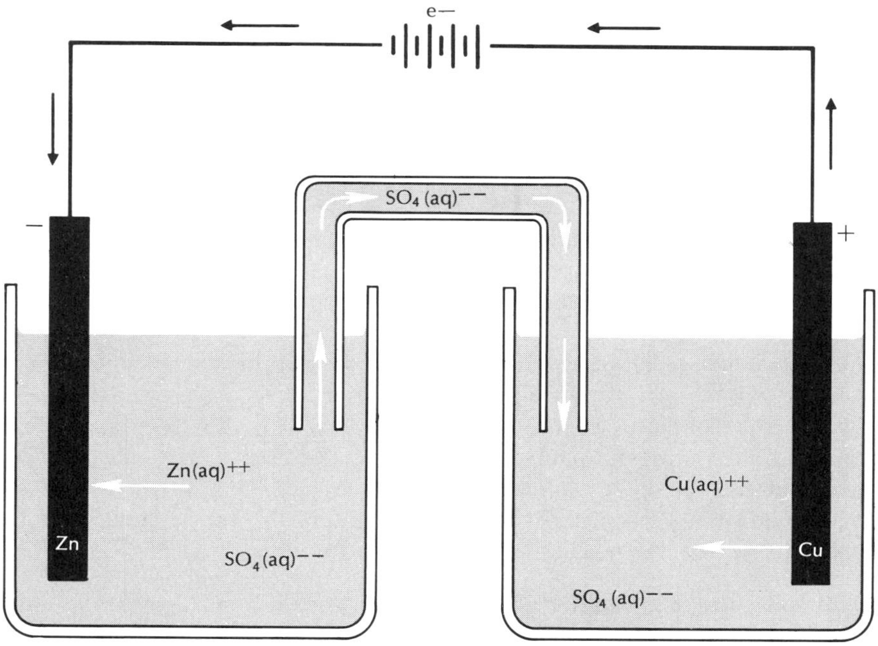

Fig. 17.3. Schematic diagram and interpretation of an electrolytic cell

sulfate reaction. An electron pump is placed in the external circuit to force electrons in the direction opposite to that in the external circuit of the voltaic cell.

In this case, the electron pump forces electrons from the copper electrode to the zinc electrode. Since the pump removes electrons from the copper electrode and deposits them on the zinc electrode, the copper electrode is positive with respect to the zinc electrode.

The half-reaction equations are the reverse of those given for the voltaic cell:

$$Zn(aq)^{++} + 2e^{-} = Zn(s) \tag{17.26}$$

$$Cu(s) = Cu(aq)^{++} + 2e^{-} \tag{17.27}$$

At the zinc electrode, the zinc ions are forced to accept electrons and are converted to zinc atoms. Since this is a reduction reaction, the zinc is the cathode. At the copper electrode, electrons are removed from copper atoms which are converted to copper(II) ions. Since this is an oxidation reaction, the copper is the anode.

The significant difference between this cell and the voltaic cell is that here the reactions are not the spontaneous ones. An external force is applied to force the electrons to move in a direction opposite to that in which they would travel if no interference was supplied. Thus an external electrical force is used to cause a nonspontaneous chemical reaction to occur. Such a cell is called an *electrolytic cell* or *secondary cell.*

Ion-Electron Equations

Ion-electron equations have just been used to describe the half-cell reactions in voltaic and electrolytic cells. It has become customary to describe some features of many redox reactions in terms of ion-electron equations in which the oxidation is considered separately from the reduction, and the over-all equation is then written as the sum of two ion-electron equations. In this section we shall look first at the mechanics of writing ion-electron equations and over-all reaction equations from the ion-electron equations and then at the stoichiometry of ion-electron equations.

CONSTRUCTION OF ION-ELECTRON EQUATIONS. Suppose that we wish to write the ion-electron equation for the reduction of nitrate ions to nitrogen(II) oxide in acidic aqueous solution. At the outset we may agree that water molecules and hydronium ions are present in abundance and may be used as before in completing these equations.

We begin with the equation skeleton indicating the species undergoing the electron change and the product of this electron transfer.

$$NO_3(aq)^{-} = NO(g) \tag{17.28}$$

Looking at the atomic balance, we note three oxygens on the left and only one on the right. *Formally*, we may consider the two extra oxygens on the left as "oxides" which will react with hydronium ions in the acidic solution to produce water. Since $4H_3O(aq)^+$ reacts with $2O(aq)^{--}$ to form $6H_2O$, we may place $4H_3O(aq)^+$ on the left and $6H_2O$ on the right to attain atomic balance.

$$NO_3(aq)^- + 4H_3O(aq)^+ = NO(g) + 6H_2O \tag{17.29}$$

Finally, charge balance may be attained by inserting $3e^-$ on the left so that the net charge on both sides is zero.

$$NO_3(aq)^- + 4H_3O(aq)^+ + 3e^- = NO(g) + 6H_2O \tag{17.30}$$

Equation (17.30) is the complete ion-electron equation for the reduction of nitrate ions to nitrogen(II) oxide in acidic solution. From the oxidation-state change it is clear that nitrate ions are reduced, and from the equation it is clear that this corresponds to the gain of electrons. Thus for electron transfer reactions, reduction corresponds to electron gain. Similarly, oxidation corresponds to electron loss for electron transfer reactions.

Equation (17.30) was derived by first considering atomic balance and then adding an appropriate number of electrons to attain charge balance. It is possible also to look at the oxidation-number change, insert the correct number of electrons first, and then add hydronium ions and water molecules to attain both atomic and charge balances. One of the advantages of the former procedure is that assignment of oxidation numbers is not necessary in order to write ion-electron equations. This is useful in cases where oxidation-number assignment might be difficult because of the complexity of the species involved.

Now let us look at this same reduction in basic solution. We may begin again with the equation skeleton.

$$NO_3(aq)^- = NO(g)$$

In basic solution the abundant species are hydroxide ions and water molecules. The "extra" two oxygens on the left this time may be absorbed by water molecules with the production of hydroxide ions. *Formally,* one oxide ion reacts with one water molecule to produce two hydroxide ions. Thus we may add $2H_2O$ to the left and $4OH(aq)^-$ to the right.

$$NO_3(aq)^- + 2H_2O = NO(g) + 4OH(aq)^- \tag{17.31}$$

Since atomic balance has been achieved, charge balance may be attained by adding $3e^-$ to the left so that the net charge on each side is $4-$.

$$NO_3(aq)^- + 2H_2O + 3e^- = NO(g) + 4OH(aq)^- \tag{17.32}$$

As a final example, consider the oxidation of phosphorus to phosphoric acid in acidic solution. Again, we begin with the equation skeleton.

$$P(s) = H_3PO_4(aq) \tag{17.33}$$

In this case there is a deficiency of four oxygens on the left which may be supplied by H_2O.

$$P(s) + 4H_2O = H_3PO_4(aq) \tag{17.34}$$

Although atomic balance for phosphorus and oxygen has been attained, it has not been attained for hydrogen. There are five more hydrogens on the left than on the right. In acidic solution we may expect these extra hydrogens to appear as hydronium ions. This requires the addition of five more waters on the left and $5H_3O(aq)^+$ on the right.

$$P(s) + 9H_2O = H_3PO_4(aq) + 5H_3O(aq)^+ \tag{17.35}$$

At this point atomic balance has been attained. To achieve charge balance, $5e^-$ must be added to the right so that the net charge on each side is zero.

$$P(s) + 9H_2O = H_3PO_4(aq) + 5H_3O(aq)^+ + 5e^- \tag{17.36}$$

Equation (17.36) is the complete ion-electron equation for the oxidation of phosphorus to phosphoric acid in acidic solution. If this oxidation is brought about with nitric acid, then Equation (17.36) may be combined with Equation (17.30) to obtain the equation for the over-all redox reaction.

$$NO_3(aq)^- + 4H_3O(aq)^+ + 3e^- = NO(g) + 6H_2O$$

$$P(s) + 9H_2O = H_3PO_4(aq) + 5H_3O(aq)^+ + 5e^-$$

Before these two equations can be added to give the over-all reaction equation, it is necessary to see that the number of electrons gained in the reduction equal the number of electrons lost in the oxidation. This is true because experimentally there is no evidence that electrons are either gained from or lost to the surroundings. This electron balance may be achieved by multiplying Equation (17.30) by 5 and Equation (17.36) by 3. These then may be added to secure over-all reaction Equation (17.37).

$$5NO_3(aq)^- + 20H_3O(aq)^+ + 15e^- = 5NO(g) + 30H_2O$$

$$3P(s) + 27H_2O = 3H_3PO_4(aq) + 15H_3O(aq)^+ + 15e^-$$

$$5NO_3(aq)^- + 3P(s) + 5H_3O(aq)^+ = 5NO(g) + 3H_3PO_4(aq) + 3H_2O \tag{17.37}$$

Compare Equation (17.37) with Equation (17.5) and note that they are the same. Thus we may achieve the same goals in equation writing by either the oxidation-state change method or the method of ion-electron equations.

Stoichiometry. Half-reaction equations may be treated stoichiometrically in the same fashion as other equations. For example, consider the reduction of nitrate to nitrogen(II) oxide in acidic solution as described by half-reaction Equation (17.30), which is rewritten here for convenience.

$$NO_3(aq)^- + 4H_3O(aq)^+ + 3e^- = NO(g) + 6H_2O$$

In words, the stoichiometric relationships may be stated as follows: 1 mole of nitrate ions reacts with 3 moles of electrons and 4 moles of hydronium ions to form 1 mole of nitrogen(II) oxide molecules and 6 moles of water molecules.

Experimentally, in working with cells, we can measure coulombs of charge passing through the cell. Thus it is convenient to know the number of coulombs of charge carried by a mole of electrons. This number can be computed in the following way:

$$1 \text{ mole e}^- \times \frac{6.023 \times 10^{23} \text{ e}^-}{1 \text{ mole e}^-} \times \frac{1 \text{ quantum}}{1 \text{ e}^-} \times \frac{1.602 \times 10^{-19} \text{ coulomb}}{1 \text{ quantum}} = 9.649 \times 10^4 \text{ coulombs}$$

In some cases it is possible to measure the coulombs of charge passing through a cell by measuring the amount of some chemical species produced or consumed. For example, suppose that the half-cell represented by Equation (17.30) is part of a voltaic cell. How many coulombs of charge have passed through the external circuit when 1.50 g NO have been produced? This may be determined in the following way:

$$1.50 \text{ g NO} \times \frac{1 \text{ mole NO}}{30.0 \text{ g NO}} \times \frac{3 \text{ moles e}^-}{1 \text{ mole NO}} \times \frac{9.65 \times 10^4 \text{ coulombs}}{1 \text{ mole e}^-} = 1.45 \times 10^4 \text{ coulombs}$$

Beginning with the mass of NO produced, the successive factors are found from the molecular weight of NO, the half-cell reaction equation, and the previously determined value for the charge carried by a mole of electrons.

In other cases, it is possible to measure the quantity of charge passing through a cell, particularly an electrolytic cell, by measuring the amperage and time. From this it is possible to compute how much of a chemical species will be consumed or produced. For example, suppose that the half-cell reaction represented by Equation (17.30) is reversed in an electrolytic cell. How many grams of NO will be oxidized by the passage of a 2.50-amp current for 20 min? This may be found as follows.

$$20 \text{ min} \times \frac{60 \text{ sec}}{1 \text{ min}} \times \frac{2.50 \text{ coul}}{\text{sec}} \times \frac{1 \text{ mole e}^-}{9.65 \times 10^4 \text{ coul}} \times \frac{1 \text{ mole NO}}{3 \text{ moles e}^-} \times \frac{30.0 \text{ g NO}}{1 \text{ mole NO}} = 0.311 \text{ g NO}$$

Beginning with the time in minutes, the successive factors are found from the number of seconds in a minute, the stated amperage, the previously determined value for the charge carried by a mole of electrons, the half-cell reaction equation, and the molecular weight of NO.

CELL POTENTIALS AND ELECTROMOTIVE FORCE

Electron transfer occurs spontaneously because one species attracts electrons from another species. Thus this transfer process involves a competition for electrons between two species. In the spontaneous process, the oxidizing agent attracts electrons more strongly than the reducing agent. This is not to suggest that the reducing agent does not attract electrons, since, in fact, it does resist the removal of electrons. Consequently, the net tendency for electrons to be transferred is a combination of the oxidizing agent's attraction for electrons and the reducing agent's resistance to having electrons removed.

In electrical terminology, the charge-attracting ability of half-cells is called the *potential* of the half-cell. The difference in charge-attracting ability between two half-cells is called the *potential difference* or *electromotive force*, *emf*, of the cell.

In an electrolytic cell, the work w required to move a charge in the external circuit is directly proportional to both the size of the charge q and the *potential difference* ΔE as shown in the following equation, where k is the proportionality constant:

$$w = kq\Delta E \tag{17.38}$$

In this proportionality, as in all others, the numerical magnitude of the proportionality constant depends upon the units in which the variables are expressed. From another point of view, if a value for the proportionality constant is chosen, then the units in which one of the variables is expressed is determined by this choice.

In the *mks* system, the work w is expressed in joules, and the charge q is expressed in coulombs. Arbitrarily, the value for k is chosen as unity so that the potential difference ΔE may be defined by rewriting Equation (17.38).

$$\Delta E = \frac{w}{kq} = \frac{w}{q} \tag{17.39}$$

In this way, the potential difference is defined as work (or energy) per unit charge. In the *mks* system this is in joules/coulomb. The unit joule/coulomb is given the name *volt* (V).

The delta notation used for the potential difference suggests that a potential, E, may be associated with each half-cell so that the potential difference is the numerical difference between these two potentials, E_2 and E_1.

$$\Delta E = E_2 - E_1 \tag{17.40}$$

However, it is not possible to measure absolute potentials for individual, isolated half-cells. Thus it is not possible to measure the electron-attracting tendency for a half-cell composed of a strip of copper metal dipping into a copper(II) sulfate solution. Although this half-cell has a tendency to attract electrons, this tendency can be measured only when the half-cell is put in competition for electrons with another half-cell. For example, if it is put in competition with the zinc-zinc sulfate half-cell as shown in Fig. 17.2, any measured potential difference indicates only that the copper-copper(II) sulfate half-cell attracts electrons more strongly than the zinc-zinc sulfate half-cell by the amount of the potential difference. For example, *if* the measured potential difference is 1 V, this means that the copper-copper(II) sulfate half-cell attracts electrons 1 V more strongly than the zinc-zinc sulfate half-cell. To state this in a slightly different way, the potentital of the copper-copper(II) sulfate half-cell is 1 V higher than the potential of the zinc-zinc sulfate half-cell.

Standards and Conventions

The previous discussion suggests that although we may not be able to measure half-cell potentials per se, we may be able to do so indirectly. We may assign half-cell potential values by choosing one specific half-cell as a standard and then comparing other half-cells to it by allowing the other half-cells to compete with it for electrons. If the potential of the standard half-cell is set at 0.00 V, the potential difference between this half-cell and another half-cell must be the potential to be assigned to that other half-cell. For those half-cells which attract electrons more strongly than the standard half-cell, the assigned potential will be positive, and for those half-cells which attract electrons less strongly than the standard half-cell, the assigned potential will be negative.

In specifying the standard half-cell, certain experimental observations must be taken into account. Experimentally, it is found that the potential difference between two half-cells depends on the temperature and the concentrations of the species which appear in the ion-electron equation for the half-cell reaction.

For example, in the case of the cell composed of the copper-copper(II) sulfate half-cell and the zinc-zinc sulfate half-cell as diagrammed in Fig. 17.2, continued operation of the cell leads to a potential difference decrease. As the reaction proceeds, the electron-attracting copper(II) ions in the copper-copper(II) sulfate half-cell decrease in concentration, so that the attraction for electrons (half-cell potential) decreases. In the zinc-zinc sulfate half-cell, the concentration of the electron-attracting zinc(II) ions increases and the half-cell potential for the zinc-zinc sulfate half-cell increases as the cell reaction proceeds. In this way, the potential difference between the two half-cells becomes smaller and smaller, approaching zero, as the reaction continues. The condition of zero potential difference should suggest the chemical equilibrium condition, which we shall discuss in more detail a little later.

The standard conditions under which potential differences are measured and half-cell potentials are assigned are the same as those given in Chapter 12 as the standard thermodynamic conditions. All dissolved species are at unit molarity; all gases are at one atmosphere of pressure; all solids are in the most stable form at 25°C. Usually the standard temperature at which measurements are made is 25°C unless otherwise specified. Furthermore, unless otherwise specified, it is understood that the half-cells contain an aqueous medium.

The specific half-cell chosen as the standard is the one represented by the following half-cell reaction equation:

$$2H_3O(aq)^+ + 2e^- = H_2(g) + 2H_2O \tag{17.41}$$

Thus the potential difference between this half-cell and any other half-cell at standard conditions is the *standard* potential of the other half-cell. Standard potentials are represented by the symbol $E°$.

To understand the assignment of standard potential values, we shall consider the copper-copper(II) sulfate, zinc-zinc sulfate, and hydrogen half-cells. The voltaic cell composed of the standard hydrogen and standard copper-copper(II) sulfate half-cells is diagrammed in Fig. 17.4. The measured potential difference between

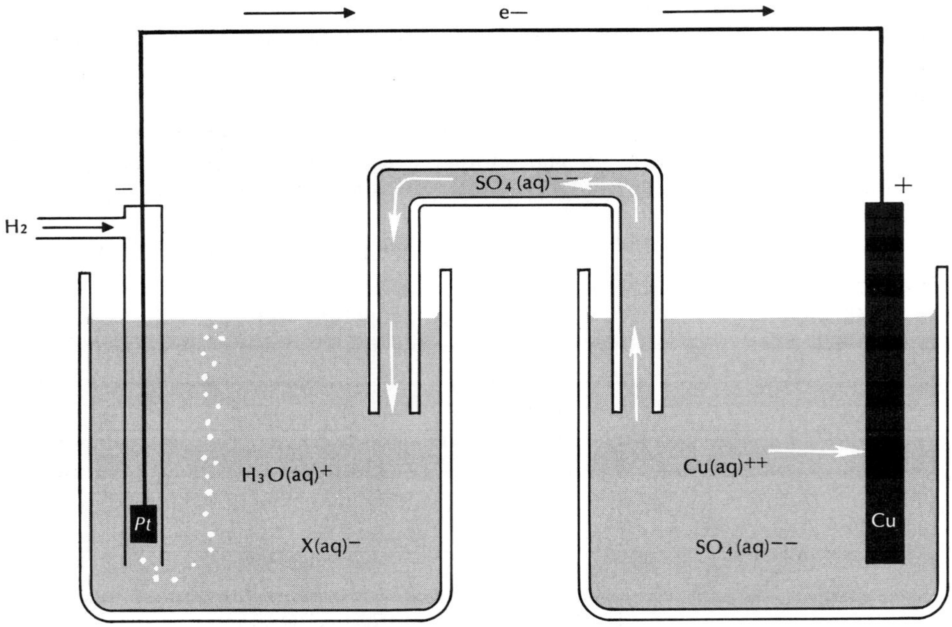

Fig. 17.4. Schematic diagram for the voltaic cell composed of the hydrogen and copper-copper sulfate half-cells

these two standard half-cells is 0.337 volts. Since the electrons flow from the hydrogen half-cell to the copper-copper(II) sulfate half-cell, the copper-copper(II) sulfate half-cell standard potential exceeds the standard potential of the hydrogen half-cell by 0.337 V. Since the standard hydrogen half-cell was assigned a potential of 0.00 V, we assign the copper-copper(II) sulfate half-cell a standard potential of +0.337 V.

Similarly, the voltaic cell composed of the standard hydrogen and standard zinc-zinc sulfate half-cells is diagrammed in Fig. 17.5. The measured potential

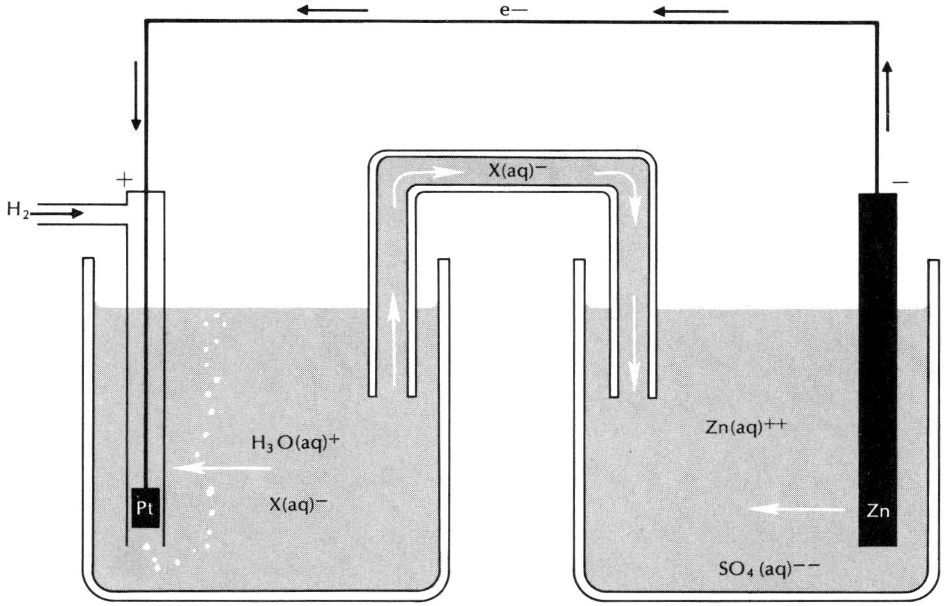

Fig. 17.5. Schematic diagram for the voltaic cell composed of the hydrogen and zinc-zinc sulfate half-cells

difference between these two standard half-cells is 0.763 V. Since the electrons flow from the zinc-zinc sulfate half-cell to the hydrogen half-cell, the hydrogen half-cell standard potential exceeds the standard potential of the zinc-zinc sulfate half-cell by 0.763 V. Since the standard hydrogen half-cell was assigned a potential of 0.00 V, we assign the zinc-zinc sulfate half-cell a standard potential of −0.763 V.

Since it is the reduction reaction in which electrons are gained, it is convenient to think of the reduction reaction as the source of electron attraction. For this reason, we represent half-cells by the reduction half-cell reaction equation, along with the standard half-cell potential, which is called the *standard reduction potential.* This representation is shown for the three half-cells discussed here in Table 17.1. In this table, the three half-cells are arranged in the order of decreasing reduction potentials.

By comparing other standard half-cells to the standard hydrogen half-cell, it is

TABLE 17.1
Standard Reduction Potentials in Volts for Three Half-cells

$Cu(aq)^{++} + 2e^- = Cu(s)$	+0.337
$2H_3O(aq)^+ + 2e^- = H_2(g) + 2H_2O$	0.000
$Zn(aq)^{++} + 2e^- = Zn(s)$	−0.763

possible to expand Table 17.1 to include many other half-cells. One such expansion is shown as Table 17.2.

TABLE 17.2
Standard Reduction Potentials in Volts at 25°C

	OXIDIZING AGENTS			REDUCING AGENTS	E°
1.	$F_2(g) + 2H_3O(aq)^+$	$+ 2e^- =$	$2H_2O$	$+ 2HF(aq)$	+3.06
2.	$F_2(g)$	$+ 2e^- =$		$2F(aq)^-$	+2.87
3.	$O_3(g) + 2H_3O(aq)^+$	$+ 2e^- =$	$3H_2O$	$+ O_2(g)$	+2.07
4.	$Co(aq)^{3+}$	$+ 1e^- =$		$Co(aq)^{++}$	+1.82
5.	$H_2O_2(aq) + 2H_3O(aq)^+$	$+ 2e^- =$		$4H_2O$	+1.77
6.	$MnO_4(aq)^- + 4H_3O(aq)^+$	$+ 3e^- =$	$6H_2O$	$+ MnO_2(s)$	+1.69
7.	$MnO_4(aq)^- + 8H_3O(aq)^+$	$+ 5e^- =$	$12H_2O$	$+ Mn(aq)^{++}$	+1.51
8.	$H_3AuO_3(s) + 3H_3O(aq)^+$	$+ 3e^- =$	$6H_2O$	$+ Au(s)$	+1.45
9.	$BrO_3(aq)^- + 6H_3O(aq)^+$	$+ 6e^- =$	$9H_2O$	$+ Br(aq)^-$	+1.44
10.	$ClO_4(aq)^- + 8H_3O(aq)^+$	$+ 8e^- =$	$12H_2O$	$+ Cl(aq)^-$	+1.39
11.	$Cl_2(g)$	$+ 2e^- =$		$2Cl(aq)^-$	+1.36
12.	$Cr_2O_7(aq)^{--} + 14H_3O(aq)^+$	$+ 6e^- =$	$21H_2O$	$+ 2Cr(aq)^{3+}$	+1.33
13.	$MnO_2(s) + 4H_3O(aq)^+$	$+ 2e^- =$	$6H_2O$	$+ Mn(aq)^{++}$	+1.23
14.	$O_2(g) + 4H_3O(aq)^+$	$+ 4e^- =$		$6H_2O$	+1.23
15.	$Br_2(l)$	$+ 2e^- =$		$2Br(aq)^-$	+1.09
16.	$NO_3(aq)^- + 4H_3O(aq)^+$	$+ 3e^- =$	$6H_2O$	$+ NO(g)$	+0.96
17.	$2Hg(aq)^{++}$	$+ 2e^- =$		$Hg_2(aq)^{++}$	+0.92
18.	$HO_2(aq)^- + H_2O$	$+ 2e^- =$		$3OH(aq)^-$	+0.88
19.	$Hg(aq)^{++}$	$+ 2e^- =$		$Hg(l)$	+0.85
20.	$2NO_3(aq)^- + 4H_3O(aq)^+$	$+ 2e^- =$	$6H_2O$	$+ N_2O_4(g)$	+0.80
21.	$Ag(aq)^+$	$+ 1e^- =$		$Ag(s)$	+0.80
22.	$Hg_2(aq)^{++}$	$+ 2e^- =$		$2Hg(l)$	+0.79
23.	$Fe(aq)^{3+}$	$+ 1e^- =$		$Fe(aq)^{++}$	+0.77
24.	$O_2(g) + 2H_3O(aq)^+$	$+ 2e^- =$	$2H_2O$	$+ H_2O_2(aq)$	+0.68
25.	$BrO_3(aq)^- + 3H_2O$	$+ 6e^- =$	$6OH(aq)^-$	$+ Br(aq)^-$	+0.61
26.	$MnO_4(aq)^- + 2H_2O$	$+ 3e^- =$	$4OH(aq)^-$	$+ MnO_2(s)$	+0.59
27.	$H_3AsO_4(aq) + 2H_3O(aq)^+$	$+ 2e^- =$	$4H_2O$	$+ As_4O_6(s)$	+0.58
28.	$I_2(s)$	$+ 2e^- =$		$2I(aq)^-$	+0.54
29.	$Cu(aq)^+$	$+ 1e^- =$		$Cu(s)$	+0.52
30.	$O_2(g) + 2H_2O$	$+ 4e^- =$		$4OH(aq)^-$	+0.40
31.	$[Fe(CN)_6](aq)^{3-}$	$+ 1e^- =$		$[Fe(CN)_6](aq)^{4-}$	+0.36
32.	$Cu(aq)^{++}$	$+ 2e^- =$		$Cu(s)$	+0.34
33.	$Cu(aq)^{++}$	$+ 1e^- =$		$Cu(aq)^+$	+0.15
34.	$S(s) + 2H_3O(aq)^+$	$+ 2e^- =$	$2H_2O$	$+ H_2S(g)$	+0.14
35.	$HSO_4(aq)^{--} + 3H_3O(aq)^+$	$+ 2e^- =$	$4H_2O$	$+ H_2SO_3(aq)$	+0.12
36.	$2H_3O(aq)^+$	$+ 2e^- =$	$2H_2O$	$+ H_2(g)$	0.00
37.	$O_2(g) + H_2O$	$+ 2e^- =$	$OH(aq)^-$	$+ HO_2(aq)^-$	−0.08
38.	$Pb(aq)^{++}$	$+ 2e^- =$		$Pb(s)$	−0.13
39.	$CrO_4(aq)^{--} + 4H_2O$	$+ 3e^- =$	$5OH(aq)^-$	$+ Cr(OH)_3(s)$	−0.13
40.	$Ni(aq)^{++}$	$+ 2e^- =$		$Ni(s)$	−0.23
41.	$Co(aq)^{++}$	$+ 2e^- =$		$Co(s)$	−0.28

TABLE 17.2 (cont.)

OXIDIZING AGENTS		REDUCING AGENTS	E°
42. $Cd(aq)^{++}$	$+ 2e^- =$	$Cd(s)$	-0.40
43. $Cr(aq)^{3+}$	$+ 1e^- =$	$Cr(aq)^{++}$	-0.41
44. $Fe(aq)^{++}$	$+ 2e^- =$	$Fe(s)$	-0.44
45. $S(s)$	$+ 2e^- =$	$S(aq)^{--}$	-0.48
46. $Fe(OH)_3(s)$	$+ 1e^- =$	$OH(aq)^- + Fe(OH)_2(s)$	-0.56
47. $AsO_4(aq)^{3-} + 2H_2O$	$+ 2e^- =$	$4OH(aq)^- + AsO_2(aq)^-$	-0.67
48. $Ni(OH)_2(s)$	$+ 2e^- =$	$2OH(aq)^- + Ni(s)$	-0.72
49. $Cr(aq)^{3+}$	$+ 3e^- =$	$Cr(s)$	-0.74
50. $Zn(aq)^{++}$	$+ 2e^- =$	$Zn(s)$	-0.76
51. $2H_2O(l)$	$+ 2e^- =$	$2OH(aq)^- + H_2(g)$	-0.83
52. $Fe(OH)_2(s)$	$+ 2e^- =$	$2OH(aq)^- + Fe(s)$	-0.88
53. $SO_4(aq)^{--} + H_2O$	$+ 2e^- =$	$2OH(aq)^- + SO_3(aq)^{--}$	-0.93
54. $Mn(aq)^{++}$	$+ 2e^- =$	$Mn(s)$	-1.18
55. $Al(aq)^{3+}$	$+ 3e^- =$	$Al(s)$	-1.66
56. $Mg(aq)^{++}$	$+ 2e^- =$	$Mg(s)$	-2.37
57. $Na(aq)^+$	$+ 1e^- =$	$Na(s)$	-2.71
58. $Ca(aq)^{++}$	$+ 2e^- =$	$Ca(s)$	-2.87
59. $Sr(aq)^{++}$	$+ 2e^- =$	$Sr(s)$	-2.89
60. $Ba(aq)^{++}$	$+ 2e^- =$	$Ba(s)$	-2.90
61. $Ra(aq)^{++}$	$+ 2e^- =$	$Ra(s)$	-2.92
62. $Cs(aq)^+$	$+ 1e^- =$	$Cs(s)$	-2.92
63. $Rb(aq)^+$	$+ 1e^- =$	$Rb(s)$	-2.92
64. $K(aq)^+$	$+ 1e^- =$	$K(s)$	-2.92
65. $Li(aq)^+$	$+ 1e^- =$	$Li(s)$	-3.04

Interpretation of Standard Reduction Potentials

Since the standard reduction potential of the copper-copper(II) sulfate half-cell is higher than the standard reduction potential of the zinc-zinc sulfate half-cell, electrons flow from the zinc-zinc sulfate half-cell to the copper-copper(II) sulfate half-cell. As a result, the two reactions that occur in the half-cells can be represented by the following equations:

$$Cu(aq)^{++} + 2e^- = Cu(s) \qquad (17.42)$$

$$Zn(s) = Zn(aq)^{++} + 2e^- \qquad (17.43)$$

Note that Equation (17.42) is written as a reduction just as it appears in Tables 17.1 and 17.2. However, Equation (17.43) is written as an oxidation, which is the reverse of the half-cell reaction equation shown in Tables 17.1 and 17.2.

This suggests the following general observation: *when two half-cells are coupled to form a voltaic cell, the half-cell with the higher reduction potential undergoes reduction, and the half-cell with the lower reduction potential undergoes oxidation.* In terms of Table 17.2, this indicates that in a voltaic cell, the half-cell appearing higher in the table undergoes the reaction indicated by the equation in the table, whereas the half-cell appearing lower in the table undergoes the reverse of the reaction shown in the table.

This also indicates that each half-cell contains an oxidizing agent and a reducing agent. For the copper-copper(II) sulfate half-cell represented by Equation (17.42), $Cu(aq)^{++}$ is the oxidizing agent and Cu(s) is the reducing agent. For the zinc-zinc sulfate half-cell represented by Equation (17.43), $Zn(aq)^{++}$ is the oxidizing agent and Zn(s) is the reducing agent. Of the two oxidizing agents, $Cu(aq)^{++}$ is the stronger corresponding to the higher reduction potential, that is to the greater tendency to gain electrons and be reduced. Of the two reducing agents, Zn(s) is the stronger. This is clear if we interpret the reduction potentials to indicate attractions for electrons. Since copper has the greater attraction for electrons, electrons will be removed from copper atoms with greater difficulty than from zinc atoms. Since a good reducing agent must allow its electrons to be removed easily, Zn(s) is a better reducing agent than Cu(s).

By generalizing this argument, we can say that those species listed on the left side of Table 17.2 are oxidizing agents and are arranged in order of decreasing strength as such. Similarly, those species listed on the right side of Table 17.2 are reducing agents and are arranged in the order of increasing strength as such. Thus the strongest oxidizing agent in Table 17.2 is elemental fluorine in an acidic medium and the strongest reducing agent is solid lithium.

The emf of a cell is the difference in reduction potentials of the two half-cells. As a matter of convention in taking differences of potential, we agree that the subtracted potential shall represent the oxidation half-reaction. This is indicated by the following equation where ΔE is the cell emf, E_{red} is the reduction potential for the half-cell undergoing reduction, and E_{ox} is the reduction potential for the half-cell undergoing oxidation.

$$\Delta E = E_{red} - E_{ox} \tag{17.44}$$

Under standard condition this equation may be written with superscripts.

$$\Delta E^\circ = E^\circ_{red} - E^\circ_{ox} \tag{17.45}$$

When done in this way, a positive cell emf indicates the spontaneous chemical reaction that occurs in a voltaic cell. A negative cell emf corresponds to the non-spontaneous chemical reaction that occurs in an electrolytic cell.

For example, in the cell composed of the copper-copper(II) sulfate and zinc-zinc sulfate half-cells, the copper-copper(II) sulfate half-cell, with $E^\circ = +0.337$ volts, undergoes reduction, and the zinc-zinc sulfate half-cell, with $E^\circ = -0.763$ volts, undergoes oxidation. Thus the standard cell emf is

$$\Delta E^\circ = E^\circ_{red} - E^\circ_{ox} = E^\circ_{Cu} - E^\circ_{Zn} = +0.337 - (-0.763) = +1.100 \text{ V}$$

If we had made the opposite assignment to indicate that the zinc-zinc sulfate half-cell undergoes reduction and the copper-copper(II) sulfate undergoes oxidation, the cell emf would have been negative.

$$\Delta E^\circ = E^\circ_{red} - E^\circ_{ox} = E^\circ_{Zn} - E^\circ_{Cu} = -0.763 - (+0.337) = -1.100 \text{ V}$$

This cell emf corresponds to the nonspontaneous reaction in which Cu(s) is oxidized to $Cu(aq)^{++}$ and $Zn(aq)^{++}$ is reduced to Zn(s).

The Nernst Equation

Standard half-cell potentials indicate that the activities of dissolved species must be unity and the gaseous substances must be at one atmosphere pressure. However, in practical laboratory work, these specific conditions are not met, and it is desirable to have some relationship between the potential of a half-cell at these standard conditions and the potential of the same half-cell at some other conditions. By comparing a half-cell at various experimental conditions to the standard hydrogen half-cell, it is possible to show that the following relationship holds, at 25°C.

$$E = E^\circ - \frac{0.0591}{n} \log Q \tag{17.46}$$

This is called the *Nernst equation;* it relates the half-cell potential at nonstandard conditions, E, to the standard half-cell potential, E°, the number of moles of electrons, n, in the equation for the half-cell reaction, and the activity quotient, Q. The activity quotient is the product of the activities of all dissolved and gaseous products raised to the powers of their coefficients in the half-cell reaction equation, divided by the product of the activities of all dissolved and gaseous reactants raised to the powers of their coefficients in the half-cell reaction equation. The form of Q resembles the equilibrium constant formulation, *except* the activities are *not* equilibrium activities.

Two examples will serve to illustrate the relationship described by the Nernst equation. Consider the half-cell reduction of dichromate in acid to chromium(III) ions.

$$Cr_2O_7(aq)^{--} + 14H_3O(aq)^+ + 6e^- = 21H_2O + 2Cr(aq)^{3+} \tag{17.47}$$

For this half-cell, the Nernst equation takes the following form:

$$E = E^\circ - \frac{0.0591}{n} \log Q = 1.33 - \frac{0.0591}{6} \log \frac{a^2_{Cr^{+++}}}{a_{Cr_2O_7^{--}} a^{14}_{H_3O^+}} \tag{17.48}$$

The value of 1.33 V for E° is found in Table 17.2 and the value for n is obtained from Equation (17.47).

As a second example, consider the following half-cell:

$$MnO_4(aq)^- + 2H_2O + 3e^- = 4OH(aq)^- + MnO_2(s) \tag{17.49}$$

For this half-cell the Nernst equation takes on the following form:

$$E = E^\circ - \frac{0.0591}{n} \log Q = 0.60 - \frac{0.0591}{3} \log \frac{a^4_{OH^-}}{a_{MnO_4^-}} \tag{17.50}$$

In this case the solid MnO_2 activity does not appear in the activity-quotient expression, unless we assign an activity of 1.00 to the solid.

In both of these examples, note that the half-cell potential depends on the acidity or basicity of the half-cell solution since the activities of hydronium or hydroxide ions are involved. This means that the half-cell potentials are changed when the pH is changed. In fact this is the basis for the construction of pH meters. An otherwise self-contained half-cell is immersed in a solution and the potential is measured against a standard half-cell, the value of the potential depending upon the pH. Meters usually are calibrated to read pH directly, but many pH meters also have potential scales, so they may serve a double function of measuring half-cell potentials as well.

It should be noted that both of these half-cells are quite sensitive to pH changes. In Equation (17.48) the hydronium ion activity appears to the fourteenth power and in Equation (17.50) the hydroxide ion activity appears to the fourth power. In each case, a small change in pH is reflected in a relatively large effect on the half-cell potential.

Using Equation (17.48) as an example, we can discern the direction of the activity-change effect on the half-cell potential. Thus an increase in the activity of $Cr(aq)^{3+}$ increases the size of Q so that log Q increases. Since this term is subtracted from the standard potential, an increase in chromium(III) ion activity leads to a decrease in reduction potential; that is, the reduction potential becomes more negative. Consequently, as the activity of $Cr(aq)^{3+}$ increases, the apparent oxidizing ability of the dichromate ion decreases, that is, it becomes more difficult to reduce the dichromate. On the other hand, as the activity of $Cr(aq)^{3+}$ increases, chromium(III) ions appear to improve as reducing agents; that is, it is easier to oxidize them to dichromate.

We might ask what effect a decrease in pH has upon the potential of the half-cell represented by Equation (17.47). The pH decrease corresponds to an increase in hydronium ion activity. Since the hydronium ion activity is in the denominator of Q, the value of Q decreases as the activity of $H_3O(aq)^+$ increases, and log Q decreases also. Since log Q is subtracted from the standard potential, the half-cell potential increases, that is, it becomes less negative. Thus, in this case, an increase in acidity increases the apparent oxidizing ability of dichromate and decreases the apparent reducing ability of chromium(III) ions.

Redox Equilibrium Constants

When two half-cell reactions are combined, whether or not they are in a cell, the cell emf, which often is called the reaction emf, is the potential difference for the two half-cell reactions. This was indicated in Equation (17.45) which is reproduced here for convenience.

$$\Delta E = E_{red} - E_{ox}$$

Both E_{red} and E_{ox} are related to their standard potentials through the Nernst equation and the Nernst relationships may be substituted into Equation (17.37).

$$E_{red} = E°_{red} - \frac{0.0591}{n} \log Q_{red} \tag{17.51}$$

$$E_{ox} = E°_{ox} - \frac{0.0591}{n} \log Q_{ox} \tag{17.52}$$

$$\Delta E = E°_{red} - E°_{ox} - \frac{0.0591}{n} \log Q_{red} + \frac{0.0591}{n} \log Q_{ox} \tag{17.53}$$

If the two n values shown in Equation (17.53) are the same as they are when the electron gain and loss have been balanced in the over-all chemical equation, then $0.0591/n$ may be factored out in the following manner:

$$\Delta E = \Delta E° - \frac{0.0591}{n}(\log Q_{red} - \log Q_{ox}) \tag{17.54}$$

$$\Delta E = \Delta E° - \frac{0.0591}{n} \log \frac{Q_{red}}{Q_{ox}} \tag{17.55}$$

If we designate Q_{red}/Q_{ox} as Q_c, the activity quotient of the over-all cell, then Equation (17.47) may be rewritten in the following form:

$$\Delta E = \Delta E° - \frac{0.0591}{n} \log Q_c \tag{17.56}$$

The activity quotient of the cell, Q_c, is the product of the activities of the over-all products of the cell reaction raised to the powers of their coefficients in the cell reaction, divided by the activities of the over-all reactants of the cell reaction raised to the powers of their coefficients in the cell reaction equation. That this is true can be seen from a specific example. Consider the cell described by the following two half-cell reaction equations, which are combined to give the over-all cell reaction equation.

$$2Hg(aq)^{++} + 2e^- = Hg_2(aq)^{++} \qquad E° = +0.92 \text{ V} \tag{17.57}$$

$$2Fe(aq)^{3+} + 1e^- = 2Fe(aq)^{++} \qquad E° = +0.77 \text{ V} \tag{17.58}$$

$$2Hg(aq)^{++} + 2Fe(aq)^{++} = Hg_2(aq)^{++} + 2Fe(aq)^{++} \quad \Delta E° = +0.15 \text{ V} \tag{17.59}$$

With the aid of these equations the respective activity quotients may be written in the following forms.

$$Q_{red} = \frac{a_{Hg_2^{++}}}{a^2_{Hg^{++}}} \tag{17.60}$$

$$Q_{ox} = \frac{a^2_{Fe^{++}}}{a^2_{Fe^{3+}}} \tag{17.61}$$

$$Q_c = \frac{Q_{\text{red}}}{Q_{\text{ox}}} = \frac{a_{\text{Hg}_2{}^{++}} \times a^2_{\text{Fe}^{3+}}}{a^2_{\text{Hg}^{++}} \times a^2_{\text{Fe}^{++}}} \tag{17.62}$$

Like those for half-cells, the activity quotient for an over-all redox reaction resembles the equilibrium constant expression in form *except* that the activities are *not* equilibrium activities. When equilibrium conditions do exist, then the two half-cell potentials are equal and $\Delta E = 0$ for the over-all reaction. Then Equation (17.56) may be rewritten as follows, where K is the equilibrium constant.

$$0 = \Delta E^\circ - \frac{0.0591}{n} \log K \tag{17.63}$$

$$\Delta E^\circ = \frac{0.0591}{n} \log K \tag{17.64}$$

$$\log K = \frac{n\Delta E^\circ}{0.0591} \tag{17.65}$$

This shows that the equilibrium constants for redox reactions are related to the standard emf values for the reactions. Thus, for the reaction represented by Equation (17.59), the equilibrium constant may be computed using Equation (17.65).

$$\log K = \frac{n\Delta E^\circ}{0.0591} = \frac{2 \times 0.15}{0.0591} = 5.08$$

$$K = 1.2 \times 10^5$$

Free Energy and Reaction EMF

Recall that the equilibrium constant is a thermodynamic function related to the standard free energy change.

$$\Delta G^\circ = -2.3RT \log K \tag{17.66}$$

$$\log K = -\frac{\Delta G^\circ}{2.3RT} \tag{17.67}$$

The expressions for log K in Equations (17.65) and (17.67) may be equated.

$$\frac{n\Delta E^\circ}{0.0591} = \frac{-\Delta G^\circ}{2.3RT} \tag{17.68}$$

$$\Delta G^\circ = \frac{-2.3nRT\Delta E^\circ}{0.0591} \tag{17.69}$$

The mks units associated with the constant, $2.3RT/0.0591$ may be determined by solving Equation (17.69) for this quantity.

$$\frac{-2.3RT}{0.0591} = \frac{\Delta G^\circ}{n\Delta E^\circ} = \frac{\text{joules}}{\text{moles e}^- \times \text{volts}} = \frac{\text{joules} \times \text{coulomb}}{\text{moles e}^- \times \text{joules}} = \frac{\text{coulombs}}{\text{moles e}^-}$$

Clearly this constant is the number of coulombs per mole of electrons, which we have already determined to be 9.65×10^4. (This constant may be evaluated from Equation (17.61) by using 8.32 joules/°K for R and 298°K for T.) The number of coulombs per mole of electrons often is called the Faraday and is given the symbol F. Using this symbol, we may rewrite Equation (17.69).

$$\Delta G^\circ = -nF\Delta E^\circ \tag{17.70}$$

From this relationship it is clear that the standard reaction emf is a thermodynamic function directly related to the standard free energy change, and as such is a measure of the tendency for redox reactions to occur.

Carefully note the signs involved in this relationship. Spontaneity is distinguished by positive values for the standard emf, and since both n and F are positive, a positive value for ΔE° corresponds to a negative value for ΔG°, which also is characteristic of spontaneous reactions.

This important relationship suggests an experimental method for the determination of standard free energy changes for redox reactions, by determining the standard reaction emf. For example, the standard emf for the reaction represented by Equation (17.59) is +0.15 V. The standard free energy change may be computed from Equation (17.70).

$$\Delta G^\circ = -nF\Delta E^\circ = -2 \text{ moles e}^- \times \frac{9.65 \times 10^4 \text{ coul}}{1 \text{ mole e}^-} \times \frac{0.15 \text{ joules}}{\text{coul}} \times \frac{1 \text{ kcal}}{4.18 \times 10^3 \text{ joules}} = -6.9 \text{ kcal}$$

Redox Titrations

Titrations in which an oxidizing agent and a reducing agent are separated, with one as titrant and the other as titree, often are carried out in the laboratory as a means of analysis. Just as acid-base titrations may be followed by pH changes, so redox titrations may be followed by potential changes.

For example, consider the titration of an aqueous solution of $Fe(aq)^{++}$ by $Cr_2O_7(aq)^{--}$ in acidic solution. The reacting system is described by the following ion-electron equations and the over-all reaction equation.

$$Cr_2O_7(aq)^{--} + 14H_3O(aq)^+ + 6e^- = 2Cr(aq)^{3+} + 21H_2O \tag{17.71}$$

$$6Fe(aq)^{++} = 6Fe(aq)^{3+} + 6e^- \tag{17.72}$$

$$Cr_2O_7(aq)^{--} + 6Fe(aq)^{++} + 14H_3O(aq)^+ = 2Cr(aq)^{3+} + 6Fe(aq)^{3+} + 21H_2O \tag{17.73}$$

The over-all reaction goes to completion as indicated by the very high value for the equilibrium constant.

$$\log K = \frac{n\Delta E^\circ}{0.0591} = \frac{6(1.33 - 0.77)}{0.0591} = 56.9$$

$$K = 8 \times 10^{56}$$

At the outset, the titree consists of an aqueous solution of some iron(II) compound and some acid. However, upon addition of the first portion of titrant, some of the iron(II) ions are converted to iron(III) ions and the added dichromate is converted to chromium(III) ions. As a result the titree effectively becomes an iron(II) ion-iron(III) ion half-cell with a measurable potential. As the titration proceeds, the concentration of iron(II) ions decreases while the concentration of iron(III) ions increases, and the half-cell potential increases accordingly as predicted by the Nernst equation.

Since the reaction is complete, at the equivalence point virtually all of the iron(II) ions have been oxidized to iron(III) ions, and virtually all of the added dichromate ions have been reduced to chromium(III) ions. Immediately beyond the equivalence point, no further oxidation or reduction occurs, and the titree becomes a chromium(III) ion-dichromate ion half-cell as the concentration of dichromate ion increases.

Data for a titration curve may be obtained for a specific case by using the Nernst equation. For example, suppose that the original titree consists of 50.00 ml of 0.0600 M iron(II) ion which also is 0.500 M in $H_3O(aq)^+$, and the titrant is 0.0100 M dichromate ion solution. We shall assume throughout that the activities may be replaced by the molarities.

The original titree consists of 50.00 ml of 0.0600 M $Fe(aq)^{++}$, which thus contains 3.00 mmoles of $Fe(aq)^{++}$. If 5.00 ml of titrant is added, this is the same as adding 0.050 mmole of dichromate. From Equation (17.73) it is clear that each mmole of dichromate consumes 6 mmoles of $Fe(aq)^{++}$. Thus the number of mmoles of $Fe(aq)^{++}$ consumed upon the addition of 5.00 ml of titrant is 0.300 mmole, leaving behind 2.700 mmoles of iron(II) ions. At the same time, 0.300 mmoles of $Fe(aq)^{3+}$ are produced and the total volume of the titree solution is 55.00 ml. Thus the molarities of $Fe(aq)^{++}$ and $Fe(aq)^{3+}$ are 2.700/55.00 = 0.0491 M and 0.300/55.00 = 0.00546 M respectively. These values may be used in the Nernst equation to compute the half-cell potential.

$$E = E^\circ - \frac{0.0591}{n}\log\frac{M_{Fe^{++}}}{M_{Fe^{3+}}} = 0.771 - \frac{0.0591}{1}\log\frac{0.0491}{0.00546}$$

$$= 0.771 - 0.056 = 0.715 \text{ V} \tag{17.74}$$

As the titration proceeds prior to the equivalence point, the molarity of $Fe(aq)^{++}$ decreases and the molarity of $Fe(aq)^{3+}$ increases. This means that the value of Q in the Nernst equation decreases so that the logarithm term decreases. However, since the logarithm term is subtracted from the standard potential, which is positive, this means that the potential of the titree increases as the titration proceeds. Some data prior to the equivalence point are recorded in Table 17.3.

In this particular titration, the equivalence point is reached when 50.00 ml of titrant have been added. The 50.00 ml of original titree contained 3.00 mmoles of $Fe(aq)^{++}$. The 50.00 ml of titrant contained 0.500 mmole of $Cr_2O_7(aq)^{--}$. However, since each mmole of $Cr_2O_7(aq)^{--}$ reacts with 6 mmoles of $Fe(aq)^{++}$, the 0.500

TABLE 17.3
Titration of 50.00 ml of 0.0600 M Fe (aq)++ in 0.500 M $H_3O(aq)^+$ with 0.0100 M $Cr_2O_7(aq)^{--}$ Before the Equivalence Point

		$Fe(aq)^{3+}$		$Fe(aq)^{++}$		
ml OF TITRANT ADDED	TOTAL TITREE (ml)	mmole IN SOLN	MOLAR CONC. $\times 10^3$	mmole LEFT	MOLAR CONC. $\times 10^3$	POTENTIAL (VOLTS)
5.00	55.00	0.300	5.46	2.70	49.1	+0.715
10.00	60.00	0.600	10.0	2.40	40.0	+0.735
20.00	70.00	1.20	17.1	1.80	25.7	+0.761
30.00	80.00	1.80	22.5	1.20	15.0	+0.781
35.00	85.00	2.10	24.7	0.90	10.6	+0.793
40.00	90.00	2.40	26.7	0.60	6.67	+0.807
45.00	95.00	2.70	28.4	0.30	3.16	+0.827
48.00	98.00	2.88	29.4	0.12	1.23	+0.853
49.00	99.00	2.94	29.0	0.06	0.61	+0.870

mmole of $Cr_2O_7(aq)^{--}$ reacts with 3.00 mmoles of $Fe(aq)^{++}$, which is exactly the amount of $Fe(aq)^{++}$ originally present.

Beyond the equivalence point, the titree is a dichromate-chromium(III) ion half-cell with the potential given by the following Nernst equation:

$$E = E^\circ - \frac{0.0591}{6} \log \frac{M^2_{Cr^{3+}}}{M_{Cr_2O_7^{--}} M^{14}_{H_3O^+}} \tag{17.75}$$

Beyond the equivalence point, the amount of $Cr(aq)^{3+}$ in the titree remains constant at 1.00 mmole. However, the titree volume increases, so the molarity decreases. After adding 1.00 ml of titrant beyond that required to reach the equivalence point, the molarity of $Cr(aq)^{3+}$ is $1.00/101.0 = 0.00990\ M$.

The dichromate ion in solution results only from the titrant which has been added beyond the equivalence point. Thus after adding 1.00 ml beyond this point, there is 0.0100 mmole of $Cr_2O_7(aq)^{--}$ in 101.0 ml of solution and the concentration is $0.0100/101.0 = 0.0000990\ M$.

Since the original titree was 0.500 M in $H_3O(aq)^+$, there were $50.00 \times 0.500 = 25.00$ mmoles of $H_3O(aq)^+$. Prior to the equivalence point, $H_3O(aq)^+$ was consumed at the rate of 14 mmoles for each mmole of $Cr_2O_7(aq)^{--}$ added. Thus the number of mmoles of $H_3O(aq)^+$ consumed was $14 \times 0.500 = 7.00$ mmoles. As a result, at the equivalence point 18.00 mmoles of $H_3O(aq)^+$ remain and the amount of $H_3O(aq)^+$ remains constant at this value after the equivalence point. However, the molarity of $H_3O(aq)^+$ decreases, since the volume of the titree increases. Thus after a total of 51.00 ml of titrant have been added, the $H_3O(aq)^+$ concentration is $18.00/101.0 = 0.178\ M$.

The values just determined may be used in Equation (17.75) to compute the potential of the titree after adding 1.00 ml of titrant beyond the equivalence point.

$$E = 1.33 - \frac{0.0591}{6} \log \frac{(9.90 \times 10^{-3})^2}{(9.90 \times 10^{-5})(1.78 \times 10^{-1})^{14}}$$

$$= 1.33 - 0.10 = 1.23 \text{ V}$$

Similarly determined data for the titration beyond the equivalence point are recorded in Table 17.4. Then the data from Tables 17.3 and 17.4 are used in Fig. 17.6, where potential is plotted against volume of titrant added. This is a typical titration curve for a redox reaction and resembles very much the acid-base titration curves discussed previously. In this case the rapid rise in potential near the equivalence point resembles the rapid rise in pH near the equivalence point in an acid-base titration.

TABLE 17.4
Titration of 50.00 ml of 0.0600 M $Fe(aq)^{++}$ in 0.500 M $H_3O(aq)^+$ with 0.0100 M $Cr_2O_7(aq)^{--}$ After the Equivalence Point

ml OF TITRANT ADDED	TOTAL TITREE (ml)	$Cr_2O_7(aq)^{--}$ mmole IN SOLN	$Cr_2O_7(aq)^{--}$ MOLAR CONC. $\times 10^4$	$H_3O(aq)^+$ mmole LEFT	$H_3O(aq)^+$ MOLAR CONC. $\times 10^1$	$Cr(aq)^{3+}$ mmole IN SOLN	$Cr(aq)^{3+}$ MOLAR CONC. $\times 10^3$	POTENTIAL (VOLTS)
51.00	101.00	0.010	0.99	18.00	1.78	1.00	9.90	+1.23
52.00	102.00	0.020	1.96	18.00	1.76	1.00	9.81	+1.23
55.00	105.00	0.050	4.76	18.00	1.71	1.00	9.52	+1.23
60.00	110.00	0.100	9.09	18.00	1.64	1.00	9.09	+1.23

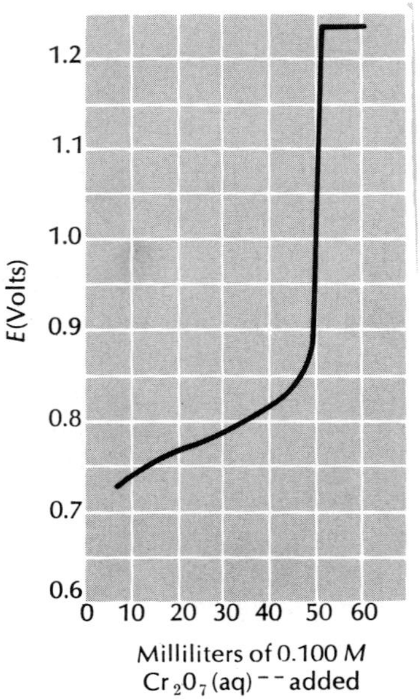

Fig. 17.6. Titration of 50.0 ml of 0.0600 *M* $Fe(aq)^{++}$ in 0.500 *M* $H_3O(aq)^+$ with 0.0100 *M* $Cr_2O_7(aq)^{--}$

It should be noted, however, that the equivalence point potential does not occur near the middle of the steeply rising portion of the curve, but rather appears near the top of this portion of the curve.

SUMMARY

Redox reactions are those reactions in which there is an oxidation number change. Electron transfer reactions are an important subclass of redox reactions, which may be used in the construction of voltaic and electrolytic cells. In addition, the thermodynamic functions of these reactions are related directly to the potentials of the reducing and oxidizing half-cells. This provides a convenient technique for the evaluation of equilibrium constants and standard free energy changes for these reactions.

Standard potentials for half-cells are defined and values are listed. From these potentials, it is possible to compute standard emf values for over-all redox reactions; it is possible to compute potentials at conditions other than standard conditions using the Nernst equation.

QUESTIONS AND PROBLEMS

17.1 Each of the following statements is a quotation from a prominent book on inorganic chemistry. Criticize each statement.

a) On a small scale, nitrogen may be obtained by the thermal decomposition of the ammonium salt of an oxidising oxyacid; the hydrogen of the ammonium radical is oxidised to water, and the nitrogen is set free, for example

$$(NH_4)_2Cr_2O_7 = Cr_2O_3 + 4H_2O + N_2$$

b) Hydrogen peroxide and hydroxylamine undergo mutual reduction:

$$HONH_2 + H_2O_2 = NH_3 + O_2 + H_2O$$

c) Sodium arsenite is reduced by hydrogen sulphide in acid solution to yield As_4S_6.

17.2 Does the following equation represent a redox reaction? Explain.

$$S_2O_8(aq)^{--} + 4H_2O = H_2O_2(aq) + 2SO_4(aq)^{--} + 2H_3O(aq)^{+}$$

17.3 Complete the following molecular-form equations by the method of oxidation number change and designate the oxidizing and reducing agents.

a) $As + HNO_3 + H_2O = H_3AsO_4 + NO$

b) $MnO_2 + HCl = MnCl_2 + Cl_2 + H_2O$

c) $CuO + NH_3 = Cu + N_2 + H_2O$

d) $Cu + HNO_3 = Cu(NO_3)_2 + NO + H_2O$

e) $AsH_3 + KClO_3 = H_3AsO_4 + KCl$

f) $KMnO_4 + H_2SO_4 + H_2S = MnSO_4 + K_2SO_4 + S + H_2O$

g) $HIO_3 + H_2SO_3 = H_2SO_4 + I_2 + H_2O$

h) $K_2Cr_2O_7 + FeSO_4 + H_2SO_4 = Cr_2(SO_4)_3 + Fe_2(SO_4)_3 + K_2SO_4 + H_2O$

i) $FeS + HNO_3 = Fe(NO_3)_3 + H_2O + NO + S$

j) $CrPO_4 + KMnO_4 + H_2O = K_2Cr_2O_7 + H_3PO_4 + MnO_2$

k) $Hg_2I_2 + S = HgS + I_2$

l) $AgClO_3 + Cl_2 = AgCl + ClO_2$

m) $K[Ag(NCS)_2] + Br_2 = AgBr + (NCS)_2 + KBr$

n) $(CN)_2 + KOH = KCN + KCNO + H_2O$
o) $Ag_2SeO_3 + H_2O + Br_2 = AgBr + H_2SeO_4$

17.4 For each of the following changes write the ion-electron equations and the complete ionic equation.

a) $Cu(s) + NO_3(aq)^- = Cu(aq)^{++} + NO(g)$ (in acid)
b) $Zn(s) + NO_3(aq)^- = Zn(aq)^{++} + NH_4(aq)^+$ (in acid)
c) $MnO_4(aq)^- + NO_2(aq)^- = MnO_4(aq)^{--} + NO_3(aq)^-$ (in base)
d) $Ag_2S(s) + NO_3(aq)^- = Ag(aq)^+ + S(s) + NO(g)$ (in acid)
e) $Al(s) + NO_3(aq)^- = [Al(OH)_4](aq)^- + NH_3(aq)$ (in base)
f) $H_2O_2(aq) + Cr_2O_7(aq)^{--} = Cr(aq)^{3+} + O_2(g)$ (in acid)
g) $HO_2(aq)^- + CrO_2(aq)^- = CrO_4(aq)^{--} + H_2O$ (in base)
h) $H_2S(g) + H_2SO_3(aq) = S(s)$ (in acid)
i) $I_2(aq) + [Ag(S_2O_3)_2](aq)^{3-} = S_4O_6(aq)^{--} + AgI(s) + I(aq)^-$ (in base)
j) $Fe(OH)_2(s) + SO_3(aq)^{--} = FeS(s) + Fe(OH)_3(s)$ (in base)

17.5 Define the terms "oxidation," "reduction," "oxidizing agent," and "reducing agent" in terms of oxidation state change and in terms of electron transfer.

17.6 The following are schematic representations of three galvanic cells. Write the ion-electron equation for each half-cell, and the ionic equation for the over-all cell reaction, and compute the standard emf for the cell.

a) $Zn(s) \mid Zn(aq)^{++} \parallel Pb(aq)^{++} \mid Pb(s)$
b) $Fe(s) \mid Fe(aq)^{++} \parallel Cu(aq)^{++} \mid Cu(s)$
c) $Pt(s) \mid I(aq)^-, I_2(aq) \parallel Cl_2(g), Cl(aq)^- \mid Pt(s)$

17.7 In the electrolysis of aqueous HCl, hydrogen gas is formed at one electrode and chlorine gas at the other. Diagram a cell suitable for this electrolysis. Label the anode and cathode, and write the two half-cell reaction equations. Indicate the direction of movement of cations and anions in the solution and electrons in the external circuit.

17.8 An electrolytic cell is constructed by immersing copper electrodes into a solution containing $CoCl_2$ and HCl. Write ion-electron equations for reactions which might occur (if there are multiple possibilities) as increasing voltage is applied to the cell. (Use standard reduction potentials.)

17.9 Design a voltaic cell to make use of the following reaction:

$$Zn(s) + Cl_2(g) = Zn(aq)^{++} + 2Cl(aq)^-$$

Label the anode and cathode, write the ion-electron equation for each half-cell reaction, and indicate the direction of motion of anions and cations in the solution and electrons in the external circuit.

17.10 Draw a diagram to show the arrangement of materials necessary to obtain an electric current from the reaction between Zn(s) and aqueous $AgNO_3$.

a) If this cell is short circuited, what is the direction of flow of electrons in the external circuit?
b) Write an ion-electron equation for each half-cell reaction and indicate whether these are reduction or oxidation processes.
c) Indicate the name and polarity (charge) for each electrode.
d) Indicate the direction of migration of ions inside the cell.
e) Compute the standard cell emf.

17.11 Criticize the following statement: When a voltaic cell is converted to an electrolytic cell by inserting an electron pump in the external circuit, the cathode of the voltaic cell becomes the anode of the electrolytic cell and the anode of the voltaic cell becomes the cathode of the electrolytic cell, but the charge signs remain the same.

17.12 Discuss the net ion flow in a salt bridge for the electrolytic cell shown in Fig. 17.3.

17.13 Write the equation for the over-all cell reaction for the cell shown in Fig. 17.3.

17.14 Is it possible to write the over-all equation for the oxidation of phosphorus to H_3PO_4 by HNO_3 by combining Equations (17.31) and (17.36)? Explain.

17.15 How many coulombs are required to deposit
a) 1.00 mole of copper atoms from $Cu(aq)^{++}$?
b) 1.00 g of copper metal from $Cu(aq)^{++}$?

17.16 An aqueous solution of sodium chloride is electrolyzed at 1.93 amps for 15 min with the evolution of hydrogen at the cathode. At standard temperature and pressure, how many moles, grams, and liters of dry hydrogen gas can be produced this way?

17.17 A beaker contains 200 ml of 0.500 *M* aqueous $CuSO_4$. Two platinum electrodes are placed in the solution, and it is electrolyzed with a constant current of 0.500 amp. Ideally, how long should it take to remove substantially all of the copper(II) ions from solution? If all of the copper(II) ions are removed, what is the solute in the remaining solution?

17.18 Using Table 17.2, arrange the following species in order of their increasing strength as oxidizing agents at standard conditions.
a) $Al(aq)^{3+}$ c) $BrO_3(aq)^-$ e) $Li(aq)^+$
b) $Fe(aq)^{3+}$ d) $Cl_2(g)$ f) $H_3O(aq)^+$

17.19 Using Table 17.2, arrange the following species in the order of their increasing strength as reducing agents at standard conditions.
a) $Cu(s)$ d) $H_2O_2(aq)$ f) $SO_3(aq)^{--}$
b) $Hg(l)$ e) $S(aq)^{--}$ g) $MnO_2(s)$
c) $N_2O_4(g)$

17.20 In Table 17.2, inspect the half-cell reaction equations numbered 1, 2, 10, 18, 23, 26, 31, and 57 and indicate for each whether the half-cell solution must be acidic, basic, or could be either or neutral.

17.21 Indicate whether each of the following species is a stronger oxidizing agent in acidic or basic media at standard conditions (use Table 17.2).
a) $F_2(g)$ b) $O_2(g)$ c) $MnO_4(aq)^-$

17.22 Indicate whether each of the following species is a stronger reducing agent in acidic or basic media at standard conditions (use Table 17.2).
a) $H_2(g)$ b) $Fe(s)$ c) $MnO_2(s)$

17.23 Using Table 17.2, predict the reaction that will occur when the following substances are brought together. Write complete ionic equations for each reaction.
a) $I_2(s)$ and $H_2S(g)$ in aqueous acid
b) $Fe(s)$ and aqueous acid
c) $MnO_4(aq)^-$ and $Cr(aq)^{3+}$ in aqueous acid
d) $H_2SO_3(aq)$ and $Br_2(l)$ in aqueous acid
e) $F_2(g)$ in water
f) $Cu(aq)^{++}$ and $Fe(aq)^{++}$ in aqueous acid
g) $Zn(s)$ and $Cr(aq)^{++}$ in aqueous acid
h) $H_2(g)$ and $NO_3(aq)^-$ in aqueous acid

17.24 For each of the following compute the reaction emf and indicate whether or not the reactions are spontaneous, at standard conditions.
a) $5H_2O_2(aq) + 2Mn(aq)^{++} + 4H_2O = 2MnO_4(aq)^- + 6H_3O(aq)^+$
b) $8NO_3(aq)^- + Cl(aq)^- + 8H_3O(aq)^+ = 4N_2O_4(g) + ClO_4(aq)^- + 12H_2O$
c) $2H_2O + S(aq)^{--} = H_2(g) + S(s) + 2OH(aq)^-$

d) $AsO_4(aq)^{3-} + SO_3(aq)^{--} + H_2O = AsO_2(aq)^- + SO_4(aq)^{--} + 2OH(aq)^-$
e) $4MnO_4(aq)^- + 2H_2O = 4MnO_2(s) + 3O_2(g) + 4OH(aq)^-$
f) $10Hg(aq)^{++} + 2Mn(aq)^{++} + 24H_2O = 5Hg_2(aq)^{++} + 2MnO_4(aq)^- + 16H_3O(aq)^+$

17.25 Using the standard half-cell potential values from Table 17.2, write the mathematical relationships for the calculation of half-cell potentials at nonstandard conditions for half-cells 10, 18, 29, 31, 32, 36, 39, and 52 in Table 17.2.

17.26 For the half-cell represented by Equation (17.49), indicate the effect of the following changes on the half-cell potential.
a) addition of more MnO_2
b) an increase in pH
c) an increase in the activity of $MnO_4(aq)^-$

17.27 Compute the half-cell potentials for the following, assuming that molar concentration may be substituted for activity.
a) a silver wire electrode dipping into 0.0010 *M* $Ag(aq)^+$
b) a metallic cadmium electrode dipping into 0.150 *M* $Cd(aq)^{++}$
c) half-cell number 33 in Table 17.2 with a platinum electrode dipping into 0.100 *M* $SO_4(aq)^{--}$, 0.0100 *M* $H_2SO_3(aq)$, at pH 2

17.28 For the following half-cells, which are indicated by number in Table 17.2, indicate the effect of the specified change on the half-cell potential.
a) half-cell 7 : Addition of potassium permanganate
b) half-cell 14: Increase the pressure on the system
c) half-cell 11: Addition of $Ag(aq)^+$
d) half-cell 40: Addition of aqueous ammonia
e) half-cell 34: Increase of pH
f) half-cell 53: Decrease of pH
g) half-cell 36: Decrease of pH

17.29 For the following half-cells, which are indicated by number in Table 17.2, indicate how you might bring about the effect indicated.
a) half-cell 9 : decrease the reduction potential
b) half-cell 20: increase the reduction potential
c) half-cell 35: increase the oxidizing strength of sulfate ion
d) half-cell 36: make the half-cell potential negative
e) half-cell 44: make the metallic ion a better reducing agent

17.30 A galvanic cell is made from the following two half-cells. Compute the potential difference between these two half-cells and indicate the reactions that take place in each of the half-cells when the over-all cell is in operation.
Half-cell A: 0.100 *M* $Ag(aq)^+$ with Ag wire electrode
Half-cell B: 0.00100 *M* $Ag(aq)^+$ with Ag wire electrode

17.31 Consider a half-cell that is 0.10 *M* in $Fe(aq)^{++}$ and 0.10 *M* in $Fe(aq)^{3+}$ with a platinum wire electrode. What is the effect upon the half-cell potential caused by the addition of
a) a strong oxidizing agent
b) water
c) soluble iron(II) salt
d) a soluble iron(III) salt
e) a strong reducing agent
f) a precipitating agent for iron(III) ions
g) a complexing agent for iron(II) ions

17.32 Criticize each of the following statements (half-cell numbers refer to Table 17.2):

a) At standard conditions reducing agents in half-cell 37 and below reduce hydronium ion and liberate $H_2(g)$.

b) At standard conditions, reducing agents in half-cell 52 and below reduce water to $H_2(g)$.

c) At standard conditions, oxidizing agents in half-cells 1 through 29 oxidize hydroxide ions to elemental oxygen.

d) At standard conditions, oxidizing agents in half-cells 1 through 14 oxidize water to elemental oxygen.

17.33 For each of the following, write the mathematical expression for computing the over-all reaction emf at non-standard activities.

a) $2Fe(aq)^{3+} + 3I(aq)^- = 2Fe(aq)^{++} + I_3(aq)^-$

b) $H_2SO_3(aq) + H_2O_2(aq) + H_2O = SO_4(aq)^{--} + 2H_3O(aq)^+$

c) $S_2O_3(aq)^{--} + 2H_3O(aq)^+ = H_2SO_3(aq) + S(s)$

d) $Br_2(l) + 2OH(aq)^- = Br(aq)^- + BrO(aq)^- + H_2O$

e) $5[Fe(CN)_6](aq)^{4-} + MnO_4(aq)^- + 8H_3O(aq)^+ = Mn(aq)^{++} + 5[Fe(CN)_6](aq)^{3-} + 12H_2O$

f) $6Fe(aq)^{++} + Cr_2O_7(aq)^{--} + 14H_3O(aq)^+ = 2Cr(aq)^{3+} + 6Fe(aq)^{3+} + 7H_2O$

17.34 Using the data in Table 17.2, compute the equilibrium constants at 25°C for the following reactions:

a) $Cl_2(g) + 2Br(aq)^- = 2Cl(aq)^- + Br_2(l)$

b) $Pb(aq)^{++} + Cu(s) = Cu(aq)^{++} + Pb(s)$

c) $5Fe(aq)^{++} + MnO_4(aq)^- + 8H_3O(aq)^+ = 5Fe(aq)^{3+} + Mn(aq)^{++} + 12H_2O$

d) $Fe(aq)^{++} + Cu(aq)^{++} = Fe(aq)^{3+} + Cu(aq)^+$

e) $3H_2O_2(aq) + 2NO(g) = 2NO_3(aq)^- + 2H_3O(aq)^+$

17.35 Compute the standard free energy change, $\Delta G°$, for each of the reactions in Problem 17.34.

17.36 What happens to Equation (17.56) under standard conditions?

17.37 Show that Equation (17.65) may be written as follows:

$$\log K = \frac{nF\Delta E°}{2.3RT}$$

17.38 Draw the titration curve for the titration of iron(II) ions with permanganate ions. Consider the titration of 50.00 ml of 0.100 *M* $Fe(aq)^{++}$ which is 1.00 *M* in $H_3O(aq)^+$ with 0.0200 *M* $MnO_4(aq)^-$.

17.39 Criticize the following statement: In the titration of $Fe(aq)^{++}$ with dichromate ions, the potential cannot exceed the standard potential of the iron(II) ion-iron(III) ion half-cell.

17.40 In the titration of $Fe(aq)^{++}$ ions with $Cr_2O_7(aq)^{--}$ ions, why is the $E°$ value for Equation (17.75) not the same as the $E°$ value for Equation (17.74)?

17.41 Why does the equivalence point potential for the titration shown in Fig. 17.6 not come near the middle of the steeply rising portion of the curve?

17.42 Verify the values in Tables 17.3 and 17.4.

17.43 A 0.1462-g sample of pure iron wire was dissolved in acid so that all of the iron was converted to $Fe(aq)^{++}$. This solution required 30.62 ml of aqueous $K_2Cr_2O_7$ for titration. What was the molarity of the $K_2Cr_2O_7$ solution?

17.44 If 0.5025 g of impure iron is converted to a solution of $Fe(aq)^{++}$ and this solution requires 35.41 ml of 0.01660 *M* $K_2Cr_2O_7$ for titration, what is the percent of iron in the original sample?

17.45 Oxalic acid is oxidized to carbon dioxide in acidic solutions by strong oxidizing agents.

a) A 0.4940-g sample of pure $H_2C_2O_4 \cdot 2H_2O$ required 41.60 ml of a $KMnO_4$ solution for titration. What is the molarity of the $KMnO_4$ solution?

b) If 27.36 ml of this $KMnO_4$ solution is required to titrate 0.4060 grams of impure $H_2C_2O_4 \cdot 2H_2O$, what is the purity of this sample of oxalic acid?

17.46 Hydrogen peroxide is oxidized to oxygen in acidic media by strong oxidizing agents. A 50.00 ml sample of aqueous hydrogen peroxide was diluted to 500.0 ml and then 25.00 ml of this solution was made acidic and titrated with 39.80 ml of 0.02660 *M* $KMnO_4$ solution. What was the concentration of the original hydrogen peroxide in mg H_2O_2/ml of solution?

17.47 Solid MnO_2 is a good oxidizing agent and oxidizes oxalates to $CO_2(g)$. A 0.6855-g sample of impure MnO_2 was reduced to $Mn(aq)^{++}$ by addition of 0.9693 g of pure $Na_2C_2O_4$. After all of the MnO_2 had reacted, 41.53 ml of 0.03100 *M* $KMnO_4$ solution was required to titrate the excess oxalate. What was the percent MnO_2 in the original sample?

18

Oxidation States of the Elements

INTRODUCTION In this final chapter, we shall look at the oxidation states of the elements from the point of view of reduction potentials to learn which states are most stable under a given set of conditions. For the most part the emphasis is on oxidation-state transitions in aqueous solutions. The chapter organization is based on the periodic table.

The discussion begins with the halogens, since relatively little good reduction potential data are available for the rare gas species. This section on the halogens is important, however, because generalizations from the techniques employed for the halogens will be useful in succeeding sections.

In Chapters 15 and 16, Bronsted-Lowry acid-base reactions were described almost completely in thermodynamic terms, that is, in terms of equilibrium constants. This was adequate, since these reactions tend to be fast, single step reactions. However, it is obvious from the equations for most redox reactions that they are not single step reactions. As a result, it should not be surprising that simple thermodynamic considerations do not always provide adequate predictions of the actual reactions which occur.

Taking into account the present state of our knowledge of redox reactions, we shall look at these reactions from the thermodynamic point of view and as appropriate point out where predictions based on thermodynamics alone fail in practice.

TABLE 18.1

Some Standard Reduction Potentials for the Halogens

	OXIDIZING AGENTS			REDUCING AGENTS	E° (VOLTS AT 25°C)
1.	$F_2(g) + 2H_3O(aq)^+$	$+\ 2e^- =$	$2H_2O$	$+\ 2HF(aq)$	+3.06
2.	$F_2(g)$	$+\ 2e^- =$		$2F(aq)^-$	+2.87
3.	$HClO_2(aq) + 2H_3O(aq)^+$	$+\ 2e^- =$	$3H_2O$	$+\ HClO(aq)$	+1.64
4.	$2HClO(aq) + 2H_3O(aq)^+$	$+\ 2e^- =$	$3H_2O$	$+\ Cl_2(g)$	+1.63
5.	$H_5IO_6(aq) + H_3O(aq)^+$	$+\ 2e^- =$	$4H_2O$	$+\ IO_3(aq)^-$	+1.6
6.	$2HBrO(aq) + 2H_3O(aq)^+$	$+\ 2e^- =$	$4H_2O$	$+\ Br_2(l)$	+1.59
7.	$2BrO_3(aq)^- + 12H_3O(aq)^+$	$+\ 10e^- =$	$18H_2O$	$+\ Br_2(l)$	+1.52
8.	$BrO_3(aq)^- + 5H_3O(aq)^+$	$+\ 4e^- =$	$7H_2O$	$+\ HBrO(aq)$	+1.49
9.	$2HIO(aq) + 2H_3O(aq)^+$	$+\ 2e^- =$	$4H_2O$	$+\ I_2(s)$	+1.45
10.	$BrO_3(aq)^- + 6H_3O(aq)^+$	$+\ 6e^- =$	$9H_2O$	$+\ Br(aq)^-$	+1.44
11.	$ClO_3(aq)^- + 5H_3O(aq)^+$	$+\ 4e^- =$	$7H_2O$	$+\ HClO(aq)$	+1.42
12.	$2ClO_4(aq)^- + 16H_3O(aq)^+$	$+\ 14e^- =$	$24H_2O$	$+\ Cl_2(g)$	+1.39
13.	$ClO_4(aq)^- + 8H_3O(aq)^+$	$+\ 8e^- =$	$12H_2O$	$+\ Cl(aq)^-$	+1.39
14.	$Cl_2(g)$	$+\ 2e^- =$		$2Cl(aq)^-$	+1.36
15.	$ClO_2(g) + H_3O(aq)^+$	$+\ 1e^- =$	H_2O	$+\ HClO_2(aq)$	+1.28
16.	$2IO_3(aq)^- + 12H_3O(aq)^+$	$+\ 10e^- =$	$18H_2O$	$+\ I_2(s)$	+1.20
17.	$ClO_4(aq)^- + 2H_3O(aq)^+$	$+\ 2e^- =$	$3H_2O$	$+\ ClO_3(aq)^-$	+1.19
18.	$ClO_2(g)$	$+\ 1e^- =$		$ClO_2(aq)^-$	+1.16
19.	$ClO_3(aq)^- + 2H_3O(aq)^+$	$+\ 1e^- =$	$3H_2O$	$+\ ClO_2(g)$	+1.14
20.	$IO_3(aq)^- + 5H_3O(aq)^+$	$+\ 4e^- =$	$7H_2O$	$+\ HIO(aq)$	+1.14
21.	$Br_2(l)$	$+\ 2e^- =$		$2Br(aq)^-$	+1.09
22.	$H_3IO_6(aq)^{--}$	$+\ 2e^- =$	$3OH(aq)^-$	$+\ IO_3(aq)^-$	+0.7
23.	$ClO_2(aq)^- + H_2O$	$+\ 2e^- =$	$2OH(aq)^-$	$+\ ClO(aq)^-$	+0.66
24.	$BrO_3(aq)^- + 3H_2O$	$+\ 6e^- =$	$6OH(aq)^-$	$+\ Br(aq)^-$	+0.61
25.	$ClO_4(aq)^- + 4H_2O$	$+\ 8e^- =$	$8OH(aq)^-$	$+\ Cl(aq)^-$	+0.56
26.	$I_2(s)$	$+\ 2e^- =$		$2I(aq)^-$	+0.54
27.	$BrO_3(aq)^- + 2H_2O$	$+\ 4e^- =$	$4OH(aq)^-$	$+\ BrO(aq)^-$	+0.54
28.	$2BrO_3(aq)^- + 6H_2O$	$+\ 10e^- =$	$12OH(aq)^-$	$+\ Br_2(l)$	+0.51
29.	$ClO_3(aq)^- + 2H_2O$	$+\ 4e^- =$	$4OH(aq)^-$	$+\ ClO(aq)^-$	+0.50
30.	$2BrO(aq)^- + 2H_2O$	$+\ 2e^- =$	$4OH(aq)^-$	$+\ Br_2(l)$	+0.45
31.	$2IO(aq)^- + 2H_2O$	$+\ 2e^- =$	$4OH(aq)^-$	$+\ I_2(s)$	+0.45
32.	$2ClO(aq)^- + 2H_2O$	$+\ 2e^- =$	$4OH(aq)^-$	$+\ Cl_2(g)$	+0.42
33.	$ClO_4(aq)^- + H_2O$	$+\ 2e^- =$	$2OH(aq)^-$	$+\ ClO_3(aq)^-$	+0.36
34.	$2IO_3(aq)^- + 6H_2O$	$+\ 10e^- =$	$12OH(aq)^-$	$+\ I_2(s)$	+0.21
35.	$IO_3(aq)^- + 2H_2O$	$+\ 4e^- =$	$4OH(aq)^-$	$+\ IO(aq)^-$	+0.14
36.	$2H_3O(aq)^+$	$+\ 2e^- =$	$2H_2O$	$+\ H_2(g)$	0.00
37.	$ClO_3(aq)^- + H_2O$	$+\ 1e^- =$	$2OH(aq)^-$	$+\ ClO_2(g)$	−0.50
38.	$H_2O(l)$	$+\ 2e^- =$	$2OH(aq)^-$	$+\ H_2(g)$	−0.83
39.	$H_2(g)$	$+\ 2e^- =$		$2H(aq)^-$	−2.25

In some respects this chapter takes on the appearance of an encyclopedia of chemical reactions. The purpose is to provide a sufficient number of actual examples which may be studied in terms of the principles which have been developed in the previous chapters of this book. This chapter should be studied from this point of view. Many of the questions at the end of the chapter ask you to look at reactions from this point of view.

FAMILY 7A

Standard Reduction Potentials

There are several ways in which standard reduction potentials may be displayed for purposes of comparisons. One is the table of ion-electron equations, arranged in the order of decreasing strength of oxidizing agents and increasing strength of reducing agents. This type of display was shown in Table 17.2 and is shown here for the halogen-containing species in Table 18.1.

From this table it is clear that most of the half-cells have positive potentials, suggesting that halogen-containing species tend to be better oxidizing agents than reducing agents. Also, it is clear that among the oxidizing agents elemental fluorine is the strongest, and among the reducing agents hydride ions are the strongest, if we include hydrogen among the halogens.

Some of the family trends may be discerned more readily from the standard potentials displayed in the diagrammatic Table 18.2. The species of each element are arranged in descending order of central element oxidation state from left to right. The standard potentials for the reduction of the form on the left to the form on the right are shown between each pair of species. Diagrams for acidic and basic conditions are given separately.

TABLE 18.2
Standard Potential Diagram for the Halogens Excluding Hydrogen

IN AQUEOUS ACID

$$F_2 \xrightarrow{+3.06} HF$$

$$ClO_3^- \xrightarrow{+1.42} HClO$$

$$ClO_4^- \xrightarrow{+1.19} ClO_3^- \xrightarrow{+1.14} ClO_2 \xrightarrow{+1.28} HClO_2 \xrightarrow{+1.64} HClO \xrightarrow{+1.63} Cl_2 \xrightarrow{+1.36} Cl^-$$

$$BrO_3^- \xrightarrow{+1.49} HBrO \xrightarrow{+1.59} Br_2 \xrightarrow{+1.09} Br^-$$

$$H_5IO_6 \xrightarrow{+1.6} IO_3^- \xrightarrow{+1.14} HIO \xrightarrow{+1.45} I_2 \xrightarrow{+0.54} I^-$$

IN AQUEOUS BASE

$$F_2 \xrightarrow{+2.87} F^-$$

$$ClO_3^- \xrightarrow{+0.50} ClO^-$$

$$ClO_4^- \xrightarrow{+0.36} ClO_3^- \xrightarrow{-0.50} ClO_2 \xrightarrow{+1.16} ClO_2^- \xrightarrow{+0.66} ClO^- \xrightarrow{+0.42} Cl_2 \xrightarrow{+1.36} Cl^-$$

$$BrO_3^- \xrightarrow{+0.53} BrO^- \xrightarrow{+0.45} Br_2 \xrightarrow{+1.09} Br^-$$

$$H_3IO_6^{--} \xrightarrow{+0.7} IO_3^- \xrightarrow{+0.14} IO^- \xrightarrow{+0.45} I_2 \xrightarrow{+0.54} I^-$$

From this table, it is possible not only to see the relationships among the several oxidation states of the same element, but also to compare similar oxidation-state transitions for the several elements in the family. For example, it is clear in looking down vertical columns in Table 18.2 that for the reduction of the elements to the halide ions in either acidic or basic medium, or for the reduction of the hypohalous acids to the halogens in acidic solution, the halogen-containing species decrease in oxidizing strength as the atomic number of the halogen increases. However, for the reduction of halate ions to hypohalite ions, the bromate ions are the best oxidizing agents, and for the reduction of hypohalite ions to the halogen molecules in basic solution, hypobromite ions and hypoiodite ions are about equally strong with both stronger than hypochlorite ions.

The standard reduction potential data in either Table 18.1 or 18.2 may be used to compute standard reduction potentials for other half-cells. For example, the reduction of chlorate ions to chlorine in acid solution is not included in either of these tables. The following equation shows the reaction for this half-cell.

$$2ClO_3(aq)^- + 12H_3O(aq)^+ + 10e^- = Cl_2(g) + 18H_2O \qquad (18.1)$$

When this half-cell is coupled to a standard hydrogen half-cell, the standard free energy change is related to the standard cell emf as shown in Chapter 17.

$$\Delta G^\circ = -nF\Delta E^\circ \qquad (18.2)$$

The standard cell emf, ΔE°, is the difference between the two standard half-cell reduction potentials.

$$\Delta E^\circ = E^\circ_{\text{red}} - E^\circ_{\text{ox}} \qquad (18.3)$$

In this case E°_{red} is the standard reduction potential for the half-cell represented by Equation (18.1) and E°_{ox} is the standard reduction potential for the hydrogen half-cell, which is zero. Thus Equation (18.2) may be rewritten as follows, where E° is the standard reduction potential for the half-cell represented by Equation (18.1).

$$\Delta G^\circ = -nFE^\circ \qquad (18.4)$$

This means that it is possible to assign free energy changes to half-cell reactions; the values of these changes are equal to the negative product of the number of moles of electrons represented in the half-cell reaction equation, the number of coulombs per mole of electrons, and the standard half-cell potential. Thus it is possible to combine half-cell reaction equations by addition or subtraction in the same way that it is possible to combine other chemical reaction equations. When this is done, the free energy change for the over-all reaction is the sum or difference of the free energy changes for each of the half-cell reactions.

For example, the half-cell reaction represented by Equation (18.1) is the sum of the half-cell reaction equations numbered 4 and 11 in Table 18.1.

$$2HClO(aq) + 2H_3O(aq)^+ + 2e^- = Cl_2(g) + 4H_2O \qquad (18.5)$$

$$2ClO_3(aq)^- + 10H_3O(aq)^+ + 8e^- = 2HClO(aq) + 14H_2O \qquad (18.6)$$

The standard free energy change, ΔG_1°, for the half-cell reaction represented by Equation (18.1) is related to its own standard reduction potential, E_1°, through Equation (18.7) and also to the sum of the standard free energy changes, ΔG_5° and ΔG_6°, for the half-cell reactions represented by Equations (18.5) and (18.6).

$$\Delta G_1^\circ = -n_1 F E_1^\circ \tag{18.7}$$

$$\Delta G_1^\circ = \Delta G_5^\circ + \Delta G_6^\circ = -n_5 F E_5^\circ - n_6 F E_6^\circ \tag{18.8}$$

The expressions in these two equations may be combined and rearranged and solved for E_1° in the following way.

$$-n_1 F E_1^\circ = -n_5 F E_5^\circ - n_6 F E_6^\circ$$

$$E_1^\circ = \frac{n_5 E_5^\circ + n_6 E_6^\circ}{n_1} \tag{18.9}$$

Using the half-cell potential values from either Table 18.1 or 18.2, the half-cell potential for the reaction represented by Equation (18.1) may be computed.

$$E_1^\circ = \frac{2(1.63) + 8(1.42)}{10} = 1.46 \text{ V}$$

This technique may be used to build a more comprehensive list of standard reduction potentials, which may be displayed to show some properties of those forms of the elements in oxidation states intermediate between the highest and lowest possible. For example, the oxidation state of chlorine in chlorate ion is 5+, which is between the maximum of 7+ and minimum of 1− exhibited by chlorine. Thus chlorate ions may serve as reducing agents when they are oxidized to perchlorate ions, or they may serve as oxidizing agents when they are reduced to chlorine(IV) oxide, chlorite ions or chlorous acid, hypochlorite ions or hypochlorous acid, chlorine, or chloride ions.

Those species which may serve as either oxidizing or reducing agents sometimes serve as both simultaneously. A reaction in which a single species is both oxidized and reduced by itself is called a *disproportionation reaction*. For example, chlorate ions may disproportionate to perchlorate ions and chlorine in acid.

$$7ClO_3(aq)^- + 2H_3O(aq)^+ = Cl_2(g) + 5ClO_4(aq)^- + 3H_2O \tag{18.10}$$

This equation is the sum of the following two half-cell reaction equations:

$$2ClO_3(aq)^- + 12H_3O(aq)^+ + 10e^- = Cl_2(g) + 18H_2O \tag{18.11}$$

$$5ClO_3(aq)^- + 15H_2O = 5ClO_4(aq)^- + 10H_3O(aq)^+ + 10e^- \tag{18.12}$$

Thus the standard emf of the over-all reaction is the potential difference between these two half-cells.

$$\Delta E^\circ = E_{red}^\circ - E_{ox}^\circ \tag{18.13}$$

Since E°_{red} is the standard reduction potential for the half-cell in which chlorate ions are serving as oxidizing agents (are being reduced) and since E°_{ox} is the standard reduction potential for the half-cell in which chlorate ions are serving as reducing agents (are being oxidized), it should be clear that the over-all reaction emf will be positive when E°_{red} is more positive than E°_{ox}. This means that chlorate ions tend to disproporationate when their reduction potential as oxidizing agents is more positive than their reduction potential as reducing agents. In this specific case, $E^\circ_{red} = +1.46$ V and $E^\circ_{ox} = +1.19$ V, and the over-all emf for the disproportionation reaction is +0.27 V. Thus we say that chlorate ions are unstable with respect to disproportionation to perchlorate ions and chlorine in acidic solution at standard conditions. It should be recognized that this is a thermodynamic instability, since the emf is related directly to the free energy change. Whether or not this disproportionation occurs in practice depends upon the mechanism of the reaction and the free energies of activation of the various steps in the mechanism.

Of course it is possible that chlorate ions may also be unstable with respect to disproportionation to other chlorine-containing species. For example, they might disproportionate to perchlorate ions and chlorine(IV) oxide, or to perchlorate ions and chloride ions. In order to observe these instabilities, it is convenient to display the standard reduction potentials for chlorine-containing species as shown in Table 18.3. In this table, all chlorine-containing species that may act as reducing agents

TABLE 18.3
Standard Half-Cell Reduction Potentials for Chlorine-containing Species (Volts at 25°C)

In Aqueous Acid

OXIDIZING AGENTS	REDUCING AGENTS					
	$Cl(aq)^-$	$Cl_2(g)$	$HClO(aq)$	$HClO_2(aq)$	$ClO_2(g)$	$ClO_3(aq)^-$
$ClO_4(aq)^-$	+1.39	+1.39	+1.35	+1.20	+1.17	+1.19
$ClO_3(aq)^-$	+1.44	+1.46	+1.42	+1.21	+1.14	
$ClO_2(g)$	+1.51	+1.55	+1.52	+1.28		
$HClO_2(aq)$	+1.54	+1.65	+1.64			
$HClO(aq)$	+1.50	+1.63				
$Cl_2(g)$	+1.36					

In Aqueous Base

OXIDIZING AGENTS	REDUCING AGENTS					
	$Cl(aq)^-$	$Cl_2(g)$	$ClO(aq)^-$	$ClO_2(aq)^-$	$ClO_2(g)$	$ClO_3(aq)^-$
$ClO_4(aq)^-$	+0.56	+0.45	+0.45	+0.34	+0.07	+0.36
$ClO_3(aq)^-$	+0.63	+0.48	+0.50	+0.33	−0.50	
$ClO_2(g)$	+0.86	+0.73	+0.83	+1.16		
$ClO_2(aq)^-$	+0.78	+0.58	+0.66			
$ClO(aq)^-$	+0.89	+0.42				
$Cl_2(g)$	+1.36					

are listed horizontally across the top. Similarly, all chlorine-containing species that may act as oxidizing agents are listed vertically at the left.

Thus, for example, the standard reduction potential for the reduction of perchlorate ions to chloride ions in acid is +1.39 V and the standard reduction potential for the reduction of perchlorate ions to chlorate ions in acid is +1.19 V. Similarly, the standard reduction potential for the oxidation of chlorine to chlorate ions in acid is +1.46 V and for the oxidation of hypochlorous acid to chlorine(IV) oxide in acid is +1.52 V.

To see how Table 18.3 may be used to check for disproportionations, consider chlorate ions in acidic solution. As reducing agents, they can be oxidized only to perchlorate ions in acidic solution, as shown in the last vertical column. Thus chlorate ions in acidic solution tend to disproportionate to perchlorate ions, and to any other species for which chlorate ion as an oxidizing agent shows a reduction potential greater than +1.19 V. Opposite chlorite ion in the second horizontal row from the top we find reduction potentials in excess of +1.19 V for the reduction of $ClO_3(aq)^-$ to $Cl(aq)^-$, $Cl_2(g)$, $HClO(aq)$, and $HClO_2(aq)$. The greatest difference in potential exists for the disproportionation of chlorate ions to perchlorate ions and chlorine in acid, but chlorate ions also are unstable with respect to disproportionation to perchlorate ions and chloride ions, perchlorate ions and hypochlorous acid, and perchlorate ions and chlorous acid.

Similarly, in basic solution, the highest reduction potential for chlorate ion as an oxidizing agent is +0.63 V to chloride ion, and the lowest potential for chlorate ion as a reducing agent is +0.36 V to perchlorate ion. Thus, in basic solution, chlorate ion has the greatest tendency to disproportionate to perchlorate ion and chloride ion, although it is unstable also toward disproportionation to $Cl_2(g)$ and $ClO_4(aq)^-$ as well as $ClO(aq)^-$ and $ClO_4(aq)^-$.

This example shows that thermodynamic instabilities often change with the pH of the solution. Note that the potentials which we have been using are standard potentials valid only when activities of dissolved species are unity. Thus in one sense, we are looking at the two extremes of pH and we can only infer what may happen at intermediate pH values. Finally we should remember that thermodynamic instability does not guarantee that the predicted reaction actually occurs, since this depends on the reaction mechanism and the free energies of activation of the various steps in that mechanism.

Factors Affecting Standard Reduction Potentials

Since standard reduction potentials are related directly to standard free energy changes, those factors which affect free energy changes must also affect the potentials. In other words, the reduction potentials may be understood in terms of enthalpy and entropy changes.

For example, consider the conversion of the halogens to the halides in basic solution, which may be represented by the following half-cell reaction, in which X stands for a halogen atom:

$$X_2 + 2e^- = 2X^- \qquad (18.14)$$

One danger in writing this type of general equation is that it makes the four half-cells under consideration look too much alike. In reality, the nature of the free elements is different; fluorine and chlorine are gases, bromine is a liquid, and iodine is a solid.

The enthalpy changes may be understood with the aid of the following Born-Haber-type cycle:

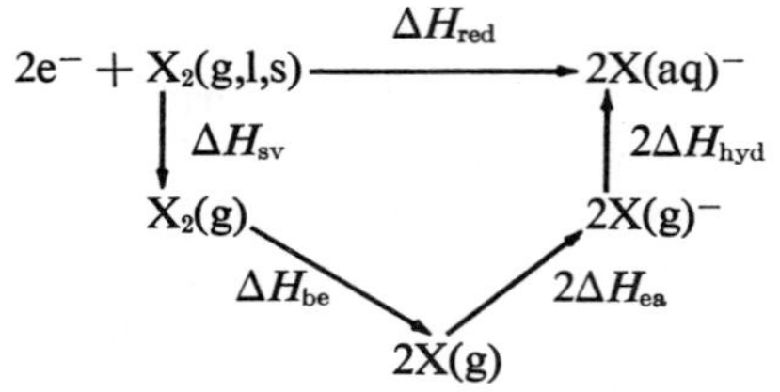

In this cycle, ΔH_{sv} is the molar enthalpy of sublimation or vaporization depending upon the original state of the halogen; ΔH_{be} is the molar enthalpy change associated with the breaking of the X—X bond; ΔH_{ea} is the molar enthalpy change associated with the absorption of electrons by the halogen atoms due to their electron affinity; ΔH_{hyd} is the molar enthalpy of hydration; ΔH_{red} is the enthalpy change associated with the over-all half-cell reaction, and it is the sum of all of the other enthalpy changes in the cycle.

$$\Delta H_{red} = \Delta H_{sv} + \Delta H_{be} + 2\Delta H_{ea} + 2\Delta H_{hyd} \qquad (18.15)$$

Some approximate values for these enthalpy changes are recorded in Table 18.4. The most negative value for ΔH_{red} corresponds to the most favorable enthalpy

TABLE 18.4

Some Standard Enthalpy Data for Half-cell Reaction Equation (18.14) in kcal

	ΔH_{sv}	ΔH_{be}	$2\Delta H_{ea}$	$2\Delta H_{hyd}$	ΔH_{red}
X = F		+37	−2 × 84	−2 × 117	−365
X = Cl		+58	−2 × 87	−2 × 85	−286
X = Br	+7	+46	−2 × 82	−2 × 78	−267
X = I	+15	+36	−2 × 72	−2 × 68	−235

change, and this occurs for fluorine. Clearly the trend in enthalpy changes is the same as the trend in standard reduction potentials. Contributing to the very high enthalpy change for the fluorine half-cell is the lack of heat absorption to change the halogen to the gaseous state, and the rather low F—F bond energy, which we have attributed to the small size of the fluorine atoms and the correspondingly great repulsions between electron pairs on the surfaces of these small atoms. Also contributing to the high value of the enthalpy change for the fluorine half-cell is the very negative enthalpy of hydration due to the small size and correspondingly high charge density on the fluoride ion.

Note that for chlorine, bromine, and iodine the bond energy enthalpy change runs counter to the over-all enthalpy change, since the smaller bond energies should lead to increasingly negative enthalpy differences. This trend is countered by both the electron affinities and the enthalpies of hydration and (for bromine and iodine) by the enthalpies of vaporization and sublimation.

Insofar as entropy trends are concerned, the over-all change from the gaseous state to the aqueous state for fluorine and chlorine is an unfavorable entropy change. In addition the high charge density on the fluoride ion leads to high attractions for water molecules, with an increase in the organization of the solvent, which also is an unfavorable entropy change. As a result, the over-all free energy change for the fluorine and chlorine half-cells is not as good as the over-all enthalpy change suggests. However, for the bromine and iodine half-cells the entropy change is small, so the enthalpy change is a close approximation to the free energy change.

The Born-Haber-type cycle is useful in pointing out some of the energy terms that may be considered in regard to half-cell changes. In more complex systems, similar changes may be considered, but the necessary data for numerical evaluations such as those given in Table 18.4 often are not available. However, in many cases it is possible to choose those particular factors which are most significant and to evaluate these factors quantitatively or qualitatively to explain the trends in reduction potential values.

For example, consider the generalized half-cell reaction in which hypohalite ions are reduced to the halogens in basic solution.

$$2XO(aq)^- + 2H_2O + 2e^- = X_2(s,l,g) + 4OH(aq)^- \qquad (18.16)$$

The followng Born-Haber cycle may be used to aid in the evaluation of the enthalpy changes.

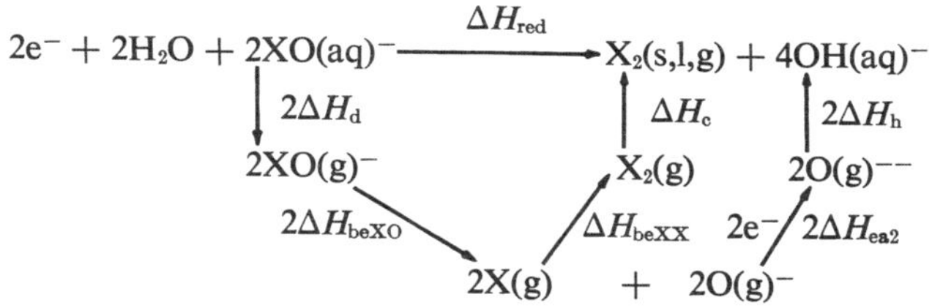

In this cycle, ΔH_d is the molar enthalpy of dehydration of the hypohalite ions; ΔH_{beXO} is the molar enthalpy change due to the breaking of X—O bonds; ΔH_{beXX} is the molar enthalpy change due to the formation of X—X bonds; ΔH_c is the molar enthalpy of condensation to a liquid or solid; ΔH_{ea2} is the molar enthalpy change due to the second electron affinity of oxygen; ΔH_h is the molar enthalpy change for the reaction of gaseous oxide ions with water; ΔH_{red} is the enthalpy change for the over-all half-cell reaction as it is represented by Equation (18.16).

Within this cycle, ΔH_{ea2} and ΔH_h have the same value regardless of the halogen involved. Furthermore, it seems reasonable to assume that ΔH_d probably does not vary greatly from one halite ion to another. Thus for comparing hypochlorite ions with hypobromite ions and hypoiodite ions, it is possible to compare the sums of $2\Delta H_{\mathrm{beXO}}$, ΔH_{beXX}, and ΔH_c. This is done in Table 18.5. It is clear that the data in the last column follow approximately the same trend as the reduction potential data in Table 18.2 for this half-cell reaction. Both sets of data indicate

TABLE 18.5

Some Standard Enthalpy Data for Half-cell Reaction Equation (18.16) in kcal

	$2\Delta H_{\mathrm{beXO}}$	ΔH_{beXX}	ΔH_c	$2\Delta H_{\mathrm{beXO}} + \Delta H_{\mathrm{beXX}} + \Delta H_c$
X = Cl	$+2 \times 60$	-58	0	$+62$
X = Br	$+2 \times 48$	-46	-7	$+43$
X = I	$+2 \times 48$	-36	-15	$+45$

that hypobromite and hypoiodite ions are about equally strong as oxidizing agents, but that both are somewhat stronger than hypochlorite ions. Considering the changes of state involved, the entropy change should be most unfavorable for the hypoiodite half-cell and the hypobromite half-cell, and least unfavorable for the hypochlorite half-cell. However, considering the magnitudes of these changes, it is not to be expected that the entropy effects will be sufficient to change the trend. Rather they will simply narrow the difference between the free energy change for the hypochlorite half-cell and the other two half-cells.

One final factor which markedly affects reduction potentials in many systems is the pH of the solution in which the reactions are carried out. This fact is evident from the data in Tables 18.2 and 18.3, for example. In general the effect of a pH change on reduction potentials can be discerned from the Nernst equation. We shall discuss these effects in the following sections, in which the species of specific halogens are discussed.

Fluorine

From the data in Table 18.1 and 18.2 it is evident that fluorine is a very strong oxidizing agent, and also that it is a stronger oxidizing agent in aqueous acid than it is in aqueous base. The reason for this is the basicity of the fluoride ion, which can be seen if we apply the Nernst equation to the half-cell reaction. If we choose the half-cell reaction in basic solution, the Nernst equation may be applied as follows:

$$F_2(g) + 2e^- = 2F(aq)^- \qquad (18.17)$$

$$E = E^\circ - \frac{0.0591}{2} \log \frac{a_{F^-}^2}{a_{F_2}} \qquad (18.18)$$

According to the data in Tables 18.1 and 18.2, when the hydroxide ion activity is 1.00, the potential of this half-cell is +2.87 V. However, as the half-cell solution is made more acidic, the activity of the fluoride ion decreases because the fluoride ion is basic and reacts with the added acid. As the fluoride ion activity decreases, the value for the logarithm term in Equation (18.18) decreases so that E becomes more positive. When the hydronium ion activity reaches 1.00, the half-cell is best described as follows:

$$F_2(g) + 2H_3O(aq)^+ + 2e^- = 2HF(aq) + 2H_2O \qquad (18.19)$$

In order to show the effect of pH changes on half-cell potentials, we find it convenient to prepare graphs in which the potential in volts is plotted against the pH of the solution. In making these graphs, we choose to keep all half-cell species at unit activity with the exception of $H_3O(aq)^+$ and $OH(aq)^-$. In this way, it is possible to see the effects of pH changes without any complications from changes in the activities of other species.

In Fig. 18.1, the stepwise construction of the plot of potential against pH for fluorine is shown. Line 1 in the graph is derived from the half-cell represented by Equation (18.19) when the half-cell solution is strongly acidic. The Nernst equation for this half-cell is as follows:

$$E = E° - \frac{0.0591}{2} \log \frac{a^2_{HF}}{a_{F_2}\, a^2_{H_3O^+}} \qquad (18.20)$$

If all activities are kept at unity except that of the hydronium ion, Equation (18.20) may be rewritten and rearranged as follows where pH is $-\log a_{H_3O^+}$:

$$E = E° - \frac{0.0591}{2} \log \frac{1}{a^2_{H_3O^+}} = 3.06 - \frac{0.0591}{2} \log a^{-2}_{H_3O^+}$$

$$= 3.06 + 0.0591 \log a_{H_3O^+}$$

$$E = 3.06 - 0.0591\ \text{pH} \qquad (18.21)$$

It is clear that if E is plotted against pH, the graph will be a straight line with a slope of -0.0591. This is line 1 in Fig. 18.1.

According to Equations (18.17) and (18.18), in highly basic solutions, the half-cell potential is not affected by pH changes if the activities of other species are kept constant at unity. Thus line 2 in Fig. 18.1b is a horizontal line at $E = E° = +2.87$ V.

Since fluoride ion is a base, the half-cell solution maintains an acid-base equilibrium described by the following equation:

$$F(aq)^- + H_2O = HF(aq) + OH(aq)^- \qquad (18.22)$$

As the acidity of the solution is increased to decrease the pH, some of the hydroxide ion is consumed and the equilibrium shifts, some of the fluoride ion being consumed in the process. Thus as we imagine the reduction of the pH we must also imagine that additional fluoride ion must be added in order to maintain the

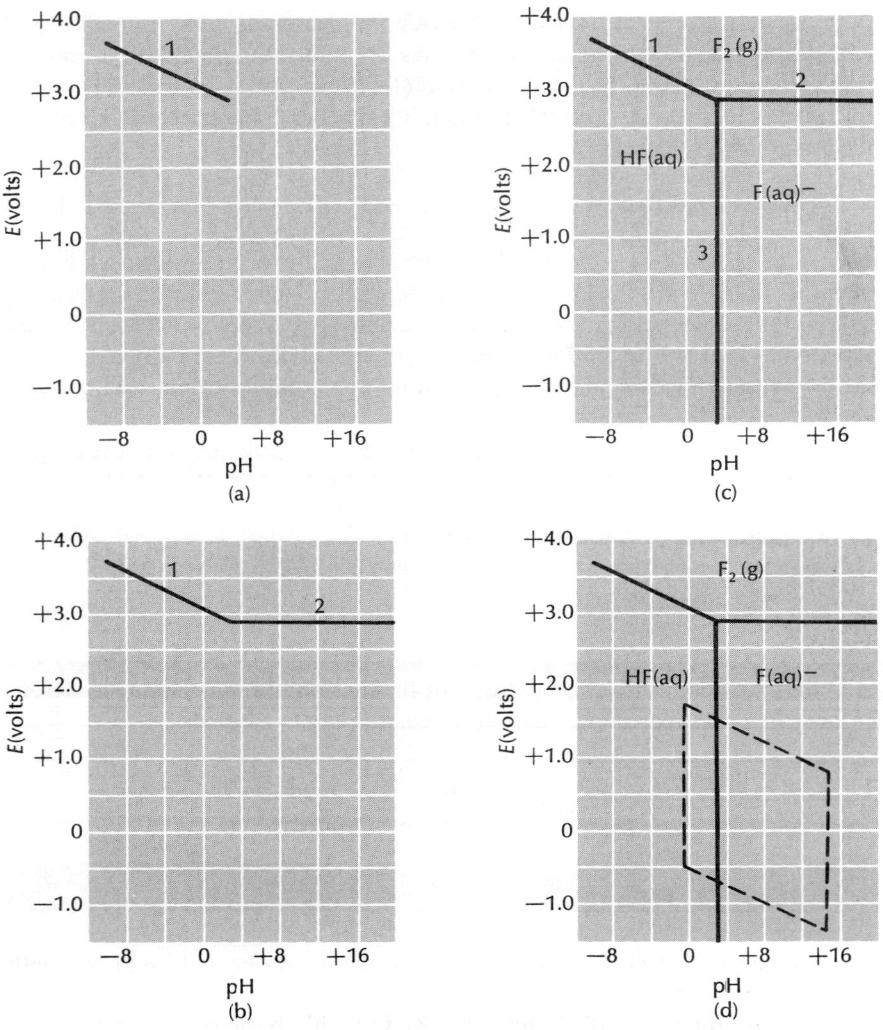

Fig. 18.1. Construction of the fluorine oxidation-state map

fluoride ion activity at 1.00. At very high pH the activity of $F(aq)^-$ is much more than the activity of HF(aq). However, as the pH is decreased and the activity of $F(aq)^-$ is maintained at 1.00, the activity of HF(aq) increases.

In plotting line 1 in Fig. 18.1, the activity of HF(aq) was maintained at 1.00. Thus when lines 1 and 2 meet in Fig. 18.1b, both activities must be 1.00 for HF(aq) and $F(aq)^-$. This means that at this pH at which the lines intersect, the activities of HF(aq) and $F(aq)^-$ must be the same. At higher pH values the predominant form of fluorine in the 1− oxidation state is $F(aq)^-$ and at lower pH values the predominant form of fluorine in the 1− oxidation state is HF(aq). The acid-base equilibrium involved in this half-cell reaction may be described by the acid constant for HF.

$$K_a = \frac{[H_3O(aq)^+][F(aq)^-]}{[HF(aq)]} \tag{18.23}$$

However, at the point of intersection of lines 1 and 2 in Fig. 18.1, $[F(aq)^-] = [HF(aq)]$, if molarities are equated to activities, and Equation (18.23) may be simplified to show that the pH is the same as pK_a.

$$K_a = [H_3O(aq)^+]$$

$$pK_a = pH \tag{18.24}$$

Thus line 3 is drawn vertically from the intersection of lines 1 and 2, and this line corresponds to a constant pH equal to pK_a for HF(aq).

Each of the areas on the graph in Fig. 18.1c may be assigned to a particular fluorine-containing species. Thus the area above lines 1 and 2 represents the combinations of pH and E values at which elemental fluorine is the thermodynamically stable form of the element. Thus, if the fluorine-fluoride half-cell could be coupled to a half-cell with a potential higher than the value on line 2 at pH 8 for example, the spontaneous reaction would be the conversion of fluoride ion to fluorine. On the other hand, if the potential is below line 2, the spontaneous reaction is the conversion of fluorine to fluoride ions, and hydrated fluoride ion is the stable form of fluorine under these conditions. Vertical line 3 indicates the pH above which fluorine in the 1− state exists predominantly as the hydrated fluoride ion and below which it exists predominantly as the hydrated hydrogen fluoride molecule.

Thus the plot of potential against pH results in a graph which may be interpreted in terms of regions in which the oxidation states of the element in question have thermodynamic stability. For this reason we shall refer to such plots as *oxidation-state maps.*

It should be clear from Fig. 18.1c that elemental fluorine is strongest as an oxidizing agent in strongly acidic solutions, that is, solutions with low pH. For as represented, it will oxidize any half-cell with a potential below the value on line 1, and since line 1 rises sharply at low pH values, elemental fluorine is stronger as an oxidizing agent as the pH decreases. Elemental fluorine exhibits its poorest oxidizing ability in basic solutions since it will only oxidize half-cells with potentials below line 2. By the same argument, fluorine in the 1− oxidation state is most easily oxidized to elemental fluorine in basic solution, where it exists predominantly as $F(aq)^-$. Fluorine in the 1− state as HF(aq) in acidic solution is oxidized to elemental fluorine only at higher potentials.

In Fig. 18.1d the area enclosed by dashed lines represents the conditions of pH and potential which one can reasonably expect to attain in aqueous solutions. The two vertical lines represent the approximate practical pH limits. As a practical matter it is difficult to prepare aqueous solutions with the hydronium ion activity much in excess of 10 M. This corresponds to a pH of −1, and thus the left-hand vertical line in Fig. 18.1d is drawn at this pH value. A similar practical limit exists on the hydroxide ion activity at about 10 *M* which corresponds to a pH of 15. Thus the right-hand vertical dotted line in Fig. 18.1d is drawn at pH 15.

The other two dashed lines in Fig. 18.1d indicate limits due to the aqueous medium. For example, in strongly acidic media, the hydronium ion is susceptible to reduction as shown by half-cell number 36 in Table 17.2, the half-cell reaction equation for which is reproduced here.

$$2H_3O(aq)^+ + 2e^- = H_2(g) + 2H_2O \tag{18.25}$$

If another half-cell with a lesser potential is coupled to this one, that other half-cell undergoes oxidation and hydrogen is evolved. Thus at any potential less than that of the hydrogen half-cell, the medium (in the form of hydronium ions) is oxidized.

The dependence of the hydrogen half-cell potential on pH is given by the Nernst equation.

$$E = E^\circ - \frac{0.0591}{2} \log \frac{a_{H_2}}{a^2_{H_3O^+}} \tag{18.26}$$

By taking the activity of $H_2(g)$ at unity, this equation may be rewritten.

$$E = 0.00 - \frac{0.0591}{2} \log \frac{1}{a^2_{H_3O^+}} = -\frac{0.0591}{2} \log a^{-2}_{H_3O^+}$$

$$= 0.0591 \log a_{H_3O^+}$$

$$E = -0.0591 \text{ pH} \tag{18.27}$$

In basic solution, the reduction of the medium is represented by half-cell number 51 in Table 17.2, the half-cell reaction equation for which is reproduced here.

$$2H_2O + 2e^- = 2OH(aq)^- + H_2(g) \tag{18.28}$$

The relationship between potential and pH given by Equation (18.27) is valid also in basic solution. A plot of potential against pH produces a line with a slope of -0.0591. Thus it is expected that any potential lower than that for the hydrogen half-cell will lead to the production of hydrogen. In practice, it is found that half-cells must have potentials somewhat lower than the anticipated before hydrogen evolution is appreciable. In effect, this gives the appearance of a lower value for the hydrogen half-cell potential than actually is the case. This apparent decrease in the hydrogen half-cell potential is called the *hydrogen overvoltage*. Although the hydrogen overvoltage can vary widely depending upon the nature of the system, for rough approximations we may estimate it to be about 0.5 V. Thus if we subtract 0.5 from Equation (18.26) we obtain the relationship which is shown as the bottom dotted line in Fig. 18.1d.

$$E = -0.0591 \text{ pH} - 0.5 \tag{18.29}$$

In a similar way, at high potentials the aqueous medium may be oxidized. In acid medium, the half-cell involved is number 14 in Table 17.2, and in basic medium, the half-cell involved is number 30 in Table 17.2. The half-cell reaction equations are reproduced here.

$$O_2(g) + 4H_3O(aq)^+ + 4e^- = 2H_2O \qquad (18.30)$$

$$O_2(g) + 2H_2O + 4e^- = 4OH(aq)^- \qquad (18.31)$$

As in the previous case, the relationship between potential and pH is the same in both acidic and basic solutions and may be derived by applying the Nernst equation to Equation (18.30).

$$E = E^\circ - \frac{0.0591}{4} \log \frac{1}{a_{O_2} \times a^4_{H_3O^+}} \qquad (18.32)$$

With the activity of oxygen kept at unity, Equation (18.31) may be rewritten and rearranged.

$$E = 1.23 - \frac{0.0591}{4} \log \frac{1}{a^4_{H_3O^+}} = 1.23 - \frac{0.0591}{4} \log a^{-4}_{H_3O^+}$$

$$= 1.23 + 0.0591 \log a_{H_3O^+}$$

$$E = 1.23 - 0.0591 \text{ pH} \qquad (18.33)$$

Theoretically, any potential above that given by Equation (18.33) should lead to the reversal of the reactions as represented in Equations (18.30) and (18.31), with the decomposition of water and evolution of oxygen. In practice, it is found that the actual potential must be somewhat higher than predicted; this increase is called the *oxygen overvoltage*. Although the oxygen overvoltage may vary widely, it may be roughly estimated to be about 0.5 V. If this is added to Equation (18.33), the following relationship results, which is used to plot the upper dashed line in Fig. 18.1d.

$$E = 1.23 - 0.0591 \text{ pH} + 0.5 = 1.7 - 0.0591 \text{ pH} \qquad (18.34)$$

The fact that the area of Fig. 18.1 in which $F_2(g)$ is stable is well above the aqueous stability region indicates that fluorine is such a strong oxidizing agent that it oxidizes the water to elemental oxygen. Thus attempts to prepare elemental fluorine by oxidizing aqueous solutions of fluoride end in failure because the water is more easily oxidized than the fluoride ion.

Chlorine

The oxidation-state map for chlorine as shown in Fig. 18.2 is quite simple despite the multiplicity of oxidation states displayed by the element. Clearly perchlorate ions and chloride ions are the two most widely stable chlorine-containing

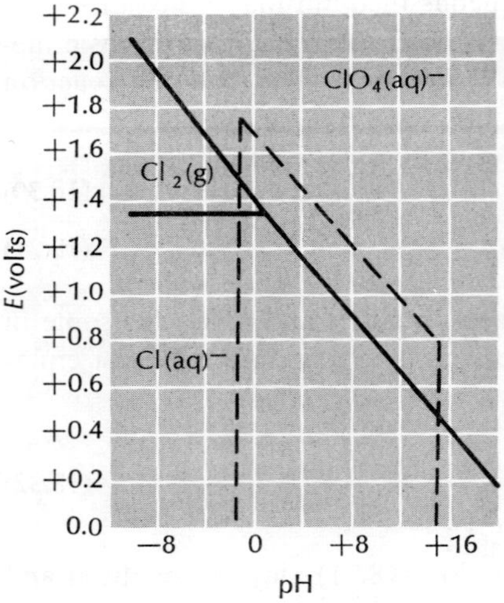

Fig. 18.2. Chlorine oxidation-state map

species. However, elemental chlorine is stable at very low pH values. Both chlorine and perchlorate ions are sufficiently strong oxidizing agents to oxidize water, neglecting overvoltage effects, but normal overvoltage is sufficient to make them appear stable in water. Perchlorate ions decrease in oxidizing strength as pH increases.

Chloride ions are poor reducing agents, serving best in this capacity in basic solutions. In strongly acidic solutions the oxidation of chloride ions leads to chlorine, and in base it leads ultimately to perchlorate ions. Why this is true can be seen from the Nernst equation.

In each of these cases, a low reduction potential for the half-cells represented by the following equations would correspond to good reducing agent properties for chloride ions.

$$Cl_2(g) + 2e^- = 2Cl(aq)^- \tag{18.35}$$

$$2ClO_4(aq)^- + 16H_3O(aq)^+ + 16e^- = 2Cl(aq)^- + 24H_2O \tag{18.36}$$

The potential for the half-cell represented by Equation (18.35) is independent of pH since chloride ions are not basic. However, the potential for the half-cell represented by Equation (18.36) is pH dependent and, according to the Nernst equation, the potential decreases as the pH increases. Thus in the oxidation of $Cl(aq)^-$ to $ClO_4(aq)^-$, chloride ion becomes a stronger reducing agent as the pH increases, that is, it is more easily oxidized.

Since the potential of the half-cell associated with Equation (18.35) remains constant and the potential of the half-cell associated with Equation (18.36) decreases as pH increases, at some sufficiently high pH, the potential of the half-cell associated with Equation (18.36) must be smaller than the potential associated with the half-cell of Equation (18.35). At pH values greater than this, when chloride ion is a reducing agent it will be oxidized to $ClO_4(aq)^-$ instead of $Cl_2(g)$.

Curiously, only three chlorine-containing species appear on the map. One might ask about such species as HClO(aq), $ClO(aq)^-$, $HClO_2(aq)$, $ClO_2(aq)^-$, $ClO_2(g)$, and $ClO_3(aq)^-$, all of which appear in half-cell reaction equations shown in Table 18.1. For example, we might ask why there is not a line on the map for the perchlorate-chlorate half-cell represented by Equations 17 and 33 in Table 18.1. Such a line has been inserted as a dashed line in the oxidation-state map for chlorine which has been reproduced as Fig. 18.3. At any potential above this line,

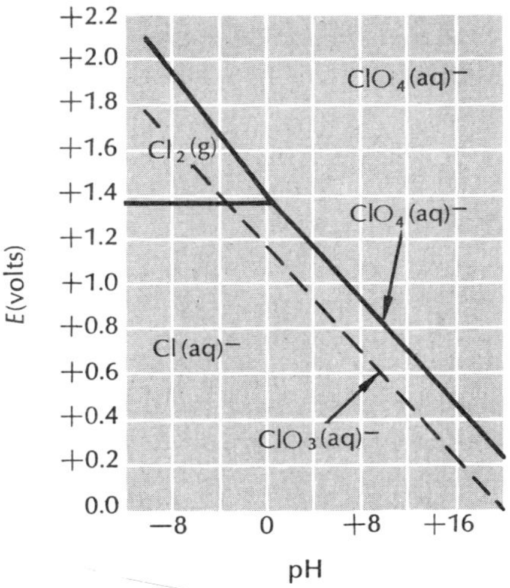

Fig. 18.3. Chlorine oxidation-state map including the perchlorate-chlorate half-cell

chlorate ions are oxidized to perchlorate ions and at any potential below this line, perchlorate ions are reduced to chlorate ions.

Now suppose that we consider the chlorine oxidation-state system at pH zero. At a potential of +1.0 V, perchlorate ions may be reduced to either chlorine or chlorate ions or chloride ions. Thus if perchlorate ions are reduced to chlorate ions, the chlorate ions may be further reduced to chloride ions. Clearly, at this potential and pH, all chlorine-containing species should be reduced to chloride ions.

Now suppose that we consider a potential of +1.3 V at pH zero. Under these conditions, chlorate ions may be oxidized to perchlorate ions, but the potential is low enough to reduce perchlorate ions to chlorine and chlorine to chloride ions. Again, all chlorine-containing species tend to be reduced to chloride ions.

Next consider a potential of +1.38 V at pH zero. Chlorate ions may be oxidized to perchlorate ions but at this potential perchlorate ions may be reduced to chlorine. Under these conditions all chlorine-containing species tend to be converted to chlorine: chloride ions by oxidation and all others by reduction.

Similarly, it can be shown that this is the case for all other combinations of species. Thus under the conditions cited, there are only three thermodynamically stable chlorine-containing species: chloride ion, chlorine, and perchlorate ion. It should be noted that thermodynamic stability may change under different conditions. Thus some species are unstable under certain sets of conditions and stable under others. It must be remembered that the oxidation state maps which we are using are prepared for rather limited conditions.

It must be noted also that not all thermodynamically feasible reactions actually occur. If a high free energy of activation appears in one of the steps of the reaction mechanism, the reaction may not occur. Under these circumstances it is possible that some less favorable but thermodynamically feasible reaction may occur. For example, at pH zero and a potential of +1.6 V, chloride ions should be oxidized to chlorine, which in turn should be oxidized to perchlorate ions. However, it may be that the oxidation to chlorine is rapid and the oxidation to perchlorate ions slow, so that the actual reaction observed is that of chloride ions oxidized to chlorine.

The disproportionation tendencies of chlorine-containing species are summarized in Table 18.3, which was discussed earlier in part. It was noted then that chlorate ions are unstable with respect to disproportionation to perchlorate ions and chlorine, chloride ions, hypochlorous acid, and chlorous acid in strongly acidic solutions. The greatest tendency is to disproportionate to perchlorate ions and chlorine.

In basic solution, chlorate ions are unstable with respect to disproportionation to perchlorate ions and chloride ions, hypochlorite ions, or chlorine; the strongest tendency is to produce perchlorate ions and chloride ions:

$$4ClO_3(aq)^- = Cl(aq)^- + 3ClO_4(aq)^- \qquad (18.37)$$

Chlorine, which shows thermodynamic stability at pH zero does not tend to disproportionate at this pH, but does tend to disproportionate in basic solution to chloride ions and hypochlorite ions, perchlorate ions, chlorate ions, chlorite ions, or chlorine(IV) oxide. It shows the greatest tendency to form chloride ions and hypochlorite ions:

$$Cl_2(g) + 2OH(aq)^- = Cl(aq)^- + ClO(aq)^- + H_2O \qquad (18.38)$$

In fact, this is the way in which hypochlorite bleaches are prepared by adding chlorine to basic aqueous solutions. It is true that the hypochlorite ions tend to disproportionate to chloride ions and perchlorate ions, chlorate ions, chlorine(IV) oxide, or chlorite ions, the greatest tendency being to form chloride ions and perchlorate ions:

$$4ClO(aq)^- = ClO_4(aq)^- + 3Cl(aq)^- \qquad (18.39)$$

SUMMARY. Chloride ion is the usual form of occurrence of chlorine in nature, as suggested by the wide range of stability in the oxidation-state map. Also chloride ion is a poor reducing agent, requiring a good oxidizing agent to convert it to higher oxidation states. With strong oxidizing agents in acidic media and at high chloride ion concentration, chloride ion is oxidized to chlorine, for example with manganese(IV) oxide:

$$MnO_2(s) + 2Cl(aq)^- + 4H_3O(aq)^+ = Cl_2(g) + Mn(aq)^{++} + 6H_2O \quad (18.40)$$

Chloride ions may be oxidized in an electrolytic cell. If an aqueous solution of sodium chloride is electrolyzed, chloride ions may be oxidized at the anode:

$$2Cl(aq)^- = Cl_2(g) + 2e^- \quad (18.41)$$

This electrolysis is carried out best at high chloride ion concentration and in acidic solutions. However, the oxidation must be carried out at a potential near the potential at which water is oxidized so there is a tendency to produce oxygen at the anode also:

$$6H_2O = O_2(g) + 4H_3O(aq)^+ + 4e^- \quad (18.42)$$

Elemental chlorine is a good oxidizing agent, although it can be either an oxidizing agent or reducing agent. As an oxidizing agent it is sufficiently strong to oxidize either bromide or iodide ions to the halogen in aqueous media:

$$Cl_2(g) + 2Br(aq)^- = 2Cl(aq)^- + Br_2(l) \quad (18.43)$$

$$Cl_2(g) + 2I(aq)^- = 2Cl(aq)^- + I_2(s) \quad (18.44)$$

Chlorine does not disproportionate readily in strongly acidic media but it does disproportionate in aqueous base. As seen earlier, it has its greatest tendency to disproportionate to chloride ions and hypochlorite ions. However, at high temperatures chlorate ions are favored over hypochlorite ions:

$$6Cl_2(g) + 12OH(aq)^- = 10Cl(aq)^- + 2ClO_3(aq)^- + 6H_2O \quad (18.45)$$

Hypochlorite ions are found mostly in solution and are prepared by the basic disproportionation of chlorine. As seen in Table 18.3, hypochlorous acid is a strong oxidizing agent and will oxidize chloride ions to chlorine in acidic solution:

$$Cl(aq)^- + HClO(aq) + H_3O(aq)^+ = Cl_2(g) + 2H_2O \quad (18.46)$$

Chlorite ions are not very common, but they may be prepared by the basic disproportionation of chlorine(IV) oxide:

$$2ClO_2(g) + 2OH(aq)^- = ClO_3(aq)^- + ClO_2(aq)^- + H_2O \quad (18.47)$$

Chlorite ions are more powerful oxidizing agents than hypochlorite ions, and thus will oxidize chloride ions to chlorine:

$$HClO_2(aq) + 3Cl(aq)^- + 3H_3O(aq)^+ = 2Cl_2(g) + 5H_2O \quad (18.48)$$

Chlorine(IV) oxide is prepared by the reduction of chlorate ions with oxalic acid in acidic solution:

$$H_2C_2O_4(aq) + 2ClO_3(aq)^- + 2H_3O(aq)^+ = 2CO_2(g) + 2ClO_2(g) + 4H_2O \quad (18.49)$$

Why this reduction stops at ClO_2 is not clear, but the reason is probably related to the mechanism of the reaction. Chlorine(IV) oxide may be produced along with oxygen by treating solid potassium chlorate with concentrated sulfuric acid. Once prepared, ClO_2 is a strong oxidizing agent. Although it disproportionates to chlorite and chlorate ions in basic media, in acids it disproportionates to chlorine and chlorate ions:

$$10ClO_2(g) + 12H_2O = Cl_2(g) + 8ClO_3(aq)^- + 8H_3O(aq)^+ \tag{18.50}$$

As indicated earlier, chlorate ions are prepared by taking advantage of the high temperature disproportionation of chlorine in base. The chlorate ion is a strong oxidizing agent in acid and also oxidizes chloride ion to chlorine:

$$ClO_3(aq)^- + 5Cl(aq)^- + 6H_3O(aq)^+ = 3Cl_2(g) + 9H_2O \tag{18.51}$$

The disproportionation reactions of chlorate ions were represented earlier by Equations (18.10) and (18.37). The basic disproportionation also occurs if dry, solid chlorates, such as $KClO_3$, are heated.

Perchlorate ions may be prepared by the basic disproportionation of chlorate ions. It should be noted that perchlorate ions are not as good oxidizing agents as chlorate ions under comparable circumstances, as can be seen in Table 18.3. In fact, the *less* highly oxygenated species tend to be the better oxidizing agents when each is considered to be reduced to the same product. This can be understood in part from a Born-Haber type cycle in which the more highly oxygenated species must have more Cl—O bonds broken in the reduction process.

Bromine

The oxidation-state map for bromine is shown as Fig. 18.4 and the summary of the standard reduction potentials for bromine-containing species is given in Table 18.6.

The oxidation-state map is similar to the one for chlorine except that the high-

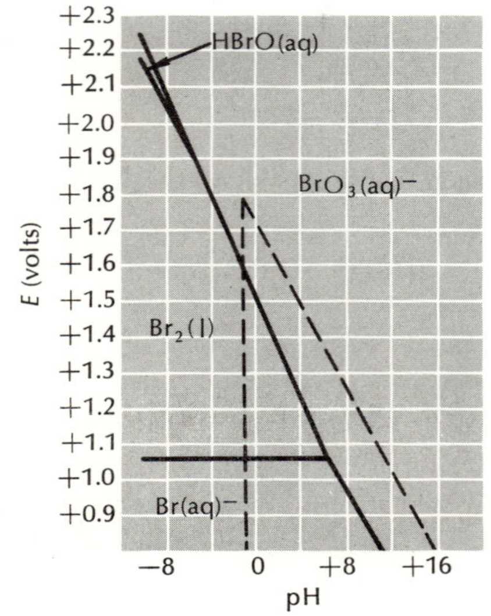

Fig. 18.4. Bromine oxidation-state map

TABLE 18.6

Standard Half-cell Reduction Potentials for Bromine-containing Species (Volts at 25°C)

In Aqueous Acid

OXIDIZING AGENTS	REDUCING AGENTS		
	$Br(aq)^-$	$Br_2(l)$	$HBrO(aq)$
$BrO_3(aq)^-$	+1.44	+1.52	+1.49
$HBrO(aq)$	+1.33	+1.59	
$Br_2(l)$	+1.09		

In Aqueous Base

OXIDIZING AGENTS	REDUCING AGENTS		
	$Br(aq)^-$	$Br_2(l)$	$BrO(aq)^-$
$BrO_3(aq)^-$	+0.61	+0.51	+0.53
$BrO(aq)^-$	+0.76	+0.45	
$Br_2(l)$	+1.09		

est oxidation state for bromine is 5+, as the bromate ion. Also hypobromous acid has a very narrow stability band in the very highly acidic region which is not accessible in practice.

As expected from the wide range of stability in the oxidation-state map, bromine occurs naturally most often as bromide ions. These occur in natural solid deposits as well as in sea water and some natural underground brine solutions. Bromide ions are better reducing agents than chloride ions and bromide ions may be oxidized by elemental chlorine. The recovery of bromine from brines and sea water depends on this reaction:

$$2Br(aq)^- + Cl_2(g) = 2Cl(aq)^- + Br_2(l) \qquad (18.52)$$

Among the enthalpy factors making bromide ions better reducing agents than chloride ions with respect to oxidation to the halogens are the lower electronegativity for bromide ions which permits easier removal of an electron and the existence of elemental bromine in the liquid state rather than the gaseous state.

Elemental bromine has a small area of stability in the highly acidic range. Like chlorine, bromine does not tend to disproportionate in strongly acidic media but does disproportionate most readily to bromide ions and hypobromite ions in bases.

Like hypochlorite ions, hypobromite ions disproportionate, although more rapidly than hypochlorite ions. In acid, hypobromous acid disproportionates to bromine and bromate ions; in base, hypobromite ions disproportionate to bromide ions and bromate ions:

$$5HBrO(aq) = 2Br_2(l) + BrO_3(aq)^- + H_3O(aq)^+ + H_2O \qquad (18.53)$$

$$3BrO(aq)^- = 2Br(aq)^- + BrO_3(aq)^- \qquad (18.54)$$

Thus bromates may be prepared by the disproportionation of hypobromite ions. Like chlorate ions, bromate ions are strong oxidizing agents, and as such sometimes are used in redox titrations.

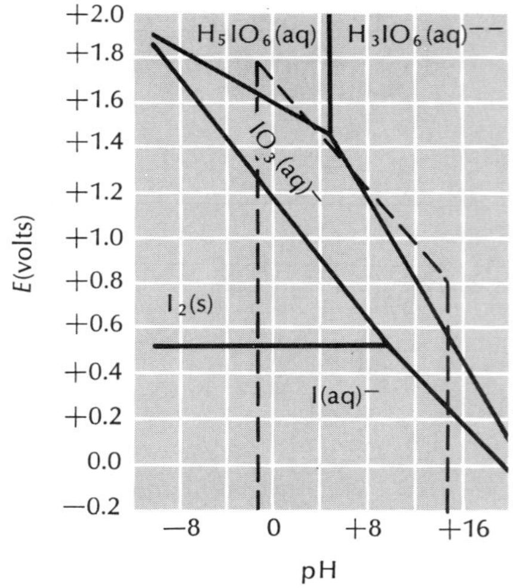

Fig. 18.5. Iodine oxidation-state map

Iodine

The oxidation-state map for iodine is shown as Fig. 18.5 and the summary of the standard reduction potentials for iodine-containing species is given in Table 18.7.

In some respects this oxidation-state map is more complex than those for the other halogens. It differs in showing ranges of stability for iodine in both the 5+ and 7+ oxidation states. In addition, the orthoperiodic acid is polyprotic, so that $H_3IO_6(aq)^{--}$ is the main species in basic solution and $H_5IO_6(aq)$ is found in acidic solution.

In nature iodine occurs as iodate ions and iodide ions. Iodide ion is the best reducing agent among the halides (excluding hydride). It is oxidized easily by

TABLE 18.7

Standard Half-cell Reduction Potentials for Iodine-containing Species (Volts at 25°C)

In Aqueous Acid

OXIDIZING AGENTS	REDUCING AGENTS			
	$I(aq)^-$	$I_2(s)$	$HIO(aq)$	$IO_3(aq)^-$
$H_5IO_6(aq)$	+1.2	+1.3	+1.3	+1.6
$IO_3(aq)^-$	+1.12	+1.20	+1.14	
$HIO(aq)$	+0.99	+1.45		
$I_2(s)$	+0.54			

In Aqueous Base

OXIDIZING AGENTS	REDUCING AGENTS			
	$I(aq)^-$	$I_2(s)$	$IO(aq)^-$	$IO_3(aq)^-$
$H_3IO_6(aq)^{--}$	+0.4	+0.3	+0.3	+0.7
$IO_3(aq)^-$	+0.29	+0.21	+0.14	
$IO(aq)^-$	+0.47	+0.45		
$I_2(s)$	+0.54			

chlorine or bromine and in acidic solutions it is oxidized by elemental oxygen to iodine.

Thus iodine is produced easily by the oxidation of iodide ions. In the oxidation-state map it has a larger stability range than the other halogens. Iodine may also be produced by the reduction of iodate ions in acid solution. Sulfurous acid is a typical reducing agent:

$$5H_2SO_3(aq) + 2IO_3(aq)^- + 7H_2O = I_2(s) + 5SO_4(aq)^{--} + 8H_3O(aq)^+ \quad (18.55)$$

Like the other halogens, iodine does not disproportionate in acidic solutions. However, in base it tends to disproportionate to iodide ions and iodate ions:

$$3I_2(s) + 6OH(aq)^- = IO_3(aq)^- + 5I(aq)^- + 3H_2O \quad (18.56)$$

Iodine is not appreciably soluble in water, but it dissolves readily in solutions containing iodide ions through the following reaction:

$$I_2(s) + I(aq)^- = I_3(aq)^- \quad (18.57)$$

Since the triiodide ion, I_3^-, is the species frequently encountered in reactions we shall use it in writing equations for those reactions in which it is commonly found.

In acids, iodine is a moderately good oxidizing agent and as such it finds use in redox titrations. One of the more commonly used reducing agents with iodine is the thiosulfate ion, which is oxidized to the tetrathionate ion:

$$2I_3(aq)^- + 2S_2O_3(aq)^{--} = 6I(aq)^- + S_4O_6(aq)^{--} \quad (18.58)$$

A technique often used is the addition of an excess of standard iodine solution to an unknown reducing agent, followed by titration of the excess iodine with standard thiosulfate ion solution.

Iodine may also be used as an oxidizing agent with arsenites and sulfides:

$$I_3(aq)^- + HAsO_2(aq) + 4H_2O = H_3AsO_4(aq) + 3I(aq)^- + 2H_3O(aq)^+ \quad (18.59)$$

$$I_3(aq)^- + H_2S(g) + 2H_2O = S(s) + 3I(aq)^- + 2H_3O(aq)^+ \quad (18.60)$$

Hypoiodite ions are very unstable and difficult to keep for any length of time. In acid they disproportionate to iodine and iodate ions, and in base to iodide ions and iodate ions.

Iodate ions are moderately stable and moderately strong oxidizing agents often used as such in redox titrations and in other analytical processes. For example, iodate ion oxidizes iodide ion to iodine:

$$IO_3(aq)^- + 8I(aq)^- + 6H_3O(aq)^+ = 3I_3(aq)^- + 9H_2O \quad (18.61)$$

Orthoperiodic acid is the strongest oxidizing agent among the oxygenated halogen-containing acids and anions. Since the reduction potential of the trihydrogen orthoperiodate ion decreases markedly as pH increases, this ion is best prepared in strongly basic solution by the oxidation of iodate ions by a strong oxidizing agent such as chlorine:

$$Cl_2(g) + IO_3(aq)^- + 3OH(aq)^- = H_3IO_6(aq)^{--} + 2Cl(aq)^- \quad (18.62)$$

FAMILY 6A

The standard reduction potentials for oxygen, sulfur, and selenium species are summarized in Tables 18.8 and 18.9. Comparisons among the chalcogens are difficult because these elements tend to form different types of species. Species containing sulfur predominate due to the tendency for sulfur to catenate and to form condensed polyanions.

It is clear that the peroxodisulfate and hydrogen peroxide are strong oxidizing agents and that they probably derive this oxidizing ability from the O—O bond. As a matter of convenience only, we are considering peroxodisulfate among the sulfur-containing species in which the oxidation state of sulfur is 7+. If we were more rigorous with oxidation states, we might want to insist that it is a species with sulfur in the 6+ state and oxygen in the 1− state. In any event, the high reduction potentials associated with these peroxo-type species probably can be attributed in part to the low O—O bond energy.

On the other hand, it appears generally that oxygenated sulfur-containing species are poorer oxidizing agents than comparable selenium-containing species. For example, sulfate ions are poorer oxidizing agents than selenate ions. This can be attributed in part to the very high bond energy for the sulfur-oxygen bonds, which probably are double bonds due to πd bonding.

Oxygen

The oxidation-state map for oxygen is shown as Fig. 18.6, and the summary of the standard reduction potentials for oxygen-containing species is given in Table 18.10.

Oxygen occurs in nature as elemental O_2 in the atmosphere and also as water in the 2− oxidation state. Both of these species have a wide stability range in the oxidation-state map.

Oxygen is a good oxidizing agent. Vigorous exothermic reactions with atmospheric oxygen are called *combustion* reactions. Less vigorous reactions with metals often are called *corrosion* reactions.

Ozone is a thermodynamically unstable modification of oxygen. It decomposes spontaneously to O_2, with an equilibrium constant of 10^{57} for the following reaction:

$$2O_3(g) = 3O_2(g) \qquad (18.63)$$

Thus ozone is a stronger oxidizing agent than oxygen.

Hydrogen peroxide is unstable with respect to disproportionation in both acidic and basic solutions:

$$2H_2O_2(aq) = 2H_2O + O_2(g) \qquad (18.64)$$

$$2HO_2(aq)^- = 2OH(aq)^- + O_2(g) \qquad (18.65)$$

TABLE 18.8

Some Standard Reduction Potentials for the Chalcogens

	OXIDIZING AGENTS			REDUCING AGENTS	E° VOLTS AT 25°C
1.	$S_2O_8(aq)^{--} + 2H_3O(aq)^+$	$+ 2e^- =$	$2H_2O$	$+ 2HSO_4(aq)^-$	+2.12
2.	$S_2O_8(aq)^{--}$	$+ 2e^- =$		$2SO_4(aq)^{--}$	+2.00
3.	$H_2O_2(aq) + 2H_3O(aq)^+$	$+ 2e^- =$		$4H_2O$	+1.77
4.	$O_3(g) + 6H_3O(aq)^+$	$+ 6e^- =$		$9H_2O$	+1.51
5.	$2H_2S_2O_4(aq)$	$+ 2e^- =$	$2H_2O$	$+ S_4O_6(aq)^{--}$	+1.39
6.	$2O_3(g) + 6H_3O(aq)^+$	$+ 6e^- =$	$6H_2O$	$+ 3H_2O_2(aq)$	+1.24
7.	$O_2(g) + 4H_3O(aq)^+$	$+ 4e^- =$		$6H_2O$	+1.23
8.	$HSeO_4(aq)^- + 3H_3O(aq)^+$	$+ 2e^- =$	$4H_2O$	$+ H_2SeO_3(aq)$	+1.09
9.	$2O_3(g) + 3H_2O$	$+ 6e^- =$	$3OH(aq)^-$	$+ 3HO_2(aq)^-$	+0.97
10.	$HO_2(aq)^- + H_2O$	$+ 2e^- =$		$3OH(aq)^-$	+0.88
11.	$H_2SeO_3(aq) + 4H_3O(aq)^+$	$+ 4e^- =$	$7H_2O$	$+ Se(s)$	+0.74
12.	$O_2(g) + 2H_3O(aq)^+$	$+ 2e^- =$	$2H_2O$	$+ H_2O_2(aq)$	+0.68
13.	$O_3(g) + 3H_2O$	$+ 6e^- =$		$6OH(aq)^-$	+0.68
14.	$H_2S_2O_3(aq) + 4H_3O(aq)^+$	$+ 4e^- =$	$7H_2O$	$+ S(s)$	+0.46
15.	$H_2S_2O_6(aq) + 2H_3O(aq)^+$	$+ 2e^- =$	$2H_2O$	$+ 2H_2SO_3(aq)$	+0.45
16.	$H_2SO_3(aq) + 4H_3O(aq)^+$	$+ 4e^- =$	$7H_2O$	$+ S(s)$	+0.45
17.	$O_2(g) + 2H_2O$	$+ 4e^- =$		$4OH(aq)^-$	+0.40
18.	$S_4O_6(aq)^{--} + 4H_3O(aq)^+$	$+ 2e^- =$	$4H_2O$	$+ 2H_2S_2O_3(aq)$	+0.22
19.	$S(s) + 2H_3O(aq)^+$	$+ 2e^- =$	$2H_2O$	$+ H_2S(g)$	+0.14
20.	$HSO_4(aq)^- + 3H_3O(aq)^+$	$+ 2e^- =$	$4H_2O$	$+ H_2SO_3(aq)$	+0.12
21.	$S_4O_6(aq)^{--}$	$+ 2e^- =$		$2S_2O_3(aq)^{--}$	+0.08
22.	$2H_2SO_3(aq) + 2H_3O(aq)^+$	$+ 2e^- =$	$4H_2O$	$+ H_2S_2O_4(aq)$	+0.06
23.	$SeO_4(aq)^{--} + H_2O$	$+ 2e^- =$	$2OH(aq)^-$	$+ SeO_3(aq)^{--}$	+0.05
24.	$S_2O_6(aq)^{--}$	$+ 2e^- =$		$2SO_3(aq)^{--}$	+0.02
25.	$O_2(g) + H_2O$	$+ 2e^- =$	$OH(aq)^-$	$+ HO_2(aq)^-$	−0.08
26.	$2S_2O_4(aq)^{--} + 2H_2O$	$+ 2e^- =$	$4OH(aq)^-$	$+ S_4O_6(aq)^{--}$	−0.12
27.	$2HSO_4(aq)^- + 4H_3O(aq)^+$	$+ 2e^- =$	$6H_2O$	$+ H_2S_2O_6(aq)$	−0.24
28.	$SeO_3(aq)^{--} + 6H_3O(aq)^+$	$+ 4e^- =$	$9H_2O$	$+ Se(s)$	−0.37
29.	$Se(s) + 2H_3O(aq)^+$	$+ 2e^- =$	$2H_2O$	$+ H_2Se(g)$	−0.40
30.	$2S(s)$	$+ 2e^- =$		$S_2(aq)^{--}$	−0.47
31.	$S(s)$	$+ 2e^- =$		$S(aq)^{--}$	−0.48
32.	$S_2(aq)^{--}$	$+ 2e^- =$		$2S(aq)^{--}$	−0.49
33.	$SO_3(aq)^{--} + 3H_2O$	$+ 4e^- =$	$6OH(aq)^-$	$+ S(s)$	−0.65
34.	$SO_4(aq)^{--} + 4H_2O$	$+ 8e^- =$	$8OH(aq)^-$	$+ S(aq)^{--}$	−0.68
35.	$S_2O_3(aq)^{--} + 3H_2O$	$+ 4e^- =$	$6OH(aq)^-$	$+ 2S(s)$	−0.74
36.	$Se(s)$	$+ 2e^- =$		$Se(aq)^{--}$	−0.92
37.	$SO_4(aq)^{--} + H_2O$	$+ 2e^- =$	$2OH(aq)^-$	$+ SO_3(aq)^{--}$	−0.93
38.	$2SO_3(aq)^{--} + 2H_2O$	$+ 2e^- =$	$4OH(aq)^-$	$+ S_2O_4(aq)^{--}$	−1.12
39.	$2SO_4(aq)^{--} + 2H_2O$	$+ 2e^- =$	$4OH(aq)^-$	$+ S_2O_6(aq)^{--}$	−1.89

TABLE **18.9**

Standard Potential Diagram for the Chalcogens *(Volts at 25°C)*

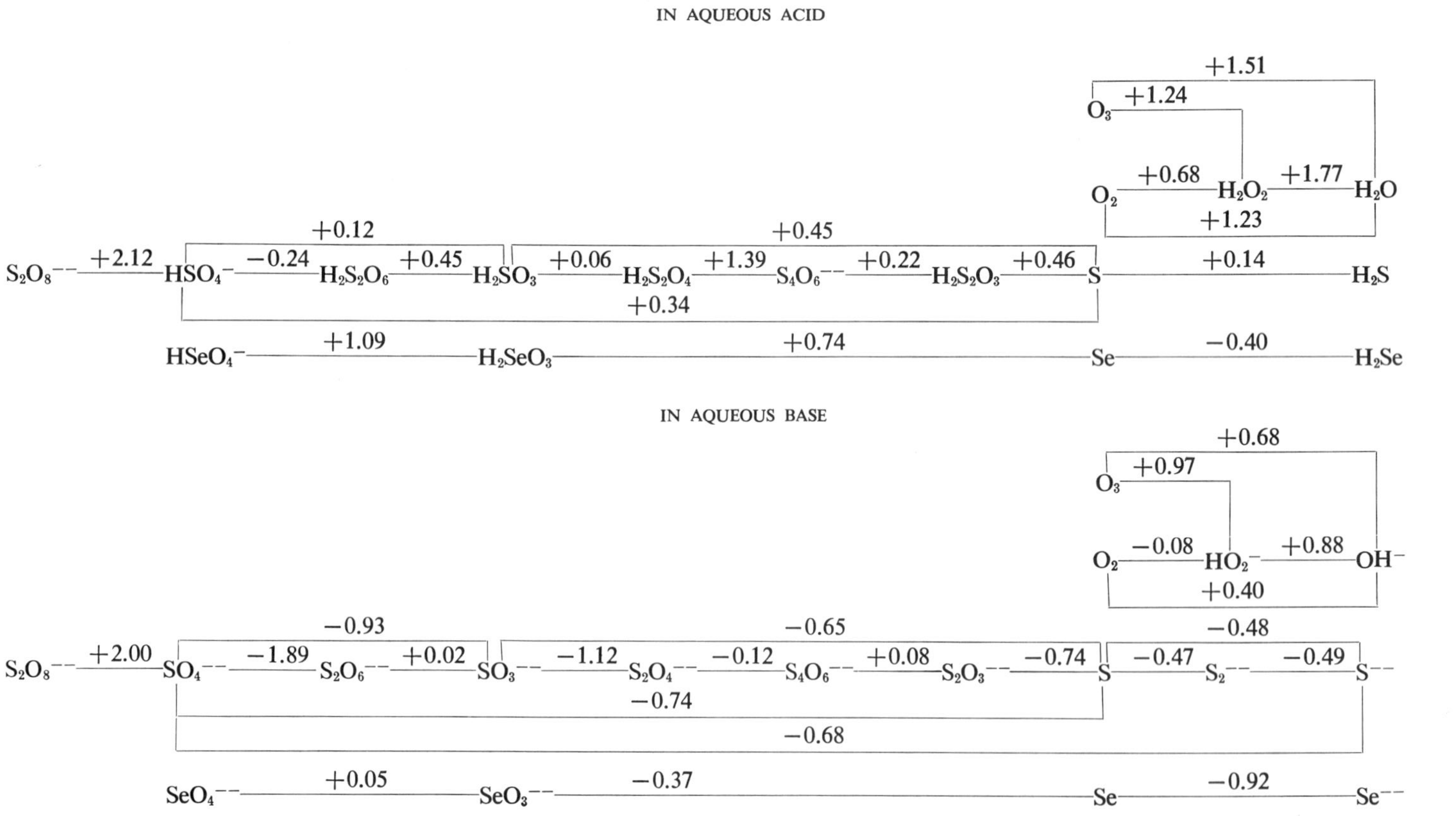

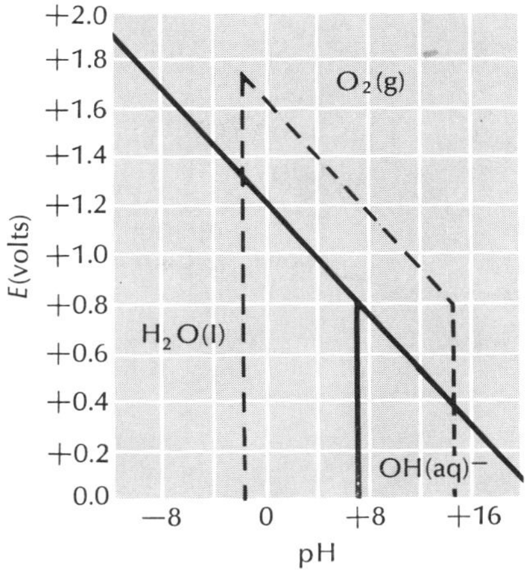

Fig. 18.6. Oxygen oxidation-state map

The basic disproportionation is more rapid than the acidic reaction. The disproportionation is catalyzed by many substances, including silver metal, manganese(IV) oxide, hydrobromic acid, and saliva. The catalytic effect of an enzyme in saliva can be demonstrated by using dilute hydrogen peroxide solution as a mouth wash and noting the vigorous effervescence.

Simple metallic peroxides may be formed by reducing atmospheric oxygen with some metals. Sodium and barium metals are examples. Hydrogen peroxide may be prepared by an acid-base reaction using sulfuric acid with barium peroxide. Another preparation depends on the decomposition of peroxodisulfate ions:

$$S_2O_8(aq)^{--} + 4H_2O = H_2O_2(aq) + 2SO_4(aq)^{--} + 2H_3O(aq)^+ \qquad (18.66)$$

Although hydrogen peroxide may be both an oxidizing agent and a reducing

TABLE 18.10

Standard Half-cell Reduction Potentials for Oxygen-containing Species (Volts at 25°C)

In Aqueous Acid

OXIDIZING AGENTS	REDUCING AGENTS	
	H_2O	$H_2O_2(aq)$
$O_3(g)$	+1.51	+1.24
$O_2(g)$	+1.23	+0.68
$H_2O_2(aq)$	+1.77	

In Aqueous Base

OXIDIZING AGENTS	REDUCING AGENTS	
	$OH(aq)^-$	$HO_2(aq)^-$
$O_3(g)$	+0.68	+0.97
$O_2(g)$	+0.40	−0.08
$HO_2(aq)^-$	+0.88	

agent, it is a stronger oxidizing agent in both acidic and basic media. In acid it oxidizes iodide ions to iodine, iron(II) ions to iron(III) ions, and sulfurous acid to sulfate ions:

$$3I(aq)^- + H_2O_2(aq) + 2H_3O(aq)^+ = I_3(aq)^- + 4H_2O \quad (18.67)$$

$$2Fe(aq)^{++} + H_2O_2(aq) + 2H_3O(aq)^+ = 2Fe(aq)^{3+} + 4H_2O \quad (18.68)$$

$$H_2SO_3(aq) + H_2O_2(aq) = HSO_4(aq)^- + H_3O(aq)^+ \quad (18.69)$$

In base hydrogen peroxide oxidizes manganese(II) hydroxide to manganese(IV) oxide:

$$Mn(OH)_2(s) + H_2O_2(aq) = MnO_2(s) + 2H_2O \quad (18.70)$$

To act as a reducing agent, hydrogen peroxide requires a moderately strong oxidizing agent. In acidic solution, permanganate ion is an example:

$$2MnO_4(aq)^- + 4H_2O_2(aq) + 6H_3O(aq)^+ = Mn(aq)^{++} + 5O_2(g) + 14H_2O \quad (18.71)$$

In basic solution hypochlorite ions and silver(I) oxide are good oxidizing agents:

$$ClO(aq)^- + H_2O_2(aq) = Cl(aq)^- + O_2(g) + H_2O \quad (18.72)$$

$$Ag_2O(s) + H_2O_2(aq) = 2Ag(s) + O_2(g) + H_2O \quad (18.73)$$

Sulfur

The oxidation-state map for sulfur is shown as Fig. 18.7, and the summary of the standard reduction potentials for sulfur-containing species is given in Table 18.11.

In nature, sulfur is found extensively in minerals as the sulfide ion and also as the disulfide ion. Neither of these is stable under normal conditions in aqueous solution, but they are stable in the solid crystalline lattice. Sulfur also is found as the free element and in the form of sulfate ions.

In the 2− state of oxidation, sulfur is easily oxidized. In acidic solutions, the oxidation of hydrogen sulfide usually leads to elemental sulfur, although sulfate ions may be produced with strong oxidizing agents. From the potentials in the column headed by $H_2S(s)$ in Table 18.11, it is evident that sulfate ions are next to the easiest species to attain by the oxidation of hydrogen sulfide in aqueous media. Thus if an oxidizing agent is strong enough to oxidize hydrogen sulfide to $H_2SO_3(aq)$, for instance, it is more than strong enough to oxidize it to sulfate ions as well. In *dilute* solution, so-called strong oxidizing agents convert hydrogen sulfide to sulfur.

$$2MnO_4(aq)^- + 5H_2S(g) + 6H_3O(aq)^+ = 5S(s) + 2Mn(aq)^{++} + 14H_2O \quad (18.74)$$

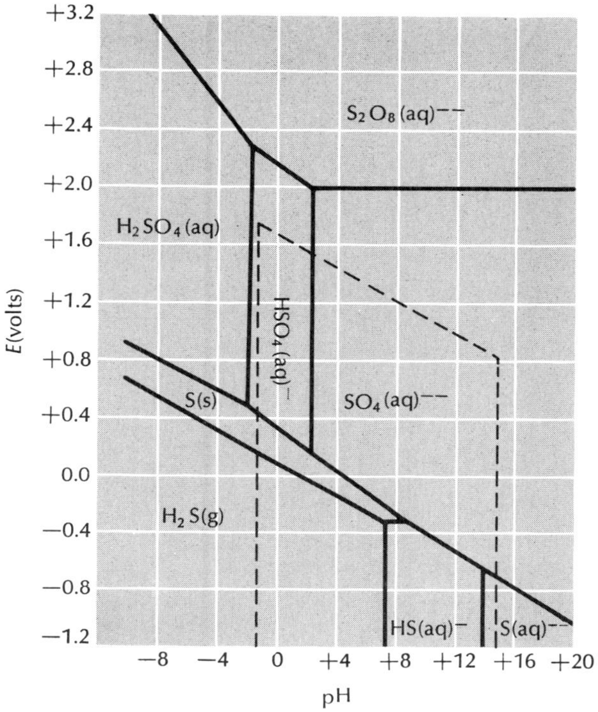

Fig. 18.7. Sulfur oxidation-state map

$$Cr_2O_7(aq)^{--} + 3H_2S(g) + 8H_3O(aq)^+ = 3S(s) + 2Cr(aq)^{3+} + 15H_2O \quad (18.75)$$

Polysulfides are formed by adding sulfur to basic aqueous solutions containing sulfide ions:

$$(x - 1)S(s) + S(aq)^{--} = S_x(aq)^{--} \quad (18.76)$$

The potential data for the polysulfides probably are unreliable. However, they do disproportionate very readily in acidic solutions:

$$H_2S_2(aq) = H_2S(g) + S(s) \quad (18.77)$$

Although the polysulfides can be both oxidizing and reducing agents, they generally find their utility as oxidizing agents:

$$S_2(aq)^{--} + SnS(s) = SnS_3(aq)^{--} \quad (18.78)$$

Although elemental sulfur is stable with respect to disproportionation in contact with acids, it is not in the presence of bases, where it tends to disproportionate mainly to sulfate and disulfide ions:

$$7S(s) + 8OH(aq)^- = 3S_2(aq)^{--} + SO_4(aq)^{--} + 4H_2O \quad (18.79)$$

TABLE 18.11

Standard Half-cell Reduction Potentials for Sulfur-containing Species
(Volts at 25°C)

In Aqueous Acid

OXIDIZING AGENTS	REDUCING AGENTS							
	$H_2S(g)$	$S(s)$	$H_2S_2O_3(aq)$	$S_4O_6(aq)^{--}$	$H_2S_2O_4(aq)$	$H_2SO_3(aq)$	$H_2S_2O_6(aq)$	$HSO_4(aq)^-$
$S_2O_8(aq)^{--}$	+0.50	+0.59	+0.65	+0.69	+0.61	+0.80	+0.94	+2.12
$HSO_4(aq)^-$	+0.30	+0.34	+0.27	+0.28	+0.10	+0.12	−0.24	
$H_2S_2O_6(aq)$	+0.37	+0.45	+0.45	+0.49	+0.26	+0.45		
$H_2SO_3(aq)$	+0.36	+0.45	+0.43	+0.51	+0.06			
$H_2S_2O_4(aq)$	+0.42	+0.58	+0.81	+1.39				
$S_4O_6(aq)^{--}$	+0.31	+0.42	+0.22					
$H_2S_2O_3(aq)$	+0.32	+0.46						
$S(s)$	+0.14							

In Aqueous Base

OXIDIZING AGENTS	REDUCING AGENTS								
	$S(aq)^{--}$	$S_2(aq)^{--}$	$S(s)$	$S_2O_3(aq)^{--}$	$S_4O_6(aq)^{--}$	$S_2O_4(aq)^{--}$	$SO_3(aq)^{--}$	$S_2O_6(aq)^{--}$	$SO_4(aq)^{--}$
$S_2O_8(aq)^{--}$	−0.38	−0.37	−0.36	−0.19	−0.23	−0.24	+0.05	+0.06	+2.00
$SO_4(aq)^{--}$	−0.68	−0.79	−0.74	−0.74	−0.86	−0.99	−0.93	−1.89	
$S_2O_6(aq)^{--}$	−0.51	−0.51	−0.52	−0.37	−0.46	−0.55	+0.02		
$SO_3(aq)^{--}$	−0.58	−0.62	−0.65	−0.57	−0.78	−1.12			
$S_2O_4(aq)^{--}$	−0.49	−0.49	−0.50	−0.02	−0.12				
$S_4O_6(aq)^{--}$	−0.51	−0.55	−0.57	+0.08					
$S_2O_3(aq)^{--}$	−0.61	−0.65	−0.74						
$S(s)$	−0.48	−0.47							
$S_2(aq)^{--}$	−0.49								

In contact with aqueous media, elemental sulfur reacts very sluggishly, probably due to its insolubility and the fact that water does not wet the surface of the sulfur and physical contact is therefore poor. The single most important reaction of sulfur is the combustion with oxygen, which is carried out without benefit of a solvent:

$$S(s) + O_2(g) = SO_2(g) \qquad (18.80)$$

The product of this reaction is sulfur(IV) oxide, not because it is the most stable possible product, but because the more stable sulfur(VI) oxide is produced very slowly. If suitable catalysts are used, oxidation to sulfur(VI) oxide occurs.

Thiosulfate ions are produced by adding sulfur to a solution of sulfite ions in basic medium:

$$S(s) + SO_3(aq)^{--} = S_2O_3(aq)^{--} \qquad (18.81)$$

The reverse of this reaction occurs in acidic solutions and thus constitutes the acidic disproportionation reaction for thiosulfate ions:

$$S_2O_3(aq)^{--} + 2H_3O(aq)^{+} = H_2SO_3(aq) + S(s) + 2H_2O \qquad (18.82)$$

Although this is a possible disproportionation according to the data in Table 18.11, it is not the most highly favored one. However, in basic solution, thiosulfate does disproportionate as predicted to sulfide ions and sulfate ions:

$$S_2O_3(aq)^{--} + 2OH(aq)^{-} = S(aq)^{--} + SO_4(aq)^{--} + H_2O \qquad (18.83)$$

From the data in Table 18.11 it is clear that thiosulfate ions are most easily oxidized to tetrathionate ions. This is one of the most important and characteristic reactions of thiosulfate ion. A number of mild oxidizing agents will bring about this conversion, including iron(III) ions and elemental iodine in the form of triiodide ions:

$$2S_2O_3(aq)^{--} + I_3(aq)^{-} = S_4O_6(aq)^{--} + 3I(aq)^{-} \qquad (18.84)$$

Once prepared, tetrathionate ions disproportionate in acid to sulfur and hydrogen sulfate ions:

$$3S_4O_6(aq)^{--} + H_2O + H_3O(aq)^{+} = 7S(s) + 5HSO_4(aq)^{-} \qquad (18.85)$$

The acid, $H_2S_2O_4$, has been named variously as dithionous acid, hyposulfurous acid, and hydrosulfurous acid. We shall use the name dithionous acid and shall call the $S_2O_4^{--}$ ion the dithionite ion. Dithionite ions usually are prepared by reduction of sulfite ions or hydrogen sulfite ions electrolytically. The dithionite ion is a rapidly acting reducing agent usually oxidized to sulfite ion by mild oxidizing agents. One of its more useful reactions is its rapid reaction with elemental oxygen:

$$2S_2O_4(aq)^{--} + O_2(g) + 4OH(aq)^{-} = 4SO_3(aq)^{--} + 2H_2O \qquad (18.86)$$

In this capacity it serves as a good absorber of oxygen in situations in which oxygen is to be removed from a system.

In acidic solutions, dithionous acid disproportionates to elemental sulfur and sulfurous acid.

Sulfur(IV) oxide dissolves in water to form what is called sulfurous acid, which might better be considered as hydrated sulfur(IV) oxide. However, for the sake of simplicity, we shall write it as $H_2SO_3(aq)$. Sulfur(IV) oxide is produced most conveniently by the combustion of sulfur or metallic sulfides. It may be produced also by the reduction of sulfate ions:

$$Cu(s) + SO_4(aq)^{--} + 4H_3O(aq)^+ = H_2SO_3(aq) + Cu(aq)^{++} + 5H_2O \quad (18.87)$$

Under conditions in which sulfurous acid is produced in high concentration, sulfur(IV) oxide is evolved as the dehydration product.

Sulfurous acid disproportionates to sulfur and sulfate ions:

$$3H_2SO_3(aq) + H_2O = S(s) + 2HSO_4(aq)^- + 2H_3O(aq)^+ \quad (18.88)$$

This is not the most highly favored disproportionation. However, the most highly favored disproportionation to tetrathionate and sulfate ions appears to be hindered kinetically and does not occur normally.

Sulfite ions may be either reducing or oxidizing agents. As a reducing agent in acid they react with strong oxidizing agents to form sulfate ions:

$$H_2O_2(aq) + H_2SO_3(aq) + H_2O = 2H_3O(aq)^+ + SO_4(aq)^{--} \quad (18.89)$$

$$I_3(aq)^- + H_2SO_3(aq) + 5H_2O = 3I(aq)^- + SO_4(aq)^{--} + 4H_3O(aq)^+ \quad (18.90)$$

Sulfurous acid usually is reduced to elemental sulfur, requiring strong reducing agents to be reduced to sulfide ions or hydrogen sulfide.

$$2H_2S(g) + H_2SO_3(aq) = 3S(s) + 3H_2O \quad (18.91)$$

Dithionate ions are produced by electrolytic oxidation of sulfite ions, or chemically with dilute aqueous or solid oxidizing agents such as MnO_2, $Fe(OH)_3$, or $Co(OH)_3$. In both acidic and basic solutions it disproportionates to the 6+ and 4+ oxidation states of sulfur.

Sulfuric acid is one of the most important chemicals commercially. Usually it is produced by the oxidation of sulfur(IV) oxide which in turn is produced by the combustion of sulfur or metallic oxides. As noted earlier, the product of these combustions should be sulfur(VI) oxide, based on thermodynamic relationships. Thus the key to sulfuric acid production lies in the discovery of catalysts to allow the faster production of sulfur(VI) oxide, according to the following reaction:

$$2SO_2 + O_2 = 2SO_3 \quad (18.92)$$

One of the first methods employed on a large scale was called the lead chamber process after the large lead chambers in which the reaction was carried out. Although the process is little used today, it is an interesting example of homogeneous catalysis. The sulfur(IV) oxide is mixed with nitrogen(II) oxide and nitrogen(IV) oxide and put into contact with steam, so that nitrogen(II) oxide is oxidized to nitrogen(IV) oxide which oxidizes sulfur(IV) oxide to nitrosyl hydrogen sulfate:

$$2NO_2(g) + 2SO_2(g) + H_2O(g) = 2NO(HSO_4)(aq) + NO(g) \tag{18.93}$$

The nitrosyl hydrogen sulfate reacts with additional water to produce sulfuric acid and reproduce the catalytic mixture of nitrogen(II) and nitrogen(IV) oxides:

$$2NO(HSO_4)(aq) + 2H_2O = 2HSO_4(aq)^- + 2H_3O(aq)^+ + NO(g) + NO_2(g) \tag{18.94}$$

This is a slightly different way to look at this catalysis which was first met in Chapter 14. See Equations (14.113)–(14.116).

The modern method for making sulfuric acid is the contact process, using a heterogeneous contact catalyst such as iron(III) oxide, vanadium(V) oxide, or platinum metal.

As indicated earlier, sulfate ions are not good oxidizing agents. Dilute sulfuric acid normally does not act as an oxidizing agent but the hot acid may oxidize some metals such as copper, silver, or mercury. Solid, dry sulfates may be reduced with hot carbon to form sulfides.

$$Na_2SO_4(s) + 2C(s) + \text{heat} = Na_2S(s) + 2CO_2(g) \tag{18.95}$$

As indicated in the oxidation-state map, peroxodisulfate ion is a powerful oxidizing agent capable of oxidizing water. However, this reaction is slow and peroxodisulfate ions may be used in aqueous solutions as oxidizing agents if they are not allowed to stand too long. The ion often is produced by the electrolysis of cold, concentrated solutions of sulfate ions. The concentrated solution favors the reaction represented by the following half-cell reaction equation.

$$2SO_4(aq)^{--} = S_2O_8(aq)^{--} + 2e^- \tag{18.96}$$

As a powerful oxidizing agent, the peroxodisulfate ion can be used to make other strong oxidizing agents such as permanganate and dichromate ions:

$$2Mn(aq)^{++} + 5S_2O_8(aq)^{--} + 14H_2O = 2MnO_4(aq)^- + 10HSO_4(aq)^- + 6H_3O(aq)^+ \tag{18.97}$$

$$2Cr(aq)^{3+} + 3S_2O_8(aq)^{--} + 15H_2O = Cr_2O_7(aq)^{--} + 6HSO_4(aq)^- + 8H_3O(aq)^+ \tag{18.98}$$

Selenium

The oxidation-state map for selenium is shown as Fig. 18.8, and the summary of the standard reduction potentials for selenium-containing species is given in Table 18.12.

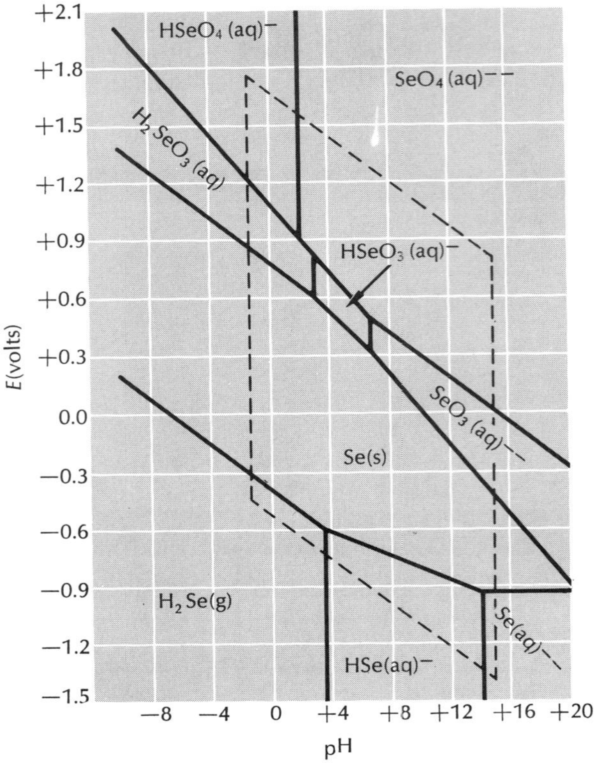

Fig. 18.8. Selenium oxidation-state map

The oxidation-state system of selenium is much simpler than that of sulfur since the diversity of species formed by selenium is less. The oxidation-state map is of interest since all of the major oxidation states are represented, 2−, 0, 4+, and 6+. Since all of the oxidation states show thermodynamic stability across the entire pH range, none of them is unstable with respect to disproportionation.

In comparison with sulfur, the selenium-containing species with selenium in the 6+ oxidation state are better oxidizing agents than the comparable species in which sulfur is in the 6+ state. In addition the species in which selenium exhibits the 4+ oxidation state have a wider range of thermodynamic stability than comparable sulfur-containing species.

Selenium occurs in nature as the free element often along with deposits of elemental sulfur. It also occurs in the form of selenide and selenite ions.

Hydrogen selenide is produced by adding acid to selenide ions. Hydrogen selenide and selenide ions are good reducing agents and usually are oxidized to elemental selenium.

Elemental selenium is both a poorer reducing agent and a poorer oxidizing agent than elemental sulfur. As a reducing agent it is usually oxidized to the 4+ oxidation state of selenium, although strong oxidizing agents may oxidize the selenium to the 6+ state. For example, nitric acid solutions oxidize selenium to selenous acid. Like sulfur, selenium undergoes combustion with oxygen:

TABLE 18.12

Standard Half-cell Reduction Potentials for Selenium-containing Species (Volts at 25°C)

In Aqueous Acid			
OXIDIZING AGENTS	REDUCING AGENTS		
	$H_2Se(g)$	Se(s)	$H_2SeO_3(aq)$
$HSeO_4(aq)^-$	+0.54	+0.86	+1.09
$H_2SeO_3(aq)$	+0.36	+0.74	
Se(s)	−0.40		

In Aqueous Base			
OXIDIZING AGENTS	REDUCING AGENTS		
	$Se(aq)^{--}$	Se(s)	$SeO_3(aq)^{--}$
$SeO_4(aq)^{--}$	−0.40	−0.23	+0.05
$SeO_3(aq)^{--}$	−0.55	−0.37	
Se(s)	−0.92		

$$Se(s) + O_2(g) = SeO_2(s) \qquad (18.99)$$

Selenous acid is a better oxidizing agent than sulfurous acid, and it is reduced easily to elemental selenium:

$$2H_2SO_3(aq) + H_2SeO_3(aq) + 3H_2O = Se(s) + 2SO_4(aq)^{--} + 4H_3O(aq)^+ \qquad (18.100)$$

$$6I(aq)^- + H_2SeO_3(aq) + 4H_3O(aq)^+ = Se(s) + 2I_3(aq)^- + 7H_2O \qquad (18.101)$$

Selenates are made from selenites by strong oxidizing agents such as chlorate or permanganate:

$$5H_2SeO_3(aq) + 2MnO_4(aq)^- + H_3O(aq)^+ = 5HSeO_4(aq)^- + 2Mn(aq)^{++} + H_2O \qquad (18.102)$$

Although selenate ions are strong oxidizing agents, their reactions tend to be slow. Dilute aqueous solutions of selenic acid quantitatively oxidize iodide ions to iodine (triiodide).

$$H_2SeO_4(aq) + 3I(aq)^- + 2H_3O(aq)^+ = I_3(aq)^- + H_2SeO_3(aq) + 3H_2O \qquad (18.103)$$

Concentrated aqueous selenic acid solutions oxidize chloride ions to chlorine, copper metal to copper(II) ions, and gold metal to gold(III) ions.

$$H_2SeO_4(aq) + 2Cl(aq)^- + 2H_3O(aq)^+ = Cl_2(g) + H_2SeO_3(aq) + 3H_2O \qquad (18.104)$$

$$H_2SeO_4(aq) + Cu(s) + 2H_3O(aq)^+ = Cu(aq)^{++} + H_2SeO_3(aq) + 3H_2O \qquad (18.105)$$

FAMILY 5A

The standard reduction potentials for nitrogen, phosphorus, and arsenic species are summarized in Tables 18.13 and 18.14. From this data, it is evident that the species which contain nitrogen in positive oxidation states tend to be the best oxidizing agents. The other species tend to be poor oxidizing agents or good reducing agents.

TABLE 18.13

Some Standard Reduction Potentials for Family 5A

	OXIDIZING AGENTS			REDUCING AGENTS	E° (VOLTS AT 25°C)
1.	$N_2O(g) + 2H_3O(aq)^+$	$+ 2e^- =$	$3H_2O$	$+ N_2(g)$	+1.77
2.	$2NO(g) + 2H_3O(aq)^+$	$+ 2e^- =$	$3H_2O$	$+ N_2O(g)$	+1.59
3.	$2HNO_2(aq) + 6H_3O(aq)^+$	$+ 6e^- =$	$10H_2O$	$+ N_2(g)$	+1.45
4.	$2NH_3OH(aq)^+ + H_3O(aq)^+$	$+ 2e^- =$	$3H_2O$	$+ N_2H_5(aq)^+$	+1.41
5.	$N_2O_4(g) + 8H_3O(aq)^+$	$+ 8e^- =$	$12H_2O$	$+ N_2(g)$	+1.35
6.	$2HNO_2(aq) + 4H_3O(aq)^+$	$+ 4e^- =$	$7H_2O$	$+ N_2O(g)$	+1.29
7.	$N_2H_5(aq)^+ + 3H_3O(aq)^+$	$+ 2e^- =$	$3H_2O$	$+ 2NH_4(aq)^+$	+1.28
8.	$2NO_3(aq)^- + 12H_3O(aq)^+$	$+ 10e^- =$	$18H_2O$	$+ N_2(g)$	+1.25
9.	$N_2O_4(g) + 2H_3O(aq)^+$	$+ 2e^- =$	$2H_2O$	$+ 2HNO_2(aq)$	+1.07
10.	$HNO_2(aq) + H_3O(aq)^+$	$+ 1e^- =$	$2H_2O$	$+ NO(g)$	+1.00
11.	$NO_3(aq)^- + 3H_3O(aq)^+$	$+ 2e^- =$	$4H_2O$	$+ HNO_2(aq)$	+0.94
12.	$N_2O(g)$	$+ 2e^- =$	$2OH(aq)^-$	$+ N_2(g)$	+0.94
13.	$N_2O_4(g)$	$+ 2e^- =$		$2NO_2(aq)^-$	+0.88
14.	$2NO_3(aq)^- + 4H_3O(aq)^+$	$+ 2e^- =$	$6H_2O$	$+ N_2O_4(g)$	+0.80
15.	$2NO(g) + H_2O$	$+ 2e^- =$	$2OH(aq)^-$	$+ N_2O(g)$	+0.76
16.	$2NH_2OH(aq)$	$+ 2e^- =$	$2OH(aq)^-$	$+ N_2H_4(aq)$	+0.73
17.	$4H_3AsO_4(aq) + 8H_3O(aq)^+$	$+ 8e^- =$	$18H_2O$	$+ As_4O_6(s)$	+0.58
18.	$N_2O_4(g) + 4H_2O$	$+ 8e^- =$	$8OH(aq)^-$	$+ N_2(g)$	+0.53
19.	$2NO_2(aq)^- + 4H_2O$	$+ 6e^- =$	$8OH(aq)^-$	$+ N_2(g)$	+0.41
20.	$N_2(g) + 8H_3O(aq)^+$	$+ 6e^- =$	$8H_2O$	$+ 2NH_4(aq)^+$	+0.27
21.	$2NO_3(aq)^- + 6H_2O$	$+ 10e^- =$	$12OH(aq)^-$	$+ N_2(g)$	+0.26
22.	$As_4O_6(s) + 12H_3O(aq)^+$	$+ 12e^- =$	$18H_2O$	$+ 4As(s)$	+0.23
23.	$2NO_2(aq)^- + 3H_2O$	$+ 4e^- =$	$6OH(aq)^-$	$+ N_2O(g)$	+0.15
24.	$N_2H_4(aq) + 2H_2O$	$+ 2e^- =$	$2OH(aq)^-$	$+ 2NH_3(aq)$	+0.15
25.	$NO_3(aq)^- + 2H_3O(aq)^+$	$+ 2e^- =$	$3H_2O$	$+ NO_2(aq)^-$	+0.01
26.	$P_2H_4(l) + 2H_3O(aq)^+$	$+ 2e^- =$	$2H_2O$	$+ 2PH_3(g)$	0.00
27.	$P(s) + 3H_3O(aq)^+$	$+ 3e^- =$	$3H_2O$	$+ 2PH_3(g)$	−0.06
28.	$2P(s) + 4H_3O(aq)^+$	$+ 4e^- =$	$4H_2O$	$+ P_2H_4(l)$	−0.1
29.	$N_2(g) + 5H_3O(aq)^+$	$+ 4e^- =$	$5H_2O$	$+ N_2H_5(aq)^+$	−0.23
30.	$H_3PO_4(aq) + 2H_3O(aq)^+$	$+ 2e^- =$	$3H_2O$	$+ H_3PO_3(aq)$	−0.28
31.	$NO_2(aq)^- + H_2O$	$+ 1e^- =$	$2OH(aq)^-$	$+ NO(g)$	−0.46

TABLE 18.13 (cont.)

	OXIDIZING AGENTS		REDUCING AGENTS	E° (VOLTS AT 25°C)
32.	$H_3PO_3(aq) + 2H_3O(aq)^+$	$+ 2e^- = 3H_2O$	$+ H_3PO_2(aq)$	−0.50
33.	$H_3PO_3(aq) + 3H_3O(aq)^+$	$+ 3e^- = 6H_2O$	$+ P(s)$	−0.50
34.	$H_3PO_2(aq) + H_3O(aq)^+$	$+ 1e^- = 3H_2O$	$+ P(s)$	−0.51
35.	$As(s) + 3H_3O(aq)^+$	$+ 3e^- = 3H_2O$	$+ AsH_3(g)$	−0.60
36.	$AsO_4(aq)^{3-} + 2H_2O$	$+ 2e^- = 4OH(aq)^-$	$+ AsO_2(aq)^-$	−0.67
37.	$AsO_4(aq)^{3-} + 4H_2O$	$+ 5e^- = 8OH(aq)^-$	$+ As(s)$	−0.68
38.	$N_2(g) + 6H_2O$	$+ 6e^- = 6OH(aq)^-$	$+ 2NH_3(aq)$	−0.72
39.	$P_2H_4(l) + 2H_2O$	$+ 2e^- = 2OH(aq)^-$	$+ 2PH_3(g)$	−0.8
40.	$2NO_3(aq)^- + 2H_2O$	$+ 2e^- = 4OH(aq)^-$	$+ N_2O_4(g)$	−0.86
41.	$P(s) + 3H_2O$	$+ 3e^- = 3OH(aq)^-$	$+ PH_3(g)$	−0.89
42.	$2P(s) + 4H_2O$	$+ 4e^- = 4OH(aq)^-$	$+ P_2H_4(l)$	−0.9
43.	$PO_4(aq)^{3-} + 2H_2O$	$+ 2e^- = 3OH(aq)^-$	$+ HPO_3(aq)^{--}$	−1.12
44.	$N_2(g) + 4H_2O$	$+ 4e^- = 4OH(aq)^-$	$+ N_2H_4(aq)$	−1.16
45.	$HN_3(aq) + 5H_3O(aq)^+$	$+ 2e^- = 2H_2O$	$+ 3NH_3OH(aq)^+$	−1.26
46.	$As(s) + 3H_2O$	$+ 3e^- = 3OH(aq)^-$	$+ AsH_3(g)$	−1.43
47.	$HPO_3(aq)^{--} + 2H_2O$	$+ 2e^- = 3OH(aq)^-$	$+ H_2PO_2(aq)^-$	−1.57
48.	$HPO_3(aq)^{--} + 2H_2O$	$+ 3e^- = 5OH(aq)^-$	$+ P(s)$	−1.73
49.	$H_2PO_2(aq)^-$	$+ 1e^- = 2OH(aq)^-$	$+ P(s)$	−2.05
50.	$N_3(aq)^- + 6H_2O$	$+ 2e^- = 3OH(aq)^-$	$+ 3NH_2OH(aq)$	−2.86
51.	$3N_2(g) + 2H_3O(aq)^+$	$+ 2e^- = 2H_2O$	$+ 2HN_3(aq)$	−3.09
52.	$3N_2(g)$	$+ 2e^- =$	$2N_3(aq)^-$	−3.4

Nitrogen

The oxidation-state map for nitrogen is shown as Fig. 18.9, and the summary of the standard reduction potentials for nitrogen-containing species is given in Table 18.15.

The simplicity of the oxidation state map is striking, considering the large number of different oxidation states which nitrogen exhibits. This is a warning that the redox chemistry of nitrogen is complex. One obvious feature of the map is the large stability area for elemental nitrogen which can be attributed to the very high bond energy for the nitrogen-nitrogen triple bond. However, it is interesting to note that despite its extreme thermodynamic stability, nitrogen is not a product of redox reactions as often as might be expected. In fact, it appears that in almost all cases in which nitrogen is produced, one of the negative oxidation states of the nitrogen has been involved. This suggests that there is a kinetic hindrance (high free energy of activation) associated with the formation of elemental nitrogen from the positive oxidation states. Thus, other less stable species are produced from the positive oxidation states. However, those reactions that do occur are still thermodynamically favorable; that is, they involve a free energy decrease.

TABLE 18.14

Standard Potential Diagram for Family 5A (Volts at 25°C)

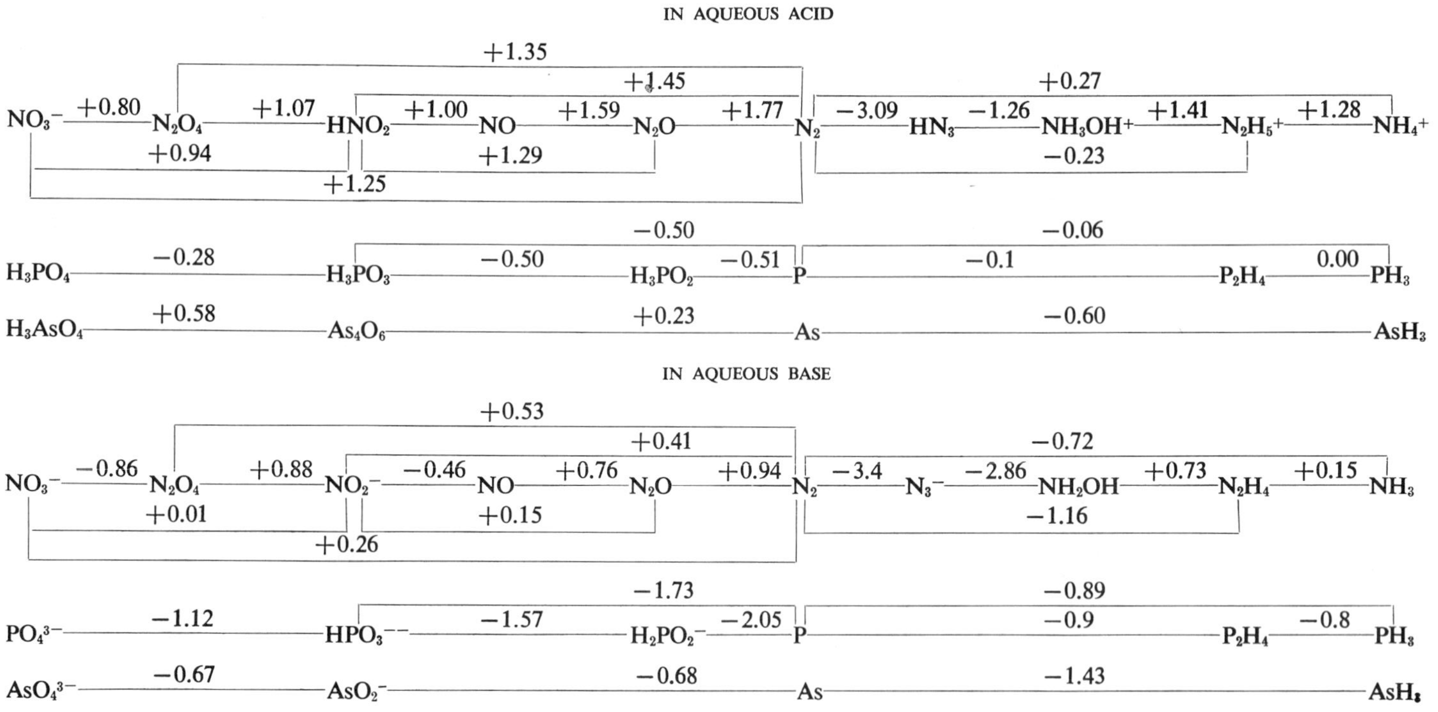

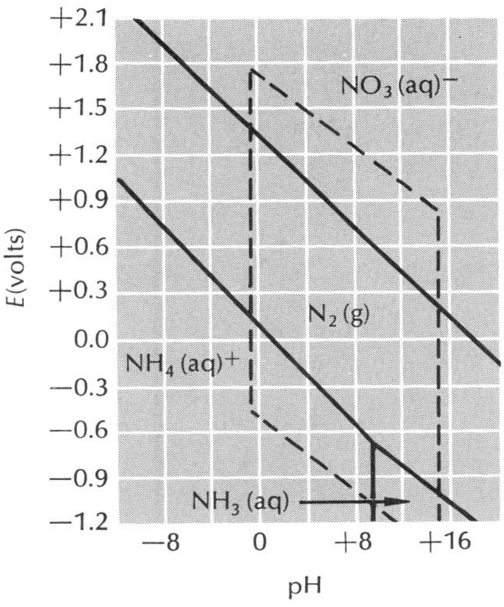

Fig. 18.9. Nitrogen oxidation-state map

Also, the wide stability band for elemental nitrogen shows that it is both a poor oxidizing agent and a poor reducing agent. Thus, nitrogen is not a common product of chemical reactions, and neither is it a common reactant, since it does not undergo redox reactions very readily. This, too, must be due to the great bond strength of the nitrogen-nitrogen triple bond.

Finally, the wide stability band for elemental nitrogen suggests that nitrogen exists predominantly in nature as the free element. However, it does occur also as nitrate ions, nitrite ions, ammonium ions, and as amino acids in animal proteins.

The commercial production of ammonia depends upon the tapping of the large reservoir of elemental nitrogen. Two major processes are now in use. The cyanamide process depends upon a high-temperature dry reaction between nitrogen and calcium carbide to form calcium cyanamide:

$$N_2(g) + CaC_2(s) = CaCN_2(s) + C(s) \qquad (18.106)$$

Ammonia is produced through reaction of the calcium cyanamide with water:

$$CaCN_2(s) + 3H_2O(l) = CaCO_3(s) + 2NH_3(g) \qquad (18.107)$$

Still more important is the Haber process, in which elemental nitrogen and hydrogen are combined in a gas-phase reaction with a heterogeneous catalyst such as iron or iron oxides at elevated temperature:

$$N_2(g) + 3H_2(g) = 2NH_3(g) \qquad (18.108)$$

Ammonia is a good reducing agent in aqueous solution as well as in the gas phase. As a gas, it burns in oxygen to form nitrogen and water:

$$4NH_3(g) + 3O_2(g) = 2N_2(g) + 6H_2O(g) \qquad (18.109)$$

TABLE 18.15

Standard Half-cell Reduction Potentials for Nitrogen-containing Species (Volts at 25°C)

In Aqueous Acid

OXIDIZING AGENTS	REDUCING AGENTS								
	$NH_4(aq)^+$	$N_2H_5(aq)^+$	$NH_3OH(aq)^+$	$HN_3(aq)$	$N_2(g)$	$N_2O(g)$	$NO(g)$	$HNO_2(aq)$	$N_2O_4(g)$
$NO_3(aq)^-$	+0.88	+0.83	+0.73	+0.10	+1.25	+1.11	+0.96	+0.94	+0.80
$N_2O_4(g)$	+0.89	+0.83	+0.71	+1.01	+1.35	+1.02	+1.03	+1.07	
$HNO_2(aq)$	+0.86	+0.78	+0.62	+1.00	+1.45	+1.29	+1.00		
$NO(g)$	+0.84	+0.73	+0.50	+1.00	+1.68	+1.59			
$N_2O(g)$	+0.65	+0.44	−0.05	+0.56	+1.77				
$N_2(g)$	+0.27	−0.23	−1.87	−3.09					
$HN_3(aq)$	+0.69	+0.34	−1.26						
$NH_3OH(aq)^+$	+1.36	+1.41							
$N_2H_5(aq)^+$	+1.28								

In Aqueous Base

OXIDIZING AGENTS	REDUCING AGENTS								
	$NH_3(aq)$	$N_2H_4(aq)$	$NH_2OH(aq)$	$N_3(aq)^-$	$N_2(g)$	$N_2O(g)$	$NO(g)$	$NO_2(aq)^-$	$N_2O_4(g)$
$NO_3(aq)^-$	−0.12	−0.15	−0.31	+0.03	+0.26	+0.08	−0.14	+0.01	−0.86
$N_2O_4(g)$	−0.02	−0.03	−0.18	+0.23	+0.53	+0.19	+0.20	+0.88	
$NO_2(aq)^-$	−0.17	−0.21	−0.45	+0.03	+0.41	+0.15	−0.46		
$NO(g)$	−0.10	−0.15	−0.45	+0.24	+0.85	+0.76			
$N_2O(g)$	−0.32	−0.46	−1.05	−0.15	+0.94				
$N_2(g)$	−0.72	−1.16	−3.04	−3.4					
$N_3(aq)^-$	−0.41	−0.71	−2.86						
$NH_2OH(aq)$	−0.46	+0.73							
$N_2H_4(aq)$	+0.15								

This is one reaction in which nitrogen is produced. Another example is the reduction of metal oxides by ammonia:

$$3CuO(s) + 2NH_3(g) = N_2(g) + 3Cu(s) + 3H_2O(g) \qquad (18.110)$$

The reaction represented by Equation (18.109) is rather slow, and with appropriate catalysts it is possible to increase the speed of a competing reaction beyond that of reaction (18.109), so that ammonia may be oxidized to nitrogen(II) oxide:

$$4NH_3(g) + 5O_2(g) = 4NO(g) + 6H_2O(g) \qquad (18.111)$$

This is an important reaction in the commercial production of nitric acid.

Sometimes it is possible to produce compounds that contain two parts, one an oxidizing agent and the other a reducing agent. Thus compounds containing ammonium ion, which is a reducing agent, and an anion which is an oxidizing agent are of this type. Such compounds tend to undergo internal redox reactions. An example is ammonium dichromate, which produces a spectacular display when heated:

$$(NH_4)_2Cr_2O_7(s) + \text{heat} = Cr_2O_3(s) + N_2(g) + 4H_2O \qquad (18.112)$$

The oxidizing agent may be another nitrogen-containing anion such as nitrate or nitrite. Usually such compounds undergo reaction to produce a single nitrogen-containing species in which the oxidation state is the average of the oxidation states in the original compound. Thus elemental nitrogen may be prepared by heating an aqueous solution of ammonium nitrite:

$$NH_4(aq)^+ + NO_2(aq)^- = N_2(g) + 2H_2O \qquad (18.113)$$

Solid ammonium nitrate decomposes very slowly. However, it is thermodynamically unstable and at sufficiently high temperatures in the presence of suitable catalysts it decomposes with explosive violence. Nitrogen(I) oxide is the main product:

$$NH_4NO_3(s) + \text{catalyst} + \text{heat} = N_2O(g) + 2H_2O(g) \qquad (18.114)$$

Amines and amides are related to ammonia. These compounds act as reducing agents toward nitrous acid to produce nitrogen in much the same way that ammonium nitrite decomposes:

$$NH_2{-}\overset{\overset{\displaystyle O}{\|}}{C}{-}NH_2(aq) + 2HNO_2(aq) = CO_2(g) + 2N_2(g) + 3H_2O \qquad (18.115)$$

This reaction finds use in the analysis of amines and amides; the measurement of the evolved nitrogen is used in quantitative determinations.

Hydrazine is produced by the basic oxidation of ammonia with hypochlorite ions:

$$2NH_3(aq) + ClO(aq)^- = N_2H_4(aq) + Cl(aq)^- + H_2O \quad (18.116)$$

Hydrazine is a good reducing agent, even in acids. In large measure, this can be attributed to the low bond energy for the N—N bond. Like ammonia, it may be oxidized to nitrogen by agents such as iodate ions, chlorine, bromine, and chlorate ions and by oxygen in basic solutions. It is not easily reduced to ammonia.

Hydroxylamine, NH_2OH, is made by reducing nitrous acid with sulfurous acid at a fairly neutral pH:

$$2H_2SO_3(aq) + HNO_2(aq) + 5H_2O = NH_2OH(aq) + 2SO_4(aq)^{--} + 4H_3O(aq)^+ \quad (18.117)$$

It may be prepared also by the reduction of either nitric or nitrous acids with strong reducing agents such as tin or zinc metals.

Hydroxylamine is best known as a reducing agent, although it may act as an oxidizing agent also. In base, it is most frequently oxidized to nitrogen(I) oxide. For example, it reduces both copper(II) oxide and iodine:

$$2NH_2OH(aq) + 4CuO(s) = N_2O(g) + 2Cu_2O(s) + 3H_2O \quad (18.118)$$

$$2NH_2OH(aq) + 2I_3(aq)^- + 4OH(aq)^- = 6I(aq)^- + N_2O(g) + 5H_2O \quad (18.119)$$

In acids it reduces iron(III) ions to iron(II) ions and nitric acid to nitrogen(II) oxide, another example of two different nitrogen oxidation states leading to the average oxidation state.

$$2NH_3OH(aq)^+ + 4Fe(aq)^{3+} + 5H_2O = 4Fe(aq)^{++} + N_2O(g) + 6H_3O(aq)^+ \quad (18.120)$$

$$NH_3OH(aq)^+ + NO_3(aq)^- = 2NO(g) + 2H_2O \quad (18.121)$$

As an oxidizing agent, hydroxylamine undergoes the following reactions.:

$$NH_2OH(aq) + H_2O_2(aq) = NH_3(aq) + O_2(g) + H_2O \quad (18.122)$$

$$NH_2OH(aq) + 2Fe(OH)_2(s) + H_2O = NH_3(aq) + 2Fe(OH)_3(s) \quad (18.123)$$

$$NH_2OH(aq) + AsO_2(aq)^- + 2OH(aq)^- = NH_3(aq) + AsO_4(aq)^{3-} + H_2O \quad (18.124)$$

$$NH_3OH(aq)^+ + 3I(aq)^- + 2H_3O(aq)^+ = I_3(aq)^- + NH_4(aq)^+ + 3H_2O \quad (18.125)$$

$$NH_3OH(aq)^+ + H_2SO_3(aq) + H_2O = HSO_4(aq)^- + NH_4(aq)^+ + H_3O(aq)^+ \quad (18.126)$$

Hydrazoic acid, HN_3, is both an oxidizing agent and a reducing agent, and it is unstable with respect to disproportionation to nitrogen and ammonium ion in acid. In fact, it can detonate with explosive violence. The salts of the acid are somewhat more stable, but azide ion still is thermodynamically unstable. The acid is made by treating hydrazine with nitrous acid in the cold.

Despite its unreactive nature, elemental nitrogen may be reduced by strong reducing agents. The Haber process is an example, and the cyanamide process is another. Similarly, nitrogen may be reduced to nitride ions by reaction with metallic magnesium:

$$N_2(g) + 3Mg(s) = Mg_3N_2(s) \quad (18.127)$$

Preparations for nitrogen(I) oxide have been shown by Equations (18.114), (18.118), (18.119), and (18.120). These all involve the oxidation of lower oxidation states of nitrogen. It may be produced also by the reduction of nitrous acid with tin(II) ions in acid.

Nitrogen(I) oxide is unstable with respect to disproportionation to elemental nitrogen and any of the higher oxidation states of nitrogen in either acidic or basic media. It is a poor reducing agent and only a fair oxidizing agent. Thus as a reducing agent it reacts only with strong oxidizing agents such as permanganate ion:

$$2MnO_4(aq)^- + 5N_2O(g) + 6H_3O(aq)^+ = 2Mn(aq)^{++} + 10NO(g) + 9H_2O \quad (18.128)$$

As an oxidizing agent it converts elemental sulfur and phosphorus to their oxides:

$$S(s) + 2N_2O(g) = SO_2(g) + 2N_2(g) \quad (18.129)$$

$$4P(s) + 6N_2O(g) = P_4O_6(s) + 6N_2(g) \quad (18.130)$$

Nitrogen(II) oxide is a common product of the reduction of nitric acid. In base, nitrogen(II) oxide disproportionates slowly to nitrogen(I) oxide and nitrite ions:

$$4NO(g) + 2OH(aq)^- = 2NO_2(aq)^- + N_2O(g) + H_2O \quad (18.131)$$

The data in Table 18.15 show that nitrogen(II) oxide is a poor oxidizing agent in acid when it is reduced to very low nitrogen oxidation states, but it is a fairly good oxidizing agent when it is reduced to the 1+ state or to elemental nitrogen. In base it is not a good oxidizing agent. It is a relatively poor reducing agent in acid but is a somewhat better one in base.

Because nitrogen(II) oxide characteristically oxidizes to nitrogen(IV) oxide with atmospheric oxygen, there is relatively little other redox chemistry of significance, for to use nitrogen(II) oxide requires the exclusion of a normal atmosphere.

$$2NO(g) + O_2(g) = N_2O_4(g) \quad (18.132)$$

Pure nitrous acid has not been isolated, but it exists in solution. In acid it is *most unstable* with respect to disproportionation to nitrogen and nitrate ions; in fact, the *most rapid* disproportionation occurs to nitrogen(II) oxide and nitrate ions:

$$3HNO_2(aq) = 2NO(g) + NO_3(aq)^- + H_3O(aq)^+ \quad (18.133)$$

Although nitrite ions are unstable with respect to disproportionation in base, in practice they do not disproportionate appreciably at moderate temperatures.

Nitrous acid and nitrite ions may be prepared through the disproportionation of nitrogen(IV) oxide, N_2O_4:

$$N_2O_4(g) + 2H_2O = HNO_2(aq) + NO_3(aq)^- + H_3O(aq)^+ \quad (18.134)$$

$$N_2O_4(g) + 2OH(aq)^- = NO_2(aq)^- + NO_3(aq)^- + H_2O \quad (18.135)$$

Nitrous acid is a moderately strong oxidizing agent (and incidentally, usually a rapid one) when it is reduced to nitrogen, nitrogen(I) oxide, or nitrogen(II) oxide. Although it is possible to reduce the acid to negative oxidation states of nitrogen with some strong reducing agents, this is not common. See Equations (18.115) and (18.117) for other reactions of nitrous acid.

Usually nitrous acid is reduced to nitrogen(II) oxide:

$$HNO_2(aq) + Fe(aq)^{++} + H_3O(aq)^+ = NO(g) + Fe(aq)^{3+} + 2H_2O \quad (18.136)$$

$$2HNO_2(aq) + 3I(aq)^- + 2H_3O(aq)^+ = I_3(aq)^- + 2NO(g) + 4H_2O \quad (18.137)$$

Nitrite ions are poor oxidizing agents in base. Although nitrous acid is a poor reducing agent, nitrite ions in base are better. Even in acidic solution, strong oxidizing agents such as permanganate ion, hydrogen peroxide, and bromate ions oxidize nitrous acid to nitrate ions:

$$5HNO_2(aq) + 2MnO_4(aq)^- + H_3O(aq)^+ = 5NO_3(aq)^- + 2Mn(aq)^{++} + 4H_2O \quad (18.138)$$

Nitrogen(IV) oxide is prepared by the oxidation of nitrogen(II) oxide with atmospheric oxygen, as shown earlier in Equation (18.132). It is produced also by the reduction of nitric acid. The standard disproportionations of nitrogen(IV) oxide have been shown as Equations (18.134) and (18.135). However, at high temperatures and high concentrations of acid, the disproportionation to nitrate ions and nitrogen(II) oxide is favored:

$$3N_2O_4(g) + 6H_2O = 4NO_3(aq)^- + 2NO(g) + 4H_3O(aq)^+ \quad (18.139)$$

Nitrogen(IV) oxide is a good oxidizing agent in acid, but a poor one in base. As a reducing agent it is very poor in acid and moderately good in base. Often it is used in the gas phase as an oxidizing agent:

$$2CO(g) + N_2O_4(g) = 2CO_2(g) + 2NO(g) \tag{18.140}$$

$$2SO_2(g) + N_2O_4(g) = 2SO_3(l) + 2NO(g) \tag{18.141}$$

$$4Cu(s) + N_2O_4(g) = Cu_2O(s) + NO(g) \tag{18.142}$$

For nitrogen(IV) oxide to act as a reducing agent, a strong oxidizing agent such as hydrogen peroxide is required:

$$N_2O_4(g) + H_2O_2(aq) + 2H_2O = 2NO_3(aq)^- + 2H_3O(aq)^+ \tag{18.143}$$

The preparation of nitric acid depends upon the catalyzed combustion of ammonia as represented by Equation (18.111). The nitrogen(II) oxide produced in this reaction is allowed to react with oxygen according to Equation (18.132) to form nitrogen(IV) oxide, which then disproportionates according to Equation (18.139) to form nitrate ions and nitrogen(II) oxide, which is recycled.

Nitric acid solutions are unstable with respect to the following decomposition:

$$4H_3O(aq)^+ + 4NO_3(aq)^- = 2N_2O_4(g) + O_2(g) + 6H_2O \tag{18.144}$$

The reduction of nitric acid is complex. However, it appears that nitrogen(II) oxide is the major reduction product from dilute nitric acid solutions and nitrogen(IV) oxide is the major reduction product from concentrated acid solution. Thus the reduction of nitric acid by copper metal may be represented by the following two equations:

$$\text{Dilute: } 2NO_3(aq)^- + 3Cu(s) + 8H_3O(aq)^+ = 3Cu(aq)^{++} + 2NO(g) + 12H_2O \tag{18.145}$$

$$\text{Conc.: } 2NO_3(aq)^- + Cu(s) + 4H_3O(aq)^+ = Cu(aq)^{++} + N_2O_4(g) + 6H_2O \tag{18.146}$$

Relatively few metals are not attacked by nitric acid. These include gold, iridium, platinum, rhodium, and tantalum. Nonmetals tend to be oxidized to their oxy-acids. With strongly reducing metals, roughly those with negative standard reduction potentials, nitric acid tends to be reduced to the lower nitrogen oxidation states exhibited by N_2O, N_2, NH_3OH^+, and NH_4^+. Often the products are mixed. However, the following equation illustrates a reaction commonly used to reduce nitrate ions to ammonia:

$$4Zn(s) + NO_3(aq)^- + 7OH(aq)^- + 6H_2O = 4[Zn(OH)_4]^{--} + NH_3(aq) \tag{18.147}$$

The presence of nitrous acid plays an important role in the reduction of nitrates and nitric acid. If nitric acid is freed of nitrous acid by the addition of urea or hydrogen peroxide, the nitric acid does not oxidize such metals as copper, silver, and bismuth. However, the presence of nitrous acid retards the reaction of such metals as magnesium and cadmium with nitric acid.

Phosphorus

The oxidation-state map for phosphorus is shown as Fig. 18.10 and the summary of the standard reduction potentials for phosphorus-containing species is given in Table 18.16.

Although the map is a bit more complex than the nitrogen map, the redox chemistry for phosphorus is somewhat simpler than that for nitrogen, since there are fewer phosphorus species that concern us. We shall not discuss the condensed phosphates here, since these are best considered as acid-base condensation products of the acids, and their redox chemistry is not illuminated significantly by the study of their reduction potentials.

From the map it is clear that none of the phosphorus-containing species is a very good oxidizing agent, but these species tend to be better reducing agents. It is also interesting to note that phosphorus in the 3+ oxidation state has its greatest stability in basic solution and that it tends to be unstable in strongly acid solutions.

Elemental phosphorus is not found on the map, indicating that it is not a

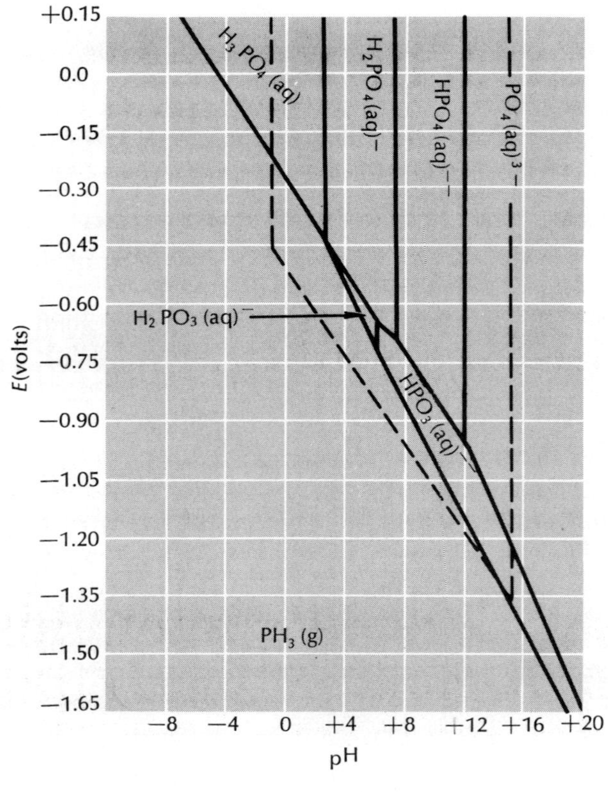

Fig. 18.10. Phosphorus oxidation-state map

TABLE 18.16

Standard Half-cell Reduction Potentials for Phosphorus-containing Species (Volts at 25°C)

In Aqueous Acid

OXIDIZING AGENTS	REDUCING AGENTS				
	$PH_3(g)$	$P_2H_4(l)$	$P(s)$	$H_3PO_2(aq)$	$H_3PO_3(aq)$
$H_3PO_4(aq)$	−0.28	−0.3	−0.41	−0.39	−0.28
$H_3PO_3(aq)$	−0.28	−0.3	−0.50	−0.50	
$H_3PO_2(aq)$	−0.18	−0.2	−0.51		
$P(s)$	−0.06	−0.1			
$P_2H_4(l)$	0.00				

In Aqueous Base

OXIDIZING AGENTS	REDUCING AGENTS				
	$PH_3(g)$	$P_2H_4(l)$	$P(s)$	$H_2PO_2(aq)^-$	$HPO_3(aq)^{--}$
$PO_4(aq)^{3-}$	−1.25	−0.3	−1.48	−1.34	−1.12
$HPO_3(aq)^{--}$	−1.31	−1.4	−1.73	−1.57	
$H_2PO_2(aq)^-$	−1.18	−0.9	−2.05		
$P(s)$	−0.89	−0.9			
$P_2H_4(l)$	−0.8				

thermodynamically stable species, which helps to account for the fact that it is not found in nature. For the most part, phosphorus occurs naturally as phosphate ions, often in the very insoluble calcium phosphate.

As evident from the data of Table 18.16, elemental phosphorus is unstable with respect to disproportionation in both acid and base. The basic disproportionation to phosphine and hypophosphite ions serves as the method for preparing phosphine:

$$4P(s) + 3OH(aq)^- + 3H_2O = 3H_2PO_2(aq)^- + PH_3(g) \quad (18.148)$$

Phosphine may be prepared also by the disproportionations of phosphorous and hypophosphorous acids:

$$4H_3PO_3(aq) = 3H_3PO_4(aq) + PH_3(g) \quad (18.149)$$

$$2H_3PO_2(aq) = H_3PO_4(aq) + PH_3(g) \quad (18.150)$$

As shown on the oxidation state map and in Table 18.16, phosphine is a good reducing agent in base. This happens to be true also of the gas in the gas phase, for it ignites spontaneously in gaseous oxygen and explodes in nitrogen(I) oxide:

$$4PH_3(g) + 8O_2(g) = P_4O_{10}(s) + 6H_2O(g) \quad (18.151)$$

$$4PH_3(g) + 16N_2O(g) = P_4O_{10}(s) + 16N_2(g) + 6H_2O(g) \quad (18.152)$$

Diphosphine is the phosphorus analog of hydrazine. It is formed along with phosphine from the basic disproportionation of elemental phosphorus. However, it is unstable and decomposes into phosphine and another complex hydride of phosphorus.

Phosphorus is prepared by the reduction of phosphates. Since phosphate ions are not good oxidizing agents, some drastic conditions are required to reduce phosphate ions to phosphorus. Usually the reaction is carried out at high temperature with carbon as the reducing agent:

$$Ca_3(PO_4)_2(s) + 2SiO_2(s) + 5C(s) = 2P(g) + 5CO(g) + Ca_3Si_2O_7(s) \quad (18.153)$$

The low values of the potentials for phosphorus in Table 18.16 indicate that elemental phosphorus is a better reducing agent than it is an oxidizing agent. Usually oxidation leads to phosphate ions or phosphoric acid:

$$5NO_3(aq)^- + 3P(s) + 5H_3O(aq)^+ = 5NO(g) + 3H_3PO_4(aq) + 3H_2O \quad (18.154)$$

$$2P(s) + 5Cu(aq)^{++} + 18H_2O = 5Cu(s) + 2H_3PO_4(aq) + 10H_3O(aq)^+ \quad (18.155)$$

Hypophosphite ions are produced along with phosphine when elemental phosphorus disproportionates in aqueous base as shown in Equation (18.148). They are made also by the mild oxidation of phosphine with iodine:

$$PH_3(g) + 2I_3(aq)^- + 6H_2O = H_3PO_2(aq) + 6I(aq)^- + 4H_3O(aq)^+ \quad (18.156)$$

Hypophosphorous acid and its salts are very poor oxidizing agents, but they are good reducing agents, especially the salts. They reduce many metallic salts to the free metals. Thus bismuth, gold, mercury, and silver metals may be produced in this way. Depending upon the strength of the oxidizing agent, the phosphorus may be oxidized to either the 3+ or 5+ state. With strong reducing agents, such as zinc metal, it may be reduced to phosphine:

$$2Zn(s) + H_3PO_2(aq) + 4H_3O(aq)^+ = PH_3(g) + 2Zn(aq)^{++} + 6H_2O \quad (18.157)$$

As a reducing agent, phosphorus burns in oxygen. The oxygen supply determines the product formed. With limited oxygen, phosphorus(III) oxide results:

$$4P(s) + 3O_2(g) = P_4O_6(s) \quad (18.158)$$

If phosphorus(III) oxide is dissolved in water, an aqueous solution of phosphorous acid results. In strong acids, phosphorous acid is unstable with respect to disproportionation to phosphine and phosphoric acid. This instability continues outside the aqueous medium, and upon heating, the acid disproportionates.

The potentials indicate that phosphorous acid is a good reducing agent and a poor oxidizing agent. However, with a strong reducing agent, it can be reduced to phosphine:

$$3Zn(s) + H_3PO_3(aq) + 6H_3O(aq)^+ = 3Zn(aq)^{++} + PH_3(g) + 9H_2O \quad (18.159)$$

As a reducing agent, it undergoes the following typical oxidations:

$$H_2SO_3(aq) + 2H_3PO_3(aq) = S(s) + 2H_3PO_4(aq) + H_2O \quad (18.160)$$

$$I_3(aq)^- + HPO_3(aq)^{--} + 3OH(aq)^- = 3I(aq)^- + PO_4(aq)^{3-} + 2H_2O \quad (18.161)$$

$$2Ag(aq)^+ + H_3PO_3(aq) + 3H_2O = 2Ag(s) + H_3PO_4(aq) + 2H_3O(aq)^+ \quad (18.162)$$

Arsenic

The oxidation-state map for arsenic is shown as Fig. 18.11, and the summary of the standard reduction potentials for arsenic-containing species is given in Table 18.17.

All of the arsenic-containing species considered here show thermodynamic stability and appear on the map. Unlike phosphorus, but like nitrogen, elemental arsenic shows thermodynamic stability across the entire pH range considered. In addition, arsenic in the 3+ state shows stability across the entire pH range, but unlike phosphorus in the 3+ state, it has a wider range of stability on the acid side than on the basic side.

Although some elemental arsenic is found in nature, arsenic is found more

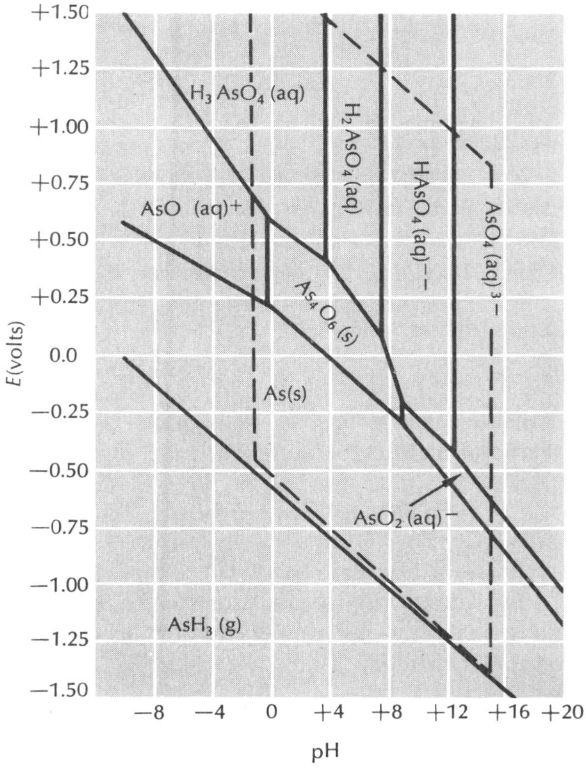

Fig. 18.11. Arsenic oxidation-state map

TABLE 18.17

Standard Half-cell Reduction Potentials for Arsenic-containing Species (Volts at 25°C)

In Aqueous Acid

OXIDIZING AGENTS	REDUCING AGENTS		
	$AsH_3(g)$	$As(s)$	$As_4O_6(s)$
$H_3AsO_4(aq)$	+0.01	+0.37	+0.58
$As_4O_6(s)$	−0.18	+0.23	
$As(s)$	−0.60		

In Aqueous Base

OXIDIZING AGENTS	REDUCING AGENTS		
	$AsH_3(g)$	$As(s)$	$AsO_2(aq)^-$
$AsO_4(aq)^{3-}$	−0.96	−0.68	−0.67
$AsO_2(aq)^-$	−1.06	−0.68	
$As(s)$	−1.43		

commonly as the insoluble sulfide, As_2S_3, and oxide, As_4O_6. In some cases it is found as the arsenide in such minerals as $FeAs_2$ or $CoAs_2$.

Whereas phosphine is almost strong enough to reduce water, arsine is actually strong enough to do so, as indicated on the oxidation state map. Thus arsine is a strong reducing agent. One of the characteristic reactions of arsine often used as a test for the presence of arsenic is the reduction of silver(I) ions to silver:

$$AsH_3(g) + 6Ag(aq)^+ + 8H_2O = 6Ag(s) + HAsO_2(aq) + 6H_3O(aq)^+ \quad (18.163)$$

Like phosphine, arsine burns vigorously in air:

$$4AsH_3(g) + 6O_2(g) = As_4O_6(s) + 6H_2O(g) \quad (18.164)$$

Since arsine is a strong reducing agent, it may be prepared by treatment of arsenic(III) oxide, or other species containing arsenic in the 3+ state, with very strong reducing agents such as zinc metal:

$$HAsO_2(aq) + 3Zn(s) + 6H_3O(aq)^+ = AsH_3(g) + 3Zn(aq)^{++} + 8H_2O \quad (18.165)$$

Elemental arsenic may be prepared by the reduction of arsenic(III) species, dry and at high temperature, with carbon as the reducing agent:

$$As_4O_6(s) + 6C(s) = 4As(g) + 6CO(g) \quad (18.166)$$

Arsenic is a better reducing agent than an oxidizing agent, and as a reducing agent it is most frequently oxidized to the 3+ state. Oxidation to the 5+ state occurs only with strong oxidizing agents such as nitric acid. Arsenic burns in air to form arsenic(III) oxide:

$$4As(s) + 3O_2(g) = As_4O_6(s) \quad (18.167)$$

This is the most commonly available compound of arsenic. Although it is insoluble in acids, it does dissolve in base, forming metaarsenite ions, $AsO_2(aq)^-$. Metaarsenious acid is a poor oxidizing agent which may be reduced to elemental arsenic by copper metal or tin(II) ions in acidic solution. It is a better reducing agent, being oxidized to arsenic acid. The arsenic content of materials often is determined by redox titrations of arsenic in the 3+ state with oxidizing agents such as permanganate ions or iodine.

$$2MnO_4(aq)^- + 5HAsO_2(aq) + 6H_3O(aq)^+ = 2Mn(aq)^{++} + 5H_3AsO_4(aq) + 4H_2O \quad (18.168)$$

$$I_3(aq)^- + HAsO_2(aq) + 4H_2O = 3I(aq)^- + H_3AsO_4(aq) + 2H_3O(aq)^+ \quad (18.169)$$

Arsenic acid is a moderately good oxidizing agent which can be prepared by the oxidation of arsenious acid. It oxidizes sulfurous acid to sulfate ions, for example:

$$H_2SO_3(aq) + H_3AsO_4(aq) + H_2O = SO_4(aq)^{--} + HAsO_2(aq) + 2H_3O(aq)^+ \quad (18.170)$$

CARBON

The study of the compounds of carbon is a broad field called organic chemistry, as we know. The important reactions of carbon chemistry often depend upon kinetic effects, and standard reduction potentials may be of little value. Nevertheless, a few standard reduction potentials are listed in Table 18.18, and a standard potential diagram for some species is given in Table 18.19. The oxidation-state map is shown as Fig. 18.12 and the oxidation state potential summary is given in Table 18.20.

From the oxidation-state map, it appears that the extremes in oxidation state are most stable, namely 4+ and 4−. Elemental carbon shows a narrow band of stability that disappears in strongly basic media.

From the potential data, methane appears to be a good reducing agent in base. In fact, in aqueous media its reducing action is very slow. It does burn in air, however, to form water and carbon dioxide, carbon monoxide, or carbon, depending upon the relative amounts of methane and oxygen used.

Methanol is a poor oxidizing agent. Its potentials suggest that it is a good reducing agent, however, particularly in base. Actually it is oxidized to formaldehyde or formic acid by strong oxidizing agents. Methanol is prepared commercially by the catalytic reduction of carbon monoxide with hydrogen:

$$CO(g) + 2H_2(g) = CH_3OH(g) \quad (18.171)$$

TABLE 18.18

Some Standard Reduction Potentials for Carbon Species (Volts at 25°C)

	OXIDIZING AGENTS		REDUCING AGENTS	E°
1.	$CH_3OH(aq) + 2H_3O(aq)^+$	$+ 2e^- = 3H_2O$	$+ CH_4(g)$	+0.59
2.	$CO(g) + 2H_3O(aq)^+$	$+ 2e^- = 3H_2O$	$+ C(s)$	+0.52
3.	$HCHO(aq) + 4H_3O(aq)^+$	$+ 4e^- = 5H_2O$	$+ CH_4(g)$	+0.41
4.	$H_2C_2O_4(aq) + 2H_3O(aq)^+$	$+ 2e^- = 4H_2O$	$+ 2CO(g)$	+0.26
5.	$HCHO(aq) + 2H_3O(aq)^+$	$+ 2e^- = 2H_2O$	$+ CH_3OH(aq)$	+0.23
6.	$CO_2(g) + 4H_3O(aq)^+$	$+ 4e^- = 6H_2O$	$+ C(s)$	+0.21
7.	$C(s) + 4H_3O(aq)^+$	$+ 4e^- = 4H_2O$	$+ CH_4(g)$	+0.13
8.	$H_2C_2O_4(aq) + 2H_3O(aq)^+$	$+ 2e^- = 2H_2O$	$+ 2HCOOH(aq)$	+0.07
9.	$HCOOH(aq) + 2H_3O(aq)^+$	$+ 2e^- = 3H_2O$	$+ HCHO(aq)$	+0.06
10.	$CO_2(g) + 2H_3O(aq)^+$	$+ 2e^- = 2H_2O$	$+ HCOOH(aq)$	−0.20
11.	$CH_3OH(aq) + H_2O$	$+ 2e^- = 2OH(aq)^-$	$+ CH_4(g)$	−0.24
12.	$CO(g) + H_2O$	$+ 2e^- = 2OH(aq)^-$	$+ C(s)$	−0.31
13.	$C(s) + 2H_3O(aq)^+$	$+ 2e^- = H_2O$	$+ CH_3OH(aq)$	−0.33
14.	$HCHO(aq) + 3H_2O$	$+ 4e^- = 4OH(aq)^-$	$+ CH_4(g)$	−0.42
15.	$2CO_2(g) + 2H_3O(aq)^+$	$+ 2e^- = 2H_2O$	$+ H_2C_2O_4(aq)$	−0.49
16.	$HCHO(aq) + 2H_2O$	$+ 2e^- = 2OH(aq)^-$	$+ CH_3OH(aq)$	−0.60
17.	$C(s) + 4H_2O$	$+ 4e^- = 4OH(aq)^-$	$+ CH_4(g)$	−0.70
18.	$CO_3(aq)^{--} + 3H_2O$	$+ 4e^- = 6OH(aq)^-$	$+ C(s)$	−0.77
19.	$C_2O_4(aq)^{--} + 2H_2O$	$+ 2e^- = 2OH(aq)^-$	$+ 2HCOO(aq)^-$	−0.82
20.	$CO_3(aq)^{--} + 2H_2O$	$+ 2e^- = 3OH(aq)^-$	$+ HCOO(aq)^-$	−1.01
21.	$HCOO(aq)^- + 2H_2O$	$+ 2e^- = 3OH(aq)^-$	$+ HCHO(aq)$	−1.07
22.	$C(s) + 3H_2O$	$+ 2e^- = 2OH(aq)^-$	$+ CH_3OH(aq)$	−1.16
23.	$2CO_3(aq)^{--} + 2H_2O$	$+ 2e^- = 4OH(aq)^-$	$+ C_2O_4(aq)^{--}$	−1.21
24.	$C_2O_4(aq)^{--} + 2H_2O$	$+ 2e^- = 4OH(aq)^-$	$+ 2CO(g)$	−1.23

Formaldehyde is made by the careful oxidation of methane or methanol. It is a fairly good reducing agent and is oxidized to formic acid or formate ions by atmospheric oxygen.

Carbon is neither a good oxidizing agent nor a good reducing agent in practice, although the potential data suggest it might be a fair reducing agent in aqueous base. It is used at very high temperatures as a cheap reducing agent. Under these conditions it will reduce steam and carbon dioxide, for instance:

TABLE 18.19

Standard Potential Diagram for Some Carbon-containing Species

IN AQUEOUS ACID

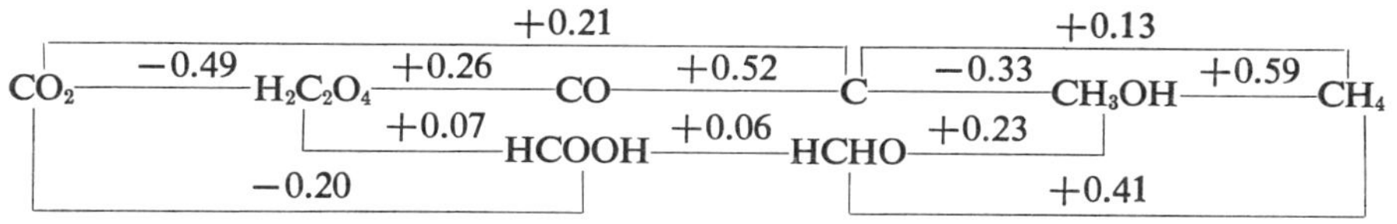

IN AQUEOUS BASE

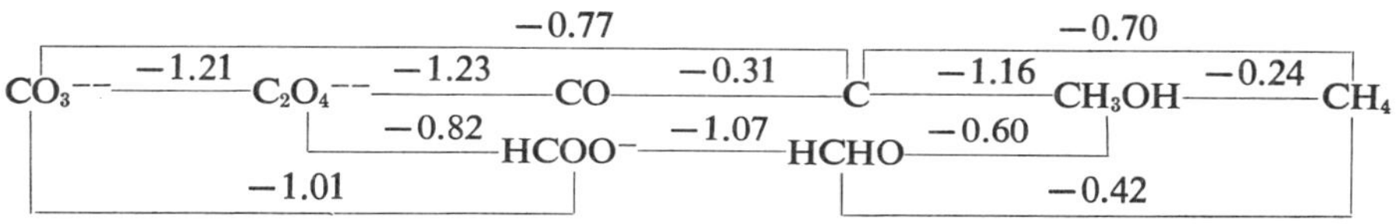

$$C(s) + H_2O(g) = CO(g) + H_2(g) \tag{18.172}$$

$$C(s) + CO_2(g) = 2CO(g) \tag{18.173}$$

Formally, formic acid is hydrated carbon monoxide, and formates may be prepared by absorption of carbon monoxide in aqueous base. In acids formic acid dehydrates, liberating carbon monoxide. Formic acid is also made by the oxida-

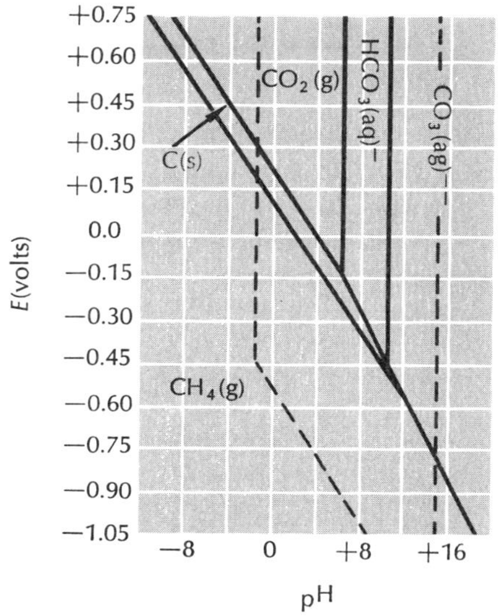

Fig. 18.12. Carbon oxidation-state map

TABLE 18.20

Standard Half-cell Reduction Potentials for Carbon-containing Species (Volts at 25°C)

In Aqueous Acid

OXIDIZING AGENTS	REDUCING AGENTS						
	CH_4(g)	CH_3OH(aq)	HCHO(aq)	C(s)	HCOOH(aq)	CO(g)	$H_2C_2O_4$(aq)
CO_2(g)	+0.17	+0.03	−0.07	+0.21	−0.20	−0.12	−0.49
$H_2C_2O_4$(aq)	+0.26	+0.13	+0.06	+0.43	+0.07	+0.26	
CO(g)	+0.26	+0.10	−0.04	+0.52			
HCOOH(aq)	+0.29	+0.14	+0.06	+0.61			
C(s)	+0.13	−0.33					
HCHO(aq)	+0.41	+0.23					
CH_3OH(aq)	+0.59						

In Aqueous Base

OXIDIZING AGENTS	REDUCING AGENTS						
	CH_4(g)	CH_3OH(aq)	HCHO(aq)	C(s)	$HCOO(aq)^-$	CO(g)	$C_2O_4(aq)^{--}$
$CO_3(aq)^{--}$	−0.73	−0.89	−1.05	−0.77	−1.01	−1.22	−1.21
$C_2O_4(aq)^{--}$	−0.66	−0.83	−0.99	−0.62	−0.82	−1.23	
CO(g)	−0.57	−0.73	−0.87	−0.31			
$HCOO(aq)^-$	−0.64	−0.84	−1.07	−0.52			
C(s)	−0.70	−1.16					
HCHO(aq)	−0.42	−0.60					
CH_3OH(aq)	−0.24						

tion of formaldehyde. Formic acid and the formates are poor oxidizing agents, as suggested by the reduction potentials, but formate ions are good reducing agents and find some use in this capacity.

Carbon monoxide is an inexpensive, easily available reducing agent for many commercial processes. The following are some typical reactions:

$$CuO(s) + CO(g) = Cu(s) + CO_2(g) \tag{18.174}$$

$$2CO(g) + O_2(g) = 2CO_2(g) \tag{18.175}$$

In acidic solutions, oxalic acid is unstable with respect to disproportionation to carbon monoxide and carbon dioxide. With a catalyst such as sulfuric acid, this reaction occurs as follows:

$$H_2C_2O_4(aq) = CO(g) + CO_2(g) + H_2O \quad (18.176)$$

Oxalic acid is a moderately good reducing agent which finds use in the analytical laboratory as a standard reducing agent for standardizing aqueous solutions of oxidizing agents, especially solutions of permanganate ions. The following equations illustrate three typical oxidations of oxalic acid or oxalate ions:

$$MnO_2(s) + H_2C_2O_4(aq) + 2H_3O(aq)^+ = Mn(aq)^{++} + 2CO_2(g) + 4H_2O \quad (18.177)$$

$$2MnO_4(aq)^- + 5H_2C_2O_4(aq) + 6H_3O(aq)^+ = 2Mn(aq)^{++} + 10CO_2(g) + 14H_2O \quad (18.178)$$

$$Cl_2(g) + C_2O_4(aq)^{--} = 2Cl(aq)^- + 2CO_2(g) \quad (18.179)$$

Although it is not a strong oxidizing agent, carbon dioxide may be reduced under some circumstances. Very strong reducing agents such as sodium, potassium, or magnesium metals actually will burn in carbon dioxide.

$$2Na(s) + CO_2(g) = CO(g) + Na_2O(s) \quad (18.180)$$

ZINC AND MERCURY

Standard reduction potentials for zinc and mercury are given in Tables 18.21 and 18.22. It is clear from these data that zinc is a good reducing agent in either acid or base, whereas mercury is not.

TABLE 18.21

Some Standard Reduction Potentials for Zinc and Mercury (Volts at 25°C)

	OXIDIZING AGENTS			REDUCING AGENTS	E°
1.	$2Hg(aq)^{++}$	$+ 2e^- =$		$Hg_2(aq)^{++}$	+0.92
2.	$Hg(aq)^{++}$	$+ 2e^- =$		$Hg(l)$	+0.85
3.	$Hg_2(aq)^{++}$	$+ 2e^- =$		$2Hg(l)$	+0.79
4.	$Hg_2SO_4(s)$	$+ 2e^- =$	$SO_4(aq)^{--}$	$+ 2Hg(l)$	+0.61
5.	$[HgCl_4](aq)^{--}$	$+ 2e^- =$	$4Cl(aq)^-$	$+ Hg(l)$	+0.48
6.	$Hg_2Cl_2(s)$	$+ 2e^- =$	$2Cl(aq)^-$	$+ 2Hg(l)$	+0.27
7.	$[HgBr_4](aq)^{--}$	$+ 2e^- =$	$4Br(aq)^-$	$+ Hg(l)$	+0.21
8.	$Hg_2Br_2(s)$	$+ 2e^- =$	$2Br(aq)^-$	$+ 2Hg(l)$	+0.14
9.	$HgO(s) + H_2O$	$+ 2e^- =$	$2OH(aq)^-$	$+ Hg(l)$	+0.10
10.	$Hg_2I_2(s)$	$+ 2e^- =$	$2I(aq)^-$	$+ 2Hg(l)$	−0.04
11.	$[HgI_4](aq)^{--}$	$+ 2e^- =$	$4I(aq)^-$	$+ Hg(l)$	−0.04
12.	$[Hg(CN)_4](aq)^{--}$	$+ 2e^- =$	$4CN(aq)^-$	$+ Hg(l)$	−0.37
13.	$HgS(s)$	$+ 2e^- =$	$S(aq)^{--}$	$+ Hg(l)$	−0.71
14.	$Zn(aq)^{++}$	$+ 2e^- =$		$Zn(s)$	−0.76
15.	$[Zn(C_2O_4)_3](aq)^{4-}$	$+ 2e^- =$	$3C_2O_4(aq)^{--}$	$+ Zn(s)$	−1.02
16.	$[Zn(NH_3)_4](aq)^{++}$	$+ 2e^- =$	$4NH_3(aq)$	$+ Zn(s)$	−1.04
17.	$Zn(OH)_2(s)$	$+ 2e^- =$	$2OH(aq)^-$	$+ Zn(s)$	−1.24
18.	$[Zn(CN)_4](aq)^{--}$	$+ 2e^- =$	$4CN(aq)^-$	$+ Zn(s)$	−1.26
19.	$ZnS(s)$	$+ 2e^- =$	$S(aq)^{--}$	$+ Zn(s)$	−1.58

TABLE 18.22
Standard Potential Diagram for Zinc- and Mercury-containing Species

IN AQUEOUS ACID

$$Zn^{++} \xrightarrow{-0.76} Zn$$

$$Hg^{++} \xrightarrow{+0.92} Hg_2^{++} \xrightarrow{+0.79} Hg$$

$$Hg^{++} \xrightarrow{+0.85} Hg$$

IN AQUEOUS BASE

$$Zn(OH)_2 \xrightarrow{-1.24} Zn$$

$$HgO \xrightarrow{+0.10} Hg$$

Zinc

The oxidation-state map for zinc is shown as Fig. 18.13. The map suggests that there may be difficulty in reducing aqueous zinc(II) ions to zinc metal by electrolysis, since water may be reduced more easily. However, an industrial technique has been developed using sulfuric acid as the electrolyte under conditions in which the hydrogen overvoltage exceeds 1.00 V.

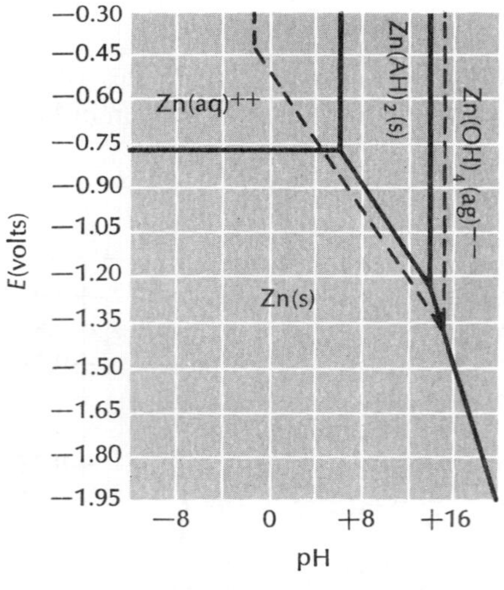

Fig. 18.13. Zinc oxidation-state map

As a strong reducing agent, zinc metal unites with atmospheric oxygen and will even reduce carbon dioxide to carbon monoxide. It also liberates hydrogen from both acidic and basic solutions:

$$Zn(s) + 2H_3O(aq)^+ = Zn(aq)^{++} + H_2(g) + 2H_2O \tag{18.181}$$

$$Zn(s) + 2OH(aq)^- = Zn(OH)_2(s) + H_2(g) \tag{18.182}$$

It is found characteristically that many precipitating or complexing species of the metal ions may be used to enhance the oxidation of the metal, that is, to sta-

bilize a positive oxidation state. To see how this comes about, consider the zinc-zinc(II) ion half-cell reaction equation:

$$Zn(aq)^{++} + 2e^- = Zn(s) \tag{18.183}$$

The Nernst equation for this half-cell is as follows:

$$E = E^\circ - \frac{0.0591}{2} \log \frac{1}{a_{zn^{++}}} \tag{18.184}$$

If in some way the activity of zinc(II) ions can be reduced, then the fraction in the logarithm term becomes larger and the logarithm value becomes more positive; this larger value must be subtracted from the standard reduction potential, so that the reduction potential becomes more negative. Under these conditions, the zinc(II) ions are poorer oxidizing agents; that is, they are reduced with greater difficulty and are thereby stabilized. The zinc metal becomes a stronger reducing agent under these circumstances and becomes less stable with respect to oxidation to zinc(II) ions.

There are two common chemical means of reducing the activity of metallic ions in solutions. One is to add a precipitating agent. When zinc(II) ions are involved, a source of sulfide ions may be introduced to precipitate zinc sulfide:

$$Zn(aq)^{++} + H_2S(g) + 2H_2O = ZnS(s) + 2H_3O(aq)^+ \tag{18.185}$$

Thus in the presence of hydrogen sulfide, which precipitates insoluble zinc sulfide, the zinc ion concentration is decreased; zinc metal then becomes a better reducing agent and zinc ions become poorer oxidizing agents. By how much the reduction potential is reduced depends upon the solubility product of zinc sulfide. If zinc sulfide is very insoluble, that is, has a very low solubility product, very few zinc ions remain in solution and the reduction potential is reduced very markedly.

If we consider half-cell 19 in Table 18.21, we may compute the standard reduction potential for this half-cell, using Equation (18.184). The standard conditions for half-cell 19 include an activity of 1.00 *M* for sulfide ion. This activity may be used with the solubility product constant value from Table 13.5 to compute the zinc ion activity.

$$K_{sp} = (Zn(aq)^{++})(S(aq)^{--}) = 1.2 \times 10^{-28}$$

$$(Zn(aq)^{++}) = \frac{1.2 \times 10^{-28}}{(S(aq)^{--})} = 1.2 \times 10^{-28}$$

The zinc ion activity may then be used in Equation (18.184) to find *E*.

$$E = 0.76 - \frac{0.0591}{2} \log \frac{1}{1.2 \times 10^{-28}} = -1.58 \text{ V}$$

In a similar way, a complexing agent may stabilize the 2+ oxidation state of

zinc by combining with zinc(II) ions. For example, ammonia forms the tetraamminezinc(II) ion:

$$Zn(aq)^{++} + 4NH_3(aq) = [Zn(NH_3)_4](aq)^{++} \tag{18.186}$$

Mercury

The oxidation-state map for mercury is shown in Fig. 18.14. Unlike zinc, mercury tends to form a 1+ oxidation state characterized by the mercury(I) ion, Hg_2^{++}. From the map it is clear that mercury in the 1+ state tends to be stable only in strongly acidic solutions. The potentials indicate that in base it is unstable toward disproportionation; this reaction occurs as follows:

$$Hg_2(aq)^{++} + 2OH(aq)^- = Hg(l) + HgO(s) + H_2O \tag{18.187}$$

This same disproportionation occurs with aqueous ammonia in the presence of chloride ions:

$$Hg_2(aq)^{++} + 2NH_3(aq) + Cl(aq)^- = Hg(l) + HgNH_2Cl(s) + NH_4(aq)^+ \tag{18.188}$$

For the most part mercury occurs in nature as the mercury(II) sulfide, which is extremely insoluble and therefore stabilizes the 2+ oxidation state. Elementary mercury may be obtained from the sulfide by heating in air:

$$HgS(s) + O_2(g) = Hg(g) + SO_2(g) \tag{18.189}$$

It is evident in the oxidation-state map, that elementary mercury is a poor reducing agent even in base. Thus mercury is oxidized only by strong oxidizing

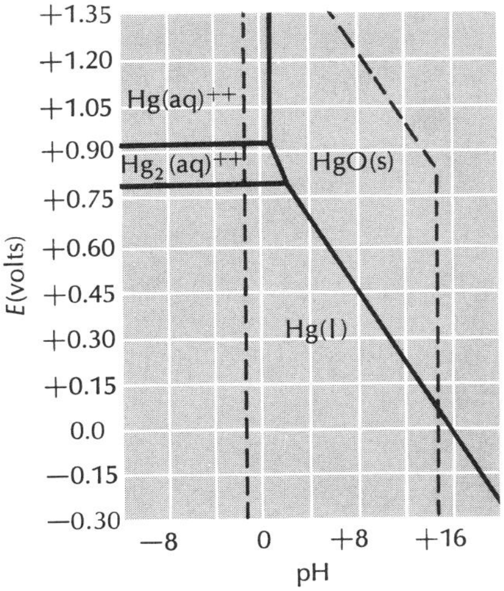

Fig. 18.14. Mercury oxidation-state map

agents. Nitric acid is used commonly. Mercury(II) ions are produced if nitric acid is in excess, and mercury(I) ions are produced if mercury is in excess:

$$\text{Excess } HNO_3\text{:}\quad 3Hg(l) + 2NO_3(aq)^- + 8H_3O(aq)^+ = 3Hg(aq)^{++} + 2NO(g) + 12H_2O \qquad (18.190)$$

$$\text{Excess Hg:}\quad 6Hg(l) + 2NO_3(aq)^- + 8H_3O(aq)^+ = 3Hg_2(aq)^{++} + 2NO(g) + 12H_2O \qquad (18.191)$$

The positive oxidation states of mercury may be stabilized by precipitation and complexation. Generally mercury(I) is stabilized as an insoluble mercury(I) halide. More frequently, mercury(II) is stabilized as a complex of mercury(II) ions. Some data are included in Table 18.21.

COPPER AND SILVER

Some standard reduction potential data are shown for copper and silver in Tables 18.23 and 18.24. Clearly, where comparable reactions occur, silver species are better oxidizing agents than copper species.

TABLE 18.23
Standard Potential Diagram for Copper and Silver

			IN AQUEOUS ACID			
				+0.34 (Cu^{++} to Cu)		
		Cu^{++}	+0.15	Cu^{+}	+0.52	Cu
Ag_2O_3	+1.57	AgO	+1.78	Ag^{+}	+0.80	Ag
			IN AQUEOUS BASE			
				−0.22 ($Cu(OH)_2$ to Cu)		
		$Cu(OH)_2$	−0.08	Cu_2O	−0.36	Cu
Ag_2O_3	+0.74	AgO	+0.57	Ag_2O	+0.34	Ag

Copper

The oxidation-state map for copper is shown as Fig. 18.15. From this map, copper in the 1+ state of oxidation is evidently stable as Cu_2O in base. In acid, copper(I) ions disproportionate:

$$2Cu(aq)^+ = Cu(s) + Cu(aq)^{++} \qquad (18.192)$$

TABLE 18.24

Some Standard Reduction Potentials for Copper and Silver (Volts at 25°C)

	OXIDIZING AGENTS				REDUCING AGENTS	E°
1.	$AgO(s) + 2H_3O(aq)^+$	$+ 1e^-$	=	$3H_2O$	$+ Ag(aq)^+$	+1.78
2.	$Ag_2O_3(s) + 2H_3O(aq)^+$	$+ 2e^-$	=	$3H_2O$	$+ 2AgO(s)$	+1.57
3.	$Cu(aq)^{++} + I(aq)^-$	$+ 1e^-$	=		$CuI(s)$	+0.88
4.	$Ag(aq)^+$	$+ 1e^-$	=		$Ag(s)$	+0.80
5.	$Ag_2O_3(s) + H_2O$	$+ 2e^-$	=	$2OH(aq)^-$	$+ 2AgO(s)$	+0.74
6.	$Ag_2SO_4(s)$	$+ 2e^-$	=	$SO_4(aq)^{--}$	$+ 2Ag(s)$	+0.65
7.	$Ag(CH_3COO)(s)$	$+ 1e^-$	=	$CH_3COO(aq)^-$	$+ Ag(s)$	+0.64
8.	$Cu(aq)^{++} + Cl(aq)^-$	$+ 1e^-$	=		$CuCl(s)$	+0.57
9.	$2AgO(s) + H_2O$	$+ 2e^-$	=	$2OH(aq)^-$	$+ Ag_2O(s)$	+0.57
10.	$Cu(aq)^+$	$+ 1e^-$	=		$Cu(s)$	+0.52
11.	$Ag_2C_2O_4(s)$	$+ 2e^-$	=	$C_2O_4(aq)^{--}$	$+ 2Ag(s)$	+0.47
12.	$AgIO_3(s)$	$+ 1e^-$	=	$IO_3(aq)^-$	$+ Ag(s)$	+0.37
13.	$[Ag(NH_3)_2](aq)^+$	$+ 1e^-$	=	$2NH_3(aq)$	$+ Ag(s)$	+0.37
14.	$Ag_2O(s) + H_2O$	$+ 2e^-$	=	$2OH(aq)^-$	$+ 2Ag(s)$	+0.34
15.	$Cu(aq)^{++}$	$+ 2e^-$	=		$Cu(s)$	+0.34
16.	$Ag_2O(s) + H_2O$	$+ 2e^-$	=	$2OH(aq)^-$	$+ 2Ag(s)$	+0.34
17.	$AgCl(s)$	$+ 1e^-$	=	$Cl(aq)^-$	$+ Ag(s)$	+0.22
18.	$Cu(aq)^{++}$	$+ 1e^-$	=		$Cu(aq)^+$	+0.15
19.	$CuCl(s)$	$+ 1e^-$	=	$Cl(aq)^-$	$+ Cu(s)$	+0.14
20.	$AgBr(s)$	$+ 1e^-$	=	$Br(aq)^-$	$+ Ag(s)$	+0.07
21.	$CuBr(s)$	$+ 1e^-$	=	$Br(aq)^-$	$+ Cu(s)$	+0.03
22.	$AgCN(s)$	$+ 1e^-$	=	$CN(aq)^-$	$+ Ag(s)$	−0.04
23.	$[Cu(NH_3)_4](aq)^{++}$	$+ 2e^-$	=	$4NH_3(aq)$	$+ Cu(s)$	−0.05
24.	$2Cu(OH)_2(s)$	$+ 2e^-$	=	$2OH(aq)^- + H_2O$	$+ Cu_2O(s)$	−0.08
25.	$AgI(s)$	$+ 1e^-$	=	$I(aq)^-$	$+ Ag(s)$	−0.15
26.	$CuI(s)$	$+ 1e^-$	=	$I(aq)^-$	$+ Cu(s)$	−0.19
27.	$Cu(OH)_2(s)$	$+ 2e^-$	=	$2OH(aq)^-$	$+ Cu(s)$	−0.22
28.	$[Ag(CN)_2](aq)^-$	$+ 1e^-$	=	$2CN(aq)^-$	$+ Ag(s)$	−0.29
29.	$Cu_2O(s) + H_2O$	$+ 2e^-$	=	$2OH(aq)^-$	$+ 2Cu(s)$	−0.36
30.	$Ag_2S(s)$	$+ 2e^-$	=	$S(aq)^{--}$	$+ 2Ag(s)$	−0.71
31.	$CuS(s)$	$+ 2e^-$	=	$S(aq)^{--}$	$+ Cu(s)$	−0.76
32.	$Cu_2S(s)$	$+ 2e^-$	=	$S(aq)^{--}$	$+ 2Cu(s)$	−0.95

Copper occurs naturally as the free element and as the very insoluble sulfides and oxides of both the 2+ and 1+ oxidation states.

As a poor reducing agent, copper requires strong oxidizing agents in order to be oxidized to copper(II) ions. Nitric acid and hot concentrated sulfuric acid are commonly used. In neutral or basic solution, copper in the 2+ state, as $Cu(aq)^{++}$

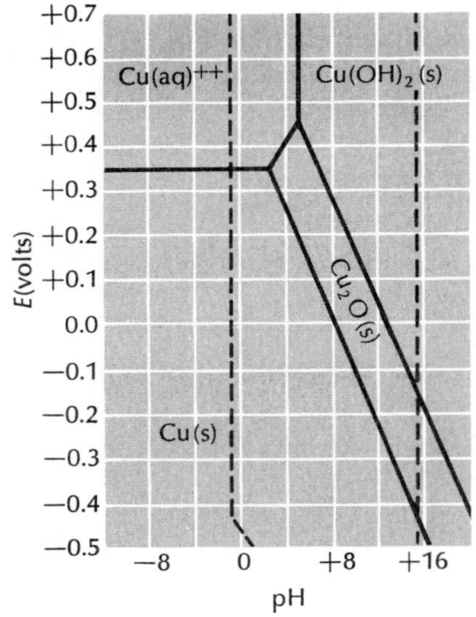

Fig. 18.15. Copper oxidation-state map

or $Cu(OH)_2(s)$, is a sufficiently good oxidizing agent to be reduced by relatively mild reducing agents to copper(I) oxide.

Both positive oxidation states of copper are stabilized by precipitating or complexing agents. A number of examples can be seen in Table 18.23. Normally the reduction of copper(II) ions in acid leads directly to metallic copper. However, if a suitable precipitating agent is used for copper(I) ions, the reduction can be carried out to the 1+ oxidation state. For example, when copper(II) ions are reduced by sulfurous acid in the presence of chloride ions, the copper(II) ions are reduced to insoluble copper(I) chloride:

$$Cu(aq)^{++} + H_2SO_3(aq) + 2Cl(aq)^- + 4H_2O = 2CuCl(s) + HSO_4(aq)^- + H_3O(aq)^+ \quad (18.193)$$

Silver

The oxidation-state map for silver is shown in Fig. 18.16. From the map, silver in the 3+ oxidation state is evidently a strong oxidizing agent. Also, silver in the 2+ state has a stability range in very strong acids as $Ag(aq)^{++}$, and in basic media as AgO, but has no thermodynamic stability between the extremes of pH. It is clear that the higher oxidation states of silver, 2+ and 3+, are easier to obtain in basic media by the oxidation of silver metal or silver(I) oxide.

Silver occurs naturally as the free metal and as the very insoluble sulfide and chloride. The purification of the metal normally depends upon the stabilization of the 1+ oxidation state as the dicyanoargentate(I) ion. If the silver exists as the free metal, then treatment with cyanide ions allows the oxidation to be carried out by atmospheric oxygen:

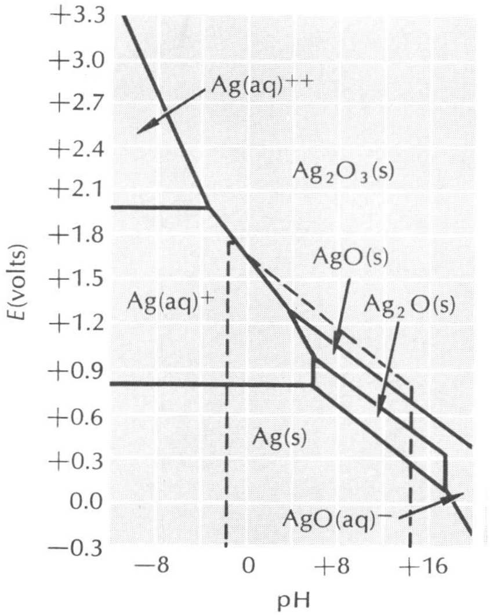

Fig. 18.16. Silver oxidation-state map

$$4Ag(s) + 8CN(aq)^- + O_2(g) + 2H_2O = 4[Ag(CN)_2](aq)^- + 2OH(aq)^- \quad (18.194)$$

If the silver is found as an insoluble substance such as silver sulfide, it can be dissolved by treatment with cyanide ions:

$$2Ag_2S(s) + 8CN(aq)^- + O_2(g) + 2H_2O = 4[Ag(CN)_2](aq)^- + 2S(s) + 4OH(aq)^- \quad (18.195)$$

Thus silver in almost any form may be put into solution as the dicyanoargentate(I) ion. Silver metal may be recovered from this solution by treatment with a strong reducing agent such as zinc metal:

$$2[Ag(CN)_2](aq)^- + Zn(s) = 2Ag(s) + [Zn(CN)_4](aq)^{--} \quad (18.196)$$

Other stabilizations of silver(I) can be seen in Table 18.24.

The +2 and +3 states of silver are obtained by oxidation of the +1 state with powerful oxidizing agents. Thus AgO is obtained by treating Ag_2O with ozone and Ag_2O_3 is obtained by treating Ag_2O with peroxodisulfate ion.

In acid, $Ag(aq)^{++}$ disproportionates to $Ag(aq)^+$ and AgO^+.

IRON

Some standard reduction potentials for iron-containing species are given in Table 18.25. The oxidation-state map is shown in Fig. 18.17. Both the 2+ and 3+ states of iron have regions of stability in this map. In strongly acidic solutions, iron(III) ions are good oxidizing agents. Thus to oxidize iron metal to iron(III)

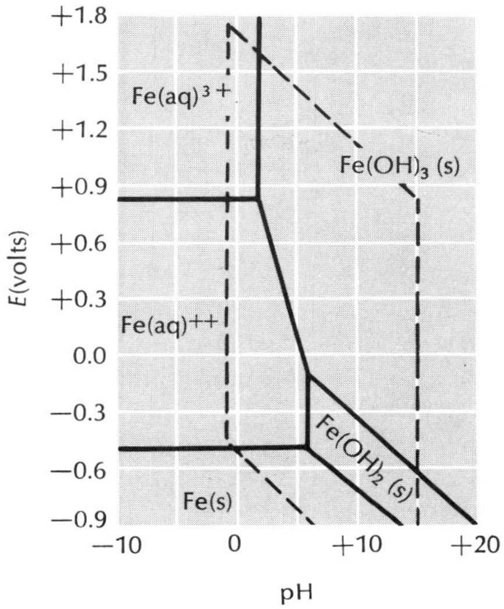

Fig. 18.17. Iron oxidation-state map

ions requires a strong oxidizing agent, whereas a weaker oxidizing agent will oxidize iron to iron(II) ions in acidic solution. It is clear, however, that the oxidation of iron to higher oxidation states is easier in base; iron(III) hydroxide is easily obtained by oxidation of iron under these conditions.

Iron occurs in nature as the insoluble silicates, carbonates, and oxides. The two most important sources of iron are Fe_2O_3 and Fe_3O_4. The metal is obtained at high temperatures by reduction of the oxides with carbon monoxide.

Since iron is a fairly good reducing agent and iron(III) ions are strong oxidizing agents, metallic iron reduces iron(III) ions to iron(II) ions:

$$Fe(s) + 2Fe(aq)^{3+} = 3Fe(aq)^{++} \quad (18.197)$$

There are several examples of the stabilization of the 2+ and 3+ oxidation states of iron in Table 18.25.

TABLE 18.25

Some Standard Reduction Potentials for Iron (Volts at 25°C)

OXIDIZING AGENTS			REDUCING AGENTS	E°
1. $Fe(aq)^{3+}$	$+ 1e^- =$		$Fe(aq)^{++}$	+0.77
2. $Fe(aq)^{3+}$	$+ 3e^- =$		$Fe(s)$	−0.04
3. $FePO_4(s)$	$+ 3e^- =$	$PO_4(aq)^{3-}$	$+ Fe(s)$	−0.39
4. $Fe(aq)^{++}$	$+ 2e^- =$		$Fe(s)$	−0.44
5. $Fe(OH)_3(s)$	$+ 1e^- =$	$OH(aq)^-$	$+ Fe(OH)_2(s)$	−0.56
6. $FeCO_3(s)$	$+ 2e^- =$	$CO_3(aq)^{--}$	$+ Fe(s)$	−0.76
7. $Fe(OH)_2(s)$	$+ 2e^- =$	$2OH(aq)^-$	$+ Fe(s)$	−0.88
8. $FeS(s)$	$+ 2e^- =$	$S(aq)^{--}$	$+ Fe(s)$	−0.97
9. $[Fe(CN)_6](aq)^{4-}$	$+ 2e^- =$	$6CN(aq)^-$	$+ Fe(s)$	−1.5
10. $Fe(OH)_3(s)$	$+ 3e^- =$	$3OH(aq)^-$	$+ Fe(s)$	−1.56
11. $[Fe(CN)_6](aq)^{3-}$	$+ 3e^- =$	$6CN(aq)^-$	$+ Fe(s)$	−2.0

MANGANESE

Some standard reduction potentials for manganese-containing species are shown in Tables 18.26, 18.27, and 18.28. The oxidation-state map for manganese is shown in Fig. 18.18.

Manganese occurs naturally as the insoluble oxides, MnO_2 and Mn_2O_3, the carbonate, $MnCO_3$, and the sulfides MnS and MnS_2. The free metal is obtained from MnO_2 at high temperature with a strong reducing agent such as aluminum metal. From the oxidation-state map it is clear that manganese metal is a strong reducing agent, sufficient to reduce water. Although this fact suggests that manganese metal cannot be deposited electrolytically from aqueous solutions of manganese(II) ions, conditions of high overvoltage may be used for that purpose.

The metal reduces hydronium ion:

$$Mn(s) + 2H_3O(aq)^+ = Mn(aq)^{++} + H_2(g) \tag{18.198}$$

In strongly basic media the metal may be oxidized to manganate ions by oxygen:

$$2Mn(s) + 4OH(aq)^- + 3O_2(g) = 2MnO_4(aq)^{--} + 2H_2O \tag{18.199}$$

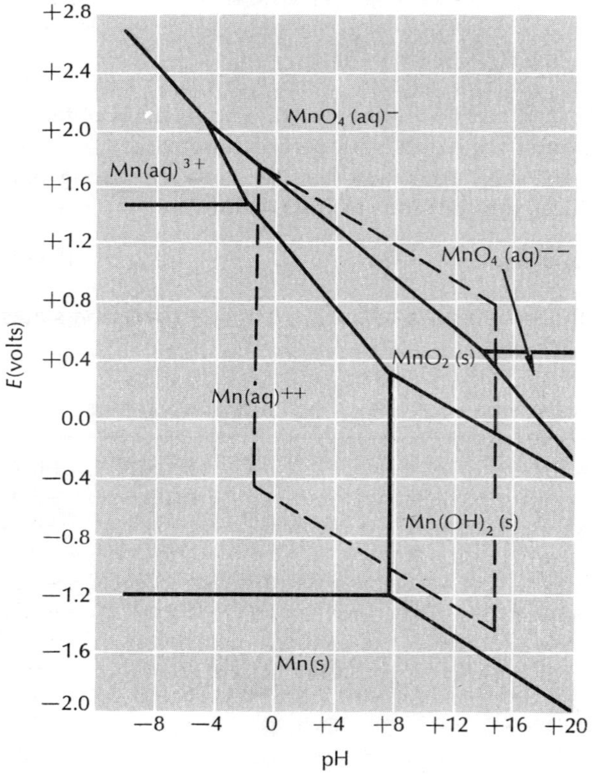

Fig. 18.18. Manganese oxidation-state map

TABLE 18.26

Some Standard Reduction Potentials for Manganese (Volts at 25°C)

	OXIDIZING AGENTS		REDUCING AGENTS	E°
1.	$MnO_4(aq)^{--} + 4H_3O(aq)^+$	$+ 2e^- =$	$6H_2O + MnO_2(s)$	+2.25
2.	$MnO_4(aq)^- + 4H_3O(aq)^+$	$+ 3e^- =$	$6H_2O + MnO_2(s)$	+1.69
3.	$MnO_4(aq)^- + 8H_3O(aq)^+$	$+ 4e^- =$	$12H_2O + Mn(aq)^{3+}$	+1.5
4.	$Mn(aq)^{3+}$	$+ 1e^- =$	$Mn(aq)^{++}$	+1.5
5.	$MnO_2(s) + 4H_3O(aq)^+$	$+ 2e^- =$	$6H_2O + Mn(aq)^{++}$	+1.23
6.	$MnO_2(s) + 4H_3O(aq)^+$	$+ 1e^- =$	$6H_2O + Mn(aq)^{3+}$	+0.95
7.	$MnO_4(aq)^{--} + 2H_2O$	$+ 2e^- =$	$4OH(aq)^- + MnO_2(s)$	+0.60
8.	$MnO_4(aq)^- + 2H_2O$	$+ 3e^- =$	$4OH(aq)^- + MnO_2(s)$	+0.59
9.	$MnO_4(aq)^-$	$+ 1e^- =$	$MnO_4(aq)^{--}$	+0.56
10.	$MnO_4(aq)^- + 4H_2O$	$+ 4e^- =$	$5OH(aq)^- + Mn(OH)_3(s)$	+0.49
11.	$Mn(OH)_3(s)$	$+ 1e^- =$	$OH(aq)^- + Mn(OH)_2(s)$	+0.2
12.	$MnO_2(s) + 2H_2O$	$+ 2e^- =$	$2OH(aq)^- + Mn(OH)_2(s)$	−0.05
13.	$MnO_2(s) + 2H_2O$	$+ 1e^- =$	$OH(aq)^- + Mn(OH)_3(s)$	−0.2
14.	$Mn(aq)^{++}$	$+ 2e^- =$	$Mn(s)$	−1.18
15.	$Mn(OH)_2(s)$	$+ 2e^- =$	$2OH(aq)^- + Mn(s)$	−1.55

TABLE 18.27

Standard Potential Diagram for Manganese

IN AQUEOUS ACID

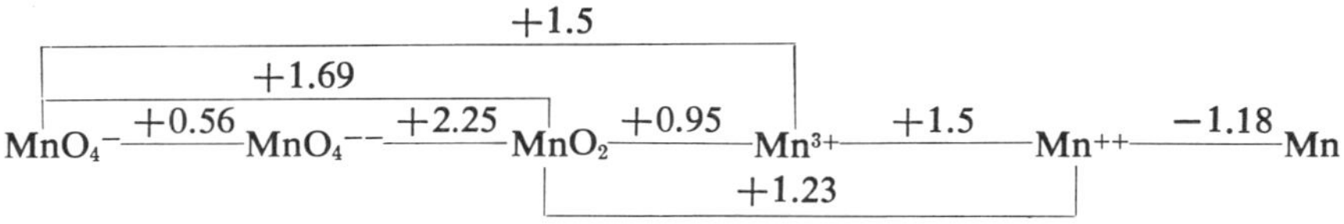

IN AQUEOUS BASE

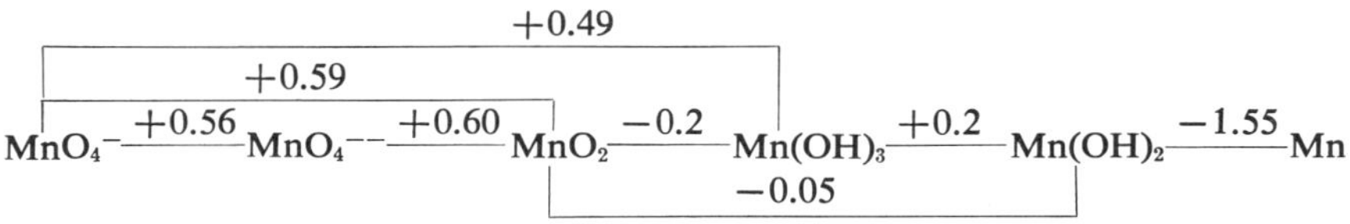

TABLE 18.28

Standard Half- cell Reduction Potentials for Manganese-containing Species (Volts at 25°C)

In Aqueous Acid

OXIDIZING AGENTS	REDUCING AGENTS				
	Mn(s)	$Mn(aq)^{++}$	$Mn(aq)^{3+}$	$MnO_2(s)$	$MnO_4(aq)^{--}$
$MnO_4(aq)^-$	+0.74	+1.51	+1.5	+1.69	+0.56
$MnO_4(aq)^{--}$	+0.87	+1.74	+1.82	+2.25	
$MnO_2(s)$	+0.03	+1.23	+0.95		
$Mn(aq)^{3+}$	−0.3	+1.5			
$Mn(aq)^{++}$	−1.18				

In Aqueous Base

OXIDIZING AGENTS	REDUCING AGENTS				
	Mn(s)	$Mn(OH)_2(s)$	$Mn(OH)_3(s)$	$MnO_2(s)$	$MnO_4(aq)^{--}$
$MnO_4(aq)^-$	−0.20	+0.33	+0.49	+0.59	+0.56
$MnO_4(aq)^{--}$	−0.23	+0.28	+0.33	+0.60	
$MnO_2(s)$	−0.80	−0.05	−0.2		
$Mn(OH)_3(s)$	−1.0	+0.2			
$Mn(OH)_2(s)$	−1.55				

The 2+ state of manganese is most easily oxidized in base. Atmospheric oxygen oxidizes manganese(II) hydroxide stepwise to manganese(III) hydroxide and then to manganese(IV) oxide. Other good oxidizing agents, such as hydrogen peroxide, also oxidize manganese(II) hydroxide to manganese(IV) oxide in base:

$$Mn(OH)_2(s) + HO_2(aq)^- = MnO_2(s) + OH(aq)^- + H_2O \qquad (18.200)$$

In acidic solutions, permanganate ions are produced by the reaction of manganese(II) ions with very strong oxidizing agents such as peroxodisulfate or metaperiodate ions:

$$2Mn(aq)^{++} + 5S_2O_8(aq)^{--} + 24H_2O = 2MnO_4(aq)^- + 10SO_4(aq)^{--} + 16H_3O(aq)^+ \qquad (18.201)$$

$$2Mn(aq)^{++} + 5IO_4(aq)^- + 9H_2O = 2MnO_4(aq)^- + 5IO_3(aq)^- + 6H_3O(aq)^+ \qquad (18.202)$$

As evident from Table 18.28, $Mn(aq)^{3+}$ is unstable in acid with respect to disproportionation to $Mn(aq)^{++}$ and MnO_2.

$$2Mn(aq)^{3+} + 6H_2O = Mn(aq)^{++} + MnO_2(s) + 4H_3O(aq)^{+} \qquad (18.203)$$

Manganese(IV) oxide is a good oxidizing agent that is normally reduced to manganese(II) ions. It is a better oxidizing agent in acid than in base, as shown on the oxidation-state map. The following equation shows a typical oxidation by MnO_2 in acid:

$$MnO_2(s) + 2Cl(aq)^- + 4H_3O(aq)^+ = Mn(aq)^{++} + Cl_2(g) + 6H_2O \qquad (18.204)$$

In basic solution, manganese(IV) oxide is oxidized by permanganate ions to manganate ions:

$$2MnO_4(aq)^- + MnO_2(s) + 4OH(aq)^- = 3MnO_4(aq)^{--} + 2H_2O \qquad (18.205)$$

Manganate ions are stable only in strongly basic solutions. In acid and even in mildly basic solutions, they disproportionate, as predicted from Table 18.26, to permanganate ions and manganese(IV) oxide:

$$3MnO_4(aq)^{--} + 4H_3O(aq)^+ = MnO_2(s) + 2MnO_4(aq)^- + 6H_2O \qquad (18.206)$$

Although manganate ion is a fairly good oxidizing agent, with strong oxidizing agents it can be oxidized to permanganate:

$$2MnO_4(aq)^{--} + ClO(aq)^- + H_2O = 2MnO_4(aq)^- + Cl(aq)^- + 2OH(aq)^- \qquad (18.207)$$

Permanganate ions are commonly employed where strong oxidizing agents are required. They are used often for redox titrations despite the fact that they slowly decompose in water by oxidizing the water. In acidic solutions they are reduced readily to manganese(II) ions but in basic solution the same reducing agents reduce the permanganate ions to manganese(IV) oxide. This can be seen in the oxidation-state map, where the MnO_2 band slopes down to the right. Thus, in strongly basic solutions permanganate ions are preferred at the same potentials that $Mn(aq)^{++}$ is preferred in much more acidic solutions.

Since the manganese(II) ion tends to form only weak coordination complexes, this oxidation state tends to be stabilized by precipitation. Some stabilizations can be seen in Table 18.26.

CHROMIUM

Some standard reduction potentials for chromium-containing species are summarized in Tables 18.29, 18.30, and 18.31. The oxidation-state map for chromium is shown in Fig. 18.19.

From the map it can be seen that chromium in the 6+ oxidation state is a strong oxidizing agent as CrO_3 in very strong acid, $Cr_2O_7(aq)^{--}$ in moderate acid, and $CrO_4(aq)^{--}$ in base. It is strongest in acids and weakest in bases.

TABLE 18.29
Some Standard Reduction Potentials for Chromium (Volts at 25°C)

	OXIDIZING AGENTS				REDUCING AGENTS	E°
1.	$Cr_2O_7(aq)^{--} + 14H_3O(aq)^+$	$+ 6e^-$	=	$21H_2O$	$+ Cr(aq)^{3+}$	+1.33
2.	$CrO_4(aq)^{--} + 4H_2O$	$+ 3e^-$	=	$5OH(aq)^-$	$+ Cr(OH)_3(s)$	−0.13
3.	$CaCrO_4(s) + 4H_2O$	$+ 3e^-$	=	$Ca(aq)^{++} + 5OH(aq)^-$	$+ Cr(OH)_3(s)$	−0.20
4.	$SrCrO_4(s) + 4H_2O$	$+ 3e^-$	=	$Sr(aq)^{++} + 5OH(aq)^-$	$+ Cr(OH)_3(s)$	−0.22
5.	$BaCrO_4(s) + 4H_2O$	$+ 3e^-$	=	$Ba(aq)^{++} + 5OH(aq)^-$	$+ Cr(OH)_3(s)$	−0.34
6.	$Cr(aq)^{3+}$	$+ 1e^-$	=		$+ Cr(aq)^{++}$	−0.41
7.	$Cr(aq)^{3+}$	$+ 3e^-$	=		$Cr(s)$	−0.74
8.	$[CrCl_2(H_2O)_4](aq)^+$	$+ 3e^-$	=	$2Cl(aq)^- + 4H_2O$	$+ Cr(s)$	−0.76
9.	$[Cr(NCS)_6](aq)^{3-}$	$+ 3e^-$	=	$6NCS(aq)^-$	$+ Cr(s)$	−0.82
10.	$Cr(aq)^{++}$	$+ 2e^-$	=		$Cr(s)$	−0.91
11.	$CrSO_4(s)$	$+ 2e^-$	=	$SO_4(aq)^{--}$	$+ Cr(s)$	−0.93
12.	$Cr(OH)_3(s)$	$+ 1e^-$	=	$OH(aq)^-$	$+ Cr(OH)_2(s)$	−1.19
13.	$Cr(OH)_3(s)$	$+ 3e^-$	=	$3OH(aq)^-$	$+ Cr(s)$	−1.34
14.	$Cr(OH)_2(s)$	$+ 2e^-$	=	$2OH(aq)^-$	$+ Cr(s)$	−1.41

TABLE 18.30
Standard Potential Diagram for Chromium

IN AQUEOUS ACID

$$Cr_2O_7^{--} \xrightarrow{+1.33} Cr^{3+} \xrightarrow{-0.41} Cr^{++} \xrightarrow{-0.91} Cr$$

IN AQUEOUS BASE

$$CrO_4^{--} \xrightarrow{-0.13} Cr(OH)_3 \xrightarrow{-1.19} Cr(OH)_2 \xrightarrow{-1.41} Cr$$

TABLE 18.31
Standard Half-cell Reduction Potentials for Chromium-containing Species (Volts at 25°C)

In Aqueous Acid

OXIDIZING AGENTS	REDUCING AGENTS: $Cr(s)$	$Cr(aq)^{++}$	$Cr(aq)^{3+}$
$Cr_2O_7(aq)^{--}$	+0.29	+0.84	+1.33
$Cr(aq)^{3+}$	−0.74	−0.41	
$Cr(aq)^{++}$	−0.91		

In Aqueous Base

OXIDIZING AGENTS	REDUCING AGENTS: $Cr(s)$	$Cr(OH)_2(s)$	$Cr(OH)_3(s)$
$CrO_4(aq)^{--}$	−0.74	−0.40	−0.13
$Cr(OH)_3(s)$	−1.34	−1.19	
$Cr(OH)_2(s)$	−1.41		

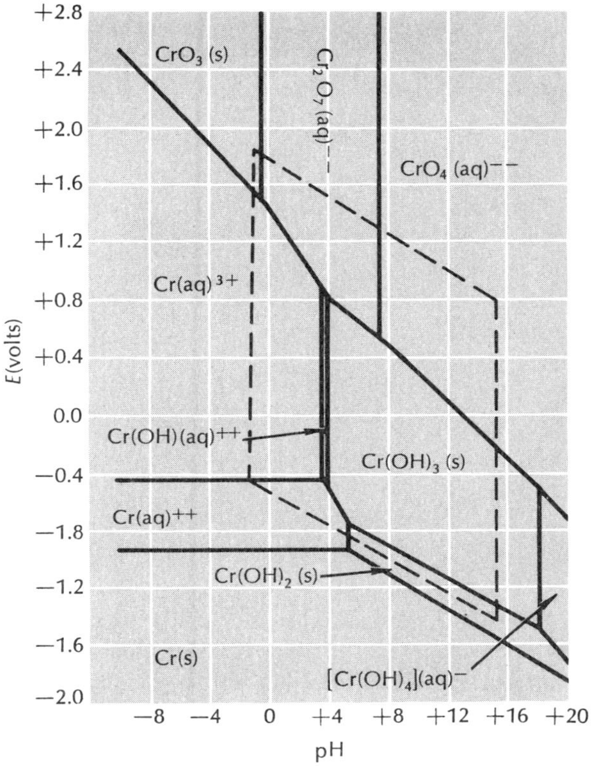

Fig. 18.19. Chromium oxidation-state map

Chromium occurs in nature as the oxide, Cr_2O_3, or as the iron(II) chromite, $Fe(CrO_2)_2$. It also occurs to some extent as insoluble lead(II) chromate.

Chromium metal is obtained from iron(II) chromite by high-temperature reduction with carbon and from chromium(III) oxide by high-temperature reduction with aluminum metal:

$$Fe(CrO_2)_2(s) + 4C(s) = Fe(s) + 2Cr(s) + 4CO(g) \tag{18.208}$$

$$Cr_2O_3(s) + 2Al(s) = Cr(s) + Al_2O_3(s) \tag{18.209}$$

Chromium metal reduces hydronium ion quickly to hydrogen, and the chromium is oxidized to chromium(II) ions:

$$Cr(s) + 2H_3O(aq)^+ = Cr(aq)^{++} + H_2(g) + 2H_2O \tag{18.210}$$

Chromium(II) ions are good reducing agents, but their oxidation to chromium(III) ions is slow. Although chromium is a good reducing agent in base, its oxidation is also slow and is not appreciable. With a strong oxidizing agent such as nitric acid, oxidation proceeds to chromium(III) ions.

Usually chromium(II) ions are produced by the reduction of chromium(III) ions. Both zinc and chromium metals are used:

$$2Cr(aq)^{3+} + Zn(s) = 2Cr(aq)^{++} + Zn(aq)^{++} \quad (18.211)$$

$$2Cr(aq)^{3+} + Cr(s) = 3Cr(aq)^{++} \quad (18.212)$$

Chromium(III) ions are poor oxidizing and reducing agents, and for this reason they tend to be the most common form of chromium, formed by the oxidation of chromium or chromium(II) ions or by the reduction of chromium in the 6+ oxidation state. However, chromium in the 3+ state can be oxidized by strong oxidizing agents, usually the oxidation is carried out in basic solution, where it is easier. Typical oxidizing agents are hypochlorite ions, hydrogen peroxide, and peroxodisulfate ions:

$$2Cr(OH)_3(s) + 3ClO(aq)^- + 4OH(aq)^- = 2CrO_4(aq)^{--} + 3Cl(aq)^- + 5H_2O \quad (18.213)$$

$$2Cr(OH)_3(s) + 3HO_2(aq)^- + OH(aq)^- = 2CrO_4(aq)^{--} + H_2O \quad (18.214)$$

$$2Cr(OH)_3(s) + 3S_2O_8(aq)^{--} + 10OH(aq)^- = 2CrO_4(aq)^{--} + 6SO_4(aq)^{--} + 8H_2O \quad (18.215)$$

In the 6+ state chromium exists in three different forms, depending on the pH of the environment. In very concentrated acid, chromium(VI) oxide precipitates as bright red needles. In a more normal acidity range, the dichromate ion, $Cr_2O_7(aq)^{--}$, is the predominant species; in base, chromate ion, $CrO_4(aq)^{--}$ predominates. Dichromates are orange and chromates are yellow.

Dichromate ion is a strong oxidizing agent often used in redox titrations. Under most conditions it does not oxidize water, so its aqueous solutions are stable. Quantitatively it is used to oxidize iron(II) ions to iron(III) ions. It is also used to oxidize iodides to iodine, which then is titrated with thiosulfate ions.

The data in Table 18.29 show some stabilizations for various oxidation states of chromium.

ALKALI AND ALKALINE EARTH METALS

The redox chemistry of the alkali and alkaline earth elements is relatively simple, since the metals all tend to be strong reducing agents. A few standard reduction potentials are given in Table 18.32 and the simple oxidation state map for sodium is given in Fig. 18.20.

All of these metals reduce water easily and none of them can be produced by the electrolysis of aqueous solutions of their salts. Thus the cathode reaction for the electrolysis of aqueous sodium chloride is as follows:

$$2H_2O + 2e^- = H_2(g) + 2OH(aq)^- \quad (18.216)$$

These elements are not found free in nature but always as the salts, often as the halides. The metals are usually produced by the electrolysis of molten salts, and halides are often used, but aqueous solutions of the salts are not used. The

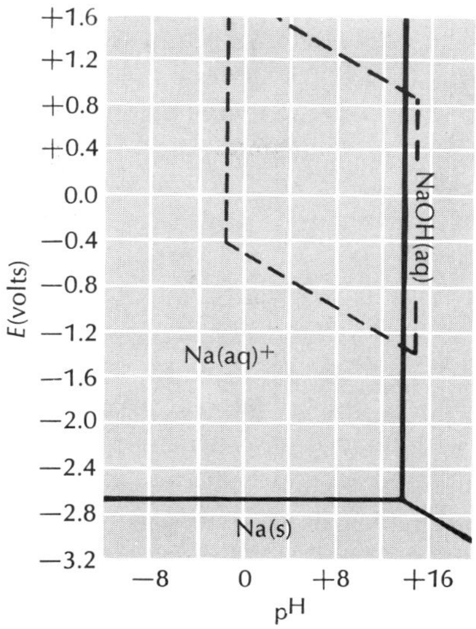

Fig. 18.20. Sodium oxidation-state map

electrolysis of molten ionic compounds must occur at very high temperatures and often this factor limits the production of the free element. It is sometimes possible to take advantage of freezing-point depression effects by mixing other compounds with the molten electrolyte so that the electrolysis may be carried out at a lower

TABLE 18.32

Some Standard Reduction Potentials of the Alkali Metals and Alkaline Earth Metals (Volts at 25°C)

	OXIDIZING AGENTS			REDUCING AGENTS	E°
1.	$Be(aq)^{++}$	$+ 2e^-$	=	Be(s)	−1.85
2.	$Mg(aq)^{++}$	$+ 2e^-$	=	Mg(s)	−2.37
3.	$Mg(OH)_2(s)$	$+ 2e^-$	=	$2OH(aq)^- +$ Mg(s)	−2.69
4.	$Na(aq)^+$	$+ 1e^-$	=	Na(s)	−2.71
5.	$Ca(aq)^{++}$	$+ 2e^-$	=	Ca(s)	−2.87
6.	$Sr(aq)^{++}$	$+ 2e^-$	=	Sr(s)	−2.89
7.	$Ba(aq)^{++}$	$+ 2e^-$	=	Ba(s)	−2.90
8.	$K(aq)^+$	$+ 2e^-$	=	K(s)	−2.92
9.	$Ra(aq)^{++}$	$+ 2e^-$	=	Ra(s)	−2.92
10.	$Rb(aq)^+$	$+ 1e^-$	=	Rb(s)	−2.92
11.	$Li(aq)^+$	$+ 1e^-$	=	Li(s)	−3.04

temperature. Thus, in the electrolysis of magnesium chloride, sodium chloride is mixed with the magnesium chloride to lower the melting point so that the electrolysis may be carried out in the molten solution at an attainable temperature.

SUMMARY

Much of the chemistry of the elements can be related to reduction potentials and the way these potentials are affected by the presence of various chemical species. The Nernst equation is a powerful tool in showing these relationships. Since the reduction potentials are thermodynamic functions, a knowledge of the potentials also reveals the tendency for reactions to occur. However, some caution must be used in predicting whether or not reactions actually occur, since kinetic effects which depend on mechanisms of reactions and the free energy of activation often control the reactions and determine which of several possibilities is actually the preferred one.

Listings of standard reduction potentials allow some predictions about the tendencies of various species to disproportionate. In addition, by using the Nernst equation, it is possible to prepare oxidation-state maps to show the thermodynamically stable species of the elements at various pH values. The Nernst equation also is valuable in interpreting the stabilization of oxidation states by relating the reduction potentials to acid-base constants and solubility product constants.

In this chapter, only a few of the elements have been reviewed for illustrative purposes to show some typical groups of reactions that do occur.

QUESTIONS AND PROBLEMS

18.1 Most half-cells involving halogen-containing species have positive standard reduction potentials. How does this fact suggest that halogen-containing species are better oxidizing agents than reducing agents?

18.2 Write the equation for the reaction of hydride ions with water.

18.3 In Table 18.1, how can you account for the fact that the acidic half-cells tend to be at the top of the table and the basic ones at the bottom?

18.4 Why is fluorine a better oxidizing agent in acidic solution than in basic solution, whereas the pH does not affect chlorine as an oxidizing agent when it is reduced to chloride ions?

18.5 For each of the following species at the stated conditions, write an equation for each of the thermodynamically feasible disproportionations. For each of these disproportionations compute the standard emf.

a) chlorate ions at pH 0
b) chlorate ions at pH 14
c) chlorine(IV) oxide at pH 0
d) hypochlorous acid at pH 0

18.6 In Table 18.3, at what pH values are the potentials measured?

18.7 What is the redox equivalent of amphoterism?

18.8 Why is it not possible to add two half-cell potentials when adding the two half-cell reaction equations to obtain the half-cell potential for a new half-cell reaction?

18.9 Of the 21 standard potentials in aqueous acid listed in Table 18.3, 9 were listed in Tables 18.1 and 18.2. Verify the values for the remaining 12 standard potentials given in Table 18.3 by using the standard potentials listed in Tables 18.1 and 18.2.

18.10 Why is the potential for the reduction of perchlorate ions to chloride ions pH dependent?

18.11 In Table 18.4, why are some enthalpy changes listed as positive and some as negative?

18.12 Criticize the following statement: In Table 18.4, the over-all half-cell enthalpy change decreases as the atomic number of the halogen increases.

18.13 Do the values in Table 18.5 indicate that the over-all enthalpy change for the half-cell represented by Equation (18.16) is unfavorable?

18.14 Criticize the following statement: The Born-Haber cycle treatment for understanding the components of the enthalpy change for a reaction constitutes the mechanism for the reaction.

18.15 Suggest how acid constants may be evaluated from measurements of half-cell potentials.

18.16 Show that Equation (18.27) is valid at pH values exceeding 7.

18.17 Write the equation for the oxidation of water by fluorine.

18.18 Elemental fluorine is made by electrolyzing molten KHF_2. Suggest the equation for the electrolysis reaction.

18.19 Write the chemical equations and mathematical equations from which all lines in Fig. 18.3 may be plotted.

18.20 Suggest why hypochlorite bleach solutions decompose upon standing.

18.21 Some laboratory instructions call for making chlorine by adding hydrochloric acid to a sodium hypochlorite bleach solution. Is this method likely to work?

18.22 Criticize the following statement: The disproportionation of chlorine(IV) oxide in acid is probably a single-step reaction.

18.23 Perchlorate ion is a poorer oxidizing agent than chlorate ion when both are reduced to chloride ions. How can this be explained in terms of a Born-Haber type cycle?

18.24 Consider the following generalized half-cell reaction where X represents a halogen atom.

$$X_2(s,l,g) + 2e^- = 2X(aq)^-$$

How does each of the following factors affect the strength of the halide ions as reducing agents?

a) The physical state of the halogen
b) The electronegativity of the halide ion
c) The X—X bond energy
d) The size of the halogen atom

18.25 Use both the Nernst equation and a Born-Haber cycle treatment to explain the following observation: When reduced to bromide ions under standard conditions, bromine is a better oxidizing agent than hypobromite ions in base, but hypobromous acid is a better oxidizing agent than bromine in acid. Does this also explain why the acidic disproportionation of hypobromous acid leads to bromine whereas the basic disproportionation of hypobromite ions leads to bromide ions?

18.26 Write chemical equations for all of the thermodynamically feasible disproportionations for elemental bromine and iodine, in both acidic and basic media.

18.27 Write an equation for each of the following reactions:

a) iodine with sulfurous acid
b) iodide ions with bromate ions in acid
c) reduction of chlorate ions to chloride ions with nitrite ions
d) bromine with iodide ions
e) oxidation of iodide ions to iodine by oxygen

18.28 A 0.5061-g sample of impure arsenic(III) oxide was dissolved in aqueous sodium hydroxide solution. Then the solution was made very slightly acid and 38.77 ml of 0.02000 *M* aqueous iodine was added. The excess iodine was titrated by 27.98 ml of 0.03333 *M* sodium thiosulfate solution. What was the percent of arsenic(III) oxide in the sample?

18.29 When 0.5560 g of KIO_3 is added to an excess of KI, the liberated I_2 requires 41.33 ml of sodium thiosulfate solution for titration. What is the molarity of the sodium thiosulfate solution?

18.30 Suggest why orthoperiodic acid is such a strong oxidizing agent.

18.31 Suggest why a moderately strong oxidizing agent such as iodine might be advantageous in redox titrations.

18.32 What are the oxidation state changes in Equation (18.58)?

18.33 In Fig. 18.6, why is the line separating elemental oxygen and oxygen in the 2− oxidation state a straight line over the entire pH range?

18.34 Write a chemical equation for each of the following:

a) hydrogen peroxide with permanganate ions in acid
b) the thermodynamically feasible disproportionations of elemental sulfur in both acid and base
c) thiosulfate ions with permanganate ions in acid
d) the thermodynamically feasible disproportionations of sulfurous acid
e) sulfurous acid with dichromate ions in acid
f) the oxidation of sulfite ions to dithionate ions by manganese(IV) oxide
g) silver with hot sulfuric acid
h) selenium with nitric acid
i) selenite ions with chlorate ions
j) oxidation of water by peroxodisulfate ions in acid

18.35 Criticize the following statement: The conversion of ozone to oxygen represents a disproportionation reaction of ozone.

18.36 How can the high oxidizing strength of peroxodisulfate ion be attributed to the low O—O bond energy?

18.37 Why does the oxidizing strength of hydrogen peroxide decrease as pH increases while the reducing strength increases as pH increases?

18.38 In Table 18.11, account for the fact that most of the standard potentials in acid are positive while most of the standard potentials in base are negative.

18.39 Account for the following two facts:

a) It is easier to oxidize sulfide ions to sulfate ions than to sulfite ions.
b) In acid, it is easier to oxidize sulfide ions to sulfur than to sulfate ions.

18.40 According to data in Table 18.11, what is the most highly favored disproportionation for thiosulfuric acid? Write an equation for this reaction and suggest why it is not the observed reaction.

18.41 Criticize the following statement: In basic solution at standard conditions, sulfite ions are stronger oxidizing agents than reducing agents.

18.42 Why are sulfate ions relatively poor oxidizing agents?

18.43 How does a high concentration of sulfate ions favor the electrolytic production of peroxodisulfate ions?

18.44 In Fig. 18.8, what does the great width of the band for elemental selenium imply about the strength of selenium as an oxidizing agent and as a reducing agent?

18.45 Use the data in Table 18.8 to compute the equilibrium constant at 25°C for each of the following reactions:

a) $2O_3(g) = 3O_2(g)$ b) $2H_2O_2(aq) = 2H_2O + O_2(g)$

18.46 In Table 18.8, which *basic* half-cell has the highest standard reduction potential?

18.47 Criticize the following statement: Equation (18.66) represents a disproportionation for peroxodisulfate ions.

18.48 Why is selenium a poorer oxidizing agent than sulfur?

18.49 In comparing the reduction of the chalcogens from the 4+ oxidation state to the elemental state, the difference between the standard acid and base conditions is virtually the same for sulfur and selenium. What does this imply about H_2SO_3 and H_2SeO_3?

18.50 Does it require a stronger oxidizing agent to oxidize selenite ions to selenate ions or sulfite ions to sulfate ions under standard basic conditions? Why?

18.51 Why does the wide stability region for elemental nitrogen in the nitrogen oxidation state map (Fig. 18.8) indicate that nitrogen is both a poor oxidizing agent and a poor reducing agent?

18.52 Why doesn't $NO_2(aq)^-$ appear on the oxidation-state map for nitrogen shown in Fig. 18.9?

18.53 Criticize the following statement: Without the overvoltage effects, nitrate ions appear to be strong enough to oxidize water.

18.54 Using the data in Table 18.13, compute the equilibrium constant for each of the following at 25°C:
a) $N_2O_4(g) = N_2(g) + 2O_2(g)$
b) $2NO(g) = N_2(g) + O_2(g)$
c) $2N_2O(g) = 2N_2(g) + O_2(g)$
d) $2NO(g) + O_2(g) = N_2O_4(g)$
e) $2As(s) + 3H_2(g) = 2AsH_3(g)$
f) $2P(s) + 3H_2(g) = 2PH_3(g)$
g) $4H_3O(aq)^+ + 4NO_3(aq)^- = 2N_2O_4(g) + O_2(g) + 6H_2O$

18.55 Why can the data in Tables 18.13 and 18.8 *not* be used to compute the equilibrium constants for the following reactions?
a) $4NH_3(g) + 5O_2(g) = 4NO(g) + 6H_2O$
b) $N_2(g) + 3H_2(g) = 2NH_3(g)$

18.56 Write a chemical equation for each of the following reactions:
a) half-cell anodic oxidation of $NH_4(aq)^+$ to $N_2H_5(aq)^+$
b) iodate ions with $N_2H_5(aq)^+$ ions in acid
c) oxygen with hydrazine in base
d) methylamine with nitrous acid
e) reduction of nitrous acid to $NH_3OH(aq)^+$ with zinc metal
f) the thermodynamically feasible disproportionations of hydrazoic acid
g) iron with nitrogen(IV) oxide
h) silver(I) ions with hypophosphorous acid
i) reduction of water by arsine

18.57 Suggest why ammonium permanganate is explosive.

18.58 Suggest a structure for the cyanamide ion, CN_2^{--}, which is found in calcium cyanamide. Suggest a reason why this ion might tend to form readily from elemental nitrogen.

18.59 Nitrogen in the 3− oxidation state is more easily oxidized to the ⅓− state in base than in acid. Why?

18.60 Nitrogen in the 3− oxidation state is more easily oxidized to the 5+ state in base than in acid. Why?

18.61 In base, why is NO more easily oxidized to nitrate ion than to nitrite ion?

18.62 Consider the reactions represented by Equations (18.109) and (18.111). Does it violate the definition of catalyst to say that a catalyst may increase the speed of one reaction more than the other? Explain.

18.63 How does the low N—N bond energy help to account for the good reducing qualities of hydrazine? Does this account also for its poor oxidizing qualities?

18.64 Is hydroxylamine a better reducing agent in acid or base? Why? Is it a better oxidizing agent in acid or base? Why?

18.65 What is the reducing agent in the cyanamide process for producing ammonia?

18.66 In enthalpy terms suggest why nitrogen(II) oxide disproportionates to nitrite ions and nitrogen(I) oxide in base.

18.67 In thermodynamic terms, at standard conditions, to what negative oxidation state should nitrous acid or nitrite ions be most readily reduced?

18.68 In general, why does the oxidizing strength of dinitrogen tetroxide decrease as pH increases whereas its reducing strength increases as pH increases?

18.69 Using the reaction represented by Equation (18.115) to analyze amines and amides by measuring the volume of evolved nitrogen, how would you go about eliminating interference from evolved carbon dioxide?

18.70 Nitrogen is present in a sample as an amine. If 0.4685 g of the sample evolves 30.12 ml of dry N_2 at 25°C, and 684.0 mm Hg when treated with excess nitrous acid, what is the percent nitrogen in the sample?

18.71 Criticize the following statement: Equation (18.144) represents a disproportionation reaction for nitric acid.

18.72 How is it possible for nitric acid to be reduced to several different products at the same time by the same reducing agent?

18.73 In Table 18.16, 14 of the 30 standard reduction potentials are given also in Table 18.14. Compute the remaining 16 from the 14 in Table 18.14.

18.74 In the nitrogen oxidation state map (Fig. 18.9), elemental nitrogen occupies a large region. In the phosphorus oxidation state map (Fig. 18.10), elemental phosphorus does not appear. Why is there this large difference?

18.75 What is the basis for saying that Equation (18.156) represents the oxidation of phosphine by iodine?

18.76 Is hypophosphorous acid unstable with respect to disproportionation? Justify your answer.

18.77 Suggest reasons why phosphate ion is a poorer oxidizing agent than nitrate ion.

18.78 What phosphorus-containing species is most easily reduced to elemental phosphorus at standard acidic conditions? Why is this the case?

18.79 Recall that phosphorous and hypophosphorous acids have structures in which hydrogen is bonded to phosphorus. Does this fact have any bearing on the fact that both of these acids tend to be good reducing agents but poor oxidizing agents? If so, how?

18.80 For the oxidation-state maps for phosphorus and arsenic, Figs. 18.10 and 18.11, write the chemical type equations for the transitions represented by each line in the map.

18.81 What factors contribute to the decreasing oxidizing strength of arsenic in the 5+ state of oxidation as basicity increases?

18.82 Why is As_4O_6(s) shown on the arsenic oxidation-state map instead of $HAsO_2$(aq) or H_3AsO_3(aq)? In view of this, what justification is there for writing $HAsO_2$(aq) in Equation (18.168), for example?

18.83 In practice, arsenic is more easily oxidized to the 3+ state than to the 5+ state. Is this in accord with the standard reduction potentials?

18.84 From the data in Tables 18.18 and 18.8 compute the equilibrium constant at 25°C for each of the following:
a) $2CO(g) + O_2(g) = 2CO_2(g)$
b) $CH_4(g) + O_2(g) = 2H_2O(l) + CO_2(g)$
c) $H_2C_2O_4(aq) = H_2(g) + 2CO_2(g)$
d) $C(s) + H_2O(l) = CO(g) + H_2(g)$
e) $C(s) + CO_2(g) = 2CO(g)$
f) $H_2C_2O_4(aq) = CO(g) + CO_2(g) + H_2O(l)$

18.85 In general, the standard potential data suggest that methane is as good a reducing agent as ammonia. Suggest a reason why the reducing action of CH_4 takes place more slowly than the reducing action of NH_3.

18.86 Write equations for the burning of methane in oxygen to form a) CO_2; b) CO; c) C. Indicate the relative proportions of methane and oxygen required in each case.

18.87 Criticize the following statement: Since many organic chemical reactions involve species not appearing on the carbon oxidation state map, these species are unstable and the reactions must occur with an increase in free energy.

18.88 From Table 18.18 combine half-cell reaction equations 4 and 8 to determine the reaction emf and equilibrium constant for:

$$HCOOH(aq) = CO(g) + H_2O$$

What is the meaning of the reaction emf? Is this a redox reaction?

18.89 Summarize the ways in which carbon monoxide may be prepared.

18.90 Write a chemical equation for each of the following:
a) oxidation of methanol to formic acid by dichromate ions
b) incomplete combustion of benzene to leave a deposit of carbon

18.91 Is oxalic acid (or oxalate ions) a better reducing agent in acid or in base?

18.92 Calcium oxalate is highly insoluble. What is the effect of calcium ions in basic solution upon the capacity of oxalate ions to act as oxidizing agents? As reducing agents?

18.93 Using the solubility product constants from Table 13.5, compute the standard reduction potentials for the following half-cells:
a) $AgBr(s) + 1e^- = Ag(s) + Br(aq)^-$
b) $MgF_2(s) + 2e^- = Mg(s) + 2F(aq)^-$
c) $Ag_2SO_4(s) + 2e^- = 2Ag(s) + SO_4(aq)^{--}$
d) $Hg_2Cl_2(s) + 2e^- = 2Hg(l) + 2Cl(aq)^-$
e) $Cu(IO_3)_2(s) + 2e^- = Cu(s) + 2IO_3(aq)^-$
f) $Cl_2(g) + 2Ag(aq)^+ + 2e^- = 2AgCl(s)$

18.94 The strength of hydride ion as a reducing agent is indicated in Table 18.1. Hydride ions form complexes with some cations and $Li[AlH_4]$ is a commonly used reducing agent.
a) What does the complexing of hydride ions do to the reduction potential for hydride ion as shown in Table 18.1?
b) Should $Li[AlH_4]$ be a stronger or weaker reducing agent than hydride ion?

18.95 What effect should the addition of mercury(II) ions have on the potential of the following half-cell?

$$S(s) + 2e^- = S(aq)^{--}$$

18.96 From the data in Table 18.23, compute the solubility product constant for each of the following: a) CuCl; b) Cu_2S; c) CuS.

18.97 From the data in Table 18.21, which of the mercury(I) halides is the least soluble? Which of the tetrahalomercurate(II) complex ions is most stable?

18.98 Compute the standard half-cell potential for each of the following:

a) $2[HgCl_4](aq)^{--} + 2e^- = Hg_2Cl_2(s) + 6Cl(aq)^-$

b) $2HgS(s) + 2e^- = Hg_2(aq)^{++} + 2S(aq)^{--}$

c) $2Hg(aq)^{++} + SO_4(aq)^{--} + 2e^- = Hg_2SO_4(s)$

18.99 Insofar as possible, verify the oxidation-state map for zinc in Fig. 18.13.

18.100 Why is it sometimes desirable to carry out electrolyses under conditions that lead to high overvoltage effects?

18.101 In enthalpy terms, why is zinc a better reducing agent than mercury?

18.102 Galvanized objects are made of iron (or steel) with a coating of zinc metal. One of the advantages of this is that if some zinc coating is chipped away, the exposed iron does not rust. Explain why this is so. If the zinc is replaced by copper, the same effect is not found. Why?

18.103 Write an equation for each of the following reactions:

a) Formaldehyde reduces copper(II) ions in neutral solution to copper(I) oxide.

b) Silver is oxidized by oxygen in the presence of thiosulfate ions which form complexes with silver(I) ions.

c) Silver(I) oxide is oxidized to silver(III) oxide by ozone.

d) Silver(I) oxide is oxidized to silver(II) oxide by peroxodisulfate ions.

18.104 Consider the data in Table 18.23. Which copper(I) halide is the most soluble?

18.105 In Table 18.24 find examples for each of the following:

a) stabilization of copper in the 1+ oxidation state in the presence of copper in the 2+ oxidation state

b) stabilization of copper in the 1+ oxidation state in the presence of metallic copper

c) stabilization of copper in the 2+ oxidation state in the presence of metallic copper

18.106 From the data in Table 18.24, compute the standard half-cell reduction potentials for the following:

a) $2CuS(s) + 2e^- = Cu_2S(s) + S(aq)^{--}$

b) $2[Cu(NH_3)_4](aq)^{++} + 2OH(aq)^- + 2e^- = Cu_2O(s) + 8NH_3(aq) + H_2O$

18.107 From the data in Table 18.24, and any other previous data required, compute the equilibrium constants for the following reactions at 25°C.

a) $4Ag(s) + O_2(g) = 2Ag_2O(s)$

b) $4Cu(s) + O_2(g) = 2Cu_2O(s)$

c) $4Ag(s) + 3O_2(g) = 2Ag_2O_3(s)$

18.108 In the oxidation state map for iron, Fig. 18.17, what factors help to account for the fact that $Fe(OH)_3(s)$ exists at lower pH values than $Fe(OH)_2(s)$?

18.109 From the data in Table 18.25, which is the more stable coordination complex, hexacyanoferrate(III) ion or hexacyanoferrate(II) ion?

18.110 In Table 18.25, find examples of the stabilization of the 2+ and 3+ oxidation states of iron.

18.111 From the data in Table 18.26, compute the equilibrium constants at 25°C for the following reactions:

a) $Mn(s) + O_2(g) = MnO_2(s)$

b) $4Mn(OH)_2(s) + O_2(g) + 2H_2O = 4Mn(OH)_3(s)$

c) $2MnO_4(aq)^- + 10Cl(aq)^- + 16H_3O(aq)^+ = 2Mn(aq)^{++} + 5Cl_2(g) + 24H_2O$

18.112 In the oxidation-state map for manganese, Fig. 18.18, why does the line separating $MnO_4(aq)^-$ and $MnO_4(aq)^{--}$ have zero slope?

18.113 How is it possible to prepare manganese(III) ions when their thermodynamic stability range lies outside the feasible conditions attainable in water, as shown in Fig. 18.18?

18.114 Write chemical equations for all of the thermodynamically feasible disproportionations of manganate ions and manganese in the 3+ oxidation state.

18.115 Manganese occurs in nature as MnS_2. Suggest two different types of compound which this might be, suggest a name for each type, and choose the one which is most likely to occur.

18.116 Suggest why manganese in the 2+ state is more easily oxidized in base when $Mn(OH)_2$ is insoluble and presumably is stable in this form.

18.117 In Chapter 10 it was pointed out that there is some resemblance between manganese-containing and chlorine-containing species. Using Tables 18.3 and 18.28, compare and contrast the redox characteristics of these species. What differences hinder comparisons? Can you make any generalization about which species are the stronger oxidizing agents or reducing agents?

18.118 Why is it that many metals are more easily oxidized in base than in acid?

18.119 How is it that chromite ions apparently exist in nature as $Fe(CrO_2)_2$, when it appears from the chromium oxidation-state map, Fig. 18.19, that they have thermodynamic stability only at very high pH values?

18.120 Write the equation for the formation of chromium(VI) oxide from dichromate ions in acid. Is this a redox reaction?

18.121 In Table 18.29, find examples of the stabilization of the 3+, 2+, and 6+ oxidation states of chromium.

18.122 Why is it easier to oxidize chromium in the 3+ oxidation state in base than it is to oxidize it in acid?

18.123 Why are the alkali metals strong reducing agents?

18.124 Suggest a Born-Haber cycle treatment to explain why lithium is a better reducing agent than sodium.

Epilogue

In this book we have barely scratched the surface in our consideration of the concepts of chemistry and the techniques employed by chemists. Nevertheless, you should now be prepared to understand many present-day chemical investigations.

We owe a great debt to the physicists, who have revealed many fundamental concepts upon which modern structural chemistry is based. The search for better understanding of atomic, molecular, and physical state structures continues, however. Many new nuclear particles have been discovered and if these can be included in a new theory, our understanding of atomic nuclei will be enhanced. New instruments and techniques are being developed to probe the structure of atoms and molecules. The instruments include microwave and infrared spectrometers, and larger accelerators with which to hurl projectiles into atomic nuclei. The techniques include electron spin resonance and nuclear magnetic resonance.

Concurrently, theoreticians are applying new techniques to provide more accurate structural models. The theory of the chemical bond continues to evolve, and there is now some deemphasis of the pairing of electrons as a prerequisite for covalent bonding. Principles of symmetry are proving to be powerful tools in the understanding of chemical bonding.

The study of molecular structure is important in many areas of chemical research. There is much interest in coordination complexes, which are proving to be important species in many chemical systems. Here there is great interest in the role of the $(n - 1)d$ electrons in the bonding of the transition elements. There is also increasing interest in the participation of the $(n - 2)f$ electrons in the bonding in species containing inner transition elements.

Structural properties play an important role in the current surge of interest in biochemistry. The structures of natural products and chemicals in living plant and animal systems are becoming better understood. This study serves as necessary background for the understanding of the thermodynamics and mechanisms of the chemical reactions that occur in living systems—another focus in biochemical studies. There is some excitement in the prospect that some simple life forms may soon be produced from nonliving matter.

Surface chemistry, the study of the boundary areas between two physical phases, is receiving much attention. Much that is being revealed about the structure of surfaces is of special interest to the geochemist, who encounters many phase boundaries within the earth.

The discovery of the compounds of the rare gases has been exciting. First, the preparation of these compounds contradicted earlier predictions that these elements cannot form compounds of the usual type. In addition, investigations of these compounds provided interesting test cases for the predictive powers of various structural theories, since the compounds were a hitherto unknown group.

In the realm of chemical reactions, extreme conditions of temperature and pressure are being studied. Properties such as superconductivity are being investigated at very low temperatures and the nature of the plasma states is being studied at very high temperatures.

The failure of thermodynamics to predict actual occurrence of reactions has focused much attention on kinetics and other techniques for revealing the nature of reaction mechanisms. The understanding of reaction mechanisms has permitted new and useful classifications for reactions, especially in organic chemistry. In inorganic chemistry, the mechanisms of electron transfer reactions and reactions involving coordination complexes have received much attention. There also is interest in photochemical reactions: that is, reactions initiated by electromagnetic radiation near the visible region of the spectrum. Similarly, research is increasing in the area of the effects of other types of radiation upon chemical reactions.

With the realization that many thermodynamically feasible reactions do not occur, catalysis research has become prolific. Both homogeneous and heterogeneous catalysis have received attention in an attempt to develop theories of catalysis that will permit the selection of catalytic materials for reactions as they are desired. This work has included studies of the actions of enzymes, which are the catalysts involved in the chemical reactions in living systems.

Some studies of solution chemistry have centered on nonaqueous solvents. Typical liquids at room temperature or other easily attainable temperatures are being used as solvents, and molten salts at high temperatures are also being used. This work is leading to more emphasis on broader concepts of acids and bases;

here the Lewis theory plays a prominent role. At the same time, this research is throwing more light on the importance of water to many of the common reactions with which we are familiar.

This brief summary is only suggestive of the broad range of interests of chemists. If you use the concepts of structure and chemical reactions to develop the theme of energy relationships in chemical changes, you should find the continued study of chemistry a rewarding experience. Even if you do not continue to study more chemistry, you should now have some understanding of the fundamental concepts that lie behind current chemical research.

APPENDIX 1 Metric Prefixes: Their Abbreviations and Numerical Equivalents

tera	T	10^{12}
giga	G	10^{9}
mega	M	10^{6}
kilo	k	10^{3}
hecto	h	10^{2}
deka	dk	10^{1}
deci	d	10^{-1}
centi	c	10^{-2}
milli	m	10^{-3}
micro	μ	10^{-6}
nano	n	10^{-9}
pico	p	10^{-12}

APPENDIX 2 Some Abbreviations

Å	angstrom unit
amp	ampere
amu	atomic mass unit
atm	atmosphere
bp	boiling point
°C	degrees Celsius
cal	calorie
cc	cubic centimeter
cgs	centimeter-gram-second
dpm	dipole moment
emf	electromotive force
esu	electrostatic unit
ev	electron volt
°F	degrees Fahrenheit
fp	freezing point
ft	foot
g	gram
hr	hour
ht	heat
in.	inch
°K	degrees Kelvin or Absolute
l	liter
lb	pound
m	meter
mi	mile
min	minute
mks	meter-kilogram-second
mp	melting point
No.	number
nt	newton
oz	ounce
pot	potential
pt	pint
qt	quart
sec	second
temp	temperature
V	volt
vol	volume
wt	weight
yd	yard
yr	year

APPENDIX 3 Values for Some Important Constants

	SYMBOL	VALUE
acceleration of gravity, terrestrial	g	$9.80665\ m/sec^2$
atmosphere of pressure		$1.01325 \times 10^5\ nt/m^2$
charge on mole of electrons	F	9.64870×10^4 coulombs
electromagnetic radiation speed	c	2.997925×10^8 m/sec
electron rest mass		9.1091×10^{-31} kg
		5.48597×10^{-4} amu
gas constant	R	82.055 atm × cm/(°K × mole)
		0.082055 1 × atm/(°K × mole)
		8.3143 joule/(°K × mole)
gravitational constant, universal	G	$6.670 \times 10^{-11}\ nt \times m^2/kg^2$
mole		6.02252×10^{23}
neutron rest mass		1.67482×10^{-27} kg
		1.0086654 amu
permeability of vacuum	μ_0	1.2566×10^{-6} mks units
permittivity of vacuum	ϵ_0	$8.85415 \times 10^{-12}\ coulomb^2/nt \times m^2$
pi	π	3.14159
Planck's constant	h	6.6256×10^{-34} joule × sec
proton rest mass		1.67252×10^{-27} kg
		1.00727663 amu
quantum of charge	e	1.60210×10^{-19} coulomb

APPENDIX 4 Some Conversion Relationships

Electric Charge

1 coulomb	$= 10^{-1}$ abcoulomb $= 2.778 \times 10^{-4}$ amp × hr
	$= 1.036 \times 10^{-6}$ F $= 2.998 \times 10^9$ statcoulomb
1 faraday	$= 9649$ abcoulombs $= 26.80$ amp × hr $= 9.649 \times 10^4$ coulomb
	$= 2.893 \times 10^{14}$ statcoulomb

Electric Current

1 amp	$= 10^{-1}$ abamp $= 2.998 \times 10^9$ statamp

Electric Dipole Moment

1 debye = 10^{-18} esu = 3.33×10^{-30} coulomb × m
= 3.33×10^{-20} coulomb × Å = 0.21 electron × Å

Electric Potential

1 volt = 10^{8} abvolt = 3.336×10^{-3} statvolt

Electric Resistance

1 ohm = 10^{9} abohm = 1.113×10^{-12} statohm

Electric Specific Resistivity

1 ohm × m = 10^{11} abohm × cm = 10^{2} ohm × cm = 1.113×10^{-10} statohm × cm

Energy and Work

(Mass units are included as energy equivalents)

1 erg = 10^{-7} joule = 2.389×10^{-8} cal = 6.242×10^{11} ev
= 1.113×10^{-24} kg = 670.5 amu

1 cal = 4.1840×10^{7} erg = 4.184 joule = 2.613×10^{19} ev
= 4.659×10^{-17} kg = 2.807×10^{10} amu

1 ev = 1.602×10^{-12} erg = 1.602×10^{-19} joule = 3.827×10^{-20} cal
= 1.783×10^{-36} kg = 1.074×10^{-9} amu

1 kg = 8.987×10^{23} erg = 8.987×10^{16} joule = 2.142×10^{16} cal
= 5.610×10^{35} ev = 6.025×10^{26} amu

1 amu = 1.492×10^{-3} erg = 1.492×10^{-10} joule = 3.564×10^{-11} cal
= 9.31×10^{8} ev = 1.660×10^{-27} kg

Force

1 nt = 10^{5} dyne = 0.2248 lb
1 lb = 4.448×10^{5} dyne = 4.448 nt

Length

1 m = 39.37 in. = 3.281 ft = 6.214×10^{-4} mi
1 in = 2.540 cm
1 Å = 10^{-10} m
1 micron = 10^{-6} m
1 light year = 9.4600×10^{12} km

Magnetic Dipole Moment

1 Bohr magnetron = 9.28×10^{-24} joule × m^2/weber

Mass and Weight

(Mass-weight equivalents are valid for terrestrial use only.)

1 g	$= 6.852 \times 10^{-5}$ slug $= 6.024 \times 10^{23}$ amu $= 3.527 \times 10^{-2}$ oz
	$= 2.205 \times 10^{-3}$ lb
1 amu	$= 1.6602 \times 10^{-24}$ g
1 lb	$= 453.6$ g
1 ton	$= 2000$ lb $= 907.2$ kg

Pressure

1 atm	$= 1.013 \times 10^{6}$ dyne/cm^2 $= 76.0$ cm Hg $= 14.70$ lb/in.^{2}
	$= 2116$ lb/ft^2
1 cm Hg	$= 1.316 \times 10^{-2}$ atm $= 1.333 \times 10^{4}$ dyne/cm^2
	$= 5.353$ in H_2O $= 0.1934$ lb/in.^{2} $= 27.85$ lb/ft^2

Temperature

b.p. of H_2O	$= 212°F = 100°C = 373.16°K$
m.p. of H_2O	$= 32°F = 0°C = 273.16°K$

Time

1 yr	$= 365.2$ day $= 8.766 \times 10^{3}$ hr $= 5.259 \times 10^{5}$ min
	$= 3.156 \times 10^{7}$ sec

Volume

1 cc	$= 3.531 \times 10^{-5}$ ft^3 $= 6.102 \times 10^{-2}$ in.^{3}
1 l	$= 1.0567$ qt $= 1.000028 \times 10^{3}$ cc

NUMERICAL ANSWERS TO PROBLEMS

Chapter 1

1.4. 1.44×10^4 dynes; 3.24×10^{-2} lb. **1.5.** 4.45×10^5 dyne/lb. **1.6.** 5.77 g. **1.7.** 5.44 slugs. **1.8.** 3.60×10^4 mi/hr^2. **1.9.** a) 90.9 kg; b) 12.2 oz.; c) 0.05 ton. **1.10.** a) 5.94×10^{24} kg; b) 2.28×10^{13} kg/mi^3. **1.11.** a) 1.63 m/sec^2; b) 0.804 lb; c) 1.53×10^{13} kg/mi^3. **1.14.** a) 2.75×10^{-28} nt; b) 1.79×10^{-9} kg; c) 3.50×10^{-30} nt. **1.17.** a) 5.91×10^{-20} dyne; b) 2.07×10^{-23} nt; c) 3.76×10^{-25} nt. **1.18.** 2. **1.21.** 1.88×10^{21} quanta. **1.22.** 2.16×10^5 coulombs; 1.35×10^{24} quanta. **1.24.** 1.26×10^{-6} mks units. **1.28.** a) 4.89×10^{12} ergs; 4.89×10^5 joules; 3.60×10^5 ft-lbs; 1.17×10^5 cal; 3.05×10^{24} ev; b) 268 mi/hr; c) 13.4 sec. **1.29.** a) 6.71×10^5 mi/hr; b) 4.10×10^{-13} erg; 4.10×10^{-20} joule; 9.80×10^{-21} cal; 0.256 ev; c) 7.74×10^{-11} mi/hr; d) 1.73×10^{-11} mi/hr; e) 1.19×10^{25}. **1.30.** a) 8.48×10^{11} ergs; 8.48×10^4 joules; 2.03×10^4 cal; 5.29×10^{23} ev; b) 7.05×10^{16} ergs; 7.05×10^9 joules; 1.68×10^9 cal; 4.40×10^{28} ev; c) 5.68×10^{-22} erg; 5.68×10^{-29} joule; 1.36×10^{-29} cal; 3.54×10^{-10} ev; d) 1.82×10^{-21} erg; 1.82×10^{-28} joule; 4.35×10^{-29} cal; 1.14×10^{-9} ev. **1.32.** a) 4.04×10^5 joules; 0 joule; -1.62×10^5 joules; b) 2.42×10^5 joules. **1.33.** a) 1.62×10^{-4} joule; b) 1.40×10^{-4} joule. **1.35.** -2.77×10^{10} joules. **1.36.** 8.4×10^9 joules.

Chapter 2

2.4. a) 1.50 tons; b) 9×10^7 mi/hr. **2.5.** a) 70.0 kg; b) 1.1×10^8 mi/hr. **2.6.** a) 9.16×10^{-28} g; b) 2.60×10^8 m/sec. **2.11.** a) 53*p*, 74*n*; b) 10*p*, 12*n*; c) 15*p*, 16*n*; d) 12*p*, 14*n*; e) 5*p*, 5*n*; f) 13*p*, 14*n*. **2.12.** a) 0.99504; b) 11.8570; c) 15.7941. **2.13.** In coulombs/g: $e = 1.769 \times 10^8$, $p = 9.583 \times 10^4$, $\alpha = 1.948 \times 10^5$, $n = 0$. **2.19.** a) ${}^1_1H = 99.98\%$, ${}^2_1H = 0.02\%$; b) ${}^7_3Li = 92.42\%$, ${}^6_3Li = 7.58\%$. **2.21.** 0.9912 amu. **2.22.** a) 1.88×10^{22} atoms; 0.0312 mole; b) 3.76×10^{22} atoms; 0.0625 mole; c) 2.53×10^{21} atoms; 0.00420 mole; d) 4.29×10^{25} atoms; 71.5 moles; e) 1.95×10^{26} atoms; 3.25×10^4 moles; f) 1.16×10^{16} atoms; 1.92×10^{-8} mole; g) 7.41×10^{20} atoms; 1.23×10^{-3} mole; h) 1.44×10^{23} atoms; 0.238 mole; i) 1.91×10^9 atoms; 3.18×10^{-15} mole; j) 2.45×10^{26} atoms; 407 moles. **2.23.** a) 9.21×10^{-4} g; b) 1.20×10^5 g; c) 738 g; d) 3.16×10^{-23} g.

Chapter 3

3.2. a) 0.023; b) 6.02×10^{23}. **3.3.** a) 4; b) 4.21×10^{24}; c) 26.6. **3.4.** 2.38×10^4 tons; 1.02×10^6 tons. **3.5.** 5000 tons, 55,000 tons. **3.14.** a) 3.96×10^{-17}; b) 3.06×10^{-14}. **3.17.** 3.56×10^{-14}. **3.18.** a) 0; b) 1.502×10^{-15}; 2.15×10^{-16}; c) 3.70×10^{-15}, 2.84×10^{-16}; d) 1.84×10^{-14}, 3.36×10^{-16}; e) 4.10×10^{-14}, 3.23×10^{-16}; f) 5.97×10^{-14}, 3.03×10^{-16}. **3.19.** a) 9.137; b) 27.4650; c) 104.804; d) 235.832. **3.20.** In the order listed: 5.0×10^{-18} yr^{-1}; 1.4×10^{-16}

yr^{-1}; 5.8×10^{-14} yr^{-1}; 1.5×10^{-10} yr^{-1}; 9×10^{-6} yr^{-1}; 4.28×10^{-4} yr^{-1}; 3.30×10^{-3} yr^{-1}; 0.010 yr^{-1}; 0.0288 $days^{-1}$; 0.0592 $days^{-1}$; 0.18 $days^{-1}$; 0.0654 $days^{-1}$; 0.113 hr^{-1}; 0.0114 min^{-1}; 0.0192 min^{-1}; 0.165 min^{-1}; 0.012 sec−1; 0.53 sec^{-1}; 4220 sec^{-1}; 2×10^6 sec^{-1}. **3.24.** 1.9×10^{10} yr. **3.25.** a) 14.6 days; c) 1.6 mg. **3.26.** 32.3 g, 19.4 g. **3.27.** a) 2370 yr; b) 5760 yr; c) 11,500 yr.

Chapter 4

4.3. a) 3.27×10^{15} sec^{-1}; b) 1.80×10^{13} sec^{-1}; c) 2.20×10^{12} sec^{-1}. **4.6.** a) 8.05×10^{-23}, 8.06×10^{-23}; b) 7.08×10^{-23}; 7.78×10^{-23}; c) 1.17×10^{-22} kcal. **4.17.** a) 1; b) 5; c) 3; d) 7. **4.24.** a) 98; b) 30; c) 2.

Chapter 5

5.1. a) 124; b) 368; c) 391. **5.3.** a) 4.31×10^{-22} kcal/atom, 259 kcal/mole; b) 2.96×10^{-22} kcal/atom, 178 kcal/mole; c) 4.59×10^{-21} kcal/atom, 2760 kcal/mole; d) 1.33×10^{-20} kcal/atom, 8010 kcal/mole; e) 3.95×10^{-21} kcal/atom; 2380 kcal/mole.

Chapter 6

6.1. a) 13; b) 7.83×10^{24}; c) 16; d) 0.562; e) 4.90×10^{23}; f) 2.25% H, 34.8% P, 62.9% O; g) 32.5. **6.2.** a) C_4H_8O; b) 66.6% C, 11.2% H, 22.2% O; c) 30.0; d) 3.34×10^{21}, 6.69×10^{21}, 8.36×10^{20}; e) 87.7 g coal, 100 g water, 96.6 g air. **6.4.** a) KI_3; b) C_3O_2; c) $CaC_4H_6O_4$; d) H_5ClO_6; e) $C_3H_8O_3$. **6.5.** PS_2, P_3S_6. **6.6.** CH_2O, $C_3H_6O_3$. **6.7.** 78.6. **6.8.** 56.1. **6.9.** 57.7. **6.31.** 109°. **6.41.** In order in which they appear in the formulas: a) 1+, 7+, 2−; b) 5+, 2−, 1−; c) 2+, 1+, 1−; d) 4+, 2−; e) 1+, 2−; f) 6+, 2−, 1−; g) 1+, 6+, 2−; h) 2−, 1+ or 2+, 1−; i) 1+, 1− or 1−, 1+; j) 1+, 1− or 1−, 1+; k) 1+, 3+, 2−; l) 1+, 7+, 2−; m) 1+, 1/3−; n) 2+, 1− or 4+, 2−; o) 8/3+, 2−; p) 1+, 5+, 2−, 1−; q) 1+, 3+, 1−.

Chapter 7

7.1. 1.01×10^6. **7.2.** 0.667. **7.4.** 1030 cm. **7.6.** a) 77.1 cm. **7.8.** 425 mm Hg. **7.9.** 424 mph. **7.10.** 43.2. **7.13.** 11.8 atm. **7.14.** a) 25.7; b) 1.81 g/l. **7.15.** 9.90 g/l. **7.16.** 229 liters. **7.17.** 10.3 atm. **7.18.** 30.5 l. **7.20.** 32.0. **7.21.** a) 112 ml; b) 0.199 g. **7.22.** 4.17. **7.23.** 3.29×10^8. **7.24.** a) S_8; b) S_2. **7.25.** $C_4H_8O_2$. **7.28.** a) 91°C; f) −2°C. **7.29.** 1437°C, −8°C. **7.31.** 26.7. **7.32.** 838 cal.

Chapter 8

8.6. 2.31 g/l. **8.19.** 59.3 kcal. **8.37.** 109° **8.69.** 2+. **8.74.** central N = 1+, other two atoms = 1/2− each

Chapter 9

9.7. 55.9. **9.17.** b) 7+. **9.34.** a) 6+; b) 6+; c) 6+ or 7+; d) 6+ or 8+. **9.36.** c) 4+. **9.47.** nitrite: N = O, O = 1/2−; nitrate: N = 1+; O = 2/3−; nitric acid: N = 1+, H = O, O = zero for one and 1/2− for other two. **9.57.** a) 5+, 3+, 1+.

Chapter 11

11.2. b) 5.0; c) 104; d) 0.33; e) 159. **11.3.** b) 47.5; c) 0.698; d) 180; e) 167,000. **11.4.** b) 0.05 mole; c) 0.15 mole potassium chloride, 0.15 mole potassium dicyanoargentate(I), 0.85 mole silver chloride, 0.20 mole potassium cyanide. **11.5.** b) II; c) same; d) II; e) 29.0% by I; f) 69.1% by I. **11.6.** c) 80.1% by I; 16.1 g; 19.9 g; d) 38.4% by I; e) 46% by I, 4.0 g. **11.7.** b) 4.00; c) 6.50; 19.5; d) 12.2 g; e) 43.4. **11.8.** b) 16.3 g; c) 5.31 g; d) 74.2. **11.9.** b) 31.4 g, 24.0 g; c) 6.6 g. **11.10.** c) 50; e) 100. **11.11.** b) 45.4 g, 15.1 g. **11.12.** b) 5.78 g; **11.14.** 19.1. **11.16.** a) 67.8; b) 3.27 g; c) 49.8% ammonium sulfate. **11.17.** b) 1.13 g; c) 85.6. **11.18.** b) 27.3% Ca, 48.4% Cl, 24.3% H_2O, $CaCl_2 \cdot 2H_2O$; c) 70.1. **11.19.** b) 6.41; c) 60.3% calcium carbonate. **11.20.** b) 71.2% potassium chlorate; c) 82.3% potassium chlorate.

Chapter 12

12.2. a) 0.0506; b) 69 cal. **12.3.** a) 5.51; b) 5.37 kcal. **12.4.** O, 240 cal, −240 cal. **12.5.** $\Delta E = 0$, $q = w =$ 841 cal. **12.8.** a) −117 kcal; b) −20 kcal; c) −34 kcal; d) −164 kcal; e) −201 kcal. **12.10.** b) −531 kcal; c) +1430 kcal; d) −174 kcal. **12.11.** a) −246 kcal; b) −269 kcal; c) −506 kcal; d) +9 kcal; e) +64.5 kcal; f) +222 kcal. **12.16.** In the order: free energy, entropy: a) +4.3 kcal, +69.8 cal; b) +0.2 kcal, +13 cal; c) −191 kcal, −1.2 cal; d) −24.59 kcal, +2.8 cal; e) −166 kcal, +755 cal; f) −0.21 kcal, +43 cal; g) −176.14 kcal, +705 cal; h) +51.31 kcal, +124 cal. **12.18.** In the order K_p, K_c: a) 8.3×10^2 atm^{-2}, 5.0×10^5 l^2/mole^2; b) 2.2×10^{13} atm^{-1}, 5.4×10^{14} l/mole; c) 1.8×10^{-24} atm, 7.4×10^{-26} mole/l; d) 2.7×10^{-56} atm^{-3}, 4.0×10^{-52} l^3/mole^3; e) 1.3×10^7 atm^{-1}; 3.2×10^8 l/mole. **12.19.** HCl: $-0.04M$, $0.04M$; O_2: $-0.01M$, $0.03M$; H_2O: $+0.02M$, $0.02M$; Cl_2: $+0.02M$; $K_c = 7.5 \times 10^{-3}$. **12.24.** 4.4. **12.26.** 20.5 mole/l. **12.27.** a) $[H_2] = [CO_2] = 0.22$, $[H_2O] = [CO] = 0.28$; b) $[H_2] = [CO_2] = 0.22$, $[H_2O] = [CO] = 0.28$; c) $[H_2] = [CO_2] = 0.44$, $[H_2O] = [CO] =$

0.56. **12.28.** $[H_2O] = 0.23$, $[CO] = 0.33$, $[H_2] = 0.17$, $[CO_2] = 0.27$. **12.29.** 9. **12.30.** 0.05 mole each of H_2 and I_2, 0.4 mole HI. **12.31.** 10^{-6}. **12.33.** $K_p = 2.81 \times 10^{-5}$; $K_c = 1.15 \times 10^{-6}$.

Chapter 13

13.2. a) 3.6; b) 4.0; c) 9.8; d) 24; e) 34.2; f) 9.8. **13.3.** a) 1.8; b) 3.3; c) 5.1; d) 10.0. **13.4.** a) 18.6; b) 42.7; c) 12.1. **13.5.** 0.180. **13.6.** 2.63. **13.7.** 12.9. **13.9.** a) 33.3 each; b) 25.5, 22.9, 51.6; c) mole fractions: 0.40, 0.36, 0.24; d) 3.3. **13.10.** a) 15.1; b) 11.7; c) 18; d) 17.6. **13.11.** 23.5. **13.12.** 6. **13.13.** a) 520; b) 0.035; c) 69.6; d) 1.6; e) 4.6. **13.14.** 0.9. **13.15.** 228 mmoles, 36.0 g. **13.16.** 85.4. **13.17.** 12.0. **13.18.** 6.71 ml diluted to 35.0 ml. **13.19.** $M_{Ca^{++}} = M_{CaCl_2} = 0.583$; $M_{Cl^-} = 1.17$; $M_{SO_4^{--}} = M_{(NH_4)_2SO_4} = 2.20$; $M_{NH_4^+} = 4.40$. **13.20.** 2.06. **13.21.** 29.9 ml diluted to 500 ml. **13.25.** Answers in order: molar solubility, solubility constant: a) 0.0404, 1.58×10^{-3}; b) 9.29×10^{-4}, 3.63×10^{-5}; c) 0.00131, 5.12×10^{-5}; d) 0.00468, 1.83×10^{-4}. **13.34.** 8.1×10^{-7}, 6.4×10^{-7}, 4.9×10^{-7}. **13.36.** a) 10^{-5}; b) 10^{-6}; c) 4×10^{-9}. **13.38.** a) 1.2×10^{-3}; b) 4.4×10^{-4}; c) 1.7×10^{-4}; d) 1.7×10^{-5}; e) 6.9×10^{-4}. **13.42.** a) 3.2×10^{-5} *M*; b) 2.6×10^{-5} *M*; c) 2.0×10^{-3} *M*. **13.43.** a) 1.4×10^{-11} *M*; b) 4.9×10^{-8} *M*;. **13.45.** a) 6.5×10^{-8}; b) 2.0×10^{-7}. **13.46.** a) 1.54×10^{-2}; b) 1.57×10^{-2}. **13.47.** a) 2.25 g; b) 0.874 g; c) 49.2 ml. **13.48.** 21.6. **13.49.** 100.26°C, −0.92°C, 760 mm Hg. **13.50.** a) 52; b) 2200; c) 180; d) 91. **13.51.** a) same; b) $C_{10}H_{16}O$; c) $C_{10}H_8$; d) C_6H_5I. **13.52.** S_6. **13.53.** ca. 61.8, $C_2H_6O_2$, 62.08. **13.56.** 1.7. **13.57.** 2.6 moles. **13.60.** 0.767.

Chapter 15

15.20. Answers in order pH, pOH: a) 2.46, 11.54; b) 3.11, 10.89; c) 11.95, 2.05; d) 10.91, 3.09; e) 7, 7. **15.21.** a) 10^{-4}; b) 10^{-7}; c) 3×10^{-3}; d) 6×10^{-9}; e) 2.7; f) 8×10^{-17}. **15.23.** 6. **15.25.** a) 1.7×10^{-4}; b) 1.0×10^{-3}; c) 2.0×10^{-12}; d) 5.3×10^{-4}; e) 2×10^{-11}. **15.31.** a) 55.5. **15.36.** a) cannot be computed; b) 2.0×10^{-21}; c) 5.2×10^{-21}. **15.37.** a) 24; b) 1.4×10^{8}; c) 1.2×10^{-9}; d) 2.46×10^{14}. **15.50.** a) 5.0; b) 6.0; c) 6.5; d) 7.0; e) 7.6.

Chapter 16

16.2. $M_{H_3O^+} = M_{NO_3^-} = 0.100$, $M_{OH^-} = 10^{-13}$, pOH = 13. **16.3.** 10^{-8} *M*. **16.6.** a) 10^{-10} *M*; b) 2×10^{-13} *M*; c) 4×10^{-9} *M*. **16.7.** a) 10^{-12} *M*; b) 6×10^{-11} *M*; c) 4×10^{-9} *M*. **16.8.** a) 2.6×10^{-11}; b) 6.1. **16.10.** 10^{-7}. **16.12.** a) 0.9; b) 1.1; c) 5.6. **16.13.** a) 1.3×10^{-5}; b) 5; c) 3.6. **16.14.** a) 1.5; b) 1.2×10^{-3} *M*; c) 11.1. **16.19.** pH = 5.5; $[OH^-] = 2.9 \times 10^{-9}$; $[NH_4^+] = [Cl^-] = 0.020$; $[NH_3] = [H_3O^+] = 3.5 \times 10^{-6}$; b) pH = 4.5; $[NO_3^-] = 0.30$; $[Ni(OH)(H_2O)_{x-1}^+] = [H_3O^+] = 3.5 \times 10^{-5}$; $[OH^-] = 2.9 \times 10^{-10}$. c) pH =

10.6; $[K^+] = [CN^-] = 6.0 \times 10^{-3}$; $[HCN] = [OH^-] = 3.6 \times 10^{-4}$; $[H_3O^+] = 2.8 \times 10^{-11}$; d) pH = 5.9; $[Na^+] = [CH_3COO^-] = 3.0 \times 10^{-3}$; $[CH_3COOH] = [H_3O^+] = 1.3 \times 10^{-6}$; $[OH^-] = 7.7 \times 10^{-9}$; e) pH = 5.3; $[CH_3COOH] = 5.5 \times 10^{-6}$; $[CH_3COO^-] = [K^+] = 5.0 \times 10^{-2}$; $[H_3O^+] = 5.5 \times 10^{-6}$; $[OH^-] = 1.8 \times 10^{-9}$. **16.21.** a) pH = 2.5; $[H_3AsO_4] = 1.8 \times 10^{-3}$; $[H_2AsO_4^-] = [H_3O^+] = 3.2 \times 10^{-3}$; $[HASO_4^{--}] = 1.1 \times 10^{-7}$; $[ASO_4^{3-}] = 10^{-16}$; $[OH^-] = 3.1 \times 10^{-12}$; b) pH = 9.9; $[Na^+] = [SO_3^{--}] = 0.040$; $[HSO_3^-] = [OH^-] = 8.0 \times 10^{-5}$; $[H_2SO_3] = 5.9 \times 10^{-13}$; $[H_3O^+] = 1.2 \times 10^{-10}$. **16.22.** a) $[H_3O^+] = [F^-] = 10^{-2}$; $[Br^-] = 3.2 \times 10^{-3}$; $[OH^-] = 10^{-12}$; $[HF] = 0.19$; b) $[H_3O^+] = [I^-] = 5.1 \times 10^{-4}$; $[CH_3COOH] = 6.5 \times 10^{-2}$; $[CH_3COO^-] = 2.2 \times 10^{-3}$; $[OH^-] = 2.0 \times 10^{-11}$; c) $[Na^+] = 0.14$; $[HO_2^-] = 0.10$; $[H_3O^+] = 1.1 \times 10^{-12}$; $[H_2O_2] = [OH^-] = 9.5 \times 10^{-3}$. **16.23.** a) 8.9; b) 4.8; c) 5.1. **16.25.** a) $[CN^-] : [HCN] = 1.6$; b) $[NH_4^+] : [NH_3] = 0.6$. **16.28.** a) pH = 4.7; $[H_3O^+] = 2.2 \times 10^{-5}$; $[Na^+] = [H_2PO_4^-] = 0.25$; $[H_3PO_4] = 7.4 \times 10^{-4}$; $[HPO_4^{--}] = 7.2 \times 10^{-4}$; $[PO_4^{3-}] = 1.5 \times 10^{-11}$; $[OH^-] = 4.7 \times 10^{-10}$; b) pH = 4.6; $[H_3O^+] = 2.7 \times 10^{-5}$; $[OH^-] = 3.7 \times 10^{-10}$; $[H_2SO_3] = 6.4 \times 10^{-5}$; $[Na^+] = [HSO_3^-] = 4 \times 10^{-2}$; $[SO_3^{--}] = 9.2 \times 10^{-5}$. **16.29.** Answers in order: % titrant, % pH: a) 40, 8; b) 80, 20; c) 90, 26; d) 98, 40. **16.30.** Answers in order: initial pH, equivalence point pH: a) 2.8, 8.7; b) 2.8, 8.3; c) 11.2, 5.1; d) 12.6, 7.0; e) 12.8, 1.7. **16.33.** 32.1, 10.5, 40.6. **16.34.** 0.357. **16.35.** 4.213. **16.36.** a) 0.283; b) 0.0103. **16.37.** $M_{H_2SO_4} = 0.176$; $M_{NaOH} = 0.352$. **16.38.** 90.1. **16.39.** RbOH.

Chapter 17

17.6. a) +0.63V; b) +0.78V; c) +0.84V. **17.10.** e) 1.56V. **17.15.** a) 1.93×10^5; b) 3040. **17.16.** 0.009 mole, 0.0182 g, 0.202 l. **17.17.** 386,000 sec. **17.24.** a) +0.26 V; b) −0.56V; c) −0.35V; d) +0.26V; e) +0.20 V; f) −0.59V. **17.27.** a) +0.62V; b) −0.42V; c) +0.15V. **17.30.** 0.10V. **17.34.** a) 10^{10}; b) 10^{-16}; c) 4×10^{62}; d) 3×10^{-11}; e) 2×10^{82}. **17.35.** In kcal: a) −13.8; b) +21.7; c) −85.4; d) +14.3; e) −112. **17.43.** 0.01424. **17.44.** 39.20. **17.45.** a) 0.03767; b) 16.01. **17.46.** 36.02. **17.47.** 50.92.

Chapter 18

18.6. 0, 14. **18.28.** 6.042. **18.29.** 0.3772. **18.45.** a) 7×10^{57}; b) 4×10^{36}. **18.54.** a) 2×10^{16}; b) 3×10^{30}; c) 4×10^{36}; d) 4×10^{13}; e) 1×10^{-61}; f) 8×10^{-7}; g) 8×10^{-30}. **18.70.** 79.18. **18.84.** a) 1×10^{90}; b) 3×10^{143}; c) 2×10^{15}; d) 3×10^{-18}; e) 1×10^{-21}; f) 1×10^{12}. **18.88.** $\Delta E° = +0.19V$, $K = 3 \times 10^6$. **18.93.** a) +0.07V; b) −2.61V; c) +0.52V; d) +0.29V; e) +0.13V; f) +1.94V. **18.96.** a) 2×10^{-7}; b) 1×10^{-50}; c) 7×10^{-38}. **18.98.** a) +0.69V; b) −1.10V; c) +0.56V. **18.106.** a) −0.57V; b) +0.26V. **18.107.** a) 1×10^4; b) 3×10^{51}; c) 1×10^{-593}. **18.111.** a) 1×10^{118}; b) 3×10^{13}; c) 2×10^{25}.

Index

(Page numbers italicized refer to illustrations. Q = Question at end of chapter.)

LOGARITHMS OF NUMBERS

n	0	1	2	3	4	5	6	7	8	9
10	0000	0043	0086	0128	0170	0212	0253	0294	0334	0374
11	0414	0453	0492	0531	0569	0607	0645	0682	0719	0755
12	0792	0828	0864	0899	0934	0969	1004	1038	1072	1106
13	1139	1173	1206	1239	1271	1303	1335	1367	1399	1430
14	1461	1492	1523	1553	1584	1614	1644	1673	1703	1732
15	1761	1790	1818	1847	1875	1903	1931	1959	1987	2014
16	2041	2068	2095	2122	2148	2175	2201	2227	2253	2279
17	2304	2330	2355	2380	2405	2430	2455	2480	2504	2529
18	2553	2577	2601	2625	2648	2672	2695	2718	2742	2765
19	2788	2810	2833	2856	2878	2900	2923	2945	2967	2989
20	3010	3032	3054	3075	3096	3118	3139	3160	3181	3201
21	3222	3243	3263	3284	3304	3324	3345	3365	3385	3404
22	3424	3444	3464	3483	3502	3522	3541	3560	3579	3598
23	3617	3636	3655	3674	3692	3711	3729	3747	3766	3784
24	3802	3820	3838	3856	3874	3892	3909	3927	3945	3962
25	3979	3997	4014	4031	4048	4065	4082	4099	4116	4133
26	4150	4166	4183	4200	4216	4232	4249	4265	4281	4298
27	4314	4330	4346	4362	4378	4393	4409	4425	4440	4456
28	4472	4487	4502	4518	4533	4548	4564	4579	4594	4609
29	4624	4639	4654	4669	4683	4698	4713	4728	4742	4757
30	4771	4786	4800	4814	4829	4843	4857	4871	4886	4900
31	4914	4928	4942	4955	4969	4983	4997	5011	5024	5038
32	5051	5065	5079	5092	5105	5119	5132	5145	5159	5172
33	5185	5198	5211	5224	5237	5250	5263	5276	5289	5302
34	5315	5328	5340	5353	5366	5378	5391	5403	5416	5428
35	5441	5453	5465	5478	5490	5502	5514	5527	5539	5551
36	5563	5575	5587	5599	5611	5623	5635	5647	5658	5670
37	5682	5694	5705	5717	5729	5740	5752	5763	5775	5786
38	5798	5809	5821	5832	5843	5855	5866	5877	5888	5899
39	5911	5922	5933	5944	5955	5966	5977	5988	5999	6010
40	6021	6031	6042	6053	6064	6075	6085	6096	6107	6117
41	6128	6138	6149	6160	6170	6180	6191	6201	6212	6222
42	6232	6243	6253	6263	6274	6284	6294	6304	6314	6325
43	6335	6345	6355	6365	6375	6385	6395	6405	6415	6425
44	6435	6444	6454	6464	6474	6484	6493	6503	6513	6522
45	6532	6542	6551	6561	6571	6580	6590	6599	6609	6618
46	6628	6637	6646	6656	6665	6675	6684	6693	6702	6712
47	6721	6730	6739	6749	6758	6767	6776	6785	6794	6803
48	6812	6821	6830	6839	6848	6857	6866	6875	6884	6893
49	6902	6911	6920	6928	6937	6946	6955	6964	6972	6981
50	6990	6998	7007	7016	7024	7033	7042	7050	7059	7067
51	7076	7084	7093	7101	7110	7118	7126	7135	7143	7152
52	7160	7168	7177	7185	7193	7202	7210	7218	7226	7235
53	7243	7251	7259	7267	7275	7284	7292	7300	7308	7316
54	7324	7332	7340	7348	7356	7364	7372	7380	7388	7396